Elaine N. Marieb, R.N., Ph.D.
Holyoke Community College

Lori A. Smith, Ph.D.
American River College

Human Anatomy & Physiology Laboratory Manual

ELEVENTH EDITION

PhysioEx™ Version 9.1 authored by

Peter Z. Zao   North Idaho College
Timothy Stabler, Ph.D.   Indiana University Northwest
Lori A. Smith, Ph.D.   American River College
Andrew Lokuta, Ph.D.   University of Wisconsin–Madison
Edwin Griff, Ph.D.   University of Cincinnati

PEARSON

Editor-in-Chief: Serina Beauparlant
Senior Acquisitions Editor: Brooke Suchomel
Program Manager: Shannon Cutt
Program Manager Team Lead: Michael Early
Director of Development: Barbara Yien
Development Editor: Lisa Clark
Editorial Assistant: Arielle Grant
Production Project and Design Manager: Michele Mangelli
Project Manager Team Lead: Nancy Tabor
Production Supervisor: Janet Vail

Art and Photo Coordinator: David Novak
Photo Researcher: Kristin Piljay
Interior and Cover Designer: tani hasegawa
Copyeditor: Sally Peyrefitte
Compositor: Cenveo® Publisher Services
Media Producers: Liz Winer and Lauren Hill
PhysioEx Developer: BinaryLabs, Inc.
Senior Manufacturing Buyer: Stacey Weinberger
Senior Marketing Manager: Allison Rona
Senior Anatomy & Physiology Specialist: Derek Perrigo

Cover photographs: Pole vaulter, Pete Saloutos/Getty Images; Sky, Gregor Schuster/Getty Images

Acknowledgments of third-party content appear on page BM-1, which constitutes an extension of this copyright page.

The Authors and Publisher believe that the lab experiments described in this publication, when conducted in conformity with the safety precautions described herein and according to the school's laboratory safety procedures, are reasonably safe for the student to whom this manual is directed. Nonetheless, many of the described experiments are accompanied by some degree of risk, including human error, the failure or misuses of laboratory or electrical equipment, mismeasurement, chemical spills, and exposure to sharp objects, heat, bodily fluids, blood, or other biologics. The Authors and Publisher disclaim any liability arising from such risks in connection with any of the experiments contained in this manual. If students have any questions or problems with materials, procedures, or instructions on any experiment, they should always ask their instructor for help before proceeding.

ISBN 10: 0-13-390238-2 (student edition)
ISBN 13: 978-0-13-390238-9 (student edition)
ISBN 10: 0-13-405733-3 (instructor's review copy)
ISBN 13: 978-0-13-405733-0 (instructor's review copy)

www.pearsonhighered.com

2 3 4 5 6 7 8 9 10—V003—18 17 16 15

Contents

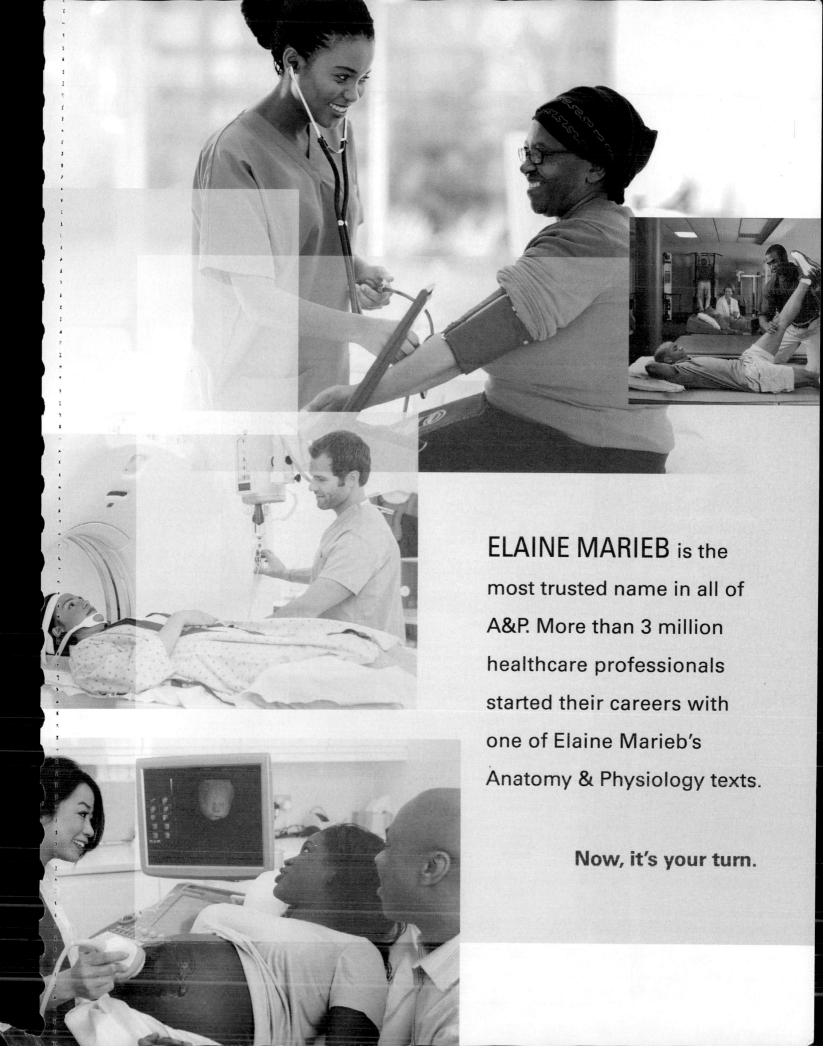

ELAINE MARIEB is the most trusted name in all of A&P. More than 3 million healthcare professionals started their careers with one of Elaine Marieb's Anatomy & Physiology texts.

Now, it's your turn.

Tools to Help You

Activity 1

Examining Epithelial Tissue Under the Microscope

Obtain slides of simple squamous, simple cuboidal, simple columnar, stratified squamous (nonkeratinized), pseudostratified ciliated columnar, stratified cuboidal, stratified columnar, and transitional epithelia. Examine each carefully, and notice how the epithelial cells fit closely together to form intact sheets of cells, a necessity for a tissue that forms linings or the coverings of membranes. Scan each epithelial type for modifications for specific functions, such as cilia (motile cell projections that help to move substances along the cell surface), and microvilli, which increase the surface area for absorption. Also be alert for goblet cells, which secrete lubricating mucus. Compare your observations with the descriptions and photomicrographs in Figure 6.3.

While working, check the questions in the Review Sheet at the end of this exercise. A number of the questions there refer to some of the observations you are asked to make during your microscopic study.

WHY THIS MATTERS | **Buccal Swabs**

A buccal, or cheek, swab is a method used to collect stratified squamous cells from the oral cavity. The cells contain DNA that can be used for DNA fingerprinting or tissue typing. DNA fingerprinting can be used in criminal investigations, and tissue typing can be used to match a recipient with a donor for organ transplant, especially a bone marrow transplant. The buccal swab procedure involves using a cotton-tipped applicator to scrape the inside of the mouth in the buccal region and remove cells at the surface. This noninvasive procedure provides an easy way to obtain the DNA profile of an individual, a unique molecular "signature." ∎

👥 Group Challenge 1

Identifying Epithelial Tissues

Following your observations of epithelial tissues under the microscope, obtain an envelope for each group that contains images of various epithelial tissues. With your lab manual closed, remove one image at a time and identify the epithelium. One member of the group will function as the verifier, whose job is to make sure that the identification is correct.

After you have correctly identified all of the images, sort them into groups to help you remember them. (*Hint:* You could sort them according to cell shape or number of layers of epithelial cells.)

Now, carefully go through each group and try to list one place in the body where the tissue is found and one function for it. After you have correctly listed the locations, take your lists and draw some general conclusions about where epithelial tissues are found in the body. Then compare and contrast the functions of the various epithelia. Finally, identify the tissues described in the **Group Challenge 1** chart, and list several locations in the body.

Group Challenge 1: Epithelial Tissue IDs

Magnified appearance	Tissue type	Locations in the body
• Apical surface has dome-shaped cells (flattened cells may also be mixed in) • Multiple layers of cells are present		
• Cells are mostly columnar • Not all cells reach the apical surface • Nuclei are located at different levels • Cilia are located at the apical surface		
• Apical surface has flattened cells with very little cytoplasm • Cells are not layered		
• Apical surface has square cells with a round nucleus • Cells are not layered		

Succeed in the Lab

NEW!

Prepare for lab by watching pre-lab videos on your smartphone, tablet, or computer. Short pre-lab videos will give you some background information on the lab, explain what materials you will be using, help you see the connections between the content you learn in the classroom and the experiment, and give you the opportunity to check your understanding—all before you even step foot in the lab.

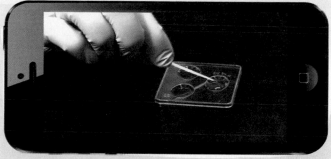

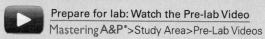
Prepare for lab: Watch the Pre-lab Video
Mastering A&P®>Study Area>Pre-Lab Videos

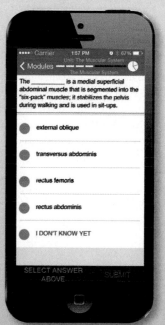

NEW!

Dynamic Study Modules offer a mobile-friendly, personalized reading experience of the chapter content. As you answer questions to master the chapter content, you receive detailed feedback with text and art from the lab manual itself. The Dynamic Study Modules help you acquire, retain, and recall information faster and more efficiently than ever before.

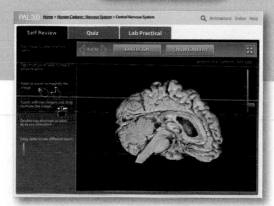

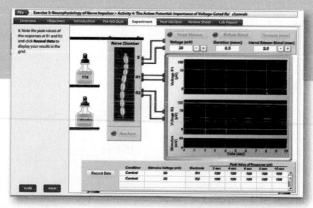

Practice Anatomy Lab™ (PAL™) 3.0 is a virtual anatomy study and practice tool that gives you 24/7 access to the most widely used lab specimens, including the human cadaver, anatomical models, histology, cat, and fetal pig. PAL 3.0 is easy to use and includes built-in audio pronunciations, rotatable bones, and simulated fill-in-the-blank lab practical exams.

PhysioEx™ 9.1 is an easy-to-use lab simulation program that allows you to repeat labs as often as you like, perform experiments without animals, and conduct experiments that are difficult to perform in a wet lab environment because of time, cost, or safety concerns. The online format with easy step-by-step instructions includes everything you need in one convenient place.

Acknowledgments

We wish to thank the following reviewers for their contribution to this edition: Matthew Abbott, Des Moines Area Community College; Lynne Anderson, Meridian Community College; Christopher W. Brooks, Central Piedmont Community College; Brandi Childress, Georgia Perimeter College; Christopher D'Arcy, Cayuga Community College; Mary E. Dawson, Kingsborough Community College; Karen Eastman, Chattanooga State Community College; Michele Finn, Monroe Community College; Lisa Flick, Monroe Community College; Abigail M. Goosie, Walters State Community College; Samuel Hirt, Auburn University; Shahdi Jalilvand, Tarrant County College—Southeast; Tiffany B. McFalls-Smith, Elizabethtown Community & Technical College; Melinda Miller, Pearl River Community College; Todd Miller, Hunter College of CUNY; Susan Mitchell, Onondaga Community College; Erin Morrey, Georgia Perimeter College; Jill Y. O'Malley, Erie Community College—City; Suzanne Oppenheimer, College of Western Idaho; Suzanne Pundt, The University of Texas at Tyler; Mark Schmidt, Clark State Community College; Teresa Stegall-Faulk, Middle Tennessee State University; Bonnie J. Tarricone, Ivy Tech Community College.

Special thanks to Susan Mitchell for her authorial contributions to this lab manual over the years.

Thanks also to Josephine Rogers of the University of Cincinnati, the original author of the pre-lab quizzes.

The excellence of PhysioEx 9.1 reflects the expertise of Peter Zao, Timothy Stabler, Lori Smith, Andrew Lokuta, Greta Peterson, Nina Zanetti, and Edwin Griff. They generated the ideas behind the activities and simulations. Credit also goes to the team at BinaryLabs, Inc., for their expert programming and design.

Continued thanks to colleagues and friends at Pearson who worked with us in the production of this edition, especially Serina Beauparlant, Editor-in-Chief; Brooke Suchomel, Senior Acquisitions Editor; Lisa Clark, Development Editor; and Shannon Cutt, Program Manager. Applause also to Lauren Hill who managed MasteringA&P, and Aimee Pavy for her work on PhysioEx 9.1. Many thanks to Amanda J.S. Kaufmann who did a super job as video production manager of the new pre-lab videos. Additional thanks go to Cheryl Chi and the group at Roaring Mouse Productions, who provided their special skills to the video production team. Many thanks to Stacey Weinberger for her manufacturing expertise. Finally, our Senior Marketing Manager, Allison Rona, has efficiently kept us in touch with the pulse of the market.

Kudos also to Michele Mangelli and her production team, who did their usual great job. Janet Vail, Production Editor for this project, got the job done in jig time. David Novak acted as Art and Photo Coordinator, and Kristin Piljay conducted photo research. Our fabulous interior and cover designs were created by tani hasegawa. Sally Peyrefitte brought her experience to copyediting the text.

We are grateful to the team at BIOPAC, especially to Jocelyn Kremer and Mike Mullins, who were extremely helpful in making sure we had the latest updates and answering all of our questions.

Elaine N. Marieb
Lori A. Smith
Anatomy and Physiology
Pearson Education
1301 Sansome Street
San Francisco, CA 94111

The Language of Anatomy

Objectives

- ☐ Describe the anatomical position, and explain its importance.
- ☐ Use proper anatomical terminology to describe body regions, orientation and direction, and body planes.
- ☐ Name the body cavities, and indicate the important organs in each.
- ☐ Name and describe the serous membranes of the ventral body cavities.
- ☐ Identify the abdominopelvic quadrants and regions on a torso model or image.

Materials

- Human torso model (dissectible)
- Human skeleton
- Demonstration: sectioned and labeled kidneys (three separate kidneys uncut or cut so that [a] entire, [b] transverse sectional, and [c] longitudinal sectional views are visible)
- Gelatin-spaghetti molds
- Scalpel
- Post-it® Notes

Pre-Lab Quiz

1. Circle True or False. In anatomical position, the body is lying down.
2. Circle the correct underlined term. With regard to surface anatomy, <u>abdominal</u> / <u>axial</u> refers to the structures along the center line of the body.
3. The term *superficial* refers to a structure that is:
   - **a.** attached near the trunk of the body
   - **b.** toward or at the body surface
   - **c.** toward the head
   - **d.** toward the midline
4. The _____ plane runs longitudinally and divides the body into right and left sides.
   - **a.** frontal
   - **b.** sagittal
   - **c.** transverse
   - **d.** ventral
5. Circle the correct underlined terms. The dorsal body cavity can be divided into the <u>cranial</u> / <u>thoracic</u> cavity, which contains the brain, and the <u>sural</u> / <u>vertebral</u> cavity, which contains the spinal cord.

MasteringA&P®

For related exercise study tools, go to the Study Area of **MasteringA&P.** There you will find:

- Practice Anatomy Lab PAL
- PhysioEx PEx
- A&PFlix *A&PFlix*
- Practice quizzes, Histology Atlas, eText, Videos, and more!

Most of us are naturally curious about our bodies. This curiosity is apparent even in infants, who are fascinated with their own waving hands or their mother's nose. Unlike an infant, however, an anatomy student must learn to observe and identify the dissectible body structures formally.

A student new to any science is often overwhelmed at first by the terminology used in that subject. The study of anatomy is no exception. But without this specialized terminology, confusion is inevitable. For example, what do *over, on top of, above,* and *behind* mean in reference to the human body? Anatomists have an accepted set of reference terms that are universally understood. These allow body structures to be located and identified precisely with a minimum of words.

This exercise presents some of the most important anatomical terminology used to describe the body and introduces you to basic concepts of **gross anatomy**, the study of body structures visible to the naked eye.

Anatomical Position

When anatomists or doctors refer to specific areas of the human body, the picture they keep in mind is a universally accepted standard position called the **anatomical position.** It is essential to understand this position because much of the directional terminology used in this book refers to the body in this position, regardless of the position the body happens to be in. In the anatomical position, the human body is erect, with the feet only slightly apart, head and toes pointed forward, and arms hanging at the sides with palms facing forward (**Figure 1.1a**).

☐ Assume the anatomical position, and notice that it is not particularly comfortable. The hands are held unnaturally forward rather than hanging with palms toward the thighs.

Check the box when you have completed this task.

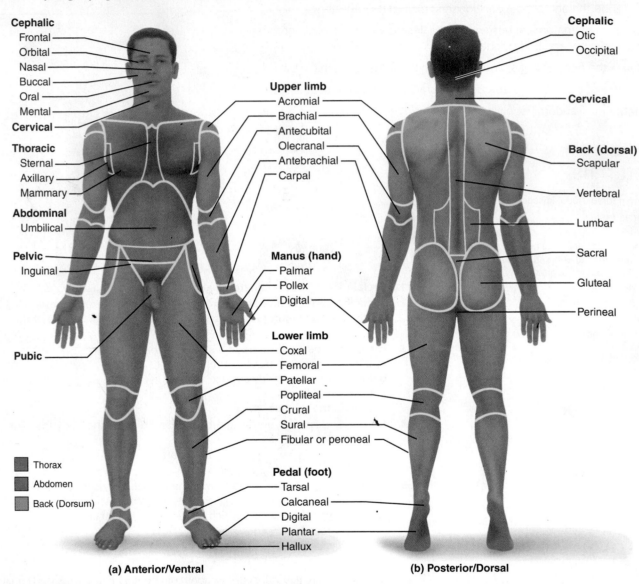

(a) Anterior/Ventral

(b) Posterior/Dorsal

Figure 1.1 Surface anatomy. (a) Anatomical position. **(b)** Heels are raised to illustrate the plantar surface of the foot.

Surface Anatomy

Body surfaces provide a wealth of visible landmarks for study. There are two major divisions of the body:

Axial: Relating to head, neck, and trunk, the axis of the body

Appendicular: Relating to limbs and their attachments to the axis

Anterior Body Landmarks

Note the following regions in Figure 1.1a:

Abdominal: Anterior body trunk region inferior to the ribs

Acromial: Point of the shoulder

Antebrachial: Forearm

Antecubital: Anterior surface of the elbow

Axillary: Armpit

Brachial: Arm

Buccal: Cheek

Carpal: Wrist

Cephalic: Head

Cervical: Neck region

Coxal: Hip

Crural: Leg

Digital: Fingers or toes

Femoral: Thigh

Fibular (peroneal): Side of the leg

Frontal: Forehead

Hallux: Great toe

Inguinal: Groin area

Mammary: Breast region

Manus: Hand

Mental: Chin

Nasal: Nose

Oral: Mouth

Orbital: Bony eye socket (orbit)

Palmar: Palm of the hand

Patellar: Anterior knee (kneecap) region

Pedal: Foot

Pelvic: Pelvis region

Pollex: Thumb

Pubic: Genital region

Sternal: Region of the breastbone

Tarsal: Ankle

Thoracic: Chest

Umbilical: Navel

Posterior Body Landmarks

Note the following body surface regions in Figure 1.1b:

Acromial: Point of the shoulder

Brachial: Arm

Calcaneal: Heel of the foot

Cephalic: Head

Dorsum: Back

Femoral: Thigh

Gluteal: Buttocks or rump

Lumbar: Area of the back between the ribs and hips; the loin

Manus: Hand

Occipital: Posterior aspect of the head or base of the skull

Olecranal: Posterior aspect of the elbow

Otic: Ear

Pedal: Foot

Perineal: Region between the anus and external genitalia

Plantar: Sole of the foot

Popliteal: Back of the knee

Sacral: Region between the hips (overlying the sacrum)

Scapular: Scapula or shoulder blade area

Sural: Calf or posterior surface of the leg

Vertebral: Area of the spinal column

Activity 1

Locating Body Regions

Locate the anterior and posterior body landmarks on yourself, your lab partner, and a human torso model.

Body Orientation and Direction

Study the terms below, referring to **Figure 1.2** on p. 4 for a visual aid. Notice that certain terms have different meanings, depending on whether they refer to a four-legged animal (quadruped) or to a human (biped).

Superior/inferior (*above/below*): These terms refer to placement of a structure along the long axis of the body. For example, the nose is superior to the mouth, and the abdomen is inferior to the chest.

Anterior/posterior *(front/back):* In humans, the most anterior structures are those that are most forward—the face, chest, and abdomen. Posterior structures are those toward the backside of the body. For instance, the spine is posterior to the heart.

Medial/lateral *(toward the midline/away from the midline or median plane):* The sternum (breastbone) is medial to the ribs; the ear is lateral to the nose.

The terms of position just described assume the person is in the anatomical position. The next four term pairs are more absolute. They apply in any body position, and they consistently have the same meaning in all vertebrate animals.

Cephalad (cranial)/caudal *(toward the head/toward the tail):* In humans, these terms are used interchangeably with *superior* and *inferior*, but in four-legged animals they are synonymous with *anterior* and *posterior*, respectively.

Ventral/dorsal *(belly side/backside):* These terms are used chiefly in discussing the comparative anatomy of animals, assuming the animal is standing. In humans, the terms *ventral* and *dorsal* are used interchangeably with the terms *anterior* and *posterior*, but in four-legged animals, *ventral* and *dorsal* are synonymous with *inferior* and *superior*, respectively.

Proximal/distal *(nearer the trunk or attached end/farther from the trunk or point of attachment):* These terms are used primarily to locate various areas of the body limbs. For example, the fingers are distal to the elbow; the knee is proximal to the toes. However, these terms may also be used to indicate regions (closer to or farther from the head) of internal tubular organs.

Superficial (external)/deep (internal) *(toward or at the body surface/away from the body surface):* For example, the skin is superficial to the skeletal muscles, and the lungs are deep to the rib cage.

Activity 2

Practicing Using Correct Anatomical Terminology

Use a human torso model, a human skeleton, or your own body to specify the relationship between the following structures when the body is in the anatomical position.

1. The wrist is _____ to the hand.

2. The trachea (windpipe) is _____ to the spine.

3. The brain is _____ to the spinal cord.

4. The kidneys are _____ to the liver.

5. The nose is _____ to the cheekbones.

6. The thumb is _____ to the ring finger.

7. The thorax is _____ to the abdomen.

8. The skin is _____ to the skeleton.

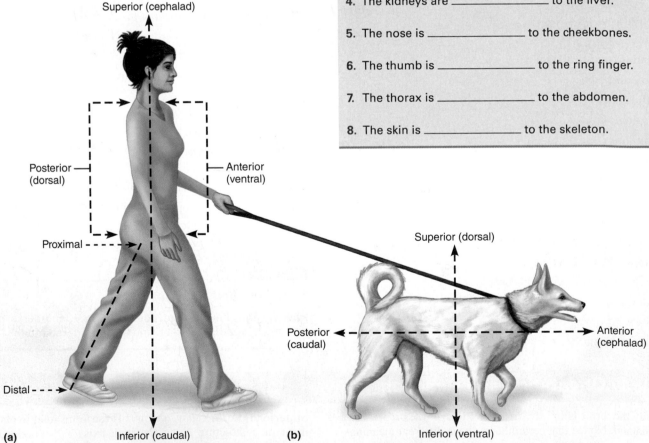

Figure 1.2 Anatomical terminology describing body orientation and direction.
(a) With reference to a human. **(b)** With reference to a four-legged animal.

Body Planes and Sections

The body is three-dimensional, and in order to observe its internal structures, it is often necessary to make a **section,** or cut. When the section is made through the body wall or through an organ, it is made along an imaginary surface or line called a **plane.** Anatomists commonly refer to three planes (**Figure 1.3**), or sections, that lie at right angles to one another.

Sagittal plane: A sagittal plane runs longitudinally and divides the body into right and left parts. If it divides the body into equal parts, right down the midline of the body, it is called a **median,** or **midsagittal, plane.**

Frontal plane: Sometimes called a **coronal plane,** the frontal plane is a longitudinal plane that divides the body (or an organ) into anterior and posterior parts.

Transverse plane: A transverse plane runs horizontally, dividing the body into superior and inferior parts. When organs are sectioned along the transverse plane, the sections are commonly called **cross sections.**

On microscope slides, the abbreviation for a longitudinal section (sagittal or frontal) is l.s. Cross sections are abbreviated x.s. or c.s.

A median or frontal plane section of any nonspherical object, be it a banana or a body organ, provides quite a different view from a cross section (**Figure 1.4**, p. 6).

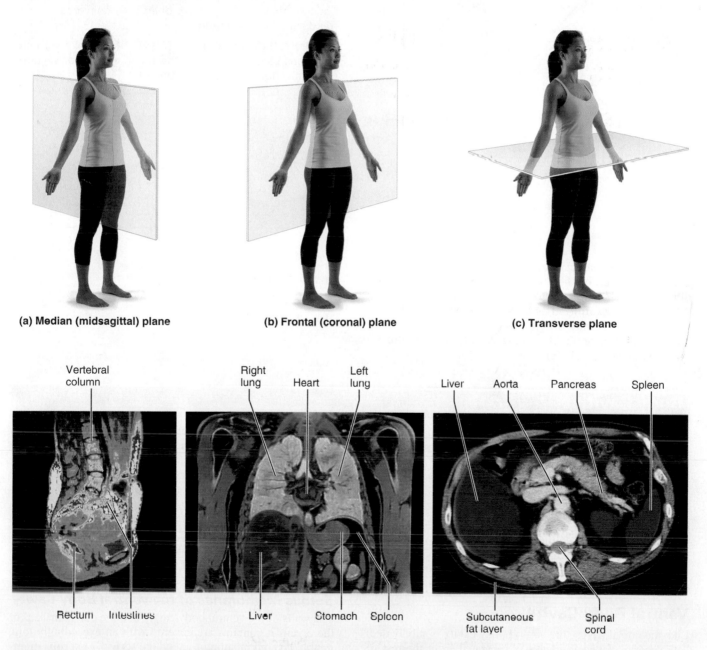

(a) Median (midsagittal) plane

(b) Frontal (coronal) plane

(c) Transverse plane

Vertebral column

Right lung   Heart   Left lung

Liver   Aorta   Pancreas   Spleen

Rectum   Intestines

Liver   Stomach   Spleen

Subcutaneous fat layer   Spinal cord

Figure 1.3 Planes of the body with corresponding magnetic resonance imaging (MRI) scans. Note the transverse section is an inferior view.

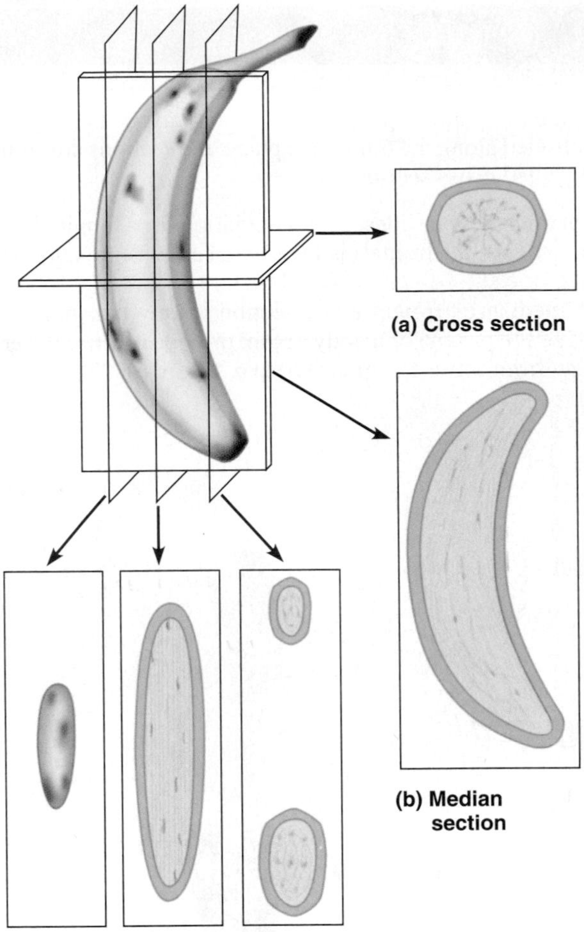

1

(a) Cross section

(b) Median section

(c) Frontal sections

Figure 1.4 Objects can look odd when viewed in section. This banana has been sectioned in three different planes **(a–c)**, and only in one of these planes **(b)** is it easily recognized as a banana. If one cannot recognize a sectioned organ, it is possible to reconstruct its shape from a series of successive cuts, as from the three serial sections in **(c)**.

Activity 3

Observing Sectioned Specimens

1. Go to the demonstration area and observe the transversely and longitudinally cut organ specimens (kidneys). Pay close attention to the different structural details in the samples; you will need to draw these views in the Review Sheet at the end of this exercise.

2. After completing instruction 1, obtain a gelatin-spaghetti mold and a scalpel, and take them to your laboratory bench. (Essentially, this is just cooked spaghetti added to warm gelatin, which is then allowed to gel.)

3. Cut through the gelatin-spaghetti mold along any plane, and examine the cut surfaces. You should see spaghetti strands that have been cut transversely (x.s.) and some cut longitudinally.

4. Draw the appearance of each of these spaghetti sections below, and verify the accuracy of your section identifications with your instructor.

Transverse cut	Longitudinal cut

Body Cavities

The axial portion of the body has two large cavities that provide different degrees of protection to the organs within them (**Figure 1.5**).

Dorsal Body Cavity

The dorsal body cavity can be subdivided into the **cranial cavity,** which lies within the rigid skull and encases the brain, and the **vertebral** (or **spinal**) **cavity,** which runs through the bony vertebral column to enclose the delicate spinal cord. Because the spinal cord is a continuation of the brain, these cavities are continuous with each other.

Ventral Body Cavity

Like the dorsal cavity, the ventral body cavity is subdivided. The superior **thoracic cavity** is separated from the rest of the ventral cavity by the dome-shaped diaphragm. The heart and lungs, located in the thoracic cavity, are protected by the bony rib cage. The cavity inferior to the diaphragm is often referred to as the **abdominopelvic cavity.** Although there is no further physical separation of the ventral cavity, some describe the abdominopelvic cavity as two areas: a superior **abdominal cavity,** the area that houses the stomach, intestines, liver, and other organs, and an inferior **pelvic cavity,** the region that is partially enclosed by the bony pelvis and contains the reproductive organs, bladder, and rectum. Notice in Figure 1.5a that the abdominal and pelvic cavities are not aligned with each other in a plane because the pelvic cavity is tipped forward.

Serous Membranes of the Ventral Body Cavity

The walls of the ventral body cavity and the outer surfaces of the organs it contains are covered with an exceedingly thin, double-layered membrane called the **serosa,** or **serous membrane.** The part of the membrane lining the cavity walls is referred to as the **parietal serosa,** and it is continuous with a

Cranial cavity
(contains brain)

Cranial
cavity

Vertebral
cavity

Superior
mediastinum

Pleural
cavity

Pericardial
cavity within
the mediastinum

Dorsal
body
cavity

Thoracic
cavity
(contains
heart and
lungs)

Vertebral cavity
(contains spinal
cord)

Diaphragm

Abdominal cavity
(contains digestive
organs)

Ventral body
cavity
(thoracic and
abdominopelvic
cavities)

Abdomino-
pelvic
cavity

Pelvic cavity
(contains urinary
bladder, reproductive
organs, and rectum)

Dorsal body cavity

Ventral body cavity

(a) Lateral view

(b) Anterior view

Figure 1.5 Dorsal and ventral body cavities and their subdivisions.

similar membrane, the **visceral serosa,** covering the external surface of the organs within the cavity. These membranes produce a thin lubricating fluid that allows the visceral organs to slide over one another or to rub against the body wall with minimal friction. Serous membranes also compartmentalize the various organs to prevent infection in one organ from spreading to others.

The specific names of the serous membranes depend on the structures they surround. The serosa lining the abdominal cavity and covering its organs is the **peritoneum,** the serosa enclosing the lungs is the **pleura,** and the serosa around the heart is the **pericardium (Figure 1.6).**

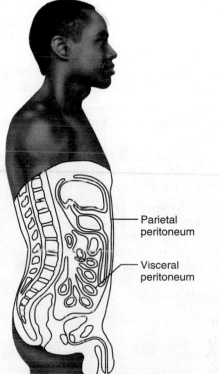

Parietal
peritoneum

Visceral
peritoneum

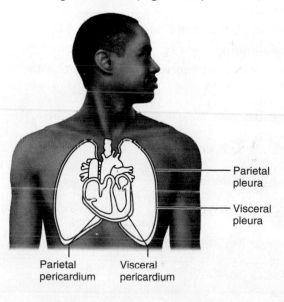

Parietal
pleura

Visceral
pleura

Parietal
pericardium

Visceral
pericardium

Figure 1.6 Serous membranes of the ventral body cavities.

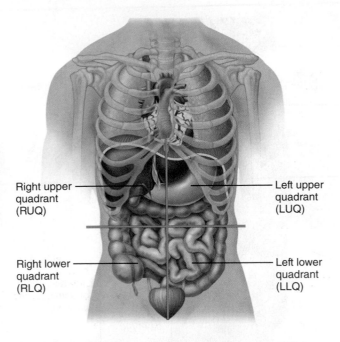

Right upper quadrant (RUQ)

Left upper quadrant (LUQ)

Right lower quadrant (RLQ)

Left lower quadrant (LLQ)

Figure 1.7 Abdominopelvic quadrants. Superficial organs all shown in each quadrant.

Abdominopelvic Quadrants and Regions

Because the abdominopelvic cavity is quite large and contains many organs, it is helpful to divide it up into smaller areas for discussion or study.

Most physicians and nurses use a scheme that divides the abdominal surface and the abdominopelvic cavity into four approximately equal regions called **quadrants**. These quadrants are named according to their relative position— that is, *right upper quadrant, right lower quadrant, left upper quadrant,* and *left lower quadrant* (**Figure 1.7**). Note that the terms *left* and *right* refer to the left and right side of

---

Activity 4

Identifying Organs in the Abdominopelvic Cavity

Examine the human torso model to respond to the following questions.

Name two organs found in the left upper quadrant.

_____ and _____

Name two organs found in the right lower quadrant.

_____ and _____

What organ (Figure 1.7) is divided into identical halves by

the median plane? _____

---

the body in the figure, not the left and right side of the art on the page.

A different scheme commonly used by anatomists divides the abdominal surface and abdominopelvic cavity into nine separate regions by four planes (**Figure 1.8**). As you read through the descriptions of these nine regions, locate them in Figure 1.8, and note the organs contained in each region.

Umbilical region: The centermost region, which includes the umbilicus (navel)

Epigastric region: Immediately superior to the umbilical region; overlies most of the stomach

Hypogastric (pubic) region: Immediately inferior to the umbilical region; encompasses the pubic area

Iliac, or inguinal, regions: Lateral to the hypogastric region and overlying the superior parts of the hip bones

Lumbar regions: Between the ribs and the flaring portions of the hip bones; lateral to the umbilical region

Hypochondriac regions: Flanking the epigastric region laterally and overlying the lower ribs

---

Activity 5

Locating Abdominal Surface Regions

Locate the regions of the abdominal surface on a human torso model and on yourself.

Other Body Cavities

Besides the large, closed body cavities, there are several types of smaller body cavities (**Figure 1.9**). Many of these are in the head, and most open to the body exterior.

Oral cavity: The oral cavity, commonly called the *mouth,* contains the tongue and teeth. It is continuous with the rest of the digestive tube, which opens to the exterior at the anus.

Nasal cavity: Located within and posterior to the nose, the nasal cavity is part of the passages of the respiratory system.

Orbital cavities: The orbital cavities (orbits) in the skull house the eyes and present them in an anterior position.

Middle ear cavities: Each middle ear cavity lies just medial to an eardrum and is carved into the bony skull. These cavities contain tiny bones that transmit sound vibrations to the hearing receptors in the inner ears.

Synovial cavities: Synovial cavities are joint cavities—they are enclosed within fibrous capsules that surround the freely movable joints of the body, such as those between the vertebrae and the knee and hip joints. Like the serous membranes of the ventral body cavity, membranes lining the synovial cavities secrete a lubricating fluid that reduces friction as the enclosed structures move across one another.

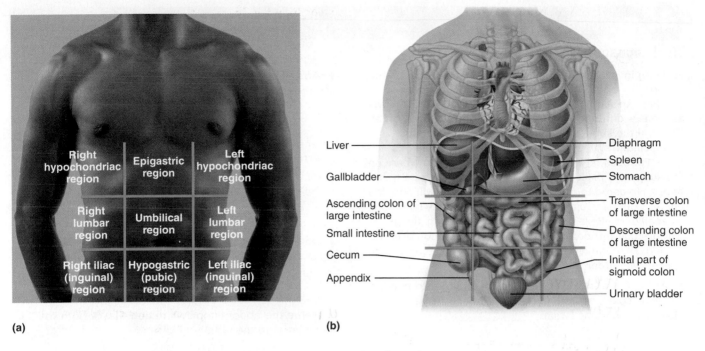

(a)

(b)

Figure 1.8 Abdominopelvic regions. Nine regions delineated by four planes.
(a) The superior horizontal plane is just inferior to the ribs; the inferior horizontal
plane is at the superior aspect of the hip bones. The vertical planes are just medial
to the nipples. (b) Superficial organs are shown in each region.

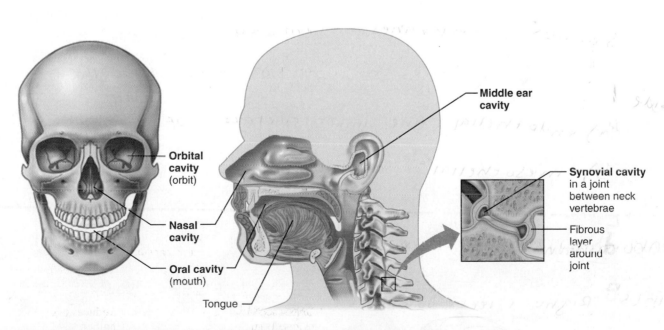

Figure 1.9 Other body cavities. The oral, nasal, orbital, and middle ear cavities
are located in the head and open to the body exterior. Synovial cavities are found
in joints between many bones, such as the vertebrae of the spine, and at the knee,
shoulder, and hip.

👥 Group Challenge

The Language of Anatomy

Working in groups of three, complete the tasks described below.

For questions 1–4, each student within a group will assume a different role: facilitator, subject, or recorder. (Remind the subject to stand in the anatomical position.) The facilitator will write each term on a separate Post-it® Note. For each term, discuss within your group where on the subject to place the Post-it®. Once your group members have come to consensus, the facilitator will stick the Post-it® on the subject on the appropriate body landmark. After all of the Post-it® Notes have been placed, the group will discuss the order in which the terms should be recorded. Then the recorder will write down the terms in the appropriate order.

1. Arrange the following terms from superior to inferior: cervical, coxal, crural, femoral, lumbar, mental, nasal, plantar, sternal, and tarsal. _____

_____

_____

_____

2. Arrange the following terms from proximal to distal: antebrachial, antecubital, brachial, carpal, digital, and palmar. _____

_____

3. Arrange the following terms from medial to lateral: acromial, axillary, buccal, otic, pollex, and umbilical.

_____

_____

4. Arrange the following terms from distal to proximal: calcaneal, femoral, hallux, plantar, popliteal, and sural.

_____

_____

5. Name a plane that you could use to section a four-legged chair and still be able to sit in the chair without falling over. _____

_____

6. Name the abdominopelvic region that is both medial and inferior to the right lumbar region.

_____

7. Name the type of inflammation (think "-itis") that is typically accompanied by pain in the lower right quadrant.

_____

Serous membranes = serosae

line body cav that do NOT open to the

outside

a) endothelial - line inner surface of blood

b) mesothelial -

abdominal cav. - peritoneal cav.

★ Pateints Right + left NOT docs right left

Organ Systems Overview

Objectives

☐ Name the human organ systems, and indicate the major functions of each.

☐ List several major organs of each system, and identify them in a dissected rat, human cadaver or cadaver image, or a dissectible human torso model.

☐ Name the correct organ system for each organ when presented with a list of organs studied in the laboratory.

Materials

- Freshly killed or preserved rat (predissected by instructor as a demonstration or for student dissection [one rat for every two to four students]) or predissected human cadaver
- Dissection trays
- Twine or large dissecting pins
- Scissors
- Probes
- Forceps
- Disposable gloves
- Human torso model (dissectible)

Pre-Lab Quiz

1. Name the structural and functional unit of all living things. _____
2. The small intestine is an example of a(n) _____, because it is composed of two or more tissue types that perform a particular function for the body.
   a. epithelial tissue   b. muscular tissue
   c. organ   d. organ system
3. The _____ system is responsible for maintaining homeostasis of the body via rapid transmission of electrical signals.
4. The kidneys are part of the _____ system.
5. The thin muscle that separates the thoracic and abdominal cavities is the _____.

The basic unit or building block of all living things is the **cell.** Cells fall into four different categories according to their structures and functions. These categories correspond to the four tissue types: epithelial, muscular, nervous, and connective. A **tissue** is a group of cells that are similar in structure and function. An **organ** is a structure composed of two or more tissue types that performs a specific function for the body. For example, the small intestine, which digests and absorbs nutrients, is made up of all four tissue types.

An **organ system** is a group of organs that act together to perform a particular body function. For example, the organs of the digestive system work together to break down foods and absorb the end products into the bloodstream in order to provide nutrients and fuel for all the body's cells. In all, there are 11 organ systems, described in **Table 2.1** on p. 16. The lymphatic system also encompasses a *functional system* called the immune system, which is composed of an army of mobile cells that protect the body from foreign substances.

Read through this summary of the body's organ systems (Table 2.1) before beginning your rat dissection or examination of the predissected human cadaver. If a human cadaver is not available, Figures 2.3–2.6 will serve as a partial replacement.

Table 2.1   Overview of Organ Systems of the Body

Organ system	Major component organs	Function
Integumentary (Skin)	Epidermal and dermal regions; cutaneous sense organs and glands	• Protects deeper organs from mechanical, chemical, and bacterial injury, and from drying out • Excretes salts and urea • Aids in regulation of body temperature • Produces vitamin D
Skeletal	Bones, cartilages, tendons, ligaments, and joints	• Body support and protection of internal organs • Provides levers for muscular action • Cavities provide a site for blood cell formation
Muscular	Muscles attached to the skeleton	• Primary function is to contract or shorten; in doing so, skeletal muscles allow locomotion (running, walking, etc.), grasping and manipulation of the environment, and facial expression • Generates heat
Nervous	Brain, spinal cord, nerves, and sensory receptors	• Allows body to detect changes in its internal and external environment and to respond to such information by activating appropriate muscles or glands • Helps maintain homeostasis of the body via rapid transmission of electrical signals
Endocrine	Pituitary, thymus, thyroid, parathyroid, adrenal, and pineal glands; ovaries, testes, and pancreas	• Helps maintain body homeostasis, promotes growth and development; produces chemical messengers called hormones that travel in the blood to exert their effect(s) on various target organs of the body
Cardiovascular	Heart, blood vessels, and blood	• Primarily a transport system that carries blood containing oxygen, carbon dioxide, nutrients, wastes, ions, hormones, and other substances to and from the tissue cells where exchanges are made; blood is propelled through the blood vessels by the pumping action of the heart • Antibodies and other protein molecules in the blood protect the body
Lymphatic/ Immunity	Lymphatic vessels, lymph nodes, spleen, thymus, tonsils, and scattered collections of lymphoid tissue	• Picks up fluid leaked from the blood vessels and returns it to the blood • Cleanses blood of pathogens and other debris • Houses lymphocytes that act via the immune response to protect the body from foreign substances
Respiratory	Nasal passages, pharynx, larynx, trachea, bronchi, and lungs	• Keeps the blood continuously supplied with oxygen while removing carbon dioxide • Contributes to the acid-base balance of the blood via its carbonic acid–bicarbonate buffer system
Digestive	Oral cavity, esophagus, stomach, small and large intestines, and accessory structures including teeth, salivary glands, liver, and pancreas	• Breaks down ingested foods to smaller particles, which can be absorbed into the blood for delivery to the body cells • Undigested residue removed from the body as feces
Urinary	Kidneys, ureters, bladder, and urethra	• Rids the body of nitrogen-containing wastes including urea, uric acid, and ammonia, which result from the breakdown of proteins and nucleic acids • Maintains water, electrolyte, and acid-base balance of blood
Reproductive	Male: testes, prostate gland, scrotum, penis, and duct system, which carries sperm to the body exterior	• Provides germ cells called sperm for perpetuation of the species
	Female: ovaries, uterine tubes, uterus, mammary glands, and vagina	• Provides germ cells called eggs; the female uterus houses the developing fetus until birth; mammary glands provide nutrition for the infant

 DISSECTION AND IDENTIFICATION

The Organ Systems of the Rat

Many of the external and internal structures of the rat are quite similar in structure and function to those of the human. So, a study of the gross anatomy of the rat should help you understand our own physical structure. The following instructions include directions for dissecting and observing a rat. In addition, the descriptions of the organs (Activity 4, Examining the Ventral Body Cavity, which begins on p. 18) also apply to superficial observations of a previously dissected human cadaver. The general instructions for observing external structures also apply to human cadaver observations. The photographs in Figures 2.3 to 2.6 will provide visual aids.

Note that four organ systems (integumentary, skeletal, muscular, and nervous) will not be studied at this time, because they require microscopic study or more detailed dissection.

Activity 1

Observing External Structures

1. If your instructor has provided a predissected rat, go to the demonstration area to make your observations. Alternatively, if you and/or members of your group will be dissecting the specimen, obtain a preserved or freshly killed rat, a dissecting tray, dissecting pins or twine, scissors, probe, forceps, and disposable gloves, and bring them to your laboratory bench.

If a predissected human cadaver is available, obtain a probe, forceps, and disposable gloves before going to the demonstration area.

⚠ **2.** Don the gloves before beginning your observations. This precaution is particularly important when handling freshly killed animals, which may harbor pathogens.

3. Observe the major divisions of the body—head, trunk, and extremities. If you are examining a rat, compare these divisions to those of humans.

Activity 2

Examining the Oral Cavity

Examine the structures of the oral cavity. Identify the teeth and tongue. Observe the extent of the hard palate (the portion underlain by bone) and the soft palate (immediately posterior to the hard palate, with no bony support). Notice that the posterior end of the oral cavity leads into the throat, or pharynx, a passageway used by both the digestive and respiratory systems.

Activity 3

Opening the Ventral Body Cavity

1. Pin the animal to the wax of the dissecting tray by placing its dorsal side down and securing its extremities to the wax with large dissecting pins as shown in **Figure 2.1a**.

Text continues on next page. →

Figure 2.1 Rat dissection: Securing for dissection and the initial incision. **(a)** Securing the rat to the dissection tray with dissecting pins. **(b)** Using scissors to make the incision on the median line of the abdominal region. **(c)** Completed incision from the pelvic region to the lower jaw. **(d)** Reflection (folding back) of the skin to expose the underlying muscles.

(a)

(b)

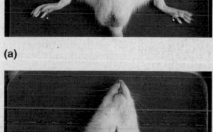

(c)

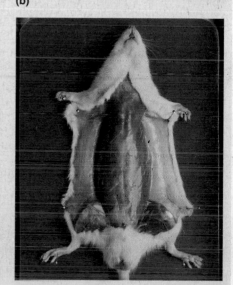

(d)

2

If the dissecting tray is not waxed, you will need to secure the animal with twine as follows. (Your instructor may prefer this method in any case.) Obtain the roll of twine. Make a loop knot around one upper limb, pass the twine under the tray, and secure the opposing limb. Repeat for the lower extremities.

2. Lift the abdominal skin with a forceps, and cut through it with the scissors (Figure 2.1b). Close the scissor blades, and insert them flat under the cut skin. Moving in a cephalad direction, open and close the blades to loosen the skin from the underlying connective tissue and muscle. Now, cut the skin along the body midline, from the pubic region to the lower jaw (Figure 2.1c). Finally, make a lateral cut about halfway down the ventral surface of each limb. Complete the job of freeing the skin with the scissor tips, and pin the flaps to the tray (Figure 2.1d). The underlying tissue that is now exposed is the skeletal musculature of the body wall and limbs. Notice that the muscles are packaged in sheets of pearly white connective tissue (fascia), which protect the muscles and bind them together.

3. Carefully cut through the muscles of the abdominal wall in the pubic region, avoiding the underlying organs. Remember, to *dissect* means "to separate"—not mutilate! Now, hold and lift the muscle layer with a forceps and cut through the muscle layer from the pubic region to the bottom of the rib cage. Make two lateral cuts at the base of the rib cage (**Figure 2.2**). A thin membrane attached to the inferior boundary of the rib cage should be obvious; this is the **diaphragm,** which separates the thoracic and abdominal cavities. Cut the diaphragm where it attaches to the ventral ribs to loosen the rib cage. Cut through the rib cage on either side. You can now lift the ribs to view the

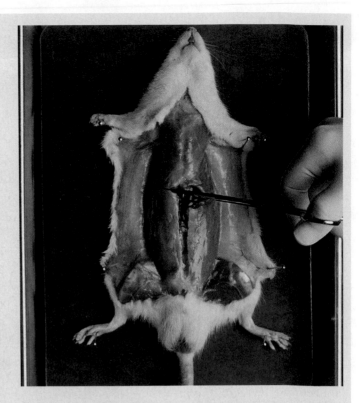

Figure 2.2 Rat dissection. Making lateral cuts at the base of the rib cage.

contents of the thoracic cavity. Cut across the flap, at the level of the neck, and remove the rib cage.

Activity 4

Examining the Ventral Body Cavity

1. Starting with the most superficial structures and working deeper, examine the structures of the thoracic cavity. Refer to **Figure 2.3**, which shows the superficial organs, as you work. Choose the appropriate view depending on whether you are examining a rat (a) or a human cadaver (b).

Thymus: An irregular mass of glandular tissue overlying the heart (not illustrated in the human cadaver photograph).

With the probe, push the thymus to the side to view the heart.

Heart: Medial oval structure enclosed within the pericardium (serous membrane sac).

Lungs: Lateral to the heart on either side.

Now observe the throat region to identify the trachea.

Trachea: Tubelike "windpipe" running medially down the throat; part of the respiratory system.

Follow the trachea into the thoracic cavity; notice where it divides into two branches. These are the bronchi.

Bronchi: Two passageways that plunge laterally into the tissue of the two lungs.

To expose the esophagus, push the trachea to one side.

Esophagus: A food chute; the part of the digestive system that transports food from the pharynx (throat) to the stomach.

Diaphragm: A thin muscle attached to the inferior boundary of the rib cage; separates the thoracic and abdominopelvic cavities.

Follow the esophagus through the diaphragm to its junction with the stomach.

Stomach: A curved organ important in food digestion and temporary food storage.

2. Examine the superficial structures of the abdominopelvic cavity. Lift the **greater omentum,** an extension of the peritoneum (serous membrane) that covers the abdominal viscera. Continuing from the stomach, trace the rest of the digestive tract (**Figure 2.4,** p. 20).

Small intestine: Connected to the stomach and ending just before the saclike cecum.

Large intestine: A large muscular tube connected to the small intestine and ending at the anus.

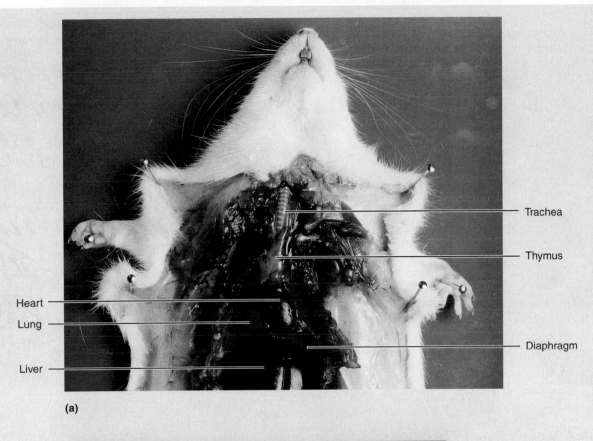

(a)

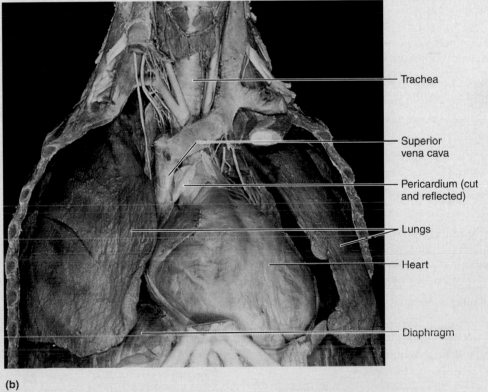

(b)

Figure 2.3 Superficial organs of the thoracic cavity. (a) Dissected rat. **(b)** Human cadaver.

Text continues on next page. →

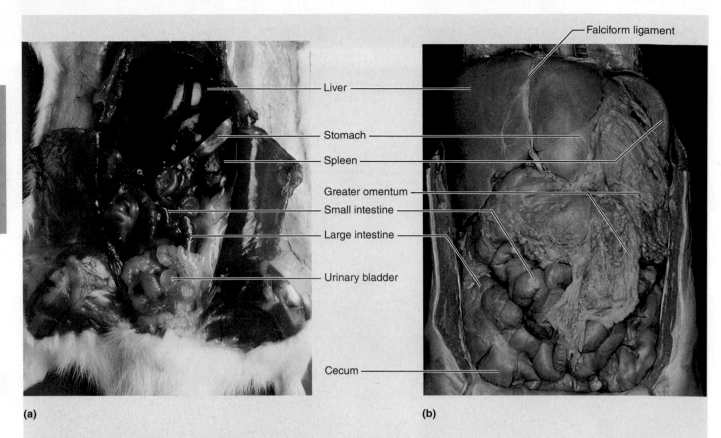

(a)   **(b)**

Figure 2.4 Abdominal organs. (a) Dissected rat, superficial view. **(b)** Human cadaver, superficial view.

Cecum: The initial portion of the large intestine.

Follow the course of the large intestine to the rectum, which is partially covered by the urinary bladder (**Figure 2.5**).

Rectum: Terminal part of the large intestine; continuous with the anal canal.

Anus: The opening of the digestive tract (through the anal canal) to the exterior.

Now lift the small intestine with the forceps to view the mesentery.

Mesentery: An apronlike serous membrane; suspends many of the digestive organs in the abdominal cavity. Notice that it is heavily invested with blood vessels and, more likely than not, riddled with large fat deposits.

Locate the remaining abdominal structures.

Pancreas: A diffuse gland; rests dorsal to and in the mesentery between the first portion of the small intestine and the stomach. You will need to lift the stomach to view the pancreas.

Spleen: A dark red organ curving around the left lateral side of the stomach; considered part of the lymphatic system and often called the red blood cell graveyard.

Liver: Large and brownish red; the most superior organ in the abdominal cavity, directly beneath the diaphragm.

3. To locate the deeper structures of the abdominopelvic cavity, move the stomach and the intestines to one side with the probe.

Examine the posterior wall of the abdominal cavity to locate the two kidneys (Figure 2.5).

Kidneys: Bean-shaped organs; retroperitoneal (behind the peritoneum).

Adrenal glands: Large endocrine glands that sit on top of each kidney; considered part of the endocrine system.

Carefully strip away part of the peritoneum with forceps and attempt to follow the course of one of the ureters to the bladder.

Ureter: Tube running from the indented region of a kidney to the urinary bladder.

Urinary bladder: The sac that serves as a reservoir for urine.

4. In the midline of the body cavity lying between the kidneys are the two principal abdominal blood vessels. Identify each.

Inferior vena cava: The large vein that returns blood to the heart from the lower body regions.

Descending aorta: Deep to the inferior vena cava; the largest artery of the body; carries blood away from the heart down the midline of the body.

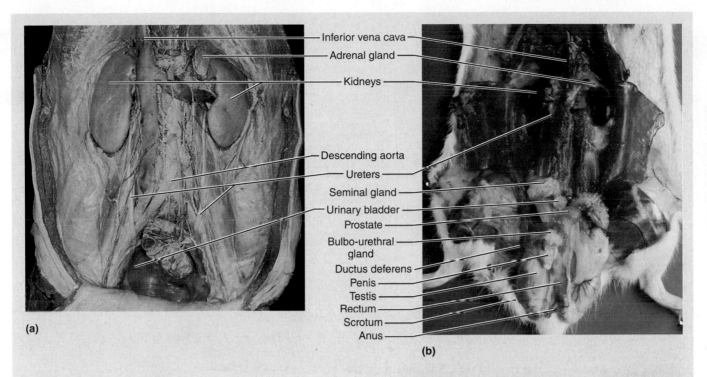

(a)

Inferior vena cava
Adrenal gland
Kidneys
Descending aorta
Ureters
Seminal gland
Urinary bladder
Prostate
Bulbo-urethral gland
Ductus deferens
Penis
Testis
Rectum
Scrotum
Anus

(b)

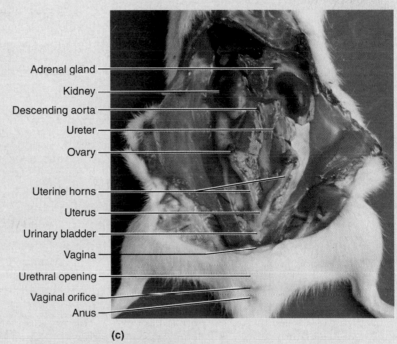

Adrenal gland
Kidney
Descending aorta
Ureter
Ovary
Uterine horns
Uterus
Urinary bladder
Vagina
Urethral opening
Vaginal orifice
Anus

(c)

Figure 2.5 Deep structures of the abdominopelvic cavity. (a) Human cadaver. **(b)** Dissected male rat. (Some reproductive structures also shown.) **(c)** Dissected female rat. (Some reproductive structures also shown.)

5. You will perform only a brief examination of reproductive organs. If you are working with a rat, first determine if the animal is a male or female. Observe the ventral body surface beneath the tail. If a saclike scrotum and an opening for the anus are visible, the animal is a male. If three body openings—urethral, vaginal, and anal—are present, it is a female.

Male Rat

Make a shallow incision into the **scrotum**. Loosen and lift out one oval **testis**. Exert a gentle pull on the testis to identify the slender **ductus deferens**, or **vas deferens**,

which carries sperm from the testis superiorly into the abdominal cavity and joins with the urethra. The urethra runs through the penis and carries both urine and sperm out of the body. Identify the **penis**, extending from the bladder to the ventral body wall. Figure 2.5b indicates other glands of the male rat's reproductive system, but they need not be identified at this time.

Female Rat

Inspect the pelvic cavity to identify the Y-shaped **uterus** lying against the dorsal body wall and superior to the

Text continues on next page. ➡

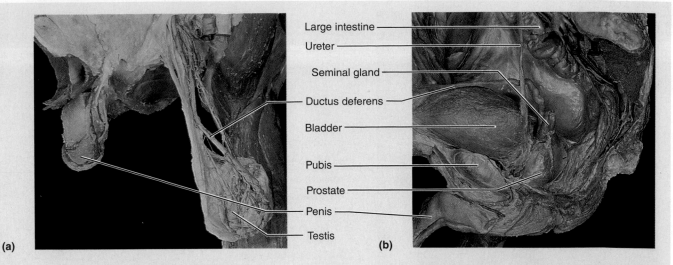

(a)

Large intestine
Ureter
Seminal gland
Ductus deferens
Bladder
Pubis
Prostate
Penis
Testis

(b)

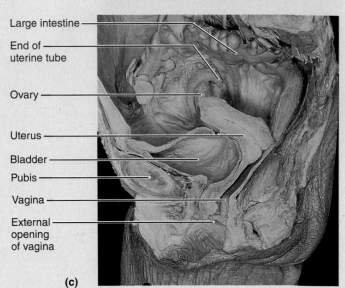

Large intestine
End of uterine tube
Ovary
Uterus
Bladder
Pubis
Vagina
External opening of vagina

(c)

Figure 2.6 Human reproductive organs. (a) Male external genitalia. **(b)** Sagittal section of the male pelvis. **(c)** Sagittal section of the female pelvis.

bladder (Figure 2.5c). Follow one of the uterine horns superiorly to identify an **ovary,** a small oval structure at the end of the uterine horn. (The rat uterus is quite different from the uterus of a human female, which is a single-chambered organ about the size and shape of a pear.) The inferior undivided part of the rat uterus is continuous with the **vagina,** which leads to the body exterior. Identify the **vaginal orifice** (external vaginal opening).

If you are working with a human cadaver, proceed as indicated next.

Male Cadaver

Make a shallow incision into the **scrotum (Figure 2.6a).** Loosen and lift out the oval **testis.** Exert a gentle pull on the testis to identify the slender **ductus (vas) deferens,** which carries sperm from the testis superiorly into the abdominopelvic cavity and joins with the urethra (Figure 2.6b). The urethra runs through the penis and carries both urine and

sperm out of the body. Identify the **penis,** extending from the bladder to the ventral body wall.

Female Cadaver

Inspect the pelvic cavity to identify the pear-shaped **uterus** lying against the dorsal body wall and superior to the bladder. Follow one of the **uterine tubes** superiorly to identify an **ovary,** a small oval structure at the end of the uterine tube (Figure 2.6c). The inferior part of the uterus is continuous with the **vagina,** which leads to the body exterior. Identify the **vaginal orifice** (external vaginal opening).

6. When you have finished your observations, rewrap or store the dissection animal or cadaver according to your instructor's directions. Wash the dissecting tools and equipment with laboratory detergent. Dispose of the gloves as instructed. Then wash and dry your hands before continuing with the examination of the human torso model.

Activity 5

Examining the Human Torso Model

1. Examine a human torso model to identify the organs listed. Some model organs will have to be removed to see the deeper organs. If a torso model is not available, the photograph of the human torso model (**Figure 2.7**) may be used for this part of the exercise.

2. Using the terms at the right of Figure 2.7, label each organ on the supplied leader line.

3. List each organ in the correct body cavity or cavities.

Dorsal body cavity _____

Thoracic cavity _____

_____

_____

Abdominopelvic cavity _____

_____

_____

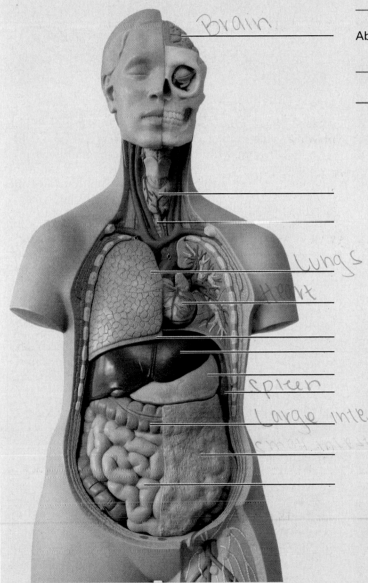

Adrenal gland	~~Lungs~~
Aortic arch	Mesentery
~~Brain~~	Pancreas
Bronchi	Rectum
Descending aorta	Small intestine
Diaphragm	Spinal cord
Esophagus	~~Spleen~~
Greater omentum	Stomach
~~Heart~~	Thyroid gland
Inferior vena cava	Trachea
Kidneys	Ureters
Large intestine	Urinary bladder
Liver	

(handwritten labels on figure: Brain, Lungs, Heart, Spleen, Large intestine, Small intestine)

Figure 2.7 Human torso model.

Text continues on next page. →

2

4. Now, assign each of the organs to one of the organ systems listed below.

Digestive: _____

_____

Urinary: _____

Cardiovascular: _____

_____

Endocrine: _____

Reproductive: _____

Respiratory: _____

Lymphatic/Immunity: _____

Nervous: _____

Group Challenge

Odd Organ Out

Each of the following sets contains four organs. One of the listed organs in each case does not share a characteristic that the other three do. Work in groups of three, and discuss the characteristics of the four organs in each set. On a separate piece of paper, one student will record the characteristics of each organ in the set. For each set of four organs, discuss the possible candidates for the "odd organ" and which characteristic it lacks, based on your recorded notes. Once you have come to a consensus among your group, circle the organ that doesn't belong with the others, and explain why it is singled out. Include as many reasons as you can think of, but make sure the "odd organ" does not have the key characteristic. Use the overview of organ systems (Table 2.1) and the pictures in your lab manual to help you select and justify your answer.

1. Which is the "odd organ"?	Why is it the odd one out?
Stomach Teeth Small intestine Oral cavity	
2. Which is the "odd organ"?	**Why is it the odd one out?**
Thyroid gland Thymus Spleen Lymph nodes	
3. Which is the "odd organ"?	**Why is it the odd one out?**
Ovaries Prostate gland Uterus Uterine tubes	
4. Which is the "odd organ"?	**Why is it the odd one out?**
Stomach Small intestine Esophagus Large intestine	

The Microscope

Objectives

☐ Identify the parts of the microscope, and list the function of each.

☐ Describe and demonstrate the proper techniques for care of the microscope.

☐ Demonstrate proper focusing technique.

☐ Define *total magnification, resolution, parfocal, field, depth of field,* and *working distance.*

☐ Measure the field diameter for one objective lens, calculate it for all the other objective lenses, and estimate the size of objects in each field.

☐ Discuss the general relationships between magnification, working distance, and field diameter.

Materials*

- Compound microscope
- Millimeter ruler
- Prepared slides of the letter *e* or newsprint
- Immersion oil
- Lens paper
- Prepared slide of grid ruled in millimeters
- Prepared slide of three crossed colored threads
- Clean microscope slide and coverslip
- Toothpicks (flat-tipped)
- Physiological saline in a dropper bottle
- Iodine or dilute methylene blue stain in a dropper bottle
- Filter paper or paper towels
- Beaker containing fresh 10% household bleach solution for wet mount disposal
- Disposable autoclave bag
- Prepared slide of cheek epithelial cells

Pre-Lab Quiz

1. The microscope slide rests on the _____ while being viewed.
   - **a.** base
   - **b.** condenser
   - **c.** iris
   - **d.** stage
2. Your lab microscope is *parfocal.* What does this mean?
   - **a.** The specimen is clearly in focus at this depth.
   - **b.** The slide should be almost in focus when changing to higher magnifications.
   - **c.** You can easily discriminate two close objects as separate.
3. If the ocular lens magnifies a specimen 10×, and the objective lens used magnifies the specimen 35×, what is the total magnification being used to observe the specimen? _____
4. How do you clean the lenses of your microscope?
   - **a.** with a paper towel
   - **b.** with soap and water
   - **c.** with special lens paper and cleaner
5. Circle True or False. You should always begin observation of specimens with the oil immersion lens.

MasteringA&P®

For related exercise study tools, go to the Study Area of **MasteringA&P.** There you will find:

- Practice Anatomy Lab PAL
- PhysioEx PEx
- A&PFlix *A&PFlix*
- Practice quizzes, Histology Atlas, eText, Videos, and more!

W ith the invention of the microscope, biologists gained a valuable tool to observe and study structures, such as cells, that are too small to be seen by the unaided eye. The knowledge they acquired helped establish many of the theories basic to the biological sciences. This exercise will familiarize you with the workhorse of microscopes—the compound microscope—and provide you with the necessary instructions for its proper use.

* Note to the Instructor: The slides and coverslips used for viewing cheek cells are to be soaked for 2 hours (or longer) in 10% bleach solution and then drained. The slides and disposable autoclave bag containing coverslips, lens paper, and used toothpicks are to be autoclaved for 15 min at 121°C and 15 pounds pressure to ensure sterility. After autoclaving, the disposable autoclave bag may be discarded in any disposal facility, and the slides and glassware washed with laboratory detergent and prepared for use. These instructions apply as well to any bloodstained glassware or disposable items used in other experimental procedures.

Care and Structure of the Compound Microscope

The **compound microscope** is a precision instrument and should always be handled with care. *At all times you must observe the following rules for its transport, cleaning, use, and storage:*

• When transporting the microscope, hold it in an upright position, with one hand on its arm and the other supporting its base. Do not swing the instrument during its transport or jar the instrument when setting it down.

• Use only special grit-free lens paper to clean the lenses. Use a circular motion to wipe the lenses, and clean all lenses before and after use.

• Always begin the focusing process with the lowest-power objective lens in position, changing to the higher-power lenses as necessary.

• Use the coarse adjustment knob only with the lowest-power objective lens.

• Always use a coverslip with wet mount preparations.

• Before putting the microscope in the storage cabinet, remove the slide from the stage, rotate the lowest-power objective lens into position, wrap the cord neatly around the base, and replace the dust cover or return the microscope to the appropriate storage area.

• Never remove any parts from the microscope; inform your instructor of any mechanical problems that arise.

Activity 1

Identifying the Parts of a Microscope

1. Using the proper transport technique, obtain a microscope and bring it to the laboratory bench.

☐ Record the number of your microscope in the **Summary chart** (p. 31).

Compare your microscope with **Figure 3.1**, and identify the microscope parts described in **Table 3.1** on p. 30.

Figure 3.1 Compound microscope and its parts.

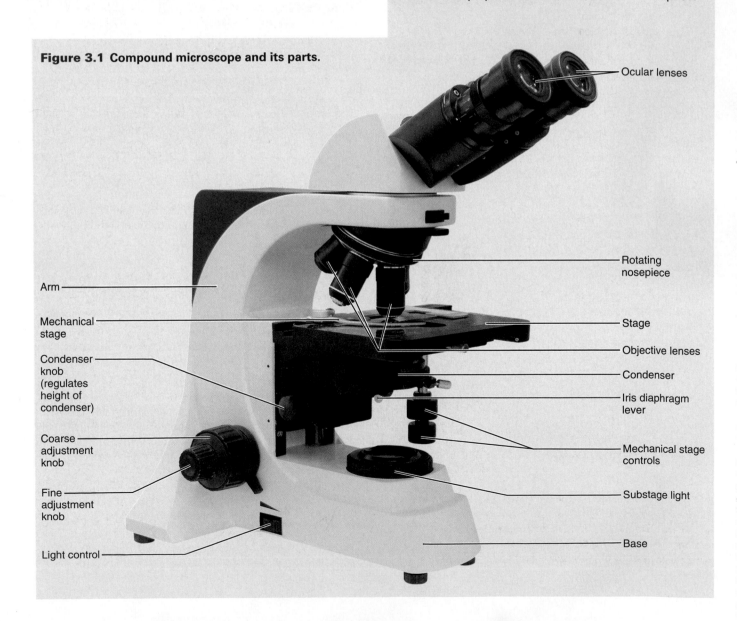

Ocular lenses

Arm

Mechanical stage

Condenser knob (regulates height of condenser)

Coarse adjustment knob

Fine adjustment knob

Light control

Rotating nosepiece

Stage

Objective lenses

Condenser

Iris diaphragm lever

Mechanical stage controls

Substage light

Base

2. Examine the objective lenses carefully; note their relative lengths and the numbers inscribed on their sides. On many microscopes, the scanning lens, with a magnification between 4× and 5×, is the shortest lens. The low-power objective lens typically has a magnification of 10×. The high-power objective lens is of intermediate length and has a magnification range from 40× to 50×, depending on the microscope. The oil immersion objective lens is usually the longest of the objective lenses and has a magnifying power of 95× to 100×. Some microscopes lack the oil immersion lens.

□ Record the magnification of each objective lens of your microscope in the first row of the Summary chart (p. 31). Also, cross out any column relating to a lens that your microscope does not have. Plan on using the same microscope for all microscopic studies.

3. Rotate the lowest-power objective lens until it clicks into position, and turn the coarse adjustment knob about 180 degrees. Notice how far the stage (or objective lens) travels during this adjustment. Move the fine adjustment knob 180 degrees, noting again the distance that the stage (or the objective lens) moves.

Magnification and Resolution

The microscope is an instrument of magnification. With the compound microscope, magnification is achieved through the interplay of two lenses—the ocular lens and the objective lens. The objective lens magnifies the specimen to produce a **real image** that is projected to the ocular. This real image is magnified by the ocular lens to produce the **virtual image** that your eye sees (**Figure 3.2**).

The **total magnification** (TM) of any specimen being viewed is equal to the power of the ocular lens multiplied by the power of the objective lens used. For example, if the ocular lens magnifies 10× and the objective lens being used magnifies 45×, the total magnification is 450× (or 10 × 45).

- Determine the total magnification you may achieve with each of the objectives on your microscope, and record the figures on the third row of the Summary chart.

The compound light microscope has certain limitations. Although the level of magnification is almost limitless, the **resolution** (or resolving power), that is, the ability to discriminate two close objects as separate, is not. The human eye can resolve objects about 100 μm apart, but the compound microscope has a resolution of 0.2 μm under ideal conditions. Objects closer than 0.2 μm are seen as a single fused image.

Resolving power is determined by the amount and physical properties of the visible light that enters the microscope. In general, the more light delivered to the objective lens, the greater the resolution. The size of the objective lens aperture (opening) decreases with increasing magnification, allowing less light to enter the objective. Thus, you will probably find it necessary to increase the light intensity at the higher magnifications.

Prepare for lab: Watch the Pre-Lab Video
Mastering A&P°>Study Area>Pre-Lab Videos

Activity 2

Viewing Objects Through the Microscope

1. Obtain a millimeter ruler, a prepared slide of the letter *e* or newsprint, a dropper bottle of immersion oil, and some lens paper. Adjust the condenser to its highest position, and switch on the light source of your microscope.

2. Secure the slide on the stage so that you can read the slide label and the letter *e* is centered over the light beam passing through the stage. On the mechanical stage of your microscope, open the jaws of its slide holder by using the control lever, typically located at the rear left corner of the mechanical stage. Insert the slide squarely within the confines of the slide holder.

3. With your lowest-power (scanning or low-power) objective lens in position over the stage, use the coarse adjustment knob to bring the objective lens and stage as close together as possible.

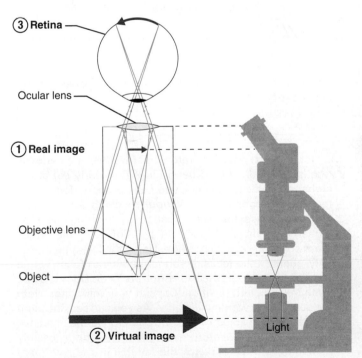

Figure 3.2 Image formation in light microscopy.
Step ① The objective lens magnifies the object, forming the real image. **Step** ② The ocular lens magnifies the real image, forming the virtual image. **Step** ③ The virtual image passes through the lens of the eye and is focused on the retina.

Text continues on next page. →

Table 3.1 Parts of the Microscope

Microscope part	Description and function
Base	The bottom of the microscope. Provides a sturdy flat surface to support and steady the microscope.
Substage light	Located in the base. The light from the lamp passes directly upward through the microscope.
Light control knob	Located on the base or arm. This dial allows you to adjust the intensity of the light passing through the specimen.
Stage	The platform that the slide rests on while being viewed. The stage has a hole in it to allow light to pass through the stage and through the specimen.
Mechanical stage	Holds the slide in position for viewing and has two adjustable knobs that control the precise movement of the slide.
Condenser	Small nonmagnifying lens located beneath the stage that concentrates the light on the specimen. The condenser may have a knob that raises and lowers the condenser to vary the light delivery. Generally, the best position is close to the inferior surface of the stage.
Iris diaphragm lever	The iris diaphragm is a shutter within the condenser that can be controlled by a lever to adjust the amount of light passing through the condenser. The lever can be moved to close the diaphragm and improve contrast. If your field of view is too dark, you can open the diaphragm to let in more light.
Coarse adjustment knob	This knob allows you to make large adjustments to the height of the stage to initially focus your specimen.
Fine adjustment knob	This knob is used for precise focusing once the initial coarse focusing has been completed.
Head	Attaches to the nosepiece to support the objective lens system. It also provides for attachment of the eyepieces which house the ocular lenses.
Arm	Vertical portion of the microscope that connects the base and the head.
Nosepiece	Rotating mechanism connected to the head. Generally, it carries three or four objective lenses and permits positioning of these lenses over the hole in the stage.
Objective lenses	These lenses are attached to the nosepiece. Usually, a compound microscope has four objective lenses: scanning (4×), low-power (10×), high-power (40×), and oil immersion (100×) lenses. Typical magnifying powers for the objectives are listed in parentheses.
Ocular lens(es)	Binocular microscopes will have two lenses located in the eyepieces at the superior end of the head. Most ocular lenses have a magnification power of 10×. Some microscopes will have a pointer and/or reticle (micrometer), which can be positioned by rotating the ocular lens.

4. Look through the ocular lens and adjust the light for comfort using the iris diaphragm lever. Now use the coarse adjustment knob to focus slowly away from the *e* until it is as clearly focused as possible. Complete the focusing with the fine adjustment knob.

5. Sketch the letter *e* in the circle on the Summary chart (p. 31) just as it appears in the **field**—the area you see through the microscope.

How far is the bottom of the objective lens from the surface of the slide? In other words, what is the **working distance**? (See Figure 3.3.) Use a millimeter ruler to make this measurement.

Record the working distance in the Summary chart.

How has the apparent orientation of the *e* changed top to bottom, right to left, and so on?

_____

_____

6. Move the slide slowly away from you on the stage as you view it through the ocular lens. In what direction does the image move?

_____

Move the slide to the left. In what direction does the image move?

_____

7. Today, most good laboratory microscopes are **parfocal**; that is, the slide should be in focus (or nearly so) at the higher magnifications once you have properly focused at the lower magnification. *Without touching the focusing knobs,* increase the magnification by rotating the next higher magnification lens into position over the stage. Make sure it clicks into position. Using the fine adjustment only, sharpen the focus. If you are unable to focus with a new lens, your microscope is not parfocal. Do not try to force the lens into position. Consult your instructor. Note the decrease in working distance. As you can see, focusing with the coarse adjustment knob could drive the objective lens through the slide, breaking the slide and possibly damaging the lens. Sketch the letter *e* in the Summary chart. What new details become clear?

_____

_____

As best you can, measure the distance between the objective and the slide.

Record the working distance in the Summary chart.

Is the image larger or smaller? _____

Approximately how much of the letter *e* is visible now?

_____

Is the field larger or smaller? _____

Why is it necessary to center your object (or the portion of the slide you wish to view) before changing to a higher power?

_____

Move the iris diaphragm lever while observing the field. What happens?

_____

Is it better to increase *or* to decrease the light when changing to a higher magnification?

_____ Why? _____

_____

8. If you have just been using the low-power objective, repeat the steps given in direction 7 using the high-power objective lens. What new details become clear?

_____

_____

Record the working distance in the Summary chart.

9. Without touching the focusing knob, rotate the high-power lens out of position so that the area of the slide over the opening in the stage is unobstructed. Place a drop of immersion oil over the *e* on the slide and rotate the oil immersion lens into position. Set the condenser at its highest point (closest to the stage), and open the diaphragm fully. Adjust the fine focus and fine-tune the light for the best possible resolution.

Note: If for some reason the specimen does not come into view after adjusting the fine focus, do not go back to the 40× lens to recenter. You do not want oil from the oil immersion lens to cloud the 40× lens. Turn the revolving nosepiece in the other direction to the low-power lens, and recenter and refocus the object. Then move the immersion lens back into position, again avoiding the 40× lens. Sketch the letter e in the Summary chart. What new details become clear?

_____

_____

Is the field again decreased in size? _____

As best you can, estimate the working distance, and record it in the Summary chart. Is the working distance less *or* greater than it was when the high-power lens was focused?

_____

Compare your observations on the relative working distances of the objective lenses with the illustration in

Summary Chart for Microscope #____

	Scanning	Low power	High power	Oil Immersion
Magnification of objective lens	_____ ×	_____ ×	_____ ×	_____ ×
Magnification of ocular lens	10 ×	10 ×	10 ×	10 ×
Total magnification	_____ ×	_____ ×	_____ ×	_____ ×
Working distance	_____ mm	_____ mm	_____ mm	_____ mm
Detail observed letter *e*				
Field diameter	____ mm ____ µm	____ mm ____ µm	____ mm ____ µm	____ mm ____ µm

Text continues on next page. ➡

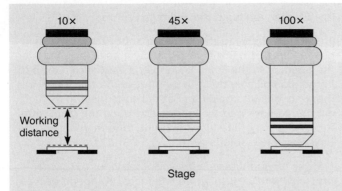

Figure 3.3 Relative working distances of the 10×, 45×, and 100× objectives.

Figure 3.3. Explain why it is desirable to begin the focusing process at the lowest power.

_____

_____

_____

10. Rotate the oil immersion lens slightly to the side, and remove the slide. Clean the oil immersion lens carefully with lens paper, and then clean the slide in the same manner with a fresh piece of lens paper.

The Microscope Field

The microscope field decreases with increasing magnification. Measuring the diameter of each of the microscope fields will allow you to estimate the size of the objects you view in any field. For example, if you have calculated the field diameter to be 4 mm and the object being observed extends across half this diameter, you can estimate that the length of the object is approximately 2 mm.

Microscopic specimens are usually measured in micrometers and millimeters, both units of the metric system. You can get an idea of the relationship and meaning of these units from **Table 3.2**. A more detailed treatment appears in the appendix.

Table 3.2	Comparison of Metric Units of Length	
Metric unit	**Abbreviation**	**Equivalent**
Meter	m	(about 39.37 in.)
Centimeter	cm	10^{-2} m
Millimeter	mm	10^{-3} m
Micrometer (or micron)	μm (μ)	10^{-6} m
Nanometer	nm (mμ)	10^{-9} m

(Refer to the Getting Started exercise on MasteringA&P for tips on metric conversions.)

Activity 3

Estimating the Diameter of the Microscope Field

1. Obtain a grid slide, which is a slide prepared with graph paper ruled in millimeters. Each of the squares in the grid is 1 mm on each side. Use your lowest-power objective to bring the grid lines into focus.

2. Move the slide so that one grid line touches the edge of the field on one side, and then count the number of squares you can see across the diameter of the field. If you can see only part of a square, as in the accompanying diagram, estimate the part of a millimeter that the partial square represents.

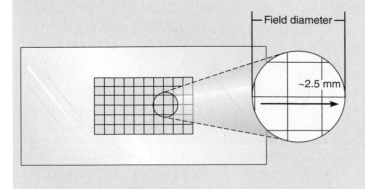

Record this figure in the appropriate space marked "field diameter" on the Summary chart (p. 31). (If you have been using the scanning lens, repeat the procedure with the low-power objective lens.)

Complete the chart by computing the approximate diameter of the high-power and oil immersion fields. The general formula for calculating the unknown field diameter is:

Diameter of field A × total magnification of field A = diameter of field B × total magnification of field B

where A represents the known or measured field and B represents the unknown field. This can be simplified to

Diameter of field B =

$$\frac{\text{diameter of field } A \times \text{total magnification of field } A}{\text{total magnification of field } B}$$

For example, if the diameter of the low-power field (field A) is 2 mm and the total magnification is 50×, you would compute the diameter of the high-power field (field B) with a total magnification of 100× as follows:

Field diameter B = (2 mm × 50)/100

Field diameter B = 1 mm

3. Estimate the length (longest dimension) of the following drawings of microscopic objects. *Base your calculations on the field sizes you have determined for your microscope and the approximate percentage of the diameter that the object occupies.*

Object seen in low-power field:

approximate length:

_____ mm

Object seen in high-power field:

approximate length:

_____ mm

or _____ μm

Object seen in oil immersion field:

approximate length:

_____ μm

3

Perceiving Depth

Any microscopic specimen has depth as well as length and width; it is rare indeed to view a tissue slide with just one layer of cells. Normally you can see two or three cell thicknesses. Therefore, it is important to learn how to determine relative depth with your microscope. In microscope work, the **depth of field** (the thickness of the plane that is clearly in focus) is greater at lower magnifications. As magnification increases, depth of field decreases.

Activity 4

Perceiving Depth

1. Obtain a slide with colored crossed threads. Focusing at low magnification, locate the point where the three threads cross each other.

2. Use the iris diaphragm lever to greatly reduce the light, thus increasing the contrast. Focus down with the coarse adjustment until the threads are out of focus, then slowly focus upward again, noting which thread comes into clear focus first. Observe: As you rotate the adjustment knob forward (away from you), does the stage rise or fall? If the stage rises, then the first clearly focused thread is the top one; the last clearly focused thread is the bottom one.

If the stage descends, how is the order affected?_____

_____

Record your observations, relative to which color of thread is uppermost, middle, or lowest:

Top thread _____

Middle thread _____

Bottom thread _____

Viewing Cells Under the Microscope

There are various ways to prepare cells for viewing under a microscope. One method is to mix the cells in physiological saline (called a *wet mount*) and stain them.

If you are not instructed to prepare your own wet mount, obtain a prepared slide of epithelial cells to make the observations in step 10 of Activity 5.

Activity 5

Preparing and Observing a Wet Mount

1. Obtain the following: a clean microscope slide and coverslip, two flat-tipped toothpicks, a dropper bottle of physiological saline, a dropper bottle of iodine or methylene blue stain, and filter paper (or paper towels). Handle only your own slides throughout the procedure.

2. Place a drop of physiological saline in the center of the slide. Using the flat end of the toothpick, *gently* scrape the inner lining of your cheek. Transfer your cheek scrapings to the slide by agitating the end of the toothpick in the drop of saline (**Figure 3.4a** on p. 34).

 Immediately discard the used toothpick in the disposable autoclave bag provided.

3. Add a tiny drop of the iodine or methylene blue stain to the preparation. (These epithelial cells are nearly transparent and thus difficult to see without the stain, which

Text continues on next page. →

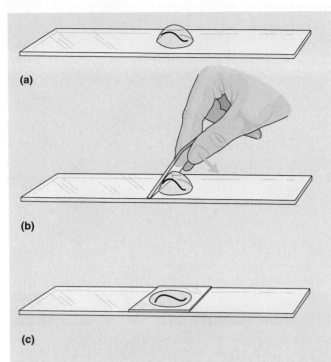

(a)

(b)

(c)

Figure 3.4 Procedure for preparation of a wet mount. **(a)** Place the object in a drop of water (or saline) on a clean slide; **(b)** hold a coverslip at a 45° angle with the fingertips; and **(c)** lower the coverslip slowly.

colors the nuclei of the cells.) Stir again, using a second toothpick.

 Immediately discard the used toothpicks in the disposable autoclave bag provided.

4. Hold the coverslip with your fingertips so that its bottom edge touches one side of the drop (Figure 3.4b), then *slowly* lower the coverslip onto the preparation (Figure 3.4c). *Do not just drop the coverslip,* or you will trap large air bubbles under it, which will obscure the cells. *Always use a coverslip with a wet mount* to protect the lens.

5. Examine your preparation carefully. The coverslip should be tight against the slide. If there is excess fluid around its edges, you will need to remove it. Obtain a piece of filter paper, fold it in half, and use the folded edge to absorb the excess fluid.

 Before continuing, discard the filter paper or paper towel in the disposable autoclave bag.

6. Place the slide on the stage, and locate the cells at the lowest power. You will probably want to dim the light to provide more contrast for viewing the lightly stained cells.

7. Cheek epithelial cells are very thin, flat cells. In the cheek, they provide a smooth, tilelike lining (**Figure 3.5**). Move to high power to examine the cells more closely.

8. Make a sketch of the epithelial cells that you observe.

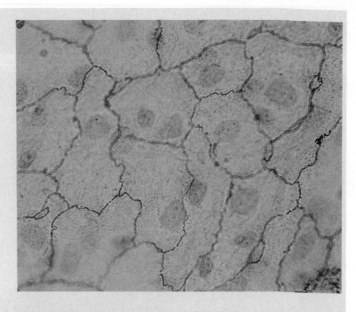

Figure 3.5 Epithelial cells of the cheek cavity (surface view, 600×).

Use information on your Summary chart (p. 31) to estimate the diameter of cheek epithelial cells. Record the total magnification (TM) used.

_____ μm   _____ × (TM)

Why do *your* cheek cells look different than those in Figure 3.5? (Hint: What did you have to *do* to your cheek to obtain them?)

_____

_____

 9. When you complete your observations of the wet mount, dispose of your wet mount preparation in the beaker of bleach solution, and put the coverslips in an autoclave bag.

10. Obtain a prepared slide of cheek epithelial cells, and view them under the microscope.

Estimate the diameter of one of these cheek epithelial cells using information from the Summary chart (p. 31).

_____ μm   _____ × (TM)

Why are these cells more similar to those in Figure 3.5 and easier to measure than those of the wet mount?

_____

_____

_____

11. Before leaving the laboratory, make sure all other materials are properly discarded or returned to the appropriate laboratory station. Clean the microscope lenses, and return the microscope to the storage cabinet.

REVIEW SHEET
The Microscope

Name _____ Lab Time/Date _____

Care and Structure of the Compound Microscope

1. Label all indicated parts of the microscope.

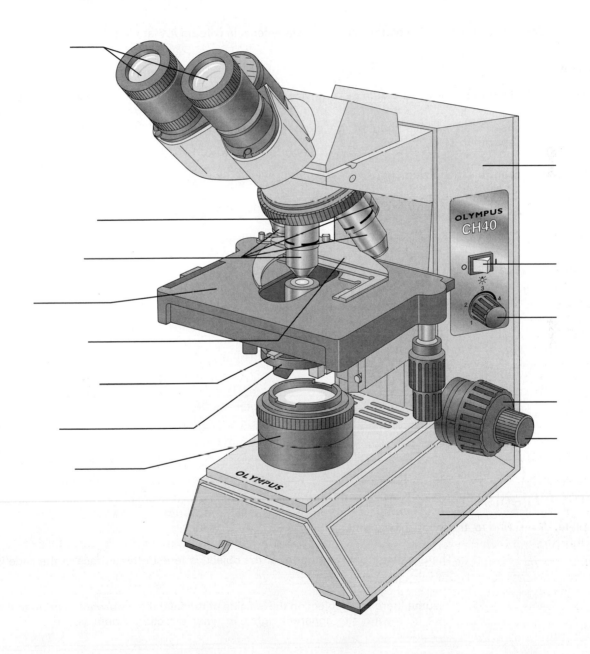

2. Explain the proper technique for transporting the microscope.

3. The following statements are true or false. If true, write *T* on the answer blank. If false, correct the statement by writing on the blank the proper word or phrase to replace the one that is underlined.

_____ 1.   The microscope lens may be cleaned <u>with any soft tissue</u>.

_____ 2.   The microscope should be stored with the <u>oil immersion</u> lens in position over the stage.

_____ 3.   When beginning to focus, use the <u>lowest-power</u> lens.

_____ 4.   When focusing on high power, always use the <u>coarse</u> adjustment knob to focus.

_____ 5.   A coverslip should always be used <u>with wet mounts</u>.

4. Match the microscope structures in column B with the statements in column A that identify or describe them.

Column A

_____ 1.   platform on which the slide rests for viewing

_____ 2.   used to adjust the amount of light passing through the specimen

_____ 3.   controls the movement of the slide on the stage

_____ 4.   delivers a concentrated beam of light to the specimen

_____ 5.   used for precise focusing once initial focusing has been done

_____ 6.   carries the objective lenses; rotates so that the different objective lenses can be brought into position over the specimen

Column B

a.   coarse adjustment knob
b.   condenser
c.   fine adjustment knob
d.   iris diaphragm lever
e.   mechanical stage
f.   nosepiece
g.   objective lenses
h.   ocular lens
i.   stage

5. Define the following terms.

virtual image: _____

_____

resolution: _____

Viewing Objects Through the Microscope

6. Complete, or respond to, the following statements:

_____ 1.   The distance from the bottom of the objective lens to the surface of the slide is called the _____.

_____ 2.   Assume there is an object on the left side of the field that you want to bring to the center (that is, toward the apparent right). In what direction would you move your slide? _____.

_____ 3.   The area of the slide seen when looking through the microscope is the _____.

_____ 4.   If a microscope has a 10× ocular lens and the total magnification at a particular time is 950×, the objective lens in use at that time is _____ ×.

_____ 5. Why should the light be dimmed when looking at living (nearly transparent) cells?

_____ 6. If, after focusing in low power, you need to use only the fine adjustment to focus the specimen at the higher powers, the microscope is said to be _____.

_____ 7. You are using a 10× ocular and a 15× objective, and the field diameter is 1.5 mm. The approximate field size with a 30× objective is _____ mm.

_____ 8. If the diameter of the high-power field is 1.2 mm, an object that occupies approximately a third of that field has an estimated diameter of _____ mm.

7. You have been asked to prepare a slide with the letter _k_ on it (as shown below). In the circle below, draw the _k_ as seen in the low-power field.

k

8. Calculate the magnification of fields 1 and 3, and the field diameter of 2. (_Hint:_ Use your ruler.) Note that the numbers for the field diameters below are too large to represent the typical compound microscope lens system, but the relationships depicted are accurate.

5 mm   _____ mm   0.5 mm

1. →O←   2. →O←   3. →o←

_____ ×   100 ×   _____ ×

9. Say you are observing an object in the low-power field. When you switch to high power, it is no longer in your field of view.

Why might this occur? _____

_____

What should you do initially to prevent this from happening? _____

_____

10. Do the following factors increase or decrease as one moves to higher magnifications with the microscope?

resolution: _____   amount of light needed: _____

working distance: _____   depth of field: _____

11. A student has the high-power lens in position and appears to be intently observing the specimen. The instructor, noting a working distance of about 1 cm, knows the student isn't actually seeing the specimen.

How so?_____

12. Describe the proper procedure for preparing a wet mount.

_____

_____

_____

13. Indicate the probable cause of the following situations during use of a microscope.

a.   Only half of the field is illuminated: _____

_____

b.   The visible field does not change as the mechanical stage is moved: _____

_____

EXERCISE 4

The Cell: Anatomy and Division

Objectives

☐ Define *cell*, *organelle*, and *inclusion*.

☐ Identify on a cell model or diagram the following cellular regions and list the major function of each: nucleus, cytoplasm, and plasma membrane.

☐ Identify the cytoplasmic organelles and discuss their structure and function.

☐ Compare and contrast specialized cells with the concept of the "generalized cell."

☐ Define *interphase*, *mitosis*, and *cytokinesis*.

☐ List the stages of mitosis, and describe the key events of each stage.

☐ Identify the mitotic phases on slides or appropriate diagrams.

☐ Explain the importance of mitotic cell division, and describe its product.

Materials

- Three-dimensional model of the "composite" animal cell or laboratory chart of cell anatomy
- Compound microscope
- Prepared slides of simple squamous epithelium, teased smooth muscle (l.s.), human blood smear, and sperm
- Animation/video of mitosis
- Three-dimensional models of mitotic stages
- Prepared slides of whitefish blastulas
- Chenille sticks (pipe cleaners), two different colors cut into 3-inch pieces, 8 pieces per group

Note to the Instructor: *See directions for handling wet mount preparations and disposable supplies (p. 34, Exercise 3). For suggestions on the animation/video of mitosis, see the Instructor's Guide.*

MasteringA&P®

For related exercise study tools, go to the Study Area of **MasteringA&P**. There you will find:

- Practice Anatomy Lab **PAL**
- A&PFlix **A&PFlix**
- PhysioEx **PEx**
- Practice quizzes, Histology Atlas, eText, Videos, and more!

Pre-Lab Quiz

1. Define *cell*. _____
   _____

2. When a cell is not dividing, the DNA is loosely spread throughout the nucleus in a threadlike form called:
   a. chromatin   c. cytosol
   b. chromosomes   d. ribosomes

3. The plasma membrane not only provides a protective boundary for the cell but also determines which substances enter or exit the cell. We call this characteristic:
   a. diffusion   c. osmosis
   b. membrane potential   d. selective permeability

4. Proteins are assembled on these organelles.
   _____

5. Because these organelles are responsible for providing most of the ATP that the cell needs, they are often referred to as the "powerhouses" of the cell. They are the:
   a. centrioles   c. mitochondria
   b. lysosomes   d. ribosomes

6. Circle the correct underlined term. During <u>cytokinesis</u> / <u>interphase</u>, the cell grows and performs its usual activities.

7. Circle True or False. The end product of mitosis is four genetically identical daughter nuclei.

8. How many stages of mitosis are there? _____

9. DNA replication occurs during:
   a. cytokinesis   c. metaphase
   b. interphase   d. prophase

10. Circle True or False. All animal cells have a cell wall.

The **cell,** the structural and functional unit of all living things, is a complex entity. The cells of the human body are highly diverse, and their differences in size, shape, and internal composition reflect their specific roles in the body. Still, cells do have many common anatomical features, and all cells must carry out certain functions to sustain life. For example, all cells can maintain their boundaries, metabolize, digest nutrients and dispose of wastes, grow and reproduce, move, and respond to a stimulus. This exercise focuses on structural similarities found in many cells and illustrated by a "composite," or "generalized," cell (**Figure 4.1a**) and considers only the function of cell reproduction (cell division).

Anatomy of the Composite Cell

In general, all animal cells have three major regions, or parts, that can readily be identified with a light microscope: the **nucleus,** the **plasma membrane,** and the **cytoplasm.** The nucleus is near the center of the cell. It is surrounded by cytoplasm, which in turn is enclosed by the plasma membrane. See the diagram (Figure 4.1a) representing the fine structure of the composite cell. An electron micrograph (Figure 4.1b) reveals the cellular structure, particularly of the nucleus.

Nucleus

The nucleus contains the genetic material, DNA, sections of which are called *genes*. Often described as the control center of the cell, the nucleus is necessary for cell reproduction. A cell that has lost or ejected its nucleus is programmed to die.

When the cell is not dividing, the genetic material is loosely dispersed throughout the nucleus in a threadlike form called **chromatin.** When the cell is in the process of dividing to form daughter cells, the chromatin coils and condenses, forming dense, rodlike bodies called **chromosomes**—much in the way a stretched spring becomes shorter and thicker when it is released.

The nucleus also contains one or more small spherical bodies, called **nucleoli,** composed primarily of proteins and ribonucleic acid (RNA). The nucleoli are assembly sites for ribosomes that are particularly abundant in the cytoplasm.

The nucleus is bound by a double-layered porous membrane, the **nuclear envelope.** The nuclear envelope is similar in composition to other cellular membranes, but it is distinguished by its large **nuclear pores.** They are spanned by protein complexes that regulate what passes through, and they permit easy passage of protein and RNA molecules.

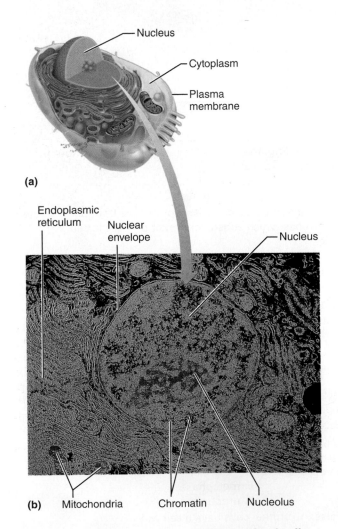

(a)

(b) Mitochondria   Chromatin   Nucleolus

Figure 4.1 Anatomy of the composite animal cell. (a) Diagram. **(b)** Transmission electron micrograph (5000×).

Activity 1

Identifying Parts of a Cell

Identify the nuclear envelope, chromatin, nucleolus, and the nuclear pores in Figure 4.1a and b and Figure 4.3.

Plasma Membrane

The **plasma membrane** separates cell contents from the surrounding environment, providing a protective barrier. Its main structural building blocks are phospholipids (fats) and globular protein molecules. Some of the externally facing proteins and lipids have sugar (carbohydrate) side chains attached to them that are important in cellular interactions (**Figure 4.2**). As described by the fluid mosaic model, the membrane is a bilayer of phospholipid molecules in which the protein molecules float. Occasional cholesterol molecules dispersed in the bilayer help stabilize it.

Because of its molecular composition, the plasma membrane is selective about what passes through it. It allows nutrients to enter the cell but keeps out undesirable substances. By the same token, valuable cell proteins and other substances are kept within the cell, and excreta, or wastes, pass to the exterior. This property is known as **selective permeability.**

Additionally, the plasma membrane maintains a resting potential that is essential to normal functioning of excitable cells, such as neurons and muscle cells, and plays a vital role

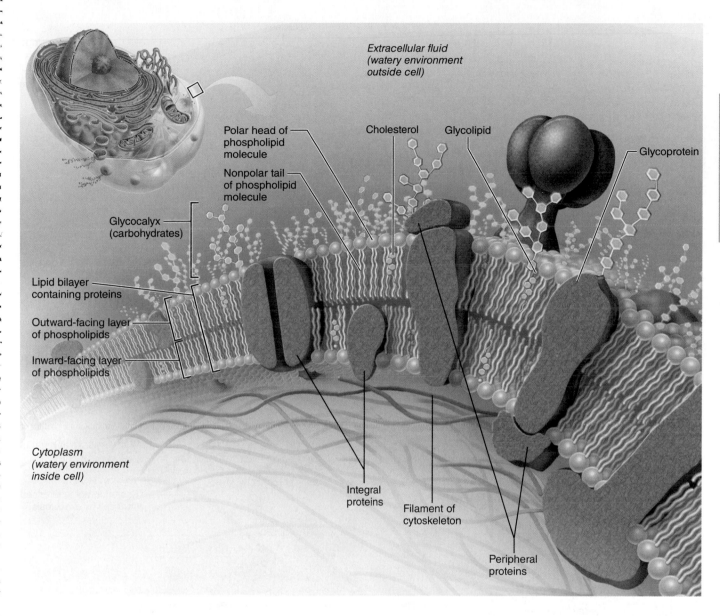

Figure 4.2 Structural details of the plasma membrane.

in cell signaling and cell to cell interactions. In some cells, the membrane is thrown into tiny fingerlike projections or folds called **microvilli** (**Figure 4.3**, p. 42). Microvilli greatly increase the surface area of the cell available for absorption or passage of materials and for the binding of signaling molecules.

Activity 2

Identifying Components of a Plasma Membrane

Identify the phospholipid and protein portions of the plasma membrane in Figure 4.2. Also locate the sugar (*glyco* = carbohydrate) side chains and cholesterol molecules. Identify the microvilli in the generalized cell diagram (Figure 4.3).

Cytoplasm and Organelles

The cytoplasm consists of the cell contents between the nucleus and plasma membrane. Suspended in the **cytosol,** the fluid cytoplasmic material, are many small structures called **organelles** (literally, "small organs"). The organelles are the metabolic machinery of the cell, and they are highly organized to carry out specific functions for the cell as a whole. The cytoplasmic organelles include the ribosomes, smooth and rough endoplasmic reticulum, Golgi apparatus, lysosomes, peroxisomes, mitochondria, cytoskeletal elements, and centrioles.

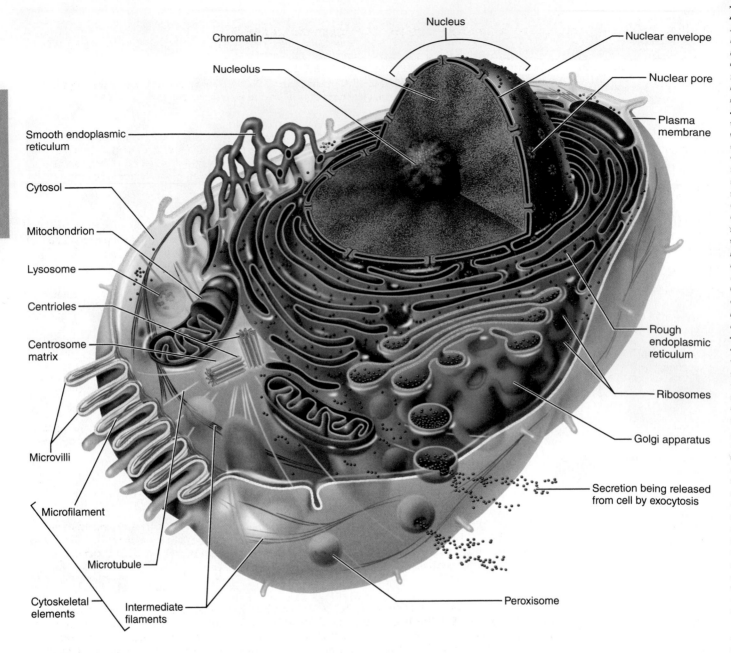

Nucleus

Chromatin

Nucleolus

Nuclear envelope

Nuclear pore

Plasma membrane

Smooth endoplasmic reticulum

Cytosol

Mitochondrion

Lysosome

Centrioles

Centrosome matrix

Microvilli

Microfilament

Microtubule

Cytoskeletal elements

Intermediate filaments

Rough endoplasmic reticulum

Ribosomes

Golgi apparatus

Secretion being released from cell by exocytosis

Peroxisome

Figure 4.3 Structure of the generalized cell. No cell is exactly like this one, but this composite illustrates features common to many human cells. Not all organelles are drawn to the same scale in this illustration.

Activity 3

Locating Organelles

Each organelle type is described in **Table 4.1**. Read through the table, and then, as best you can, locate the organelles in Figures 4.1b and 4.3.

The cell cytoplasm may or may not contain **inclusions.** Examples of inclusions are stored foods (glycogen granules and lipid droplets), pigment granules, crystals of various types, water vacuoles, and ingested foreign materials.

Activity 4

Examining the Cell Model

Once you have located all of these structures in the art (Figures 4.1b and 4.3), examine the cell model (or cell chart) to repeat and reinforce your identifications.

Table 4.1	Summary of Structure and Function of Cytoplasmic Organelles
Organelle	**Location and function**
Ribosomes	Tiny spherical bodies composed of RNA and protein; floating free or attached to a membranous structure (the rough ER) in the cytoplasm. Actual sites of protein synthesis.
Endoplasmic reticulum (ER)	Membranous system of tubules that extends throughout the cytoplasm; two varieties: rough and smooth. Rough ER is studded with ribosomes; tubules of the rough ER provide an area for storage and transport of the proteins made on the ribosomes to other cell areas. Smooth ER, which has no function in protein synthesis, is a site of steroid and lipid synthesis, lipid metabolism, and drug detoxification.
Golgi apparatus	Stack of flattened sacs with bulbous ends and associated small vesicles; found close to the nucleus. Plays a role in packaging proteins or other substances for export from the cell or incorporation into the plasma membrane and in packaging lysosomal enzymes.
Lysosomes	Various-sized membranous sacs containing digestive enzymes including acid hydrolases; function to digest worn-out cell organelles and foreign substances that enter the cell. Have the capacity of total cell destruction if ruptured and are for this reason referred to as "suicide sacs."
Peroxisomes	Small lysosome-like membranous sacs containing oxidase enzymes that detoxify alcohol, free radicals, and other harmful chemicals. They are particularly abundant in liver and kidney cells.
Mitochondria	Generally rod-shaped bodies with a double-membrane wall; inner membrane is shaped into folds, or cristae; contain enzymes that oxidize foodstuffs to produce cellular energy (ATP); often referred to as "powerhouses of the cell."
Centrioles	Paired, cylindrical bodies that lie at right angles to each other, close to the nucleus. Internally, each centriole is composed of nine triplets of microtubules. As part of the centrosome, they direct the formation of the mitotic spindle during cell division; form the bases of cilia and flagella and in that role are called *basal bodies*.
Cytoskeletal elements: microfilaments, intermediate filaments, and microtubules	Form an internal scaffolding called the *cytoskeleton*. Provide cellular support; function in intracellular transport. Microfilaments are formed largely of actin, a contractile protein, and thus are important in cell mobility, particularly in muscle cells. Intermediate filaments are stable elements composed of a variety of proteins and resist mechanical forces acting on cells. Microtubules form the internal structure of the centrioles and help determine cell shape.

Differences and Similarities in Cell Structure

Activity 5

Observing Various Cell Structures

1. Obtain a compound microscope and prepared slides of simple squamous epithelium, smooth muscle cells (teased), human blood, and sperm.

2. Observe each slide under the microscope, carefully noting similarities and differences in the cells. See

photomicrographs for simple squamous epithelium (Figure 3.5 in Exercise 3) and teased smooth muscle (Figure 6.7c in Exercise 6). The oil immersion lens will be needed to observe blood and sperm. Distinguish the boundaries of the individual cells, and notice the shape

Text continues on next page. ➡

(four circles for sketches)

**Simple squamous
epithelium**
Diameter _____

Sperm cells
Length _____

**Human
red blood cells**
Diameter _____

**Teased smooth
muscle cells**
Length _____

and position of the nucleus in each case. When you look at the human blood smear, direct your attention to the red blood cells, the pink-stained cells that are most numerous. The color photomicrographs illustrating a blood smear (Figure 29.3 in Exercise 29) and sperm (Figure 43.3 in Exercise 43) may be helpful in this cell structure study. Sketch your observations in the circles provided above.

3. Measure the length or diameter of each cell, and record below the appropriate sketch.

4. How do these four cell types differ in shape and size?

_____

_____

_____

_____

How might cell shape affect cell function?

_____

_____

_____

_____

_____

Which cells have visible projections? _____

_____

How do these projections relate to the function of these cells?

_____

_____

Do any of these cells lack a plasma membrane? _____

A nucleus? _____

In the cells with a nucleus, can you discern nucleoli?

_____

Were you able to observe any of the organelles in these

cells? _____ Why or why not? _____

_____

_____

_____

_____

Cell Division: Mitosis and Cytokinesis

A cell's *life cycle* is the series of changes it goes through from the time it is formed until it reproduces. It consists of two stages—**interphase,** the longer period during which the cell grows and carries out its usual activities (Figure 4.4a), and **cell division,** when the cell reproduces itself by dividing. In an interphase cell about to divide, the genetic material (DNA) is copied exactly via DNA replication. Once this important event has occurred, cell division ensues.

Cell division in all cells other than bacteria consists of two events called mitosis and cytokinesis. **Mitosis** is the division of the copied DNA of the mother cell to two daughter nuclei. **Cytokinesis** is the division of the cytoplasm, which begins when mitosis is nearly complete. Although mitosis is usually accompanied by cytokinesis, in some instances cytoplasmic division does not occur, leading to the formation of binucleate or multinucleate cells.

The product of **mitosis** is two daughter nuclei that are genetically identical to the mother nucleus. This distinguishes mitosis from **meiosis,** a specialized type of nuclear division (covered in Exercise 43) that occurs only in the reproductive organs (testes or ovaries). Meiosis, which yields four daughter nuclei that differ genetically in composition from the mother nucleus, is used only for the production of gametes (eggs and sperm) for sexual reproduction. The function of cell division, including mitosis and cytokinesis in the body, is to increase the number of cells for growth and repair.

The phases of mitosis include **prophase, metaphase, anaphase,** and **telophase.** The detailed events of interphase, mitosis, and cytokinesis are described and illustrated in **Figure 4.4** on pp. 46–47.

Mitosis is essentially the same in all animal cells, but depending on the tissue, it takes from 5 minutes to several hours to complete. In most cells, centriole replication occurs during interphase of the next cell cycle.

At the end of cell division, two daughter cells exist—each with a smaller cytoplasmic mass than the mother cell but genetically identical to it. The daughter cells grow and carry out the normal spectrum of metabolic processes until it is their turn to divide.

Cell division is extremely important during the body's growth period. Most cells divide until puberty, when adult body size is achieved and overall body growth ceases. After this time in life, only certain cells carry out cell division

routinely—for example, cells subjected to abrasion (epithelium of the skin and lining of the gut). Other cell populations—such as liver cells—stop dividing but retain this ability should some of them be removed or damaged. Skeletal muscle, cardiac muscle, and most mature neurons almost completely lose this ability to divide and thus are severely handicapped by injury.

Activity 6

Identifying the Mitotic Stages

1. Watch an animation or video presentation of mitosis (if available).

2. Using the three-dimensional models of dividing cells provided, identify each of the mitotic phases illustrated and described in Figure 4.4.

3. Obtain a prepared slide of whitefish blastulas to study the stages of mitosis. The cells of each *blastula* (a stage of embryonic development consisting of a hollow ball of cells) are at approximately the same mitotic stage, so it may be necessary to observe more than one blastula to view all the mitotic stages. A good analogy for a blastula is a soccer ball in which each leather piece making up the ball's surface represents an embryonic cell. The exceptionally high rate of mitosis observed in this tissue is typical of embryos, but if it occurs in specialized tissues it can indicate cancerous cells, which also have an extraordinarily high mitotic rate. Examine the slide carefully, identifying the four mitotic phases and the process of cytokinesis. Compare your observations with the photomicrographs (Figure 4.4), and verify your identifications with your instructor.

Activity 7

"Chenille Stick" Mitosis

1. Obtain a total of eight 3-inch pieces of chenille stick, four of one color and four of another color (e.g., four green and four purple).

2. Assemble the chenille sticks into a total of four chromosomes (each with two sister chromatids) by twisting two sticks of the same color together at the center with a single twist.

What does the twist at the center represent? _____

_____

3. Arrange the chromosomes as they appear in early prophase.

Name the structure that assembles during this phase.

_____

Draw early prophase in the space provided in the Review Sheet (question 10, p. 51).

4. Arrange the chromosomes as they appear in late prophase.

What structure on the chromosome centromere do

the growing spindle microtubules attach to? _____

What structure is now present as fragments? _____

_____

Draw late prophase in the space provided on the Review Sheet (question 10, p. 51).

5. Arrange the chromosomes as they appear in metaphase.

What is the name of the imaginary plane that the

chromosomes align along? _____

Draw metaphase in the space provided on the Review Sheet (question 10, p. 51).

Text continues on page 48. →

Interphase	Early Prophase	Late Prophase
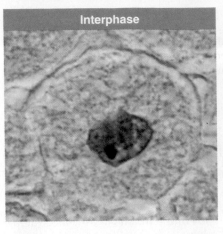	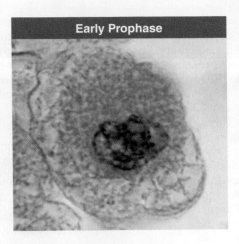	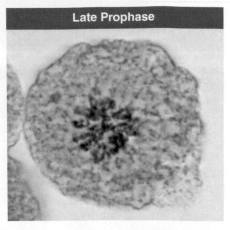

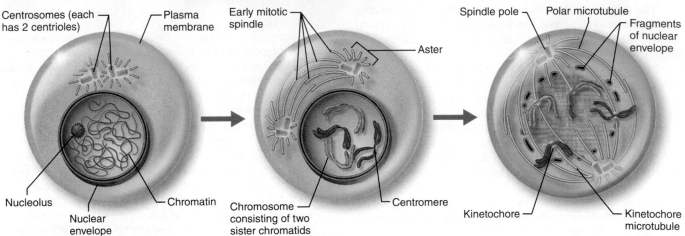

Centrosomes (each has 2 centrioles) — Plasma membrane

Nucleolus — Nuclear envelope — Chromatin

Early mitotic spindle — Aster

Chromosome consisting of two sister chromatids — Centromere

Spindle pole — Polar microtubule — Fragments of nuclear envelope

Kinetochore — Kinetochore microtubule

Interphase

Interphase is the period when the cell carries out its normal metabolic activities and grows. Interphase is not part of mitosis.

• During interphase, the DNA-containing material is in the form of chromatin. The nuclear envelope and one or more nucleoli are intact and visible.

• There are three distinct periods of interphase:
  G₁: The centrioles begin replicating.
  S: DNA is replicated.
  G₂: Final preparations for mitosis are completed, and centrioles finish replicating.

Prophase—first phase of mitosis

Early Prophase
• The chromatin condenses, forming barlike chromosomes.

• Each duplicated chromosome consists of two identical threads, called **sister chromatids**, held together at the **centromere**. (Later when the chromatids separate, each will be a new chromosome.)

• As the chromosomes appear, the nucleoli disappear, and the two centrosomes separate from one another.

• The centrosomes act as focal points for growth of a microtubule assembly called the **mitotic spindle**. As the microtubules lengthen, they propel the centrosomes toward opposite ends (poles) of the cell.

• Microtubule arrays called **asters** ("stars") extend from the centrosome matrix.

Late Prophase
• The nuclear envelope breaks up, allowing the spindle to interact with the chromosomes.

• Some of the growing spindle microtubules attach to **kinetochores**, special protein structures at each chromosome's centromere. Such microtubules are called **kinetochore microtubules**.

• The remaining spindle microtubules (not attached to any chromosomes) are called **polar microtubules**. The microtubules slide past each other, forcing the poles apart.

• The kinetochore microtubules pull on each chromosome from both poles in a tug-of-war that ultimately draws the chromosomes to the center, or equator, of the cell.

Figure 4.4 The interphase cell and the events of cell division. The cells shown are from an early embryo of a whitefish. Photomicrographs are above; corresponding diagrams are below. (Micrographs approximately 1600×.)

| Metaphase | Anaphase | Telophase | Cytokinesis |

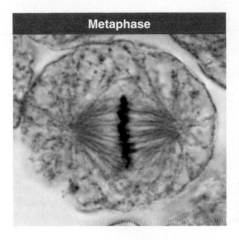 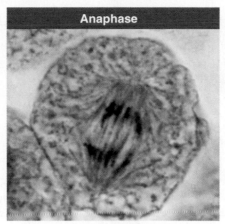 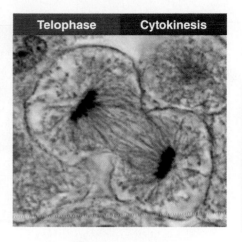

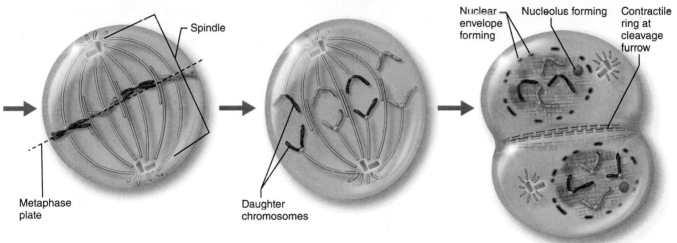

Spindle

Metaphase plate

Daughter chromosomes

Nuclear envelope forming   Nucleolus forming   Contractile ring at cleavage furrow

Metaphase—second phase of mitosis

• The two centrosomes are at opposite poles of the cell.

• The chromosomes cluster at the midline of the cell, with their centromeres precisely aligned at the **equator** of the spindle. This imaginary plane midway between the poles is called the **metaphase plate**.

• Enzymes act to separate the chromatids from each other.

Anaphase—third phase of mitosis

The shortest phase of mitosis, anaphase begins abruptly as the centromeres of the chromosomes split simultaneously. Each chromatid now becomes a chromosome in its own right.

• The kinetochore microtubules, moved along by motor proteins in the kinetochores, gradually pull each chromosome toward the pole it faces.

• At the same time, the polar microtubules slide past each other, lengthen, and push the two poles of the cell apart.

• The moving chromosomes look V shaped. The centromeres lead the way, and the chromosomal "arms" dangle behind them.

• The fact that the chromosomes are short, compact bodies makes it easier for them to move and separate. Diffuse threads of chromatin would trail, tangle, and break, resulting in imprecise "parceling out" to the daughter cells.

Telophase—final phase of mitosis

Telophase begins as soon as chromosomal movement stops. This final phase is like prophase in reverse.

• The identical sets of chromosomes at the opposite poles of the cell uncoil and resume their threadlike chromatin form.

• A new nuclear envelope forms around each chromatin mass, nucleoli reappear within the nuclei, and the spindle breaks down and disappears.

• Mitosis is now ended. The cell, for just a brief period, is binucleate (has two nuclei), and each new nucleus is identical to the original mother nucleus.

Cytokinesis—division of cytoplasm

Cytokinesis begins during late anaphase and continues through and beyond telophase. A contractile ring of actin microfilaments forms the **cleavage furrow** and pinches the cell apart.

Figure 4.4 *(continued)* **The events of cell division.**

4

6. Arrange the chromosomes as they appear in anaphase.

What does untwisting of the chenille sticks represent?

_____

_____

Each sister chromatid has now become a _____.

Draw anaphase in the space provided on the Review Sheet (question 10, p. 51).

7. Arrange the chromosomes as they appear in telophase.

Briefly list four reasons why telophase is like the reverse of prophase.

_____

_____

_____

Draw telophase in the space provided on the Review Sheet (question 10, p. 51).

REVIEW SHEET
The Cell: Anatomy and Division

Name _____ Lab Time/Date _____

Anatomy of the Composite Cell

1. Define the following terms:

   organelle: _____

   _____

   cell: _____

2. Cells have differences that reflect their specific functions in the body, but what functions do they have in common?

   _____

3. Identify the following cell structures:

   _____ 1. external boundary of cell; regulates flow of materials into and out of the cell; site of cell signaling

   _____ 2. contains digestive enzymes of many varieties; "suicide sac" of the cell

   _____ 3. scattered throughout the cell; major site of ATP synthesis

   _____ 4. slender extensions of the plasma membrane that increase its surface area

   _____ 5. stored glycogen granules, crystals, pigments; present in some cell types

   _____ 6. membranous system consisting of flattened sacs and vesicles; packages proteins for export

   _____ 7. control center of the cell; necessary for cell division and cell life

   _____ 8. two rod-shaped bodies near the nucleus; associated with the formation of the mitotic spindle

   _____ 9. dense nuclear body; packaging site for ribosomes

   _____ 10. contractile elements of the cytoskeleton

   _____ 11. membranous tubules covered with ribosomes; involved in intracellular transport of proteins

   _____ 12. attached to membrane systems or scattered in the cytoplasm; site of protein synthesis

   _____ 13. threadlike structures in the nucleus; contain genetic material (DNA)

   _____ 14. site of free radical detoxification

4. In the following diagram, label all parts provided with a leader line.

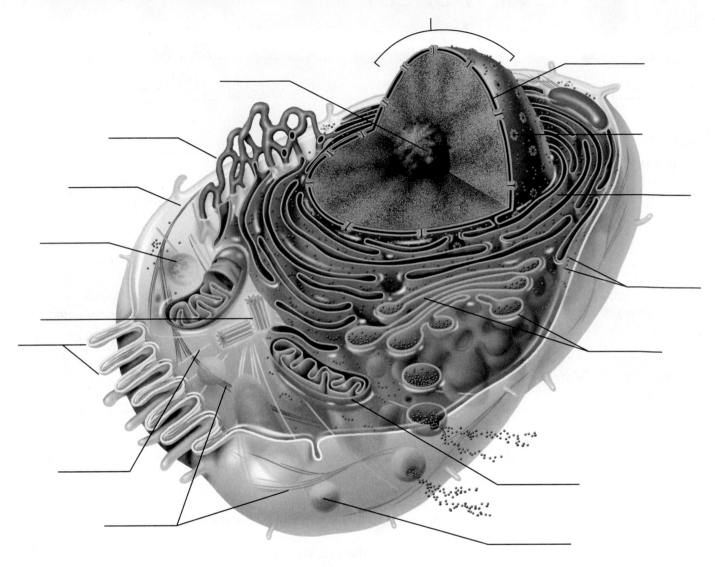

Differences and Similarities in Cell Structure

5. For each of the following cell types, list (a) *one* important structural characteristic observed in the laboratory, and (b) the function that the structure complements or ensures.

squamous epithelium   a. _____

                       b. _____

sperm   a. _____

                       b. _____

smooth muscle   a. _____

                       b. _____

red blood cells   a. _____

                       b. _____

6. What is the consequence of the red blood cell being anucleate (without a nucleus)? _____

_____

   Did it ever have a nucleus? (Use an appropriate reference.) _____ If so, when? _____

7. Of the four cells observed microscopically (squamous epithelial cells, red blood cells, smooth muscle cells, and sperm),

   which has the smallest diameter? _____ Which is longest? _____

Cell Division: Mitosis and Cytokinesis

8. Identify the three phases of mitosis in the following photomicrographs.

   a. _____   b. _____   c. _____

9. What is the function of mitotic cell division? _____

_____

10. Draw the phases of mitosis for a cell that contains four chromosomes as its diploid, or 2*n*, number.

11. Complete or respond to the following statements:

Division of the __1__ is referred to as mitosis. Cytokinesis is division of the __2__. The major structural difference between chromatin and chromosomes is that the latter are __3__. Chromosomes attach to the spindle fibers by undivided structures called __4__. If a cell undergoes mitosis but not cytokinesis, the product is __5__. The structure that acts as a scaffolding for chromosomal attachment and movement is called the __6__. __7__ is the period of cell life when the cell is not involved in division. Three cell populations in the body that do not routinely undergo cell division are __8__, __9__ and __10__.

1. _____

2. _____

3. _____

4. _____

5. _____

6. _____

7. _____

8. _____

9. _____

10. _____

12. Using the key, categorize each of the events described below according to the phase in which it occurs.

Key: a. anaphase   b. interphase   c. metaphase   d. prophase   e. telophase

_____ 1. Chromatin coils and condenses, forming chromosomes.

_____ 2. The chromosomes are V shaped.

_____ 3. The nuclear envelope re-forms.

_____ 4. Chromosomes stop moving toward the poles.

_____ 5. Chromosomes line up in the center of the cell.

_____ 6. The nuclear envelope fragments.

_____ 7. The mitotic spindle forms.

_____ 8. DNA replication occurs.

_____ 9. Centrioles replicate.

_____ 10. Chromosomes first appear to be duplex structures.

_____ 11. Cleavage furrow forms.

_____ and _____ 12. The nuclear envelope is completely absent.

13. What is the physical advantage of the chromatin coiling and condensing to form short chromosomes at the onset of mitosis?

_____

_____

The Cell: Transport Mechanisms and Cell Permeability

Objectives

☐ Define *selective permeability*, and explain the difference between active and passive transport processes.

☐ Define *diffusion*, and explain how simple diffusion and facilitated diffusion differ.

☐ Define *osmosis*, and explain the difference between isotonic, hypotonic, and hypertonic solutions.

☐ Define *filtration*, and discuss where it occurs in the body.

☐ Define *vesicular transport*, and describe phagocytosis, pinocytosis, receptor-mediated endocytosis, and exocytosis.

☐ List the processes that account for the movement of substances across the plasma membrane, and indicate the driving force for each.

☐ Name one substance that uses each membrane transport process.

☐ Determine which way substances will move passively through a selectively permeable membrane when given appropriate information about their concentration gradients.

Materials

Passive Processes

Diffusion of Dye Through Agar Gel

- Petri dish containing 12 ml of 1.5% agar-agar
- Millimeter-ruled graph paper
- Wax marking pencil
- 3.5% methylene blue solution (approximately 0.1 *M*) in dropper bottles
- 1.6% potassium permanganate solution (approximately 0.1 *M*) in dropper bottles
- Medicine dropper

Text continues on next page. →

MasteringA&P®

For related exercise study tools, go to the Study Area of **MasteringA&P**. There you will find:

- Practice Anatomy Lab **PAL**
- PhysioEx **PEx**
- A&PFlix **A&P Flix**
- Practice quizzes, Histology Atlas, eText, Videos, and more!

Pre-Lab Quiz

1. Circle the correct underlined term. A passive process, <u>diffusion</u> / <u>osmosis</u> is the movement of solute molecules from an area of greater concentration to an area of lesser concentration.

2. A solution surrounding a cell is *hypertonic* if:
   a. it contains fewer nonpenetrating solute particles than the interior of the cell
   b. it contains more nonpenetrating solute particles than the interior of the cell
   c. it contains the same amount of nonpenetrating solute particles as the interior of the cell

3. Which of the following would require an input of energy?
   a. diffusion
   b. filtration
   c. osmosis
   d. vesicular transport

4. Circle the correct underlined term. In <u>pinocytosis</u> / <u>phagocytosis</u>, parts of the plasma membrane and cytoplasm extend and engulf a relatively large or solid material.

5. Circle the correct underlined term. In <u>active</u> / <u>passive</u> processes, the cell provides energy in the form of ATP to power the transport process.

(Materials list continued.)

Diffusion and Osmosis Through Nonliving Membranes

- Four dialysis sacs
- Small funnel
- 25-ml graduated cylinder
- Wax marking pencil
- Fine twine or dialysis tubing clamps
- 250-ml beakers
- Distilled water
- 40% glucose solution
- 10% sodium chloride (NaCl) solution
- 40% sucrose solution colored with Congo red dye
- Laboratory balance
- Paper towels
- Hot plate and large beaker for hot water bath
- Benedict's solution in dropper bottle
- Silver nitrate ($AgNO_3$) in dropper bottle
- Test tubes in rack, test tube holder

Experiment 1

- Deshelled eggs
- 400-ml beakers
- Wax marking pencil
- Distilled water
- 30% sucrose solution

- Laboratory balance
- Paper towels
- Graph paper
- Weigh boat

Experiment 2

- Clean microscope slides and coverslips
- Medicine dropper
- Compound microscope
- Vials of mammalian blood obtained from a biological supply house or veterinarian—at option of instructor
- Freshly prepared physiological (mammalian) saline solution in dropper bottle
- 5% sodium chloride solution in dropper bottle
- Distilled water
- Filter paper
- Disposable gloves
- Basin and wash bottles containing 10% household bleach solution
- Disposable autoclave bag
- Paper towels

Diffusion Demonstrations

1. Diffusion of a dye through water

Prepared the morning of the laboratory session with setup time noted. Potassium permanganate crystals are placed in a 1000-ml graduated cylinder, and distilled water

is added slowly and with as little turbulence as possible to fill to the 1000-ml mark.

2. Osmometer

Just before the laboratory begins, the broad end of a thistle tube is closed with a selectively permeable dialysis membrane, and the tube is secured to a ring stand. Molasses is added to approximately 5 cm above the thistle tube bulb, and the bulb is immersed in a beaker of distilled water. At the beginning of the lab session, the level of the molasses in the tube is marked with a wax pencil.

Filtration

- Ring stand, ring, clamp
- Filter paper, funnel
- Solution containing a mixture of uncooked starch, powdered charcoal, and copper sulfate ($CuSO_4$)
- 10-ml graduated cylinder
- 100-ml beaker
- Lugol's iodine in a dropper bottle

Active Processes

- Video showing phagocytosis (if available)
- Video viewing system

Note to the Instructor: See directions for handling wet mount preparations and disposable supplies (p. 34, Exercise 3).

PEx PhysioEx™ 9.1 Computer Simulation Ex.1 on p. PEx-3.

Because of its molecular composition, the plasma membrane is selective about what passes through it. It allows nutrients to enter the cell but keeps out undesirable substances. By the same token, valuable cell proteins and other substances are kept within the cell, and excreta or wastes pass to the exterior. This property is known as **selective,** or **differential, permeability.** Transport through the plasma membrane occurs in two basic ways. In **passive processes,** concentration or pressure differences drive the movement. In **active processes,** the cell provides energy (ATP) to power the transport process.

Passive Processes

The two important passive processes of membrane transport are *diffusion* and *filtration*. Diffusion is an important transport process for every cell in the body. By contrast, filtration usually occurs only across capillary walls.

Molecules possess **kinetic energy** and are in constant motion. As molecules move about randomly at high speeds, they collide and ricochet off one another, changing direction with each collision (**Figure 5.1**). The driving force for diffusion is kinetic energy of the molecules themselves, and the speed of diffusion depends on molecular size and temperature. Smaller molecules move faster, and molecules move faster as temperature increases.

Diffusion

When a **concentration gradient** (difference in concentration) exists, the net effect of this random molecular movement is that the molecules eventually become evenly distributed throughout the environment. **Diffusion** is the movement of molecules from a region of their higher concentration to a region of their lower concentration.

There are many examples of diffusion in nonliving systems. For example, if a bottle of ether was uncorked at the front of the laboratory, very shortly thereafter you would be nodding off as the molecules became distributed throughout the room. The ability to smell a friend's fragrance shortly after he or she has entered the room is another example.

In general, molecules diffuse passively through the plasma membrane if they can dissolve in the lipid portion of the membrane, as CO_2 and O_2 can. The unassisted diffusion of solutes (dissolved substances) through a selectively permeable membrane is called **simple diffusion.**

Certain molecules, glucose for example, are transported across the plasma membrane with the assistance of a protein carrier molecule. The substances move by a passive transport process called **facilitated diffusion.** As with simple diffusion, the substances move from an area of higher concentration to

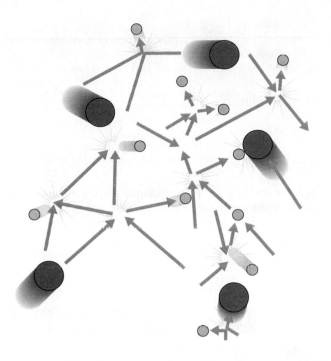

Figure 5.1 Random movement and numerous collisions cause molecules to become evenly distributed. The small spheres represent water molecules; the large spheres represent glucose molecules.

one of lower concentration, that is, down their concentration gradients.

Osmosis

The flow of water across a selectively permeable membrane is called **osmosis.** During osmosis, water moves down its concentration gradient. The concentration of water is inversely related to the concentration of solutes. If the solutes can diffuse across the membrane, both water and solutes will move down their concentration gradients through the membrane. If the particles in solution are nonpenetrating solutes (prevented from crossing the membrane), water alone will move by osmosis and in doing so will cause changes in the volume of the compartments on either side of the membrane.

Diffusion of Dye Through Agar Gel and Water

The relationship between molecular weight and the rate of diffusion can be examined easily by observing the diffusion of two different types of dye molecules through an agar gel. The dyes used in this experiment are methylene blue, which has a molecular weight of 320 and is deep blue in color, and potassium permanganate, a purple dye with a molecular weight of 158. Although the agar gel appears quite solid, it is primarily (98.5%) water and allows free movement of the dye molecules through it.

Activity 1

Observing Diffusion of Dye Through Agar Gel

1. Work with members of your group to formulate a hypothesis about the rates of diffusion of methylene blue and potassium permanganate through the agar gel. Justify your hypothesis.

2. Obtain a petri dish containing agar gel, a piece of millimeter-ruled graph paper, a wax marking pencil, dropper bottles of methylene blue and potassium permanganate, and a medicine dropper.

3. Using the wax marking pencil, draw a line on the bottom of the petri dish dividing it into two sections. Place the petri dish on the ruled graph paper.

4. Create a well in the center of each section using the medicine dropper. To do this, squeeze the bulb of the medicine dropper, and push it down into the agar. Release the bulb as you slowly pull the dropper vertically out of the agar. This should remove an agar plug, leaving a well in the agar. (See **Figure 5.2a.**)

5. Carefully fill one well with the methylene blue solution and the other well with the potassium permanganate solution (Figure 5.2b).

Record the time. _____

6. At 15-minute intervals, measure the distance the dye has diffused from each well by measuring the radius of the dye. Continue these observations for 1 hour, and record the results in the **Activity 1 chart** (p. 56).

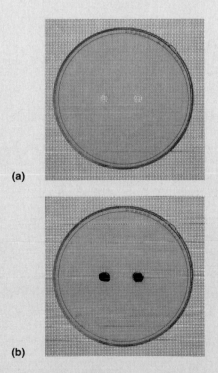

(a)

(b)

Figure 5.2 Comparing diffusion rates. Agar plated petri dish as it appears after the placement of 0.1 *M* methylene blue in one well and 0.1 *M* potassium permanganate in another.

Text continues on next page. ➡

Activity 1: Dye Diffusion Results		
Time (min)	Diffusion of methylene blue (mm)	Diffusion of potassium permanganate (mm)
15		
30		
45		
60		

Which dye diffused more rapidly? _____

What is the relationship between molecular weight and rate of molecular movement (diffusion)?

_____

Why did the dye molecules move? _____

_____

Compute the rate of diffusion of the potassium permanganate molecules in millimeters per minute (mm/min) and record.

_____ mm/min

Compute the rate of diffusion of the methylene blue molecules in mm/min and record.

_____ mm/min

7. Prepare a lab report for these experiments. (See Getting Started, on MasteringA&P.)

Make a mental note to yourself to go to the demonstration area at the end of the laboratory session to observe the extent of diffusion of the potassium permanganate dye through water. At that time, follow the next directions given.

Activity 2

Observing Diffusion of Dye Through Water

1. Go to the diffusion demonstration area, and observe the cylinder containing dye crystals and water set up at the beginning of the lab.

2. Measure the number of millimeters the dye has diffused from the bottom of the graduated cylinder, and record.

_____ mm

3. Record the time the demonstration was set up and the time of your observation. Then compute the rate of the dye's diffusion through water and record below.

Time of setup _____

Time of observation _____

Rate of diffusion _____ mm/min

4. Does the potassium permanganate dye diffuse more rapidly through water or agar gel? Explain your answer.

_____

_____

 Prepare for lab: Watch the Pre-Lab Video
MasteringA&P®>Study Area>Pre-Lab Videos

Activity 3

Investigating Diffusion and Osmosis Through Nonliving Membranes

The following experiment provides information on the movement of water and solutes through selectively permeable membranes called dialysis sacs. Dialysis sacs have pores of a particular size. The selectivity of living membranes depends on more than just pore size, but using the dialysis sacs will allow you to examine selectivity due to this factor.

1. Read through the experiments in this activity, and develop a hypothesis for each part.

2. Obtain four dialysis sacs, a small funnel, a 25-ml graduated cylinder, a wax marking pencil, fine twine or dialysis tubing clamps, and four beakers (250 ml). Number the beakers 1 to 4 with the wax marking pencil, and half fill all of them with distilled water except beaker 2, to which you should add 125 ml of the 40% glucose solution.

3. Prepare the dialysis sacs one at a time. Using the funnel, half fill each with 20 ml of the specified liquid (see Activity 3 chart). Press out the air, fold over the open end of the sac, and tie it securely with fine twine or clamp it. Before proceeding to the next sac, rinse it under the tap, and quickly and carefully blot the sac dry by rolling it on a paper towel. Weigh it with a laboratory balance. Record the weight in the **Activity 3 chart**, and then drop the sac into the corresponding beaker. Be sure the sac is completely covered by the beaker solution, adding more solution if necessary.

- Sac 1: 40% glucose solution
- Sac 2: 40% glucose solution
- Sac 3: 10% NaCl solution
- Sac 4: Congo red dye in 40% sucrose solution

				Activity 3: Experimental Data on Diffusion and Osmosis Through Nonliving Membranes		
Beaker	**Contents of sac**	**Initial weight**	**Final weight**	**Weight change**	**Tests— beaker fluid**	**Tests— sac fluid**
Beaker 1 ½ filled with distilled water	Sac 1, 20 ml of 40% glucose solution				Benedict's test:	Benedict's test:
Beaker 2 ½ filled with 40% glucose solution	Sac 2, 20 ml of 40% glucose solution					
Beaker 3 ½ filled with distilled water	Sac 3, 20 ml of 10% NaCl solution				$AgNO_3$ test:	
Beaker 4 ½ filled with distilled water	Sac 4, 20 ml of 40% sucrose solution containing Congo red dye				Benedict's test:	

Allow the sacs to remain undisturbed in the beakers for 1 hour. Use this time to continue with other experiments.

4. After an hour, boil a beaker of water on the hot plate. Obtain the supplies you will need to determine your experimental results: dropper bottles of Benedict's solution and silver nitrate solution, a test tube rack, four test tubes, and a test tube holder.

5. Quickly and gently blot sac 1 dry and weigh it. (**Note:** Do not squeeze the sac during the blotting process.) Record the weight in the data chart.

Was there any change in weight? _____

Conclusions: _____

_____

_____

Place 5 drops of Benedict's solution in each of two test tubes. Put 4 ml of the beaker fluid into one test tube and 4 ml of the sac fluid into the other. Mark the tubes for identification, and then place them in a beaker containing boiling water. Boil 2 minutes. Cool slowly. If a green, yellow, or rusty red precipitate forms, the test is positive, meaning that glucose is present. If the solution remains the original blue color, the test is negative. Record results in the data chart.

Was glucose still present in the sac? _____

Was glucose present in the beaker? _____

Conclusions: _____

_____

_____

6. Blot gently and weigh sac 2. Record the weight in the data chart.

Was there an *increase* or *decrease* in weight? _____

With 40% glucose in the sac and 40% glucose in the beaker, would you expect to see any net movement of water (osmosis) or of glucose molecules (simple diffusion)?

_____ Why or why not? _____

_____

7. Blot gently and weigh sac 3. Record the weight in the data chart.

Was there any change in weight? _____

Conclusions: _____

_____

Take a 5-ml sample of beaker 3 solution and put it in a clean test tube. Add a drop of silver nitrate ($AgNO_3$). The appearance of a white precipitate or cloudiness indicates the presence of silver chloride (AgCl), which is formed by the reaction of $AgNO_3$ with NaCl (sodium chloride). Record results in the data chart.

Text continues on next page. →

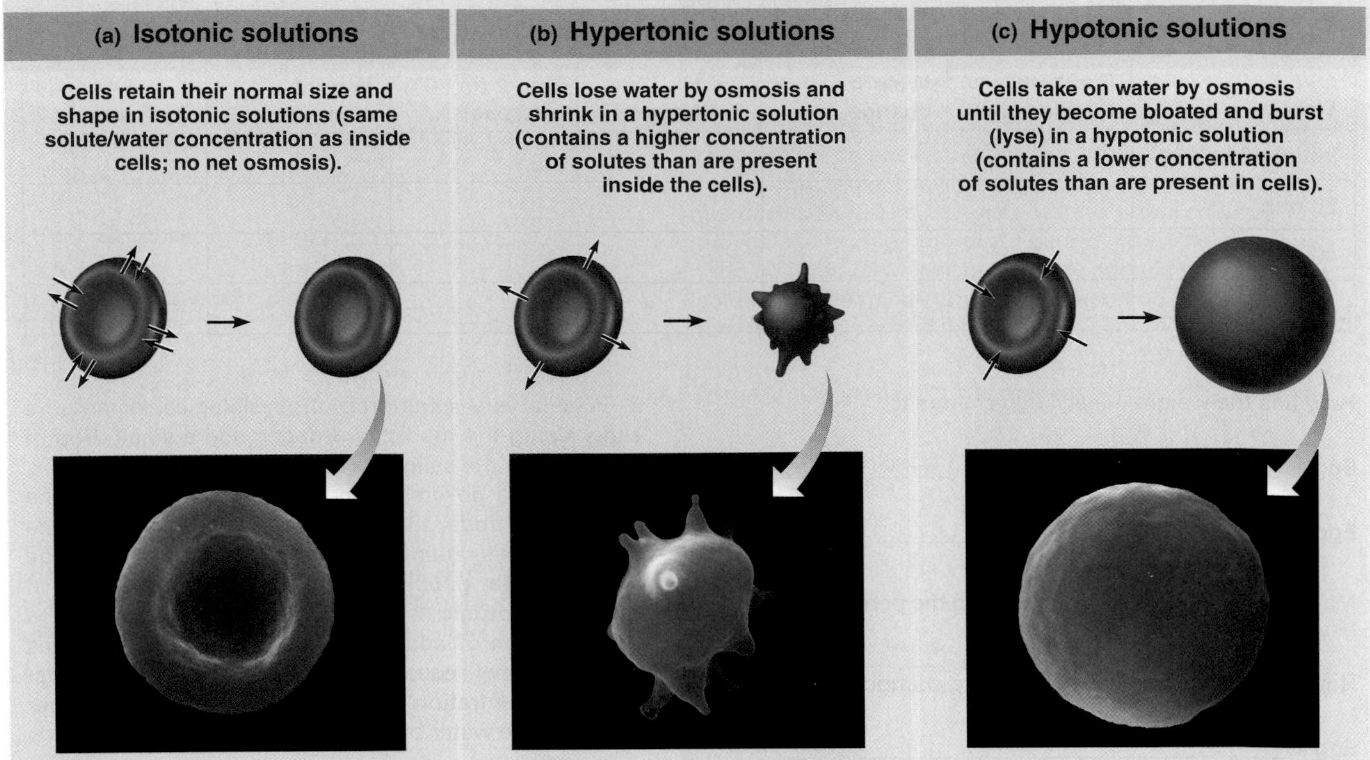

5

(a) Isotonic solutions	(b) Hypertonic solutions	(c) Hypotonic solutions
Cells retain their normal size and shape in isotonic solutions (same solute/water concentration as inside cells; no net osmosis).	Cells lose water by osmosis and shrink in a hypertonic solution (contains a higher concentration of solutes than are present inside the cells).	Cells take on water by osmosis until they become bloated and burst (lyse) in a hypotonic solution (contains a lower concentration of solutes than are present in cells).

Figure 5.3 Influence of isotonic, hypertonic, and hypotonic solutions on red blood cells.

blood cells first "plump up" (Figure 5.3c), but then they suddenly start to disappear. The red blood cells burst as the water floods into them, leaving "ghosts" in their wake—a phenomenon called **hemolysis.**

⚠ **5.** Place the blood-soiled slides and test tube in the bleach-containing basin. Put the coverslips you used into the disposable autoclave bag. Obtain a wash (squirt) bottle containing 10% bleach solution, and squirt the bleach liberally over the bench area where blood was handled. Wipe the bench down with a paper towel wet with the bleach solution, and allow it to dry before continuing. Remove gloves, and discard in the autoclave bag.

6. Prepare a lab report for experiments 1 and 2. (See Getting Started, on MasteringA&P.) Be sure to include in the discussion answers to the questions proposed in this activity.

WHY THIS MATTERS | **Isotonic Sports Drinks**

You have just completed your daily run or fitness class. Do you need to drink a specialized post-workout beverage? A body fluid loss of as little as 2% of your body weight can affect physiological function, so this is an important question. For an exercise session lasting 90 minutes or less, water should suffice. Exercise sessions lasting longer than 90 minutes, however, can result in a significant loss of fluid and electrolytes. After exercising that long, you should drink an isotonic sports drink. Isotonic sports drinks are formulated to contain the same concentration of nonpenetrating solutes as our cells, hence the name *isotonic*. They are designed to restore hydration, replace electrolytes, and replenish carbohydrates. ■

Filtration

Filtration is a passive process in which water and solutes are forced through a membrane by hydrostatic (fluid) pressure. For example, fluids and solutes filter out of the capillaries in the kidneys and into the kidney tubules because the blood pressure in the capillaries is greater than the fluid pressure in the tubules. Filtration is not selective. The amount of filtrate (fluids and solutes) formed depends almost entirely on the pressure gradient (difference in pressure on the two sides of the membrane) and on the size of the membrane pores.

Activity 6

Observing the Process of Filtration

1. Obtain the following equipment: a ring stand, ring, and ring clamp; a funnel; a piece of filter paper; a beaker; a 10-ml graduated cylinder; a solution containing uncooked starch, powdered charcoal, and copper sulfate; and a dropper bottle of Lugol's iodine. Attach the ring to the ring stand with the clamp.

2. Fold the filter paper in half twice, open it into a cone, and place it in a funnel. Place the funnel in the ring of the ring stand and place a beaker under the funnel. Shake the starch solution, and fill the funnel with it to just below the top of the filter paper. When the steady stream of filtrate changes to countable filtrate drops, count the number of drops formed in 10 seconds and record.

_____ drops

When the funnel is half empty, again count the number of drops formed in 10 seconds, and record the count.

_____ drops

3. After all the fluid has passed through the filter, check the filtrate and paper to see which materials were retained by the paper. If the filtrate is blue, the copper sulfate passed. Check both the paper and filtrate for black particles to see whether the charcoal passed. Finally, using a 10-ml graduated cylinder, put a 2-ml filtrate sample into a test tube. Add several drops of Lugol's iodine. If the sample turns blue/black when iodine is added, starch is present in the filtrate.

Passed: _____

Retained: _____

What does the filter paper represent? _____

During which counting interval was the filtration rate

greatest? _____

Explain: _____

_____

_____

What characteristic of the three solutes determined whether or not they passed through the filter paper?

_____

Active Processes

Whenever a cell uses the bond energy of ATP to move substances across its boundaries, the process is an *active process*. Substances moved by active means are generally unable to pass by diffusion. They may not be lipid soluble; they may be too large to pass through the membrane channels; or they may have to move against rather than with a concentration gradient. There are two types of active processes: *active transport* and *vesicular transport.*

Active Transport

Like carrier-mediated facilitated diffusion, **active transport** requires carrier proteins that combine specifically with the transported substance. Active transport may be primary, driven directly by hydrolysis of ATP, or secondary, driven indirectly by energy stored in ionic gradients. In most cases, the substances move against concentration or electrochemical gradients or both. These substances are insoluble in lipid and too large to pass through membrane channels but are necessary for cell life.

Vesicular Transport

In **vesicular transport,** fluids containing large particles and macromolecules are transported across cellular membranes inside membranous sacs called *vesicles*. Like active transport, vesicular transport moves substances into the cell (**endocytosis)** and out of the cell (**exocytosis).** Vesicular transport requires energy, usually in the form of ATP, and all forms of vesicular transport involve protein-coated vesicles to some extent.

There are three types of endocytosis: phagocytosis, pinocytosis, and receptor-mediated endocytosis. In **phagocytosis** ("cell eating"), the cell engulfs some relatively large or solid material such as a clump of bacteria, cell debris, or inanimate particles (**Figure 5.4a**, p. 62). When a particle binds to receptors on the cell's surface, cytoplasmic extensions called pseudopods form and flow around the particle. This produces a vesicle called a *phagosome*. In most cases, the phagosome then fuses with a lysosome and its contents are digested. Indigestible contents are ejected from the cell by exocytosis. In the human body, only macrophages and certain other white blood cells perform phagocytosis. These cells help protect the body from disease-causing microorganisms and cancer cells.

In **pinocytosis** ("cell drinking"), also called **fluid-phase endocytosis,** the cell "gulps" a drop of extracellular fluid containing dissolved molecules (Figure 5.4b). Since no receptors are involved, the process is nonspecific. Unlike phagocytosis, pinocytosis is a routine activity of most cells, allowing them a way of sampling the extracellular fluid. It is particularly important in cells that absorb nutrients, such as cells that line the intestines.

The main mechanism for *specific* endocytosis of most macromolecules is **receptor-mediated endocytosis** (Figure 5.4c). The receptors for this process are plasma membrane proteins that bind only certain substances. This exquisitely selective mechanism allows cells to concentrate material that is present only in small amounts in the extracellular fluid. The ingested vesicle may fuse with a lysosome that either digests or releases its contents, or it may be transported across the cell to release its contents by exocytosis. The latter case is common in endothelial cells lining blood vessels because it provides a quick means to get substances from blood to extracellular fluid. Substances taken up by receptor-mediated endocytosis include enzymes, insulin and some other hormones, cholesterol (attached to a transport protein), and iron.

facilitated diffusion: _____

_____

osmosis: _____

_____

filtration: _____

_____

vesicular transport: _____

_____

endocytosis: _____

_____

exocytosis: _____

_____

Slide box 10

6

Classification of Tissues

Objectives

☐ Name the four primary tissue types in the human body, and state a general function of each.

☐ Name the major subcategories of the primary tissue types, and identify the tissues of each subcategory microscopically or in an appropriate image.

☐ State the locations of the various tissues in the body.

☐ List the general function and structural characteristics of each of the tissues studied.

Materials

- Compound microscope
- Immersion oil
- Prepared slides of simple squamous, simple cuboidal, simple columnar, stratified squamous (nonkeratinized), stratified cuboidal, stratified columnar, pseudostratified ciliated columnar, and transitional epithelium
- Prepared slides of mesenchyme; of adipose, areolar, reticular, and dense (both regular and irregular connective tissues); of hyaline and elastic cartilage; of fibrocartilage; of bone (x.s.); and of blood
- Prepared slide of nervous tissue (spinal cord smear)
- Prepared slides of skeletal, cardiac, and smooth muscle (l.s.)
- Envelopes containing index cards with color photomicrographs of tissues

MasteringA&P®

For related exercise study tools, go to the Study Area of **MasteringA&P**. There you will find:

- Practice Anatomy Lab **PAL**
- A&PFlix **A&PFlix**
- PhysioEx **PEx**
- Practice quizzes, Histology Atlas, eText, Videos, and more!

Pre-Lab Quiz

1. Groups of cells that are anatomically similar and share a function are called:
   a. organ systems
   b. organisms
   c. organs
   d. tissues

2. How many primary tissue types are found in the human body? _____

3. Circle True or False. Endocrine and exocrine glands are classified as epithelium because they usually develop from epithelial membranes.

4. Epithelial tissues can be classified according to cell shape. _____ epithelial cells are scalelike and flattened.
   a. Columnar
   b. Cuboidal
   c. Squamous
   d. Transitional

5. All connective tissue is derived from an embryonic tissue known as:
   a. cartilage
   b. ground substance
   c. mesenchyme
   d. reticular

6. All the following are examples of connective tissue *except*:
   a. bones
   b. ligaments
   c. neurons
   d. tendons

7. Circle True or False. Blood is a type of connective tissue.

8. Circle the correct underlined term. Of the two major cell types found in nervous tissue, <u>neurons</u> / <u>neuroglial cells</u> are highly specialized to generate and conduct electrical signals.

9. How many basic types of muscle tissue are there? _____

10. This type of muscle tissue is found in the walls of hollow organs. It has no striations, and its cells are spindle shaped. It is:
    a. cardiac muscle
    b. skeletal muscle
    c. smooth muscle

C ells are the building blocks of life and the all-inclusive functional units of unicellular organisms. However, in higher organisms, cells do not usually operate as isolated, independent entities. In humans and other multicellular organisms, cells depend on one another and cooperate to maintain homeostasis in the body.

With a few exceptions, even the most complex animal starts out as a single cell, the fertilized egg, which divides almost endlessly. The trillions of cells that result become specialized for a particular function; some become supportive bone, others the transparent lens of the eye, still others skin cells, and so on. Thus a division of labor exists, with certain groups of cells highly specialized to perform functions that benefit the organism as a whole.

Groups of cells that are similar in structure and function are called **tissues.** The four primary tissue types—epithelium, connective tissue, nervous tissue, and muscle—have distinctive structures, patterns, and functions. The four primary tissues are further divided into subcategories, as described shortly.

To perform specific body functions, the tissues are organized into **organs** such as the heart, kidneys, and lungs. Most organs contain several representatives of the primary tissues, and the arrangement of these tissues determines the organ's structure and function. Thus **histology,** the study of tissues, complements a study of gross anatomy and provides the structural basis for a study of organ physiology.

The main objective of this exercise is to familiarize you with the major similarities and differences of the primary tissues, so that when the tissue composition of an organ is described, you will be able to more easily understand (and perhaps even predict) the organ's major function.

Epithelial Tissue

Epithelial tissue, or an **epithelium,** is a sheet of cells that covers a body surface or lines a body cavity. It occurs in the body as (1) covering and lining epithelium and (2) glandular epithelium. Covering and lining epithelium forms the outer layer of the skin and lines body cavities that open to the outside. It covers the walls and organs of the closed ventral body cavity. Since glands almost invariably develop from epithelial sheets, glands are also classed as epithelium.

Epithelial functions include protection, absorption, filtration, excretion, secretion, and sensory reception. For example, the epithelium covering the body surface protects against bacterial invasion and chemical damage. Epithelium specialized to absorb substances lines the stomach and small intestine. In the kidney tubules, the epithelium absorbs, secretes, and filters. Secretion is a specialty of the glands.

The following characteristics distinguish epithelial tissues from other types:

- Polarity. The membranes always have one free surface, called the *apical surface,* and typically that surface is significantly different from the *basal surface.*

- Specialized contacts. Cells fit closely together to form membranes, or sheets of cells, and are bound together by specialized junctions.

- Supported by connective tissue. The cells are attached to and supported by an adhesive **basement membrane,** which is an amorphous material secreted partly by the epithelial cells (*basal lamina*) and connective tissue cells (*reticular lamina*) that lie next to each other.

- Avascular but innervated. Epithelial tissues are supplied by nerves but have no blood supply of their own (are avascular). Instead they depend on diffusion of nutrients from the underlying connective tissue.

- Regeneration. If well nourished, epithelial cells can easily divide to regenerate the tissue. This is an important characteristic because many epithelia are subjected to a good deal of friction.

The covering and lining epithelia are classified according to two criteria—arrangement or relative number of layers and cell shape (**Figure 6.1**). On the basis of arrangement, epithelia are classified as follows:

- **Simple** epithelia consist of one layer of cells attached to the basement membrane.

- **Stratified** epithelia consist of two or more layers of cells.

Based on cell shape, epithelia are classified into three categories:

- **Squamous** (scalelike)
- **Cuboidal** (cubelike)
- **Columnar** (column-shaped)

The terms denoting shape and arrangement of the epithelial cells are combined to describe the epithelium fully. *Stratified epithelia are named according to the cells at the apical surface of the epithelial sheet,* not those resting on the basement membrane.

There are, in addition, two less easily categorized types of epithelia.

- **Pseudostratified epithelium** is actually a simple columnar epithelium (one layer of cells), but because its cells vary in height and the nuclei lie at different levels above the basement membrane, it gives the false appearance of being stratified. This epithelium is often ciliated.

- **Transitional epithelium** is a rather peculiar stratified squamous epithelium formed of rounded, or "plump," cells with the ability to slide over one another to allow the organ to be stretched. Transitional epithelium is found only in urinary system organs subjected to stretch, such as the bladder. The superficial cells are flattened (like true squamous cells) when the organ is full and rounded when the organ is empty.

Epithelial cells forming glands are highly specialized to remove materials from the blood and to manufacture them into new materials, which they then secrete. There are two types of glands, *endocrine* and *exocrine,* shown in **Figure 6.2. Endocrine glands** lose their surface connection (duct) as they develop; thus they are referred to as ductless glands. They secrete hormones into the extracellular fluid, and from there the hormones enter the blood or the lymphatic vessels that weave through the glands. **Exocrine glands** retain their ducts, and their secretions empty through these ducts either to the body surface or into body cavities. The exocrine glands include the sweat and oil glands, liver, and pancreas.

The most common types of epithelia, their characteristic locations in the body, and their functions are described in **Figure 6.3**, p. 70.

(*Text continues on page 74.*)

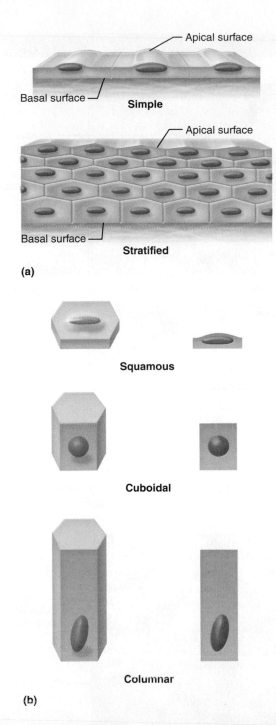

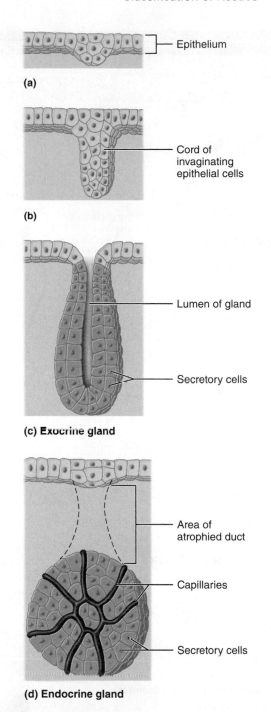

Figure 6.1 Classification of epithelia. (a) Classification based on number of cell layers. **(b)** Classification based on cell shape. For each category, a whole cell is shown on the left, and a longitudinal section is shown on the right.

Figure 6.2 Formation of endocrine and exocrine glands from epithelial sheets. (a) Epithelial cells grow and push into the underlying tissue. **(b)** A cord of epithelial cells forms. **(c)** In an exocrine gland, a lumen (cavity) forms. The inner cells form the duct, the outer cells produce the secretion. **(d)** In a forming endocrine gland, the connecting duct cells atrophy, leaving the secretory cells with no connection to the epithelial surface. However, they do become heavily invested with blood and lymphatic vessels that receive the secretions.

6

(a) Simple squamous epithelium

Description: Single layer of flattened cells with disc-shaped central nuclei and sparse cytoplasm; the simplest of the epithelia.

Function: Allows materials to pass by diffusion and filtration in sites where protection is not important; secretes lubricating substances in serosae.

Location: Kidney glomeruli; air sacs of lungs; lining of heart, blood vessels, and lymphatic vessels; lining of ventral body cavity (serosae).

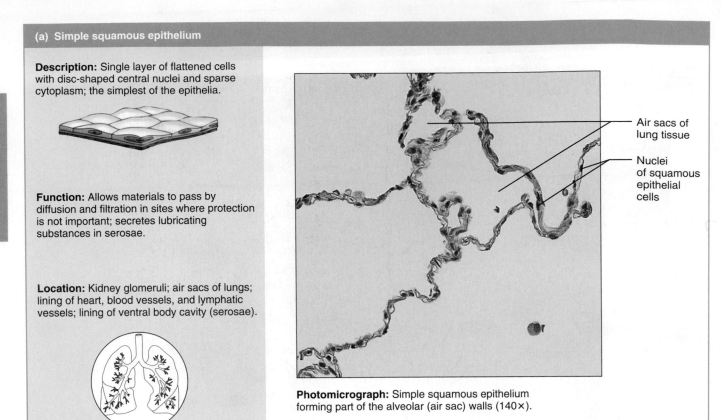

Air sacs of lung tissue

Nuclei of squamous epithelial cells

Photomicrograph: Simple squamous epithelium forming part of the alveolar (air sac) walls (140×).

(b) Simple cuboidal epithelium

Description: Single layer of cubelike cells with large, spherical central nuclei.

Function: Secretion and absorption.

Location: Kidney tubules; ducts and secretory portions of small glands; ovary surface.

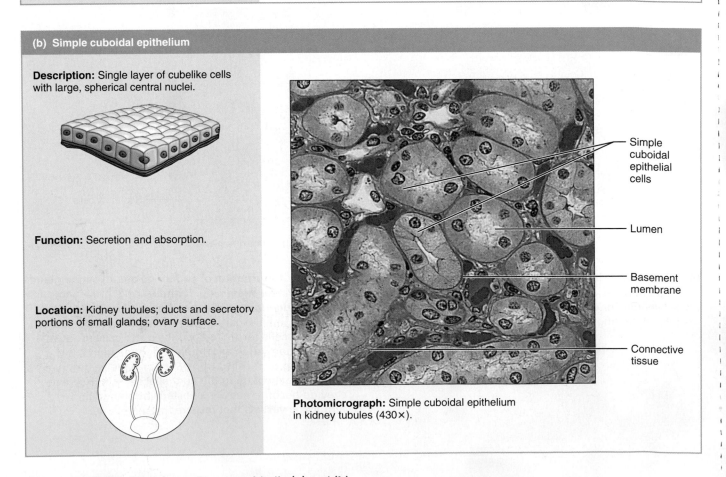

Simple cuboidal epithelial cells

Lumen

Basement membrane

Connective tissue

Photomicrograph: Simple cuboidal epithelium in kidney tubules (430×).

Figure 6.3 Epithelial tissues. Simple epithelia **(a)** and **(b)**.

(c) Simple columnar epithelium

Description: Single layer of tall cells with *round* to *oval* nuclei; some cells bear cilia; layer may contain mucus-secreting unicellular glands (goblet cells).

Function: Absorption; secretion of mucus, enzymes, and other substances; ciliated type propels mucus (or reproductive cells) by ciliary action.

Location: Nonciliated type lines most of the digestive tract (stomach to rectum), gallbladder, and excretory ducts of some glands; ciliated variety lines small bronchi, uterine tubes, and some regions of the uterus.

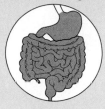

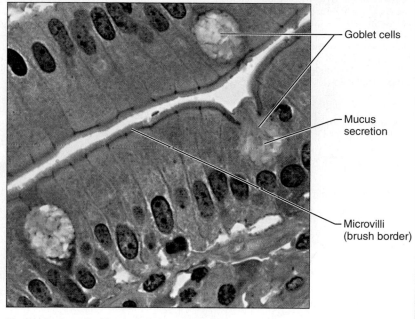

Goblet cells

Mucus secretion

Microvilli (brush border)

Photomicrograph: Simple columnar epithelium containing goblet cells from the small intestine (640×).

(d) Pseudostratified columnar epithelium

Description: Single layer of cells of differing heights, some not reaching the free surface; nuclei seen at different levels; may contain mucus-secreting goblet cells and bear cilia.

Function: Secretes substances, particularly mucus; propulsion of mucus by ciliary action.

Location: Nonciliated type in male's sperm-carrying ducts and ducts of large glands; ciliated variety lines the trachea, most of the upper respiratory tract.

Trachea

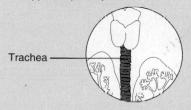

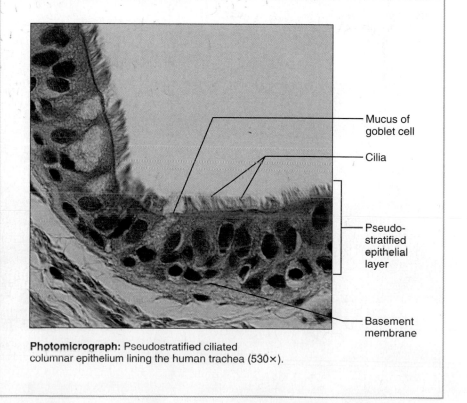

Mucus of goblet cell

Cilia

Pseudo-stratified epithelial layer

Basement membrane

Photomicrograph: Pseudostratified ciliated columnar epithelium lining the human trachea (530×).

Figure 6.3 *(continued)* Simple epithelia (c) and (d).

6

(e) Stratified squamous epithelium

Description: Thick membrane composed of several cell layers; basal cells are cuboidal or columnar and metabolically active; surface cells are flattened (squamous); in the keratinized type, the surface cells are full of keratin and dead; basal cells are active in mitosis and produce the cells of the more superficial layers.

Function: Protects underlying tissues in areas subjected to abrasion.

Location: Nonkeratinized type forms the moist linings of the esophagus, mouth, and vagina; keratinized variety forms the epidermis of the skin, a dry membrane.

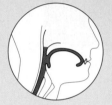

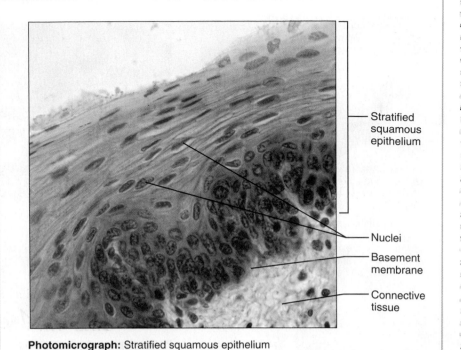

Stratified squamous epithelium

Nuclei

Basement membrane

Connective tissue

Photomicrograph: Stratified squamous epithelium lining the esophagus (280×).

(f) Stratified cuboidal epithelium

Description: Generally two layers of cubelike cells.

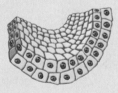

Function: Protection.

Location: Largest ducts of sweat glands, mammary glands, and salivary glands.

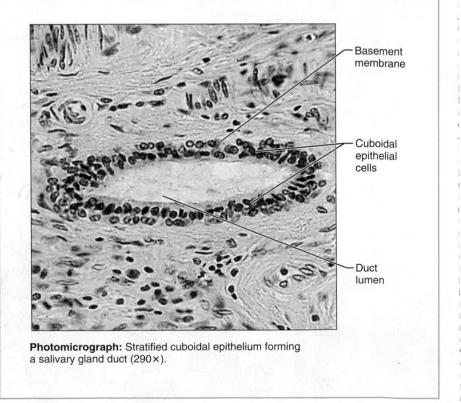

Basement membrane

Cuboidal epithelial cells

Duct lumen

Photomicrograph: Stratified cuboidal epithelium forming a salivary gland duct (290×).

Figure 6.3 *(continued)* **Epithelial tissues.** Stratified epithelia **(e)** and **(f)**.

(g) Stratified columnar epithelium

Description: Several cell layers; basal cells usually cuboidal; superficial cells elongated and columnar.

Function: Protection; secretion.

Location: Rare in the body; small amounts in male urethra and in large ducts of some glands.

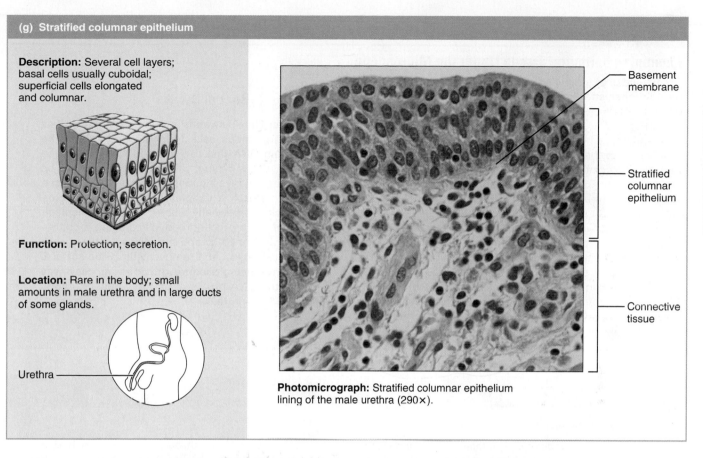

Urethra

Basement membrane

Stratified columnar epithelium

Connective tissue

6

Photomicrograph: Stratified columnar epithelium lining of the male urethra (290×).

(h) Transitional epithelium

Description: Resembles both stratified squamous and stratified cuboidal; basal cells cuboidal or columnar; surface cells dome shaped or squamouslike, depending on degree of organ stretch.

Function: Stretches readily and permits distension of urinary organ by contained urine.

Location: Lines the ureters, urinary bladder, and part of the urethra.

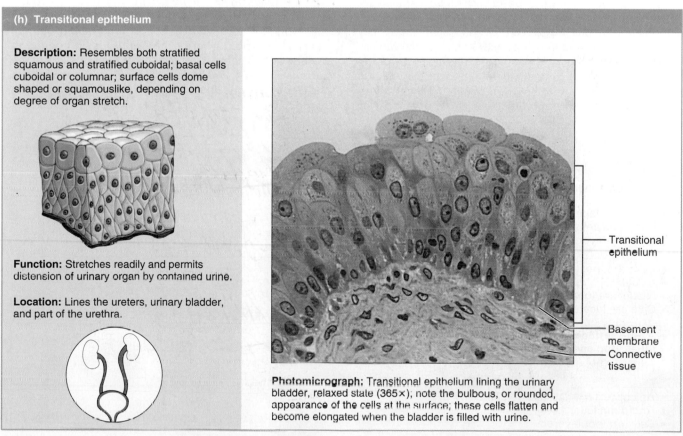

Transitional epithelium

Basement membrane

Connective tissue

Photomicrograph: Transitional epithelium lining the urinary bladder, relaxed state (365×); note the bulbous, or rounded, appearance of the cells at the surface; these cells flatten and become elongated when the bladder is filled with urine.

Figure 6.3 *(continued)* Stratified epithelia **(g)** and **(h)**.

Activity 1

Examining Epithelial Tissue Under the Microscope

Obtain slides of simple squamous, simple cuboidal, simple columnar, stratified squamous (nonkeratinized), pseudostratified ciliated columnar, stratified cuboidal, stratified columnar, and transitional epithelia. Examine each carefully, and notice how the epithelial cells fit closely together to form intact sheets of cells, a necessity for a tissue that forms linings or the coverings of membranes. Scan each epithelial type for modifications for specific functions, such as cilia (motile cell projections that help to move substances along the cell surface), and microvilli, which increase the surface area for absorption. Also be alert for goblet cells, which secrete lubricating mucus. Compare your observations with the descriptions and photomicrographs in Figure 6.3.

While working, check the questions in the Review Sheet at the end of this exercise. A number of the questions there refer to some of the observations you are asked to make during your microscopic study.

WHY THIS MATTERS | Buccal Swabs

A buccal, or cheek, swab is a method used to collect stratified squamous cells from the oral cavity. The cells contain DNA that can be used for DNA fingerprinting or tissue typing. DNA fingerprinting can be used in criminal investigations, and tissue typing can be used to match a recipient with a donor for organ transplant, especially a bone marrow transplant. The buccal swab procedure involves using a cotton-tipped applicator to scrape the inside of the mouth in the buccal region and remove cells at the surface. This noninvasive procedure provides an easy way to obtain the DNA profile of an individual, a unique molecular "signature." ■

Group Challenge 1

Identifying Epithelial Tissues

Following your observations of epithelial tissues under the microscope, obtain an envelope for each group that contains images of various epithelial tissues. With your lab manual closed, remove one image at a time and identify the epithelium. One member of the group will function as the verifier, whose job is to make sure that the identification is correct.

After you have correctly identified all of the images, sort them into groups to help you remember them. (*Hint:* You could sort them according to cell shape or number of layers of epithelial cells.)

Now, carefully go through each group and try to list one place in the body where the tissue is found and one function for it. After you have correctly listed the locations, take your lists and draw some general conclusions about where epithelial tissues are found in the body. Then compare and contrast the functions of the various epithelia. Finally, identify the tissues described in the **Group Challenge 1** chart, and list several locations in the body.

Group Challenge 1: Epithelial Tissue IDs		
Magnified appearance	**Tissue type**	**Locations in the body**
• Apical surface has dome-shaped cells (flattened cells may also be mixed in) • Multiple layers of cells are present		
• Cells are mostly columnar • Not all cells reach the apical surface • Nuclei are located at different levels • Cilia are located at the apical surface		
• Apical surface has flattened cells with very little cytoplasm • Cells are not layered		
• Apical surface has square cells with a round nucleus • Cells are not layered		

Connective Tissue

Connective tissue is found in all parts of the body as discrete structures or as part of various body organs. It is the most abundant and widely distributed of the tissue types.

There are four main types of adult connective tissue. These are **connective tissue proper, cartilage, bone,** and **blood.** All of these derive from an embryonic tissue called *mesenchyme.* Connective tissue proper has two subclasses: **loose connective tissues** (areolar, adipose, and reticular) and **dense connective tissues** (dense regular, dense irregular, and elastic). **Connective tissues** perform a variety of functions, but they primarily protect, support, insulate, and bind together other tissues of the body. For example, bones are composed of connective tissue (**bone,** or **osseous tissue**), and they protect and support other body tissues and organs. The ligaments and tendons (**dense regular connective tissue**) bind the bones together or connect skeletal muscles to bones.

Areolar connective tissue (**Figure 6.4**) is a soft packaging material that cushions and protects body organs. **Adipose** (fat) tissue provides insulation for the body tissues and a source of stored energy. Connective tissue also serves a vital function in the repair of all body tissues, since many wounds are repaired by connective tissue in the form of scar tissue.

The characteristics of connective tissue include the following:

• With a few exceptions (cartilages, tendons, and ligaments, which are poorly vascularized), connective tissues have a rich supply of blood vessels.

• Connective tissues are composed of many types of cells.

• There is a great deal of noncellular, nonliving material (matrix) between the cells of connective tissue.

The nonliving material between the cells—the **extracellular matrix**—deserves a bit more explanation because it distinguishes connective tissue from all other tissues. It is produced by the cells and then extruded. The matrix is primarily responsible for the strength associated with connective tissue, but there is variation in the amount of matrix. At one extreme, adipose tissue is composed mostly of cells. At the opposite extreme, bone and cartilage have few cells and large amounts of matrix.

The matrix has two components—ground substance and fibers. The **ground substance** is composed chiefly of interstitial fluid, cell adhesion proteins, and proteoglycans. Depending on its specific composition, the ground substance may be liquid, semisolid, gel-like, or very hard. When the matrix is firm, as in cartilage and bone, the connective tissue cells reside in cavities in the matrix called *lacunae.* The fibers, which provide support, include **collagen** (white) **fibers, elastic** (yellow) **fibers,** and **reticular** (fine collagen) **fibers.** Of these, the collagen fibers are most abundant.

The connective tissues have a common structural plan seen best in *areolar connective tissue* (Figure 6.4). Since all other connective tissues are variations of areolar, it is considered the model, or prototype, of the connective tissues. Notice that areolar tissue has all three varieties of fibers, but they are sparsely

6

Cell types

Extracellular matrix

Ground substance

Fibers

• Collagen fiber

• Elastic fiber

• Reticular fiber

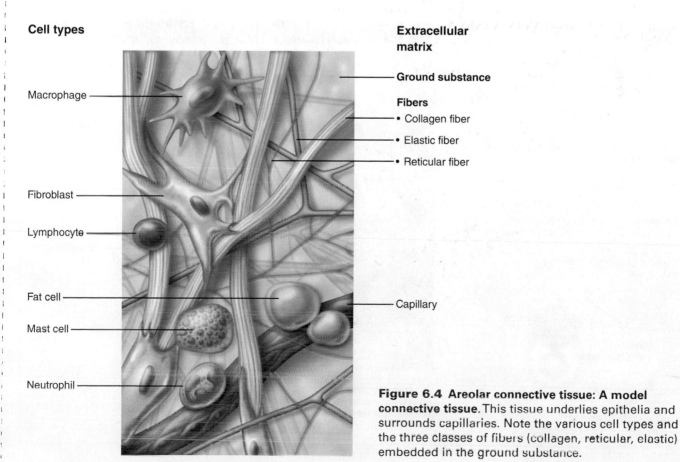

Macrophage

Fibroblast

Lymphocyte

Fat cell

Mast cell

Neutrophil

Capillary

Figure 6.4 Areolar connective tissue: A model connective tissue. This tissue underlies epithelia and surrounds capillaries. Note the various cell types and the three classes of fibers (collagen, reticular, elastic) embedded in the ground substance.

arranged in its transparent gel-like ground substance (Figure 6.4). The cell type that secretes its matrix is the *fibroblast*, but a wide variety of other cells (including phagocytic cells such as macrophages and certain white blood cells and mast cells that act in the inflammatory response) are present as well. The more durable connective tissues, such as bone, cartilage, and the dense connective tissues, characteristically have a firm ground substance and many more fibers.

Figure 6.5 lists the general characteristics, location, and function of some of the connective tissues found in the body.

Activity 2

Examining Connective Tissue Under the Microscope

Obtain prepared slides of mesenchyme; of adipose, areolar, reticular, dense regular, elastic, and dense irregular connective tissue; of hyaline and elastic cartilage and fibrocartilage; of osseous connective tissue (bone); and of blood. Compare your observations with the views illustrated in Figure 6.5.

Distinguish the living cells from the matrix. Pay particular attention to the denseness and arrangement of the matrix. For example, notice how the matrix of the dense regular and dense irregular connective tissues, respectively making up tendons and the dermis of the skin, is packed with collagen fibers. Note also that in the *regular* variety (tendon), the fibers are all running in the same direction, whereas in the dermis they appear to be running in many directions.

While examining the areolar connective tissue, notice how much empty space there appears to be (*areol* = small empty space), and distinguish the collagen fibers from the coiled elastic fibers. Identify the starlike fibroblasts. Also, try to locate a **mast cell,** which has large, darkly staining granules in its cytoplasm (*mast* = stuffed full of granules). This cell type

releases histamine, which makes capillaries more permeable during inflammation and allergies and thus is partially responsible for that "runny nose" of some allergies.

In adipose tissue, locate a "signet ring" cell, a fat cell in which the nucleus can be seen pushed to one side by the large, fat-filled vacuole that appears to be a large empty space. Also notice how little matrix there is in adipose (fat) tissue. Distinguish the living cells from the matrix in the dense connective tissue, bone, and hyaline cartilage preparations.

Scan the blood slide at low and then high power to examine the general shape of the red blood cells. Then, switch to the oil immersion lens for a closer look at the various types of white blood cells. How does the matrix of blood differ from all other connective tissues?

_____

_____

(a) Embryonic connective tissue: Mesenchyme

Description: Embryonic connective tissue; gel-like ground substance containing fibers; star-shaped mesenchymal cells.

Function: Gives rise to all other connective tissue types.

Location: Primarily in embryo.

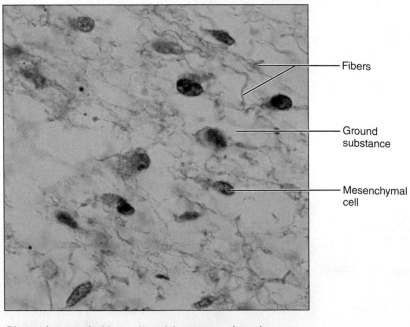

— Fibers

— Ground substance

— Mesenchymal cell

Photomicrograph: Mesenchymal tissue, an embryonic connective tissue (627×); the clear-appearing background is the fluid ground substance of the matrix; notice the fine, sparse fibers.

Figure 6.5 Connective tissues. Embryonic connective tissue **(a).**

(b) Connective tissue proper: loose connective tissue, areolar

Description: Gel-like matrix with all three fiber types; cells: fibroblasts, macrophages, mast cells, and some white blood cells.

Function: Wraps and cushions organs; its macrophages phagocytize bacteria; plays important role in inflammation; holds and conveys tissue fluid.

Location: Widely distributed under epithelia of body, e.g., forms lamina propria of mucous membranes; packages organs; surrounds capillaries.

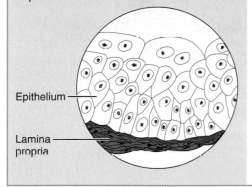

Epithelium

Lamina propria

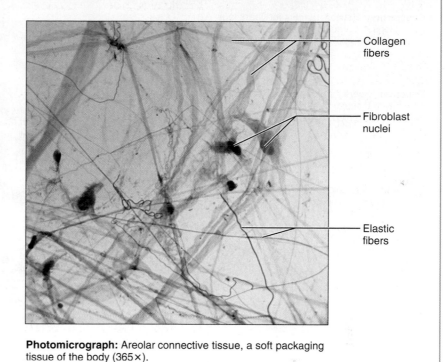

Collagen fibers

Fibroblast nuclei

Elastic fibers

Photomicrograph: Areolar connective tissue, a soft packaging tissue of the body (365×).

(c) Connective tissue proper: loose connective tissue, adipose

Description: Matrix as in areolar, but very sparse; closely packed adipocytes, or fat cells, have nucleus pushed to the side by large fat droplet.

Function: Provides reserve fuel; insulates against heat loss; supports and protects organs.

Location: Under skin; around kidneys and eyeballs; within abdomen; in breasts.

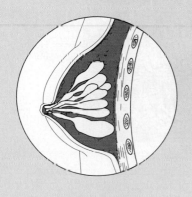

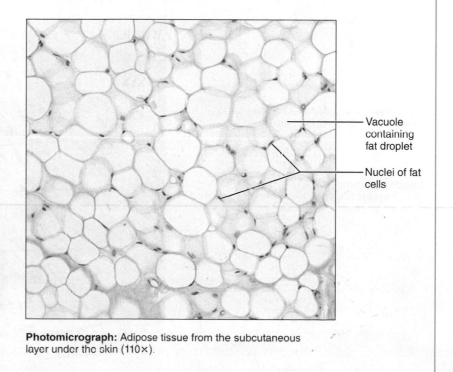

Vacuole containing fat droplet

Nuclei of fat cells

Photomicrograph: Adipose tissue from the subcutaneous layer under the skin (110×).

Figure 6.5 (continued) Connective tissue proper (b) and (c).

(d) Connective tissue proper: loose connective tissue, reticular

Description: Network of reticular fibers in a typical loose ground substance; reticular cells lie on the network.

Function: Fibers form a soft internal skeleton (stroma) that supports other cell types, including white blood cells, mast cells, and macrophages.

Location: Lymphoid organs (lymph nodes, bone marrow, and spleen).

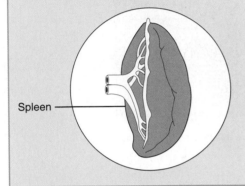

Spleen

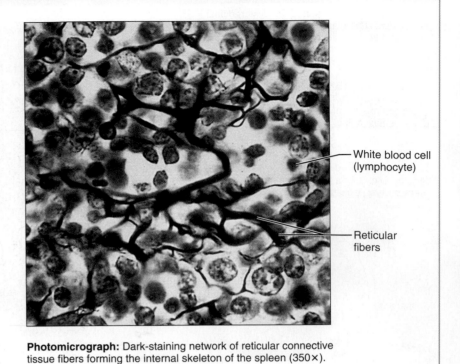

White blood cell (lymphocyte)

Reticular fibers

Photomicrograph: Dark-staining network of reticular connective tissue fibers forming the internal skeleton of the spleen (350×).

(e) Connective tissue proper: dense regular connective tissue

Description: Primarily parallel collagen fibers; a few elastic fibers; major cell type is the fibroblast.

Function: Attaches muscles to bones or to other muscles; attaches bones to bones; withstands great tensile stress when pulling force is applied in one direction.

Location: Tendons, most ligaments, aponeuroses.

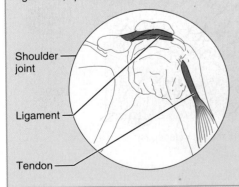

Shoulder joint

Ligament

Tendon

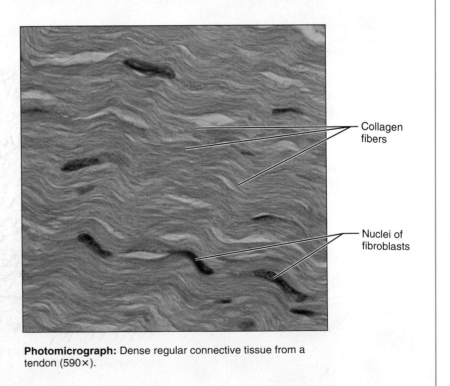

Collagen fibers

Nuclei of fibroblasts

Photomicrograph: Dense regular connective tissue from a tendon (590×).

Figure 6.5 *(continued)* **Connective tissues.** Connective tissue proper **(d)** and **(e)**.

(f) Connective tissue proper: elastic connective tissue

Description: Dense regular connective tissue containing a high proportion of elastic fibers.

Function: Allows recoil of tissue following stretching; maintains pulsatile flow of blood through arteries; aids passive recoil of lungs following inspiration.

Location: Walls of large arteries; within certain ligaments associated with the vertebral column; within the walls of the bronchial tubes.

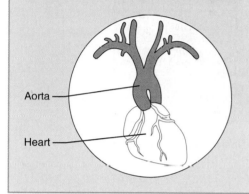

Aorta

Heart

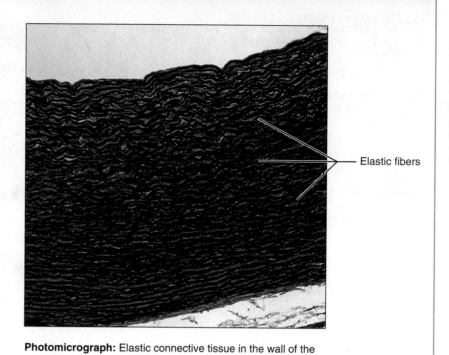

Elastic fibers

Photomicrograph: Elastic connective tissue in the wall of the aorta (250×).

(g) Connective tissue proper: dense irregular connective tissue

Description: Primarily irregularly arranged collagen fibers; some elastic fibers; major cell type is the fibroblast.

Function: Able to withstand tension exerted in many directions; provides structural strength.

Location: Fibrous capsules of organs and of joints; dermis of the skin; submucosa of digestive tract.

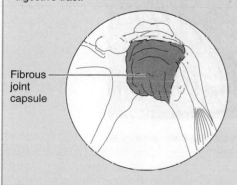

Fibrous joint capsule

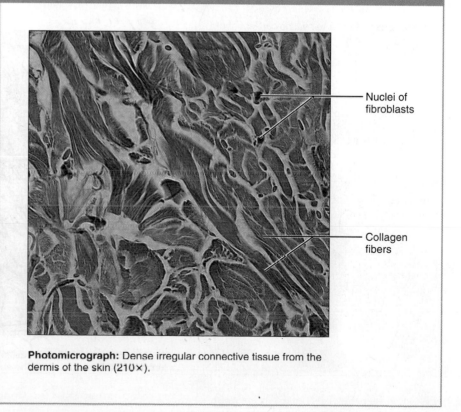

Nuclei of fibroblasts

Collagen fibers

Photomicrograph: Dense irregular connective tissue from the dermis of the skin (210×).

Figure 6.5 *(continued)* Connective tissue proper (f) and (g).

6

(h) Cartilage: hyaline

Description: Amorphous but firm matrix; collagen fibers form an imperceptible network; chondroblasts produce the matrix and, when mature (chondrocytes), lie in lacunae.

Function: Supports and reinforces; serves as resilient cushion; resists compressive stress.

Location: Forms most of the embryonic skeleton; covers the ends of long bones in joint cavities; forms costal cartilages of the ribs; cartilages of the nose, trachea, and larynx.

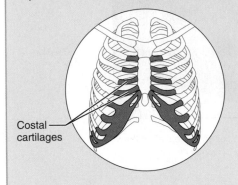

Costal cartilages

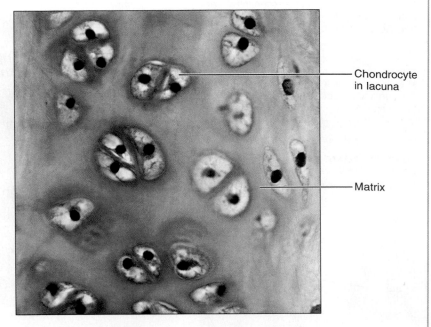

Chondrocyte in lacuna

Matrix

Photomicrograph: Hyaline cartilage from a costal cartilage of a rib (470×).

(i) Cartilage: elastic

Description: Similar to hyaline cartilage, but more elastic fibers in matrix.

Function: Maintains the shape of a structure while allowing great flexibility.

Location: Supports the external ear (auricle); epiglottis.

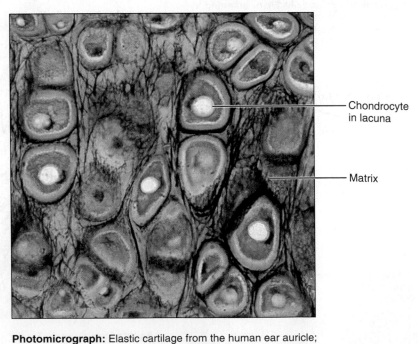

Chondrocyte in lacuna

Matrix

Photomicrograph: Elastic cartilage from the human ear auricle; forms the flexible skeleton of the ear (510×).

Figure 6.5 *(continued)* **Connective tissues.** Cartilage **(h)** and **(i)**.

(j) Cartilage: fibrocartilage

Description: Matrix similar to but less firm than matrix in hyaline cartilage; thick collagen fibers predominate.

Function: Tensile strength with the ability to absorb compressive shock.

Location: Intervertebral discs; pubic symphysis; discs of knee joint.

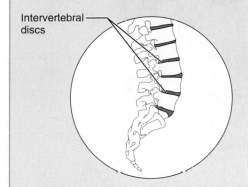

Intervertebral discs

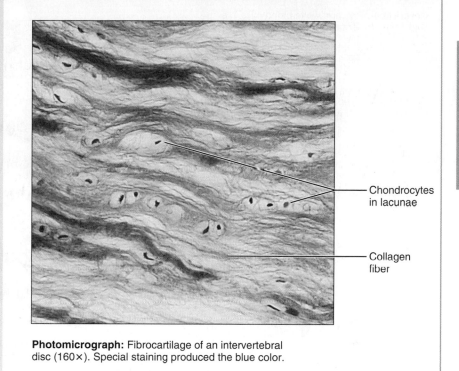

Chondrocytes in lacunae

Collagen fiber

Photomicrograph: Fibrocartilage of an intervertebral disc (160×). Special staining produced the blue color.

(k) Bones (osseous tissue)

Description: Hard, calcified matrix containing many collagen fibers; osteocytes lie in lacunae. Very well vascularized.

Function: Bone supports and protects (by enclosing); provides levers for the muscles to act on; stores calcium and other minerals and fat; marrow inside bones is the site for blood cell formation (hematopoiesis).

Location: Bones

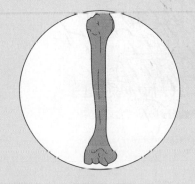

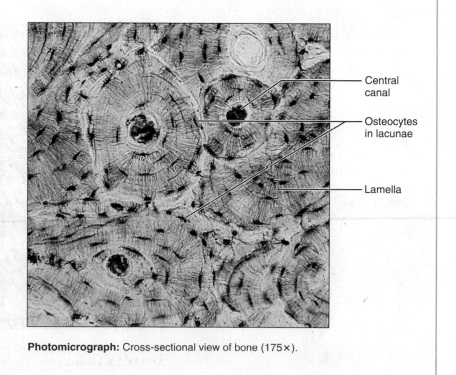

Central canal

Osteocytes in lacunae

Lamella

Photomicrograph: Cross-sectional view of bone (175×).

Figure 6.5 *(continued)* Cartilage (j) and bone (k).

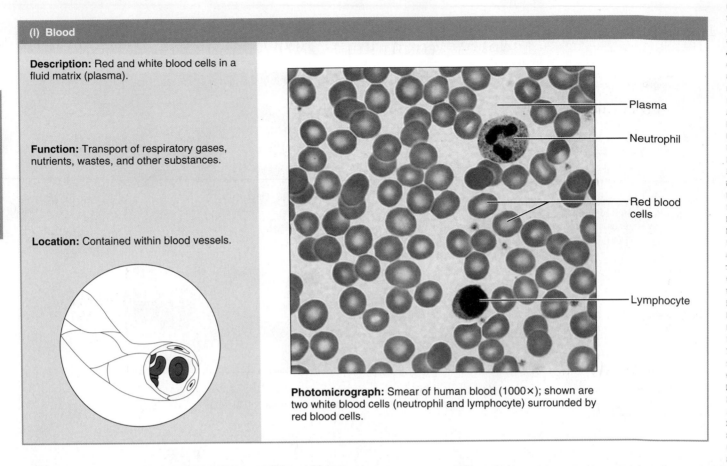

(I) Blood

Description: Red and white blood cells in a fluid matrix (plasma).

Function: Transport of respiratory gases, nutrients, wastes, and other substances.

Location: Contained within blood vessels.

Plasma

Neutrophil

Red blood cells

Lymphocyte

Photomicrograph: Smear of human blood (1000×); shown are two white blood cells (neutrophil and lymphocyte) surrounded by red blood cells.

Figure 6.5 *(continued)* **Connective tissues.** Blood **(I)**.

Nervous Tissue

Nervous tissue is made up of two major cell populations. The **neuroglia** are special supporting cells that protect, support, and insulate the more delicate neurons. The **neurons** are highly specialized to receive stimuli (excitability) and to generate electrical signals that may be sent to all parts of the body (conductivity).

The structure of neurons is markedly different from that of all other body cells. They have a nucleus-containing cell body, and their cytoplasm is drawn out into long extensions (cell processes)—sometimes as long as 1 m (about 3 feet), which allows a single neuron to conduct an electrical signal over relatively long distances. (More detail about the anatomy of the different classes of neurons and neuroglia appears in Exercise 15.)

Activity 3

Examining Nervous Tissue Under the Microscope

Obtain a prepared slide of a spinal cord smear. Locate a neuron and compare it to **Figure 6.6**. Keep the light dim—this will help you see the cellular extensions of the neurons. (See also Figure 15.2 in Exercise 15.)

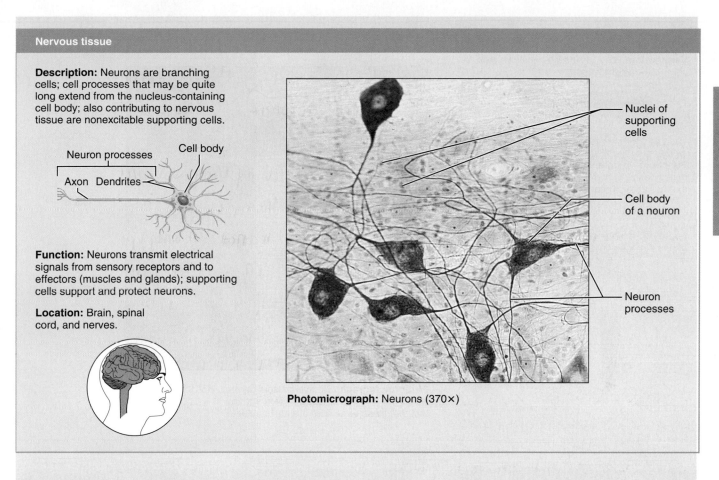

Nervous tissue

Description: Neurons are branching cells; cell processes that may be quite long extend from the nucleus-containing cell body; also contributing to nervous tissue are nonexcitable supporting cells.

Neuron processes

Cell body

Axon   Dendrites

Function: Neurons transmit electrical signals from sensory receptors and to effectors (muscles and glands); supporting cells support and protect neurons.

Location: Brain, spinal cord, and nerves.

Nuclei of supporting cells

Cell body of a nouron

Neuron processes

Photomicrograph: Neurons (370×)

Figure 6.6 Nervous tissue.

Muscle Tissue

Muscle tissue (**Figure 6.7**, p. 84) is highly specialized to contract and produces most types of body movement. As you might expect, muscle cells tend to be elongated, providing a long axis for contraction. The three basic types of muscle tissue are described briefly here.

Skeletal muscle, the "meat," or flesh, of the body, is attached to the skeleton. It is under voluntary control (consciously controlled), and its contraction moves the limbs and other external body parts. The cells of skeletal muscles are long, cylindrical, nonbranching, and multinucleate (several nuclei per cell), with the nuclei pushed to the periphery of the cells; they have obvious *striations* (stripes).

Cardiac muscle is found only in the heart. As it contracts, the heart acts as a pump, propelling the blood into the blood vessels. Cardiac muscle, like skeletal muscle, has striations, but cardiac cells are branching uninucleate cells that interdigitate (fit together) at junctions called **intercalated discs.** These structural modifications allow the cardiac muscle to act as a unit. Cardiac muscle is under involuntary control, which means that we cannot voluntarily or consciously control the operation of the heart.

Smooth muscle is found mainly in the walls of hollow organs (digestive and urinary tract organs, uterus, blood vessels). Typically it has two layers that run at right angles to each other; consequently its contraction can constrict or dilate the lumen (cavity) of an organ and propel substances along predetermined pathways. Smooth muscle cells are quite different in appearance from those of skeletal or cardiac muscle. No striations are visible, and the uninucleate smooth muscle cells are spindle-shaped (tapered at the ends, like a candle). Like cardiac muscle, smooth muscle is under involuntary control.

Activity 4

Examining Muscle Tissue Under the Microscope

Obtain and examine prepared slides of skeletal, cardiac, and smooth muscle. Notice their similarities and dissimilarities in your observations and in the illustrations and photomicrographs in Figure 6.7.

6

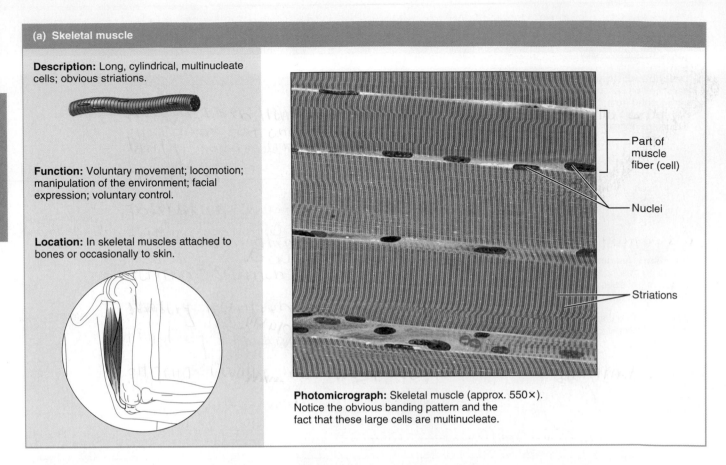

(a) Skeletal muscle

Description: Long, cylindrical, multinucleate cells; obvious striations.

Function: Voluntary movement; locomotion; manipulation of the environment; facial expression; voluntary control.

Location: In skeletal muscles attached to bones or occasionally to skin.

Part of muscle fiber (cell)

Nuclei

Striations

Photomicrograph: Skeletal muscle (approx. 550×). Notice the obvious banding pattern and the fact that these large cells are multinucleate.

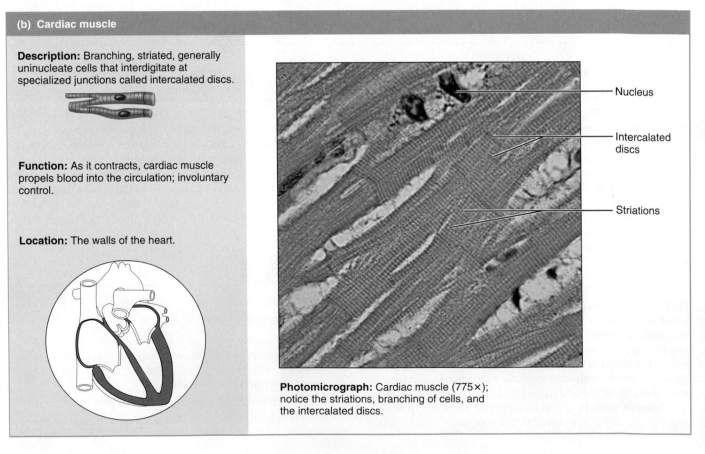

(b) Cardiac muscle

Description: Branching, striated, generally uninucleate cells that interdigitate at specialized junctions called intercalated discs.

Function: As it contracts, cardiac muscle propels blood into the circulation; involuntary control.

Location: The walls of the heart.

Nucleus

Intercalated discs

Striations

Photomicrograph: Cardiac muscle (775×); notice the striations, branching of cells, and the intercalated discs.

Figure 6.7 Muscle tissues. Skeletal muscle **(a)** and cardiac muscle **(b).**

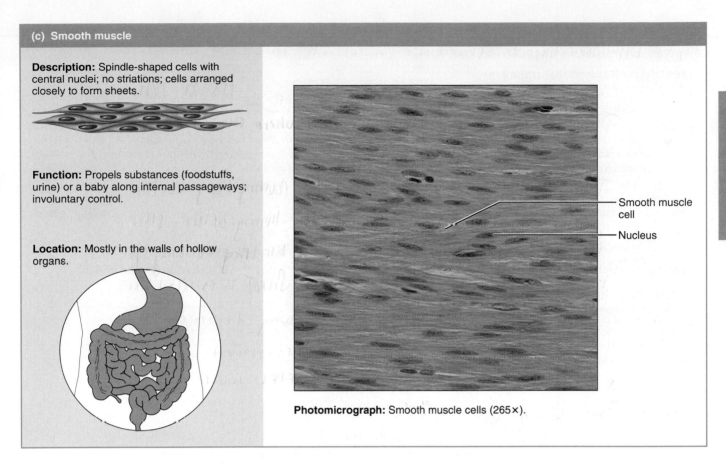

(c) Smooth muscle

Description: Spindle-shaped cells with central nuclei; no striations; cells arranged closely to form sheets.

Function: Propels substances (foodstuffs, urine) or a baby along internal passageways; involuntary control.

Location: Mostly in the walls of hollow organs.

Smooth muscle cell

Nucleus

Photomicrograph: Smooth muscle cells (265×).

Figure 6.7 *(continued)* Smooth muscle **(c)**.

Group Challenge 2

Identifying Connective Tissue

Following your observations of connective tissues under the microscope, obtain an envelope for each group that contains images of some of the tissues you have studied. With your lab manual closed, remove one image at a time and identify the tissue. One member of the group will function as the verifier, whose job is to make sure that the identification is correct.

After you have correctly identified all of the images, sort them into groups according to their primary tissue type and subcategory (if appropriate).

Now, carefully go through each group. List one place in the body where the tissue is found and one function for it.

Next, obtain an envelope from your instructor that contains an image of a section through an organ. Identify all of the tissues that you see in this section, and use it to review the relationship between the location and function of the tissue types that you have studied.

Finally, identify the tissues described in the **Group Challenge 2 chart**, and list several locations in the body.

Group Challenge 2: Connective Tissue IDs		
Magnified appearance	**Tissue type**	**Locations in the body**
• Large, round cells are densely packed • Nucleus is pushed to one side		
• Lacunae (small cavities within the tissue) are present • Lacunae are not arranged in a concentric circle • No visible fibers in the matrix		
• Fibers and cells are loosely packed, with visible space between fibers • Fibers overlap but do not form a network		
• Extracellular fibers run parallel to each other • Nuclei of fibroblasts are visible		
• Lacunae are sparsely distributed • Lacunae are not arranged in a concentric circle • Fibers are visible and fairly organized		
• Tapered cells with darkly stained nucleus centrally located • No striations • Cells layered to form a sheet		

Objectives

- ☐ List several important functions of the skin, or integumentary system.
- ☐ Identify the following skin structures on a model, image, or microscope slide: epidermis, dermis (papillary and reticular layers), hair follicles and hair, sebaceous glands, and sweat glands.
- ☐ Name and describe the layers of the epidermis.
- ☐ List the factors that determine skin color, and describe the function of melanin.
- ☐ Identify the major regions of nails.
- ☐ Describe the distribution and function of hairs, sebaceous glands, and sweat glands.
- ☐ Discuss the difference between eccrine and apocrine sweat glands.
- ☐ Compare and contrast the structure and functions of the epidermis and the dermis.

Materials

- Skin model (three-dimensional, if available)
- Compound microscope
- Prepared slide of human scalp
- Prepared slide of skin of palm or sole
- Sheet of 20# bond paper ruled to mark off cm² areas
- Scissors
- Betadine® swabs, or Lugol's iodine and cotton swabs
- Adhesive tape
- Disposable gloves
- Data collection sheet for plotting distribution of sweat glands
- Porelon® fingerprint pad or portable inking foils
- Ink cleaner towelettes
- Index cards (4 in. × 6 in.)
- Magnifying glasses

MasteringA&P®

For related exercise study tools, go to the Study Area of **MasteringA&P**. There you will find:

- Practice Anatomy Lab PAL
- PhysioEx PEx
- A&PFlix **A&PFlix**
- Practice quizzes, Histology Atlas, eText, Videos, and more!

Pre-Lab Quiz

1. All the following are functions of the skin *except*:
   a. excretion of body wastes
   b. insulation
   c. protection from mechanical damage
   d. site of vitamin A synthesis
2. The skin has two distinct regions. The superficial layer is the _____, and the underlying connective tissue is the _____.
3. The most superficial layer of the epidermis is the:
   a. stratum basale   c. stratum granulosum
   b. stratum spinosum   d. stratum corneum
4. Thick skin of the epidermis contains _____ layers.
5. _____ is a yellow-orange pigment found in the stratum corneum and the hypodermis.
   a. Keratin   c. Melanin
   b. Carotene   d. Hemoglobin
6. These cells produce a brown-to-black pigment that colors the skin and protects DNA from ultraviolet radiation damage. The cells are:
   a. dendritic cells   c. melanocytes
   b. keratinocytes   d. tactile cells
7. Circle True or False. Nails originate from the epidermis.
8. The portion of a hair that projects from the surface of the skin is known as the:
   a. bulb   c. root
   b. matrix   d. shaft
9. Circle the correct underlined term. The ducts of <u>sebaceous</u> / <u>sweat</u> glands usually empty into a hair follicle but may also open directly on the skin surface.
10. Circle the correct underlined term. <u>Eccrine</u> / <u>Apocrine</u> sweat glands are found primarily in the genital and axillary areas.

The **integument** is considered an organ system because it consists of multiple organs, the **skin** and its accessory organs. It is much more than an external body covering; architecturally, the skin is a marvel. It is tough yet pliable, a characteristic that enables it to withstand constant insult from outside agents.

The skin has many functions, most concerned with protection. It insulates and cushions the underlying body tissues and protects the entire body from abrasion, exposure to harmful chemicals, temperature extremes, and bacterial invasion. The hardened uppermost layer of the skin prevents water loss from the body surface. The skin's abundant capillary network (under the control of the nervous system) plays an important role in temperature regulation by regulating heat loss from the body surface.

The skin has other functions as well. For example, it acts as an excretory system; urea, salts, and water are lost through the skin pores in sweat. The skin also has important metabolic duties. For example, like liver cells, it carries out some chemical conversions that activate or inactivate certain drugs and hormones, and it is the site of vitamin D synthesis for the body. Vitamin D plays a role in calcium absorption in the digestive system. Finally, the sense organs for touch, pressure, pain, and temperature are located here.

Basic Structure of the Skin

The skin has two distinct regions—the superficial *epidermis* composed of epithelium and an underlying connective tissue, the *dermis* (**Figure 7.1**). These layers are firmly "cemented" together along a wavy border. But friction, such as the rubbing of a poorly fitting shoe, may cause them to separate, resulting in a blister. Immediately deep to the dermis is the **hypodermis,** or **superficial fascia,** which is not considered part of the skin. It consists primarily of adipose tissue. The main skin areas and structures are described below.

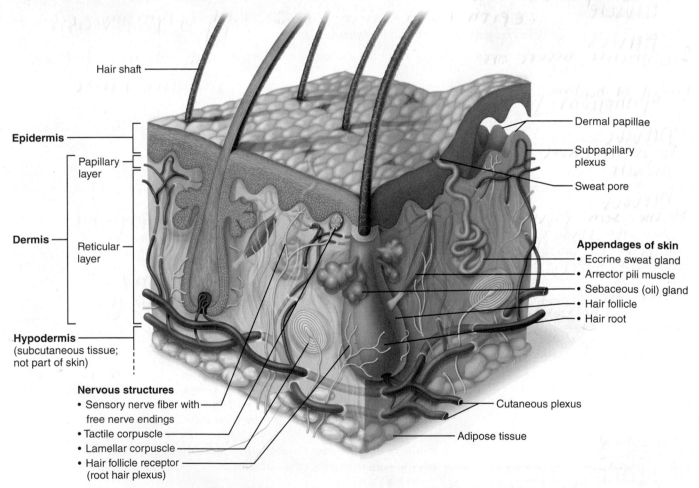

Figure 7.1 Skin structure. Three-dimensional view of the skin and the underlying hypodermis. The epidermis and dermis have been pulled apart at the right corner to reveal the dermal papillae. Tactile corpuscles are not common in hairy skin, but are included here for illustrative purposes.

Figure 7.2 The main structural features in epidermis of thin skin. (a) Photomicrograph depicting the four major epidermal layers (430×). **(b)** Diagram showing the layers and relative distribution of the different cell types. Keratinocytes (orange), melanocytes (gray), dendritic cells (purple), and tactile (Merkel) cells (blue). A sensory nerve ending (yellow) extending from the dermis is associated with a tactile cell, forming a tactile disc (touch receptor). Notice that the keratinocytes are joined by numerous desmosomes. The stratum lucidum, present in thick skin, is not illustrated here.

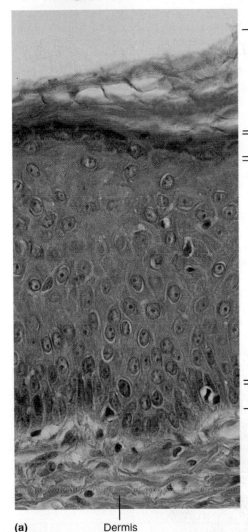

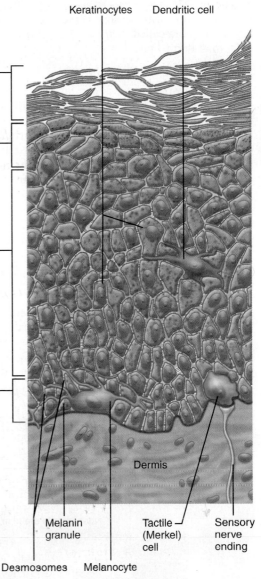

Stratum corneum
Most superficial layer; 20–30 layers of dead cells, essentially flat membranous sacs filled with keratin. Glycolipids in extracellular space.

Stratum granulosum
One to five layers of flattened cells, organelles deteriorating; cytoplasm full of lamellar granules (release lipids) and keratohyaline granules.

Stratum spinosum
Several layers of keratinocytes joined by desmosomes. Cells contain thick bundles of intermediate filaments made of pre-keratin.

Stratum basale
Deepest epidermal layer; one row of actively mitotic stem cells; some newly formed cells become part of the more superficial layers.

Keratinocytes   Dendritic cell

Dermis

Melanin granule   Tactile (Merkel) cell   Sensory nerve ending

Desmosomes   Melanocyte

(a) Dermis

(b)

Activity 1

Locating Structures on a Skin Model

As you read, locate the following structures in the diagram (Figure 7.1) and on a skin model.

Epidermis

Structurally, the avascular epidermis is a keratinized stratified squamous epithelium consisting of four distinct cell types and four or five distinct layers.

Cells of the Epidermis

- **Keratinocytes** (literally, keratin cells): The most abundant epidermal cells, their main function is to produce keratin fibrils. **Keratin** is a fibrous protein that gives the epidermis its durability and protective capabilities. Keratinocytes are tightly connected to each other by desmosomes.

Far less numerous are the following types of epidermal cells (**Figure 7.2**):

- **Melanocytes:** Spidery black cells that produce the brown-to-black pigment called **melanin.** The skin tans because melanin production increases when the skin is exposed to sunlight. The melanin provides a protective pigment umbrella over the nuclei of the cells in the deeper epidermal layers, thus shielding their genetic material (DNA) from the damaging effects of ultraviolet radiation. A concentration of melanin in one spot is called a *freckle.*

- **Dendritic cells:** Also called *Langerhans cells,* these cells play a role in immunity by performing phagocytosis.

- **Tactile (Merkel) cells:** Occasional spiky hemispheres that, in combination with sensory nerve endings, form sensitive touch receptors called *tactile* or *Merkel discs* located at the epidermal-dermal junction.

Layers of the Epidermis

The epidermis consists of four layers in thin skin, which covers most of the body. Thick skin, found on the palms of the hands and soles of the feet, contains an additional layer, the stratum lucidum. From deep to superficial, the layers of the epidermis are the stratum basale, stratum spinosum, stratum granulosum, stratum lucidum, and stratum corneum (Figure 7.2). The layers of the epidermis are summarized in **Table 7.1.**

Dermis

The dense irregular connective tissue making up the dermis consists of two principal regions—the papillary and reticular areas (Figure 7.1). Like the epidermis, the dermis varies in thickness.

- **Papillary layer:** The more superficial dermal region composed of areolar connective tissue. It is very uneven and has fingerlike projections from its superior surface, the **dermal papillae,** which attach it to the epidermis above. These projections lie on top of the larger dermal ridges. In the palms of

the hands and soles of the feet, they produce the *fingerprints,* unique patterns of *epidermal ridges* that remain unchanged throughout life. Abundant capillary networks in the papillary layer furnish nutrients for the epidermal layers and allow heat to radiate to the skin surface. The pain (free nerve endings) and touch receptors (*tactile corpuscles* in hairless skin) are also found here.

- **Reticular layer:** The deepest skin layer. It is composed of dense irregular connective tissue and contains many arteries and veins, sweat and sebaceous glands, and pressure receptors (*lamellar corpuscles*).

Both the papillary and reticular layers are abundant in collagen and elastic fibers. The elastic fibers give skin its exceptional elasticity in youth. With age, the number of elastic fibers decreases, and the subcutaneous layer loses fat, which leads to wrinkling and inelasticity of the skin. Fibroblasts, adipose cells, various types of macrophages, and other cell types are found throughout the dermis.

The abundant dermal blood supply allows the skin to play a role in the regulation of body temperature. When body temperature is high, the arterioles serving the skin dilate, and the capillary network of the dermis becomes engorged with the heated blood. Thus body heat is allowed to radiate from the skin surface.

Any restriction of the normal blood supply to the skin results in cell death and, if severe enough, skin ulcers (**Figure 7.3**). Bedsores (**decubitus ulcers**) occur in bedridden patients who are not turned regularly enough. The weight of the body puts pressure on the skin, especially over bony projections (hips, heels, etc.), which leads to restriction of the blood supply and tissue death. ✚

The dermis is also richly provided with lymphatic vessels and nerve fibers. Many of the nerve endings bear highly specialized receptor organs that, when stimulated by environmental changes, transmit messages to the central nervous system for interpretation. Some of these receptors—

Table 7.1	**Layers of the Epidermis**
Epidermal layer	**Description**
Stratum basale (basal layer)	A single row of cells immediately above the dermis. Its cells are constantly undergoing mitosis to form new cells, hence its alternate name, *stratum germinativum.* Some 10–25% of the cells in this layer are melanocytes, which thread their processes through this and adjacent layers of keratinocytes. Occasional tactile cells are also present in this layer.
Stratum spinosum (spiny layer)	Several layers of cells that contain thick, weblike bundles of intermediate filaments made of a pre-keratin protein. The cells in this layer appear spiky because when the tissue is prepared, the cells shrink, but their desmosomes hold tight to adjacent cells. Cells in this layer and the basal layer are the only ones to receive adequate nourishment from diffusion of nutrients from the dermis.
Stratum granulosum (granular layer)	A thin layer named for the abundant granules its cells contain. These granules are (1) *lamellar granules,* which contain a waterproofing glycolipid that is secreted into the extracellular space; and (2) *keratohyaline granules,* which help to form keratin in the more superficial layers. At the upper border of this layer, the cells are beginning to die.
Stratum lucidum (clear layer)	Present only in thick skin. A very thin transparent band of flattened, dead keratinocytes with indistinct boundaries.
Stratum corneum (horny layer)	The outermost layer consisting of 20–30 layers of dead, scalelike keratinocytes. They are constantly being exfoliated and replaced by the division of the deeper cells.

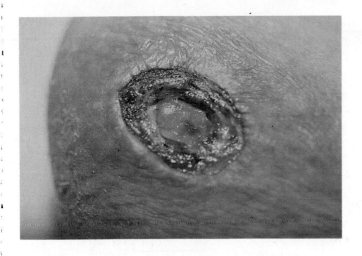

Figure 7.3 Photograph of a deep (stage III) decubitus ulcer.

free nerve endings (pain receptors), a lamellar corpuscle, and a hair follicle receptor (also called a *root hair plexus*)—are shown in Figure 7.1. (These receptors are discussed in depth in Exercise 22.)

Accessory Organs of the Skin

The accessory organs of the skin—cutaneous glands, hair, and nails—are all derivatives of the epidermis, but they reside primarily in the dermis. They originate from the stratum basale and grow downward into the deeper skin regions.

Nails

Nails are hornlike derivatives of the epidermis (**Figure 7.4**). Their named parts are:

- **Body:** The visible attached portion.
- **Free edge:** The portion of the nail that grows out away from the body.
- **Hyponychium:** The region beneath the free edge of the nail.
- **Root:** The part that is embedded in the skin and adheres to an epithelial nail bed.
- **Nail folds:** Skin folds that overlap the borders of the nail.
- **Eponychium:** Projection of the thick proximal nail fold commonly called the cuticle.
- **Nail bed:** Extension of the stratum basale beneath the nail.
- **Nail matrix:** The thickened proximal part of the nail bed containing germinal cells responsible for nail growth. As the matrix produces the nail cells, they become heavily keratinized and die. Thus nails, like hairs, are mostly nonliving material.
- **Lunule:** The proximal region of the thickened nail matrix, which appears as a white crescent moon. Everywhere else, nails are transparent and nearly colorless, but they appear pink because of the blood supply in the underlying dermis. When someone is cyanotic because of a lack of oxygen in the blood, the nail beds take on a blue cast.

Skin Color

Skin color is a result of the relative amount of melanin in skin, the relative amount of carotene in skin, and the degree of oxygenation of the blood. *Carotene* is a yellow-orange pigment present primarily in the stratum corneum and in the adipose tissue of the hypodermis. Its presence is most noticeable when large amounts of carotene-rich foods (carrots, for instance) are eaten.

Skin color may be an important diagnostic tool. For example, flushed skin may indicate hypertension or fever, whereas pale skin is typically seen in anemic individuals. When the blood is inadequately oxygenated, as during asphyxiation and serious lung disease, both the blood and the skin take on a bluish cast, a condition called **cyanosis. Jaundice,** in which the tissues become yellowed, is almost always diagnostic for liver disease, whereas a bronzing of the skin hints that a person's adrenal cortex is hypoactive (**Addison's disease**). ✚

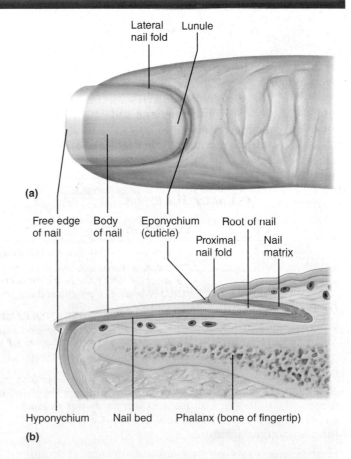

Figure 7.4 Structure of a nail. (a) Surface view of the distal part of a finger showing nail parts. The nail matrix that forms the nail lies beneath the lunule; the epidermis of the nail bed underlies the nail. **(b)** Sagittal section of the fingertip.

Identifying Nail Structures

Identify the parts of a nail (as shown in Figure 7.4) on yourself or your lab partner.

Hairs and Associated Structures

Hairs, enclosed in hair follicles, are found all over the entire body surface, except for thick-skinned areas (the palms of the hands and the soles of the feet), parts of the external genitalia, the nipples, and the lips.

Hair consists of two primary regions: the **hair shaft,** the region projecting from the surface of the skin, and the **hair root,** which is beneath the surface of the skin and is embedded within the **hair follicle.** The **hair bulb** is a collection of well-nourished epithelial cells at the base of the hair follicle (**Figure 7.5**). The hair shaft and the hair root have three layers of keratinized cells: the *medulla* in the center, surrounded by the *cortex,* and the protective *cuticle.* Abrasion of the cuticle at the tip of the hair shaft results in split ends. Hair color depends on the amount and type of melanin pigment found in the hair cortex.

- **Hair follicle:** A structure formed from both epidermal and dermal cells (Figure 7.5). Its inner epithelial root sheath, with two parts (internal and external), is enclosed by a thickened basement membrane, the glassy membrane, and by a peripheral connective tissue (or fibrous) sheath, which is essentially dermal tissue. A small nipple of dermal tissue protrudes into the hair bulb from the peripheral connective tissue sheath and provides nutrition to the growing hair. It is called the **hair papilla.** A layer of actively dividing epithelial cells called the **hair matrix** is located on top of the hair papilla.

- **Arrector pili muscle:** Small bands of smooth muscle cells connect each hair follicle to the papillary layer of the dermis (Figures 7.1 and 7.5). When these muscles contract (during cold or fright), the slanted hair follicle is pulled upright, dimpling the skin surface with goose bumps. This phenomenon is especially dramatic in a scared cat, whose fur actually stands on end to increase its apparent size.

Comparing Hairy and Relatively Hair-Free Skin Microscopically

Whereas thick skin has no hair follicles or sebaceous (oil) glands, thin skin typical of most of the body usually has both. The scalp, of course, has the highest density of hair follicles.

1. Obtain a prepared slide of the human scalp, and study it carefully under the microscope. Compare your tissue slide to **Figure 7.6a**, and identify as many as possible of the diagrammed structures in Figure 7.1.

How is this stratified squamous epithelium different from that observed in the esophagus (Exercise 6)?

_____

_____

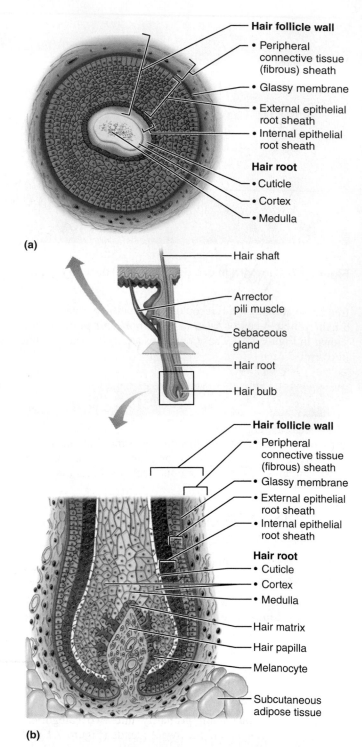

(a)

(b)

Figure 7.5 Structure of a hair and hair follicle.
(a) Diagram of a cross section of a hair within its follicle. **(b)** Diagram of a longitudinal view of the expanded bulb of the hair follicle, which encloses the matrix, the actively dividing epithelial cells that produce the hair.

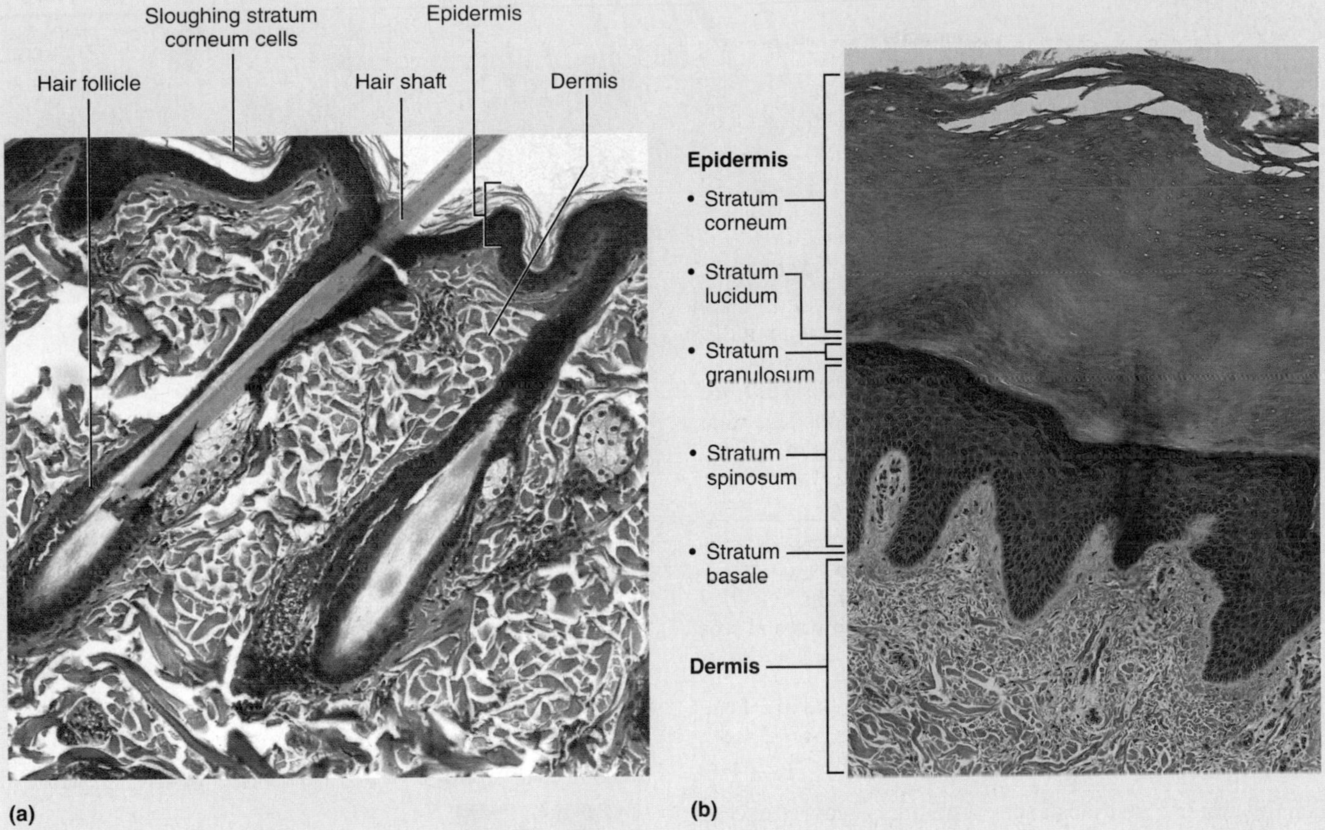

(a)

(b)

Figure 7.6 Photomicrographs of skin. (a) Thin skin with hairs (120×). **(b)** Thick hairless skin (75×).

How do these differences relate to the functions of these two similar epithelia?

_____

_____

_____

2. Obtain a prepared slide of hairless skin of the palm or sole (Figure 7.6b). Compare the slide to the previous photomicrograph (Figure 7.6a). In what ways does the thick skin of the palm or sole differ from the thin skin of the scalp?

_____

_____

_____

Cutaneous Glands

The cutaneous glands fall primarily into two categories: the sebaceous glands and the sweat glands (Figure 7.1 and **Figure 7.7**, p. 100).

Sebaceous (Oil) Glands

The sebaceous glands are found nearly all over the skin, except for the palms of the hands and the soles of the feet. Their ducts usually empty into a hair follicle, but some open directly on the skin surface.

Sebum is the product of sebaceous glands. It is a mixture of oily substances and fragmented cells that acts as a lubricant to keep the skin soft and moist (a natural skin cream) and keeps the hair from becoming brittle. The sebaceous glands become particularly active during puberty, when more male hormones (androgens) begin to be produced for both genders; thus the skin tends to become oilier during this period of life.

Blackheads are accumulations of dried sebum, bacteria, and melanin from epithelial cells in the oil duct. **Acne** is an active infection of the sebaceous glands. ✚

Sweat (Sudoriferous) Glands

Sweat, or sudoriferous, glands are exocrine glands that are widely distributed all over the skin. Outlets for the glands are epithelial openings called _pores_. Sweat glands are categorized by the composition of their secretions.

• **Eccrine sweat glands:** Also called **merocrine sweat glands,** these glands are distributed all over the body. They produce clear perspiration consisting primarily of water, salts (mostly NaCl), and urea. Eccrine sweat glands, under the control of the nervous system, are an important part of the body's heat-regulating apparatus. They secrete perspiration when the external temperature or body temperature is high.

7

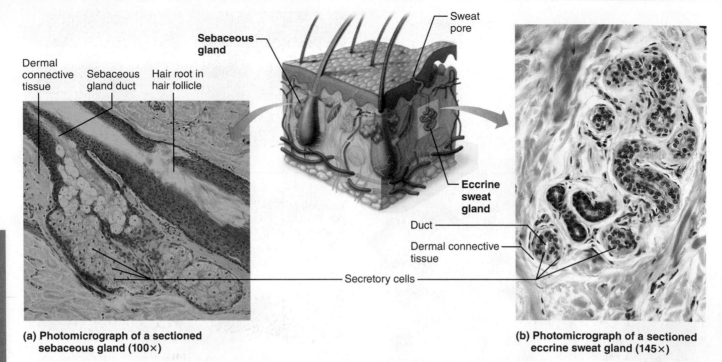

(a) Photomicrograph of a sectioned sebaceous gland (100×)

(b) Photomicrograph of a sectioned eccrine sweat gland (145×)

Figure 7.7 Cutaneous glands.

When this water-based substance evaporates, it carries excess body heat with it.

- **Apocrine sweat glands:** Found predominantly in the axillary and genital areas, these glands secrete the basic components of eccrine sweat plus proteins and fat-rich substances. Apocrine sweat is an excellent nutrient medium for the microorganisms typically found on the skin. This sweat is initially odorless, but when bacteria break down its organic components, it begins to smell unpleasant.

Activity 4

Differentiating Sebaceous and Sweat Glands Microscopically

Using the slide *thin skin with hairs* and the photomicrographs of cutaneous glands (Figure 7.7) as a guide, identify sebaceous and eccrine sweat glands. What characteristics relating to location or gland structure allow you to differentiate these glands?

_____

_____

_____

Activity 5

Plotting the Distribution of Sweat Glands

1. Form a hypothesis about the relative distribution of sweat glands on the palm and forearm. Justify your hypothesis.

2. The bond paper for this simple experiment has been preruled in cm² — put on disposable gloves, and cut along the lines to obtain the required squares. You will need two squares of bond paper (each 1 cm × 1 cm), adhesive tape, and a Betadine (iodine) swab *or* Lugol's iodine and a cotton-tipped swab.

3. Paint an area of the medial aspect of your left palm (avoid the deep crease lines) and a region of your left forearm with the iodine solution, and allow it to dry thoroughly. The painted area in each case should be slightly larger than the paper squares to be used.

4. Have your lab partner *securely* tape a square of bond paper over each iodine-painted area, and leave the paper squares in place for 20 minutes. (If it is very warm in the laboratory while this test is being conducted, you can obtain good results within 10 to 15 minutes.)

5. After 20 minutes, remove the paper squares, and count the number of blue-black dots on each square. The presence of a blue-black dot on the paper indicates an active sweat gland. The iodine in the pore is dissolved in the sweat and reacts chemically with the starch in the bond paper to produce the blue-black color. You have produced "sweat maps" for the two skin areas.

6. Which skin area tested has the greater density of sweat glands?

_____

_____

_____

7. Tape your results (bond paper squares) to a data collection sheet labeled "palm" and "forearm" at the front of the lab. Be sure to put your paper squares in the correct columns on the data sheet.

8. Once all the data have been collected, review the class results.

9. Prepare a lab report for the experiment. (See Getting Started, on MasteringA&P).

Dermography: Fingerprinting

As noted previously, each of us has a unique genetically determined set of fingerprints. Because fingerprinting is useful for identifying and apprehending criminals, most people associate this craft solely with criminal investigations. However, fingerprints are also invaluable for quick identification of amnesia victims, missing persons, and unknown deceased, such as people killed in major disasters.

The friction ridges responsible for fingerprints appear in several patterns, which are clearest when the fingertips are inked and then pressed against white paper. Impressions are also made when perspiration or any foreign material such as blood, dirt, or grease adheres to the ridges and the fingers are then pressed against a smooth, nonabsorbent surface. The three most common patterns are *arches, loops,* and *whorls* (**Figure 7.8**).

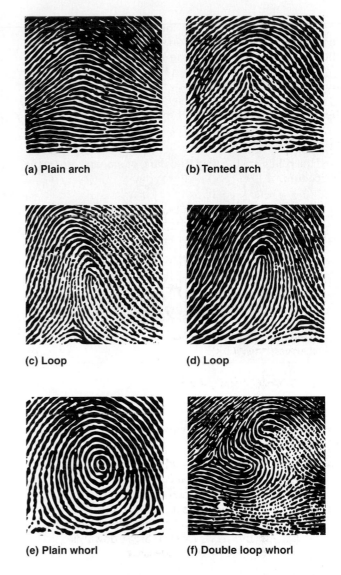

(a) Plain arch   **(b) Tented arch**

(c) Loop   **(d) Loop**

(e) Plain whorl   **(f) Double loop whorl**

Figure 7.8 Main types of fingerprint patterns.
(a, b) Arches. **(c, d)** Loops. **(e, f)** Whorls.

Activity 6

Taking and Identifying Inked Fingerprints

For this activity, you will be working as a group with your lab partners. Though the equipment for professional fingerprinting is fairly basic, consisting of a glass or metal inking plate, printer's ink (a heavy black paste), ink roller, and standard 8 in. × 8 in. cards, you will be using supplies that are even easier to handle. Each student will prepare two index cards, each bearing his or her thumbprint and index fingerprint of the right hand.

1. Obtain the following supplies and bring them to your bench: two 4 in. × 6 in. index cards per student, Porelon fingerprint pad or portable inking foils, ink cleaner towelettes, and a magnifying glass.

2. The subject should wash and dry the hands. Open the ink pad or peel back the covering over the ink foil, and

position it close to the edge of the laboratory bench. The subject should position himself or herself at arm's length from the bench edge and inking object.

3. A second student, called the *operator,* stands to the left of the subject and with two hands holds and directs movement of the subject's fingertip. During this process, the subject should look away, try to relax, and refrain from trying to help the operator.

4. The thumbprint is to be placed on the left side of the index card, the index fingerprint on the right. The operator should position the subject's right thumb or index finger on the side of the bulb of the finger in such a way that the area to be inked spans the distance from the fingertip to just beyond the first joint, and then roll

Text continues on next page →

the finger lightly across the inked surface until its bulb faces in the opposite direction. To prevent smearing, the thumb is rolled away from the body midline (from left to right as the subject sees it; see **Figure 7.9**), and the index finger is rolled toward the body midline (from right to left). The same ink foil can be reused for all the students at the bench; the ink pad is good for thousands of prints. Repeat the procedure (still using the subject's right hand) on the second index card.

5. If the prints are too light, too dark, or smeary, repeat the procedure.

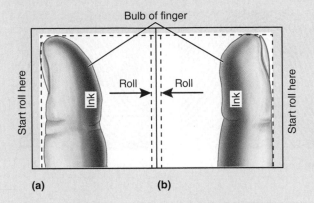

(a)   (b)

Figure 7.9 Fingerprinting. Method of inking and printing **(a)** the thumb and **(b)** the index finger of the right hand.

6. While other students in the group are making clear prints of their thumb and index finger, those who have completed that activity should clean their inked fingers with a towelette and attempt to classify their own prints as arches, loops, or whorls. Use the magnifying glass as necessary to see ridge details.

7. When all group members at a bench have completed the above steps, they are to write their names on the backs of their index cards. The students combine their cards, shuffle them, and transfer them to the opposite bench. Finally, students classify the patterns and identify which prints were made by the same individuals.

How difficult was it to classify the prints into one of the three categories given?

_____

_____

Why do you think this is so?

_____

_____

Was it easy or difficult to identify the prints made by the same individual?

_____

_____

Why do you think this was so?

_____

_____

Overview of the Skeleton: Classification and Structure of Bones and Cartilages

Objectives

☐ Name the two tissue types that form the skeleton.

☐ List the functions of the skeletal system.

☐ Locate and identify the three major types of skeletal cartilages.

☐ Name the four main groups of bones based on shape.

☐ Identify surface bone markings and list their functions.

☐ Identify the major anatomical areas on a longitudinally cut long bone or on an appropriate image.

☐ Explain the role of inorganic salts and organic matrix in providing flexibility and hardness to bone.

☐ Locate and identify the major parts of an osteon microscopically, or on a histological model or appropriate image of compact bone.

Materials

- Long bone sawed longitudinally (beef bone from a slaughterhouse, if possible, or prepared laboratory specimen)
- Disposable gloves
- Long bone soaked in 10% hydrochloric acid (HCl) (or vinegar) until flexible
- Long bone baked at 250°F for more than 2 hours
- Compound microscope
- Prepared slide of ground bone (x.s.)
- Three-dimensional model of microscopic structure of compact bone
- Prepared slide of a developing long bone undergoing endochondral ossification
- Articulated skeleton

MasteringA&P®

For related exercise study tools, go to the Study Area of **MasteringA&P**. There you will find:

- Practice Anatomy Lab **PAL**
- **PhysioEx** **PEx**
- A&PFlix **A&PFlix**
- Practice quizzes, Histology Atlas, eText, Videos, and more!

Pre-Lab Quiz

1. All the following are functions of the skeleton *except*:
   a. attachment for muscles
   b. production of melanin
   c. site of red blood cell formation
   d. storage of lipids

2. Circle the correct underlined term. The <u>axial</u> / <u>appendicular</u> skeleton consists of bones that surround the body's center of gravity.

3. The type of cartilage that has the greatest strength and is found in the knee joint and intervertebral discs is:
   a. elastic   b. fibrocartilage   c. hyaline

4. Circle the correct underlined term. <u>Compact</u> / <u>Spongy</u> bone looks smooth and homogeneous.

5. _____ bones are generally thin and have a layer of spongy bone between two layers of compact bone.
   a. Flat   b. Irregular   c. Long   d. Short

6. The femur is an example of a(n) _____ bone.
   a. flat   b. irregular   c. long   d. short

7. Circle the correct underlined term. The shaft of a long bone is known as the <u>epiphysis</u> / <u>diaphysis</u>.

8. The structural unit of compact bone is the:
   a. osteon   b. canaliculus   c. lacuna

9. Circle True or False. Embryonic skeletons consist primarily of elastic cartilage, which is gradually replaced by bone during development and growth.

10. Circle True or False. Cartilage has a covering made of dense irregular connective tissue called a periosteum.

The **skeleton,** the body's framework, is constructed of two of the most supportive tissues found in the human body—cartilage and bone. In embryos, the skeleton is predominantly made up of hyaline cartilage, but in the adult, most of the cartilage is replaced by more rigid bone. Cartilage remains only in such isolated areas as the external ear, bridge of the nose, larynx, trachea, joints, and parts of the rib cage (see Figure 8.2).

Besides supporting and protecting the body as an internal framework, the skeleton provides a system of levers with which the skeletal muscles work to move the body. In addition, the bones store lipids and many minerals (the most important of which is calcium). Finally, the red marrow cavities of bones provide a site for hematopoiesis (blood cell formation).

The skeleton is made up of bones that are connected at *joints,* or *articulations*. The skeleton is subdivided into two divisions: the **axial skeleton** (the bones that lie around the body's center of gravity) and the **appendicular skeleton** (bones of the limbs, or appendages) **(Figure 8.1)**.

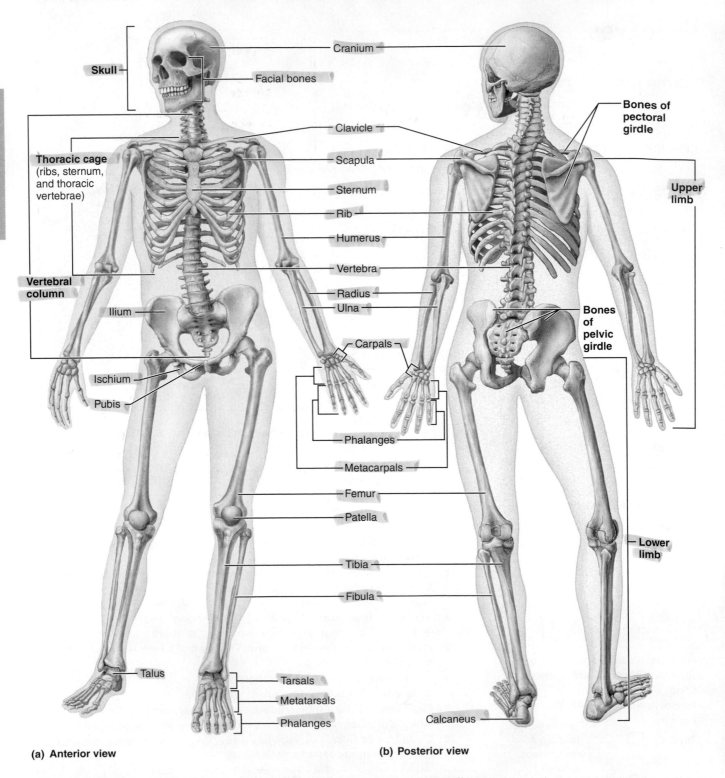

(a) Anterior view

(b) Posterior view

Figure 8.1 The human skeleton. The bones of the axial skeleton are colored green to distinguish them from the bones of the appendicular skeleton.

Cartilages of the Skeleton

The most important of the adult skeletal cartilages are (1) **articular cartilages,** which cover the bone ends at movable joints; (2) **costal cartilages,** which connect the ribs to the sternum (breastbone); (3) **laryngeal cartilages,** which largely construct the larynx (voice box); (4) **tracheal** and **bronchial cartilages,** which reinforce other passageways of the respiratory system; (5) **nasal cartilages,** which support the external nose; (6) **intervertebral discs,** which separate and cushion the vertebrae; and (7) the cartilage supporting the external ear (**Figure 8.2**).

Cartilage tissues are distinguished by the fact that they contain no nerves and very few blood vessels. Like bones, each cartilage is surrounded by a covering of dense irregular connective tissue, called a *perichondrium*, which acts to resist

8

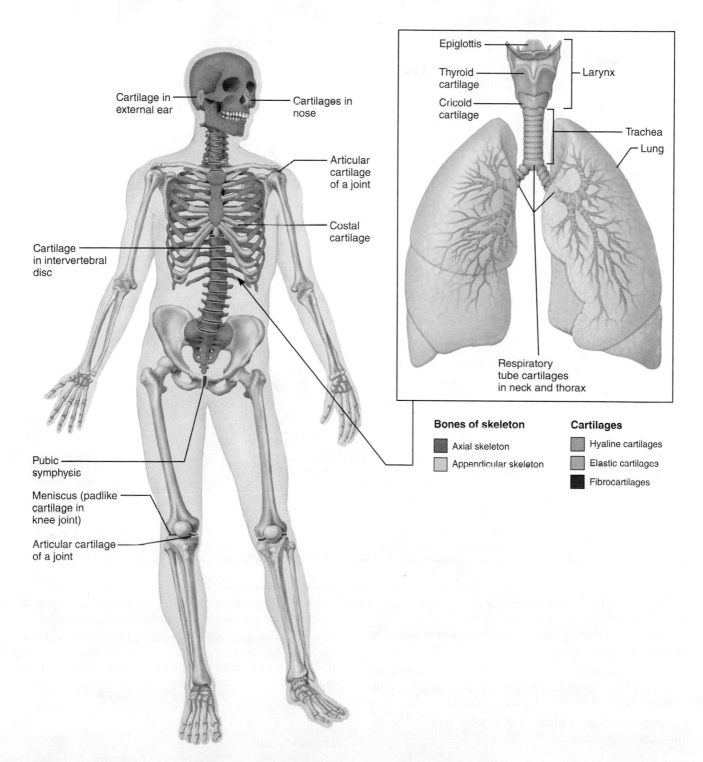

Figure 8.2 Cartilages in the adult skeleton and body. Additional cartilages that support the respiratory tubes and larynx are shown separately at the upper right.

Table 8.1	Bone Markings	
Name of bone marking	**Description**	**Illustration**

Projections That Are Sites of Muscle and Ligament Attachment

Tuberosity	Large rounded projection; may be roughened	
Crest	Narrow ridge of bone; usually prominent	
Trochanter	Very large, blunt, irregularly shaped process (the only examples are on the femur)	
Line	Narrow ridge of bone; less prominent than a crest	
Tubercle	Small rounded projection or process	
Epicondyle	Raised area on or above a condyle	
Spine	Sharp, slender, often pointed projection	
Process	Any bony prominence	

Projections That Help Form Joints

Head	Bony expansion carried on a narrow neck	
Facet	Smooth, nearly flat articular surface	
Condyle	Rounded articular projection	
Ramus	Armlike bar of bone	

Depressions and Openings for Passage of Blood Vessels and Nerves

Groove	Furrow	
Fissure	Narrow, slitlike opening	
Foramen	Round or oval opening through a bone	
Notch	Indentation at the edge of a structure	

Others

Meatus	Canal-like passageway	
Sinus	Bone cavity, filled with air and lined with mucous membrane	
Fossa	Shallow basinlike depression in a bone, often serving as an articular surface	

distortion of the cartilage when it is subjected to pressure and plays a role in cartilage growth and repair.

The skeletal cartilages have representatives from each of the three cartilage tissue types—hyaline, elastic, and fibrocartilage.

- **Hyaline cartilage** provides sturdy support with some flexibility. Most skeletal cartilages are composed of hyaline cartilage (Figure 8.2).

- **Elastic cartilage** is much more flexible than hyaline cartilage, and it tolerates repeated bending. Only the cartilages

of the external ear and the epiglottis (which flops over and covers the larynx when we swallow) are elastic cartilage.

- **Fibrocartilage** consists of rows of chondrocytes alternating with rows of thick collagen fibers. Fibrocartilage, which has great tensile strength and can withstand heavy compression, is used to construct the intervertebral discs and the cartilages within the knee joint (see Figure 8.2).

Classification of Bones

The 206 bones of the adult skeleton are composed of two basic kinds of osseous tissue that differ in their texture. **Compact bone** looks smooth and homogeneous; **spongy** (or *cancellous*) **bone** is composed of small *trabeculae* (columns) of bone and lots of open space.

Bones may be classified further on the basis of their gross anatomy into four groups: long, short, flat, and irregular bones.

Long bones, such as the femur and phalanges (Figure 8.1), are much longer than they are wide, generally consisting of a shaft with heads at either end. Long bones are composed mostly of compact bone. **Short bones** are typically cube shaped, and they contain more spongy bone than compact bone. The tarsals and carpals are examples (see Figure 8.1).

Flat bones are generally thin, with two waferlike layers of compact bone sandwiching a thicker layer of spongy bone between them. Although the name "flat bone" implies a structure that is straight, many flat bones are curved (for example, the bones of the skull). Bones that do not fall into one of the preceding categories are classified as **irregular bones.** The vertebrae are irregular bones (see Figure 8.1).

Some anatomists also recognize two other subcategories of bones. **Sesamoid bones** are special types of short bones formed within tendons. The patellas (kneecaps) are sesamoid bones. **Sutural bones** are tiny bones between cranial bones.

Bone Markings

Even a casual observation of the bones will reveal that bone surfaces are not featureless smooth areas. They have an array of bumps, holes, and ridges. These **bone markings** reveal where bones form joints with other bones, where muscles, tendons, and ligaments were attached, and where blood vessels and nerves passed. Bone markings fall into two main categories: projections that grow out from the bone and serve as sites of muscle attachment or help form joints; and depressions or openings in the bone that often serve as conduits for nerves and blood vessels. The bone markings are summarized in **Table 8.1**.

Gross Anatomy of the Typical Long Bone

Activity 1

Examining a Long Bone

1. Obtain a long bone that has been sawed along its longitudinal axis. If a cleaned dry bone is provided, no special preparations need be made.

⚠ Note: If the bone supplied is a fresh beef bone, don disposable gloves before beginning your observations.

Identify the **diaphysis** or shaft (**Figure 8.3** on p. 112 may help). Observe its smooth surface, which is composed of compact bone. If you are using a fresh specimen, carefully pull away the **periosteum,** a fibrous membrane covering made up of dense irregular connective tissue, to view the bone surface. Notice that many fibers of the periosteum penetrate into the bone. These collagen fibers are called **perforating (Sharpey's) fibers.** Blood vessels and nerves travel through the periosteum and invade the bone. *Osteoblasts* (bone-forming cells) and *osteogenic cells* are found on the inner, or osteogenic, layer of the periosteum.

2. Now inspect the **epiphysis,** the end of the long bone. Notice that it is composed of a thin layer of compact bone that encloses spongy bone.

3. Identify the **articular cartilage,** which covers the epiphysis in place of the periosteum. The glassy hyaline cartilage provides a smooth surface to minimize friction at joints.

4. If the animal was still young and growing, you will be able to see the **epiphyseal plate,** a thin area of hyaline cartilage that provides for longitudinal growth of the bone during youth. Once the long bone has stopped growing,

WHY THIS MATTERS | **Shin Splints**

Runners, gymnasts, basketball players, and other athletes often complain of a dull ache in the shin. "Shin splints" is a term for any pain in the leg. Most cases of shin pain are classified as medial tibial stress syndrome (MTSS), a more scientific term than "shin splints" but still not very explanatory. Assuming there is no stress fracture in the tibia, the "stress" can be on muscles, tendons, the periosteum, perforating fibers, or any combination of these structures. Most commonly, though, the pain is caused by inflammation of the periosteum (periostitis) and perforating fibers. A variety of mechanical factors, including flat feet and turning the feet inward with impact exercise, can contribute to MTSS. ■

these areas are replaced with bone and appear as thin, barely discernible remnants—the **epiphyseal lines.**

5. In an adult animal, the central cavity of the shaft (*medullary cavity*) is essentially a storage region for adipose, or **yellow marrow.** In the infant, this area is involved in forming blood cells, and so **red marrow** is found in the marrow cavities. In adult bones, the red marrow is confined to the interior of the epiphyses, where it occupies the spaces between the trabeculae of spongy bone.

Text continues on next page. →

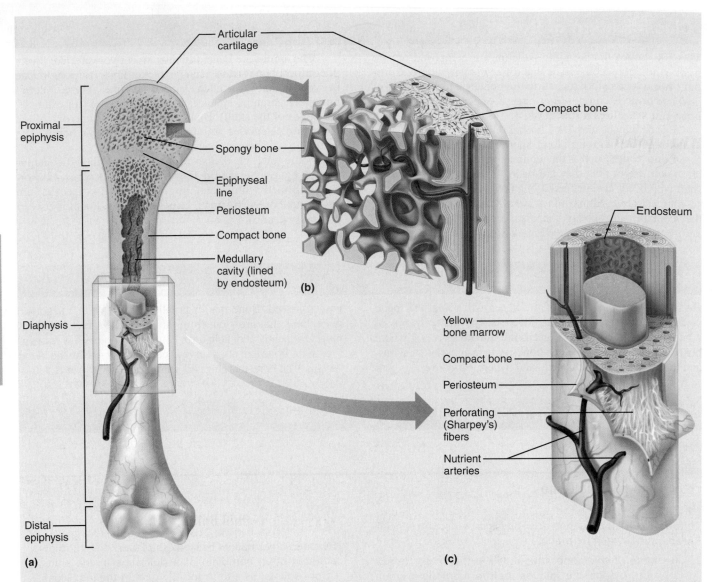

Figure 8.3 The structure of a long bone (humerus of the arm). **(a)** Anterior view with longitudinal section cut away at the proximal end. **(b)** Pie-shaped, three-dimensional view of spongy bone and compact bone of the epiphysis. **(c)** Cross section of diaphysis (shaft). Note that the external surface of the diaphysis is covered by a periosteum but that the articular surface of the epiphysis is covered with hyaline cartilage.

6. If you are examining a fresh bone, look carefully to see if you can distinguish the delicate **endosteum** lining the shaft. The endosteum also covers the trabeculae of spongy bone and lines the central and perforating canals of compact bone. Like the periosteum, the endosteum contains osteogenic cells that differentiate into osteoblasts. As the bone grows in diameter on its external surface, it is constantly being broken down on its inner surface. Thus the thickness of the compact bone layer composing the shaft remains relatively constant.

7. If you have been working with a fresh bone specimen, return it to the appropriate area and properly dispose of your gloves, as designated by your instructor.

Longitudinal bone growth at epiphyseal plates (growth plates) follows a predictable sequence and provides a reliable indicator of the age of children exhibiting normal growth. If problems of long-bone growth are suspected (for example, pituitary dwarfism), X-ray films are taken to view the width of the growth plates. An abnormally thin epiphyseal plate indicates growth retardation. ✚

Chemical Composition of Bone

Bone is one of the hardest materials in the body. Although relatively light, bone has a remarkable ability to resist tension and shear forces that continually act on it. An engineer would tell you that a cylinder (like a long bone) is one of the strongest structures for its mass.

The hardness of bone is due to the inorganic calcium salts deposited in its ground substance. Its flexibility comes from the organic elements of the matrix, particularly the collagen fibers.

Activity 2

Examining the Effects of Heat and Hydrochloric Acid on Bones

Obtain a bone sample that has been soaked in hydrochloric acid (HCl) (or in vinegar) and one that has been baked. Heating removes the organic part of bone, whereas acid dissolves the minerals.

Gently apply pressure to each bone sample. What happens to the heated bone?

_____

_____

_____

What happens to the bone treated with acid?

_____

What does the acid appear to remove from the bone?

_____

What does baking appear to do to the bone?

_____

_____

In rickets, the bones are not properly calcified. Which of the demonstration specimens would more closely resemble the bones of a child with rickets?

_____

Microscopic Structure of Compact Bone

Spongy bone has a spiky, open-work appearance, resulting from the arrangement of the **trabeculae** that compose it, whereas compact bone appears to be dense and homogeneous. However, microscopic examination of compact bone reveals that it is riddled with passageways carrying blood vessels, nerves, and lymphatic vessels that provide the living bone cells with needed substances and a way to eliminate wastes.

Activity 3

Examining the Microscopic Structure of Compact Bone

1. Obtain a prepared slide of ground bone, and examine it under low power. Focus on a central canal (**Figure 8.4** on p. 114). The **central (Haversian) canal** runs parallel to the long axis of the bone and carries blood vessels, nerves, and lymphatic vessels through the bony matrix. Identify the **osteocytes** (mature bone cells) in **lacunae** (chambers), which are arranged in concentric circles called **concentric lamellae** around the central canal. Because bone remodeling is going on all the time, you will also see some _interstitial lamellae,_ remnants of osteons that have been broken down (Figure 8.4c).

A central canal and all the concentric lamellae surrounding it are referred to as an **osteon,** or **Haversian system.** Also identify **canaliculi,** tiny canals radiating outward from a central canal to the lacunae of the first lamella and then from lamella to lamella. The canaliculi form a dense transportation network through the hard bone matrix, connecting all the living cells of the osteon to the nutrient supply. You may need a higher-power magnification to see the fine canaliculi.

2. Also note the **perforating (Volkmann's) canals** (Figure 8.4). These canals run at right angles to the shaft and connect the blood and nerve supply of the medullary cavity to the central canals.

3. If a model of bone histology is available, identify the same structures on the model.

Text continues on next page. →

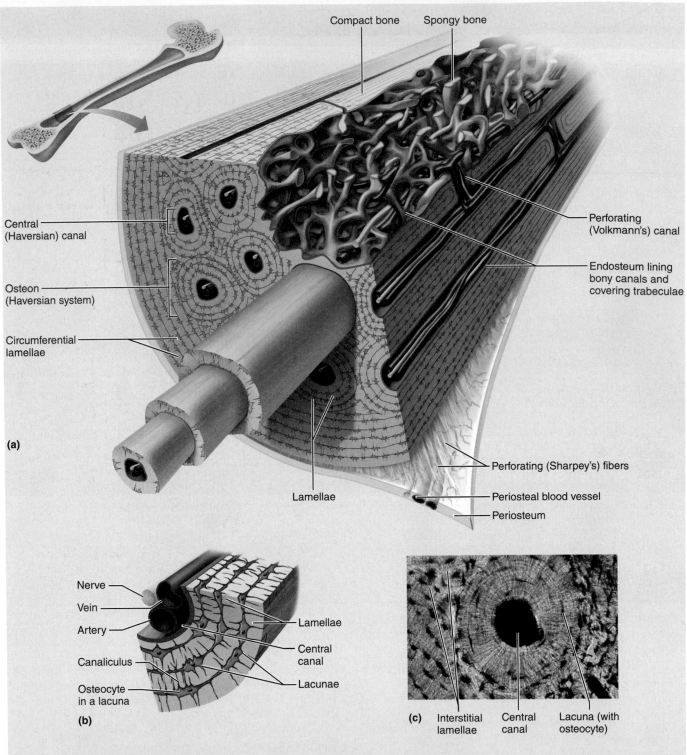

Figure 8.4 Microscopic structure of compact bone. (a) Diagram of a pie-shaped segment of compact bone, illustrating its structural units (osteons). **(b)** A portion of one osteon. **(c)** Photomicrograph of a cross-sectional view of an osteon (320×).

Ossification: Bone Formation and Growth in Length

Except for the collarbones, all bones of the body inferior to the skull form by the process of **endochondral ossification,** which uses hyaline cartilage as a model for bone formation. The major events of this process are as follows:

• Blood vessels invade the perichondrium covering the hyaline cartilage model and convert it to a periosteum.

• Osteoblasts at the inner surface of the periosteum secrete bone matrix around the hyaline cartilage model, forming a bone collar.

• Cartilage in the shaft center calcifies and then hollows out, forming an internal cavity.

• A *periosteal bud* (blood vessels, nerves, red marrow elements, osteoblasts, and osteoclasts) invades the cavity and forms spongy bone, which is removed by osteoclasts, producing the medullary cavity. This process proceeds in both directions from the *primary ossification center.*

As bones grow longer, the medullary cavity gets larger and longer. Chondroblasts lay down new cartilage matrix on the epiphyseal face of the epiphyseal plate, and it is eroded away and replaced by bony spicules on the diaphyseal face (**Figure 8.5**). This process continues until late adolescence when the entire epiphyseal plate is replaced by bone.

Activity 4

Examining the Osteogenic Epiphyseal Plate

Obtain a slide depicting endochondral ossification (cartilage bone formation), and bring it to your bench to examine under the microscope. Identify the proliferation, hypertrophic, calcification, and ossification zones of the epiphyseal plate (Figure 8.5). Then, also identify the area of resting cartilage cells, some hypertrophied chondrocytes, and bony spicules.

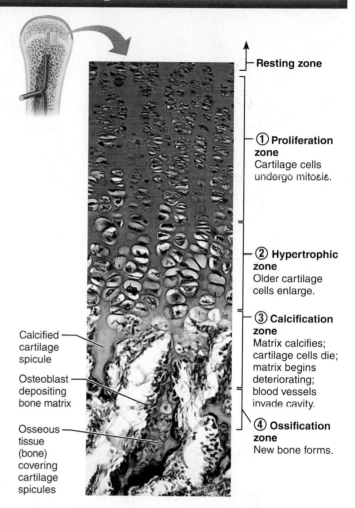

Resting zone

① **Proliferation zone**
Cartilage cells undergo mitosis.

② **Hypertrophic zone**
Older cartilage cells enlarge.

③ **Calcification zone**
Matrix calcifies; cartilage cells die; matrix begins deteriorating; blood vessels invade cavity.

④ **Ossification zone**
New bone forms.

Calcified cartilage spicule

Osteoblast depositing bone matrix

Osseous tissue (bone) covering cartilage spicules

Figure 8.5 Growth in length of a long bone occurs at the epiphyseal plate. The side of the epiphyseal plate facing the epiphysis (distal face) contains resting cartilage cells. The cells of the epiphyseal plate proximal to the resting cartilage area are arranged in four zones—proliferation, hypertrophic, calcification, and ossification—from the region of the earliest stage of growth ① to the region where bone is replacing the cartilage ④(125×).

8

The Axial Skeleton

Objectives

☐ Name the three parts of the axial skeleton.

☐ Identify the bones of the axial skeleton, either by examining isolated bones or by pointing them out on an articulated skeleton or skull, and name the important bone markings on each.

☐ Name and describe the different types of vertebrae.

☐ Discuss the importance of intervertebral discs and spinal curvatures.

☐ Identify three abnormal spinal curvatures.

☐ List the components of the thoracic cage.

☐ Identify the bones of the fetal skull by examining an articulated skull or image.

☐ Define *fontanelle*, and discuss the function and fate of fontanelles.

☐ Discuss important differences between the fetal and adult skulls.

Materials

- Intact skull and Beauchene skull
- X-ray images of individuals with scoliosis, lordosis, and kyphosis (if available)
- Articulated skeleton, articulated vertebral column, removable intervertebral discs
- Isolated cervical, thoracic, and lumbar vertebrae, sacrum, and coccyx
- Isolated fetal skull

MasteringA&P®

For related exercise study tools, go to the Study Area of **MasteringA&P**. There you will find:

- Practice Anatomy Lab PAL
- PhysioEx PEx
- A&PFlix *A&PFlix*
- Practice quizzes, Histology Atlas, eText, Videos, and more!

Pre-Lab Quiz

1. The axial skeleton can be divided into the skull, the vertebral column, and the:
   - **a.** thoracic cage
   - **b.** femur
   - **c.** hip bones
   - **d.** humerus

2. Eight bones make up the _____, which encloses and protects the brain.
   - **a.** cranium
   - **b.** face
   - **c.** skull

3. How many bones of the skull are considered facial bones? _____

4. Circle the correct underlined term. The lower jawbone, or <u>maxilla</u> / <u>mandible</u>, articulates with the temporal bones in the only freely movable joints in the skull.

5. Circle the correct underlined term. The <u>body</u> / <u>spinous process</u> of a typical vertebra forms the rounded, central portion that faces anteriorly in the human vertebral column.

6. The seven bones of the neck are called _____ vertebrae.
   - **a.** cervical
   - **b.** lumbar
   - **c.** spinal
   - **d.** thoracic

7. The _____ vertebrae articulate with the corresponding ribs.
   - **a.** cervical
   - **b.** lumbar
   - **c.** spinal
   - **d.** thoracic

8. The _____, commonly referred to as the breastbone, is a flat bone formed by the fusion of three bones: the manubrium, the body, and the xiphoid process.
   - **a.** coccyx
   - **b.** sacrum
   - **c.** sternum

9. Circle True or False. The first seven pairs of ribs are called floating ribs because they have only indirect cartilage attachments to the sternum.

10. A fontanelle:
    - **a.** is found only in the fetal skull
    - **b.** is a fibrous membrane
    - **c.** allows for compression of the skull during birth
    - **d.** all of the above

The **axial skeleton** (the green portion of Figure 8.1 on p. 108) can be divided into three parts: the skull, the vertebral column, and the thoracic cage. This division of the skeleton forms the longitudinal axis of the body and protects the brain, spinal cord, heart, and lungs.

The Skull

The **skull** is composed of two sets of bones. Those of the **cranium** (8 bones) enclose and protect the fragile brain tissue. The **facial bones** (14 bones) support the eyes and position them anteriorly. They also provide attachment sites for facial muscles. All but one of the bones of the skull are joined by interlocking fibrous joints called *sutures*. The mandible is attached to the rest of the skull by a freely movable joint.

read through this material, identify each bone on an intact and/or Beauchene skull (see Figure 9.10).

Note: *Important bone markings are listed in the tables for the bones on which they appear, and each bone name is colored to correspond to the bone color in the figures.*

Activity 1

Identifying the Bones of the Skull

The bones of the skull (**Figures 9.1–9.10**, pp. 123–131) are described in **Tables 9.1** and **9.2** on p. 128. As you

The Cranium

The cranium may be divided into two major areas for study—the **cranial vault,** or **calvaria,** forming the superior, lateral, and posterior walls of the skull; and the **cranial base,** forming

(Text continues on page 128.)

Table 9.1A	The Axial Skeleton: Cranial Bones and Important Bone Markings	
Cranial bone	**Important markings**	**Description**
Frontal (1) Figures 9.1, 9.3, 9.7, 9.9, and 9.10	N/A	Forms the forehead, superior part of the orbit, and the floor of the anterior cranial fossa.
	Supraorbital margin	Thick margin of the eye socket that lies beneath the eyebrows.
	Supraorbital foramen (notch)	Opening above each orbit allowing blood vessels and nerves to pass.
	Glabella	Smooth area between the eyes.
Parietal (2) Figures 9.1, 9.3, 9.6, 9.7, and 9.10	N/A	Form the superior and lateral aspects of the skull.
Temporal (2) Figures 9.1, 9.2, 9.3, 9.6, 9.7, and 9.10	N/A	Form the inferolateral aspects of the skull and contribute to the middle cranial fossa; each has squamous, tympanic, and petrous parts.
	Squamous part	Located inferior to the squamous suture. The next two markings are located in this part.
	Zygomatic process	A bridgelike projection that articulates with the zygomatic bone to form the zygomatic arch.
	Mandibular fossa	Located on the inferior surface of the zygomatic process; receives the condylar process of the mandible to form the temporomandibular joint.
	Tympanic part	Surrounds the external ear opening. The next two markings are located in this part.
	External acoustic meatus	Canal leading to the middle ear and eardrum.
	Styloid process	Needlelike projection that serves as an attachment point for ligaments and muscles of the neck. (This process is often missing from demonstration skulls because it has broken off.)
	Petrous part	Forms a bony wedge between the sphenoid and occipital bones and contributes to the cranial base. The remaining temporal markings are located in this part.
	Jugular foramen	Located where the petrous part of the temporal bone joins the occipital bone. Forms an opening which the internal jugular vein and cranial nerves IX, X, and XI pass.
	Carotid canal	Opening through which the internal carotid artery passes into the cranial cavity.
	Foramen lacerum	Almost completely closed by cartilage in the living person but forms a jagged opening in dried skulls.
	Stylomastoid foramen	Tiny opening between the mastoid and styloid processes through which cranial nerve VII leaves the cranium.
	Mastoid process	Located posterior to the external acoustic meatus; serves as an attachment point for neck muscles.

(Table continues on page 126.)

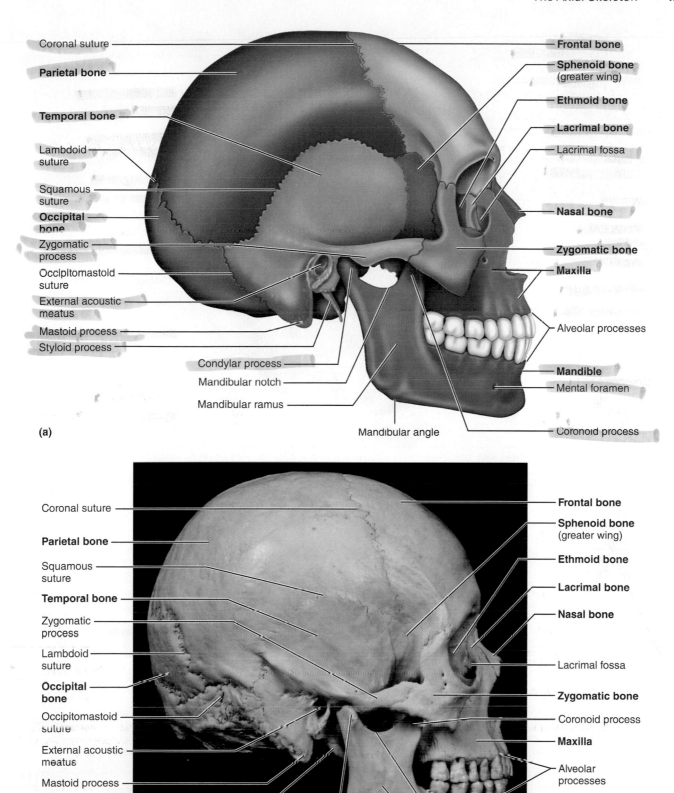

Coronal suture

Parietal bone

Temporal bone

Lambdoid
suture

Squamous
suture

**Occipital
bone**

Zygomatic
process

Occipitomastold
suture

External acoustic
meatus

Mastoid process

Styloid process

Condylar process

Mandibular notch

Mandibular ramus

Mandibular angle

Frontal bone

Sphenoid bone
(greater wing)

Ethmoid bone

Lacrimal bone

Lacrimal fossa

Nasal bone

Zygomatic bone

Maxilla

Alveolar processes

Mandible

Mental foramen

Coronoid process

(a)

9

Coronal suture

Parietal bone

Squamous
suture

Temporal bone

Zygomatic
process

Lambdoid
suture

**Occipital
bone**

Occipitomastoid
suture

External acoustic
meatus

Mastoid process

Styloid process

Condylar
process

Mandibular angle

Frontal bone

Sphenoid bone
(greater wing)

Ethmoid bone

Lacrimal bone

Nasal bone

Lacrimal fossa

Zygomatic bone

Coronoid process

Maxilla

Alveolar
processes

Mandible

Mental foramen

Mandibular notch

Mandibular ramus

(b)

Figure 9.1 External anatomy of the right lateral aspect of the skull. (a) Diagram. **(b)** Photograph.

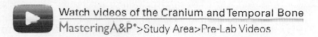

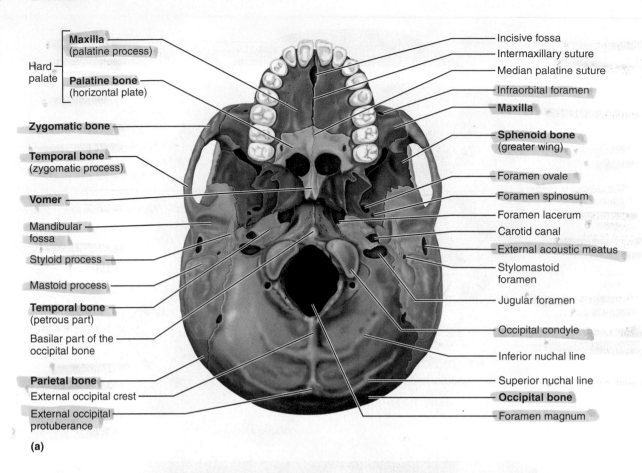

(a)

Maxilla (palatine process)
Hard palate
Palatine bone (horizontal plate)
Zygomatic bone
Temporal bone (zygomatic process)
Vomer
Mandibular fossa
Styloid process
Mastoid process
Temporal bone (petrous part)
Basilar part of the occipital bone
Parietal bone
External occipital crest
External occipital protuberance

Incisive fossa
Intermaxillary suture
Median palatine suture
Infraorbital foramen
Maxilla
Sphenoid bone (greater wing)
Foramen ovale
Foramen spinosum
Foramen lacerum
Carotid canal
External acoustic meatus
Stylomastoid foramen
Jugular foramen
Occipital condyle
Inferior nuchal line
Superior nuchal line
Occipital bone
Foramen magnum

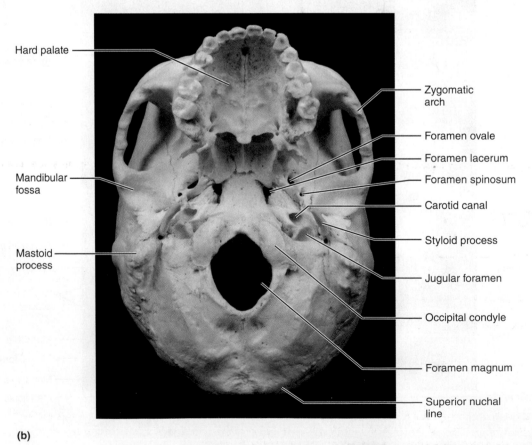

(b)

Hard palate
Mandibular fossa
Mastoid process

Zygomatic arch
Foramen ovale
Foramen lacerum
Foramen spinosum
Carotid canal
Styloid process
Jugular foramen
Occipital condyle
Foramen magnum
Superior nuchal line

Figure 9.2 Inferior view of the skull, mandible removed.

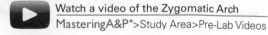

Watch a video of the Zygomatic Arch
MasteringA&P*>Study Area>Pre-Lab Videos

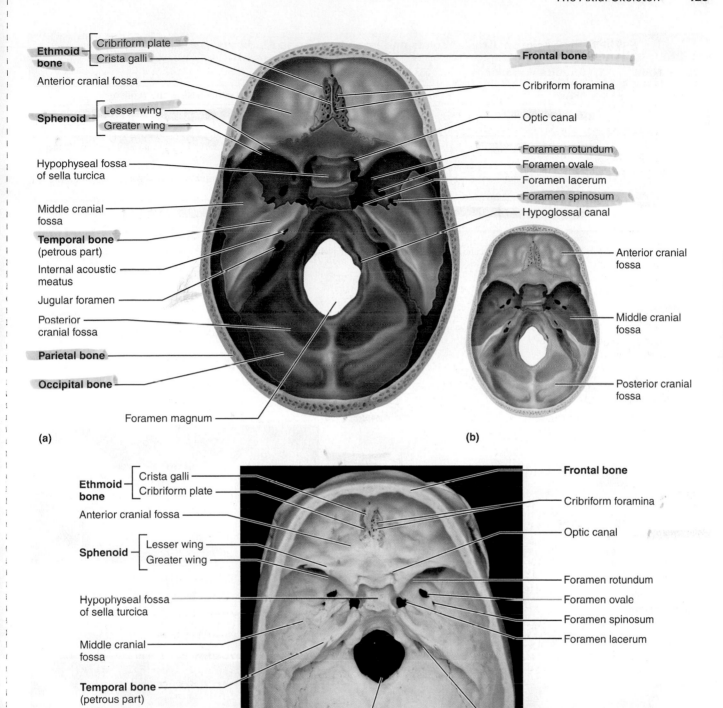

(a) Superior view of the base of the cranial cavity, calvaria removed.

(b) Diagram of the cranial base showing the extent of its major fossae.

(c) Photograph of superior view of the base of the cranial cavity, calvaria removed.

Figure 9.3 Internal anatomy of the inferior portion of the skull.

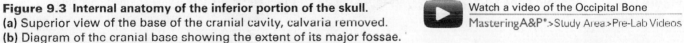

Watch a video of the Occipital Bone
MasteringA&P®>Study Area>Pre-Lab Videos

Table 9.1A The Axial Skeleton: Cranial Bones and Important Bone Markings *(continued)*

Cranial bone	Important markings	Description
Occipital (1) Figures 9.1, 9.2, 9.3, and 9.6	N/A	Forms the posterior aspect and most of the base of the skull.
	Foramen magnum	Large opening in the base of the bone, which allows the spinal cord to join with the brain stem.
	Occipital condyles	Rounded projections lateral to the foramen magnum that articulate with the first cervical vertebra (atlas).
	Hypoglossal canal	Opening medial and superior to the occipital condyle through which cranial nerve XII (the hypoglossal nerve) passes.
	External occipital protuberance	Midline prominence posterior to the foramen magnum.

The number in parentheses ( ) following the bone name indicates the total number of such bones in the body.

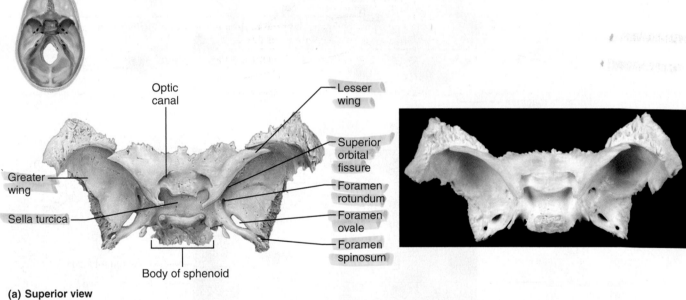

(a) Superior view

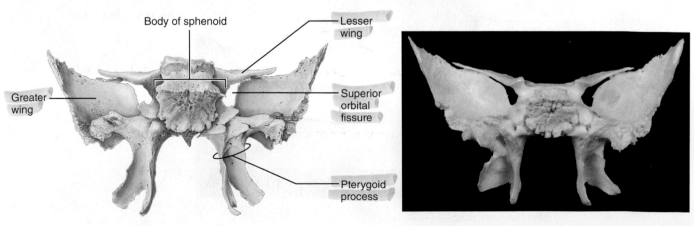

(b) Posterior view

Figure 9.4 The sphenoid bone.

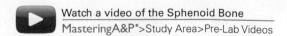

Watch a video of the Sphenoid Bone
MasteringA&P®>Study Area>Pre-Lab Videos

Table 9.1B	The Axial Skeleton: Cranial Bones and Important Bone Markings	
Cranial bone	**Important markings**	**Description**
Sphenoid bone (1) Figures 9.1, 9.2, 9.3, 9.4, 9.7, and 9.10	N/A	Bat-shaped bone that is described as the keystone bone of the cranium because it articulates with all other cranial bones.
	Greater wings	Project laterally from the sphenoid body, forming parts of the middle cranial fossa and the orbits.
	Pterygoid processes	Project inferiorly from the greater wings; attachment site for chewing muscles (pterygoid muscles).
	Superior orbital fissures	Slits in the orbits providing passage of cranial nerves that control eye movements (III, IV, VI, and the ophthalmic division of V).
	Sella turcica	"Turkish saddle" located on the superior surface of the body; the seat of the saddle, called the *hypophyseal fossa*, holds the pituitary gland.
	Lesser wings	Form part of the floor of the anterior cranial fossa and part of the orbit.
	Optic canals	Openings in the base of the lesser wings; cranial nerve II (optic nerve) passes through to serve the eye.
	Foramen rotundum	Openings located in the medial part of the greater wing; a branch of cranial nerve V (maxillary division) passes through.
	Foramen ovale	Openings located posterolateral to the foramen rotundum; a branch of cranial nerve V (mandibular division) passes through.
	Foramen spinosum	Openings located posterolateral to the foramen spinosum; provides passageway for the middle meningeal artery.

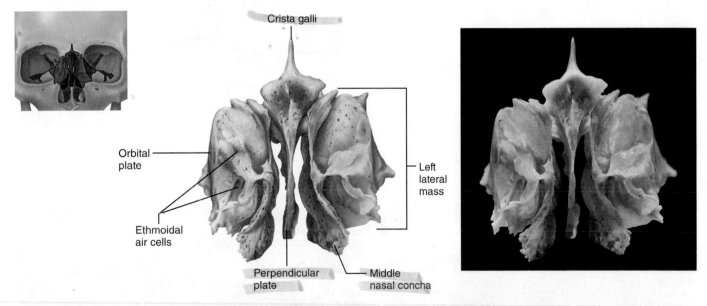

Figure 9.5 The ethmoid bone. Anterior view. The superior nasal conchae are located posteriorly and are therefore not visible in the anterior view.

Watch a video of the Ethmoid Bone
MasteringA&P®>Study Area>Pre-Lab Videos

Table 9.1C	The Axial Skeleton: Cranial Bones and Important Bone Markings	
Cranial bone	**Important markings**	**Description**
Ethmoid (1) Figures 9.1, 9.3, 9.5, 9.7, and 9.10	N/A	Contributes to the anterior cranial fossa; forms part of the nasal septum and the nasal cavity; contributes to the medial wall of the orbit.
	Crista galli	"Rooster's comb"; a superior projection that attaches to the dura mater, helping to secure the brain within the skull.
	Cribriform plates	Located lateral to the crista galli; form a portion of the roof of the nasal cavity and the floor of the anterior cranial fossa.

(Table continues on page 128.)

Table 9.1C

Cranial bone	Important markings	Description
Ethmoid (1) *(continued)*	Cribriform foramina	Tiny holes in the cribriform plates that allow for the passage of filaments of cranial nerve I (olfactory nerve).
	Perpendicular plate	Inferior projection that forms the superior portion of the nasal septum.
	Lateral masses	Flank the perpendicular plate on each side and are filled with sinuses called *ethmoidal air cells.*
	Orbital plates	Lateral surface of the lateral masses that contribute to the medial wall of the orbits.
	Superior and middle nasal conchae	Extend medially from the lateral masses; act as turbinates to improve airflow through the nasal cavity.

Table 9.1C The Axial Skeleton: Cranial Bones and Important Bone Markings *(continued)*

the skull bottom. Internally, the cranial base has three distinct depressions: the **anterior, middle,** and **posterior cranial fossae** (see Figure 9.3). The brain sits in these fossae, completely enclosed by the cranial vault. Overall, the brain occupies the *cranial cavity.*

Major Sutures

The four largest sutures are located where the parietal bones articulate with each other and where the parietal bones articulate with other cranial bones:

- **Sagittal suture:** Occurs where the left and right parietal bones meet superiorly in the midline of the cranium (Figure 9.6).
- **Coronal suture:** Running in the frontal plane, occurs anteriorly where the parietal bones meet the frontal bone (Figure 9.1).
- **Squamous suture:** Occurs where each parietal bone meets the temporal bone, on each lateral aspect of the skull (Figure 9.1).
- **Lambdoid suture:** Occurs where the parietal bones meet the occipital bone posteriorly (Figure 9.6).

Facial Bones

Of the 14 bones composing the face, 12 are paired. *Only the mandible and vomer are single bones.* An additional bone,

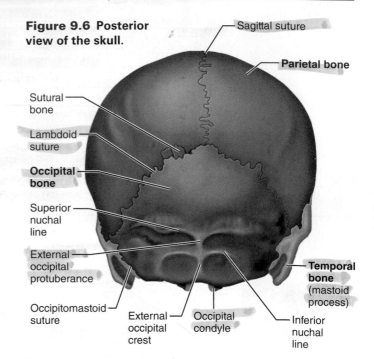

Figure 9.6 Posterior view of the skull.

the hyoid bone, although not a facial bone, is considered here because of its location.

Table 9.2 The Axial Skeleton: Facial Bones and Important Bone Markings (Figures 9.1, 9.7, 9.9, and 9.10, with additional figures listed for specific bones)

Facial bone	Important markings	Description
■ Nasal (2)	N/A	Small rectangular bones forming the bridge of the nose.
■ Lacrimal (2)	N/A	Each forms part of the medial orbit in between the maxilla and ethmoid bone.
	Lacrimal fossa	Houses the lacrimal sac, which helps to drain tears from the nasal cavity.
Zygomatic (2) (also Figure 9.2)	N/A	Commonly called the cheekbones; each forms part of the lateral orbit
■ Inferior nasal concha (2)	N/A	Inferior turbinate; each forms part of the lateral walls of the nasal cavities; improves the airflow through the nasal cavity
Palatine (2) (also Figure 9.2)	N/A	Forms the posterior hard palate, a small part of the nasal cavity, and part of the orbit.
	Horizontal plate	Forms the posterior portion of the hard palate.
	Median palatine suture	Median fusion point of the horizontal plates of the palatine bones.
■ Vomer (1)	N/A	Thin, blade-shaped bone that forms the inferior nasal septum.
Maxilla (2) (also Figures 9.2 and 9.8)	N/A	Keystone facial bones because they articulate with all other facial bones except the mandible; form the upper jaw and parts of the hard palate, orbits, and nasal cavity.
	Frontal process	Forms part of the lateral aspect of the bridge of the nose.
	Infraorbital foramen	Opening under the orbit that forms a passageway for the infraorbital artery and nerve.

Table 9.2	(continued)	
Facial bone	**Important markings**	**Description**
◼ Maxilla (2) *(continued)*	Palatine process	Forms the anterior hard palate; meet anteriorly in the intermaxillary suture (Note: Seen in inferior view).
	Zygomatic process	Articulation process for zygomatic bone.
	Alveolar process	Inferior margin of the maxilla; contains sockets in which the teeth lie
◼ Mandible (1) (also Figures 9.2 and 9.8)	N/A	The lower jawbone, which articulates with the temporal bone to form the only freely movable joints in the skull (the temporomandibular joint).
	Condylar processes	Articulate with the mandibular fossae of the temporal bones.
	Coronoid processes	"Crown-shaped" portion of the ramus for muscle attachment.
	Mandibular notches	Separate the condylar process and the coronoid process.
	Body	Horizontal portion that forms the chin.
	Ramus	Vertical extension of the body.
	Mandibular angles	Posterior points where the ramus meets the body.
	Mental foramina	Paired openings on the body (lateral to the midline); transmit blood vessels and nerves to the lower lip and skin of the chin.
	Alveolar process	Superior margin of the mandible; contains sockets in which the teeth lie.
	Mandibular foramina	Located on the medial surface of each ramus; passageway for the nerve involved in tooth sensation. (Dentists inject anesthetic into this foramen before working on the lower teeth.)

9

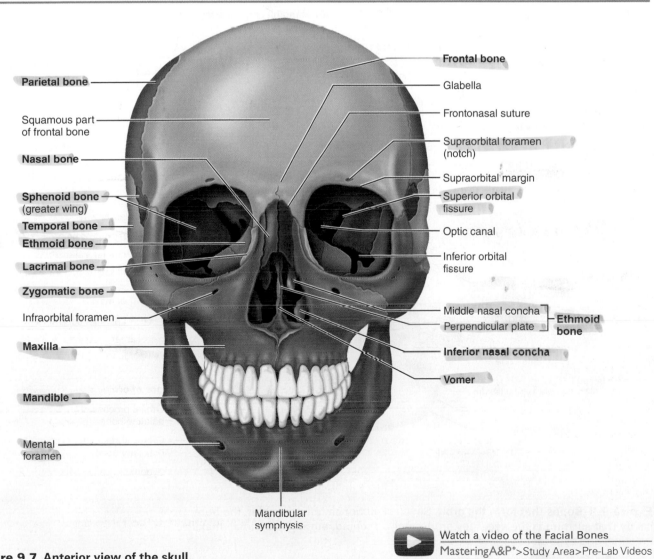

Figure 9.7 Anterior view of the skull.

Watch a video of the Facial Bones
MasteringA&P®>Study Area>Pre-Lab Videos

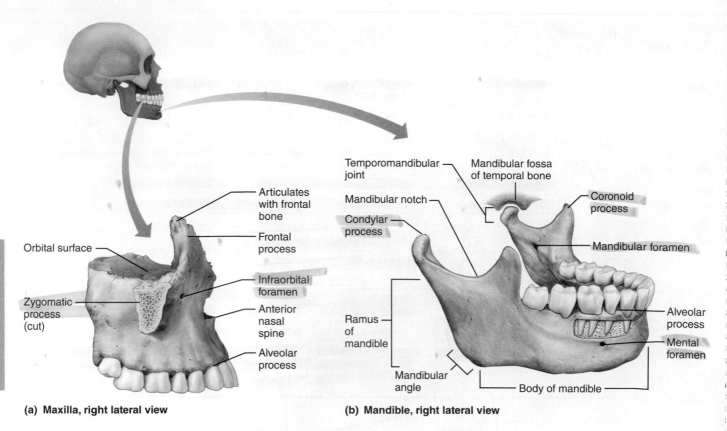

(a) Maxilla, right lateral view

Articulates with frontal bone
Frontal process
Orbital surface
Infraorbital foramen
Anterior nasal spine
Zygomatic process (cut)
Alveolar process

(b) Mandible, right lateral view

Temporomandibular joint
Mandibular notch
Condylar process
Mandibular fossa of temporal bone
Coronoid process
Mandibular foramen
Ramus of mandible
Alveolar process
Mental foramen
Mandibular angle
Body of mandible

Figure 9.8 Detailed anatomy of the maxilla and mandible.

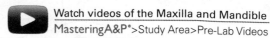
Watch videos of the Maxilla and Mandible
MasteringA&P*>Study Area>Pre-Lab Videos

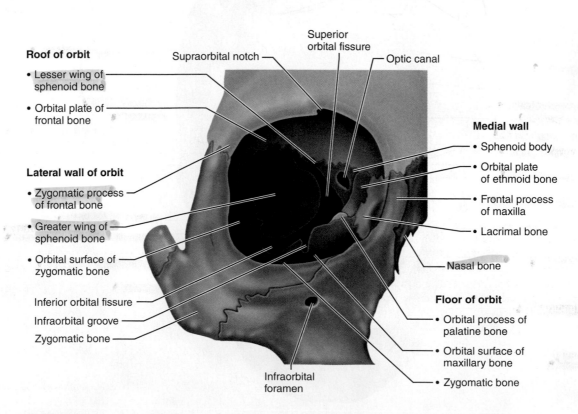

Roof of orbit
• Lesser wing of sphenoid bone
• Orbital plate of frontal bone

Lateral wall of orbit
• Zygomatic process of frontal bone
• Greater wing of sphenoid bone
• Orbital surface of zygomatic bone

Inferior orbital fissure
Infraorbital groove
Zygomatic bone

Supraorbital notch
Superior orbital fissure
Optic canal

Medial wall
• Sphenoid body
• Orbital plate of ethmoid bone
• Frontal process of maxilla
• Lacrimal bone

Nasal bone

Floor of orbit
• Orbital process of palatine bone
• Orbital surface of maxillary bone
• Zygomatic bone

Infraorbital foramen

Figure 9.9 Bones that form the orbit. Seven skull bones form the orbit, the bony cavity that surrounds the eye. They are frontal, sphenoid, ethmoid, lacrimal, maxilla, palatine, and zygomatic.

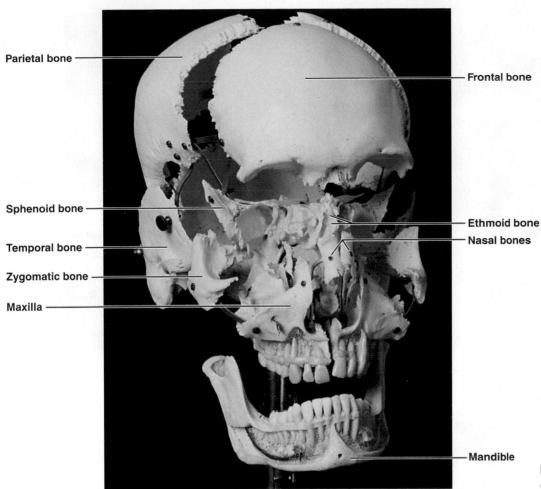

Figure 9.10 Frontal view of the Beauchene skull.

Parietal bone

Frontal bone

Sphenoid bone

Ethmoid bone

Nasal bones

Temporal bone

Zygomatic bone

Maxilla

Mandible

9

Group Challenge

Odd Bone Out

Each of the following sets contains four bones. One of the listed bones does not share a characteristic that the other three do. Work in groups of three, and discuss the characteristics of the bones in each group. On a separate piece of paper, one student will record the characteristics of each bone. For each set of bones, discuss the possible candidates for the "odd bone out" and which characteristic it lacks, based on your notes. Once your group has come to a consensus, circle the bone that doesn't belong with the others, and explain why it is singled out. What characteristic is it missing? Include as many characteristics as you can think of, but make sure your choice does not have the key characteristic. Use an articulated skull, disarticulated skull bones, and the pictures in your lab manual to help you select and justify your answer.

1. Which is the "odd bone"?	Why is it the odd one out?
Zygomatic bone	
Maxilla	
Vomer	
Nasal bone	

2. Which is the "odd bone"?	Why is it the odd one out?
Parietal bone	
Sphenoid bone	
Frontal bone	
Occipital bone	

3. Which is the "odd bone"?	Why is it the odd one out?
Lacrimal bone	
Nasal bone	
Zygomatic bone	
Maxilla	

9

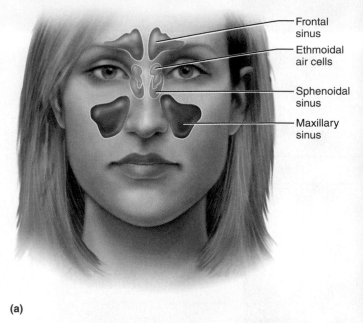

(a)

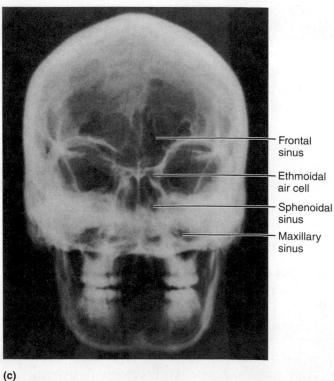

(c)

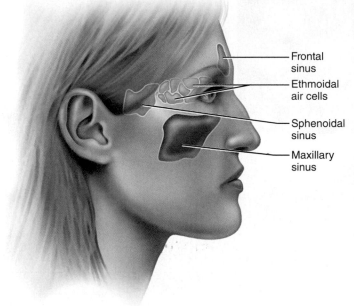

(b)

Figure 9.11 Paranasal sinuses. (a) Anterior aspect.
(b) Medial aspect. **(c)** Skull X-ray image showing the
paranasal sinuses, anterior view.

Paranasal Sinuses

Four skull bones—maxillary, sphenoid, ethmoid, and frontal—
contain sinuses (mucosa-lined air cavities) that lead into the
nasal passages (see Figure 9.5 and **Figure 9.11**). These
paranasal sinuses lighten the skull and may act as resonance
chambers for speech. The maxillary sinus is the largest of the
sinuses found in the skull.

Sinusitis, or inflammation of the sinuses, sometimes occurs as a result of an allergy or bacterial
invasion of the sinus cavities. In such cases, some of the
connecting passageways between the sinuses and nasal
passages may become blocked with thick mucus or infectious
material. Then, as the air in the sinus cavities is absorbed, a
partial vacuum forms. The result is a sinus headache localized
over the inflamed sinus area. Severe sinus infections may
require surgical drainage to relieve this painful condition. ✚

Hyoid Bone

Not really considered or counted as a skull bone, the hyoid
bone is located in the throat above the larynx. It serves as a
point of attachment for many tongue and neck muscles. It
does not articulate with any other bone and is thus unique. It
is horseshoe shaped with a body and two pairs of **horns,** or
cornua (**Figure 9.12**).

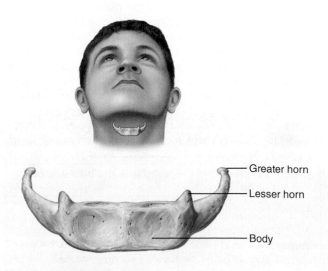

Figure 9.12 Hyoid bone.

Activity 2

Palpating Skull Markings

Palpate the following areas on yourself. Place a check mark in the boxes as you locate the skull markings. Ask your instructor for help with any markings that you are unable to locate.

☐ Zygomatic bone and arch. (The most prominent part of your cheek is your zygomatic bone. Follow the posterior course of the zygomatic arch to its junction with your temporal bone.)

☐ Mastoid process (the rough area behind your ear).

☐ Temporomandibular joints. (Open and close your jaws to locate these.)

☐ Greater wing of sphenoid. (Find the indentation posterior to the orbit and superior to the zygomatic arch on your lateral skull.)

☐ Supraorbital foramen. (Apply firm pressure along the superior orbital margin to find the indentation resulting from this foramen.)

☐ Infraorbital foramen. (Apply firm pressure just inferior to the inferomedial border of the orbit to locate this large foramen.)

☐ Mandibular angle (most inferior and posterior aspect of the mandible).

☐ Mandibular symphysis (midline of chin).

☐ Nasal bones. (Run your index finger and thumb along opposite sides of the bridge of your nose until they "slip" medially at the inferior end of the nasal bones.)

☐ External occipital protuberance. (This midline projection is easily felt by running your fingers up the furrow at the back of your neck to the skull.)

☐ Hyoid bone. (Place a thumb and index finger beneath the chin just anterior to the mandibular angles, and squeeze gently. Exert pressure with the thumb, and feel the horn of the hyoid with the index finger.)

 Watch a video of the Thoracic Vertebra
MasteringA&P®>Study Area>Pre-Lab Video

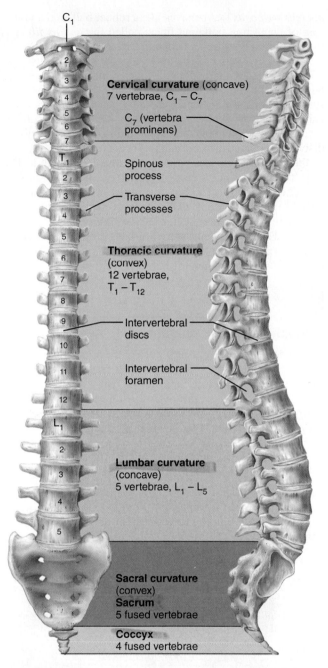

Cervical curvature (concave)
7 vertebrae, $C_1 – C_7$

C_7 (vertebra prominens)

Spinous process

Transverse processes

Thoracic curvature (convex)
12 vertebrae, $T_1 – T_{12}$

Intervertebral discs

Intervertebral foramen

Lumbar curvature (concave)
5 vertebrae, $L_1 – L_5$

Sacral curvature (convex)
Sacrum
5 fused vertebrae

Coccyx
4 fused vertebrae

Anterior view   Right lateral view

Figure 9.13 The vertebral column. Notice the curvatures in the lateral view. (The terms *convex* and *concave* refer to the curvature of the posterior aspect of the vertebral column.)

The Vertebral Column

The **vertebral column,** extending from the skull to the pelvis, forms the body's major axial support. Additionally, it surrounds and protects the delicate spinal cord while allowing the spinal nerves to emerge from the cord via openings between adjacent vertebrae. The term *vertebral column* might suggest a rigid supporting rod, but this is far from the truth. The vertebral column consists of 24 single bones called **vertebrae** and two composite, or fused, bones (the sacrum and coccyx) that are connected in such a way as to provide a flexible curved structure (**Figure 9.13**). Of the 24 single vertebrae, the seven bones of the neck are called *cervical vertebrae;* the next 12 are *thoracic vertebrae;* and the 5 supporting the lower back are *lumbar vertebrae.* Remembering common mealtimes for breakfast, lunch, and dinner (7 A.M., 12 noon, and 5 P.M.) may help you to remember the number of bones in each region.

The vertebrae are separated by pads of fibrocartilage, **intervertebral discs,** that cushion the vertebrae and absorb shocks. Each disc has two major regions, a central gelatinous *nucleus pulposus* that behaves like a rubber ball, and an outer ring of encircling collagen fibers called the *anulus fibrosus* that stabilizes the disc and contains the nucleus pulposus.

As a person ages, the water content of the discs decreases (as it does in other tissues throughout the body), and the discs become thinner and less compressible. This situation, along with other degenerative changes such as weakening of the ligaments and tendons of the vertebral column, predisposes older people to a ruptured disc, called a **herniated disc.** In a herniated disc, the anulus fibrosus commonly ruptures and the nucleus pulposus protrudes (herniates) through it. This event typically compresses adjacent nerves, causing pain. ✚

The thoracic and sacral curvatures of the spine are referred to as *primary curvatures* because they are present and well developed at birth. Later the *secondary curvatures* are formed. The cervical curvature becomes prominent when the baby begins to hold its head up independently, and the lumbar curvature develops when the baby begins to walk.

Activity 3

Examining Spinal Curvatures

1. Observe the normal curvature of the vertebral column in the articulated vertebral column or laboratory skeleton and compare it to Figure 9.13. Note the differences between normal curvature and three abnormal spinal curvatures seen in the figure—*scoliosis, kyphosis,* and *lordosis* (**Figure 9.14**). These abnormalities may result from disease or poor posture. Also examine X-ray images, if they are available, showing these same conditions in a living patient.

2. Then, using the articulated vertebral column (or an articulated skeleton), examine the freedom of movement between two lumbar vertebrae separated by an intervertebral disc.

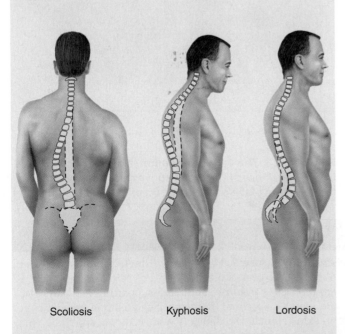

Scoliosis   Kyphosis   Lordosis

Figure 9.14 Abnormal spinal curvatures.

When the fibrous disc is properly positioned, are the spinal cord or peripheral nerves impaired in any way?

_____

_____

Remove the disc, and put the two vertebrae back together. What happens to the nerve?

_____

_____

What would happen to the spinal nerves in areas of malpositioned or "slipped" discs?

_____

_____

Structure of a Typical Vertebra

Although they differ in size and specific features, all vertebrae have some features in common (**Figure 9.15**).

- **Body** (or **centrum**): Rounded central portion of the vertebra, which faces anteriorly in the human vertebral column.

- **Vertebral arch:** Composed of pedicles, laminae, and a spinous process, it represents the junction of all posterior extensions from the vertebral body.

- **Vertebral (spinal) foramen:** Opening enclosed by the body and vertebral arch; a passageway for the spinal cord.

- **Transverse processes:** Two lateral projections from the vertebral arch.

- **Spinous process:** Single medial and posterior projection from the vertebral arch.

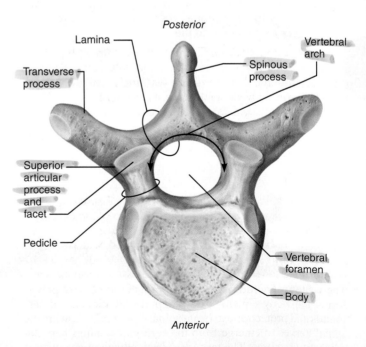

Figure 9.15 A typical vertebra, superior view. Inferior articulating surfaces not shown.

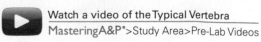

Watch a video of the Typical Vertebra
MasteringA&P®>Study Area>Pre-Lab Videos

- **Superior and inferior articular processes:** Paired projections lateral to the vertebral foramen that enable articulation with adjacent vertebrae. The superior articular processes typically face toward the spinous process (posteriorly), whereas the inferior articular processes face (anteriorly) away from the spinous process.

- **Intervertebral foramina:** The right and left pedicles have notches (see Figure 9.17) on their inferior and superior surfaces that create openings, the intervertebral foramina, for spinal nerves to leave the spinal cord between adjacent vertebrae.

Figures 9.16–9.18 and **Table 9.3** on p. 137 show how specific vertebrae differ; refer to them as you read the following sections.

Cervical Vertebrae

The seven cervical vertebrae (referred to as C_1 through C_7) form the neck portion of the vertebral column. The first two cervical vertebrae (atlas and axis) are highly modified to perform special functions (**Figure 9.16**). The **atlas** (C_1) lacks a body, and its lateral processes contain large concave depressions on their superior surfaces that receive the occipital condyles of the skull. This joint enables you to nod "yes." The **axis** (C_2) acts as a pivot for the rotation of the atlas (and skull) above. It bears a large vertical process, the **dens,** that serves as the pivot point. The articulation between C_1 and C_2 allows you to rotate your head from side to side to indicate "no."

The more typical cervical vertebrae (C_3 through C_7) are distinguished from the thoracic and lumbar vertebrae by several features (see Table 9.3 and **Figure 9.17** on p. 136). They are the smallest, lightest vertebrae, and the vertebral foramen is triangular. The spinous process is short and often bifurcated (divided into two branches). The spinous process of C_7 is not branched, however, and is substantially longer than that of the other cervical vertebrae. Because the spinous process of C_7 is visible through the skin at the base of the neck, it is called the *vertebra prominens* (Figure 9.13) and is used as a landmark for counting the vertebrae. Transverse processes of the cervical vertebrae are wide, and they contain foramina through which the vertebral arteries pass superiorly on their way to the brain. Any time you see these foramina in a vertebra, you can be sure that it is a cervical vertebra.

☐ Palpate your vertebra prominens. Place a check mark in the box when you locate the structure.

Thoracic Vertebrae

The 12 thoracic vertebrae (referred to as T_1 through T_{12}) may be recognized by the following structural characteristics. They have a larger body than the cervical vertebrae (see Figure 9.17). The body is somewhat heart shaped, with two small articulating surfaces, or **costal facets,** on each side (one superior, the other inferior) close to the origin of the vertebral arch. These facets articulate with the heads of the corresponding ribs. The vertebral foramen is oval or round, and the spinous process is long, with a sharp downward hook. The closer the thoracic vertebra is to the lumbar region, the less sharp and shorter the spinous process. Articular facets on the transverse processes articulate with the tubercles of the ribs. Besides forming the thoracic part of the spine, these vertebrae form

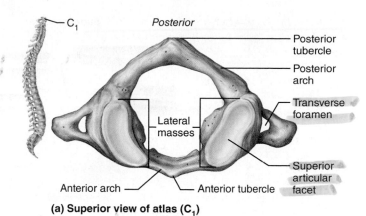

(a) Superior view of atlas (C_1)

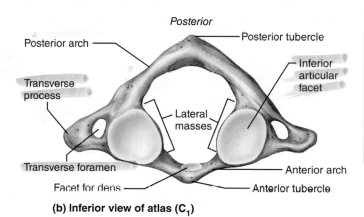

(b) Inferior view of atlas (C_1)

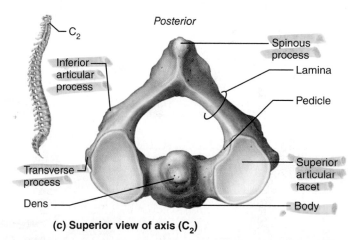

(c) Superior view of axis (C_2)

Figure 9.16 The first and second cervical vertebrae.

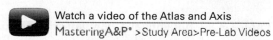

Watch a video of the Atlas and Axis
MasteringA&P® > Study Area > Pre-Lab Videos

the posterior aspect of the thoracic cage. Indeed, they are the only vertebrae that articulate with the ribs.

Lumbar Vertebrae

The five lumbar vertebrae (L_1 through L_5) have massive blocklike bodies and short, thick, hatchet-shaped spinous processes extending directly backward (see Table 9.3 and

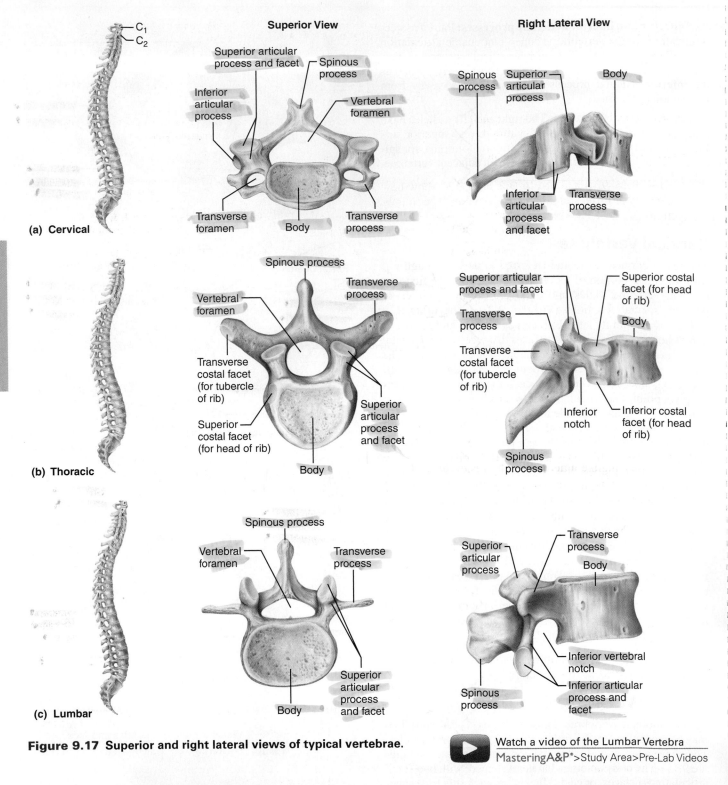

Superior View | Right Lateral View

(a) Cervical

(b) Thoracic

(c) Lumbar

Figure 9.17 Superior and right lateral views of typical vertebrae.

Watch a video of the Lumbar Vertebra
MasteringA&P®>Study Area>Pre-Lab Videos

Figure 9.17). The superior articular facets face posteromedially; the inferior ones are directed anterolaterally. These structural features reduce the mobility of the lumbar region of the spine. Since most stress on the vertebral column occurs in the lumbar region, these are also the sturdiest of the vertebrae.

The spinal cord ends at the superior edge of L_2, but the outer covering of the cord, filled with cerebrospinal fluid,

extends an appreciable distance beyond. Thus a *lumbar puncture* (for examination of the cerebrospinal fluid) or the administration of "saddle block" anesthesia for childbirth is normally done between L_3 and L_4 or L_4 and L_5, where there is little or no chance of injuring the delicate spinal cord.

Table 9.3	Regional Characteristics of Cervical, Thoracic, and Lumbar Vertebrae		
Characteristic	**(a) Cervical (C$_3$–C$_7$)**	**(b) Thoracic**	**(c) Lumbar**
Body	Small, wide side to side	Larger than cervical; heart shaped; bears costal facets	Massive; kidney shaped
Spinous process	Short; bifid; projects directly posteriorly	Long; sharp; projects inferiorly	Short; blunt; projects directly posteriorly
Vertebral foramen	Triangular	Circular	Triangular
Transverse processes	Contain foramina	Bear facets for ribs (except T$_{11}$ and T$_{12}$)	Thin and tapered
Superior and inferior articulating processes	Superior facets directed superoposteriorly	Superior facets directed posteriorly	Superior facets directed posteromedially (or medially)
	Inferior facets directed inferoanteriorly	Inferior facets directed anteriorly	Inferior facets directed anterolaterally (or laterally)
Movements allowed	Flexion and extension; lateral flexion; rotation; the spine region with the greatest range of movement	Rotation; lateral flexion possible but limited by ribs; flexion and extension prevented	Flexion and extension; some lateral flexion; rotation prevented

The Sacrum

The **sacrum** (**Figure 9.18**) is a composite bone formed from the fusion of five vertebrae. Superiorly it articulates with L$_5$, and inferiorly it connects with the coccyx. The **median sacral crest** is a remnant of the spinous processes of the fused vertebrae. The winglike **alae,** formed by fusion of the transverse processes, articulate laterally with the hip bones. The sacrum is concave anteriorly and forms the posterior border of the pelvis. Four ridges (lines of fusion) cross the anterior part of the sacrum, and **sacral foramina** are located at either end of these ridges. These foramina allow blood vessels and nerves to pass. The vertebral canal continues inside the sacrum as the **sacral canal** and terminates near the coccyx via an enlarged opening called the **sacral hiatus**. The **sacral promontory** (anterior border of the body of S$_1$) is an important anatomical landmark for obstetricians.

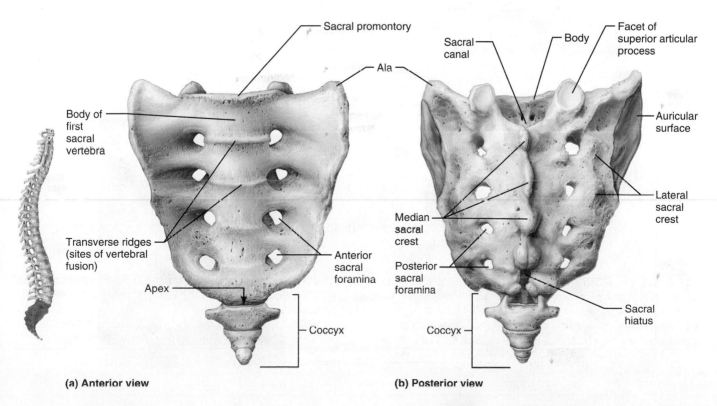

(a) Anterior view

(b) Posterior view

Figure 9.18 Sacrum and coccyx.

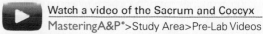

Watch a video of the Sacrum and Coccyx
MasteringA&P®>Study Area>Pre-Lab Videos

□ Attempt to palpate the median sacral crest of your sacrum. (This is more easily done by thin people and obviously in privacy.) Place a check mark in the box when you locate the structure.

The Coccyx

The **coccyx** (see Figure 9.18) is formed from the fusion of three to five small irregularly shaped vertebrae. It is literally the human tailbone, a vestige of the tail that other vertebrates have. The coccyx is attached to the sacrum by ligaments.

The Thoracic Cage

The **thoracic cage** consists of the bony thorax, which is composed of the sternum, ribs, and thoracic vertebrae, plus the costal cartilages (**Figure 9.19**). Its cone-shaped, cagelike structure protects the organs of the thoracic cavity, including the critically important heart and lungs.

The Sternum

The **sternum** (breastbone), a typical flat bone, is a result of the fusion of three bones—the manubrium, body, and xiphoid process. It is attached to the first seven pairs of ribs. The superiormost **manubrium** looks like the knot of a tie;

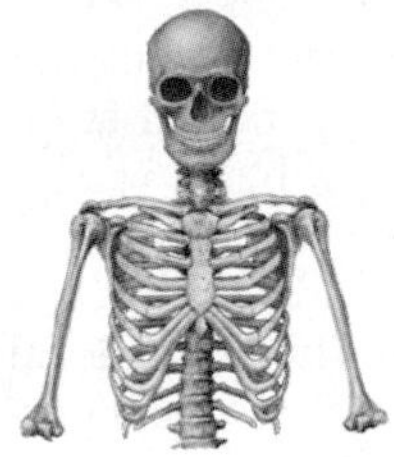

Figure 9.19 The thoracic cage. (a) Anterior view with costal cartilages shown in blue. **(b)** Median section of the thorax, illustrating the relationship of the surface anatomical landmarks of the thorax to the thoracic portion of the vertebral column.

▶ Watch a video of the Sternum
MasteringA&P®>Study Area>Pre-Lab Videos

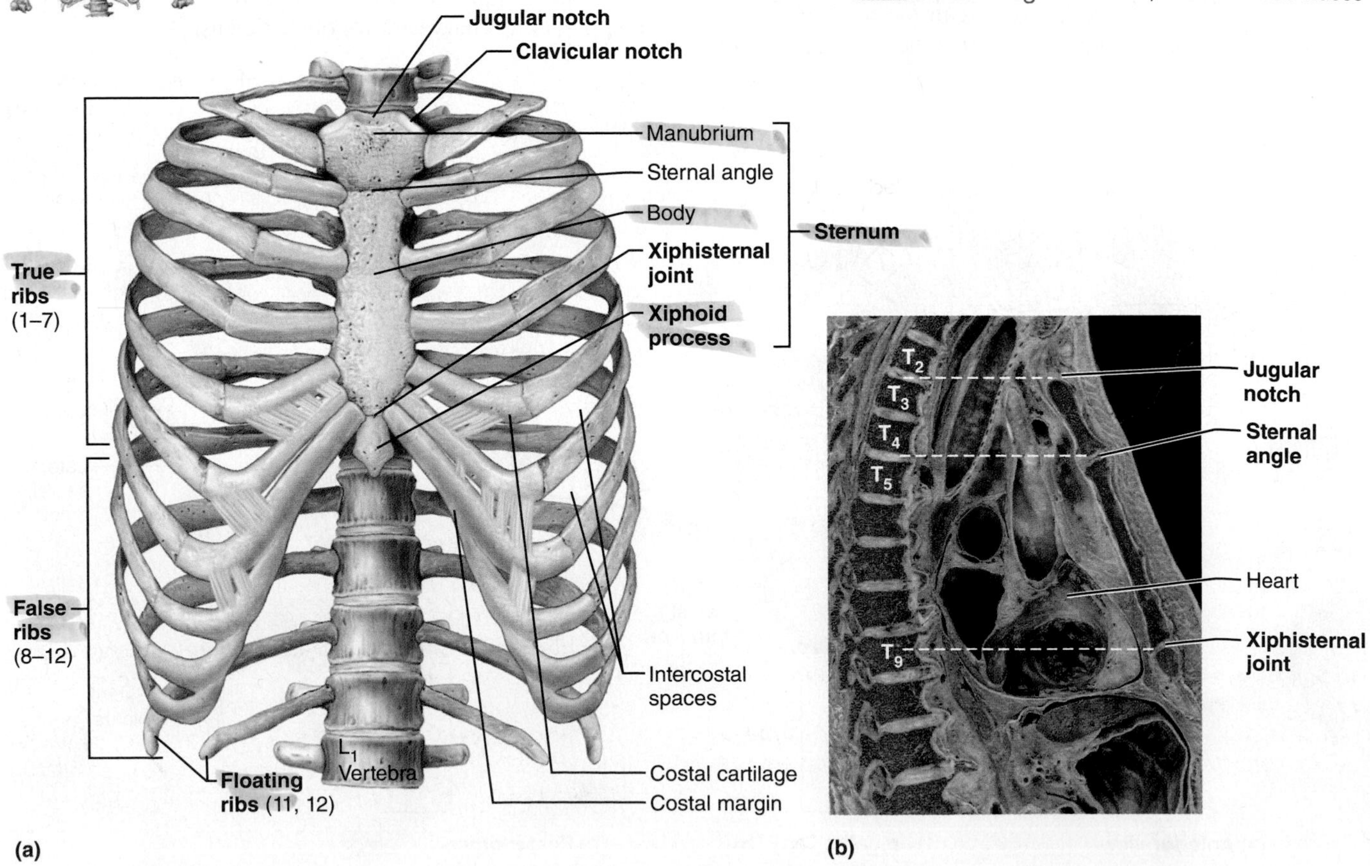

(a)

(b)

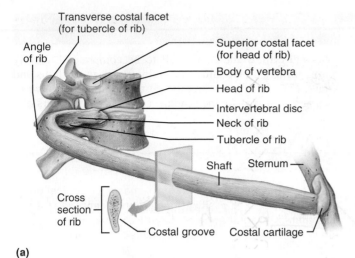

(a)

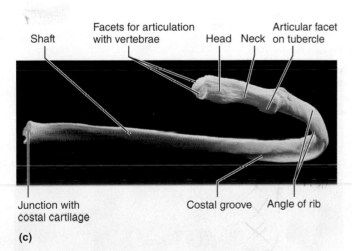

(b)

Shaft
Facets for articulation with vertebrae
Head Neck
Articular facet on tubercle

Junction with costal cartilage
Costal groove Angle of rib

(c)

Figure 9.20 Structure of a typical true rib and its articulations. (a) Vertebral and sternal articulations of a typical true rib. **(b)** Superior view of the articulation between a rib and a thoracic vertebra, with costovertebral ligaments. **(c)** Right rib 6, posterior view.

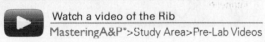

Watch a video of the Rib
MasteringA&P®>Study Area>Pre-Lab Videos

it articulates with the clavicle (collarbone) laterally. The **body** forms the bulk of the sternum. The **xiphoid process** constructs the inferior end of the sternum and lies at the level of the fifth intercostal space. Although it is made of hyaline cartilage in children, it is usually ossified in adults over the age of 40.

In some people, the xiphoid process projects dorsally. This may present a problem because physical trauma to the chest can push such a xiphoid into the underlying heart or liver, causing massive hemorrhage. ✚

The sternum has three important bony landmarks—the jugular notch, the sternal angle, and the xiphisternal joint. The **jugular notch** (concave upper border of the manubrium) can be palpated easily; generally it is at the level of the disc in between the second and third thoracic vertebrae. The **sternal angle** is a result of the manubrium and body meeting at a slight angle to each other, so that a transverse ridge is formed at the level of the second ribs. It provides a handy reference point for counting ribs to locate the second intercostal space for listening to certain heart valves, and it is an important anatomical landmark for thoracic surgery. The **xiphisternal joint,** the point where the sternal body and xiphoid process fuse, lies at the level of the ninth thoracic vertebra.

☐ Palpate your sternal angle and jugular notch. Place a check mark in the box when you locate the structures.

Because of its accessibility, the sternum is a favored site for obtaining samples of blood-forming (hematopoietic) tissue for the diagnosis of suspected blood diseases. A needle is inserted into the marrow of the sternum, and the sample is withdrawn (sternal puncture).

The Ribs

The 12 pairs of **ribs** form the walls of the thoracic cage (see Figure 9.19 and **Figure 9.20**). All of the ribs articulate posteriorly with the vertebral column via their heads and tubercles and then curve downward and toward the anterior body surface. The first seven pairs, called the *true,* or *vertebrosternal, ribs,* attach directly to the sternum by their "own" costal cartilages. The next five pairs are called *false ribs;* they attach indirectly to the sternum or entirely lack a sternal attachment. Of these, rib pairs 8–10, which are also called *vertebrochondral ribs,* have indirect cartilage attachments to the sternum via the costal cartilage of rib 7. The last two pairs, called *floating,* or *vertebral, ribs,* have no sternal attachment.

Activity 5

Examining the Relationship Between Ribs and Vertebrae

First take a deep breath to expand your chest. Notice how your ribs seem to move outward and how your sternum rises. Then examine an articulated skeleton to observe the relationship between the ribs and the vertebrae.

(Refer to Activity 3, Palpating Landmarks of the Trunk, section on The Thorax: Bones, steps 1 and 3, in Exercise 46, Surface Anatomy Roundup.)

9

The Fetal Skull

One of the most obvious differences between fetal and adult skeletons is the huge size of the fetal skull relative to the rest of the skeleton. Skull bones are incompletely formed at birth and connected by fibrous membranes called **fontanelles.** The fontanelles allow the fetal skull to be compressed slightly during birth and also allow for brain growth in the fetus and infant. They ossify (become bone) as the infant ages, completing the process by the time the child is 1½ to 2 years old.

Activity 6

Examining a Fetal Skull

1. Obtain a fetal skull, and study it carefully.

- Does it have the same bones as the adult skull?

- How does the size of the fetal face relate to the cranium?

- How does this compare to what is seen in the adult?

2. Locate the following fontanelles on the fetal skull (refer to **Figure 9.21**): *anterior* (or *frontal*) *fontanelle, mastoid fontanelle, sphenoidal fontanelle,* and *posterior* (or *occipital) fontanelle.*

3. Notice that some of the cranial bones have conical protrusions. These are **ossification (growth) centers.** Notice also that the frontal bone is still in two parts and that the temporal bone is incompletely ossified, little more than a ring of bone.

4. Before completing this study, check the questions on the Review Sheet at the end of this exercise to ensure that you have made all of the necessary observations.

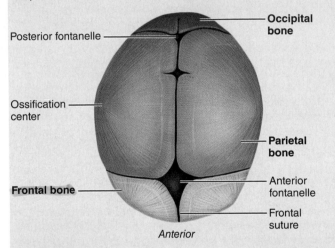

(a) Superior view

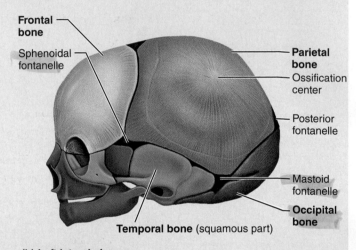

(b) Left lateral view

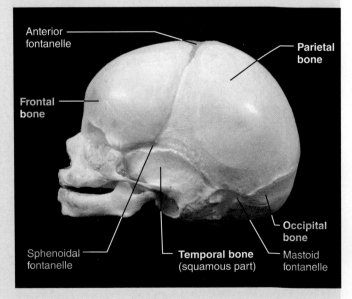

(c) Anterior view

(d) Left lateral view

Figure 9.21 Skull of a newborn.

The Appendicular Skeleton

Objectives

☐ Identify the bones of the pectoral and pelvic girdles and their attached limbs by examining isolated bones or an articulated skeleton, and name the important bone markings on each.

☐ Describe the differences between a male and a female pelvis, and explain why these differences are important.

☐ Compare the features of the human pectoral and pelvic girdles, and discuss how their structures relate to their specialized functions.

☐ Arrange unmarked, disarticulated bones in their proper places to form an entire skeleton.

Materials

- Articulated skeletons
- Disarticulated skeletons (complete)
- Articulated pelves (male and female for comparative study)
- X-ray images of bones of the appendicular skeleton

Pre-Lab Quiz

1. The _____ skeleton is made up of 126 bones of the limbs and girdles.
2. Circle the correct underlined term. The <u>pectoral</u> / <u>pelvic</u> girdle attaches the upper limb to the axial skeleton.
3. The _____, on the posterior thorax, are roughly triangular in shape. They have no direct attachment to the axial skeleton but are held in place by trunk muscles.
4. The arm consists of one long bone, the:
   a. femur   c. tibia
   b. humerus   d. ulna
5. The hand consists of three groups of bones. The carpals make up the wrist. The _____ make up the palm, and the phalanges make up the fingers.
6. You are studying a pelvis that is wide and shallow. The acetabula are small and far apart. The pubic arch is rounded and greater than 90°. It appears to be tilted forward, with a wide, short sacrum. Is this a male or a female pelvis? _____
7. The strongest, heaviest bone of the body is in the thigh. It is the
   a. femur   b. fibula   c. tibia
8. The _____, or "knee cap," is a sesamoid bone that is found within the quadriceps tendon.
9. Circle True or False. The fingers of the hand and the toes of the foot—with the exception of the great toe and the thumb—each have three phalanges.
10. Each foot has a total of _____ bones.

MasteringA&P®

For related exercise study tools, go to the Study Area of **MasteringA&P**. There you will find:

- Practice Anatomy Lab **PAL**
- PhysioEx **PEx**
- A&PFlix **A&PFlix**
- Practice quizzes, Histology Atlas, eText, Videos, and more!

The **appendicular skeleton** (the gold-colored portion of Figure 8.1) is composed of the 126 bones of the appendages and the pectoral and pelvic girdles, which attach the limbs to the axial skeleton. Although the upper and lower limbs differ in their functions and mobility, they have the same fundamental plan, with each limb made up of three major segments connected by freely movable joints.

Activity 1

Examining and Identifying Bones of the Appendicular Skeleton

Carefully examine each of the bones described in this exercise, and identify the characteristic bone markings of each (see Tables 10.1–10.5). The markings aid in determining whether a bone is the right or left member of its pair; for example, the glenoid cavity is on the lateral aspect of the scapula, and the spine is on its posterior aspect.

This is a very important instruction because you will be constructing your own skeleton to finish this laboratory exercise. Additionally, when corresponding X-ray images are available, compare the actual bone specimen to its X-ray image.

Bones of the Pectoral Girdle and Upper Limb

The Pectoral (Shoulder) Girdle

The paired **pectoral,** or **shoulder, girdles** (**Figures 10.1** and **10.2** and **Table 10.1**, p. 152) each consist of two bones—the anterior clavicle and the posterior scapula. The clavicle articulates with the axial skeleton via the sternum. The scapula does not articulate with the axial skeleton. The shoulder girdles attach the upper limbs to the axial skeleton and provide attachment points for many trunk and neck muscles.

The shoulder girdle is exceptionally light and allows the upper limb a degree of mobility not seen anywhere else in the body. This is due to the following factors:

• The sternoclavicular joints are the *only* site of attachment of the shoulder girdles to the axial skeleton.

• The relative looseness of the scapular attachment allows it to slide back and forth against the thorax with muscular activity.

• The glenoid cavity is shallow and does little to stabilize the shoulder joint.

However, this exceptional flexibility exacts a price: the arm bone (humerus) is very susceptible to dislocation, and fracture of the clavicle disables the entire upper limb. This is because the clavicle acts as a brace to hold the scapula and arm away from the top of the thoracic cage.

The Arm

The arm, or brachium (**Figure 10.3**, p. 153 and **Table 10.2**, p. 153), contains a single bone—the **humerus.** Proximally its rounded *head* fits into the shallow glenoid cavity of the scapula. The head is separated from the shaft by the *anatomical neck* and the more constricted *surgical neck,* which is a common site of fracture.

The Forearm

Two bones, the radius and the ulna, compose the skeleton of the forearm, or antebrachium (**Figure 10.4**, p. 154). When the body is in the anatomical position, the **radius** is lateral, the **ulna** is medial, and the radius and ulna are parallel.

The Hand

The skeleton of the hand, or manus (**Figure 10.5**, p. 155), includes three groups of bones, those of the carpus, the metacarpals, and the phalanges.

The wrist is the proximal portion of the hand. It is referred to anatomically as the **carpus;** the eight bones

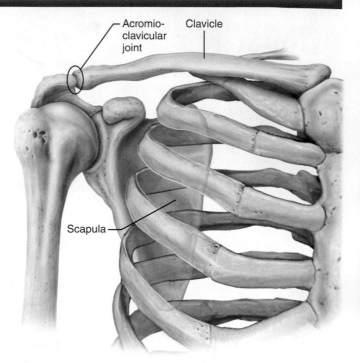

Figure 10.1 Articulated bones of the pectoral (shoulder) girdle. The right pectoral girdle is articulated to show the relationship of the girdle to the bones of the thorax and arm.

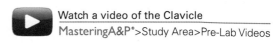
Watch a video of the Clavicle
MasteringA&P®>Study Area>Pre-Lab Videos

composing it are the **carpals.** The carpals are arranged in two irregular rows of four bones each, illustrated in Figure 10.5. In the proximal row (lateral to medial) are the *scaphoid, lunate, triquetrum,* and *pisiform bones;* the scaphoid and lunate articulate with the distal end of the radius. In the distal row (lateral to medial) are the *trapezium, trapezoid, capitate,* and *hamate.*

The **metacarpals,** numbered I to V from the thumb side of the hand toward the little finger, radiate out from the wrist like spokes to form the palm of the hand. The *bases* of the metacarpals articulate with the carpals; their more bulbous *heads* articulate with the phalanges distally. When the fist is clenched, the heads of the metacarpals become prominent as the knuckles.

The fingers are numbered from 1 to 5, beginning from the thumb *(pollex)* side of the hand. The 14 bones of the fingers, or

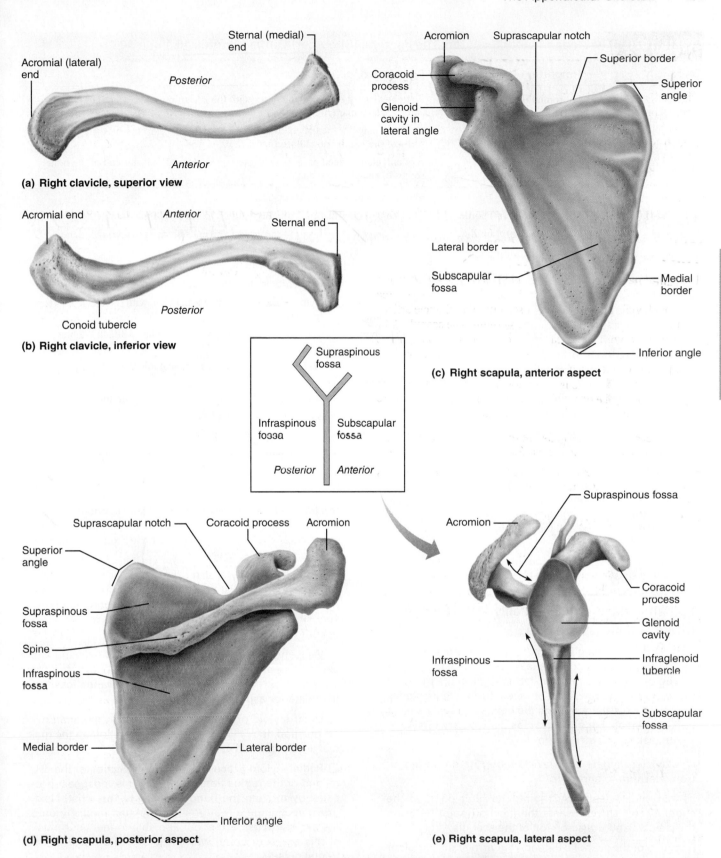

(a) **Right clavicle, superior view**

(b) **Right clavicle, inferior view**

(c) **Right scapula, anterior aspect**

(d) **Right scapula, posterior aspect**

(e) **Right scapula, lateral aspect**

Figure 10.2 Individual bones of the pectoral (shoulder) girdle. View (e) is accompanied by a schematic representation of its orientation.

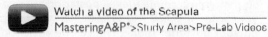

Watch a video of the Scapula
MasteringA&P*>Study Area>Pre-Lab Videos

Table 10.1	The Appendicular Skeleton: The Pectoral (Shoulder) Girdle (Figures 10.1 and 10.2)	
Bone	**Important markings**	**Description**
Clavicle ("collarbone") (Figures 10.1 and 10.2)	Acromial (lateral) end	Flattened lateral end that articulates with the acromion of the scapula to form the acromioclavicular (AC) joint
	Sternal (medial) end	Oval or triangular medial end that articulates with the sternum to form the lateral walls of the jugular notch (see Figure 9.19, p. 138)
	Conoid tubercle	A small, cone-shaped projection located on the lateral, inferior end of the bone; serves to anchor ligaments
Scapula ("shoulder blade") (Figures 10.1 and 10.2)	Superior border	Short, sharp border located superiorly
	Medial (vertebral) border	Thin, long border that runs roughly parallel to the vertebral column
	Lateral (axillary) border	Thick border that is closest to the armpit and ends superiorly with the glenoid cavity
	Glenoid cavity	A shallow socket that articulates with the head of the humerus
	Spine	A ridge of bone on the posterior surface that is easily felt through the skin
	Acromion	The lateral end of the spine of the scapula that articulates with the clavicle to form the AC joint
	Coracoid process	Projects above the glenoid cavity as a hooklike process; helps attach the biceps brachii muscle
	Suprascapular notch	Small notch located medial to the coracoid process that allows for the passage of blood vessels and a nerve
	Subscapular fossa	A large shallow depression that forms the anterior surface of the scapula
	Supraspinous fossa	A depression located superior to the spine of the scapula
	Infraspinous fossa	A broad depression located inferior to the spine of the scapula

digits, are miniature long bones, called **phalanges** (singular: *phalanx*). Each finger contains three phalanges (proximal, middle, and distal) except the thumb, which has only two (proximal and distal).

Activity 2

Palpating the Surface Anatomy of the Pectoral Girdle and the Upper Limb

Identify the following bone markings on the skin surface of the upper limb. It is usually preferable to palpate the bone markings on your lab partner since many of these markings can only be seen from the posterior aspect. Place a check mark in the boxes as you locate the bone markings. For any markings that you are unable to locate, ask your instructor for help.

☐ Clavicle: Palpate the clavicle along its entire length from sternum to shoulder.

☐ Acromioclavicular (AC) joint: The high point of the shoulder, which represents the junction point between the clavicle and the acromion of the scapula.

☐ Spine of the scapula: Extend your arm at the shoulder so that your scapula moves posteriorly. As you do this, your scapular spine will be seen as a winglike protrusion on your posterior thorax and can be easily palpated by your lab partner.

☐ Lateral epicondyle of the humerus: After you have located the epicondyle, run your finger posteriorly into the hollow immediately dorsal to the epicondyle. This is the site where the extensor muscles of the forearm are attached and is a common site of the excruciating pain of tennis elbow.

☐ Medial epicondyle of the humerus: Feel this medial projection at the distal end of the humerus. The ulnar nerve, which runs behind the medial epicondyle, is responsible for the tingling, painful sensation felt when you hit your "funnybone."

☐ Olecranon of the ulna: Work your elbow—flexing and extending—as you palpate its posterior aspect to feel the olecranon moving into and out of the olecranon fossa on the posterior aspect of the humerus.

☐ Ulnar styloid process: With the hand in the anatomical position, feel this small inferior projection on the medial aspect of the distal end of the ulna.

☐ Radial styloid process: Find this projection at the distal end of the radius (lateral aspect). It is most easily located by moving the hand medially at the wrist. Next, move your fingers (not your thumb) just medially onto the anterior wrist. You should be able to feel your pulse at this pressure point, which lies over the radial artery (radial pulse).

☐ Pisiform: Just distal to the ulnar styloid process, feel the rounded, pealike pisiform bone.

☐ Metacarpophalangeal joints (knuckles): Clench your fist and locate your knuckles—these are your metacarpophalangeal joints.

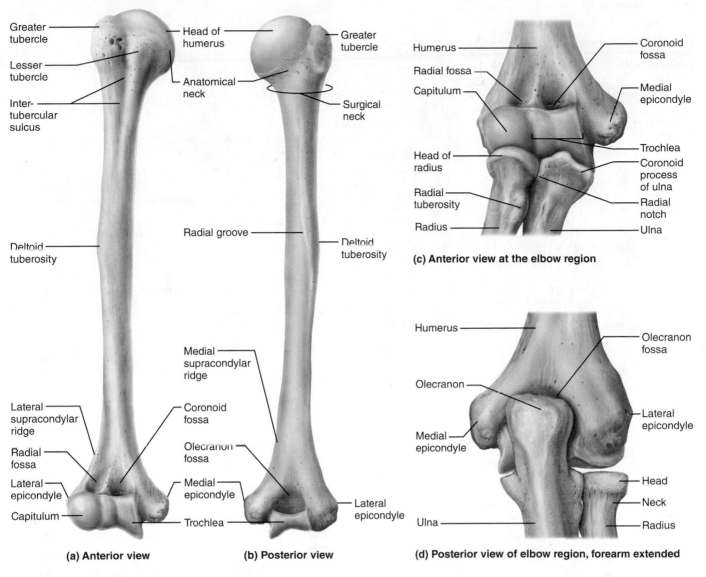

(a) Anterior view

(b) Posterior view

(c) Anterior view at the elbow region

(d) Posterior view of elbow region, forearm extended

Figure 10.3 Bone of the right arm. (a, b) Humerus. **(c, d)** Detailed views of elbow region.

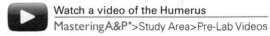

Watch a video of the Humerus
MasteringA&P®>Study Area>Pre-Lab Videos

Table 10.2A	The Appendicular Skeleton: The Upper Limb (Figure 10.3)	
Bone	**Important markings**	**Description**
Humerus (Figure 10.3) (only bone of the arm)	Greater tubercle	Large lateral prominence; site of the attachment of rotator cuff muscles
	Lesser tubercle	Small medial prominence; site of attachment of rotator cuff muscles
	Intertubercular sulcus	A groove separating the greater and lesser tubercles; the tendon of the biceps brachii lies in this groove
	Deltoid tuberosity	A roughened area about midway down the shaft of the lateral humerus; site of attachment of the deltoid muscle
	Radial fossa	Small lateral depression; receives the head of the radius when the forearm is flexed
	Coronoid fossa	Small medial anterior depression; receives the coronoid process of the ulna when the forearm is flexed
	Capitulum	A rounded lateral condyle that articulates with the radius
	Trochlea	A flared medial condyle that articulates with the ulna
	Lateral epicondyle	Small condyle proximal to the capitulum
	Medial epicondyle	Rough condyle proximal to the trochlea
	Radial groove	Small posterior groove, marks the course of the radial nerve
	Olecranon fossa	Large distal posterior depression that accommodates the olecranon of the ulna

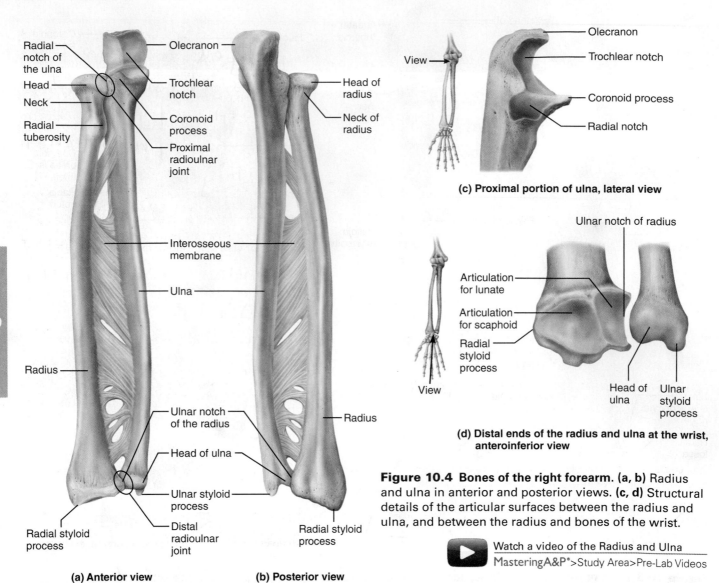

Figure 10.4 Bones of the right forearm. (a, b) Radius and ulna in anterior and posterior views. **(c, d)** Structural details of the articular surfaces between the radius and ulna, and between the radius and bones of the wrist.

▶ Watch a video of the Radius and Ulna
MasteringA&P°>Study Area>Pre-Lab Videos

(a) Anterior view

(b) Posterior view

(c) Proximal portion of ulna, lateral view

(d) Distal ends of the radius and ulna at the wrist, anteroinferior view

Table 10.2B	The Appendicular Skeleton: The Upper Limb (Figures 10.3 and 10.4)	
Bone	**Important markings**	**Description**
Radius (lateral bone of the forearm in the anatomical position) (Figures 10.4 and 10.3c, d)	Head	Proximal end of the radius that forms part of the proximal radioulnar joint and articulates with the capitulum of the humerus
	Radial tuberosity	Medial prominence just below the head of the radius; site of attachment of the biceps brachii
	Radial styloid process	Distal prominence; site of attachment for ligaments that travel to the wrist
	Ulnar notch	Small distal depression that accommodates the head of the ulna, forming the distal radioulnar joint
Ulna (medial bone of the forearm in the anatomical position) (Figures 10.4 and 10.3c, d)	Olecranon	Prominent process on the posterior proximal ulna; articulates with the olecranon fossa of the humerus when the forearm is extended
	Trochlear notch	Deep notch that separates the olecranon and the coronoid process; articulates with the trochlea of the humerus
	Coronoid process	Shaped like a point on a crown; articulates with the trochlea of the humerus
	Radial notch	Small proximal lateral notch that articulates with the head of the radius; forms part of the proximal radioulnar joint
	Head	Slim distal end of the ulna; forms part of the distal radioulnar joint
	Ulnar styloid process	Distal pointed projection; located medial to the head of the ulna

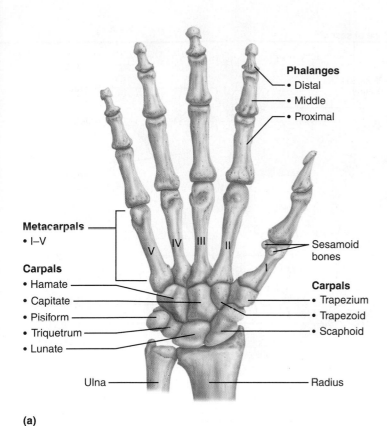

Phalanges
• Distal
• Middle
• Proximal

Metacarpals
• I–V

Carpals
• Hamate
• Capitate
• Pisiform
• Triquetrum
• Lunate

Sesamoid bones

Carpals
• Trapezium
• Trapezoid
• Scaphoid

Ulna Radius

(a)

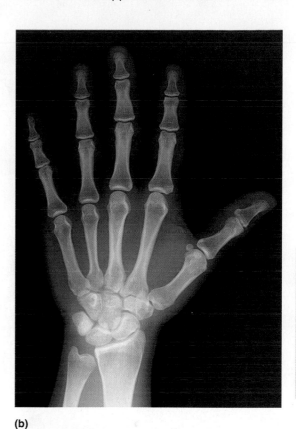

(b)

Figure 10.5 Bones of the right hand. (a) Anterior view showing the relationships of the carpals, metacarpals, and phalanges. **(b)** X-ray image of the right hand in the anterior view.

Watch a video of the Hand
MasteringA&P®>Study Area>Pre-Lab Videos

Bones of the Pelvic Girdle and Lower Limb

The Pelvic (Hip) Girdle

As with the bones of the pectoral girdle and upper limb, pay particular attention to bone markings needed to identify right and left bones.

The **pelvic girdle**, or **hip girdle** (**Figure 10.6**, p. 156 and **Table 10.3**, p. 157), is formed by the two **hip bones** (also called the **ossa coxae**, or *coxal bones*) and the sacrum. The deep structure formed by the hip bones, sacrum, and coccyx is called the **pelvis** or *bony pelvis*. In contrast to the bones of the shoulder girdle, those of the pelvic girdle are heavy and massive, and they attach securely to the axial skeleton. The sockets for the heads of the femurs (thigh bones) are deep and heavily reinforced by ligaments to ensure a stable, strong limb attachment. The ability to bear weight is more important here than mobility and flexibility. The combined weight of the upper body rests on the pelvic girdle.

Each hip bone is a result of the fusion of three bones—the **ilium, ischium,** and **pubis**—which are distinguishable in the young child.

The ilium, ischium, and pubis fuse at the deep socket called the **acetabulum** (literally, "vinegar cup"), which receives the head of the thigh bone. The rami of the pubis and ischium form a bar of bone enclosing the **obturator foramen,** through which a few blood vessels and nerves pass.

Activity 3

Observing Pelvic Articulations

Take time to examine an articulated pelvis. Notice how each hip bone articulates with the sacrum posteriorly and how the two hip bones join at the pubic symphysis. The sacroiliac joint is a common site of lower back problems because of the pressure it must bear.

Comparison of the Male and Female Pelves

The female pelvis reflects differences for childbearing—it is wider, shallower, lighter, and rounder than that of the male. Her pelvis not only must support the increasing size of a fetus, but also must be large enough to allow the infant's head (its largest dimension) to descend through the birth canal at birth.

The **pelvic brim** is a continuous oval ridge of bone that runs along the pubic symphysis, pubic crests, arcuate lines, sacral alae, and sacral promontory. The **false pelvis** is that portion superior to the pelvic brim; it is bounded by the alae of the ilia laterally and the sacral promontory and lumbar vertebrae posteriorly. The false pelvis supports the abdominal viscera, but it does not restrict childbirth in any way. The **true**

10

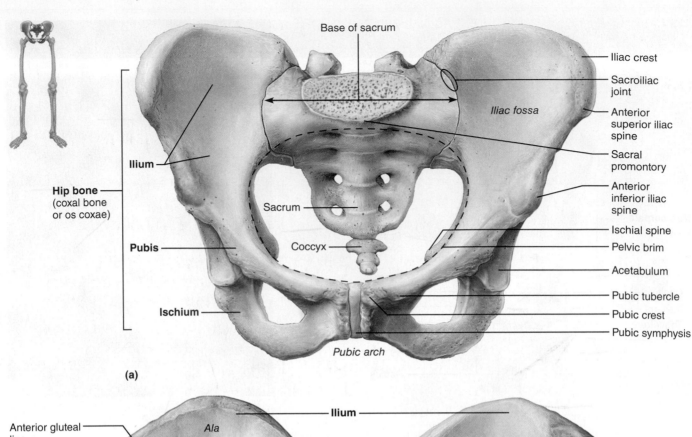

Hip bone
(coxal bone
or os coxae)

Base of sacrum

Iliac crest

Sacroiliac
joint

Iliac fossa

Anterior
superior iliac
spine

Sacral
promontory

Ilium

Anterior
inferior iliac
spine

Sacrum

Ischial spine

Pubis

Coccyx

Pelvic brim

Acetabulum

Pubic tubercle

Pubic crest

Ischium

Pubic symphysis

Pubic arch

(a)

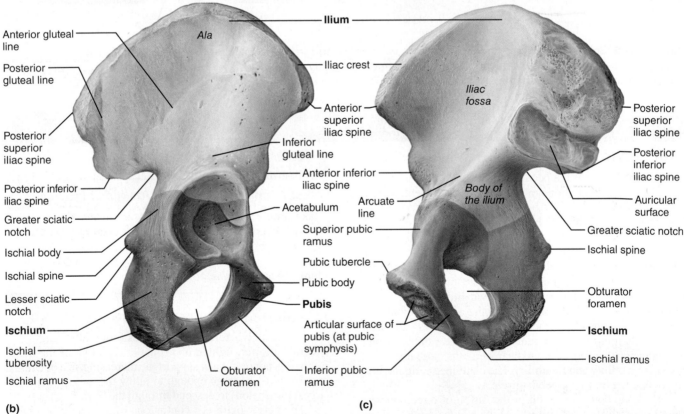

Anterior gluteal
line

Ala

Ilium

Posterior
gluteal line

Iliac crest

*Iliac
fossa*

Posterior
superior
iliac spine

Posterior
inferior iliac
spine

Anterior
superior
iliac spine

Posterior
superior
iliac spine

Posterior
inferior
iliac spine

Inferior
gluteal line

Auricular
surface

Posterior inferior
iliac spine

Anterior inferior
iliac spine

*Body of
the ilium*

Greater sciatic
notch

Acetabulum

Arcuate
line

Greater sciatic notch

Ischial body

Superior pubic
ramus

Ischial spine

Ischial spine

Pubic tubercle

Obturator
foramen

Lesser sciatic
notch

Pubic body

Ischium

Pubis

Ischium

Ischial
tuberosity

Articular surface of
pubis (at pubic
symphysis)

Ischial ramus

Ischial ramus

Obturator
foramen

Inferior pubic
ramus

(b)

(c)

Figure 10.6 Bones of the pelvic girdle. (a) Articulated pelvis,
showing the two hip bones (coxal bones), which together with the
sacrum comprise the pelvic girdle, and the coccyx. **(b)** Right hip bone,
lateral view, showing the point of fusion of the ilium, ischium, and
pubis. **(c)** Right hip bone, medial view.

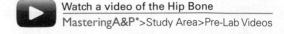

Watch a video of the Hip Bone
MasteringA&P®>Study Area>Pre-Lab Videos

Table 10.3	The Appendicular Skeleton: The Pelvic (Hip) Girdle (Figure 10.6)	
Bone	**Important markings**	**Description**
Ilium (Figure 10.6)	Iliac crest	Thick superior margin of bone
	Anterior superior iliac spine	The blunt anterior end of the iliac crest
	Posterior superior iliac spine	The sharp posterior end of the iliac crest
	Anterior inferior iliac spine	Small projection located just below the anterior superior iliac spine
	Posterior inferior iliac spine	Small projection located just below the posterior superior iliac spine
	Greater sciatic notch	Deep notch located inferior to the posterior inferior iliac spine; allows the sciatic nerve to enter the thigh
	Iliac fossa	Shallow depression below the iliac crest; forms the internal surface of the ilium
	Auricular surface	Rough medial surface that articulates with the auricular surface of the sacrum, forming the sacroiliac joint
	Arcuate line	A ridge of bone that runs inferiorly and anteriorly from the auricular surface
Ischium ("sit-down" bone) (Figure 10.6)	Ischial tuberosity	Rough projection that receives the weight of our body when we are sitting
	Ischial spine	Located superior to the ischial tuberosity and projects medially into the pelvic cavity
	Lesser sciatic notch	A small notch located inferior to the ischial spine
	Ischial ramus	Narrow portion of the bone that articulates with the pubis
Pubis (Figure 10.6)	Superior pubic ramus	Superior extension of the body of the pubis
	Inferior pubic ramus	Inferior extension of the body of the pubis; articulates with the ischium
	Pubic crest	Thick anterior border
	Pubic tubercle	Lateral end of the pubic crest; lateral attachment for the inguinal ligament
	Articular surface	Surface of each pubis that combines with fibrocartilage to form the pubic symphysis

10

pelvis is the region inferior to the pelvic brim that is almost entirely surrounded by bone. Its posterior boundary is formed by the sacrum. The ilia, ischia, and pubic bones define its limits laterally and anteriorly.

The **pelvic inlet** is the opening delineated by the pelvic brim. The widest dimension of the pelvic inlet is from left to right, that is, along the frontal plane. The **pelvic outlet** is the inferior margin of the true pelvis. It is bounded anteriorly by the pubic arch, laterally by the ischia, and posteriorly by the sacrum and coccyx. Since both the coccyx and the ischial spines protrude into the outlet opening, a sharply angled coccyx or large, sharp ischial spines can dramatically narrow the outlet. The largest dimension of the outlet is the anterior-posterior diameter.

Activity 4

Comparing Male and Female Pelves

Examine male and female pelves for the following differences. For *female* pelves:

- The pelvic inlet is larger and more circular.
- The pelvis as a whole is shallower, and the bones are lighter and thinner.
- The sacrum is broader and less curved, and the pubic arch is more rounded and broader.

- The acetabula are smaller and farther apart, and the ilia flare more laterally.
- The ischial spines are shorter, farther apart, and everted, thus enlarging the pelvic outlet.

The major differences between the male and female pelves are summarized in **Table 10.4** on p. 158.

The Thigh

The **femur,** or thigh bone (**Figure 10.7b**, p. 159 and **Table 10.5**, pp. 159–160), is the only bone of the thigh. It is the heaviest, strongest bone in the body. The ball-like head of the femur articulates with the hip bone via the deep, secure socket of the acetabulum.

The femur angles medially as it runs downward to the leg bones; this brings the knees in line with the body's center of gravity. The medial course of the femur is more noticeable in females because of the wider female pelvis.

The **patella** (Figure 10.7a) or kneecap is a triangular sesamoid bone enclosed in the quadriceps tendon that secures the anterior thigh muscles to the tibia.

(Text continues on page 161.)

10

Table 10.4	Comparison of the Male and Female Pelves	
Characteristic	**Female**	**Male**
General structure and functional modifications	Tilted forward; adapted for childbearing; true pelvis defines the birth canal; cavity of the true pelvis is broad, shallow, and has a greater capacity	Tilted less forward; adapted for support of a male's heavier build and stronger muscles; cavity of the true pelvis is narrow and deep
Bone thickness	Bones lighter, thinner, and smoother	Bones heavier and thicker, and markings are more prominent
Acetabula	Smaller; farther apart	Larger; closer together
Pubic arch	Broader angle (80°–90°); more rounded	Angle is more acute (50°–60°)
Anterior view	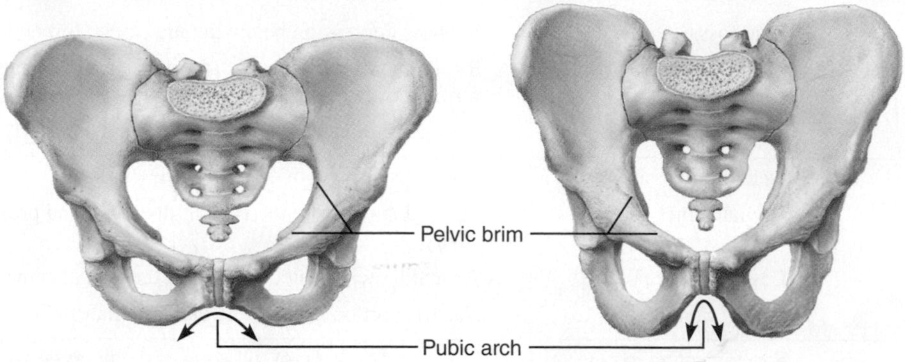	
Sacrum	Wider; shorter; sacrum is less curved	Narrow; longer; sacral promontory projects anteriorly
Coccyx	More movable; straighter; projects inferiorly	Less movable; curves and projects anteriorly
Left lateral view	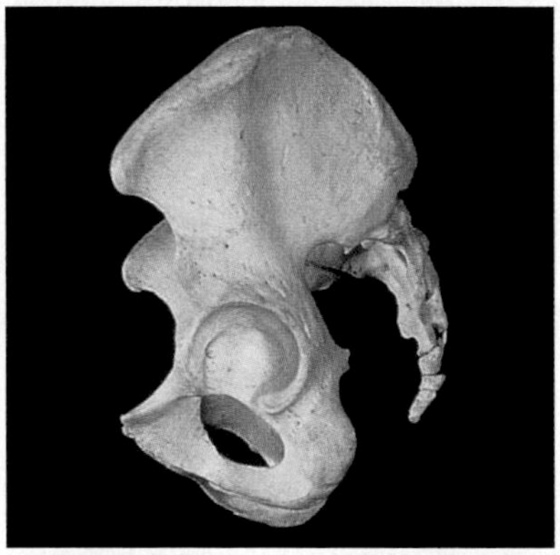	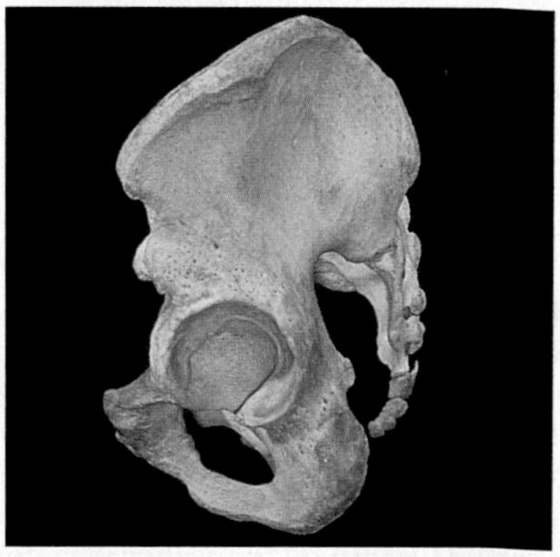
Pelvic inlet	Wider; oval from side to side	Narrow; basically heart shaped
Pelvic outlet	Wider; ischial spines shorter, farther apart, and everted	Narrower; ischial spines longer, sharper, and point more medially
Posteroinferior view	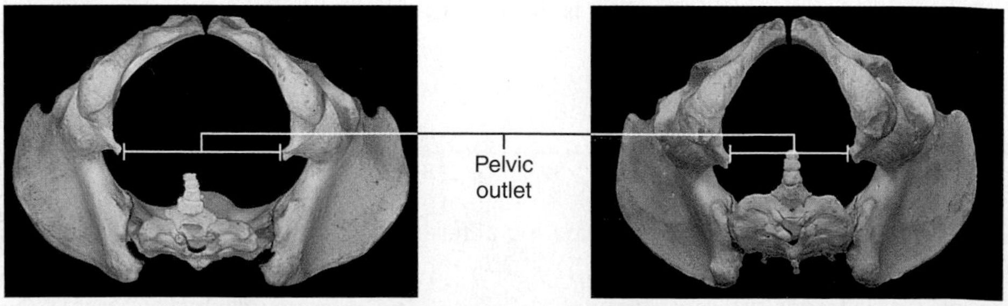	

Pelvic brim

Pubic arch

Pelvic outlet

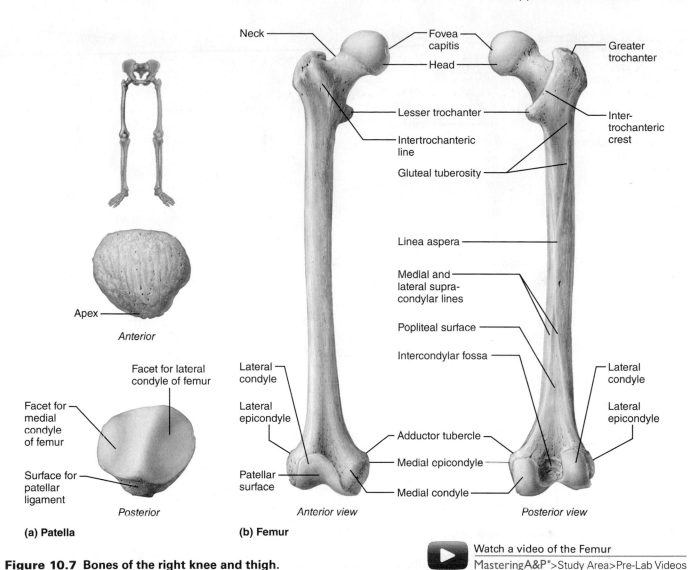

Figure 10.7 Bones of the right knee and thigh.

Watch a video of the Femur
MasteringA&P®>Study Area>Pre-Lab Videos

Table 10.5A	The Appendicular Skeleton: The Lower Limb (Figure 10.7)	
Bone	**Important markings**	**Description**
Femur (thigh bone) (Figure 10.7)	Fovea capitis	A small pit in the head of the femur for the attachment of a short ligament that runs to the acetabulum
	Neck	Weakest part of the femur, the usual fracture site of a "broken hip"
	Greater trochanter	Large lateral projection; serves a site for muscle attachment on the proximal femur
	Lesser trochanter	Large posteromedial projection; serves a site for muscle attachment on the proximal femur
	Intertrochanteric line	Thin ridge of bone that connects the two trochanters anteriorly
	Intertrochanteric crest	Prominent ridge of bone that connects the two trochanters posteriorly
	Gluteal tuberosity	Thin ridge of bone located posteriorly; serves as a site for muscle attachment on the proximal femur
	Linea aspera	Long vertical ridge of bone on the posterior shaft of the femur
	Medial and lateral supracondylar lines	Two lines that diverge from the linea aspera and travel to their respective condyles
	Medial and lateral condyles	Distal "wheel shaped" projections that articulate with the tibia, each condyle has a corresponding epicondyle
	Intercondylar fossa	Deep depression located between the condyles and beneath the popliteal surface
	Adductor tubercle	A small bump on the superior portion of the medial epicondyle; attachment site for the large adductor magnus muscle
	Patellar surface	Smooth distal anterior surface between the condyles; articulates with the patella

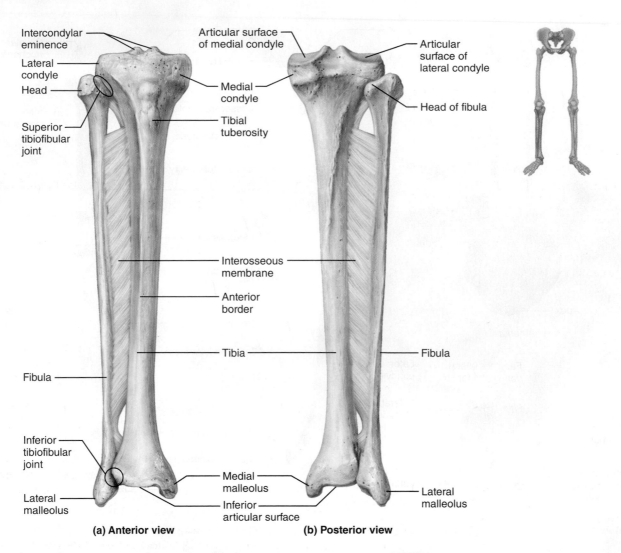

Figure 10.8 Bones of the right leg. Tibia and fibula, anterior and posterior views.

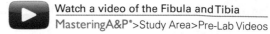

Watch a video of the Fibula and Tibia
MasteringA&P°>Study Area>Pre-Lab Videos

Table 10.5B	The Appendicular Skeleton: The Lower Limb (Figure 10.8)	
Bone	**Important markings**	**Description**
Tibia (shin bone, medial bone of the leg) (Figure 10.8)	Lateral condyle	Slightly concave surface that articulates with the lateral condyle of the femur; the inferior region of this condyle articulates with the fibula to form the superior tibiofibular joint
	Medial condyle	Slightly concave surface that articulates with the medial condyle of the femur
	Intercondylar eminence	Irregular projection located between the two condyles
	Tibial tuberosity	Roughened anterior surface; site of patellar ligament attachment
	Anterior border	Sharp ridge of bone easily palpated because it is close to the surface
	Medial malleolus	Forms the medial bulge of the ankle
	Inferior articular surface	Distal surface of the tibia that articulates with the talus
	Fibular notch	Smooth lateral surface that articulates with the fibula to form the inferior tibiofibular joint
Fibula (lateral bone of the leg) (Figure 10.8)	Head	Proximal end of the fibula that articulates with the tibia to form the superior tibiofibular joint
	Lateral malleolus	Forms the lateral bulge of the ankle and articulates with the talus

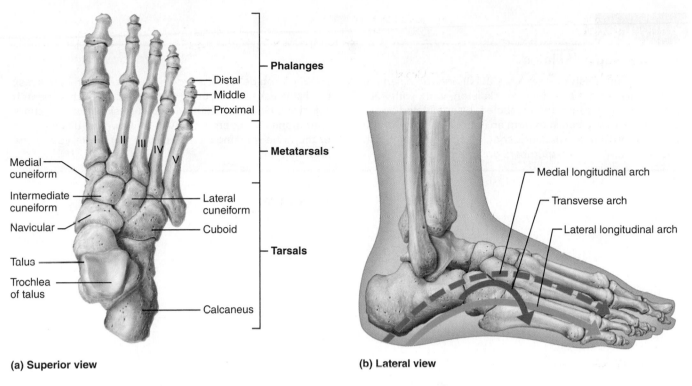

(a) Superior view

(b) Lateral view

Figure 10.9 Bones of the right foot. Arches of the right foot are diagrammed in **(b)**.

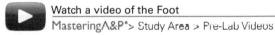

Watch a video of the Foot
MasteringA&P®> Study Area > Pre-Lab Videos

10

The Leg

Two bones, the tibia and the fibula, form the skeleton of the leg (**Figure 10.8**). The **tibia,** or *shinbone,* is the larger, medial, weight-bearing bone of the leg.

The **fibula,** which lies parallel to the tibia, takes no part in forming the knee joint. Its proximal head articulates with the lateral condyle of the tibia.

The Foot

The bones of the foot include the 7 **tarsal** bones, 5 **metatarsals,** which form the instep, and 14 **phalanges,** which form the toes (**Figure 10.9**). Body weight is concentrated on the two largest tarsals, which form the posterior aspect of the foot. These are the larger *calcaneus* (heel bone) and the *talus,* which lies between the tibia and the calcaneus. (The other tarsals are named and identified in Figure 10.9). The metatarsals are numbered I through V, medial to lateral. Like the fingers of the hand, each toe has three phalanges except the great toe, which has two.

The bones in the foot are arranged to produce three strong arches—two longitudinal arches (medial and lateral) and one transverse arch (Figure 10.9b). Ligaments bind the foot bones together, and tendons of the foot muscles hold the bones firmly in the arched position but still allow a certain degree of give. Weakened tendons and ligaments supporting the arches are referred to as *fallen arches* or *flat feet.*

Activity 5

Palpating the Surface Anatomy of the Pelvic Girdle and Lower Limb

Locate and palpate the following bone markings on yourself and/or your lab partner. Place a check mark in the

boxes as you locate the bone markings. Ask your instructor for help with any markings that you are unable to locate.

☐ Iliac crest and anterior superior iliac spine: Rest your hands on your hips—they will be overlying the iliac crests. Trace the crest as far posteriorly as you can, and then follow it anteriorly to the anterior superior iliac spine. This latter bone marking is clearly visible through the skin of very slim people.

☐ Greater trochanter of the femur: This is easier to locate in females than in males because of the wider female pelvis. Try to locate it on yourself as the most lateral point of the proximal femur. It typically lies about 6 to 8 inches below the iliac crest.

☐ Patella and tibial tuberosity: Feel your kneecap and palpate the ligaments attached to its borders. Follow the inferior patellar ligament to the tibial tuberosity.

☐ Medial and lateral condyles of the femur and tibia: As you move from the patella inferiorly on the medial (and then the lateral) knee surface, you will feel first the femoral and then the tibial condyle.

☐ Medial malleolus: Feel the medial protrusion of your ankle, the medial malleolus of the distal tibia.

☐ Lateral malleolus: Feel the bulge of the lateral aspect of your ankle, the lateral malleolus of the fibula.

☐ Calcaneus: Attempt to follow the extent of your calcaneus, or heel bone.

Activity 6

Constructing a Skeleton

1. When you finish examining yourself and the disarticulated bones of the appendicular skeleton, work with your lab partner to arrange the unlabeled, disarticulated bones on the laboratory bench to form an entire skeleton. Careful observation of bone markings should help you distinguish between right and left members of bone pairs.

2. When you believe that you have accomplished this task correctly, ask the instructor to check your arrangement. If it is not correct, go to the articulated skeleton and check your bone arrangements. Also review the descriptions of the bone markings as necessary to correct your bone arrangement.

Articulations and Body Movements

Objectives

☐ Name and describe the three functional categories of joints.

☐ Name and describe the three structural categories of joints, and discuss how their structure is related to mobility.

☐ Identify the types of synovial joints; indicate whether they are nonaxial, uniaxial, biaxial, or multiaxial; and describe the movements each makes.

☐ Define *origin* and *insertion* of muscles.

☐ Demonstrate or describe the various body movements.

☐ Compare and contrast the structure and function of the shoulder and hip joints.

☐ Describe the structure and function of the knee and temporomandibular joints.

Materials

- Skull
- Articulated skeleton
- X-ray image of a child's bone showing the cartilaginous growth plate (if available)
- Anatomical chart of joint types (if available)
- Diarthrotic joint (fresh or preserved), preferably a beef knee joint sectioned sagittally (Alternatively, pig's feet with phalanges sectioned frontally could be used)
- Disposable gloves
- Water balloons and clamps
- Functional models of hip, knee, and shoulder joints (if available)
- X-ray images of normal and arthritic joints (if available)

MasteringA&P®

For related exercise study tools, go to the Study Area of **MasteringA&P**. There you will find:

- Practice Anatomy Lab **PAL**
- A&PFlix *A&PFlix*
- PhysioEx **PEx**
- Practice quizzes, Histology Atlas, eText, Videos, and more!

Pre-Lab Quiz

1. Name one of the two functions of an articulation, or joint. _____
   _____

2. The functional classification of joints is based on:
   a. a joint cavity
   b. amount of connective tissue
   c. amount of movement allowed by the joint

3. Structural classification of joints includes fibrous, cartilaginous, and _____, which have a fluid-filled cavity between articulating bones.

4. Circle the correct underlined term. Sutures, which have their irregular edges of bone joined by short fibers of connective tissue, are an example of <u>fibrous</u> / <u>cartilaginous</u> joints.

5. Circle True or False. All synovial joints are diarthroses, or freely movable joints.

6. Circle the correct underlined term. Every muscle of the body is attached to a bone or other connective tissue structure at two points. The <u>origin</u> / <u>insertion</u> is the more movable attachment.

7. The hip joint is an example of a _____ synovial joint.
   a. ball-and-socket   **c.** pivot
   b. hinge   **d.** plane

8. Movement of a limb *away* from the midline or median plane of the body in the frontal plane is known as:
   a. abduction   **c.** extension
   b. eversion   **d.** rotation

9. Circle the correct underlined term. This type of movement is common in ball-and-socket joints and can be described as the movement of a bone around its longitudinal axis. It is <u>rotation</u> / <u>flexion</u>.

10. Circle True or False. The knee joint is the most freely movable joint in the body.

W ith rare exceptions, every bone in the body is connected to, or forms a joint with, at least one other bone. **Articulations,** or joints, perform two functions for the body. They (1) hold the bones together and (2) allow the rigid skeletal system some flexibility so that gross body movements can occur.

Classification of Joints

Joints may be classified structurally or functionally. The structural classification is based on the presence of connective tissue fiber, cartilage, or a joint cavity between the articulating bones. Structurally, there are *fibrous, cartilaginous,* and *synovial joints.*

The functional classification focuses on the amount of movement allowed at the joint. On this basis, there are **synarthroses,** or immovable joints; **amphiarthroses,** or slightly movable joints; and **diarthroses,** or freely movable joints.

A summary of the structural and functional classifications of joints can be found in **Table 11.1**. A sample of the types of joints can be found in **Figure 11.1**.

Activity 1

Identifying Fibrous Joints

Examine a human skull. Notice that adjacent bone surfaces do not actually touch but are separated by fibrous connective tissue. Also examine a skeleton and anatomical chart of joint types and **Table 11.3**, pp. 183–184, for examples of fibrous joints.

Activity 2

Identifying Cartilaginous Joints

Identify the cartilaginous joints on a human skeleton, Table 11.3, and an anatomical chart of joint types. View an X-ray image of the cartilaginous growth plate (epiphyseal plate) of a child's bone if one is available.

Table 11.1	Summary of Joint Classifications (Figure 11.1)			
Structural class	**Structural characteristics**	**Structural types**	**Examples**	**Functional classification**
Fibrous	Adjoining bones connected by dense regular connective tissue; no joint cavity	Suture (short fibers)	Squamous suture between the parietal and temporal bones	Synarthrosis (immovable)
		Syndesmosis (longer fibers)	Between the tibia and fibula	Amphiarthrosis (slightly movable)
		Gomphosis (periodontal ligament)	Tooth in a bony socket (Figure 38.12)	Synarthrosis (immovable)
Cartilaginous	Adjoining bones united by cartilage; no joint cavity	Synchondrosis (hyaline cartilage)	Between the costal cartilage of rib 1 and the sternum and the epiphyseal plate in growing long bones (Figure 8.5)	Synarthrosis (immovable)
		Symphysis (fibrocartilage)	Intervertebral discs between adjacent vertebrae and the anterior connection between the pubic bones	Amphiarthrosis (slightly movable)
Synovial	Adjoining bones covered in articular cartilage; separated by a joint cavity and enclosed in an articular capsule lined with a synovial membrane	Plane joint	Between the carpals of the wrist	Diarthrosis (freely movable)
		Hinge joint	Elbow joint	
		Pivot joint	Proximal radioulnar joint	
		Condylar joint	Between the metacarpals and the proximal phalanx	
		Saddle joint	Between the trapezium (carpal) and metatarsal I	
		Ball-and-socket joint	Shoulder joint	

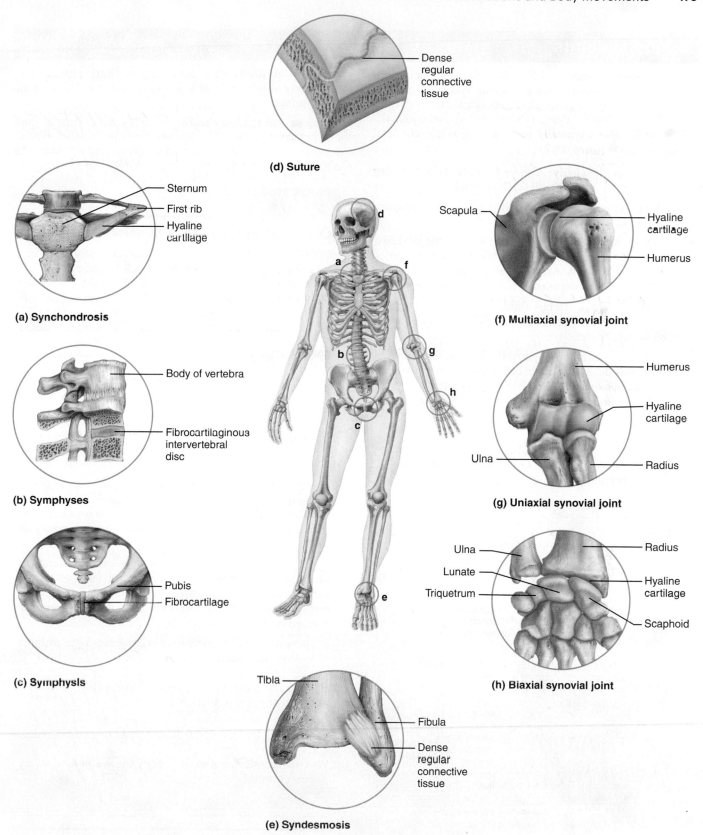

(d) Suture — Dense regular connective tissue

(a) Synchondrosis — Sternum, First rib, Hyaline cartilage

(b) Symphyses — Body of vertebra, Fibrocartilaginous intervertebral disc

(c) Symphysis — Pubis, Fibrocartilage

(e) Syndesmosis — Tibia, Fibula, Dense regular connective tissue

(f) Multiaxial synovial joint — Scapula, Hyaline cartilage, Humerus

(g) Uniaxial synovial joint — Humerus, Hyaline cartilage, Ulna, Radius

(h) Biaxial synovial joint — Ulna, Lunate, Triquetrum, Radius, Hyaline cartilage, Scaphoid

Figure 11.1 Types of joints. Joints to the left of the skeleton are cartilaginous joints; joints above and below the skeleton are fibrous joints; joints to the right of the skeleton are synovial joints. **(a)** Joint between costal cartilage of rib 1 and the sternum. **(b)** Intervertebral discs of fibrocartilage connecting adjacent vertebrae. **(c)** Fibrocartilaginous pubic symphysis connecting the pubic bones anteriorly. **(d)** Dense regular connective tissue connecting interlocking skull bones. **(e)** Ligament of dense regular connective tissue connecting the interior ends of the tibia and fibula. **(f)** Shoulder joint. **(g)** Elbow joint. **(h)** Wrist joint.

11

Synovial Joints

Synovial joints are those in which the articulating bone ends are separated by a joint cavity containing synovial fluid (see Figure 11.1f–h). All synovial joints are diarthroses, or freely movable joints. Most joints in the body are synovial joints.

Synovial joints typically have the following structural characteristics (**Figure 11.2**):

- **Joint (articular) cavity:** A space between the articulating bones. The cavity is filled with synovial fluid.

- **Articular cartilage:** Hyaline cartilage that covers the surfaces of the bones forming the joint.

- **Articular capsule:** Two layers that enclose the joint cavity. The tough external layer is the *fibrous layer* composed of dense irregular connective tissue. The inner layer is the *synovial membrane* composed of loose connective tissue.

- **Synovial fluid:** A viscous fluid, the consistency of egg whites, located in the joint cavity. This fluid acts as a lubricant, reducing friction.

- **Reinforcing ligaments:** Synovial joints are reinforced by ligaments outside the articular capsule (extracapsular ligaments) and ligaments found within the articular capsule (intracapsular ligaments). Intracapsular ligaments are covered with the synovial membrane, which separates them from the joint cavity.

- **Nerves and blood vessels:** Sensory nerve fibers detect pain and joint stretching. Most of the blood vessels supply the synovial membrane.

- **Articular discs:** Pads composed of fibrocartilage (menisci) may also be present to minimize wear and tear on the bone surfaces.

- **Bursa and tendon sheath:** Sac filled with synovial fluid that reduces friction where tendons cross the bone. A tendon sheath is an elongated bursa that wraps around a tendon like a bun around a hot dog.

Activity 3

Examining Synovial Joint Structure

Examine a beef or pig joint to identify the general structural features of diarthrotic joints as listed above.

⚠ If the joint is freshly obtained from the slaughterhouse and you will be handling it, don disposable gloves before beginning your observations.

Activity 4

Demonstrating the Importance of Friction-Reducing Structures

1. Obtain a small water balloon and clamp. Partially fill the balloon with water (it should still be flaccid), and clamp it closed.

2. Position the balloon atop one of your fists and press down on its top surface with the other fist. Push on the balloon until your two fists touch and move your fists back and forth over one another. Assess the amount of friction generated.

3. Unclamp the balloon and add more water. The goal is to get just enough water in the balloon so that your fists cannot come into contact with one another, but instead remain separated by a thin water layer when pressure is applied to the balloon.

4. Repeat the movements in step 2 to assess the amount of friction generated.

How does the presence of a cavity containing fluid influence the amount of friction generated?

_____

What anatomical structure(s) does the water-containing balloon mimic?

_____

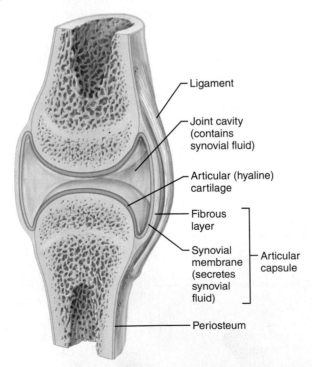

Figure 11.2 General structure of a synovial joint. The articulating bone ends are covered with articular cartilage, and enclosed within an articular capsule that is typically reinforced by ligaments externally. Internally the fibrous layer is lined with a smooth synovial membrane that secretes synovial fluid.

Labels:
- Ligament
- Joint cavity (contains synovial fluid)
- Articular (hyaline) cartilage
- Fibrous layer
- Synovial membrane (secretes synovial fluid)
- Articular capsule
- Periosteum

Types of Synovial Joints

The many types of synovial joints can be subdivided according to their function and structure. The shapes of the articular surfaces determine the types of movements that can occur at the joint, and they also determine the structural classification of the joints (**Table 11.2** and **Figure 11.3**).

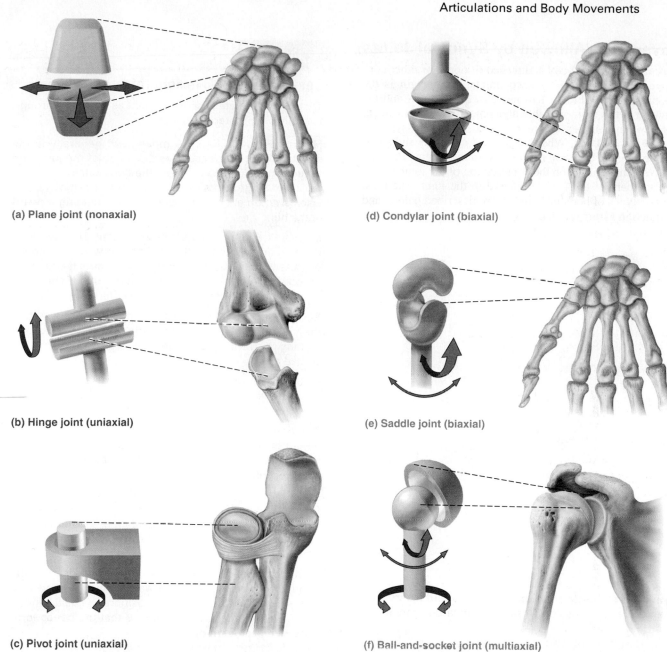

(a) Plane joint (nonaxial)

(d) Condylar joint (biaxial)

(b) Hinge joint (uniaxial)

(e) Saddle joint (biaxial)

(c) Pivot joint (uniaxial)

(f) Ball-and-socket joint (multiaxial)

11

Figure 11.3 Types of synovial joints. Dashed lines indicate the articulating bones. **(a)** Intercarpal joint. **(b)** Elbow. **(c)** Proximal radioulnar joint. **(d)** Metacarpo-phalangeal joint. **(e)** Carpometacarpal joint of the thumb. **(f)** Shoulder.

Table 11.2	**Types of Synovial Joints (Figure 11.3)**		
Synovial joint	**Description of articulating surfaces**	**Movement**	**Examples**
Plane	Flat or slightly curved bones	Nonaxial: gliding	Intertarsal, intercarpal joints
Hinge	A rounded or cylindrical bone fits into a concave surface on the other bone	Uniaxial: flexion and extension	Elbow, interphalangeal joints
Pivot	A rounded bone fits into a sleeve (a concave bone plus a ligament)	Uniaxial: rotation	Proximal radioulnar, atlantoaxial joints
Condylar	An oval condyle fits into an oval depression on the other bone	Biaxial: flexion, extension, adduction, and abduction	Metacarpophalangeal (knuckle) and radiocarpal joints
Saddle	Articulating surfaces are saddle shaped; one surface is concave, the other surface is convex	Biaxial: flexion, extension, adduction, and abduction	Carpometacarpal joint of the thumb
Ball-and-socket	The ball-shaped head of one bone fits into the cuplike depression of the other bone	Multiaxial: flexion, extension, adduction, abduction, and rotation	Shoulder, hip joints

Movements Allowed by Synovial Joints

Every muscle of the body is attached to bone (or other connective tissue structures) at two points. The **origin** is the stationary, immovable, or less movable attachment, and the **insertion** is the more movable attachment. Body movement occurs when muscles contract across diarthrotic synovial joints (**Figure 11.4**). When the muscle contracts and its fibers shorten, the insertion moves toward the origin. The type of movement depends on the construction of the joint and on the placement of the muscle relative to the joint. The most common types of body movements are described below (and illustrated in **Figure 11.5**).

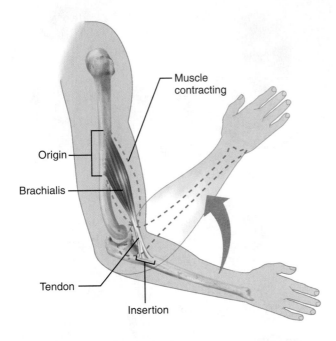

Figure 11.4 Muscle attachments (origin and insertion). When a skeletal muscle contracts, its insertion moves toward its origin.

Demonstrating Movements of Synovial Joints

Try to demonstrate each movement as you read through the following material:

Flexion (Figure 11.5a–c): A movement, generally in the sagittal plane, that decreases the angle of the joint and reduces the distance between the two bones. Flexion is typical of hinge joints (bending the knee or elbow) but is also common at ball-and-socket joints (bending forward at the hip).

Extension (Figure 11.5a–c): A movement that increases the angle of a joint and the distance between two bones or parts of the body; the opposite of flexion. If extension proceeds beyond anatomical position (bends the trunk backward), it is termed *hyperextension.*

Abduction (Figure 11.5d): Movement of a limb away from the midline of the body, along the frontal plane, or the fanning movement of fingers or toes when they are spread apart.

Adduction (Figure 11.5d): Movement of a limb toward the midline of the body or drawing the fingers or toes together; the opposite of abduction.

Rotation (Figure 11.5e): Movement of a bone around its longitudinal axis without lateral or medial displacement. Rotation, a common movement of ball-and-socket joints, also describes the movement of the atlas around the dens of the axis.

Circumduction (Figure 11.5d): A combination of flexion, extension, abduction, and adduction commonly observed in ball-and-socket joints like the shoulder. The limb as a whole outlines a cone.

Pronation (Figure 11.5f): Movement of the palm of the hand from an anterior or upward-facing position to a posterior or downward-facing position. The distal end of the radius rotates over the ulna so that the bones form an X with pronation of the forearm.

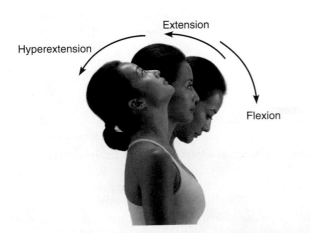

(a) Flexion, extension, and hyperextension of the neck

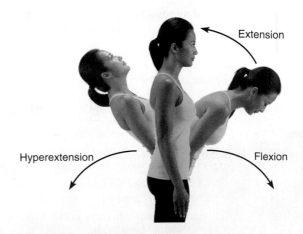

(b) Flexion, extension, and hyperextension of the vertebral column

Figure 11.5 Movements occurring at synovial joints of the body.

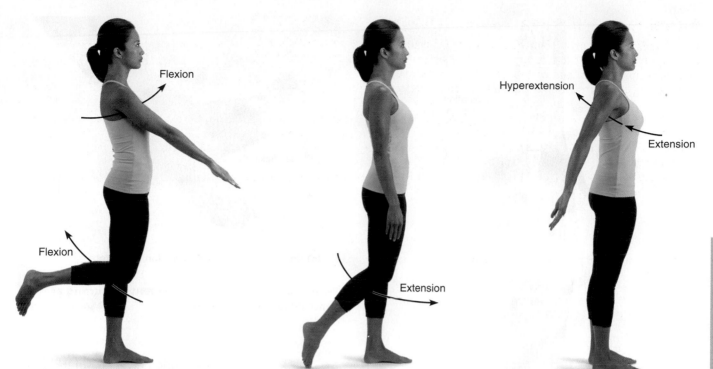

(c) Flexion and extension at the shoulder and knee, and hyperextension of the shoulder

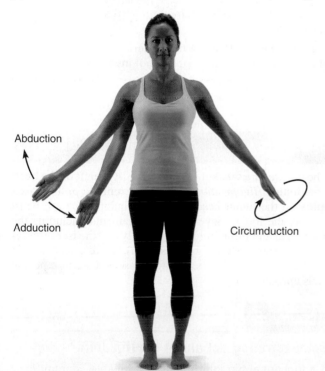

(d) Abduction, adduction, and circumduction of the upper limb at the shoulder

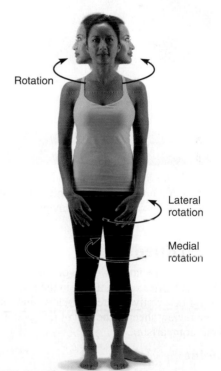

(e) Rotation of the head and lower limb

Figure 11.5 *(continued)*

Supination (Figure 11.5f): Movement of the palm from a posterior position to an anterior position (the anatomical position); the opposite of pronation. During supination, the radius and ulna are parallel.

The last four terms refer to movements of the foot:

Dorsiflexion (Figure 11.5g): A movement of the ankle joint that lifts the foot so that its superior surface approaches the shin.

Text continues on next page. ➔

11

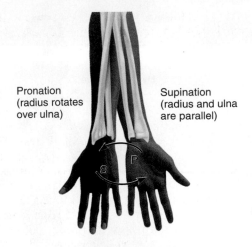

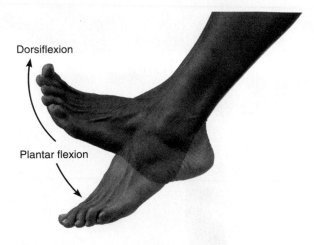

(f) Supination (S) and pronation (P) of the forearm

(g) Dorsiflexion and plantar flexion of the foot

Figure 11.5 *(continued)* **Movements occurring at synovial joints of the body.**

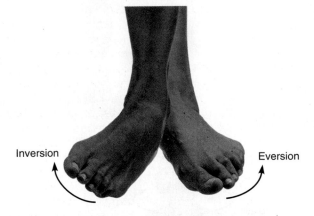

(h) Inversion and eversion of the foot

Plantar flexion (Figure 11.5g): A movement of the ankle joint in which the foot is flexed downward as if standing on one's toes or pointing the toes.

Inversion (Figure 11.5h): A movement that turns the sole of the foot medially.

Eversion (Figure 11.5h): A movement that turns the sole of the foot laterally; the opposite of inversion.

Selected Synovial Joints

Now you will have the opportunity to compare and contrast the structure of the hip and knee joints and to investigate the structure and movements of the temporomandibular joint and shoulder joint.

The Hip and Knee Joints

Both of these joints are large weight-bearing joints of the lower limb, but they differ substantially in their security. Read through the brief descriptive material below, and look at the questions in the review sheet at the end of this exercise before beginning your comparison.

The Hip Joint

The hip joint is a ball-and-socket joint, so movements can occur in all possible planes. However, its movements are definitely limited by its deep socket and strong reinforcing ligaments, the two factors that account for its exceptional stability (**Figure 11.6**).

The deeply cupped acetabulum that receives the head of the femur is enhanced by a circular rim of fibrocartilage called the **acetabular labrum.** Because the diameter of the labrum is smaller than that of the femur's head, dislocations

of the hip are rare. A short ligament, the **ligament of the head of the femur** *(ligamentum teres)* runs from the pitlike **fovea capitis** on the femur head to the acetabulum where it helps to secure the femur. Several strong ligaments, including the **iliofemoral** and **pubofemoral** anteriorly and the **ischiofemoral** that spirals posteriorly (not shown), are arranged so that they "screw" the femur head into the socket when a person stands upright.

Activity 6

Demonstrating Actions at the Hip Joint

If a functional hip joint model is available, identify the joint parts and manipulate it to demonstrate the following movements: flexion, extension, abduction, and medial and lateral rotation that can occur at this joint.

Reread the information on what movements the associated ligaments restrict, and verify that information during your joint manipulations.

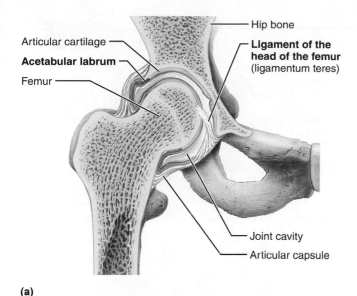

Hip bone

Articular cartilage

Acetabular labrum

Femur

Ligament of the head of the femur (ligamentum teres)

Joint cavity

Articular capsule

(a)

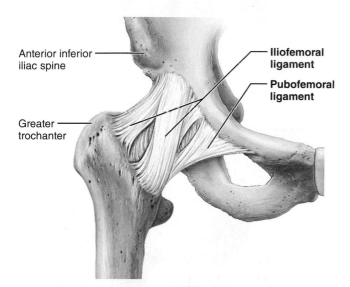

Anterior inferior iliac spine

Greater trochanter

Iliofemoral ligament

Pubofemoral ligament

(b)

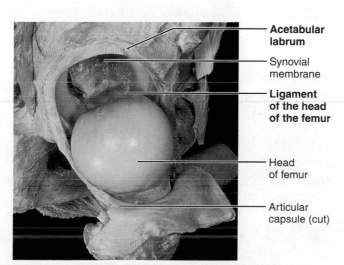

Acetabular labrum

Synovial membrane

Ligament of the head of the femur

Head of femur

Articular capsule (cut)

(c)

Figure 11.6 Hip joint relationships. (a) Frontal section through the right hip joint. **(b)** Anterior superficial view of the right hip joint. **(c)** Photograph of the interior of the hip joint, lateral view.

The Knee Joint

The knee is the largest and most complex joint in the body. Three joints in one (**Figure 11.7**, p.180), it allows extension, flexion, and a little rotation. The **tibiofemoral joint,** actually a bicondyloid joint between the femoral condyles above and the **menisci** (semilunar cartilages) of the tibia below, is functionally a hinge joint, a very unstable one made slightly more secure by the menisci (Figure 11.7b and d). Some rotation occurs when the knee is partly flexed, but during extension, the menisci and ligaments counteract rotation and side-to-side movements. The other joint is the **femoropatellar joint,** the intermediate joint anteriorly (Figure 11.7a and c).

The knee is unique in that it is only partly enclosed by an articular capsule. Anteriorly, where the capsule is absent, are three broad ligaments, the **patellar ligament** and the **medial** and **lateral patellar retinacula** (retainers), which run from the patella to the tibia below and merge with the capsule on either side.

Extracapsular ligaments including the **fibular** and **tibial collateral ligaments** (which prevent rotation during extension) and the **oblique popliteal** and **arcuate popliteal ligaments** are crucial in reinforcing the knee. The knees have a built-in locking device that must be "unlocked" by the popliteus muscles (Figure 11.7e) before the knees can be flexed again. The **anterior** and **posterior cruciate ligaments** are intracapsular ligaments that cross (*cruci* = cross) in the notch between the femoral condyles. They help to prevent anterior-posterior displacement of the joint and overflexion and hyperextension of the joint.

Activity 7

Demonstrating Actions at the Knee Joint

If a functional model of a knee joint is available, identify the joint parts and manipulate it to illustrate the following movements: flexion, extension, and medial and lateral rotation.

Reread the information on what movements the various associated ligaments restrict, and verify that information during your joint manipulations.

The Shoulder Joint

The shoulder joint, or **glenohumeral joint,** is the most freely moving joint of the body. The rounded head of the humerus fits the shallow glenoid cavity of the scapula (**Figure 11.8**, p.181). A rim of fibrocartilage, the **glenoid labrum,** deepens the cavity slightly.

The articular capsule enclosing the joint is thin and loose, contributing to ease of movement. The **coracohumeral ligament** helps support the weight of the upper limb, and three weak **glenohumeral ligaments** strengthen the front of the capsule. Muscle tendons from the biceps brachii and **rotator cuff** muscles contribute most to shoulder stability.

11

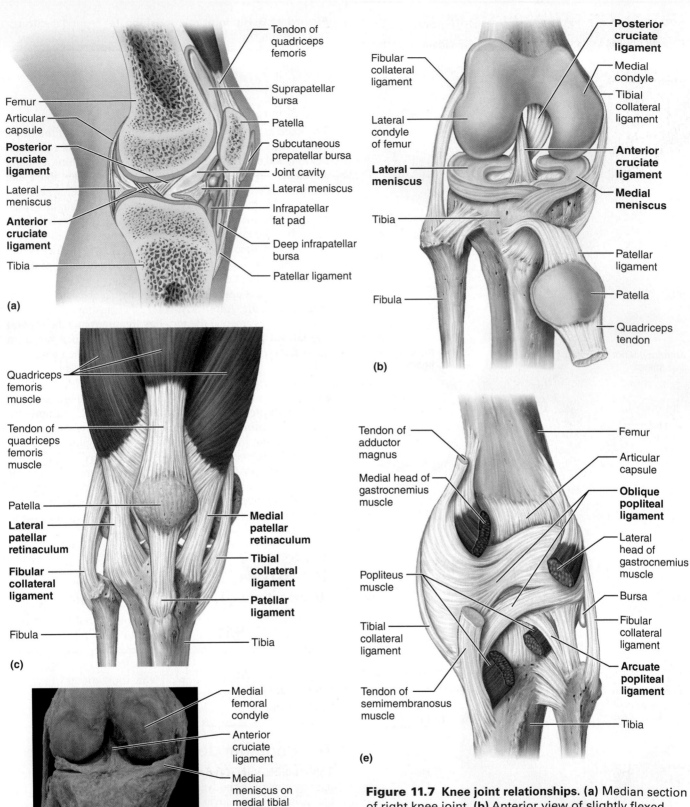

(a)

Femur
Articular capsule
Posterior cruciate ligament
Lateral meniscus
Anterior cruciate ligament
Tibia

Tendon of quadriceps femoris
Suprapatellar bursa
Patella
Subcutaneous prepatellar bursa
Joint cavity
Lateral meniscus
Infrapatellar fat pad
Deep infrapatellar bursa
Patellar ligament

(b)

Fibular collateral ligament
Lateral condyle of femur
Lateral meniscus
Tibia
Fibula

Posterior cruciate ligament
Medial condyle
Tibial collateral ligament
Anterior cruciate ligament
Medial meniscus
Patellar ligament
Patella
Quadriceps tendon

(c)

Quadriceps femoris muscle
Tendon of quadriceps femoris muscle
Patella
Lateral patellar retinaculum
Fibular collateral ligament
Fibula

Medial patellar retinaculum
Tibial collateral ligament
Patellar ligament
Tibia

(d)

Medial femoral condyle
Anterior cruciate ligament
Medial meniscus on medial tibial condyle
Patella

(e)

Tendon of adductor magnus
Medial head of gastrocnemius muscle
Popliteus muscle
Tibial collateral ligament
Tendon of semimembranosus muscle

Femur
Articular capsule
Oblique popliteal ligament
Lateral head of gastrocnemius muscle
Bursa
Fibular collateral ligament
Arcuate popliteal ligament
Tibia

Figure 11.7 Knee joint relationships. (a) Median section of right knee joint. **(b)** Anterior view of slightly flexed right knee joint showing the cruciate ligaments. Articular capsule has been removed; the quadriceps tendon has been cut and reflected distally. **(c)** Anterior superficial view of the right knee. **(d)** Photograph of an opened knee joint corresponds to view in (b). **(e)** Posterior superficial view of the ligaments clothing the knee joint.

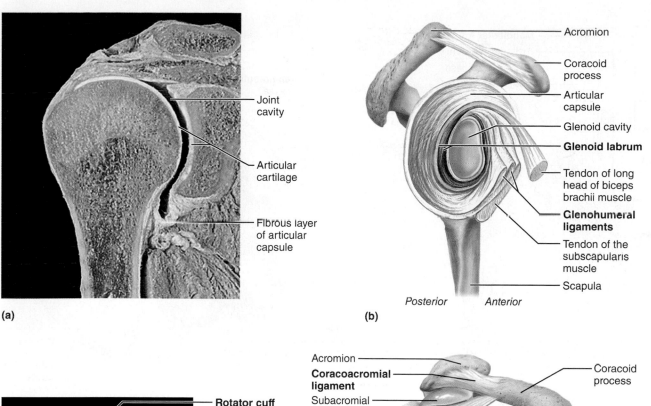

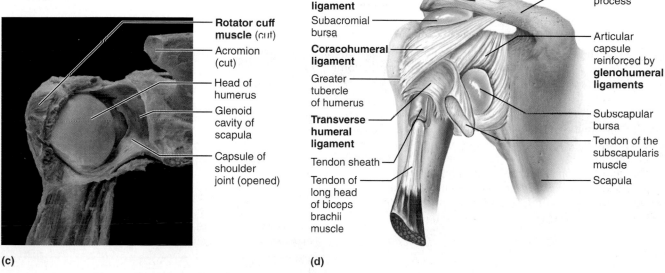

Figure 11.8 Shoulder joint relationships. (a) Frontal section through the shoulder. **(b)** Right shoulder joint, cut open and viewed from the lateral aspect; humerus removed. **(c)** Photograph of the interior of the shoulder joint, anterior view. **(d)** Anterior superficial view of the right shoulder.

Activity 8

Demonstrating Actions at the Shoulder Joint

If a functional shoulder joint model is available, identify the joint parts and manipulate the model to demonstrate the following movements: flexion, extension, abduction, adduction, circumduction, and medial and lateral rotation.

Note where the joint is weakest, and verify the most common direction of a dislocated humerus.

The Temporomandibular Joint

The **temporomandibular joint (TMJ)** lies just anterior to the ear (**Figure 11.9**, p. 182), where the egg-shaped condylar process of the mandible articulates with the inferior surface of the squamous part of the temporal bone. The temporal bone joint surface has a complicated shape: posteriorly is the **mandibular fossa** and anteriorly is a bony knob called the **articular tubercle.** The joint's articular capsule, though strengthened by the **lateral ligament,** is loose; an articular disc divides the joint cavity into superior and inferior compartments.

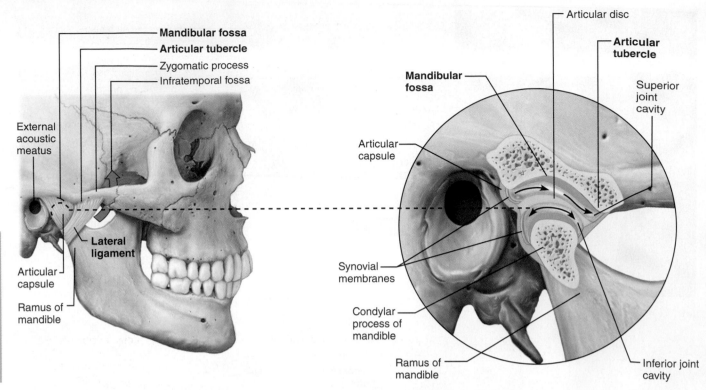

(a) Location of the joint in the skull

(b) Enlargement of a sagittal section through the joint

Figure 11.9 The temporomandibular (jaw) joint relationships. Note that the superior and inferior compartments of the joint cavity allow different movements indicated by arrows.

▶ Watch a video of the Temporomandibular Joint
MasteringA&P®>Study Area>Pre-Lab Videos

Typically, the condylar process–mandibular fossa connection allows the familiar hingelike movements of elevating and depressing the mandible to open and close the mouth. However, when the mouth is opened wide, the condylar process glides anteriorly and is braced against the dense bone of the articular tubercle so that the mandible is not forced superiorly through the thin mandibular fossa when we bite hard foods.

Activity 9

Examining the Action at the TMJ

While placing your fingers over the area just anterior to the ear, open and close your mouth to feel the hinge action at the TMJ. Then, keeping your fingers on the TMJ, yawn to demonstrate the anterior gliding of the condylar process of the mandible.

Joint Disorders

Most of us don't think about our joints until something goes wrong with them. Joint pains and malfunctions have a variety of causes. For example, a hard blow to the knee can cause a painful bursitis, known as "water on the knee," due to damage to, or inflammation of, the patellar bursa.

Sprains and dislocations are other types of joint problems. In a **sprain,** the ligaments reinforcing a joint are damaged by overstretching or are torn away from the bony attachment. Since both ligaments and tendons are cords of dense regular connective tissue with a poor blood supply, sprains heal slowly and are quite painful.

Dislocations occur when bones are forced out of their normal position in the joint cavity. They are normally accompanied by torn or stressed ligaments and considerable inflammation. The process of returning the bone to its proper position, called reduction, should be done only by a physician. Attempts by the untrained person to "snap the bone back into its socket" are often more harmful than helpful.

Advancing years also take their toll on joints. Weight-bearing joints in particular eventually begin to degenerate. *Adhesions* (fibrous bands) may form between the surfaces where bones join, and extraneous bone tissue *(spurs)* may grow along the joint edges. Such degenerative changes lead to the complaint so often heard from the elderly: "My joints are getting so stiff. . . ."

• If possible, compare an X-ray image of an arthritic joint to one of a normal joint. ✚

Table 11.3	Structural and Functional Characteristics of Body Joints			
Illustration	**Joint**	**Articulating bones**	**Structural type***	**Functional type; movements allowed**
	Skull	Cranial and facial bones	Fibrous; suture	Synarthrotic; no movement
	Temporo-mandibular	Temporal bone of skull and mandible	Synovial; modified hinge† (contains articular disc)	Diarthrotic; gliding and uniaxial rotation; slight lateral movement, elevation, depression, protraction, and retraction of mandible
	Atlanto-occipital	Occipital bone of skull and atlas	Synovial; condylar	Diarthrotic; biaxial; flexion, extension, lateral flexion, circumduction of head on neck
	Atlantoaxial	Atlas (C_1) and axis (C_2)	Synovial; pivot	Diarthrotic; uniaxial; rotation of the head
	Intervertebral	Between adjacent vertebral bodies	Cartilaginous; symphysis	Amphiarthrotic; slight movement
	Intervertebral	Between articular processes	Synovial; plane	Diarthrotic; gliding
	Costovertebral	Vertebrae (transverse processes or bodies) and ribs	Synovial; plane	Diarthrotic; gliding of ribs
	Sternoclavicular	Sternum and clavicle	Synovial; shallow saddle (contains articular disc)	Diarthrotic; multiaxial (allows clavicle to move in all axes)
	Sternocostal (first)	Sternum and rib 1	Cartilaginous; synchondrosis	Synarthrotic; no movement
	Sternocostal	Sternum and ribs 2–7	Synovial; double plane	Diarthrotic; gliding
	Acromio-clavicular	Acromion of scapula and clavicle	Synovial; plane (contains articular disc)	Diarthrotic; gliding and rotation of scapula on clavicle
	Shoulder (glenohumeral)	Scapula and humerus	Synovial; ball and socket	Diarthrotic; multiaxial; flexion, extension, abduction, adduction, circumduction, rotation of humerus
	Elbow	Ulna (and radius) with humerus	Synovial; hinge	Diarthrotic; uniaxial; flexion, extension of forearm
	Proximal radioulnar	Radius and ulna	Synovial; pivot	Diarthrotic; uniaxial; pivot (head of radius rotates in radial notch of ulna)
	Distal radioulnar	Radius and ulna	Synovial; pivot (contains articular disc)	Diarthrotic; uniaxial; rotation of radius to allow pronation and supination
	Wrist	Radius and proximal carpals	Synovial; condylar	Diarthrotic; biaxial; flexion, extension, abduction, adduction, circumduction of hand
	Intercarpal	Adjacent carpals	Synovial; plane	Diarthrotic; gliding
	Carpometacarpal of digit 1 (thumb)	Carpal (trapezium) and metacarpal I	Synovial; saddle	Diarthrotic; biaxial; flexion, extension, abduction, adduction, circumduction, opposition of metacarpal I
	Carpometacarpal of digits 2–5	Carpal(s) and metacarpal(s)	Synovial; plane	Diarthrotic; gliding of metacarpals
	Metacarpo-phalangeal (knuckle)	Metacarpal and proximal phalanx	Synovial; condylar	Diarthrotic; biaxial; flexion, extension, abduction, adduction, circumduction of fingers
	Interphalangeal (finger)	Adjacent phalanges	Synovial; hinge	Diarthrotic; uniaxial; flexion, extension of fingers

Table continues on next page. →

Table 11.3   Structural and Functional Characteristics of Body Joints (continued)

Illustration	Joint	Articulating bones	Structural type*	Functional type; movements allowed
	Sacroiliac	Sacrum and hip bone	Synovial; plane	Diarthrotic; little movement, slight gliding possible (more during pregnancy)
	Pubic symphysis	Pubic bones	Cartilaginous; symphysis	Amphiarthrotic; slight movement (enhanced during pregnancy)
	Hip (coxal)	Hip bone and femur	Synovial; ball and socket	Diarthrotic; multiaxial; flexion, extension, abduction, adduction, rotation, circumduction of thigh
	Knee (tibiofemoral)	Femur and tibia	Synovial; modified hinge† (contains articular discs)	Diarthrotic; biaxial; flexion, extension of leg, some rotation allowed
	Knee (femoropatellar)	Femur and patella	Synovial; plane	Diarthrotic; gliding of patella
	Superior tibiofibular	Tibia and fibula (proximally)	Synovial; plane	Diarthrotic; gliding of fibula
	Inferior tibiofibular	Tibia and fibula (distally)	Fibrous; syndesmosis	Synarthrotic; slight "give" during dorsiflexion of foot
	Ankle	Tibia and fibula with talus	Synovial; hinge	Diarthrotic; uniaxial; dorsiflexion, and plantar flexion of foot
	Intertarsal	Adjacent tarsals	Synovial; plane	Diarthrotic; gliding; inversion and eversion of foot
	Tarsometatarsal	Tarsal(s) and metatarsal(s)	Synovial; plane	Diarthrotic; gliding of metatarsals
	Metatarso-phalangeal	Metatarsal and proximal phalanx	Synovial; condylar	Diarthrotic; biaxial; flexion, extension, abduction, adduction, circumduction of great toe
	Interphalangeal (toe)	Adjacent phalanges	Synovial; hinge	Diarthrotic; uniaxial; flexion, extension of toes

*Fibrous joint indicated by orange circles; cartilaginous joints, by blue circles; synovial joints, by purple circles.
†These modified hinge joints are structurally bicondylar.

Group Challenge

Articulations: "Simon Says"

Working in groups of three or four, play a game of "Simon Says" using the movements defined in the exercise (see pp. 176–178). One student will play the role of "Simon" while the others perform the movement. For example, when "Simon" says, "Simon says, perform flexion at the elbow," the remaining students would flex their forearm. Take turns playing the role of Simon. As you perform the movements, discuss whether the joint is uniaxial, biaxial, or multiaxial. (Use Table 11.3 as a guide.) After playing for 15–20 minutes, complete the following tables.

	Name of joint	Movements allowed
1. List one uniaxial joints, and describe the movements at each.		
2. List one biaxial joints, and describe the movements at each.		
3. List two multiaxial joints, and describe the movements at each.		

12

Microscopic Anatomy and Organization of Skeletal Muscle

Objectives

☐ Define *muscle fiber*, *myofibril*, and *myofilament*, and describe the structural relationships among them.

☐ Describe thick (myosin) and thin (actin) filaments and their relationship to the sarcomere.

☐ Discuss the structure and location of T tubules and terminal cisterns.

☐ Define *endomysium*, *perimysium*, and *epimysium*, and relate them to muscle fibers, fascicles, and entire muscles.

☐ Define *tendon* and *aponeurosis*, and describe the difference between them.

☐ Describe the structure of skeletal muscle from gross to microscopic levels.

☐ Explain the connection between motor neurons and skeletal muscle, an and function of the neuromuscular junction.

Materials

- Three-dimensional model of skeletal muscle fibers (if available)
- Forceps
- Dissecting needles
- Clean microscope slides and coverslips
- 0.9% saline solution in dropper bottles
- Chicken breast or thigh muscle (fresh from the market)
- Compound microscope
- Prepared slides of skeletal muscle (l.s. and x.s. views) and skeletal muscle showing neuromuscular junctions
- Three-dimensional model of skeletal muscle showing neuromuscular junction (if available)

Pre-Lab Quiz

1. Which is *false* of skeletal muscle?
   a. It enables you to manipulate your environment.
   b. It influences the body's contours and shape.
   c. It is one of the major components of hollow organ
   d. It provides a means of locomotion.

2. Circle the correct underlined term. Because the cells keletal muscle are relatively large and cylindrical in shape, they are known as fibers / tubules.

3. Circle True or False. Skeletal muscle cells have more t n one nucleus.

4. The two contractile proteins that make up the myofila ents of skeletal muscle are _____ and _____.

5. Each muscle fiber is surrounded by thin connective tisue called the
   a. aponeurosis   c. endomysium
   b. epimysium   d. perimysium

6. A cordlike structure that connects a muscle to anothor muscle or bone is:
   a. a fascicle
   b. a tendon
   c. deep fascia

7. The junction between an axon and a muscle fiber is called a _____.

8. Circle True or False. The neuron and muscle fiber membranes do not actually touch but are separated by a fluid-filled gap.

9. Circle the correct underlined term. The contractile unit of muscle is the sarcolemma / sarcomere.

10. Circle True or False. Larger, more powerful muscles have relatively less connective tissue than smaller muscles.

MasteringA&P®

For related exercise study tools, go to the Study Area of **MasteringA&P**. There you will find:

- Practice Anatomy Lab **PAL**
- PhysioEx **PEx**
- A&PFlix **A&PFlix**
- Practice quizzes, Histology Atlas, eText, Videos, and more!

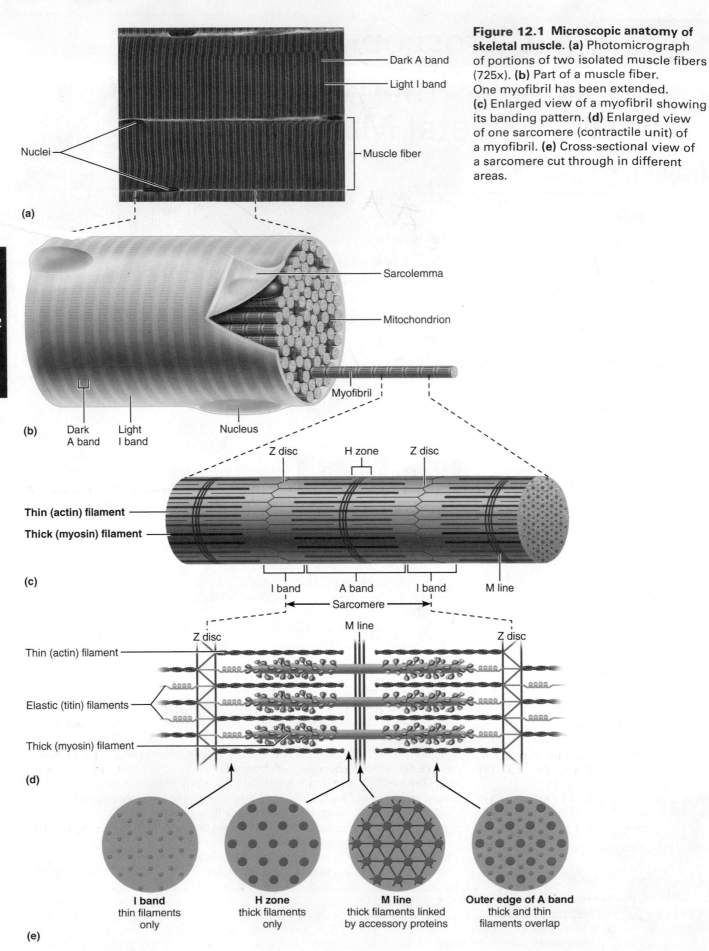

Figure 12.1 Microscopic anatomy of skeletal muscle. (a) Photomicrograph of portions of two isolated muscle fibers (725x). **(b)** Part of a muscle fiber. One myofibril has been extended. **(c)** Enlarged view of a myofibril showing its banding pattern. **(d)** Enlarged view of one sarcomere (contractile unit) of a myofibril. **(e)** Cross-sectional view of a sarcomere cut through in different areas.

(a)

Dark A band
Light I band
Muscle fiber
Nuclei

(b)

Sarcolemma
Mitochondrion
Myofibril
Dark A band   Light I band   Nucleus

(c)

Z disc   H zone   Z disc
Thin (actin) filament
Thick (myosin) filament
I band   A band   I band   M line
Sarcomere

(d)

M line
Z disc   Z disc
Thin (actin) filament
Elastic (titin) filaments
Thick (myosin) filament

(e)

I band
thin filaments only

H zone
thick filaments only

M line
thick filaments linked by accessory proteins

Outer edge of A band
thick and thin filaments overlap

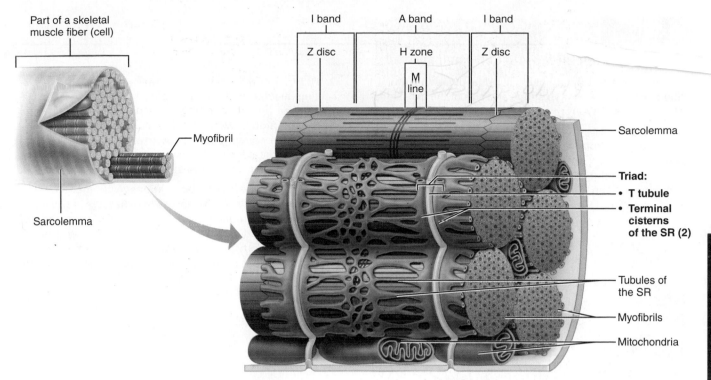

Part of a skeletal muscle fiber (cell)

I band   A band   I band

Z disc   H zone   Z disc

M line

Myofibril

Sarcolemma

Sarcolemma

Triad:
- **T tubule**
- **Terminal cisterns of the SR (2)**

Tubules of the SR

Myofibrils

Mitochondria

12

Figure 12.2 Relationship of the sarcoplasmic reticulum and T tubules to the myofibrils of skeletal muscle.

Most of the muscle tissue in the body is **skeletal muscle,** which attaches to the skeleton or associated connective tissue. Skeletal muscle shapes the body and gives you the ability to move—to walk, run, jump, and dance; to draw, paint, and play a musical instrument; and to smile and frown. The remaining muscle tissue of the body consists of smooth muscle that forms the walls of hollow organs and cardiac muscle that forms the walls of the heart. Smooth and cardiac muscle move materials within the body. For example, smooth muscle moves digesting food through the gastrointestinal system, and urine from the kidneys to the exterior of the body. Cardiac muscle moves blood through the blood vessels.

Each of the three muscle types has a structure and function uniquely suited to its function in the body. Our focus here is to investigate the structure of skeletal muscle.

Skeletal muscle is also known as *voluntary muscle* because it can be consciously controlled, and as striated muscle because it appears to be striped.

The Cells of Skeletal Muscle

Skeletal muscle is made up of relatively large, long cylindrical cells, called **muscle fibers.** These cells range from 10 to 100 μm in diameter. Some are up to 30 cm long.

Since hundreds of embryonic cells fuse to produce each muscle fiber, the cells (**Figure 12.1a** and **b**) are multinucleate; multiple oval nuclei can be seen just beneath the plasma membrane (called the *sarcolemma* in these cells). The nuclei are pushed peripherally by the longitudinally arranged **myofibrils,** long, rod-shaped organelles that nearly fill the sarcoplasm, the cytoplasm of the muscle cell. Alternating light (I) and dark (A) bands along the length of the perfectly aligned myofibrils give the muscle fiber its striped appearance.

Electron microscope studies have revealed that the myofibrils are made up of even smaller threadlike structures called **myofilaments** (Figure 12.1d). The myofilaments are composed largely of two varieties of contractile proteins—**actin** and **myosin**—which slide past each other during muscle activity to bring about shortening or contraction of the muscle cells. The actual contractile units of muscle, called **sarcomeres,** extend from the middle of one I band (its Z disc) to the middle of the next along the length of the myofibrils (Figure 12.1c and d.) Cross sections of the sarcomere in areas where **thick** and **thin filaments** overlap show that each thick filament is surrounded by six thin filaments; each thin filament is surrounded by three thick filaments (Figure 12.1e).

At each junction of the A and I bands, the sarcolemma indents into the muscle cell, forming a **transverse tubule (T tubule).** These tubules run deep into the muscle fiber between cross channels, or **terminal cisterns,** of the elaborate smooth endoplasmic reticulum called the **sarcoplasmic reticulum (SR) (Figure 12.2).** Regions where the SR terminal cisterns border a T tubule on each side are called **triads.**

Activity 1

Examining Skeletal Muscle Cell Anatomy

1. Look at the three-dimensional model of skeletal muscle fibers, noting the relative shape and size of the cells. Identify the nuclei, myofibrils, and light and dark bands.

2. Obtain forceps, two dissecting needles, slide and coverslip, and a dropper bottle of saline solution. With forceps, remove a very small piece of muscle (about 1 mm diameter) from a fresh chicken breast or thigh. Place the tissue on a clean microscope slide, and add a drop of the saline solution.

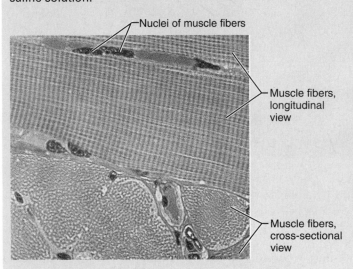

Figure 12.3 **Photomicrograph of muscle fibers, longitudinal and cross sections (800×).**

3. Pull the muscle fibers apart (tease them) with the dissecting needles until you have a fluffy-looking mass of tissue. Cover the teased tissue with a coverslip, and observe under the high-power lens of a compound microscope. Look for the banding pattern by examining muscle fibers isolated at the edge of the tissue mass. Regulate the light carefully to obtain the highest possible contrast.

4. Now compare your observations with the photomicrograph (**Figure 12.3**) and with what can be seen in professionally prepared muscle tissue. Obtain a slide of skeletal muscle (longitudinal section), and view it under high power. From your observations, draw a small section of a muscle fiber in the space provided below. Label the nuclei, sarcolemma, and A and I bands.

What structural details become apparent with the prepared slide?

Organization of Skeletal Muscle Cells into Muscles

Muscle fibers are soft and surprisingly fragile. Thousands of muscle fibers are bundled together with connective tissue to form the organs we refer to as skeletal muscles (**Figure 12.4**). Each muscle fiber is enclosed in a delicate, areolar connective tissue sheath called the **endomysium.** Several sheathed muscle fibers are wrapped by a collagenic membrane called the **perimysium,** forming a bundle of muscle fibers called a **fascicle.** A large number of fascicles are bound together by a much coarser "overcoat" of dense irregular connective tissue called the **epimysium,** which sheathes the entire muscle. These epimysia blend into strong cordlike **tendons** or sheetlike **aponeuroses,** which attach muscles to each other or indirectly to bones. A muscle's more movable attachment is called its *insertion* whereas its fixed (or immovable) attachment is the *origin* (Exercise 11).

Tendons perform several functions, two of the most important being to provide durability and to conserve space. Because tendons are tough dense regular connective tissue, they can span rough bony projections that would destroy the more delicate muscle tissues. Because of their relatively small size, more tendons than fleshy muscles can pass over a joint.

In addition to supporting and binding the muscle fibers, and providing strength to the muscle as a whole, the connective tissue wrappings provide a route for the entry and exit of nerves and blood vessels that serve the muscle fibers. The larger, more powerful muscles have relatively more connective tissue than muscles involved in fine or delicate movements.

As we age, the mass of the muscle fibers decreases, and the amount of connective tissue increases; thus the skeletal muscles gradually become more sinewy, or "stringier." ✚

Activity 2

Observing the Histological Structure of a Skeletal Muscle

Identify the muscle fibers, their peripherally located nuclei, and their connective tissue wrappings—the endomysium, perimysium, and epimysium, if visible (use Figure 12.4 as a reference).

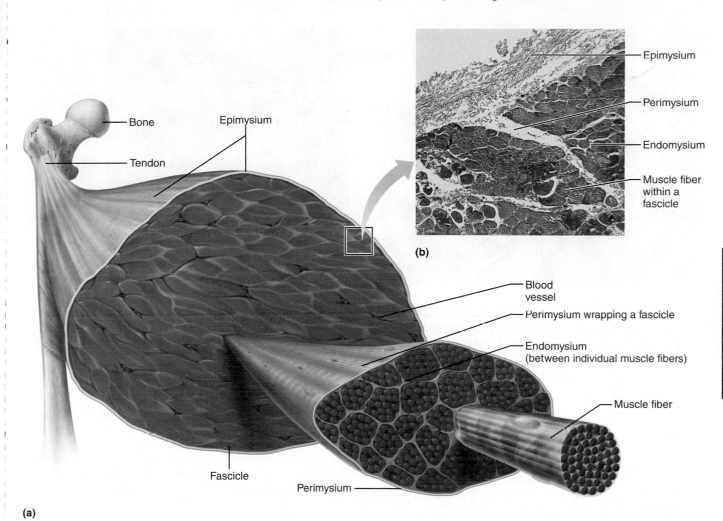

Figure 12.4 Connective tissue coverings of skeletal muscle. (a) Diagram. **(b)** Photomicrograph of a cross section of skeletal muscle (90×).

The Neuromuscular Junction

The voluntary skeletal muscle cells must be stimulated by motor neurons via nerve impulses. The junction between an axon of a motor neuron and a muscle fiber is called a **neuromuscular junction (Figure 12.5**, p. 194).

Each axon of the motor neuron usually divides into many branches called *terminal branches* as it approaches the muscle. Each of these branches ends in an axon terminal that participates in forming a neuromuscular junction with a single muscle fiber. Thus a single neuron may stimulate many muscle fibers. Together, a neuron and all the muscle fibers it stimulates make up the functional structure called the **motor unit.** Part of a motor unit, showing two neuromuscular junctions, is shown in **Figure 12.6**, p. 194. The neuron and muscle fiber membranes, close as they are, do not actually touch. They are separated by a small fluid-filled gap called the **synaptic cleft** (see Figure 12.5).

Within the axon terminals are many mitochondria and vesicles containing a neurotransmitter called acetylcholine (ACh). When an action potential reaches the axon terminal, some of these vesicles release their contents into the synaptic cleft. The ACh rapidly diffuses across the junction and combines with the receptors on the sarcolemma. When receptors bind ACh, a change in the permeability of the sarcolemma occurs. Ion channels open briefly, depolarizing the sarcolemma, and subsequent contraction of the muscle fiber occurs.

Activity 3

Studying the Structure of a Neuromuscular Junction

1. If possible, examine a three-dimensional model of skeletal muscle fibers that illustrates the neuromuscular junction. Identify the structures just described.

2. Obtain a slide of skeletal muscle stained to show a portion of a motor unit. Examine the slide under high power to identify the axon branches that extend like a leash to the muscle fibers.

Text continues on next page. →

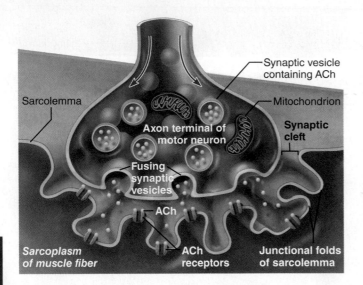

Figure 12.5 The neuromuscular junction. Red arrows indicate arrival of the action potential, which ultimately causes vesicles to release ACh. The ACh receptor is part of the ion channel that opens briefly, causing depolarization of the sarcolemma.

Follow one of the axons to its terminal branch to identify the oval-shaped axon terminal. Compare your observations to the photomicrograph (Figure 12.6). Sketch a small section in the space provided below. Label the axon of the motor neuron, its terminal branches, and muscle fibers.

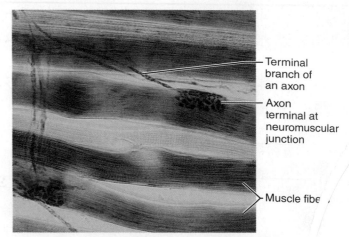

Figure 12.6 Photomicrograph of neuromuscular junctions (750×).

WHY THIS MATTERS | Botox—The Good, the Bad, and the Wrinkle-Free

A diluted form of the botulinum toxin, Botox, received FDA approval to treat uncontrollable blinking (blepharospasm) in 1989. Prior to this time, the botulinum toxin was better known for causing respiratory failure with the foodborne illness known as botulism. The bacterium *Clostridium botulinum* secretes botulinum toxin, which is a neurotoxin. The toxin causes paralysis by blocking the release of the neurotransmitter acetylcholine (ACh) from the axon terminal of a motor neuron. If the neurotoxin blocks axon terminals serving the diaphragm, death is a common result. If the ACh receptors are located in a facial muscle, such as the corrugator supercilii, a brow is unfurrowed, and a wrinkle-free glabella results. ∎

Gross Anatomy of the Muscular System

Objectives

☐ Define *prime mover (agonist)*, *antagonist*, *synergist*, and *fixator*.

☐ List the criteria used in naming skeletal muscles.

☐ Identify the major muscles of the human body on a torso model, a human cadaver, lab chart, or image, and state the action of each.

☐ Name muscle origins and insertions as required by the instructor.

☐ Explain how muscle actions are related to their location.

☐ List antagonists for the prime movers listed.

Materials

- Human torso model or large anatomical chart showing human musculature
- Human cadaver for demonstration (if available)
- Disposable gloves
- *Human Musculature* video on the Cadaver Dissection Video Series for Human Anatomy and Physiology DVD (available from Pearson Education to qualified adopters).
- Tubes of body (or face) paint
- 1" wide artist's brushes

✂ For instructions on animal dissections, see the dissection exercises (starting on p. 705) in the cat and fetal pig editions of this manual.

MasteringA&P®

For related exercise study tools, go to the Study Area of **MasteringA&P**. There you will find:

- Practice Anatomy Lab **PAL**
- A&PFlix **A&PFlix**
- PhysioEx **PEx**
- Practice quizzes, Histology Atlas, eText, Videos, and more!

Pre-Lab Quiz

1. A prime mover, or _____, produces a particular type of movement.
   a. agonist   c. fixator
   b. antagonist   d. synergist

2. Skeletal muscles are named on the basis of many criteria. Name one.

   _____

3. Circle True or False. Muscles of facial expression differ from most skeletal muscles because they usually do not insert into a bone.

4. The _____ musculature includes muscles that move the vertebral column, the ribs, and the abdominal wall.
   a. head and neck   b. lower limb   c. trunk

5. Muscles that act on the _____ cause movement at the hip, knee, and foot joints.
   a. lower limb   b. trunk   c. upper limb

6. This two-headed muscle bulges when the forearm is flexed. It is the most familiar muscle of the anterior humerus. It is the:
   a. biceps brachii   c. extensor digitorum
   b. flexor carpii radialis   d. triceps brachii

7. These abdominal muscles are responsible for giving me my "six-pack." They also stabilize my pelvis when walking. They are the _____ muscles.
   a. internal intercostals   c. quadriceps
   b. rectus abdominis   d. triceps femoris

8. Circle the correct underlined term. This lower limb muscle, which attaches to the calcaneus via the calcaneal tendon and plantar flexes the foot when the leg is extended, is the <u>tibialis anterior</u> / gastrocnemius.

9. The _____ is the largest and most superficial of the gluteal muscles.
   a. gluteus internus   c. gluteus maximus
   b. gluteus medius   d. gluteus minimus

10. Circle True or False. The biceps femoris is located in the anterior compartment of the thigh.

Skeletal muscles enable movement. Smiling, frowning, speaking, singing, breathing, dancing, running, and playing a musical instrument are just a few examples. Most often, purposeful movements require the coordinated action of several skeletal muscles.

Classification of Skeletal Muscles

Types of Muscles

Muscles that are most responsible for producing a particular movement are called **prime movers,** or **agonists.** Muscles that oppose or reverse a movement are called **antagonists.** When a prime mover is active, the fibers of the antagonist are stretched and in the relaxed state. Antagonists can be prime movers in their own right. For example, the biceps muscle of the arm (a prime mover of flexion at the elbow) is antagonized by ___ ___ceps (a prime mover of extension at the elbow).

Synergists aid the action of agonists either by assisting with the same movement or by reducing undesirable or unnecessary movement. Contraction of a muscle crossing two or more joints would cause movement at all joints spanned if the synergists were not there to stabilize them. For example, the muscles that flex the fingers cross both the wrist and finger joints, but you can make a fist without bending at the wrist because synergist muscles stabilize the wrist joint.

Fixators, or fixation muscles, are specialized synergists. They immobilize the origin of a prime mover so that all the tension is exerted at the insertion. Muscles that help maintain posture are fixators; so too are muscles of the back that stabilize or "fix" the scapula during arm movements.

Naming Skeletal Muscles

Remembering the names of the skeletal muscles is a monumental task, but certain clues help. Muscles are named on the basis of the following criteria:

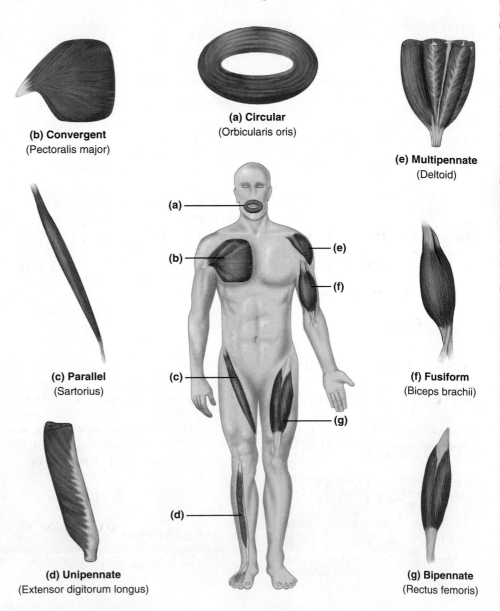

(b) Convergent
(Pectoralis major)

(a) Circular
(Orbicularis oris)

(e) Multipennate
(Deltoid)

(c) Parallel
(Sartorius)

(f) Fusiform
(Biceps brachii)

(d) Unipennate
(Extensor digitorum longus)

(g) Bipennate
(Rectus femoris)

Figure 13.1 Patterns of fascicle arrangement in muscles.

13

- **Direction of muscle fibers:** Some muscles are named in reference to some imaginary line, usually the midline of the body. A muscle with fibers running parallel to that imaginary line will have the term *rectus* (straight) in its name. For example, the rectus abdominis is the straight muscle of the abdomen. Likewise, the terms *transverse* and *oblique* indicate that the muscle fibers run at right angles and obliquely (respectively) to the imaginary line. Muscle structure is determined by fascicle arrangement (**Figure 13.1**).

- **Relative size of the muscle:** Terms such as *maximus* (largest), *minimus* (smallest), *longus* (long), and *brevis* (short) are often used in naming muscles—as in gluteus maximus and gluteus minimus.

- **Location of the muscle:** Some muscles are named for the bone with which they are associated. For example, the temporalis muscle overlies the temporal bone.

- **Number of origins:** When the term *biceps, triceps,* or *quadriceps* forms part of a muscle name, you can generally assume that the muscle has two, three, or four origins (respectively). For example, the biceps brachii has two origins.

- **Location of the muscle's origin and insertion:** For example, the sternocleidomastoid muscle has its origin on the sternum *(sterno)* and clavicle *(cleido)*, and inserts on the mastoid process of the temporal bone.

- **Shape of the muscle:** For example, the deltoid muscle is roughly triangular *(deltoid* = triangle), and the trapezius muscle resembles a trapezoid.

- **Action of the muscle:** For example, all the adductor muscles of the anterior thigh bring about its adduction, and all the extensor muscles of the wrist extend the hand.

Identification of Human Muscles

While reading the tables and identifying the various human muscles in the figures, try to visualize what happens when the muscle contracts. Since muscles often have many actions, we have indicated the primary action of each muscle in blue type in the tables. Then, use a torso model or an anatomical chart to identify as many muscles as possible. If a human cadaver is available, your instructor will provide specific instructions. Then, carry out the instructions for demonstrating and palpating muscles. **Figures 13.2** and **13.3**, pp. 202–203, are summary figures illustrating the superficial musculature of the body.

Muscles of the Head and Neck

The muscles of the head serve many specific functions. For instance, the muscles of facial expression differ from most skeletal muscles because they insert into the skin or other muscles rather than into bone. As a result, they move the facial skin, allowing a wide range of emotions to be expressed. Other muscles of the head are the muscles of mastication, which move the mandible during chewing, and the six extrinsic eye muscles located within the orbit, which aim the eye. (Orbital muscles are studied in Exercise 23.) Neck muscles provide for the movement of the head and shoulder girdle.

Activity 1

Identifying Head and Neck Muscles

Read the descriptions of specific head and neck muscles and identify the various muscles in the figures (**Tables 13.1** and **13.2**, pp. 204–208 and **Figures 13.4** and **13.5**, pp. 205–207), trying to visualize their action when they contract. Then identify them on a torso model or anatomical chart.

Demonstrating Operations of Head Muscles

1. Raise your eyebrow to wrinkle your forehead. You are using the *frontal belly* of the *epicranius* muscle.

2. Blink your eyes; wink. You are contracting *orbicularis oculi.*

3. Close your lips and pucker up. This requires contraction of *orbicularis oris.*

4. Smile. You are using *zygomaticus.*

5. To demonstrate the *temporalis,* place your hands on your temples and clench your teeth. The *masseter* can also be palpated now at the angle of the jaw.

Muscles of the Trunk

The trunk musculature includes muscles that move the vertebral column; anterior thorax muscles that act to move ribs, head, and arms; and muscles of the abdominal wall that play a role in the movement of the vertebral column but more importantly form the "natural girdle," or the major portion of the abdominal body wall.

Activity 2

Identifying Muscles of the Trunk

Read the descriptions of specific trunk muscles and identify them in the figures (**Tables 13.3**, pp. 209–210 and **13.4**, pp. 212–214 and **Figures 13.6–13.9**, pp. 209–215), visualizing their action when they contract. Then identify them on a torso model or anatomical chart.

Demonstrating Operations of Trunk Muscles

Now, work with a partner to demonstrate the operation of the following muscles. One of you can demonstrate the movement; the following steps are addressed to this partner. The other can supply resistance and palpate the muscle being tested.

1. Fully abduct the arm and extend at the elbow. Now adduct the arm against resistance. You are using the *latissimus dorsi.*

2. To observe the action of the *deltoid,* try to abduct your arm against resistance. Now attempt to elevate your shoulder against resistance; you are contracting the upper portion of the *trapezius.*

3. The *pectoralis major* is used when you press your hands together at chest level with your elbows widely abducted.

13

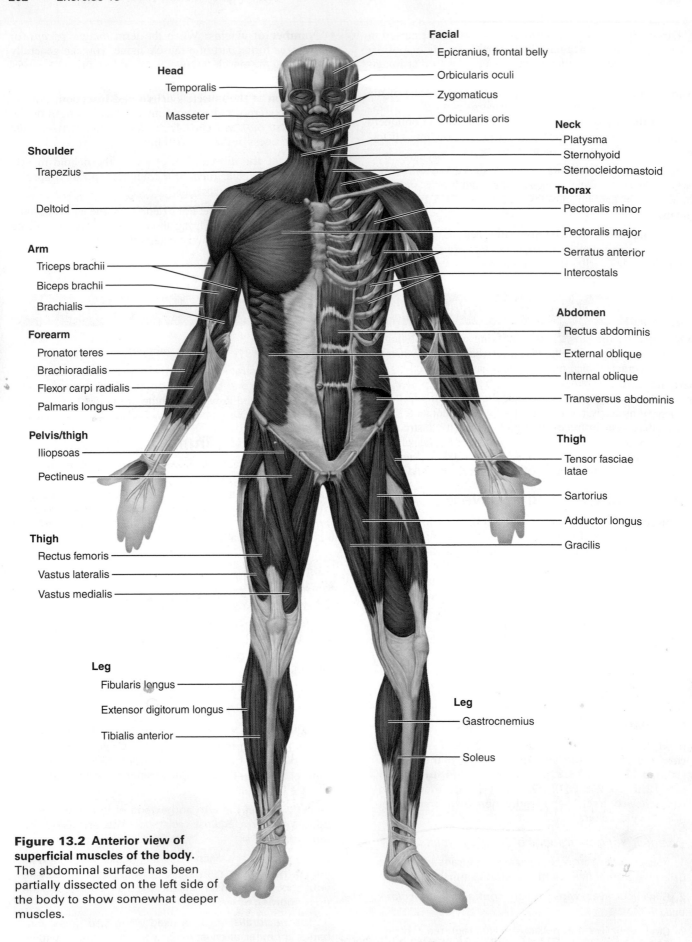

Facial
Epicranius, frontal belly
Orbicularis oculi
Zygomaticus
Orbicularis oris

Head
Temporalis
Masseter

Neck
Platysma
Sternohyoid
Sternocleidomastoid

Shoulder
Trapezius
Deltoid

Thorax
Pectoralis minor
Pectoralis major
Serratus anterior
Intercostals

Arm
Triceps brachii
Biceps brachii
Brachialis

Forearm
Pronator teres
Brachioradialis
Flexor carpi radialis
Palmaris longus

Abdomen
Rectus abdominis
External oblique
Internal oblique
Transversus abdominis

Pelvis/thigh
Iliopsoas
Pectineus

Thigh
Tensor fasciae latae
Sartorius
Adductor longus
Gracilis

Thigh
Rectus femoris
Vastus lateralis
Vastus medialis

Leg
Fibularis longus
Extensor digitorum longus
Tibialis anterior

Leg
Gastrocnemius
Soleus

Figure 13.2 Anterior view of superficial muscles of the body. The abdominal surface has been partially dissected on the left side of the body to show somewhat deeper muscles.

13

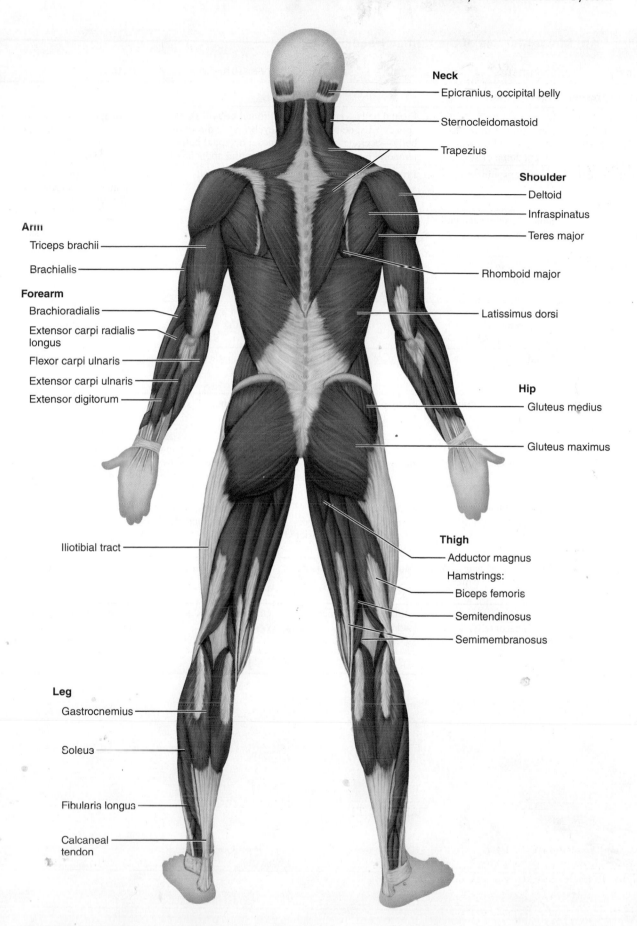

Figure 13.3 Posterior view of superficial muscles of the body.

Table 13.1   Major Muscles of the Human Head (Figure 13.4)

Muscle	Comments	Origin	Insertion	Action
Facial Expression (Figure 13.4a)				
Epicranius—frontal and occipital bellies	Bipartite muscle consisting of frontal and occipital bellies, which covers dome of skull	Frontal belly—epicranial aponeurosis; occipital belly—occipital and temporal bones	Frontal belly—skin of eyebrows and root of nose; occipital belly—epicranial aponeurosis	With aponeurosis fixed, frontal belly raises eyebrows; occipital belly fixes aponeurosis and pulls scalp posteriorly
Orbicularis oculi	Tripartite sphincter muscle of eyelids	Frontal and maxillary bones and ligaments around orbit	Tissue of eyelid	Closes eye, produces blinking, squinting, and draws eyebrows inferiorly
Corrugator supercilii	Small muscle; acts with orbicularis oculi	Arch of frontal bone above nasal bone	Skin of eyebrow	Draws eyebrows medially and inferiorly; wrinkles skin of forehead vertically
Levator labii superioris	Thin muscle between orbicularis oris and inferior eye margin	Zygomatic bone and infraorbital margin of maxilla	Skin and muscle of upper lip	Raises and furrows upper lip; opens lips
Zygomaticus—major and minor	Extends diagonally from corner of mouth to cheekbone	Zygomatic bone	Skin and muscle at corner of mouth	Raises lateral corners of mouth upward (smiling muscle)
Risorius	Slender muscle; runs inferior and lateral to zygomaticus	Fascia of masseter muscle	Skin at angle of mouth	Draws corner of lip laterally; tenses lip; zygomaticus synergist
Depressor labii inferioris	Small muscle running from lower lip to mandible	Body of mandible lateral to its midline	Skin and muscle of lower lip	Draws lower lip inferiorly
Depressor anguli oris	Small muscle lateral to depressor labii inferioris	Body of mandible below incisors	Skin and muscle at angle of mouth below insertion of zygomaticus	Zygomaticus antagonist; draws corners of mouth downward and laterally
Orbicularis oris	Multilayered muscle of lips with fibers that run in many different directions; most run circularly	Arises indirectly from maxilla and mandible; fibers blended with fibers of other muscles associated with lips	Encircles mouth; inserts into muscle and skin at angles of mouth	Closes lips; purses and protrudes lips (kissing and whistling muscle)
Mentalis	One of muscle pair forming V-shaped muscle mass on chin	Mandible below incisors	Skin of chin	Protrudes lower lip; wrinkles chin
Buccinator	Principal muscle of cheek; runs horizontally, deep to the masseter	Molar region of maxilla and mandible	Orbicularis oris	Draws corner of mouth laterally; compresses cheek (as in whistling); holds food between teeth during chewing

(Table continues on page 206.)

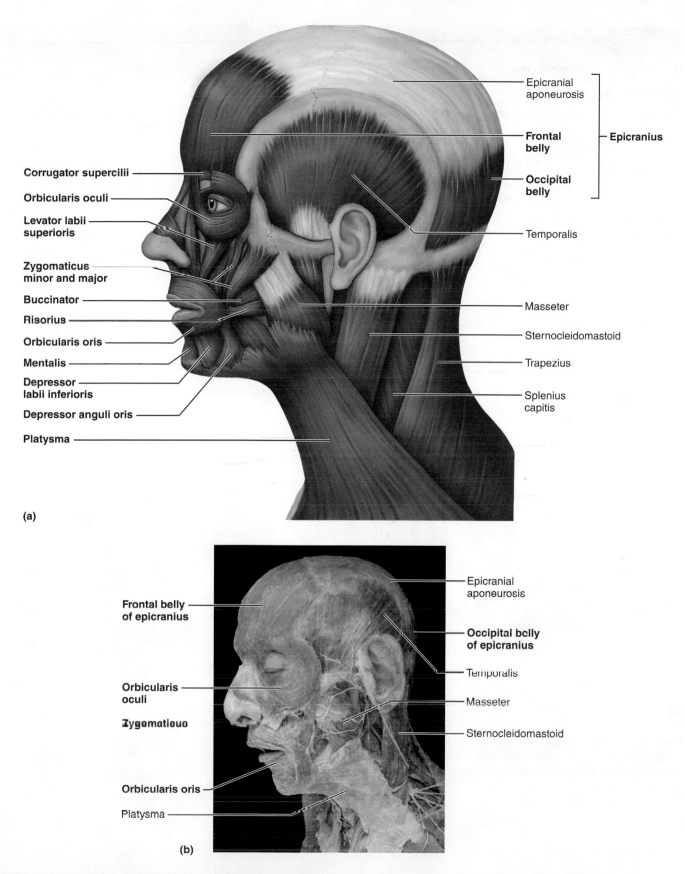

Corrugator supercilii

Orbicularis oculi

Levator labii superioris

Zygomaticus minor and major

Buccinator

Risorius

Orbicularis oris

Mentalis

Depressor labii inferioris

Depressor anguli oris

Platysma

Epicranial aponeurosis

Frontal belly

Occipital belly

Epicranius

Temporalis

Masseter

Sternocleidomastoid

Trapezius

Splenius capitis

13

(a)

Frontal belly of epicranius

Orbicularis oculi

Zygomaticus

Orbicularis oris

Platysma

Epicranial aponeurosis

Occipital belly of epicranius

Temporalis

Masseter

Sternocleidomastoid

(b)

Figure 13.4 Muscles of the head (left lateral view). (a) Superficial muscles.
(b) Photo of superficial structures of head and neck.

Table 13.1	**Major Muscles of the Human Head** *(continued)*			
Muscle	**Comments**	**Origin**	**Insertion**	**Action**
Mastication (Figure 13.4c, d)				
Masseter	Covers lateral aspect of mandibular ramus; can be palpated on forcible closure of jaws	Zygomatic arch and maxilla	Angle and ramus of mandible	Prime mover of jaw closure; elevates mandible
Temporalis	Fan-shaped muscle lying over parts of frontal, parietal, and temporal bones	Temporal fossa	Coronoid process of mandible	Closes jaw; elevates and retracts mandible
Buccinator	(See muscles of facial expression.)			
Medial pterygoid	Runs along internal (medial) surface of mandible (thus largely concealed by that bone)	Sphenoid, palatine, and maxillary bones	Medial surface of mandible, near its angle	Synergist of temporalis and masseter; elevates mandible; in conjunction with lateral pterygoid, aids in grinding movements of teeth
Lateral pterygoid	Superior to medial pterygoid	Greater wing of sphenoid bone	Condylar process of mandible	Protracts jaw (moves it anteriorly); in conjunction with medial pterygoid, aids in grinding movements of teeth

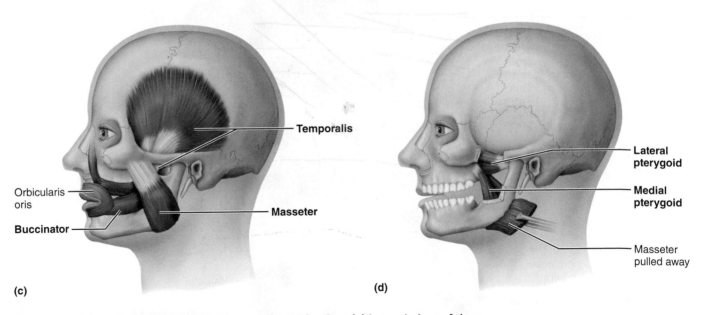

(c)

(d)

Figure 13.4 *(continued)* **Muscles of the head mastication. (c)** Lateral view of the temporalis, masseter, and buccinator muscles. **(d)** Lateral view of the deep chewing muscles, the medial and lateral pterygoid muscles.

Table 13.2	**Anterolateral Muscles of the Human Neck (Figure 13.5)**			
Muscle	**Comments**	**Origin**	**Insertion**	**Action**
Superficial				
Platysma (Figure 13.4a)	Unpaired, thin, sheetlike superficial neck muscle, plays role in facial expression	Fascia of chest (over pectoral muscles) and deltoid	Lower margin of mandible, skin, and muscle at corner of mouth	Tenses skin of neck; depresses mandible; pulls lower lip back and down
Sternocleidomastoid	Two-headed muscle located deep to platysma on anterolateral surface of neck; indicate limits of anterior and posterior triangles of neck	Manubrium of sternum and medial portion of clavicle	Mastoid process of temporal bone and superior nuchal line of occipital bone	Simultaneous contraction of both muscles of pair causes flexion of neck, acting independently, rotate head toward shoulder on opposite side
Scalenes—anterior, middle, and posterior (Figure 13.5c)	Located more on lateral than anterior neck; deep to platysma and sternocleidomastoid	Transverse processes of cervical vertebrae	Anterolaterally on ribs 1–2	Flex and slightly rotate neck; elevate ribs 1–2 (aid in inspiration)

(Table continues on page 208.)

13

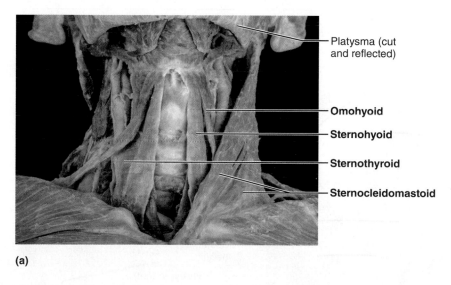

Platysma (cut and reflected)

Omohyoid

Sternohyoid

Sternothyroid

Sternocleidomastoid

(a)

Figure 13.5 Muscles of the anterolateral neck and throat. (a) Photo of the anterior and lateral regions of the neck.

Table 13.2	Anterolateral Muscles of the Human Neck *(continued)*			
Muscle	**Comments**	**Origin**	**Insertion**	**Action**
Deep (Figure 13.5a, b)				
Digastric	Consists of two bellies united by an intermediate tendon; forms a V-shape under chin	Lower margin of mandible (anterior belly) and mastoid process (posterior belly)	By a connective tissue loop to hyoid bone	Acting together, elevate hyoid bone; open mouth and depress mandible
Stylohyoid	Slender muscle parallels posterior border of digastric; below angle of jaw	Styloid process of temporal bone	Hyoid bone	Elevates and retracts hyoid bone
Mylohyoid	Just deep to digastric; forms floor of mouth	Medial surface of mandible	Hyoid bone and median raphe	Elevates hyoid bone and base of tongue during swallowing
Sternohyoid	Runs most medially along neck; straplike	Manubrium and medial end of clavicle	Lower margin of hyoid bone	Acting with sternothyroid and omohyoid, depresses larynx and hyoid bone if mandible is fixed; may also flex skull
Sternothyroid	Lateral and deep to sternohyoid	Posterior surface of manubrium	Thyroid cartilage of larynx	(See Sternohyoid above)
Omohyoid	Straplike with two bellies; lateral to sternohyoid	Superior surface of scapula	Hyoid bone; inferior border	(See Sternohyoid above)
Thyrohyoid	Appears as a superior continuation of sternothyroid muscle	Thyroid cartilage	Hyoid bone	Depresses hyoid bone; elevates larynx if hyoid is fixed

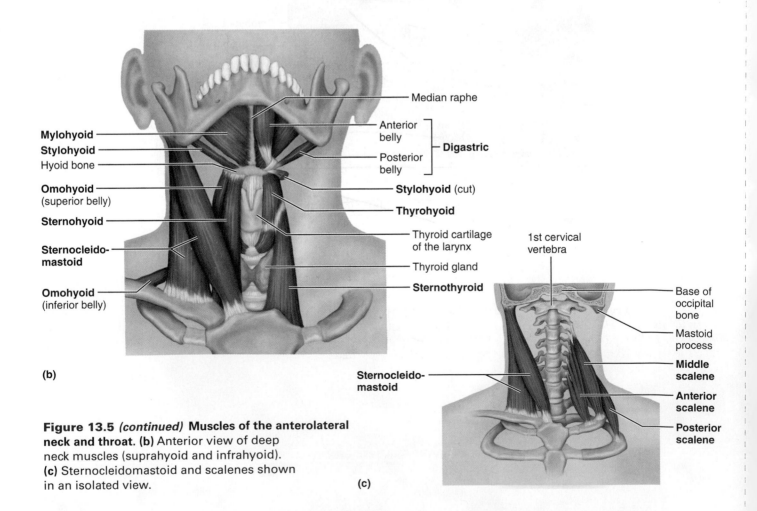

Figure 13.5 *(continued)* Muscles of the anterolateral neck and throat. (b) Anterior view of deep neck muscles (suprahyoid and infrahyoid). **(c)** Sternocleidomastoid and scalenes shown in an isolated view.

Table 13.3	**Anterior Muscles of the Human Thorax, Shoulder, and Abdominal Wall (Figures 13.6, 13.7, and 13.8)**			
Muscle	**Comments**	**Origin**	**Insertion**	**Action**
Thorax and Shoulder, Superficial (Figure 13.6)				
Pectoralis major	Large fan-shaped muscle covering upper portion of chest	Clavicle, sternum, cartilage of ribs 1–6 (or 7), and aponeurosis of external oblique muscle	Fibers converge to insert by short tendon into intertubercular sulcus of humerus	Prime mover of arm flexion; adducts, medially rotates arm; with arm fixed, pulls chest upward (thus also acts in forced inspiration)
Serratus anterior	Fan-shaped muscle deep to scapula; deep and inferior to pectoral muscles on lateral rib cage	Lateral aspect of ribs 1–8 (or 9)	Vertebral border of anterior surface of scapula	Prime mover to protract and hold scapula against chest wall; rotates scapula, causing inferior angle to move laterally and upward; essential to raising arm; fixes scapula for arm abduction
Deltoid (see also Figure 13.9)	Fleshy triangular muscle forming shoulder muscle mass; intramuscular injection site	Lateral ¹/₃ of clavicle; acromion and spine of scapula	Deltoid tuberosity of humerus	Acting as a whole, prime mover of arm abduction; when only specific fibers are active, can act as a synergist in flexion, extension, and rotation of arm

13

(Table continues on page 210.)

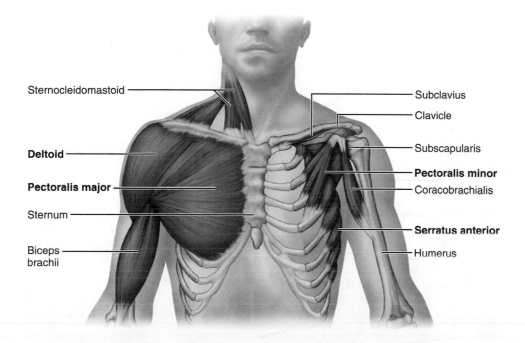

Figure 13.6 Muscles of the thorax and shoulder acting on the scapula and arm (anterior view). The superficial muscles, which effect arm movements, are shown on the left side of the figure. These muscles have been removed on the right side of the figure to show the muscles that stabilize or move the pectoral girdle.

Table 13.3	Anterior Muscles of the Human Thorax, Shoulder, and Abdominal Wall *(continued)*			
Muscle	**Comments**	**Origin**	**Insertion**	**Action**
Thorax and Shoulder, Superficial *(continued)*				
Pectoralis minor	Flat, thin muscle deep to pectoralis major	Anterior surface of ribs 3–5, near their costal cartilages	Coracoid process of scapula	With ribs fixed, draws scapula forward and inferiorly; with scapula fixed, draws rib cage superiorly
Thorax, Deep: Muscles of Respiration (Figure 13.7)				
External intercostals	11 pairs lie between ribs; fibers run obliquely downward and forward toward sternum	Inferior border of rib above (not shown in figure)	Superior border of rib below	Pull ribs toward one another to elevate rib cage; aid in inspiration
Internal intercostals	11 pairs lie between ribs; fibers run deep and at right angles to those of external intercostals	Superior border of rib below	Inferior border of rib above (not shown in figure)	Draw ribs together to depress rib cage; aid in forced expiration; antagonistic to external intercostals
Diaphragm	Broad muscle; forms floor of thoracic cavity; dome-shaped in relaxed state; fibers converge from margins of thoracic cage toward a central tendon	Inferior border of rib and sternum, costal cartilages of last six ribs and lumbar vertebrae	Central tendon	Prime mover of inspiration flattens on contraction, increasing vertical dimensions of thorax; increases intra-abdominal pressure
Abdominal Wall (Figure 13.8a and b)				
Rectus abdominis	Medial superficial muscle, extends from pubis to rib cage; ensheathed by aponeuroses of oblique muscles; segmented	Pubic crest and symphysis	Xiphoid process and costal cartilages of ribs 5–7	Flexes and rotates vertebral column; increases abdominal pressure; fixes and depresses ribs; stabilizes pelvis during walking; used in sit-ups and curls
External oblique	Most superficial lateral muscle; fibers run downward and medially; ensheathed by an aponeurosis	Anterior surface of lower eight ribs	Linea alba,* pubic crest and tubercles, and iliac crest	See rectus abdominis, above; compresses abdominal wall; also aids muscles of back in trunk rotation and lateral flexion; used in oblique curls
Internal oblique	Most fibers run at right angles to those of external oblique, which it underlies	Lumbar fascia, iliac crest, and inguinal ligament	Linea alba, pubic crest, and costal cartilages of last three ribs	As for external oblique
Transversus abdominis	Deepest muscle of abdominal wall; fibers run horizontally	Inguinal ligament, iliac crest, cartilages of last five or six ribs, and lumbar fascia	Linea alba and pubic crest	Compresses abdominal contents

*The linea alba (white line) is a narrow, tendinous sheath that runs along the middle of the abdomen from the sternum to the pubic symphysis. It is formed by the fusion of the aponeurosis of the external oblique and transversus muscles.

Figure 13.7 Deep muscles of the thorax: muscles of respiration. (a) The external intercostals (inspiratory muscles) are shown on the left and the internal intercostals (expiratory muscles) are shown on the right. These two muscle layers run obliquely and at right angles to each other. **(b)** Inferior view of the diaphragm, the prime mover of inspiration. Notice that its muscle fibers converge toward a central tendon, an arrangement that causes the diaphragm to flatten and move inferiorly as it contracts. The diaphragm and its tendon are pierced by the great vessels (aorta and inferior vena cava) and the esophagus.

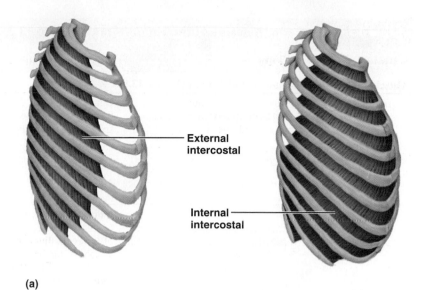

External intercostal

Internal intercostal

(a)

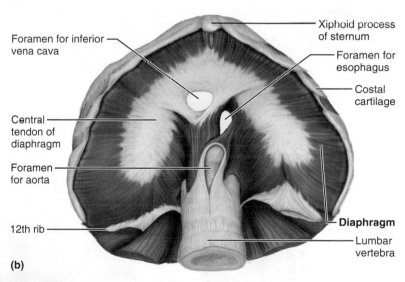

Foramen for inferior vena cava

Central tendon of diaphragm

Foramen for aorta

12th rib

Xiphoid process of sternum

Foramen for esophagus

Costal cartilage

Diaphragm

Lumbar vertebra

(b)

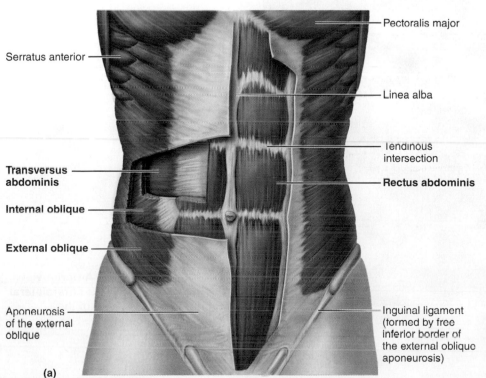

Serratus anterior

Transversus abdominis

Internal oblique

External oblique

Aponeurosis of the external oblique

Pectoralis major

Linea alba

Tendinous intersection

Rectus abdominis

Inguinal ligament (formed by free inferior border of the external oblique aponeurosis)

(a)

Figure 13.8 Anterior view of the muscles forming the anterolateral abdominal wall. (a) The superficial muscles have been partially cut away on the left side of the diagram to reveal the deeper internal oblique and transversus abdominis muscles.

Table 13.4 Posterior Muscles of the Human Trunk (Figure 13.9)

Muscle	Comments	Origin	Insertion	Action
Muscles of the Neck, Shoulder, and Thorax (Figure 13.9a)				
Trapezius	Most superficial muscle of posterior thorax; very broad origin and insertion	Occipital bone; ligamentum nuchae; spines of C_7 and all thoracic vertebrae	Acromion and spinous process of scapula; lateral third of clavicle	Extends head; raises, rotates, and retracts (adducts) scapula and stabilizes it; superior fibers elevate scapula (as in shrugging the shoulders); inferior fibers depress scapula
Latissimus dorsi	Broad flat muscle of lower back (lumbar region); extensive superficial origins	Indirect attachment to spinous processes of lower six thoracic vertebrae, lumbar vertebrae, last three to four ribs, and iliac crest	Floor of intertubercular sulcus of humerus	Prime mover of arm extension; adducts and medially rotates arm; brings arm down in power stroke, as in striking a blow
Infraspinatus	A rotator cuff muscle; partially covered by deltoid and trapezius	Infraspinous fossa of scapula	Greater tubercle of humerus	Lateral rotation of arm; helps hold head of humerus in glenoid cavity; stabilizes shoulder
Teres minor	A rotator cuff muscle; small muscle inferior to infraspinatus	Lateral margin of posterior scapula	Greater tubercle of humerus	Same as for infraspinatus

13

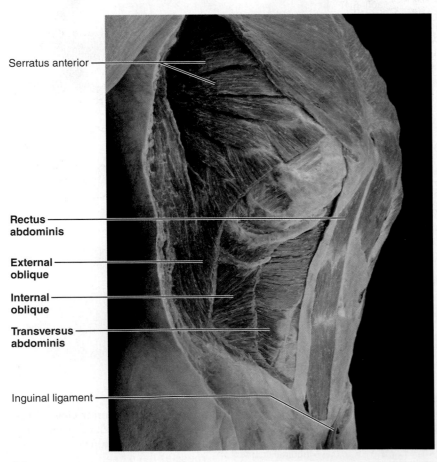

Serratus anterior

Rectus abdominis

External oblique

Internal oblique

Transversus abdominis

Inguinal ligament

(b)

Figure 13.8 *(continued)* **Anterior view of the muscles forming the anterolateral abdominal wall. (b)** Photo of the anterolateral abdominal wall.

Table 13.4	(continued)			
Muscle	**Comments**	**Origin**	**Insertion**	**Action**
Teres major	Located inferiorly to teres minor	Posterior surface at inferior angle of scapula	Intertubercular sulcus of humerus	Extends, medially rotates, and adducts arm; synergist of latissimus dorsi
Supraspinatus	A rotator cuff muscle; obscured by trapezius	Supraspinous fossa of scapula	Greater tubercle of humerus	Initiates abduction of arm; stabilizes shoulder joint
Levator scapulae	Located at back and side of neck, deep to trapezius	Transverse processes of C_1–C_4	Medial border of scapula superior to spine	Elevates and adducts scapula; with fixed scapula, laterally flexes neck to the same side
Rhomboids—major and minor	Beneath trapezius and inferior to levator scapulae; rhomboid minor is the more superior muscle	Spinous processes of C_7 and T_1–T_5	Medial border of scapula	Pull scapulae medially (retraction); stabilize scapulae; rotate glenoid cavity downward
Muscles Associated with the Vertebral Column (Figure 13.9b)				
Semispinalis	Deep composite muscle of the back—thoracis, cervicis, and capitis portions	Transverse processes of C_7–T_{12}	Occipital bone and spinous processes of cervical vertebrae and T_1–T_4	Acting together, extend head and vertebral column; acting independently (right vs. left), causes rotation toward the opposite side

(Table continues on page 214.)

13

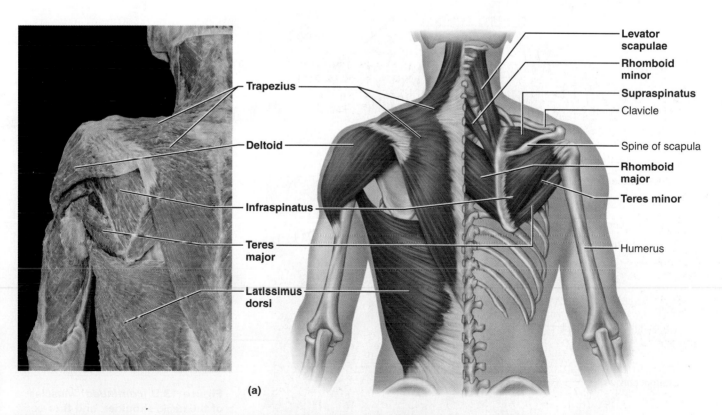

(a)

Figure 13.9 Muscles of the neck, shoulder, and thorax (posterior view).
(a) The superficial muscles of the back are shown for the left side of the body, with a corresponding photograph. The superficial muscles are removed on the right side of the illustration to reveal the deeper muscles acting on the scapula and the rotator cuff muscles that help to stabilize the shoulder joint.

Table 13.4	Posterior Muscles of the Human Trunk *(continued)*			
Muscle	**Comments**	**Origin**	**Insertion**	**Action**
Muscles Associated with the Vertebral Column *(continued)*				
Erector spinae	A long tripartite muscle composed of iliocostalis (lateral), longissimus, and spinalis (medial) muscle columns; superficial to semispinalis muscles; extends from pelvis to head	Iliac crest, transverse processes of lumbar, thoracic, and cervical vertebrae, and/or ribs 3–6 depending on specific part	Ribs and transverse processes of vertebrae about six segments above origin; longissimus also inserts into mastoid process	Extend and bend the vertebral column laterally; fibers of the longissimus also extend and rotate head

13

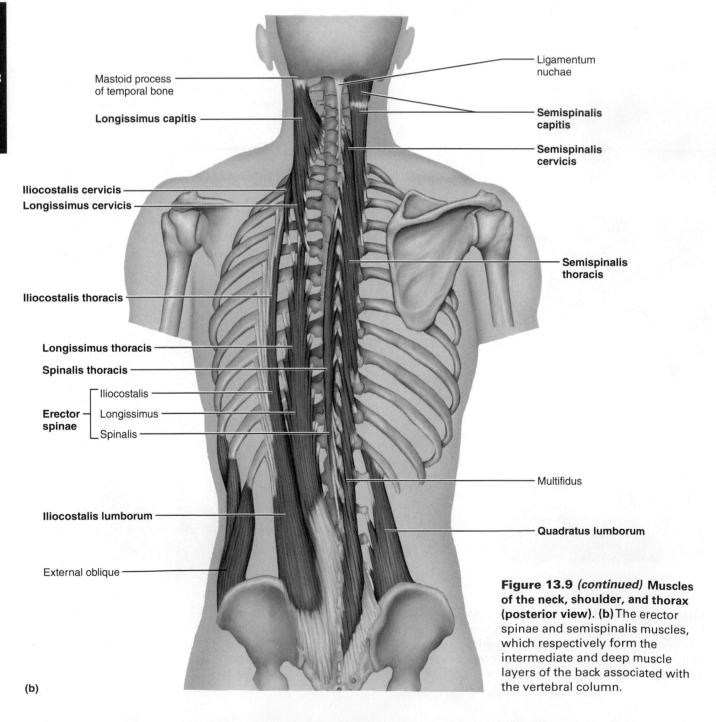

Figure 13.9 *(continued)* **Muscles of the neck, shoulder, and thorax (posterior view). (b)** The erector spinae and semispinalis muscles, which respectively form the intermediate and deep muscle layers of the back associated with the vertebral column.

(b)

Table 13.4 *(continued)*				
Muscle	**Comments**	**Origin**	**Insertion**	**Action**
Splenius (Figure 13.9c)	Superficial muscle (capitis and cervicis parts) extending from upper thoracic region to skull	Ligamentum nuchae and spinous processes of C_7–T_6	Mastoid process, occipital bone, and transverse processes of C_2–C_4	As a group, extend or hyperextend head; when only one side is active, head is rotated and bent toward the same side
Quadratus lumborum	Forms greater portion of posterior abdominal wall	Iliac crest and lumbar fascia	Inferior border of rib 12; transverse processes of lumbar vertebrae	Each flexes vertebral column laterally; together extend the lumbar spine and fix rib 12; maintains upright posture

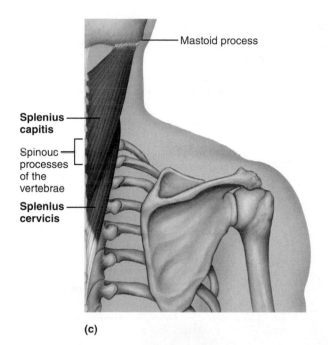

Mastoid process

Splenius capitis

Spinous processes of the vertebrae

Splenius cervicis

(c)

Figure 13.9 *(continued)* **Muscles of the neck, shoulder, and thorax (posterior view). (c)** Deep (splenius) muscles of the posterior neck. Superficial muscles have been removed.

Muscles of the Upper Limb

The muscles that act on the upper limb fall into three groups: those that move the arm, those causing movement of the forearm, and those moving the hand and fingers.

The muscles that cross the shoulder joint to insert on the humerus and move the arm (subscapularis, supraspinatus and infraspinatus, deltoid, and so on) are primarily trunk muscles that originate on the axial skeleton or shoulder girdle. These muscles are included with the trunk muscles.

The second group of muscles, which cross the elbow joint and move the forearm, consists of muscles forming the musculature of the humerus. These muscles arise mainly from the humerus and insert in forearm bones. They are responsible for flexion, extension, pronation, and supination.

The third group forms the musculature of the forearm. For the most part, these muscles cross the wrist to insert on the digits and produce movements of the hand and fingers.

Activity 3

Identifying Muscles of the Upper Limb

Study the origins, insertions, and actions of muscles that move the forearm, and identify them (**Table 13.5**, p. 216 and **Figure 13.10**, p. 216).

Do the same for muscles acting on the wrist and hand (**Table 13.6**, pp. 217–218 and **Figure 13.11**, pp. 217–219). They are more easily identified if you locate their insertion tendons first.

Then see if you can identify the upper limb muscles on a torso model, anatomical chart, or cadaver. Complete this portion of the exercise with palpation demonstrations as outlined next.

Demonstrating Operations of Upper Limb Muscles

1. To observe the *biceps brachii,* attempt to flex your forearm (hand supinated) against resistance. The insertion tendon of this biceps muscle can also be felt in the lateral aspect of the cubital fossa (where it runs toward the radius to attach).

2. If you acutely flex at your elbow and then try to extend the forearm against resistance, you can demonstrate the action of your *triceps brachii.*

3. Strongly flex your hand, and make a fist. Palpate your contracting wrist flexor muscles' origins at the medial epicondyle of the humerus and their insertion tendons, at the anterior aspect of the wrist.

4. Flare your fingers to identify the tendons of the *extensor digitorum* muscle on the dorsum of your hand.

(Text continues on page 220.)

13

Table 13.5   Muscles of the Human Humerus That Act on the Forearm (Figure 13.10)

Muscle	Comments	Origin	Insertion	Action
Triceps brachii	Large fleshy muscle of posterior humerus; three-headed origin	Long head—inferior margin of glenoid cavity; lateral head—posterior humerus; medial head—distal radial groove on posterior humerus	Olecranon of ulna	Powerful forearm extensor; antagonist of forearm flexors (brachialis and biceps brachii)
Anconeus	Short triangular muscle blended with triceps	Lateral epicondyle of humerus	Lateral aspect of olecranon of ulna	Abducts ulna during forearm pronation; synergist of triceps brachii in forearm extension
Biceps brachii	Most familiar muscle of anterior humerus because this two-headed muscle bulges when forearm is flexed	Short head: coracoid process; long head: supraglenoid tubercle and lip of glenoid cavity; tendon of long head runs in intertubercular sulcus and within capsule of shoulder joint	Radial tuberosity	Flexion (powerful) and supination of forearm; "it turns the corkscrew and pulls the cork"; weak arm flexor
Brachioradialis	Superficial muscle of lateral forearm; forms lateral boundary of cubital fossa	Lateral ridge at distal end of humerus	Base of radial styloid process	Synergist in forearm flexion
Brachialis	Immediately deep to biceps brachii	Distal portion of anterior humerus	Coronoid process of ulna	Flexor of forearm

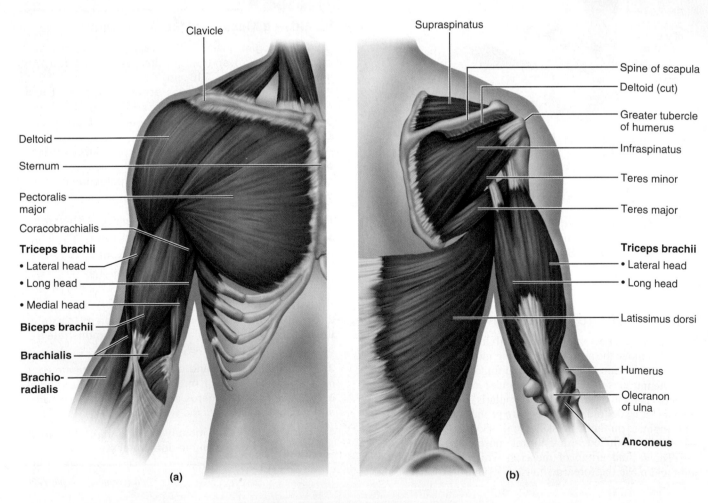

(a)   (b)

Figure 13.10 Muscles acting on the arm and forearm. (a) Superficial muscles of
the anterior thorax, shoulder, and arm, anterior view. **(b)** Posterior aspect of the arm
showing the lateral and long heads of the triceps brachii muscle.

Table 13.6	Muscles of the Human Forearm That Act on the Hand and Fingers (Figure 13.11)			
Muscle	**Comments**	**Origin**	**Insertion**	**Action**
Anterior Compartment, Superficial (Figure 13.11a, b, c)				
Pronator teres	Seen in a superficial view between proximal margins of brachioradialis and flexor carpi radialis	Medial epicondyle of humerus and coronoid process of ulna	Midshaft of radius	Acts synergistically with pronator quadratus to pronate forearm; weak forearm flexor
Flexor carpi radialis	Superficial; runs diagonally across forearm	Medial epicondyle of humerus	Base of metacarpals II and III	Powerful flexor and abductor of the hand
Palmaris longus	Small fleshy muscle with a long tendon; medial to flexor carpi radialis	Medial epicondyle of humerus	Palmar aponeurosis; skin and fascia of palm	Flexes hand (weak); tenses skin and fascia of palm

(Table continues on page 218.)

13

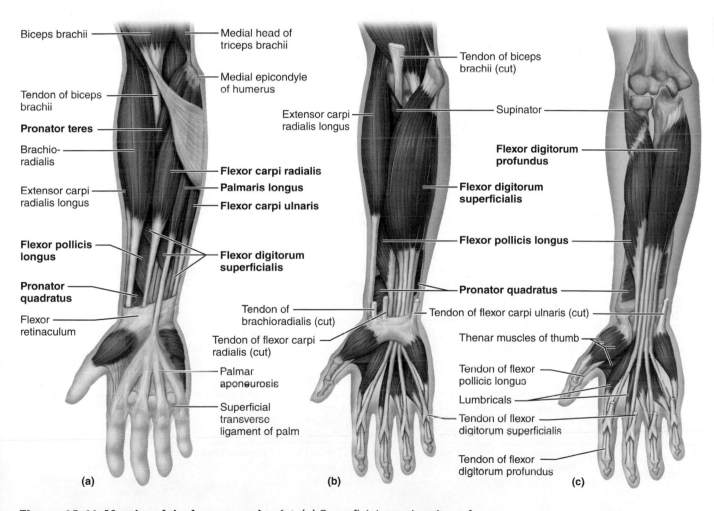

(a)   (b)   (c)

Figure 13.11 Muscles of the forearm and wrist. (a) Superficial anterior view of right forearm and hand. **(b)** The brachioradialis, flexors carpi radialis and ulnaris, and palmaris longus muscles have been removed to reveal the position of the somewhat deeper flexor digitorum superficialis. **(c)** Deep muscles of the anterior compartment. Superficial muscles have been removed. (*Note:* The thenar muscles of the thumb and the lumbricals that help move the fingers are illustrated here but are not described in Table 13.6.)

Table 13.6   Muscles of the Human Forearm That Act on the Hand and Fingers *(continued)*

Muscle	Comments	Origin	Insertion	Action
Anterior Compartment, Superficial *(continued)*				
Flexor carpi ulnaris	Superficial; medial to palmaris longus	Medial epicondyle of humerus and olecranon and posterior surface of ulna	Base of metacarpal V; pisiform and hamate bones	Flexes and adducts hand
Flexor digitorum superficialis	Deeper muscle (deep to muscles named above); visible at distal end of forearm	Medial epicondyle of humerus, coronoid process of ulna, and shaft of radius	Middle phalanges of fingers 2–5	Flexes hand and middle phalanges of fingers 2–5
Anterior Compartment, Deep (Figure 13.11a, b, c)				
Flexor pollicis longus	Deep muscle of anterior forearm; distal to and paralleling lower margin of flexor digitorum superficialis	Anterior surface of radius, and interosseous membrane	Distal phalanx of thumb	Flexes thumb (*pollex* is Latin for "thumb")
Flexor digitorum profundus	Deep muscle; overlain entirely by flexor digitorum superficialis	Anteromedial surface of ulna, interosseous membrane, and coronoid process	Distal phalanges of fingers 2–5	Sole muscle that flexes distal phalanges; assists in hand flexion
Pronator quadratus	Deepest muscle of distal forearm	Distal portion of anterior ulnar surface	Anterior surface of radius, distal end	Pronates forearm
Posterior Compartment, Superficial (Figure 13.11d, e, f)				
Extensor carpi radialis longus	Superficial; parallels brachioradialis on lateral forearm	Lateral supracondylar ridge of humerus	Base of metacarpal II	Extends and abducts hand
Extensor carpi radialis brevis	Deep to extensor carpi radialis longus	Lateral epicondyle of humerus	Base of metacarpal III	Extends and abducts hand; steadies wrist during finger flexion
Extensor digitorum	Superficial; medial to extensor carpi radialis brevis	Lateral epicondyle of humerus	By four tendons into distal phalanges of fingers 2–5	Prime mover of finger extension; extends hand; can flare (abduct) fingers
Extensor carpi ulnaris	Superficial; medial posterior forearm	Lateral epicondyle of humerus; posterior border of ulna	Base of metacarpal V	Extends and adducts hand
Posterior Compartment, Deep (Figure 13.11d, e, f)				
Extensor pollicis longus and brevis	Muscle pair with a common origin and action; deep to extensor carpi ulnaris	Dorsal shaft of ulna and radius, interosseous membrane	Base of distal phalanx of thumb (longus) and proximal phalanx of thumb (brevis)	Extend thumb
Abductor pollicis longus	Deep muscle; lateral and parallel to extensor pollicis longus	Posterior surface of radius and ulna; interosseous membrane	Metacarpal I and trapezium	Abducts and extends thumb
Supinator	Deep muscle at posterior aspect of elbow	Lateral epicondyle of humerus; proximal ulna	Proximal end of radius	Synergist of biceps brachii to supinate forearm; antagonist of pronator muscles

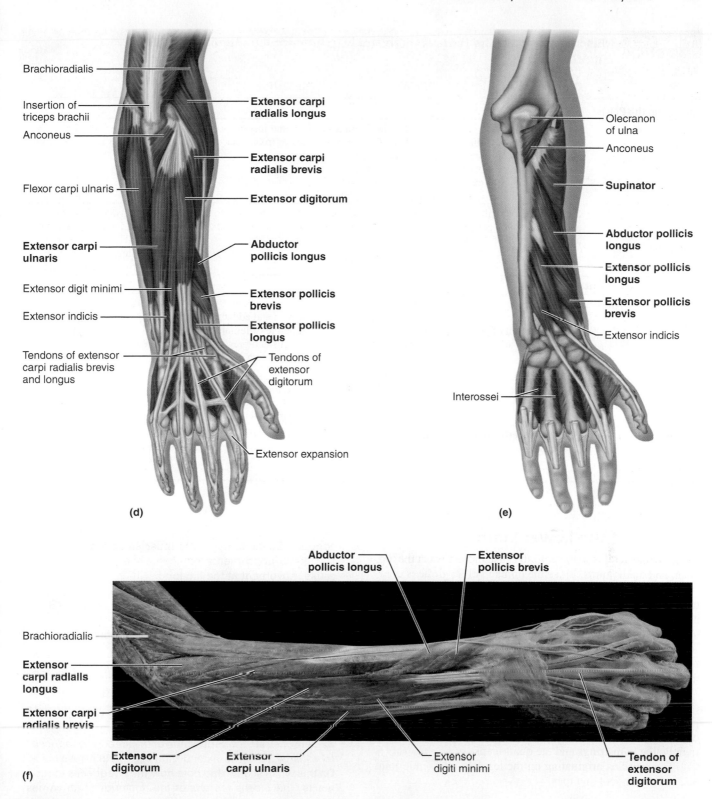

Brachioradialis

Insertion of triceps brachii

Anconeus

Flexor carpi ulnaris

Extensor carpi ulnaris

Extensor digit minimi

Extensor indicis

Tendons of extensor carpi radialis brevis and longus

Extensor carpi radialis longus

Extensor carpi radialis brevis

Extensor digitorum

Abductor pollicis longus

Extensor pollicis brevis

Extensor pollicis longus

Tendons of extensor digitorum

Extensor expansion

(d)

Olecranon of ulna

Anconeus

Supinator

Abductor pollicis longus

Extensor pollicis longus

Extensor pollicis brevis

Extensor indicis

Interossei

(e)

13

Abductor pollicis longus

Extensor pollicis brevis

Brachioradialis

Extensor carpi radialis longus

Extensor carpi radialis brevis

Extensor digitorum

Extensor carpi ulnaris

Extensor digiti minimi

Tendon of extensor digitorum

(f)

Figure 13.11 *(continued)* **Muscles of the forearm and wrist. (d)** Superficial muscles, posterior view. **(e)** Deep posterior muscles; superficial muscles have been removed. The interossei, the deepest layer of intrinsic hand muscles, are also illustrated. **(f)** Photo of posterior muscles of the right forearm.

Table 13.7	Muscles Acting on the Human Thigh and Leg, Anterior and Medial Aspects (Figure 13.12)			
Muscle	**Comments**	**Origin**	**Insertion**	**Action**
Origin on the Pelvis				
Iliopsoas—iliacus and psoas major	Two closely related muscles; fibers pass under inguinal ligament to insert into femur via a common tendon; iliacus is more lateral	Iliacus—iliac fossa and crest, lateral sacrum; psoas major—transverse processes, bodies, and discs of T_{12} and lumbar vertebrae	On and just below lesser trochanter of femur	Flex trunk at hip joint, flex thigh; lateral flexion of vertebral column (psoas)
Sartorius	Straplike superficial muscle running obliquely across anterior surface of thigh to knee	Anterior superior iliac spine	By an aponeurosis into medial aspect of proximal tibia	Flexes, abducts, and laterally rotates thigh; flexes leg; known as "tailor's muscle" because it helps effect cross-legged position in which tailors are often depicted
Medial Compartment				
Adductors—magnus, longus, and brevis	Large muscle mass forming medial aspect of thigh; arise from front of pelvis and insert at various levels on femur	Magnus—ischial and pubic rami and ischial tuberosity; longus—pubis near pubic symphysis; brevis—body and inferior pubic ramus	Magnus—linea aspera and adductor tubercle of femur; longus and brevis—linea aspera	Adduct and medially rotate and flex thigh; posterior part of magnus is also a synergist in thigh extension
Pectineus	Overlies adductor brevis on proximal thigh	Pectineal line of pubis (and superior pubic ramus)	Inferior from lesser trochanter to linea aspera of femur	Adducts, flexes, and medially rotates thigh
Gracilis	Straplike superficial muscle of medial thigh	Inferior ramus and body of pubis	Medial surface of tibia just inferior to medial condyle	Adducts thigh; flexes and medially rotates leg, especially during walking

(Table continues on page 222.)

Muscles of the Lower Limb

Muscles that act on the lower limb cause movement at the hip, knee, and ankle joints. Since the human pelvic girdle is composed of heavy, fused bones that allow very little movement, no special group of muscles is necessary to stabilize it. This is unlike the shoulder girdle, where several muscles (mainly trunk muscles) are needed to stabilize the scapulae.

Muscles acting on the thigh (femur) cause various movements at the multiaxial hip joint. These include the iliopsoas, the adductor group, and others.

Muscles acting on the leg form the major musculature of the thigh. (Anatomically, the term *leg* refers only to that portion between the knee and the ankle.) The thigh muscles cross the knee to allow flexion and extension of the leg. They include the hamstrings and the quadriceps.

The muscles originating on the leg cross the ankle joint and act on the foot and toes.

Activity 4

Identifying Muscles of the Lower Limb

Read the descriptions of specific muscles acting on the thigh and leg and identify them (**Tables 13.7**, pp. 220–222 and **13.8**, pp. 222–223 and **Figures 13.12** and **13.13**, p. 223), trying to visualize their action when they contract. Since some of the muscles acting on the leg also have attachments on the pelvic girdle, they can cause movement at the hip joint.

Do the same for muscles acting on the foot and toes (**Table 13.9**, p. 224 and **Figures 13.14**, p. 225 and **13.15**, pp. 227–228).

Then identify all the muscles on a model or anatomical chart.

Demonstrating Operations of Lower Limb Muscles

1. Go into a deep knee bend and palpate your own *gluteus maximus* muscle as you extend your thigh to resume the upright posture.

2. Demonstrate the contraction of the anterior *quadriceps femoris* by trying to extend your leg against resistance. Do this while seated and note how the quadriceps tendon reacts. The *biceps femoris* of the posterior thigh comes into play when you flex your leg against resistance.

3. Now stand on your toes. Have your partner palpate the lateral and medial heads of the *gastrocnemius* and follow it to its insertion in the calcaneal tendon.

4. Dorsiflex and invert your foot while palpating your *tibialis anterior* muscle (which parallels the sharp anterior crest of the tibia laterally).

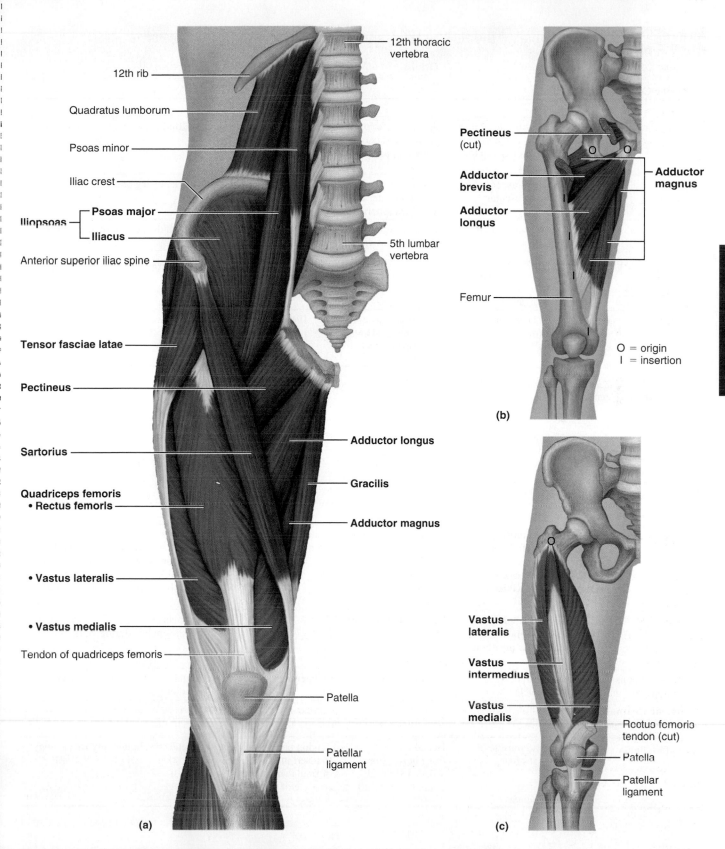

12th thoracic vertebra

12th rib

Quadratus lumborum

Psoas minor

Iliac crest

Iliopsoas { **Psoas major**
Iliacus

Anterior superior iliac spine

5th lumbar vertebra

Tensor fasciae latae

Pectineus

Sartorius

Quadriceps femoris
• Rectus femoris

• Vastus lateralis

• Vastus medialis

Tendon of quadriceps femoris

Patella

Patellar ligament

(a)

Adductor longus

Gracilis

Adductor magnus

Pectineus (cut)

Adductor brevis

Adductor longus

Adductor magnus

Femur

O = origin
I = insertion

(b)

13

Vastus lateralis

Vastus intermedius

Vastus medialis

Rectus femoris tendon (cut)

Patella

Patellar ligament

(c)

Figure 13.12 Anterior and medial muscles promoting movements of the thigh and leg. (a) Anterior view of the deep muscles of the pelvis and superficial muscles of the right thigh. **(b)** Adductor muscles of the medial compartment of the thigh. **(c)** The vastus muscles (isolated) of the quadriceps group.

Table 13.7   Muscles Acting on the Human Thigh and Leg, Anterior and Medial Aspects *(continued)*

Muscle	Comments	Origin	Insertion	Action
Anterior Compartment				
Quadriceps femoris*				
Rectus femoris	Superficial muscle of thigh; runs straight down thigh; only muscle of group to cross hip joint	Anterior inferior iliac spine and superior margin of acetabulum	Tibial tuberosity and patella	Extends leg and flexes thigh
Vastus lateralis	Forms lateral aspect of thigh; intramuscular injection site	Greater trochanter, intertrochanteric line, and linea aspera	Tibial tuberosity and patella	Extends leg and stabilizes knee
Vastus medialis	Forms inferomedial aspect of thigh	Linea aspera and intertrochanteric line	Tibial tuberosity and patella	Extends leg; stabilizes patella
Vastus intermedius	Obscured by rectus femoris; lies between vastus lateralis and vastus medialis on anterior thigh	Anterior and lateral surface of femur	Tibial tuberosity and patella	Extends leg
Tensor fasciae latae	Enclosed between fascia layers of thigh	Anterior aspect of iliac crest and anterior superior iliac spine	Iliotibial tract (lateral portion of fascia lata)	Flexes, abducts, and medially rotates thigh; steadies trunk

*The quadriceps form the flesh of the anterior thigh and have a common insertion in the tibial tuberosity via the patellar ligament. They are powerful leg extensors, enabling humans to kick a football, for example.

Table 13.8   Muscles Acting on the Human Thigh and Leg, Posterior Aspect (Figure 13.13)

Muscle	Comments	Origin	Insertion	Action
Origin on the Pelvis				
Gluteus maximus	Largest and most superficial of gluteal muscles (which form buttock mass); intramuscular injection site	Dorsal ilium, sacrum, and coccyx	Gluteal tuberosity of femur and iliotibial tract*	Complex, powerful thigh extensor (most effective when thigh is flexed, as in climbing stairs—but not as in walking); antagonist of iliopsoas; laterally rotates and abducts thigh
Gluteus medius	Partially covered by gluteus maximus; intramuscular injection site	Upper lateral surface of ilium	Greater trochanter of femur	Abducts and medially rotates thigh; steadies pelvis during walking
Gluteus minimus (not shown in figure)	Smallest and deepest gluteal muscle	External inferior surface of ilium	Greater trochanter of femur	Abducts and medially rotates thigh; steadies pelvis
Posterior Compartment				
Hamstrings†				
Biceps femoris	Most lateral muscle of group; arises from two heads	Ischial tuberosity (long head); linea aspera and distal femur (short head)	Tendon passes laterally to insert into head of fibula and lateral condyle of tibia	Extends thigh; laterally rotates leg; flexes leg

(Table continues on page 223.)

Table 13.8	(continued)			
Muscle	**Comments**	**Origin**	**Insertion**	**Action**
Semitendinosus	Medial to biceps femoris	Ischial tuberosity	Medial aspect of upper tibial shaft	Extends thigh; flexes leg; medially rotates leg
Semimembranosus	Deep to semitendinosus	Ischial tuberosity	Medial condyle of tibia; lateral condyle of femur	Extends thigh; flexes leg; medially rotates leg

*The iliotibial tract, a thickened lateral portion of the fascia lata, ensheathes all the muscles of the thigh. It extends as a tendinous band from the iliac crest to the knee.

†The hamstrings are the fleshy muscles of the posterior thigh. As a group, they are strong extensors of the thigh; they counteract the powerful quadriceps by stabilizing the knee joint when standing.

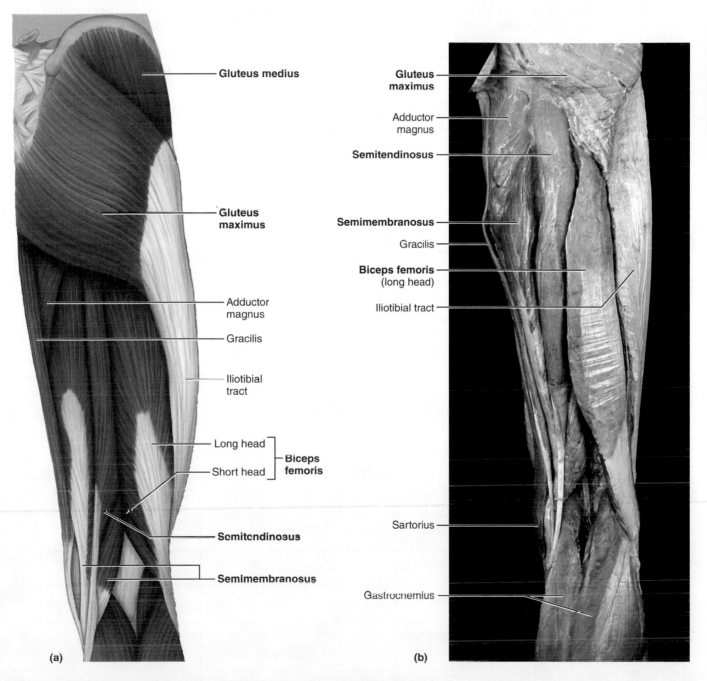

Figure 13.13 Muscles of the posterior aspect of the right hip and thigh.
(a) Superficial view showing the gluteus muscles of the buttock and hamstring muscles of the thigh. (b) Photo of muscles of the posterior thigh.

Table 13.9 Muscles Acting on the Human Foot and Ankle (Figures 13.14 and 13.15)

Muscle	Comments	Origin	Insertion	Action
Lateral Compartment (Figure 13.14a, b and Figure 13.15b)				
Fibularis (peroneus) longus	Superficial lateral muscle; overlies fibula	Head and upper portion of fibula	By long tendon under foot to metatarsal I and medial cuneiform	Plantar flexes and everts foot
Fibularis (peroneus) brevis	Smaller muscle; deep to fibularis longus	Distal portion of fibula shaft	By tendon running behind lateral malleolus to insert on proximal end of metatarsal V	Plantar flexes and everts foot
Anterior Compartment (Figure 13.14a, b)				
Tibialis anterior	Superficial muscle of anterior leg; parallels sharp anterior margin of tibia	Lateral condyle and upper $^2/_3$ of tibia; interosseous membrane	By tendon into inferior surface of first cuneiform and metatarsal I	Prime mover of dorsiflexion; inverts foot; supports longitudinal arch of foot
Extensor digitorum longus	Anterolateral surface of leg; lateral to tibialis anterior	Lateral condyle of tibia; proximal ¾ of fibula; interosseous membrane	Tendon divides into four parts; inserts into middle and distal phalanges of toes 2–5	Prime mover of toe extension; dorsiflexes foot
Fibularis (peroneus) tertius	Small muscle; often fused to distal part of extensor digitorum longus	Distal anterior surface of fibula and interosseous membrane	Tendon inserts on dorsum of metatarsal V	Dorsiflexes and everts foot
Extensor hallucis longus	Deep to extensor digitorum longus and tibialis anterior	Anteromedial shaft of fibula and interosseous membrane	Tendon inserts on distal phalanx of great toe	Extends great toe; dorsiflexes foot

(Table continues on page 226.)

Group Challenge

Name That Muscle

Work in groups of three or four to fill out the Group Challenge chart for muscle IDs. Refrain from looking back at the tables. Use the "brain power" of your group and the appropriate muscle models. All members of the group discuss where the origins and insertions in the table are located. One student in the group acts as the "muscle model." The "model" points to the appropriate attachment locations on his or her body and performs the primary action as advised by the group.

To assist in this task, recall that when a muscle contracts, the muscle's insertion moves toward the muscle's origin. Also, in the muscles of the limbs, the origin typically lies proximal to the insertion. Sometimes the origin and insertion are even part of the muscle's name!

Group Challenge: Muscle IDs

Origin	Insertion	Muscle	Primary action
Zygomatic arch and maxilla	Angle and ramus of the mandible		
Anterior surface of ribs 3–5	Coracoid process of the scapula		
Inferior border of rib above	Superior border of rib below		
Distal portion of anterior humerus	Coronoid process of the ulna		
Anterior inferior iliac spine and superior margin of acetabulum	Tibial tuberosity and patella		
By two heads from medial and lateral condyles of femur	Calcaneus via calcaneal tendon		

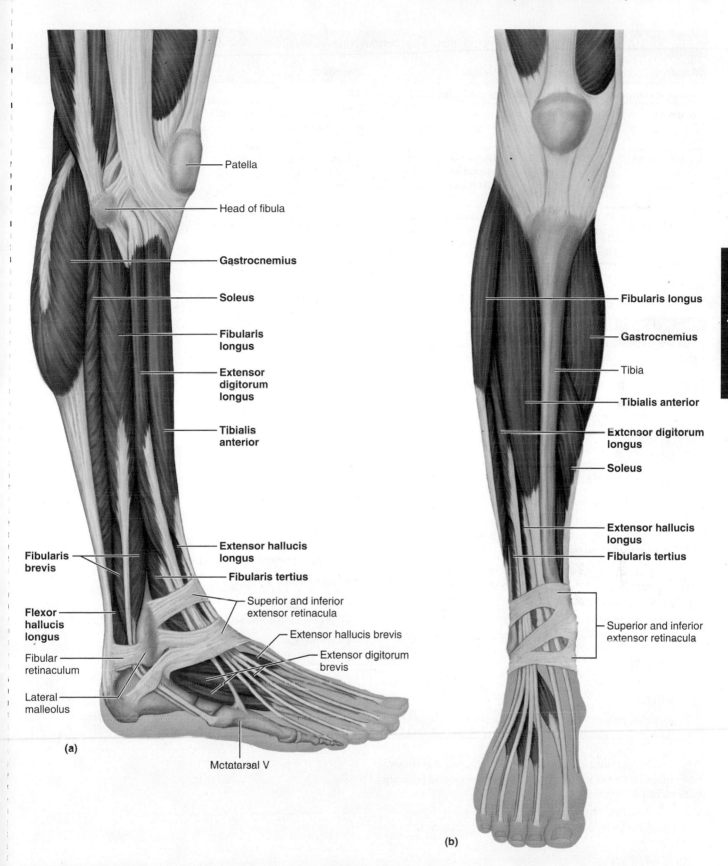

- Patella
- Head of fibula
- **Gastrocnemius**
- **Soleus**
- **Fibularis longus**
- **Extensor digitorum longus**
- **Tibialis anterior**
- **Fibularis longus**
- **Gastrocnemius**
- Tibia
- **Tibialis anterior**
- **Extensor digitorum longus**
- **Soleus**
- **Extensor hallucis longus**
- **Fibularis brevis**
- **Fibularis tertius**
- **Extensor hallucis longus**
- **Fibularis tertius**
- Superior and inferior extensor retinacula
- **Flexor hallucis longus**
- Extensor hallucis brevis
- Fibular retinaculum
- Extensor digitorum brevis
- Superior and inferior extensor retinacula
- Lateral malleolus

(a)

Metatarsal V

(b)

Figure 13.14 Muscles of the anterolateral aspect of the right leg. (a) Superficial view of lateral aspect of the leg, illustrating the positioning of the lateral compartment muscles (fibularis longus and brevis) relative to anterior and posterior leg muscles. **(b)** Superficial view of anterior leg muscles.

Table 13.9	Muscles Acting on the Human Foot and Ankle *(continued)*			
Muscle	**Comments**	**Origin**	**Insertion**	**Action**
Posterior Compartment, Superficial (Figure 13.15a; also Figure 13.14)				
Triceps surae	Refers to muscle pair below that shapes posterior calf		Via common tendon (calcaneal) into calcaneus of the heel	Plantar flex foot
Gastrocnemius	Superficial muscle of pair; two prominent bellies	By two heads from medial and lateral condyles of femur	Calcaneus via calcaneal tendon	Plantar flexes foot when leg is extended; crosses knee joint; thus can flex leg (when foot is dorsiflexed)
Soleus	Deep to gastrocnemius	Proximal portion of tiba and fibula; interosseous membrane	Calcaneus via calcaneal tendon	Plantar flexion; an important muscle for locomotion

(Table continues on page 228.)

Activity 5

Review of Human Musculature

Review the muscles by watching the *Human Musculature* video.

Activity 6

Making a Muscle Painting

1. Choose a male student to be "muscle painted."

2. Obtain brushes and water-based paints from the supply area while the "volunteer" removes his shirt and rolls up his pant legs (if necessary).

3. Using different colored paints, identify the muscles listed below by painting his skin. If a muscle covers a large body area, you may opt to paint only its borders.

- biceps brachii
- deltoid
- erector spinae
- pectoralis major
- rectus femoris
- tibialis anterior
- triceps brachii
- vastus lateralis
- biceps femoris
- extensor carpi radialis longus
- latissimus dorsi
- rectus abdominis
- sternocleidomastoid
- trapezius
- triceps surae
- vastus medialis

4. Check your "human painting" with your instructor before cleaning your bench and leaving the laboratory.

For instructions on animal dissections, see the dissection exercises (starting on p. 705) in the cat and fetal pig editions of this manual.

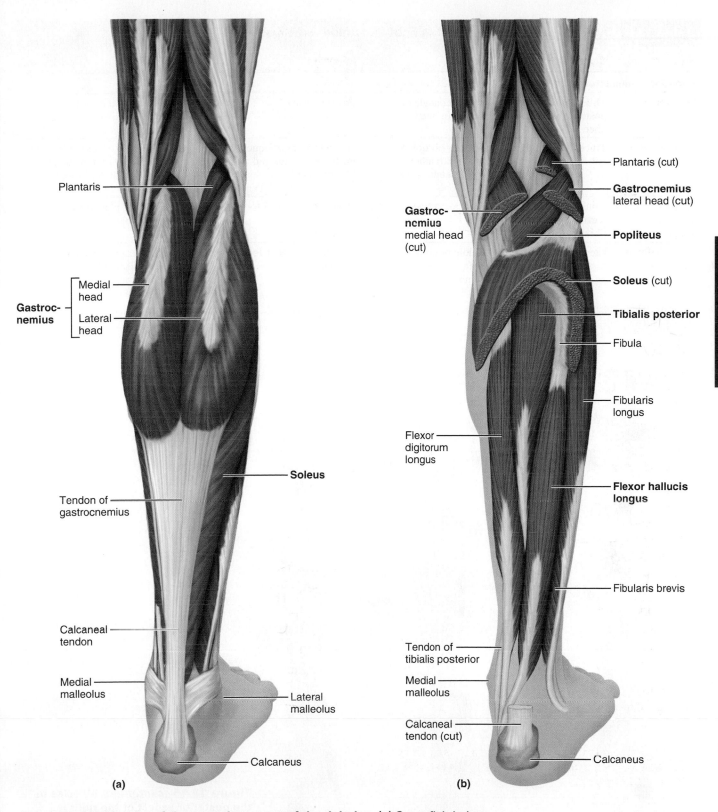

Plantaris

Gastroc-
nemius

Medial
head

Lateral
head

Soleus

Tendon of
gastrocnemius

Calcaneal
tendon

Medial
malleolus

Lateral
malleolus

Calcaneus

(a)

Plantaris (cut)

Gastrocnemius
lateral head (cut)

Gastroc-
nemius
medial head
(cut)

Popliteus

Soleus (cut)

Tibialis posterior

Fibula

Fibularis
longus

Flexor
digitorum
longus

**Flexor hallucis
longus**

Fibularis brevis

Tendon of
tibialis posterior

Medial
malleolus

Calcaneal
tendon (cut)

Calcaneus

(b)

13

Figure 13.15 Muscles of the posterior aspect of the right leg. (a) Superficial view
of the posterior leg. **(b)** The triceps surae has been removed to show the deep
muscles of the posterior compartment.

Table 13.9	Muscles Acting on the Human Foot and Ankle *(continued)*			
Muscle	**Comments**	**Origin**	**Insertion**	**Action**
Posterior Compartment, Deep (Figure 13.15b–e)				
Popliteus	Thin muscle at posterior aspect of knee	Lateral condyle of femur and lateral meniscus	Proximal tibia	Flexes and rotates leg medially to "unlock" knee when leg flexion begins
Tibialis posterior	Thick muscle deep to soleus	Superior portion of tibia and fibula and interosseous membrane	Tendon passes obliquely behind medial malleolus and under arch of foot; inserts into several tarsals and metatarsals II–IV	Prime mover of foot inversion; plantar flexes foot; stabilizes longitudinal arch of foot
Flexor digitorum longus	Runs medial to and partially overlies tibialis posterior	Posterior surface of tibia	Distal phalanges of toes 2–5	Flexes toes; plantar flexes and inverts foot
Flexor hallucis longus (see also Figure 13.14a)	Lies lateral to inferior aspect of tibialis posterior	Middle portion of fibula shaft; interosseous membrane	Tendon runs under foot to distal phalanx of great toe	Flexes great toe (*hallux* = great toe); plantar flexes and inverts foot; the "push-off muscle" during walking

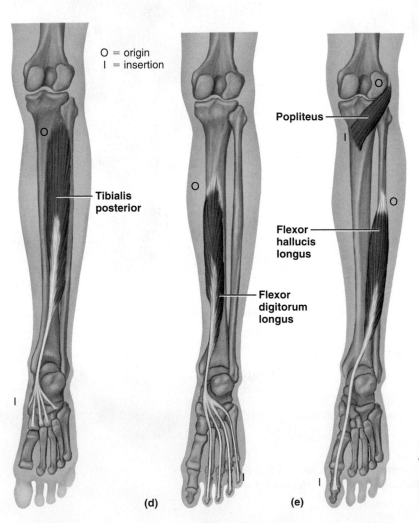

O = origin
I = insertion

Popliteus

Tibialis posterior

Flexor hallucis longus

Flexor digitorum longus

(c)   (d)   (e)

Figure 13.15 *(continued)* **Muscles of the posterior aspect of the right leg. (c–e)** Individual muscles are shown in isolation so that their origins and insertions may be observed.

13

Skeletal Muscle Physiology: Frogs and Human Subjects

Objectives

☐ Define the terms *resting membrane potential*, *depolarization*, *repolarization*, *action potential*, and *absolute* and *relative refractory periods*, and explain the physiological basis of each.

☐ Observe muscle contraction microscopically, and describe the role of ATP and various ions in muscle contraction.

☐ Define *muscle twitch*, and describe its three phases.

☐ Differentiate between subthreshold stimulus, threshold stimulus, and maximal stimulus.

☐ Define *tetanus* in skeletal muscle, and explain how it comes about.

☐ Define *muscle fatigue*, and describe the effects of load on muscle fatigue.

☐ Define *motor unit*, and relate recruitment of motor units and temporal summation to production of a graded contraction.

☐ Demonstrate how a physiograph or a computer with a data acquisition unit can be used to record skeletal muscle activity.

☐ Explain the significance of recordings obtained during experimentation.

Materials

- ATP muscle kits (glycerinated rabbit psoas muscle;* ATP and salt solutions obtainable from Carolina Biological Supply)
- Petri dishes
- Microscope slides
- Coverslips
- Millimeter ruler
- Compound microscope
- Stereomicroscope
- Pointed glass probes (teasing needles)
- Small beakers (50 ml)
- Distilled water

Text continues on next page. →

Pre-Lab Quiz

1. Circle the correct underlined term. The potential difference, or voltage, across the plasma membrane is the result of the difference in membrane permeability to <u>anions</u> / <u>cations</u>, most importantly Na^+ and K^+.

2. The _____ wave follows the depolarization wave across the sarcolemma.
   a. hyperpolarization
   b. refraction
   c. repolarization

3. Circle True or False. The voltage at which the first noticeable contractile response is achieved is called the threshold stimulus.

4. Circle True or False. A single muscle is made up of many motor units, and the gradual activation of these motor units results in a graded contraction of the whole muscle.

5. A sustained, smooth, muscle contraction that is a result of high-frequency stimulation is:
   a. tetanus   b. tonus   c. twitch

MasteringA&P®

For related exercise study tools, go to the Study Area of **MasteringA&P**. There you will find:

- Practice Anatomy Lab **PAL**
- A&PFlix **A&PFlix**
- PhysioEx **PEx**
- Practice quizzes, Histology Atlas, eText, Videos, and more!

The contraction of skeletal and cardiac muscle fibers can be considered in terms of three events—electrical excitation of the muscle fiber, excitation contraction coupling, and shortening of the muscle fiber due to sliding of the myofilaments within it.

At rest, all cells maintain a potential difference, or voltage, across their plasma membrane; the inner face of the membrane is approximately −60 to −90 millivolts (mV) compared with the cell exterior. This potential difference

- Glass-marking pencil
- Textbooks or other heavy books
- Watch or timer
- Ringer's solution (frog)
- Scissors
- Metal needle probes
- Medicine dropper
- Cotton thread
- Forceps
- Disposable gloves
- Glass or porcelain plate
- Pithed bullfrog[†]
- Physiograph or BIOPAC® equipment:[‡] Physiograph apparatus, physiograph paper and ink, force transducer, pin electrodes, stimulator, stimulator output extension cable, transducer stand and cable, straight pins, frog board

BIOPAC® BIOPAC® BSL System with BSL software version 3.7.5 to 3.7.7 (for Windows 7/Vista/XP or Mac OS X 10.4-10.6), data acquisition unit MP36/35 or MP45, PC or Mac computer, Biopac Student Lab electrode lead set, hand dynamometer, headphones, metric tape measure, disposable vinyl electrodes, and conduction gel.

Instructors using the MP36/35/30 data acquisition unit with BSL software versions earlier than 3.7.5 (for Windows or Mac) will need slightly different channel settings and collection strategies. Instructions for using the older data acquisition unit can be found on MasteringA&P.

Notes to the Instructor: *At the beginning of the lab, the muscle bundle should be removed from the test tube and cut into approximately 2-cm lengths. Both the cut muscle segments and the entubed glycerol should be put into a petri dish. One muscle segment is sufficient for every two to four students making observations.*

[†]Bullfrogs to be pithed by lab instructor as needed for student experimentation. (If instructor prefers that students pith their own specimens, an instructional sheet on that procedure suitable for copying for student handouts is provided in the Instructor's Guide.)

[‡]Additionally, instructions for Activity 3 using a kymograph can be found in the Instructor Guide. Instructions for using PowerLab® equipment can be found on MasteringA&P.

PEx PhysioEx™ 9.1 Computer Simulation Ex.2 on p. PEx-17.

is a result of differences in membrane permeability to cations, most importantly sodium (Na^+) and potassium (K^+) ions. Intracellular potassium concentration is much greater than its extracellular concentration, and intracellular sodium concentration is considerably less than its extracellular concentration. Hence, steep concentration gradients across the membrane exist for both cations. Because the plasma membrane is more permeable to K^+ than to Na^+, the cell's **resting membrane potential** is more negative inside than outside. The resting membrane potential is of particular interest in excitable cells, like muscle fibers and neurons, because changes in that voltage underlie their ability to do work (to contract and/or issue electrical signals).

Action Potential

When a muscle fiber is stimulated, the sarcolemma becomes temporarily more permeable to Na^+, which enters the cell. This sudden influx of Na^+ alters the membrane potential. That is, the cell interior becomes less negatively charged at that point, an event called **depolarization.** When depolarization reaches a certain level and the sarcolemma momentarily changes its polarity, a depolarization wave travels along the sarcolemma. Even as the influx of Na^+ occurs, the sarcolemma becomes less permeable to Na^+ and more permeable to K^+. Consequently, K^+ ions move out of the cell, restoring the resting membrane potential (but not the original ionic conditions), an event called **repolarization.** The repolarization wave follows the depolarization wave across the sarcolemma. This rapid depolarization and repolarization of the membrane that is propagated along the entire membrane from the point of stimulation is called the **action potential.**

The **absolute refractory period** is the period of time required for the Na^+ channel gates to reset to their resting positions. During this period there is no possibility of generating another action potential. As Na^+ permeability is gradually restored to resting levels during repolarization, an especially strong stimulus to the muscle fiber may provoke another action potential. This period of time is the **relative refractory period.** Repolarization restores the muscle fiber's normal excitability. The sodium-potassium pump, which actively transports K^+ into the cell and Na^+ out of the cell, must be "revved up" (become more active) to reestablish the ionic concentrations of the resting state.

Contraction

Propagation of the action potential down the T tubules of the sarcolemma causes the release of calcium ions (Ca^{2+}) from storage in the sarcoplasmic reticulum within the muscle fiber. When the calcium ions bind to the regulatory protein troponin on the actin myofilaments, they act as an ionic trigger that initiates contraction, and the actin and myosin filaments slide past each other. Once the action potential ends, the calcium ions are almost immediately transported back into the sarcoplasmic reticulum. Instantly the muscle fiber relaxes.

Activity 1

Prepare for lab: Watch the Pre-Lab Video
MasteringA&P®>Study Area>Pre-Lab Videos

Observing Muscle Fiber Contraction

In this simple experiment, you will have the opportunity to review your understanding of muscle fiber anatomy and to watch fibers respond to the presence of ATP and/or a solution of K^+ and magnesium ions (Mg^{2+}).

This experiment uses preparations of glycerinated muscle. The glycerination process denatures troponin and tropomyosin. Consequently, calcium, so critical for contraction in vivo, is not necessary here. The role of magnesium and potassium salts as cofactors in the contraction process is not well understood, but magnesium and potassium salts seem to be required for ATPase activity in this system.

1. Talk with other members of your lab group to develop a hypothesis about requirements for muscle fiber contraction for this experiment. The hypothesis should have three parts: (1) salts and ATP, (2) ATP only, and (3) salts only.

2. Obtain the following materials from the supply area: two glass teasing needles or forceps; six glass microscope slides and six coverslips; millimeter ruler; dropper bottles containing the following solutions: (a) 0.25% ATP plus 0.05 M KCl plus 0.001 M MgCl$_2$ in distilled water, (b) 0.25% ATP in triply distilled water, and (c) 0.05 M KCl plus 0.001 M MgCl$_2$ in distilled water; a petri dish; a beaker of distilled water; a glass-marking pencil; and a small portion of a previously cut muscle bundle segment. While you are at the supply area, place the muscle fibers in the petri dish, and pour a small amount of glycerol (the fluid in the supply petri dish) over your muscle fibers. Also obtain both a compound and a stereomicroscope and bring them to your laboratory bench.

3. Using clean fine glass needles or forceps, tease the muscle segment to separate its fibers. The objective is to isolate *single* muscle fibers for observation. Be patient and work carefully so that the fibers do not get torn during this isolation procedure.

4. Transfer one or more of the fibers (or the thinnest strands you have obtained) onto a clean microscope slide with a glass needle, and cover with a coverslip. Examine the fiber under the compound microscope at low- and then high-power magnifications to observe the striations and the smoothness of the fibers when they are in the relaxed state.

5. Clean three microscope slides well and rinse in distilled water. Label the slides A, B, and C.

6. Transfer three or four fibers to microscope slide A with a glass needle. Using the needle as a prod, carefully position the fibers so that they are parallel to one another and as straight as possible. Place this slide under a *stereomicroscope,* and measure the length of each fiber by holding a millimeter ruler adjacent to it. Alternatively, you can rest the microscope slide *on* the millimeter ruler to make your length determinations. Record the data in the **Activity 1 chart.**

7. Flood the fibers (situated under the stereomicroscope) with several drops of the solution containing ATP, K$^+$, and Mg^{2+}. Watch the reaction of the fibers after adding the solution. After 30 seconds (or slightly longer), remeasure each fiber and record the observed ending lengths on the chart. Calculate the percentage of contraction by using the simple formula below, and record this data on the chart.

$$\begin{array}{c}\text{Initial} \\ \text{length (mm)}\end{array} - \begin{array}{c}\text{ending} \\ \text{length (mm)}\end{array} = \begin{array}{c}\text{net} \\ \text{change (mm)}\end{array}$$

then:

$$\frac{\text{net change (mm)}}{\text{initial length (mm)}} \times 100 = \underline{\hspace{1cm}}\% \text{ contraction}$$

8. Carefully transfer one of the fibers from step 7 to a clean, unmarked microscope slide, cover with a coverslip, and observe with the compound microscope. Mentally compare your initial observations with the view you are observing now. What differences do you see? (Be specific.)

_____

_____

_____

What zones (or bands) have disappeared?

_____

9. Repeat steps 6 through 8 twice more, using clean slides and fresh muscle fibers. On slide B, use the solution of ATP in distilled water (no salts). Then on slide C, use the solution containing only salts (no ATP) for the third series. Record data on the Activity 1 chart.

Activity 1: Observations of Muscle Fiber Contraction				
	Muscle fiber 1	**Muscle fiber 2**	**Muscle fiber 3**	**Average**
Salts and ATP, slide A				
Initial length (mm)				
Ending length (mm)				
% Contraction				
ATP only, slide B				
Initial length (mm)				
Ending length (mm)				
% Contraction				
Salts only, slide C				
Initial length (mm)				
Ending length (mm)				
% Contraction				

Text continues on next page. →

10. Calculate the averages for each slide. Collect the average data from all the groups in your laboratory and use these data to prepare a lab report. (See Getting Started, on MasteringA&P.) Include in your discussion the following questions:

What % contraction was observed when ATP was applied in the absence of K^+ and Mg^{2+}?

_____

What % contraction was observed when the muscle fibers were flooded with a solution containing K^+ and Mg^{2+}, and lacking ATP?

_____

What conclusions can you draw about the importance of ATP, K^+, and Mg^{2+} to the contractile process?

_____

_____

Can you draw exactly the same conclusions from the data provided by each group? List some variables that might have been introduced into the procedure and that might account for any differences.

_____

_____

Activity 2

Inducing Contraction in the Frog Gastrocnemius Muscle

Physiologists have learned a great deal about the way muscles function by isolating muscles from laboratory animals and then stimulating these muscles to observe their responses. Various stimuli—electrical shock, temperature changes, extremes of pH, certain chemicals—elicit muscle activity, but laboratory experiments of this type typically use electrical shock. This is because it is easier to control the onset and cessation of electrical shock, as well as the strength of the stimulus.

Various types of apparatus are used to record muscle contraction. All include a way to mark time intervals, a way to indicate exactly when the stimulus was applied, and a way to measure the magnitude of the contractile response. Instructions are provided here for setting up a physiograph apparatus (**Figure 14.1**). Specific instructions for use of recording apparatus during recording will be provided by your instructor.

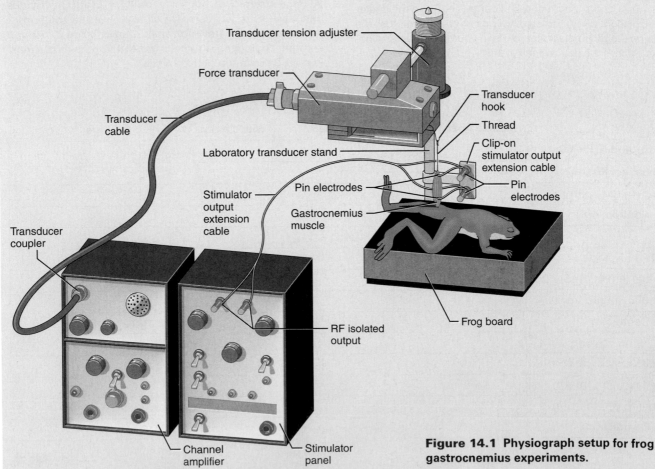

Figure 14.1 Physiograph setup for frog gastrocnemius experiments.

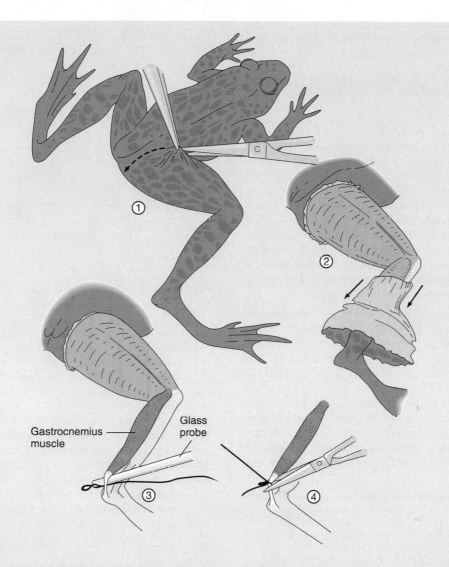

14

Figure 14.2 Preparation of the frog gastrocnemius muscle. Numbers indicate the sequence of steps for preparing the muscle.

Preparing a Muscle for Experimentation

The preparatory work that precedes the recording of muscle activity tends to be quite time-consuming. If you work in teams of two or three, the work can be divided. While one of you is setting up the recording apparatus, one or two students can dissect the frog leg (**Figure 14.2**). Experimentation should begin as soon as the dissection is completed.

Materials

Channel amplifier and transducer cable

Stimulator panel and stimulator output extension cable

Force transducer

Transducer tension adjuster

Transducer stand

Two pin electrodes

Frog board and straight pins

Prepared frog (gastrocnemius muscle freed and calcaneal tendon ligated)

Frog Ringer's solution

Procedure

1. Connect transducer to transducer stand and attach frog board to stand.

2. Attach transducer cable to transducer and to input connection on channel amplifier.

3. Attach stimulator output extension cable to output on stimulator panel (red to red, black to black).

4. Using clip at opposite end of extension cable, attach cable to bottom of transducer stand adjacent to frog board.

5. Attach two pin electrodes securely to electrodes on clip.

6. Place knee of prepared frog in clip-on frog board and secure by inserting a straight pin through tissues of frog. Keep frog muscle moistened with Ringer's solution.

7. Attach thread from the calcaneal tendon of frog to transducer spring hook.

8. Adjust position of transducer on stand to produce a constant tension on thread attached to tendon (taut but not tight). Gastrocnemius muscle should hang vertically directly below hook.

Text continues on next page. →

9. Insert free ends of pin electrodes into the muscle, one at proximal end and the other at distal end.

DISSECTION

Frog Hind Limb

1. Before beginning the frog dissection, have the following supplies ready at your laboratory bench: a small beaker containing 20 to 30 ml of frog Ringer's solution, scissors, a metal needle probe, a glass probe with a pointed tip, a medicine dropper, cotton thread, forceps, a glass or porcelain plate, and disposable gloves. While these supplies are being gathered, one member of your team should notify the instructor that you are ready to begin experimentation, so that a frog can be prepared (pithed). Preparation of a frog in this manner renders it unable to feel pain and prevents reflex movements (like hopping) that would interfere with the experiments.

⚠ **2.** All students who will be handling the frog should don disposable gloves. Obtain a pithed frog and place it ventral surface down on the glass plate. Make an incision into the skin approximately midthigh (Figure 14.2), and then continue the cut completely around the thigh. Grasp the skin with the forceps, and strip it from the leg and hindfoot. The skin adheres more at the joints, but a careful, persistent pulling motion—somewhat like pulling off a stocking—will enable you to remove it in one piece. _From this point on, keep the exposed muscle tissue moistened with the Ringer's solution_ to prevent spontaneous twitches.

3. Identify the gastrocnemius muscle (the fleshy muscle of the posterior calf) and the calcaneal tendon that secures it to the heel.

4. Slip a glass probe under the gastrocnemius muscle, and run it along the entire length and under the calcaneal tendon to free them from the underlying tissues.

5. Cut a piece of thread about 10 inches long, and use the glass probe to slide the thread under the calcaneal tendon. Knot the thread firmly around the tendon and then sever the tendon distal to the thread. Alternatively, you can bend a common pin into a Z shape and insert the pin securely into the tendon. The thread is then attached to the opposite end of the pin. Once the tendon has been tied or pinned, the frog is ready for experimentation (see Figure 14.2).

Recording Muscle Activity

Skeletal muscles consist of thousands of muscle fibers and react to stimuli with graded responses. Thus muscle contractions can be weak or vigorous, depending on the requirements of the task. Graded responses (different degrees of shortening) of a skeletal muscle depend on the number of muscle fibers being stimulated. In the intact organism, the number of motor units firing at any one time determines how many muscle fibers will be stimulated. In this experiment, the frequency and strength of an electrical current determines the response.

A single contraction of skeletal muscle is called a **muscle twitch**. A tracing of a muscle twitch (**Figure 14.3**) shows

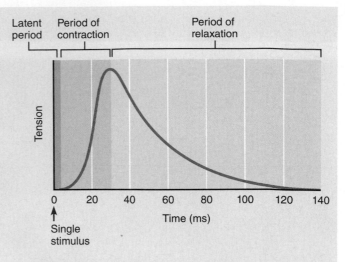

Figure 14.3 Tracing of a muscle twitch.

three distinct phases: latent, contraction, and relaxation. The **latent period** is the interval from stimulus application until the muscle begins to shorten. Although no activity is indicated on the tracing during this phase, excitation-contraction coupling is occurring within the muscle. During the **period of contraction,** the muscle fibers shorten; the tracing shows an increasingly higher needle deflection and the tracing peaks. During the **period of relaxation,** represented by a downward curve of the tracing, the muscle fibers relax and lengthen. On a slowly moving recording surface, the single muscle twitch appears as a spike rather than a bell-shaped curve, as in Figure 14.3, but on a rapidly moving recording surface, the three distinct phases just described become recognizable.

Determining the Threshold Stimulus

1. Assuming that you have already set up the recording apparatus, set the time marker to deliver one pulse per second and set the paper speed at a slow rate, approximately 0.1 cm per second.

2. Set the duration control on the stimulator between 7 and 15 milliseconds (msec), multiplier ×1. Set the voltage control at 0 V, multiplier ×1. Turn the sensitivity control knob of the stimulator fully clockwise (lowest value, greatest sensitivity).

3. Administer single stimuli to the muscle at 1-minute intervals, beginning with 0.1 V and increasing each successive stimulus by 0.1 V until a contraction is obtained (shown by a spike on the paper).

At what voltage did contraction occur? _____ V

The voltage at which the first perceptible contractile response is obtained is called the **threshold stimulus.** All stimuli applied prior to this point are termed **subthreshold stimuli,** because at those voltages no response was elicited.

4. Stop the recording and mark the record to indicate the threshold stimulus, voltage, and time. _Do not remove the record from the recording surface;_ continue with the next experiment. _Remember: keep the muscle preparation moistened with Ringer's solution at all times._

Observing Graded Muscle Response to Increased Stimulus Intensity

1. Follow the previous setup instructions, but set the voltage control at the threshold voltage (as determined in the first experiment).

2. Deliver single stimuli at 1-minute intervals. Initially increase the voltage between shocks by 0.5 V; then increase the voltage by 1 to 2 V between shocks as the experiment continues, until contraction height increases no further. Stop the recording apparatus.

What voltage produced the highest spike (and thus the

maximal strength of contraction)? _____ V

This voltage, called the **maximal stimulus** (for *your* muscle specimen), is the weakest stimulus at which all muscle fibers are being stimulated.

3. Mark the record *maximal stimulus.* Record the maximal stimulus voltage and the time you completed the experiment.

4. What is happening to the muscle as the voltage is increased?

_____

_____

What is another name for this phenomenon? (Use an appropriate reference if necessary.)

_____

5. Explain why the strength of contraction does not increase once the maximal stimulus is reached.

_____

_____

Timing the Muscle Twitch

1. Follow the previous setup directions, but set the voltage for the maximal stimulus (as determined in the preceding experiment) and set the paper advance or recording speed at maximum. Record the paper speed setting:

_____mm/sec

2. Determine the time required for the paper to advance 1 mm by using the formula

$$\frac{1 \text{ mm}}{\text{mm/sec (paper speed)}}$$

(Thus, if your paper speed is 25 mm/sec, each mm on the chart equals 0.04 sec.) Record the computed value:

1 mm = _____ sec

3. Deliver single stimuli at 1-minute intervals to obtain several "twitch" curves. Stop the recording.

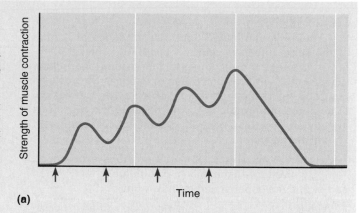

(a)

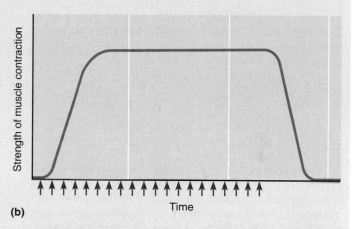

(b)

Figure 14.4 Muscle response to stimulation. Arrows represent stimuli. **(a)** Wave summation at low-frequency stimulation. **(b)** Fused tetanus occurs as stimulation rate is increased.

4. Determine the duration of the latent, contraction, and relaxation phases of the twitches, and record the data below.

Duration of latent period: _____ sec

Duration of period of contraction: _____ sec

Duration of period of relaxation: _____ sec

5. Label the record to indicate the point of stimulus, the beginning of contraction, the end of contraction, and the end of relaxation.

6. Allow the muscle to rest (but keep it moistened with frog Ringer's solution) before continuing with the next experiment.

Observing Graded Muscle Response to Increased Stimulus Frequency

Muscles subjected to frequent stimulation, without a chance to relax, exhibit two kinds of responses—wave summation and tetanus—depending on the stimulus frequency (**Figure 14.4**).

Text continues on next page. ➡

Wave Summation

If a muscle is stimulated with a rapid series of stimuli of the same intensity before it has had a chance to relax completely, the response to the second and subsequent stimuli will be greater than to the first stimulus (see Figure 14.4a). This phenomenon, called **wave summation,** or **temporal summation,** occurs because the muscle is already in a partially contracted state when subsequent stimuli are delivered.

1. Set up the apparatus as in the previous experiment, setting the voltage to the maximal stimulus as determined earlier and the paper speed to maximum.

2. With the stimulator in single mode, deliver two successive stimuli as rapidly as possible.

3. Shut off the recorder and label the record as *wave summation.* Note also the time, the voltage, and the frequency. What did you observe?

_____

_____

Tetanus

Stimulation of a muscle at an even higher frequency will produce a "fusion" (complete tetanization) of the summated twitches. In effect, a single sustained contraction is achieved in which no evidence of relaxation can be seen (see Figure 14.4b). **Fused tetanus,** or **complete tetanus,** demonstrates the maximum force generated by a skeletal muscle; the single muscle twitch is primarily a laboratory phenomenon.

1. To demonstrate fused tetanus, maintain the conditions used for wave summation except for the frequency of stimulation. Set the stimulator to deliver 60 stimuli per second.

2. As soon as you obtain a single smooth, sustained contraction (with no evidence of relaxation), discontinue stimulation and shut off the recorder.

3. Label the tracing with the conditions of experimentation, the time, and the area of *fused* or *complete tetanus.*

Inducing Muscle Fatigue

Muscle fatigue is a reversible physiological condition in which a muscle is unable to contract even though it is being stimulated. Fatigue can occur with short-duration maximal contraction or long-duration submaximal contraction. Although the phenomenon of muscle fatigue is not completely understood, several factors appear to contribute to it. Most affect excitation-contraction coupling. One theory involves the buildup of inorganic phosphate (P_i) from ATP and creatine phosphate breakdown, which may block calcium release from the sarcoplasmic reticulum (SR). Another theory suggests that potassium accumulation in the T tubules may block calcium release from the SR and alter the membrane potential of the muscle fiber. Lactic acid buildup, long implicated as a cause of fatigue, does not appear to play a role.

1. To demonstrate muscle fatigue, set up an experiment like the tetanus experiment but continue stimulation until the muscle completely relaxes and the contraction curve returns to the baseline.

2. Measure the time interval between the beginning of complete tetanus and the beginning of fatigue (when the tracing begins its downward curve). Mark the record appropriately.

3. Determine the time required for complete fatigue to occur (the time interval from the beginning of fatigue until the return of the curve to the baseline). Mark the record appropriately.

4. Allow the muscle to rest (keeping it moistened with Ringer's solution) for 10 minutes, and then repeat the experiment.

What was the effect of the rest period on the fatigued muscle?

_____

_____

What might be the physiological basis for this reaction?

_____

_____

Determining the Effect of Load on Skeletal Muscle

When the fibers of a skeletal muscle are slightly stretched by a weight or tension, the muscle responds by contracting more forcibly and thus is capable of doing more work. When the actin and myosin barely overlap, sliding can occur along nearly the entire length of the actin filaments. If the load is increased beyond the optimum, the latent period becomes longer, contractile force decreases, and relaxation (fatigue) occurs more quickly. With excessive stretching, the muscle is unable to develop any active tension and no contraction occurs. Since the filaments no longer overlap at all with this degree of stretching, the sliding force cannot be generated.

If your equipment allows you to add more weights to the muscle specimen or to increase the tension on the muscle, perform the following experiment to determine the effect of loading on skeletal muscle and to develop a work curve for the frog's gastrocnemius muscle.

1. Set the stimulator to deliver the maximal voltage as previously determined.

2. Stimulate the unweighted muscle with single shocks at 1– to 2– second intervals to achieve three or four muscle twitches.

3. Stop the recording apparatus and add 10 g of weight or tension to the muscle. Restart and advance the recording about 1 cm, and then stimulate again to obtain three or four spikes.

4. Repeat the previous step seven more times, increasing the weight by 10 g each time until the total load on the muscle is 80 g or the muscle fails to respond. If the calcaneal tendon tears, the weight will drop, which ends the trial. In such cases, you will need to prepare another frog's leg to continue the experiments and the maximal

stimulus will have to be determined for the new muscle preparation.

5. When these "loading" experiments are completed, discontinue recording. Mark the curves on the record to indicate the load (in grams).

6. Measure the height of contraction (in millimeters) for each sequence of twitches obtained with each load, and insert this information in the Activity 2 chart.

7. Compute the work done by the muscle for each twitch (load) sequence.

 Weight of load (g) × distance load lifted (mm) = work done

Enter these calculations into the chart in the column labeled Work done, Trial 1.

8. Allow the muscle to rest for 5 minutes. Then conduct a second trial in the same manner (i.e., repeat steps 2 through 7). Record this second set of measurements and calculations in the columns labeled Trial 2. Be sure to keep the muscle well moistened with Ringer's solution during the resting interval.

9. Using two different colors, plot a line graph of work done against the weight on the grid accompanying the chart for each trial. Label each plot appropriately.

10. Dismantle all apparatus and prepare the equipment for storage. Dispose of the frog remains in the appropriate container. Discard the gloves as instructed and wash and dry your hands.

11. Inspect your records of the experiments, and make sure each is fully labeled with the experimental conditions, the date, and the names of those who conducted the experiments. For future reference, attach a tracing (or a copy of the tracing) for each experiment to this page.

Activity 2: Results for Effect of Load on Skeletal Muscle

Load (g)	Distance load lifted (mm)		Work done	
	Trial 1	Trial 2	Trial 1	Trial 2
0				
10				
20				
30				
40				
50				
60				
70				
80				

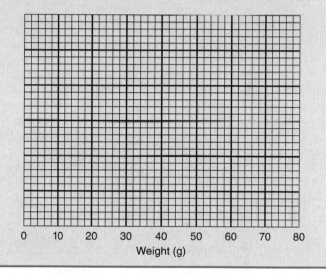

Weight (g)

14

Activity 3

Demonstrating Muscle Fatigue in Humans

1. Work in small groups. In each group select a subject, a timer, and a recorder.

2. Obtain a copy of the laboratory manual and a copy of the textbook. Weigh each book separately, and then record the weight of each in the Activity 3 chart in step 6.

3. The subject is to extend an upper limb straight out in front of him or her, holding the position until the arm shakes or the muscles begin to ache. Record the time to fatigue on the chart.

4. Allow the subject to rest for several minutes. Now ask the subject to hold the laboratory manual while keeping the arm and forearm in the same position as in step 3 above. Record the time to fatigue on the chart.

5. Allow the subject to rest again for several minutes. Now ask the subject to hold the textbook while keeping the upper limb in the same position as in steps 3 and 4 above. Record the time to fatigue on the chart.

6. Each person in the group should take a turn as the subject, and all data should be recorded in the chart.

Activity 3: Results for Human Muscle Fatigue

Load	Weight of object	Time elapsed until fatigue		
		Subject 1	Subject 2	Subject 3
Appendage	N/A			
Lab Manual				
Textbook				

7. What can you conclude about the effect of load on muscle fatigue? Explain.

_____

_____

_____

Activity 4

Electromyography in a Human Subject Using BIOPAC®

Part 1: Temporal and Multiple Motor Unit Summation

This activity is an introduction to a procedure known as **electromyography,** the recording of skin-surface voltage that indicates underlying skeletal muscle contraction. The actual visible recording of the resulting voltage waveforms is called an **electromyogram (EMG).**

A single skeletal muscle consists of numerous elongated *skeletal muscle fibers* (Exercise 12). These muscle fibers are excited by *motor neurons* of the central nervous system whose *axons* terminate at the muscle. An axon of a motor neuron branches profusely at the muscle. Each branch produces multiple **axon terminals,** each of which innervates a single muscle fiber. The number of muscle fibers controlled by a single motor neuron can vary greatly, from five (for fine control needed in the hand) to 500 (for gross control, such as in the buttocks). The most important organizational concept in the physiology of muscle contraction is the **motor unit,** a single motor neuron and all of the fibers within a muscle that it activates (see Figure 12.6 p. 194). Understanding gross muscular contraction depends on realizing that a single muscle consists of multiple motor units, and that the gradual and coordinated activation of these motor units results in **graded contraction** of the whole muscle.

The nervous system controls muscle contraction by two mechanisms:

- **Recruitment (multiple motor unit summation):** The gradual activation of more and more motor units

- **Wave summation:** An increase in the *frequency* of stimulation for each active motor unit

Thus, increasing the force of contraction of a muscle arises from gradually increasing the number of motor units being activated and increasing the frequency of nerve impulses delivered by those active motor units.

Setting Up the Equipment

1. Connect the BIOPAC® unit to the computer, then turn the computer **ON.**

2. Make sure the BIOPAC® unit is **OFF.**

3. Plug in the equipment (as shown in **Figure 14.5**).

- Electrode lead set—CH 1

- Headphones—back of MP36/35 unit or top of MP45 unit

4. Turn the BIOPAC® unit **ON.**

5. Attach three electrodes to the subject's dominant forearm as shown in **Figure 14.6** and attach the electrode leads according to the colors indicated.

6. Start the Biopac Student Lab program by double-clicking the icon on the desktop or by following your instructor's guidance.

7. Select lesson **L01-EMG-1** from the menu and click **OK.**

8. Type in a filename that will save this subject's data on the computer hard drive. You may want to use the subject's last name followed by EMG-1 (for example, SmithEMG-1). Then click **OK.**

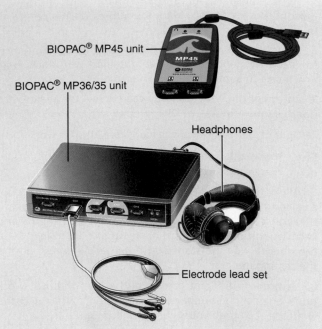

BIOPAC® MP45 unit

BIOPAC® MP36/35 unit

Headphones

Electrode lead set

Figure 14.5 Setting up the BIOPAC® equipment to observe recruitment and temporal summation. Plug the headphones into the back of the MP36/35 data acquisition unit or into the top of the MP45 unit, and the electrode lead set into Channel 1. Electrode leads and headphones are shown connected to the MP36/35 unit.

Calibrating the Equipment

1. With the subject in a still position, click **Calibrate.** This initiates the process by which the computer automatically establishes parameters to record the data properly for the subject.

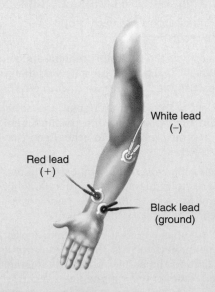

White lead (−)

Red lead (+)

Black lead (ground)

Figure 14.6 Placement of electrodes and the appropriate attachment of electrode leads by color.

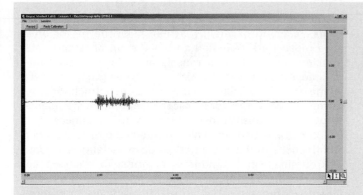

Figure 14.7 Example of waveform during the calibration procedure.

2. After you click **OK,** have the subject wait for 2 seconds, clench the fist tightly for 2 seconds, then release the fist and relax. The computer then automatically stops the recording.

3. Observe the recording of the calibration data, which should look like the waveform in the example (**Figure 14.7**).

- If the data look very different, click **Redo Calibration** and repeat the steps above.

- If the data look similar, proceed to the next section.

Recording the Data

1. Tell the subject that the recording will be of a series of four fist clenches, with the following instructions: First clench the fist softly for 2 seconds, then relax for 2 seconds, then clench harder for 2 seconds, and relax for 2 seconds, then clench even harder for 2 seconds, and relax for 2 seconds, and finally clench with maximum strength, then relax. The result should be a series of four clenches of increasing intensity.

When the subject is ready to do this, click **Record;** then click **Suspend** when the subject is finished.

2. Observe the recording of the data, which should look like the waveforms in the example (**Figure 14.8**).

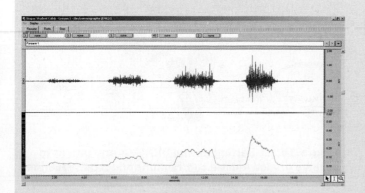

Figure 14.8 Example of waveforms during the recording of data. Note the increased signal strength with the increasing force of the clench.

- If the data look very different, click **Redo** and repeat the steps above.

- If the data look similar, click **STOP.** Click **YES** in response to the question, "Are you finished with both forearm recordings?"

Optional: Anyone can use the headphones to listen to an "auditory version" of the electrical activity of contraction by clicking **Listen** and having the subject clench and relax. Note that the frequency of the auditory signal corresponds with the frequency of action potentials stimulating the muscles. The signal will continue to run until you click **STOP.**

3. Click **Done,** and then remove all electrodes from the forearm.

- If you wish to record from another subject, choose the **Record from another subject** option and return to step 5 under Setting Up the Equipment.

- If you are finished recording, choose **Analyze current data file** and click **OK.** Proceed to Data Analysis, step 2.

Data Analysis

1. If you are just starting the BIOPAC® program to perform data analysis, enter **Review Saved Data** mode and choose the file with the subject's EMG data (for example, SmithEMG-1).

2. Observe the **Raw EMG** recording and computer-calculated **Integrated EMG.** The raw EMG is the actual recording of the voltage (in mV) at each instant in time, while the integrated EMG reflects the absolute intensity of the voltage from baseline at each instant in time.

3. To analyze the data, set up the first four pairs of channel/measurement boxes at the top of the screen by selecting the following channels and measurement types from the drop-down menus.

Channel	Measurement	Data
CH 1	min	raw EMG
CH 1	max	raw EMG
CH 1	p-p	raw EMG
CH 40	mean	integrated EMG

4. Use the arrow cursor and click the I-beam cursor box at the lower right of the screen to activate the "area calculation" function. Using the activated I-beam cursor, highlight the first EMG cluster, representing the first fist clenching (**Figure 14.9**, p. 248).

5. Notice that the computer automatically calculates the **min, max, p-p,** and **mean** for the selected area. These measurements, calculated from the data by the computer, represent the following:

min: Displays the *minimum* value in the selected area

max: Displays the *maximum* value in the selected area

p-p (peak-to-peak): Measures the difference in value between the highest and lowest values in the selected area

mean: Displays the *average* value in the selected area

Text continues on next page. →

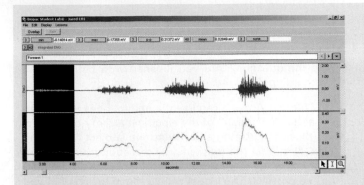

Figure 14.9 Using the I-beam cursor to highlight a cluster of data for analysis.

6. Write down the data for clench 1 in the chart in step 7 (round to the nearest 0.01 mV).

7. Using the I-beam cursor, highlight the clusters for clenches 2, 3, and 4, and record the data in the **EMG Cluster Results chart.**

EMG Cluster Results			
	Min	**Max**	**P-P**	**Mean**
Clench 1				
Clench 2				
Clench 3				
Clench 4				

From the data recorded in the chart, what trend do you observe for each of these measurements as the subject gradually increases the force of muscle contraction?

_____

_____

What is the relationship between maximum voltage for each clench and the number of motor units in the forearm that are being activated?

_____

_____

Part 2: Force Measurement and Fatigue

In this set of activities, you will be observing graded muscle contractions and fatigue in a subject. **Graded muscle contractions,** which represent increasing levels of force generated by a muscle, depend on (1) the gradual activation of more motor units, and (2) increasing the frequency of action potentials (stimulation) for each active motor unit. This permits a range of forces to be generated by any given muscle or group of muscles, all the way up to the maximum force.

For example, the biceps muscle will have more active motor units and exert more force when lifting a 10-kg object than when lifting a 2-kg object. In addition, the motor neuron of each active motor unit will increase the frequency of action potentials delivered to the motor units, resulting in tetanus. When all of the motor units of a muscle are activated and in a state of tetanus, the maximum force of that muscle is achieved. Recall that fatigue is a condition in which the muscle gradually loses some or all of its ability to contract after contracting for an extended period of time. Recent experimental evidence suggests that this is mostly due to ionic imbalances.

In this exercise, you will observe and measure graded contractions of the fist, and then observe fatigue, in both the dominant and nondominant arms. To measure the force generated during fist contraction, you will use a **hand dynamometer** (*dynamo* = force; *meter* = measure). The visual recording of force is called a **dynagram,** and the procedure of measuring the force itself is called **dynamometry.**

You will first record data from the subject's **dominant arm** (forearm 1) indicated by his or her "handedness," then repeat the procedures on the subject's **nondominant arm** (forearm 2) for comparison.

Setting Up the Equipment

1. Connect the BIOPAC® unit to the computer and turn the computer **ON.**

2. Make sure the BIOPAC® unit is **OFF.**

3. Plug in the equipment (as shown in **Figure 14.10**).

- Electrode lead set—CH1

- Hand dynamometer—CH2

- Headphones—back of MP36/35 unit or top of MP45 unit

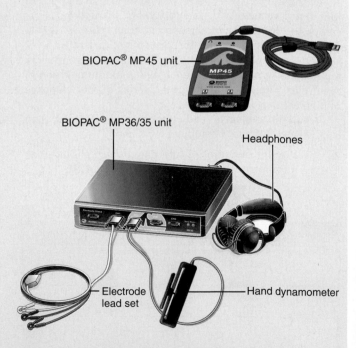

Figure 14.10 Setting up the BIOPAC® equipment to observe graded muscle contractions and muscle fatigue. Plug the headphones into the back of the MP36/35 data acquisition unit or into the top of the MP45 unit, the electrode lead set into Channel 1, and the hand dynamometer into Channel 2. Electrode leads and dynamometer are shown connected to the MP36/35 unit.

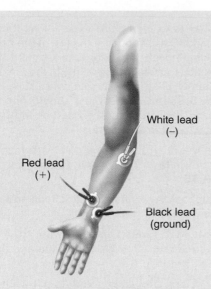

Figure 14.11 Placement of electrodes and the appropriate attachment of electrode leads by color.

4. Turn the BIOPAC® unit **ON**.

5. Attach three electrodes to the subject's *dominant* forearm (forearm 1; see the attachments in **Figure 14.11**), and attach the electrode leads according to the colors indicated.

6. Start the Biopac Student Lab program on the computer by double-clicking the icon on the desktop or by following your instructor's guidance.

7. Select lesson **L02-EMG-2** from the menu and click **OK**.

8. Type in a filename that will save this subject's data on the computer hard drive. You may want to use subject's last name followed by EMG-2 (for example, SmithEMG-2). Then click **OK**.

Calibrating the Equipment

1. With the hand dynamometer at rest on the table, click **Calibrate**. This initiates the process by which the computer automatically establishes parameters to record the data properly for the subject.

2. A pop-up window prompts the subject to remove any grip force. This is to ensure that the dynamometer has been at rest on the table and that no force is being applied. When this is so, click **OK**.

3. As instructed by the pop-up window, the subject is to grasp the hand dynamometer with the dominant hand. With model SS25LA, grasp the dynamometer with the palm of the hand against the short grip bar (as shown in **Figure 14.12a**). Hold model SS25LA vertically. With model SS56L, wrap the hand around the bulb (as shown in Figure 14.12b). Do not curl the fingers into the bulb. (The older SS25L hand dynamometer may also be used with any of the data acquisition units.) Then, click **OK**. The instructions that follow apply to this model of dynamometer.

4. Tell the subject that the instructions will be to wait 2 seconds, then squeeze the hand dynamometer as hard as possible for 2 seconds, and then relax. The computer will automatically stop the calibration.

5. When the subject is ready to proceed, click **OK** and follow the instructions in step 4, which are also in the pop-up window. The calibration will stop automatically.

6. Observe the recording of the calibration data, which should look like the waveforms in the calibration example (**Figure 14.13**, p. 250).

- If the data look very different, click **Redo Calibration** and repeat the steps above.

- If the data look similar, proceed to the next section.

Recording Incremental Force Data for the Forearm

1. Using the force data from the calibration procedure, estimate the **maximum force** that the subject generated (kg).

2. Divide that maximum force by four. In the following steps, the subject will gradually increase the force in approximately these increments. For example, if the

Text continues on next page. →

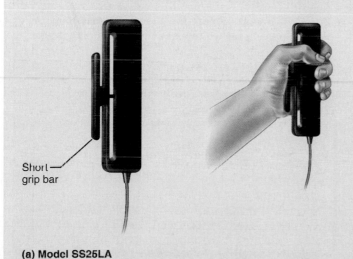

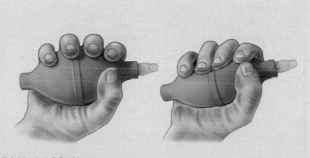

(a) Model SS25LA

(b) Model SS56L

Figure 14.12 Proper grasp of the hand dynamometer using either model SS25LA or model SS56L.

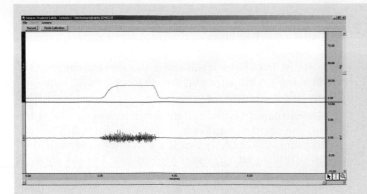

Figure 14.13 Example of calibration data. Force is measured in kilograms or kgf/m^2 at the top, and electromyography is measured in millivolts at the bottom.

maximum force generated was 20 kg, the increment will be 20/4 = 5 kg. The subject will grip at 5 kg, then 10 kg, then 15 kg, and then 20 kg. The subject should watch the tracing on the computer screen and compare it to the scale on the right to determine each target force. Click **Continue**.

3. After you click **Record,** have the subject wait 2 seconds, clench at the first force increment 2 seconds (for example, 5 kg), then relax 2 seconds, clench at the second force increment 2 seconds (10 kg), then relax 2 seconds, clench at the third force increment 2 seconds (15 kg), then relax 2 seconds, then clench with the maximum force 2 seconds (20 kg), and then relax. When the subject relaxes after the maximum clench, click **Suspend** to stop the recording.

4. Observe the recording of the data, which should look similar to the data in the example (**Figure 14.14**).

- If the data look very different, click **Redo** and repeat the steps above.
- If the data look similar, click **Continue** and proceed to observation and recording of muscle fatigue.

Recording Muscle Fatigue Data for the Forearm

Continuing from the end of the incremental force recording, the subject will next record muscle fatigue.

1. After you click **Resume,** the recording will continue from where it stopped and the subject will clench the

dynamometer with maximum force for as long as possible. A "marker" will appear at the top of the data, denoting the beginning of this recording segment. When the subject's clench force falls below 50% of the maximum (for example, below 10 kg for a subject with 20 kg maximum force), click **Suspend**. The subject should not watch the screen during this procedure; those helping can inform the subject when it is time to relax.

2. Observe the recording of the data, which should look similar to the data in the muscle fatigue data example (**Figure 14.15**).

- If the data look very different, click **Redo** and repeat the steps above.
- If the data look similar, and you want to record from the nondominant arm, click **Continue** and proceed to Recording from the Nondominant Arm.
- If the data look similar, and you do not want to record from the nondominant arm or you have just finished recording the nondominant arm, click **STOP**. A dialog box comes up asking if you are sure you want to stop recording. Click **NO** to return to the **Resume** or **Stop** options, providing one last chance to redo the fatigue recording. Click **YES** to end the recording session and automatically save your data.

Optional: Anyone can use the headphones to listen to an "auditory version" of the electrical activity of contraction by clicking **Listen** and having the subject clench and relax. Note that the frequency of the auditory signal corresponds to the frequency of action potentials stimulating the muscles. The signal will continue to run until you click **STOP**.

3. Click **Done**. Choose **Analyze current data file** and proceed to Data Analysis, step 2.

4. Remove all electrodes from the arm of the subject.

Recording from the Nondominant Arm

1. To record from the nondominant forearm, attach three electrodes to the subject's forearm (as shown in Figure 14.11) and attach the electrode leads according to the colors indicated.

2. Click **Resume**. Repeat the Clench-Release-Wait cycles with increasing clench force as performed with the dominant arm.

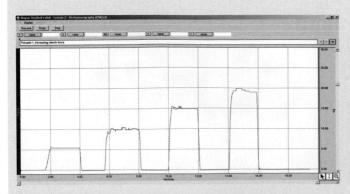

Figure 14.14 Example of incremental force data.

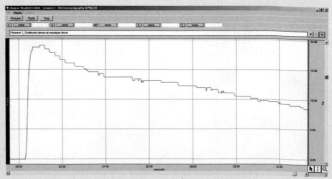

Figure 14.15 Example of muscle fatigue data.

3. Observe the recording of the data, which should look similar to the data in the muscle fatigue data example (Figure 14.15). If the data look very different, click **Redo** and repeat.

4. End the session by repeating the steps for muscle fatigue. Click **Continue** and repeat steps 1 and 2 of the muscle fatigue section, recording the nondominant arm.

Data Analysis

1. In **Review Saved Data** mode, select the file that is to be analyzed (for example, SmithEMG-2-1-L02).

2. Observe the recordings of the clench **Force** (kg or kgf/m^2), **Raw EMG** (mV), and computer-calculated **Integrated EMG** (mV). The force is the actual measurement of the strength of clench in kilograms at each instant in time. The raw EMG is the actual recording of the voltage (in mV) at each instant in time, and the integrated EMG indicates the absolute intensity of the voltage from baseline at each instant in time.

3. To analyze the data, set up the first three pairs of channel/measurement boxes at the top of the screen. Select the following channels and measurement types:

Channel	Measurement	Data
CH 41	mean	clench force
CH 40	mean	integrated EMG

4. Use the arrow cursor and click the I-beam cursor box on the lower right side of the screen to activate the "area selection" function. Using the activated I-beam cursor, highlight the "plateau phase" of the first clench cluster. The plateau should be a relatively flat force in the middle of the cluster (**Figure 14.16**).

5. Observe that the computer automatically calculates the **p-p** and **mean** values for the selected area. These measurements, calculated from the data by the computer, represent the following:

p-p (peak-to-peak): Measures the difference in value between the highest and lowest values in the selected area

mean: Displays the average value in the selected area

6. In the chart in step 7, record the data for clench 1 (for example, 5-kg clench) to the nearest 0.01.

7. Using the I-beam cursor, highlight the clusters for the subsequent clenches, and record the data in the **Dominant Forearm Clench Increments chart**.

Dominant Forearm Clench Increments		
	Force at plateau Mean (kg or kgf/m^2)	Integrated EMG Mean (mV-sec)
Clench 1		
Clench 2		
Clench 3		
Clench 4		

8. Scroll along the bottom of the data page to the segment that includes the recording of muscle fatigue (it should begin after the "marker" that appears at the top of the data).

9. Change the channel/measurement boxes so that the first two selected appear as follows (the third should be set to "none"):

Channel	Measurement	data
CH 41	value	force
CH 40	delta T	integrated EMG

value: Measures the highest value in the selected area (force measured in kg with SS25LA or kgf/m^2 with the SS56L.)

delta T: Measures the time elapsed in the selected area (time measured in seconds.)

10. Use the arrow cursor and click on the I-beam cursor box on the lower right side of the screen to activate the "area selection" function.

11. Using the activated I-beam cursor, select just the single point of maximum clench strength at the start of this data segment as shown in **Figure 14.17**.

12. In the **Dominant Forearm Fatigue Measurement chart** on p. 252, note the maximum force measurement for this point.

Text continues on next page. →

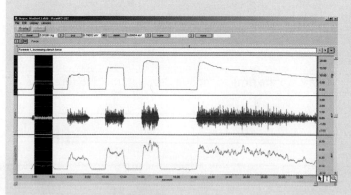

Figure 14.16 Highlighting the "plateau" of clench cluster 1.

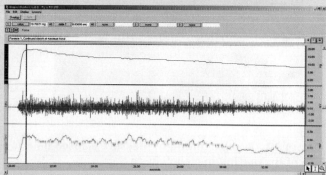

Figure 14.17 Selection of single point of maximum clench.

14

Dominant Forearm Fatigue Measurement		
Maximum clench force (kg or kgf/m²)	50% of the maximum clench force (divide maximum clench force by 2)	Time to fatigue (seconds)

13. Calculate the value of 50% of the maximum clench force, and record this in the data chart.

14. Using a metric tape measure, measure the circumference of the subject's dominant forearm at its greatest diameter:

_____ cm

15. Measure the amount of time that elapsed between the initial maximum force and the point at which the subject fatigued to 50% of this level. Using the activated I-beam cursor, highlight the area from the point of 50% clench force back to the point of maximal clench force, as shown in the example (**Figure 14.18**).

16. Note the time it took the subject to reach this point of fatigue (CH 40 delta T), and record this data in the Dominant Forearm Fatigue Measurement chart.

Repeat Data Analysis for the Nondominant Forearm

1. Return to step 1 of the Data Analysis section, and repeat the same measurements for the nondominant forearm (forearm 2).

2. Record your data in the **Nondominant Forearm Clench Increments chart** and the **Nondominant Forearm Fatigue Measurement chart**. These data will be used for comparison.

3. Using a tape measure, measure the circumference of the subject's nondominant forearm at its greatest:

_____ cm

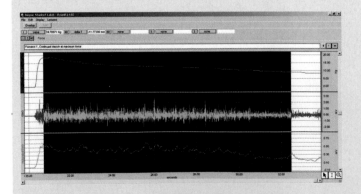

Figure 14.18 Highlighting to measure elapsed time to 50% of the maximum clench force.

4. When finished, exit the program by going to the **File** menu at the top of the screen and clicking **Quit**.

Is there a difference in maximal force that was generated between the dominant and nondominant forearms? If so, how much?

_____

Calculate the percentage difference in force between the dominant maximal force and nondominant maximal force.

_____

_____

Nondominant Forearm Clench Increments		
	Force at plateau Mean (kg or kgf/m²)	Integrated EMG Mean (mV-sec)
Clench 1		
Clench 2		
Clench 3		
Clench 4		

Nondominant Forearm Fatigue Measurement		
Maximum Clench Force (kg or kgf/m²)	50% of the maximum clench force (divide maximum clench force by 2)	Time to fatigue (seconds)

Is there a difference in the circumference between the dominant and nondominant forearms? If so, how much?

_____

If there is a difference in circumference, is this difference likely to be due to a difference in the *number* of muscle fibers in each forearm or in the *diameter* of each muscle fiber in the forearms? Explain. Use an appropriate reference if needed.

_____

_____

Compare the time to fatigue between the two forearms.

_____

_____

REVIEW SHEET
Skeletal Muscle Physiology: Frogs and Human Subjects

Name _____ Lab Time/Date _____

Muscle Activity

1. The following group of incomplete statements begins with a description of a muscle fiber in the resting state just before stimulation. Complete each statement by choosing the correct response from the key items.

   Key: a. Na⁺ diffuses out of the cell
         b. K⁺ diffuses out of the cell
         c. Na⁺ diffuses into the cell
         d. K⁺ diffuses into the cell
         e. inside the cell
         f. outside the cell
         g. relative ionic concentrations on the two sides of the membrane
         h. electrical conditions
         i. activation of the sodium-potassium pump, which moves K⁺ into the cell and Na⁺ out of the cell
         j. activation of the sodium-potassium pump, which moves Na⁺ into the cell and K⁺ out of the cell

   There is a greater concentration of Na⁺ _____; there is a greater concentration of K⁺ _____.

   When the stimulus is delivered, the permeability of the membrane at that point is changed; and _____, initiating

   the depolarization of the membrane. Almost as soon as the depolarization wave has begun, a repolarization wave follows

   it across the membrane. This occurs as _____. Repolarization restores the _____ of the resting

   cell membrane. The _____ is (are) reestablished by _____.

2. Number the following statements in the proper sequence to describe the contraction mechanism in a skeletal muscle fiber. Number 1 has already been designated.

   _____1_____ Depolarization occurs, and the action potential is generated.

   _____ The muscle fiber relaxes and lengthens.

   _____ The calcium ion concentrations at the myofilaments increase; the myofilaments slide past one another, and the cell shortens.

   _____ The action potential, carried deep into the cell by the T tubules, triggers the release of calcium ions from the sarcoplasmic reticulum.

   _____ The concentration of the calcium ions at the myofilaments decreases as they are actively transported into the sarcoplasmic reticulum.

3. Refer to your observations of muscle fiber contraction in Activity 1 to answer the following questions.

   a. Did your data support your hypothesis? _____

   b. *Explain* your observations fully. _____

   _____

   _____

c. Draw a relaxed and a contracted sarcomere below. Label the Z discs, thick filaments, and thin filaments.

Relaxed	**Contracted**

Induction of Contraction in the Frog Gastrocnemius Muscle

4. Why is it important to destroy the brain and spinal cord of a frog before conducting physiological experiments on

    muscle contraction? _____

    _____

5. What kind of stimulus (electrical or chemical) travels from the motor neuron to skeletal muscle? _____

    What kind of stimulus (electrical or chemical) travels from the axon terminal to the sarcolemma? _____

6. Give the name and duration of each of the three phases of the muscle twitch, and describe what is happening during
    each phase.

    a. _____, _____ msec, _____

        _____

    b. _____, _____ msec, _____

    c. _____, _____ msec, _____

7. Use the items in the key to identify the conditions described.

    Key:

    a. maximal stimulus   d. tetanus
    b. recruitment   e. threshold stimulus
    c. subthreshold stimulus   f. wave summation

    _____ 1. sustained contraction without any   _____ 4. increasingly stronger contractions
                       evidence of relaxation   owing to stimulation at a rapid rate

    _____ 2. stimulus that results in no   _____ 5. increasingly stronger contractions
                       perceptible contraction   owing to increased stimulus strength

    _____ 3. stimulus at which the muscle first   _____ 6. weakest stimulus at which all muscle
                       contracts perceptibly   fibers in the muscle are contracting

8. Complete the following statements by writing the appropriate words on the corresponding numbered blanks at the right.

     When a weak but smooth muscle contraction is desired, a few motor units are stimulated at a _1_ rate. Within limits, as the load on a muscle is increased, the muscle contracts _2_ (more/less) strongly.

1. _____

2. _____

9. During the frog experiment on muscle fatigue, how did the muscle contraction pattern change as the muscle began to fatigue?

_____

_____

How long was stimulation continued before fatigue was apparent? _____

If the sciatic nerve that stimulates the living frog's gastrocnemius muscle had been left attached to the muscle and the stimulus had been applied to the nerve rather than the muscle, would fatigue have become apparent sooner, later, or at the same time?

_____

10. What will happen to a muscle in the body when its nerve supply is destroyed or badly damaged? _____

_____

_____

11. Explain the relationship between the load on a muscle and its strength of contraction. _____

_____

_____

12. The skeletal muscles are maintained in a slightly stretched condition for optimal contraction. How is this accomplished?

_____

Why does stretching a muscle beyond its optimal length reduce its ability to contract? (Include an explanation of the

events at the level of the myofilaments.) _____

_____

_____

13. If the length but not the tension of a muscle is changed, the contraction is called an isotonic contraction. In an isometric contraction, the tension is increased but the muscle does not shorten. Which type of contraction did you observe most often during the laboratory experiments?

_____

Electromyography in a Human Subject Using BIOPAC®

14. If you were a physical therapist applying a constant voltage to the forearm, what might you observe if you gradually increased the *frequency* of stimulatory impulses, keeping the voltage constant each time?

_____

_____

15. Describe what is meant by the term *recruitment.* _____

_____

16. Describe the physiological processes occurring in the muscle fibers that account for the gradual onset of muscle fatigue.

_____

_____

_____

17. Most subjects use their dominant forearm far more than their nondominant forearm. What does this indicate about degree of activation of motor units and these factors: muscle fiber diameter, maximum muscle fiber force, and time to muscle fatigue? (You may need to use your textbook for help with this one.)

_____

_____

_____

_____

18. Define *dynamometry.* _____

_____

19. How might dynamometry be used to assess patients in a clinical setting? _____

_____

_____

EXERCISE 15

Histology of Nervous Tissue

Objectives

- ☐ Discuss the functional differences between neurons and neuroglia.
- ☐ List six types of neuroglia, and indicate where each is found in the nervous system.
- ☐ Identify the important anatomical features of a neuron on an appropriate image.
- ☐ List the functions of dendrites, axons, and axon terminals.
- ☐ Explain how a nerve impulse is transmitted from one neuron to another.
- ☐ State the function of myelin sheaths, and explain how Schwann cells myelinate nerve fibers (axons) in the peripheral nervous system.
- ☐ Classify neurons structurally and functionally.
- ☐ Differentiate between a nerve and a tract, and between a ganglion and a CNS nucleus.
- ☐ Identify an endoneurium, perineurium, and epineurium microscopically or in an appropriate image and cite their functions.

Materials

- Model of a "typical" neuron (if available)
- Compound microscope
- Immersion oil
- Prepared slides of an ox spinal cord smear and teased myelinated nerve fibers
- Prepared slides of Purkinje cells (cerebellum), pyramidal cells (cerebrum), and a dorsal root ganglion
- Prepared slide of a nerve (x.s.)

MasteringA&P®

For related exercise study tools, go to the Study Area of **MasteringA&P**. There you will find:

- Practice Anatomy Lab PAL
- A&PFlix A&PFlix
- PhysioEx PEx
- Practice quizzes, Histology Atlas, eText, Videos, and more!

Pre-Lab Quiz

1. Circle the correct underlined term. Nervous tissue is made up of <u>two</u> / <u>three</u> main cell types.

2. Neuroglia of the peripheral nervous system include:
   - **a.** ependymal cells and satellite cells
   - **b.** oligodendrocytes and astrocytes
   - **c.** satellite cells and Schwann cells

3. _____ are the functional units of nervous tissue.

4. These branching neuron processes serve as receptive regions and transmit electrical signals toward the cell body. They are:
   - **a.** axons
   - **c.** dendrites
   - **b.** collaterals
   - **d.** neuroglia

5. Circle True or False. Axons are the neuron processes that generate and conduct nerve impulses.

6. Most axons are covered with a fatty material called _____, which insulates the fibers and increases the speed of neurotransmission.

7. Circle the correct underlined term. Axons running through the central nervous system form <u>tracts</u> / <u>nerves</u> of white matter.

8. Neurons can be classified according to structure. _____ neurons have many processes that issue from the cell body.
   - **a.** Bipolar
   - **b.** Multipolar
   - **c.** Unipolar

9. Circle the correct underlined term. Neurons can be classified according to function. <u>Afferent</u> / <u>Efferent</u>, or motor neurons, carry electrical signals from the central nervous system primarily to muscles or glands.

10. Within a nerve, each axon is surrounded by a covering called the:
    - **a.** endoneurium
    - **b.** epineurium
    - **c.** perineurium

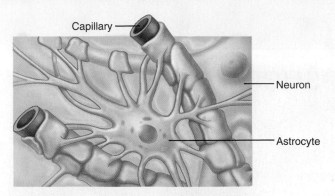

(a) Astrocytes are the most abundant CNS neuroglia.

Capillary

Neuron

Astrocyte

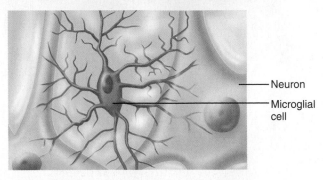

(b) Microglial cells are defensive cells in the CNS.

Neuron

Microglial cell

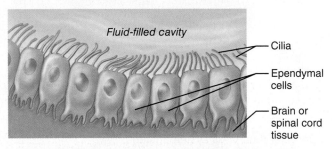

(c) Ependymal cells line cerebrospinal fluid–filled cavities.

Fluid-filled cavity

Cilia

Ependymal cells

Brain or spinal cord tissue

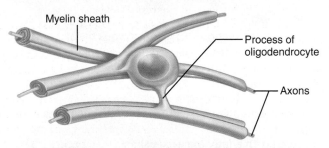

(d) Oligodendrocytes have processes that form myelin sheaths around CNS axons.

Myelin sheath

Process of oligodendrocyte

Axons

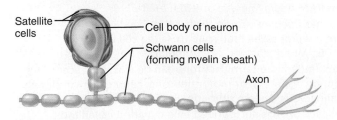

(e) Satellite cells and Schwann cells (which form myelin) surround neurons in the PNS.

Satellite cells

Cell body of neuron

Schwann cells (forming myelin sheath)

Axon

15

The nervous system is the master integrating and coordinating system, continuously monitoring and processing sensory information both from the external environment and from within the body. Every thought, action, and sensation is a reflection of its activity. Like a computer, it processes and integrates new "inputs" with information previously fed into it to produce a response. However, no computer can possibly compare in complexity and scope to the human nervous system.

Two primary divisions make up the nervous system: the central nervous system, or CNS, consisting of the brain and spinal cord, and the peripheral nervous system, or PNS, which includes all the nervous elements located outside the central nervous system. PNS structures include nerves, sensory receptors, and some clusters of neuron cell bodies.

Despite its complexity, nervous tissue is made up of just two principal cell types: neurons and neuroglia.

Neuroglia

The **neuroglia** ("nerve glue"), or **glial cells,** of the CNS include *astrocytes, oligodendrocytes, microglial cells,* and *ependymal cells* (**Figure 15.1**). The neuroglia found in the PNS include *Schwann cells,* also called neurolemmocytes, and *satellite cells.*

Neuroglia serve the needs of the delicate neurons by supporting and protecting them. In addition, they act as phagocytes (microglial cells), myelinate the cytoplasmic extensions of the neurons (oligodendrocytes and Schwann cells), play a role in capillary-neuron exchanges, and control the chemical environment around neurons (astrocytes). Although neuroglia resemble neurons in some ways (many have branching cellular extensions), they are not capable of generating and transmitting nerve impulses. In this exercise, we focus on the highly excitable neurons.

Neurons

Neurons, or nerve cells, are the basic functional units of nervous tissue. They are highly specialized to transmit messages from one part of the body to another in the form of nerve impulses. Although neurons differ structurally, they have many identifiable features in common (**Figure 15.2a and b**). All have a **cell body** from which slender processes extend. The cell body is both the *biosynthetic center* of the neuron and part of its *receptive region.* Neuron cell bodies make up the gray matter of the CNS, and form clusters there that are called **nuclei.** In the PNS, clusters of neuron cell bodies are called **ganglia.**

The neuron cell body contains a large round nucleus surrounded by cytoplasm. Two prominent structures are found in the cytoplasm: One is cytoskeletal elements called **neurofibrils,** which provide support for the cell and a means to transport substances throughout the neuron. The second is darkly staining structures called **chromatophilic substance,** clusters of rough endoplasmic reticulum and ribosomes involved in the metabolic activities of the cell.

Figure 15.1 Neuroglia. (a–d) The four types of neuroglia in the central nervous system. **(e)** Neuroglia of the peripheral nervous system.

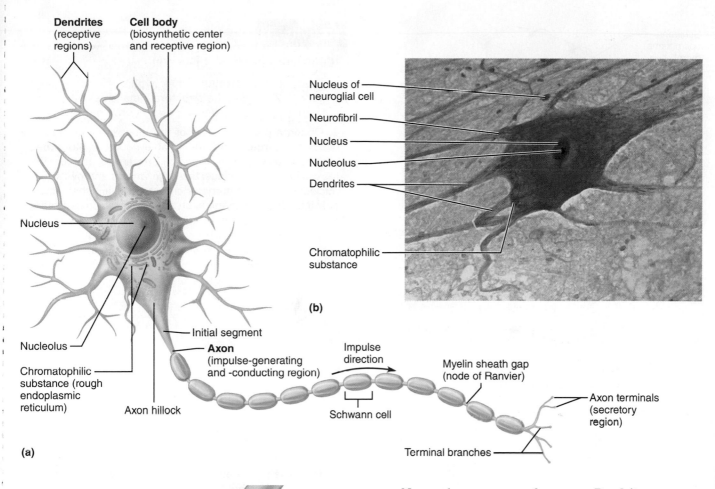

Dendrites (receptive regions)

Cell body (biosynthetic center and receptive region)

Nucleus of neuroglial cell

Neurofibril

Nucleus

Nucleolus

Dendrites

Chromatophilic substance

(b)

Nucleus

Nucleolus

Chromatophilic substance (rough endoplasmic reticulum)

Axon hillock

Initial segment

Axon (impulse-generating and -conducting region)

Impulse direction

Myelin sheath gap (node of Ranvier)

Axon terminals (secretory region)

Schwann cell

Terminal branches

(a)

15

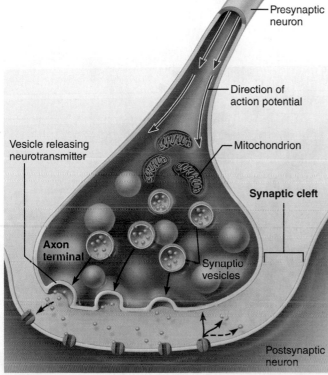

Presynaptic neuron

Direction of action potential

Mitochondrion

Synaptic cleft

Vesicle releasing neurotransmitter

Axon terminal

Synaptic vesicles

Postsynaptic neuron

(c)

Figure 15.2 Structure of a typical motor neuron.
(a) Diagram. **(b)** Photomicrograph (450×) **(c)** Enlarged diagram of a synapse.

Neurons have two types of processes. **Dendrites** are *receptive regions* that bear receptors for neurotransmitters released by the axon terminals of other neurons. **Axons,** also called *nerve fibers* when they are long, form the *impulse generating and conducting region* of the neuron. The white matter of the nervous system is made up of axons. Neurons may have many dendrites, but they have only a single axon. The axon may branch, forming one or more processes called **axon collaterals.**

In general, a neuron is excited by other neurons when their axons release neurotransmitters close to its dendrites or cell body. The electrical signal produced travels across the cell body and if it is great enough, it elicits a regenerative electrical signal, a *nerve impulse* or *action potential,* that travels down the axon. The axon in motor neurons begins just distal to a slightly enlarged cell body structure called the **axon hillock** (Figure 15.2a). The axon ends in many small structures called **axon terminals,** or *terminal boutons,* which form **synapses** with neurons or effector cells. These terminals store the neurotransmitter chemical in tiny vesicles. Each axon terminal of the neuron is separated from the cell body or dendrites of the next neuron by a tiny gap called the **synaptic cleft** (Figure 15.2c).

Most long nerve fibers are covered with a fatty material called *myelin,* and such fibers are referred to as **myelinated fibers.** Nerve fibers in the peripheral nervous system are typically heavily myelinated by special supporting cells called **Schwann cells,** which wrap themselves tightly around the axon in jelly roll fashion (**Figure 15.3,** p. 260). The wrapping is the **myelin sheath.** Since the myelin sheath is formed by many individual Schwann cells, it is a discontinuous sheath. The gaps, or indentations, in the sheath are called **myelin sheath gaps** or **nodes of Ranvier** (see Figure 15.2a).

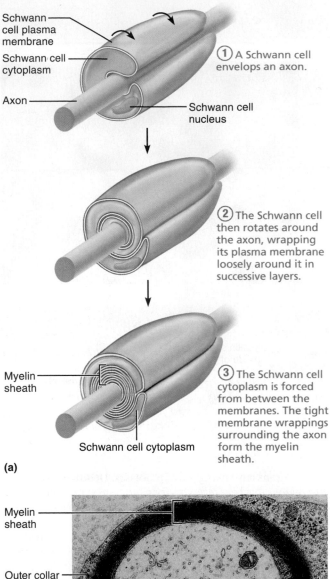

(1) A Schwann cell envelops an axon.

(2) The Schwann cell then rotates around the axon, wrapping its plasma membrane loosely around it in successive layers.

(3) The Schwann cell cytoplasm is forced from between the membranes. The tight membrane wrappings surrounding the axon form the myelin sheath.

Schwann cell plasma membrane

Schwann cell cytoplasm

Axon

Schwann cell nucleus

Myelin sheath

Schwann cell cytoplasm

(a)

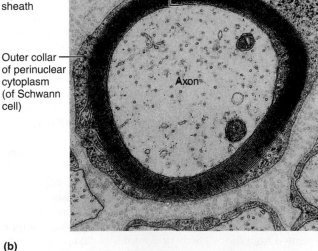

Myelin sheath

Outer collar of perinuclear cytoplasm (of Schwann cell)

Axon

(b)

Figure 15.3 Myelination of a nerve fiber (axon) by Schwann cells. (a) Nerve fiber myelination. **(b)** Electron micrograph of cross section through a myelinated axon (7500×).

Within the CNS, myelination is accomplished by neuroglia called **oligodendrocytes** (see Figure 15.1d). Because of its chemical composition, myelin electrically insulates the fibers and greatly increases the transmission speed of nerve impulses.

Activity 1

Identifying Parts of a Neuron

1. Study the illustration of a typical motor neuron (Figure 15.2), noting the structural details described and then identify these structures on a neuron model.

2. Obtain a prepared slide of the ox spinal cord smear, which has large, easily identifiable neurons. Study one representative neuron under oil immersion and identify the cell body; the nucleus; the large, prominent "owl's eye" nucleolus; and the granular chromatophilic substance. If possible, distinguish the axon from the many dendrites.

Sketch the neuron in the space provided below, and label the important anatomical details you have observed. Compare your sketch to Figure 15.2b.

3. Obtain a prepared slide of teased myelinated nerve fibers. Identify the following (use **Figure 15.4** as a guide): myelin sheath gaps, axon, Schwann cell nuclei, and myelin sheath.

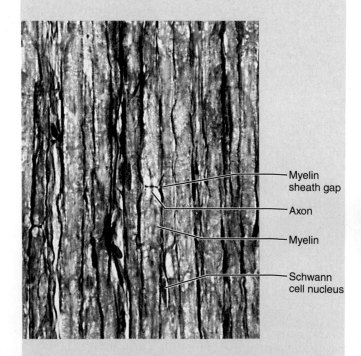

Myelin sheath gap

Axon

Myelin

Schwann cell nucleus

Figure 15.4 Photomicrograph of a small portion of a peripheral nerve in longitudinal section (400×).

Text continues on next page. →

Sketch a portion of a myelinated nerve fiber in the space provided below, illustrating a myelin sheath gap. Label the axon, myelin sheath, myelin sheath gap, and Schwann cell nucleus.

Do the gaps seem to occur at consistent intervals, or are they irregularly distributed? _____

Explain the functional significance of this finding: _____

_____

_____

Neuron Classification

Neurons may be classified on the basis of structure or of function.

Classification by Structure

Structurally, neurons may be differentiated by the number of processes attached to the cell body (**Figure 15.5a**). In **unipolar neurons,** one very short process, which divides into *peripheral* and *central processes,* extends from the cell body. Functionally, only the most distal parts of the peripheral process act as receptive endings; the rest acts as an axon along with the central process. Unipolar neurons are more accurately called **pseudounipolar neurons** because they are

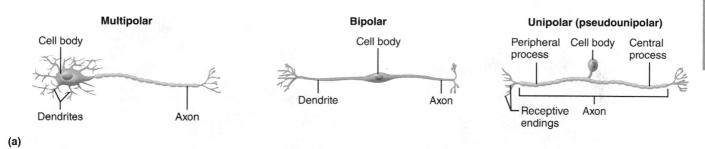

(a)

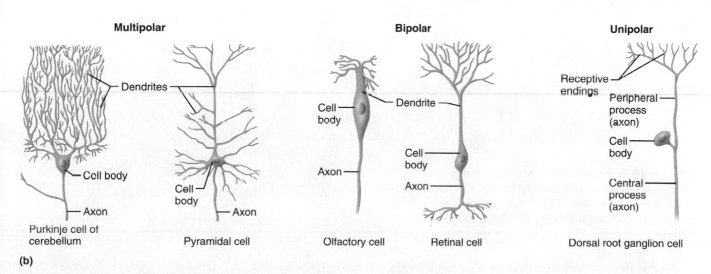

(b)

Figure 15.5 Classification of neurons according to structure. (a) Classification of neurons based on structure (number of processes extending from the cell body). **(b)** Structural variations within the classes.

derived from bipolar neurons. Nearly all neurons that conduct impulses toward the CNS are unipolar.

Bipolar neurons have two processes attached to the cell body. This neuron type is quite rare, typically found only as part of the receptor apparatus of the eye, ear, and olfactory mucosa.

Many processes issue from the cell body of **multipolar neurons,** all classified as dendrites except for a single axon. Most neurons in the brain and spinal cord and those whose axons carry impulses away from the CNS fall into this last category.

Studying the Microscopic Structure of Selected Neurons

Obtain prepared slides of pyramidal cells of the cerebral cortex, Purkinje cells of the cerebellar cortex, and a dorsal root ganglion. As you observe them under the microscope, try to pick out the anatomical details and compare the cells to Figure 15.5b and **Figure 15.6**. Notice that the neurons of the cerebral and cerebellar tissues (both brain tissues) are extensively branched; in contrast, the neurons of the dorsal root ganglion are more rounded. The many small nuclei visible surrounding the neurons are those of bordering neuroglia.

Which of these neuron types would be classified as multipolar neurons?

_____

As unipolar? _____

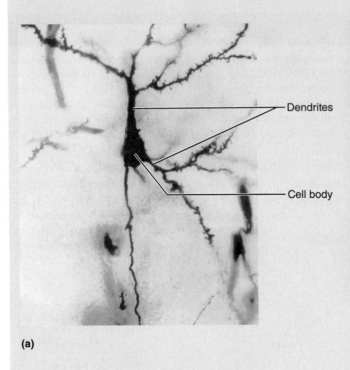

(a)

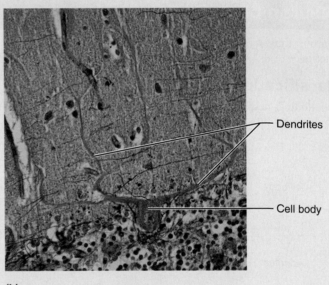

(b)

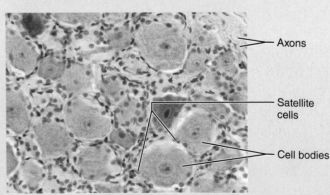

(c)

Figure 15.6 Photomicrographs of neurons.
(a) Pyramidal neuron from the cerebral cortex (600×).
(b) Purkinje cell from the cerebellar cortex (600×).
(c) Dorsal root ganglion cells (245×).

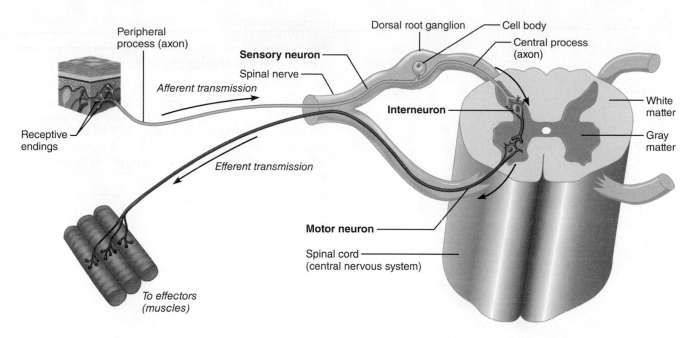

Figure 15.7 Classification of neurons on the basis of function. Sensory (afferent) neurons conduct impulses from the body's sensory receptors to the central nervous system; most are unipolar neurons with their cell bodies in ganglia in the peripheral nervous system (PNS). Motor (efferent) neurons transmit impulses from the CNS to effectors (muscles). Interneurons complete the communication line between sensory and motor neurons. They are typically multipolar, and their cell bodies reside in the CNS.

Classification by Function

In general, neurons carrying impulses from sensory receptors in the internal organs (viscera), the skin, skeletal muscles, joints, or special sensory organs are termed **sensory,** or **afferent, neurons** (**Figure 15.7**). The receptive endings of sensory neurons are often equipped with specialized receptors that are stimulated by specific changes in their immediate environment. (The structure and function of these receptors are considered separately in Exercise 22, General Sensation.) The cell bodies of sensory neurons are always found in a ganglion outside the CNS, and these neurons are typically unipolar.

Neurons carrying impulses from the CNS to the viscera and/or body muscles and glands are termed **motor,** or **efferent, neurons.** Motor neurons are most often multipolar, and their cell bodies are almost always located in the CNS.

The third functional category of neurons is **interneurons,** which are situated between and contribute to pathways that connect sensory and motor neurons. Their cell bodies are always located within the CNS, and they are multipolar neurons structurally.

Structure of a Nerve

In the CNS, bundles of axons are called **tracts.** In the PNS, bundles of axons are called **nerves.** Wrapped in connective tissue coverings, nerves extend to and/or from the CNS and visceral organs or structures of the body periphery such as skeletal muscles, glands, and skin.

Like neurons, nerves are classified according to the direction in which they transmit impulses. **Sensory (afferent) nerves** conduct impulses only toward the CNS. A few of the cranial nerves are pure sensory nerves. **Motor (efferent) nerves** carry impulses only away from the CNS. The ventral roots of the spinal cord are motor nerves. Nerves carrying both sensory (afferent) and motor (efferent) fibers are called **mixed nerves;** most nerves of the body, including all spinal nerves, are mixed nerves.

Within a nerve, each axon is surrounded by a delicate connective tissue sheath called an **endoneurium,** which insulates it from the other neuron processes adjacent to it. The endoneurium is often mistaken for the myelin sheath; it is instead an additional sheath that surrounds the myelin sheath. Groups of axons are bound by a coarser connective tissue, called the **perineurium,** to form bundles of fibers called **fascicles.** Finally, all the fascicles are bound together by a white, fibrous connective tissue sheath called the **epineurium,** forming the cordlike nerve (**Figure 15.8,** p. 264). In addition to the connective tissue wrappings, blood vessels and lymphatic vessels serving the fibers also travel within a nerve.

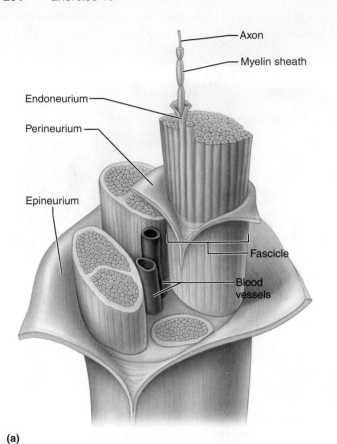

Axon

Myelin sheath

Endoneurium

Perineurium

Epineurium

Fascicle

Blood vessels

(a)

15

Activity 3

Examining the Microscopic Structure of a Nerve

Use the compound microscope to examine a prepared cross section of a peripheral nerve. Using the photomicrograph (Figure 15.8b) as an aid, identify axons, myelin sheaths, fascicles, and endoneurium, perineurium, and epineurium sheaths. If desired, sketch the nerve in the space below.

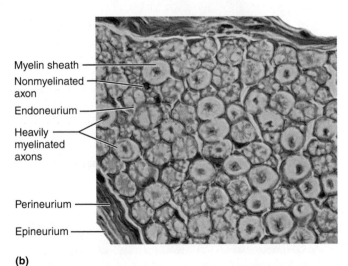

Myelin sheath

Nonmyelinated axon

Endoneurium

Heavily myelinated axons

Perineurium

Epineurium

(b)

Figure 15.8 Structure of a nerve showing connective tissue wrappings. (a) Three-dimensional view of a portion of a nerve. **(b)** Photomicrograph of a cross-sectional view of part of a peripheral nerve (510×).

REVIEW SHEET
Histology of Nervous Tissue

Name_____ Lab Time/Date_____

1. The basic functional unit of the nervous system is the neuron. What is the major function of this cell type?

_____

2. Name four types of neuroglia in the CNS, and list a function for each of these cells. (You will need to consult your textbook for this.)

Types   **Functions**

a. _____   a. _____

                                                              _____

b. _____   b. _____

                                                              _____

c. _____   c. _____

                                                              _____

d. _____   d. _____

                                                              _____

Name the PNS neuroglial cell that forms myelin. _____

Name the PNS neuroglial cell that surrounds neuron cell bodies in ganglia. _____

3. Match each description with a term from the key.

   Key: a. afferent neuron   e. interneuron   i. nuclei
         b. central nervous system   f. neuroglia   j. peripheral nervous system
         c. efferent neuron   g. neurotransmitters   k. synaptic cleft
         d. ganglion   h. nerve   l. tract

   _____ 1. the brain and spinal cord collectively

   _____ 2. specialized supporting cells in the nervous system

   _____ 3. junction or point of close contact between neurons

   _____ 4. a bundle of axons inside the CNS

   _____ 5. neuron serving as part of the conduction pathway between sensory and motor neurons

   _____ 6. ganglia and spinal and cranial nerves

   _____ 7. collection of nerve cell bodies found outside the CNS

_____ 8. neuron that conducts impulses away from the CNS to muscles and glands

_____ 9. neuron that conducts impulses toward the CNS from the body periphery

_____ 10. chemicals released by neurons that stimulate or inhibit other neurons or effectors

Neuron Anatomy

4. Match the following anatomical terms (column B) with the appropriate description or function (column A).

Column A

_____ 1. region of the cell body from which the axon originates

_____ 2. secretes neurotransmitters

_____ 3. receptive regions of a neuron (2 terms)

_____ 4. insulates the nerve fibers

_____ 5. site of the nucleus and most important metabolic area

_____ 6. involved in the transport of substances within the neuron

_____ 7. essentially rough endoplasmic reticulum, important metabolically

_____ 8. impulse generator and transmitter

Column B

a. axon

b. axon terminal

c. axon hillock

d. cell body

e. chromatophilic substance

f. dendrite

g. myelin sheath

h. neurofibril

5. Draw a "typical" multipolar neuron in the space below. Include and label the following structures on your diagram: cell body, nucleus, nucleolus, chromatophilic substance, dendrites, axon, myelin sheath, myelin sheath gaps, and axon terminals.

6. What substance is found in synaptic vesicles of the axon terminal? _____

7. What anatomical characteristic determines whether a particular neuron is classified as unipolar, bipolar, or multipolar?

_____

Make a simple line drawing of each type here.

Unipolar neuron	Bipolar neuron	Multipolar neuron

8. Correctly identify the sensory (afferent) neuron, interneuron, and motor (efferent) neuron in the figure below.

Which of these neuron types is/are unipolar? _____

Which is/are most likely multipolar? _____

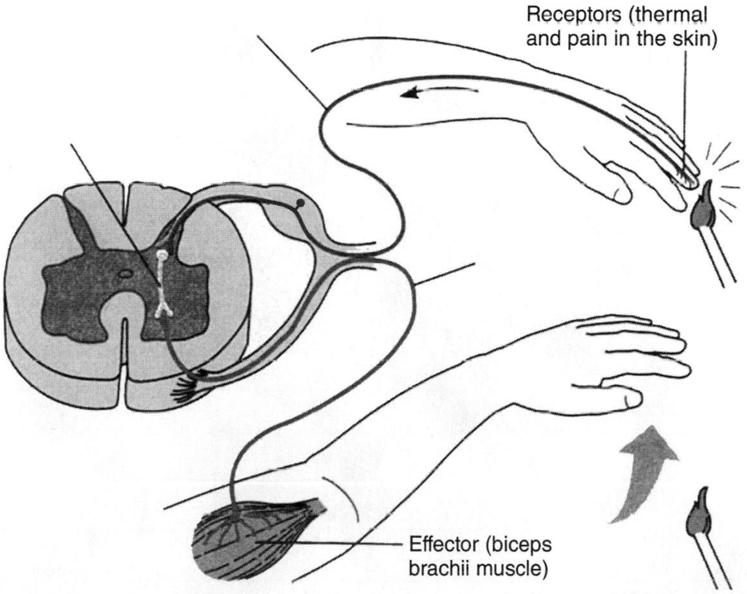

Receptors (thermal and pain in the skin)

Effector (biceps brachii muscle)

9. Describe how the Schwann cells form the myelin sheath encasing the nerve fibers.

_____

_____

Structure of a Nerve

10. What is a nerve? _____

_____

11. State the location of each of the following connective tissue coverings.

endoneurium: _____

perineurium: _____

epineurium: _____

12. What is the function of the connective tissue wrappings found in a nerve? _____

_____

13. Define *mixed nerve*. _____

_____

14. Identify all indicated parts of the nerve section.

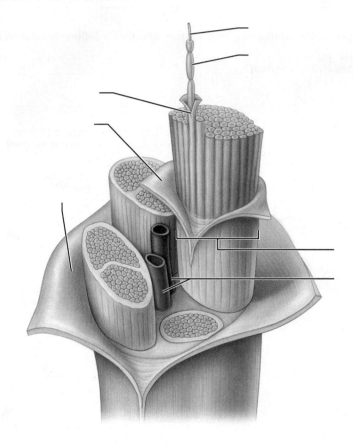

16

Neurophysiology of Nerve Impulses: Frog Subjects

Objectives

☐ Describe the resting membrane potential in neurons.

☐ Define *depolarization, repolarization, action potential,* and *relative refractory period* and *absolute refractory period.*

☐ Describe the events that lead to the generation and conduction of an action potential.

☐ Explain briefly how a nerve impulse is transmitted from one neuron to another, and how a neurotransmitter may be either excitatory or inhibitory to the recipient cell.

☐ Define *compound action potential,* and discuss how it differs from an action potential in a single neuron.

☐ Describe the preparation used to examine contraction of the frog gastrocnemius muscle.

☐ List various substances and factors that can stimulate neurons.

☐ State the site of action of the blocking agents ether and curare.

☐ Describe the experimental setup used to record compound action potentials in the frog sciatic nerve.

Materials

- *Rana pipiens**
- Dissecting instruments and tray
- Disposable gloves
- Ringer's solution (frog) in dropper bottles, some at room temperature and some in an ice bath
- Thread
- Glass rods or probes
- Glass plates or slides
- Ring stand and clamp
- Stimulator; platinum electrodes
- Forceps
- Filter paper
- 0.01% hydrochloric acid (HCl) solution
- Sodium chloride (NaCl) crystals
- Heat-resistant mitts
- Bunsen burner

Text continues on next page. →

MasteringA&P®

For related exercise study tools, go to the Study Area of **MasteringA&P.** There you will find:

- Practice Anatomy Lab PAL
- A&PFlix **A&PFlix**
- PhysioEx **PEx**
- Practice quizzes, Histology Atlas, eText, Videos, and more!

Pre-Lab Quiz

1. Circle the correct underlined term. <u>Excitability</u> / <u>Conductivity</u> is the ability to transmit nerve impulses to other neurons.

2. When a neuron is stimulated, the membrane becomes more permeable to Na^+ ions, which diffuse into the cell and cause:
   a. depolarization
   b. hyperpolarization
   c. repolarization

3. As an action potential progresses, the permeability to Na^+ decreases, and the permeability to this ion increases:
   a. Ca^{2+} b. K^+ c. Na^+

4. The period of time when the neuron is totally insensitive to further stimulation and cannot generate another action potential is:
   a. absolute refractory period   c. repolarization
   b. membrane potential   d. threshold

5. What muscle and nerve will you need to isolate to study the physiology of nerve fibers?
   a. gastrocnemius and sciatic
   b. sartorius and femoral
   c. triceps brachii and radial

N eurons are **excitable;** they respond to stimuli by producing an electrical signal. Excited neurons communicate—they transmit electrical signals to neurons, muscles, glands, and other tissues of the body, a property called **conductivity.** In a resting neuron, the interior of the cell membrane is slightly

- Safety goggles
- Absorbent cotton
- Ether
- Pipettes
- 1-cc syringe with small-gauge needle
- 0.5% tubocurarine solution
- Frog board
- Oscilloscope
- Nerve chamber

*Instructor to provide freshly pithed frogs (*Rana pipiens*) for student experimentation.

PEx PhysioEx™ 9.1 Computer Simulation Ex. 3 on p. PEx-35.

more negatively charged than the exterior surface (**Figure 16.1**). The difference in electrical charge produces a **resting membrane potential** across the membrane that is measured in millivolts. As in most cells, the predominant intracellular cation is K^+; Na^+ is the predominant cation in the extracellular fluid. In a resting neuron, Na^+ leaks into the cell and K^+ leaks out. The resting membrane potential is maintained by the sodium-potassium pump, which transports Na^+ back out of the cell and K^+ back into the cell.

The Action Potential

When a neuron receives an excitatory stimulus, the membrane becomes more permeable to sodium ions, and Na^+ diffuses down its electrochemical gradient into the cell. As a result, the interior of the membrane becomes less negative

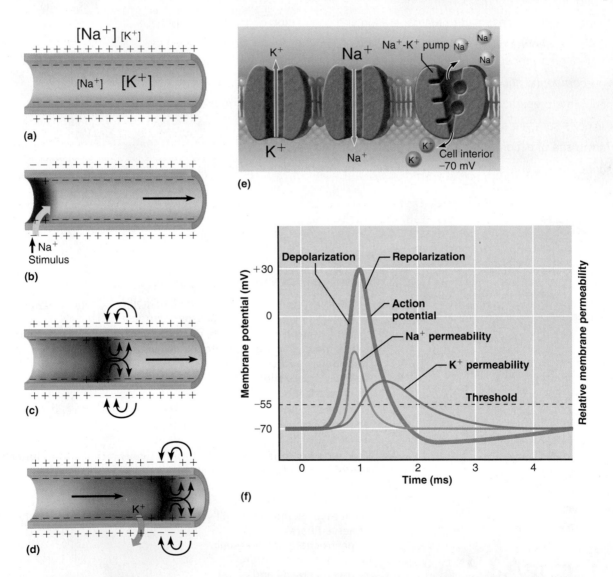

Figure 16.1 The action potential.
(a) Resting membrane potential (RMP). There is an excess of positive ions at the external cell surface, with Na^+ the predominant extracellular fluid ion and K^+ the predominant intracellular ion. The plasma membrane has a low permeability to Na^+. **(b)** Depolarization—reversal of the RMP. Application of a stimulus changes the membrane permeability, and Na^+ ions are allowed to diffuse rapidly into the cell. **(c)** Generation of the action potential. If the stimulus is of adequate intensity, the depolarization wave spreads rapidly along the entire length of the membrane. **(d)** Repolarization— reestablishment of the RMP. The negative charge on the internal plasma membrane surface and the positive charge on its external surface are reestablished by diffusion of K^+ ions out of the cell, proceeding in the same direction as in depolarization. **(e)** In the resting state, Na^+ ions leak into the cell and K^+ ions leak out. The RMP is maintained by the active sodium-potassium pump. **(f)** The action potential is caused by permeability changes in the plasma membrane.

(Figure 16.1b), an event called **depolarization.** If the stimulus is great enough to depolarize the initial segment of the axon to **threshold,** an **action potential** is generated. The initial segment of the axon in multipolar neurons is at the axon hillock of the cell body. In peripheral sensory neurons, the initial segment is just proximal to the sensory receptor, far from the cell body located in the dorsal root ganglion.

When the threshold voltage is reached, the membrane permeability to Na^+ increases rapidly (Figure 16.1f). As the neuron depolarizes, the polarity of the membrane reverses: the interior surface now becomes more positive than the exterior (Figure 16.1c). As the membrane permeability to Na^+ falls, the permeability to K^+ increases, and K^+ diffuses down its electrochemical gradient and out of the cell (Figure 16.1d). Once again the interior of the membrane becomes more negative than the exterior. This event is called **repolarization.** As you can see, the action potential is a brief reversal of the neuron's membrane potential.

The period of time when Na^+ permeability is rapidly changing and maximal, and the period immediately following when Na^+ permeability becomes restricted, together correspond to a time when the neuron is insensitive to further stimulation and cannot generate another action potential. This period is called the **absolute refractory period.** As Na^+ permeability is gradually restored to resting levels during repolarization, an especially strong stimulus to the neuron may provoke another action potential. This period of time is the **relative refractory period.** Restoration of the resting membrane potential restores the neuron's normal excitability.

Once generated, the action potential propagates along the entire length of the axon. It is never partially transmitted. Furthermore, it retains a constant amplitude and duration; the action potential is not small when a stimulus is small and large when a stimulus is large. Since the action potential of a given neuron is always the same, it is said to be an all-or-none response. When an action potential reaches the axon terminals, it causes neurotransmitter to be released. The neurotransmitter may be excitatory or inhibitory to the next cell in the transmission chain, depending on the receptor types on that cell. (The experiments in this exercise consider only excitatory neurotransmitters.)

Physiology of Nerves

The sciatic nerve is a bundle of axons that vary in diameter. An electrical signal recorded from a nerve represents the summed electrical activity of all the axons in the nerve. This summed activity is called a **compound action potential.** Unlike an action potential in a single axon, the compound action potential varies in shape according to which axons are producing action potentials. When a nerve is stimulated by external electrodes, as in our experiments, the largest axons reach threshold first and generate action potentials. Higher-intensity stimuli are required to produce action potentials in smaller axons.

In this laboratory session, you will investigate the functioning of a nerve by subjecting the sciatic nerve of a frog to various types of stimuli and blocking agents. Work in groups of two to four to lighten the workload.

16

✂ DISSECTION

Isolating the Gastrocnemius Muscle and Sciatic Nerve

⚠ **1. Don** gloves to protect yourself from any parasites the frogs might have. Obtain a pithed frog from your instructor, and bring it to your laboratory bench. Also obtain dissecting instruments, a tray, thread, two glass rods or probes, and frog Ringer's solution at room temperature from the supply area.

2. Prepare the sciatic nerve as illustrated (**Figure 16.2**). Place the pithed frog on the dissecting tray, dorsal side down. Make a cut through the skin around the circumference of the frog approximately halfway down the trunk, and then pull the skin down over the muscles of the legs.

Text continues on next page. →

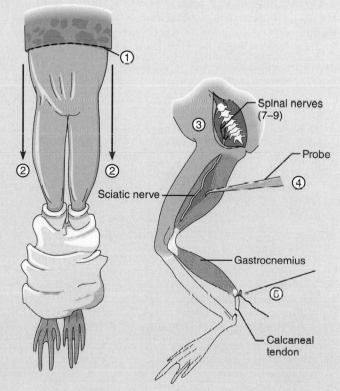

Figure 16.2 Removal of the sciatic nerve and gastrocnemius muscle. ① Cut through the frog's skin around the circumference of the trunk. ② Pull the skin down over the trunk and legs. ③ Make a longitudinal cut through the abdominal musculature, and expose the roots of the sciatic nerve (arising from spinal nerves 7–9). Ligate the nerve and cut the roots proximal to the ligature. ④ Use a glass probe to expose the sciatic nerve beneath the posterior thigh muscles. ⑤ Ligate the calcaneal tendon, and cut it free distal to the ligature. Release the gastrocnemius muscle from the connective tissue of the knee region.

Open the abdominal cavity, and push the abdominal organs to one side to expose the origin of the glistening white sciatic nerve, which arises from the last three spinal nerves. _Once the sciatic nerve has been exposed, keep it continually moist with room-temperature frog Ringer's solution._

3. Using a glass probe, slip a piece of thread moistened with frog Ringer's solution under the sciatic nerve close to its origin at the vertebral column. Make a single ligature (tie it firmly with the thread), and then cut through the nerve roots to free the proximal end of the sciatic nerve from its attachments. Using a glass rod or probe, carefully separate the posterior thigh muscles to locate and then free the sciatic nerve, which runs down the posterior aspect of the thigh.

4. Tie a piece of thread around the calcaneal tendon of the gastrocnemius muscle, and then cut through the tendon distal to the ligature to free the gastrocnemius muscle from the heel. Using a scalpel, carefully release the gastrocnemius muscle from the connective tissue in the knee region. At this point you should have completely freed both the gastrocnemius muscle and the sciatic nerve, which innervates it.

Activity 1

Stimulating the Nerve

In this first set of experiments, stimulation of the nerve and generation of the compound action potential will be indicated by contraction of the gastrocnemius muscle. Because you will make no mechanical recording (unless your instructor asks you to), you must keep complete and accurate records of all experimental procedures and results.

1. Obtain a glass slide or plate, ring stand and clamp, stimulator, electrodes, salt (NaCl), forceps, filter paper, 0.01% hydrochloric acid (HCl) solution, Bunsen burner, and heat-resistant mitts. With glass rods, transfer the isolated muscle-nerve preparation to a glass plate or slide, and then attach the slide to a ring stand with a clamp. Allow the end of the sciatic nerve to hang over the free edge of the glass slide, so that it is easily accessible for stimulation. _Remember to keep the nerve moist at all times._

2. You are now ready to investigate the response of the sciatic nerve to various stimuli, beginning with electrical stimulation. Using the stimulator and platinum electrodes, stimulate the sciatic nerve with single shocks, gradually increasing the intensity of the stimulus until the threshold stimulus is determined.

The muscle as a whole will just barely contract at the threshold stimulus. Record the voltage of this stimulus:

Threshold stimulus: _____ V

Continue to increase the voltage until you find the point beyond which no further increase occurs in the strength of muscle contraction—that is, the point at which the maximal contraction of the muscle is obtained. Record this voltage below.

Maximal stimulus: _____ V

Delivering multiple or repeated shocks to the sciatic nerve causes volleys of impulses in the nerve. Shock the nerve with multiple stimuli. Observe the response of the muscle. How does this response compare with the response to the single electrical shocks?

_____

_____

3. To investigate mechanical stimulation, pinch the free end of the nerve by firmly pressing it between two glass rods or by pinching it with forceps. What is the result?

_____

4. Test chemical stimulation by applying a small piece of filter paper saturated with HCl solution to the free end of the nerve. What is the result?

_____

Drop a few crystals of salt (NaCl) on the free end of the nerve. What is the result?

_____

5. Now test thermal stimulation. Wearing the heat-resistant mitts, heat a glass rod for a few moments over a Bunsen burner. Then touch the rod to the free end of the nerve. What is the result?

_____

What do these muscle reactions say about the excitability and conductivity of neurons?

_____

_____

Most neurons within the body are stimulated to the greatest degree by a particular stimulus (in many cases, a chemical neurotransmitter), but a variety of other stimuli may trigger nerve impulses, as seen in the experimental series just conducted. Generally, no matter what type of stimulus is present, if the affected part responds by becoming activated, it will always react in the same way. Familiar examples are the well-known phenomenon of "seeing stars" when you receive a blow to the head or press on your eyeball, both of which trigger impulses in your optic nerves.

Activity 2

Inhibiting the Nerve

Numerous physical factors and chemical agents can impair the ability of nerve fibers to function. For example, deep pressure and cold temperature both block nerve impulse transmission by preventing the local blood supply from reaching the nerve fibers. Local anesthetics, alcohol, and numerous other chemicals are also very effective at blocking nerve transmission. Ether, one such chemical blocking agent, will be investigated first.

! Since ether is extremely volatile and explosive, perform this experiment in a vented hood. *Don safety goggles before beginning this procedure.*

1. Obtain another glass slide or plate, absorbent cotton, ether, and a pipette. Clamp the new glass slide to the ring stand slightly below the first slide of the apparatus setup for the previous experiment. With glass rods, gently position the sciatic nerve on this second slide, allowing a small portion of the nerve's distal end to extend over the edge. Place a piece of absorbent cotton soaked with ether under the midsection of the nerve on the slide, prodding it into position with a glass rod. Using a voltage slightly above the threshold stimulus, stimulate the distal end of the nerve at 2-minute intervals until the muscle fails to respond. (If the cotton dries before this, re-wet it with ether using a pipette.) How long did it take for anesthesia to occur?

_____ sec

2. Once anesthesia has occurred, stimulate the nerve beyond the anesthetized area, between the ether-soaked pad and the muscle. What is the result?

_____

3. Remove the ether-soaked pad and flush the nerve fibers with Ringer's solution. Again stimulate the nerve at its distal end at 2-minute intervals. How long does it take for recovery?

_____

Does ether exert its blocking effect on the nerve *or* on the muscle fibers?

_____ Explain your reasoning.

_____

_____

If sufficient frogs are available and time allows, you may do the following experiment. *Curare* was used by

some South American Indian tribes to tip their arrows. Victims struck with these arrows were paralyzed, but the paralysis was not accompanied by loss of sensation.

1. Prepare another frog as described in steps 1 through 3 of the dissection instructions. However, in this case position the frog ventral side down on a frog board. In exposing the sciatic nerve, take care not to damage the blood vessels in the thigh region, as the success of the experiment depends on maintaining the blood supply to the muscles of the leg.

2. Expose and gently tie the left sciatic nerve so that it can be lifted away from the muscles of the leg for stimulation. Slip another length of thread under the nerve, and then tie the thread tightly around the thigh muscles to cut off circulation to the leg. The sciatic nerve should be above the thread and *not in* the ligated tissue. Expose and ligate the sciatic nerve of the *right* leg in the same manner, but this time do *not* ligate the thigh muscles.

! 3. *Take great care in handling tubocurarine, because it is extremely poisonous.* Do not get any on your skin. Get a syringe and needle, and a vial of 0.5% tubocurarine, the main toxin in curare. Obtain 1 cc of the tubocurarine by injecting 1 cc of air into the vial through the rubber membrane, and then drawing up 1 cc of the chemical into the syringe. Slowly and carefully inject 1 cc of the tubocurarine into the dorsal lymph sac of the frog. The dorsal lymph sacs are located dorsally at the level of the scapulae, so introduce the needle of the syringe just beneath the skin between the scapulae and toward one side of the spinal column.

4. Wait 15 minutes after injection of the tubocurarine to allow it to be distributed throughout the body in the blood and lymphatic stream. Then electrically stimulate the left sciatic nerve. Be careful not to touch any of the other tissues with the electrode. Gradually increasing the voltage, deliver single shocks until the threshold stimulus is determined for this specimen.

Threshold stimulus: _____ V

Now stimulate the right sciatic nerve with the same voltage intensity. Is there any difference in the reaction of the two muscles?

_____ If so, explain. _____

_____

_____

Text continues on next page. →

16

If you did not find any difference, wait an additional 10 to 15 minutes and restimulate both sciatic nerves.

What is the result? _____

_____

5. To determine the site at which tubocurarine acts, directly stimulate each gastrocnemius muscle. What is the result?

_____

_____

_____

Explain the difference between the responses of the right and left sciatic nerves.

_____

_____

Explain the results when the muscles were stimulated directly.

_____

_____

_____

_____

At what site does tubocurarine (or curare) act?

_____

_____

The Oscilloscope: An Experimental Tool

In this exercise, a frog's sciatic nerve will be electrically stimulated, and the compound action potential generated will be observed on the oscilloscope. The **oscilloscope** is an instrument that visually displays the rapid changes in voltage that occur during an action potential. The dissected nerve will be placed in contact with two pairs of electrodes—*stimulating* and *recording*. The stimulating electrodes will be used to deliver a pulse of electricity to a point on the sciatic nerve. At another point on the nerve, a pair of recording electrodes connected to the oscilloscope will deliver the signal to the oscilloscope, and the electrical pulse will be visible on the screen as a vertical deflection, or a *stimulus artifact* (**Figure 16.3**). As the nerve is stimulated with increasingly higher voltage, the stimulus artifact increases in amplitude as well. When the stimulus voltage reaches a high enough level, a compound action potential will be generated by the nerve, and a *second* vertical deflection will appear on the screen, approximately 2 milliseconds after the stimulus artifact (Figure 16.3b–d). This second deflection reports the potential difference that represents the compound action potential.

Activity 3

Visualizing the Compound Action Potential with an Oscilloscope

1. Obtain a nerve chamber, an oscilloscope, a stimulator, frog Ringer's solution (room temperature), a dissecting needle, and glass probes. Set up the experimental apparatus as illustrated (**Figure 16.4**). Connect the two stimulating electrodes to the output terminals of the stimulator and the two recording electrodes to the oscilloscope.

2. Obtain another pithed frog, and prepare one of its sciatic nerves for experimentation as indicated in steps 1 through 3 of the dissection instructions (p. 271). While

(a)

(b)

(c)

(d)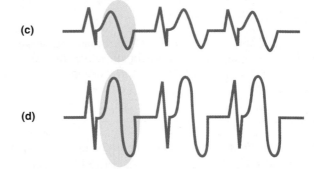

Figure 16.3 Recordings of the compound action potential from the sciatic nerve. The first compound action potential in each scan is circled. **(a)** Stimulus artifacts only. **(b–d)** Increasing stimulus strengths reveal the graded nature of the compound action potential.

working, be careful not to touch the nerve with your fingers, and do not allow the nerve to touch the frog's skin.

3. When you have freed the sciatic nerve to the knee region with the glass probe, slip another thread length beneath that end of the nerve and make a ligature. Cut the

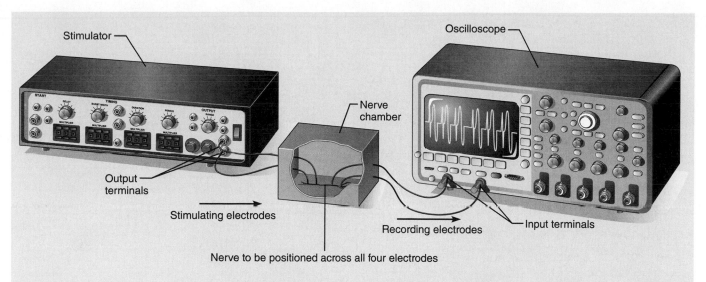

Figure 16.4 Setup for oscilloscope visualization of action potentials in a nerve.

nerve distal to this tied thread, and then carefully lift the cut nerve away from the thigh of the frog by holding the threads at the nerve's proximal and distal ends. Place the nerve in the nerve chamber so that it rests across all four electrodes—the two stimulating and two recording electrodes (see Figure 16.4). Flush the nerve with room-temperature frog Ringer's solution.

4. Adjust the horizontal sweep according to the instructions given in the manual or by your instructor, and set the stimulator duration, frequency, and amplitude to their lowest settings.

5. Begin to stimulate the nerve with single stimuli, slowly increasing the voltage until a threshold stimulus is achieved. The compound action potential will appear as a small rounded "hump" immediately following the stimulus artifact. Record the voltage of the threshold stimulus:

Threshold stimulus: _____ V

6. Flush the nerve with the frog Ringer's solution and continue to increase the voltage, watching as the vertical deflections produced by the compound action potential become diphasic (show both upward and downward vertical deflections). Record the voltage at which the compound action potential reaches its maximal amplitude; this is the maximal stimulus.

Maximal stimulus: _____ V

7. Set the stimulus voltage at a level just slightly lower than the maximal stimulus, and gradually increase the

frequency of stimulation. What is the effect on the amplitude of the compound action potential?

_____

8. Flush the nerve with room-temperature frog Ringer's solution once again, and allow the nerve to sit for a few minutes while you obtain a bottle of frog Ringer's solution from the ice bath. Repeat steps 5 and 6 while your partner continues to flush the nerve preparation with the cold frog Ringer's solution. Record the threshold and maximal stimuli, and watch the oscilloscope pattern carefully to detect any differences in the velocity or speed of conduction from what was seen previously.

Threshold stimulus: _____ V

Maximal stimulus: _____ V

9. Flush the nerve preparation with room-temperature frog Ringer's solution again, and then gently lift the nerve by its attached threads. Then turn the nerve around so that the end formerly resting on the stimulating electrodes now rests on the recording electrodes and vice versa. Stimulate the nerve. Is the impulse conducted in the opposite direction?

_____

10. Dispose of the frog remains and gloves in the appropriate containers, clean the lab bench and equipment, and return your equipment to the proper supply area.

REVIEW SHEET
Neurophysiology of Nerve Impulses:
Frog Subjects

Name _____ Lab Time/Date _____

The Action Potential

1. Match the terms in column B to the appropriate definition in column A.

Column A

_____ 1. period of depolarization of the neuron membrane during which it cannot respond to a second stimulus

_____ 2. reversal of the resting potential due to an influx of sodium ions

_____ 3. period during which potassium ions diffuse out of the neuron because of a change in membrane permeability

_____ 4. period of repolarization when only a strong stimulus will elicit an action potential

_____ 5. mechanism in which ATP is used to move sodium out of the cell and potassium into the cell; restores the resting membrane voltage and intracellular ionic concentrations

Column B

a. absolute refractory period

b. action potential

c. depolarization

d. relative refractory period

e. repolarization

f. sodium-potassium pump

2. Define the term *depolarization*. _____

_____

How does an action potential differ from simple depolarization? _____

_____

3. Would a substance that decreases membrane permeability to sodium increase or decrease the probability of generating an action potential? Why?

_____

_____

4. The diagram here represents a section of an axon. Complete the figure by illustrating an area of resting membrane potential, an area of depolarization, and local current flow. Indicate the direction of the depolarization wave.

$[Na^+]$ $[K^+]$

$[Na^+]$ $[K^+]$

Physiology of Nerves Stimulating and Inhibiting the Nerve

5. Respond appropriately to each question posed below. Insert your responses in the corresponding numbered blanks to the right.

1–3.   Name three types of stimuli that resulted in action potential generation in the sciatic nerve of the frog.

1. _____

2. _____

4.   Which of the stimuli resulted in the most effective nerve stimulation?

3. _____

5.   Which of the stimuli employed in that experiment might represent
types of stimuli to which nerves in the human body are subjected?

4. _____

6.   What is the usual mode of stimulus transfer in neuron-to-neuron
interactions?

5. _____

6. _____

7.   Since the action potentials themselves were not visualized with an
oscilloscope during this initial set of experiments, how did you recognize
that impulses were being transmitted?

7. _____

6. How did the site of action of ether and tubocurarine differ?

_____

In the tubocurarine experiment, why was one of the frog's legs ligated? _____

_____

Visualizing the Compound Action Potential with an Oscilloscope

7. Explain why the amplitude of the compound action potential recorded from the frog sciatic nerve increased when the

voltage of the stimulus was increased above the threshold value. _____

_____

8. What was the effect of cold temperature (flooding the nerve with iced frog Ringer's solution) on the functioning of the

sciatic nerve tested? _____

_____

9. When the nerve was reversed in position, was the impulse conducted in the opposite direction? _____

How can this result be reconciled with the concept of one-way conduction in neurons? _____

_____

_____

_____

Gross Anatomy of the Brain and Cranial Nerves

Objectives

☐ List the elements of the central and peripheral divisions of the nervous system.

☐ Discuss the difference between the sensory and motor portions of the nervous system, and name the two divisions of the motor portion.

☐ Recognize the terms that describe the development of the human brain, and discuss the relationships between the terms.

☐ As directed by your instructor, identify the bold terms associated with the cerebral hemispheres, diencephalon, brain stem, and cerebellum on a dissected human brain, brain model, or appropriate image. State the functions of these structures.

☐ State the difference between gyri, fissures, and sulci.

☐ Describe the composition of gray matter and white matter in the nervous system.

☐ Name and describe the three meninges that cover the brain, state their functions, and locate the falx cerebri, falx cerebelli, and tentorium cerebelli.

☐ Discuss the formation, circulation, and drainage of cerebrospinal fluid.

☐ Identify the cranial nerves by number and name on a model or image, stating the origin and function of each.

☐ Identify at least four key anatomical differences between the human and sheep brain.

Materials

- Human brain model (dissectible)
- Preserved human brain (if available)
- Three-dimensional model of ventricles
- Frontally sectioned human brain slice (if available)
- Materials as needed for cranial nerve testing (see Table 17.1): aromatic oils (e.g., vanilla and cloves); eye chart; ophthalmoscope; penlight; safety pin; blunt probe; cotton; solutions of sugar, salt, vinegar, and quinine; ammonia; tuning fork; and tongue depressor

Text continues on next page. →

MasteringA&P®

For related exercise study tools, go to the Study Area of **MasteringA&P.** There you will find:

- Practice Anatomy Lab **PAL**
- PhysioEx **PEx**
- A&PFlix **A&PFlix**
- Practice quizzes, Histology Atlas, eText, Videos, and more!

Pre-Lab Quiz

1. Circle the correct underlined term. The <u>central nervous system</u> / <u>peripheral nervous system</u> consists of the brain and spinal cord.

2. Circle the correct underlined term. The most superior portion of the brain includes the <u>cerebral hemispheres</u> / <u>brain stem</u>.

3. Circle True or False. Deep grooves within the cerebral hemispheres are known as gyri.

4. On the ventral surface of the brain, you can observe the optic nerves and chiasma, the pituitary gland, and the mammillary bodies. These externally visible structures form the floor of the:
   a. brain stem   b. diencephalon   c. frontal lobe   d. occipital lobe

5. Circle the correct underlined term. The inferior region of the brain stem, the <u>medulla oblongata</u> / <u>cerebellum</u> houses many vital autonomic centers involved in the control of heart rate, respiratory rhythm, and blood pressure.

6. Directly under the occipital lobe of the cerebrum is a large cauliflower-like structure known as the _____.
   a. brain stem   b. cerebellum   c. diencephalon

7. Circle the correct underlined term. The outer cortex of the brain contains the cell bodies of cerebral neurons and is known as <u>white matter</u> / <u>gray matter</u>.

8. The brain and spinal cord are covered and protected by three connective tissue layers called:
   a. lobes   b. meninges   c. sulci   d. ventricles

Text continues on next page. →

- Preserved sheep brain (meninges and cranial nerves intact)
- Dissecting instruments and tray
- Disposable gloves

9. Circle True or False. Cerebrospinal fluid is produced by the frontal lobe of the cerebrum and is unlike any other body fluid.
10. How many pairs of cranial nerves are there? _____

When viewed alongside all Earth's animals, humans are indeed unique, and the key to their uniqueness is found in the brain. Each of us is a reflection of our brain's experience. If all past sensory input could mysteriously and suddenly be "erased," we would be unable to walk, talk, or communicate in any manner. Spontaneous movement would occur, as in a fetus, but no voluntary integrated function of any type would be possible. Clearly we would cease to be the same individuals.

For convenience, the nervous system is considered in terms of two principal divisions: the central nervous system and the peripheral nervous system. The **central nervous system (CNS)** consists of the brain and spinal cord, which primarily interpret incoming sensory information and issue instructions based on that information and on past experience. The **peripheral nervous system (PNS)** consists of the cranial and spinal nerves, ganglia, and sensory receptors.

The PNS has two major subdivisions: the **sensory portion,** which consists of nerve fibers that conduct impulses from sensory receptors toward the CNS, and the **motor portion,** which contains nerve fibers that conduct impulses away from the CNS. The motor portion, in turn, consists of the **somatic division** (sometimes called the *voluntary system*), which controls the skeletal muscles, and the **autonomic nervous system (ANS),** which controls smooth and cardiac muscles and glands.

In this exercise the brain (CNS) and cranial nerves (PNS) will be studied together because of their close anatomical relationship.

17   The Human Brain

During embryonic development of all vertebrates, the CNS first makes its appearance as a simple tubelike structure, the **neural tube,** that extends down the dorsal median plane. By the fourth week, the human brain begins to form as an expansion of the anterior or rostral end of the neural tube (the end toward the head). Shortly thereafter, constrictions appear, dividing the developing brain into three major regions— **forebrain, midbrain,** and **hindbrain (Figure 17.1).** The remainder of the neural tube becomes the spinal cord.

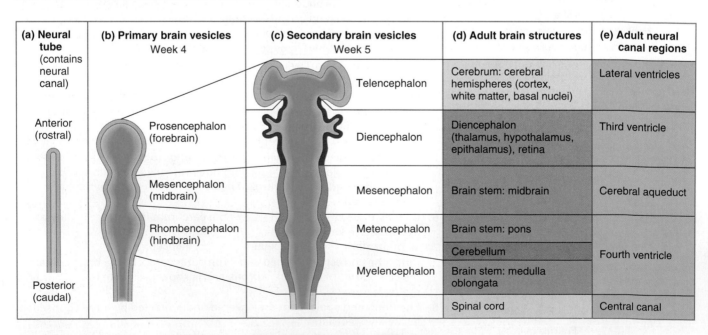

(a) Neural tube (contains neural canal)	(b) Primary brain vesicles Week 4	(c) Secondary brain vesicles Week 5	(d) Adult brain structures	(e) Adult neural canal regions
Anterior (rostral)	Prosencephalon (forebrain)	Telencephalon	Cerebrum: cerebral hemispheres (cortex, white matter, basal nuclei)	Lateral ventricles
		Diencephalon	Diencephalon (thalamus, hypothalamus, epithalamus), retina	Third ventricle
	Mesencephalon (midbrain)	Mesencephalon	Brain stem: midbrain	Cerebral aqueduct
	Rhombencephalon (hindbrain)	Metencephalon	Brain stem: pons	Fourth ventricle
			Cerebellum	Fourth ventricle
Posterior (caudal)		Myelencephalon	Brain stem: medulla oblongata	
			Spinal cord	Central canal

Figure 17.1 Embryonic development of the human brain. (a) The neural tube subdivides into **(b)** the primary brain vesicles, which subsequently form **(c)** the secondary brain vesicles, which differentiate into **(d)** the adult brain structures. **(e)** The adult structures derived from the neural canal.

The central canal of the neural tube, which remains continuous throughout the brain and cord, enlarges in four regions of the brain, forming chambers called **ventricles** (see Figure 17.8a and b, p. 289).

Activity 1

Identifying External Brain Structures

Identify external brain structures using the figures cited. Also use a model of the human brain and other learning aids as they are mentioned.

Generally, the brain is studied in terms of four major regions: the cerebral hemispheres, diencephalon, brain stem, and cerebellum.

Cerebral Hemispheres

The **cerebral hemispheres** are the most superior portion of the brain (**Figure 17.2**, p. 282). Their entire surface is thrown into elevated ridges of tissue called **gyri** that are separated by shallow grooves called **sulci** or deeper grooves called **fissures**. Many of the fissures and gyri are important anatomical landmarks.

The cerebral hemispheres are divided by a single deep fissure, the **longitudinal fissure**. The **central sulcus** divides the **frontal lobe** from the **parietal lobe,** and the **lateral sulcus** separates the **temporal lobe** from the parietal lobe. The **parieto-occipital sulcus** on the medial surface of each hemisphere divides the **occipital lobe** from the parietal lobe. The sulcus is not visible externally. A fifth lobe of each cerebral hemisphere, the **insula,** is buried deep within the lateral sulcus, and is covered by portions of the temporal, parietal, and frontal lobes. Notice that most cerebral hemisphere lobes are named for the cranial bones that lie over them.

Some important functional areas of the cerebral hemispheres have also been located (Figure 17.2d).

The cell bodies of cerebral neurons involved in these functions are found only in the outermost gray matter of the cerebrum, the **cerebral cortex**. Most of the balance of cerebral tissue—the deeper **cerebral white matter**—is composed of myelinated fibers bundled into tracts carrying impulses to or from the cortex. Some of these functional areas are summarized in **Table 17.1**.

Using a model of the human brain (and a preserved human brain, if available), identify the areas and structures of the cerebral hemispheres described above.

Then continue using the model and preserved brain along with the figures as you read about other structures.

Diencephalon

The **diencephalon** is embryologically part of the forebrain, along with the cerebral hemispheres (see Figure 17.1).

Turn the brain model so the ventral surface of the brain can be viewed. Identify the externally visible structures that mark the position of the floor of the diencephalon using **Figure 17.3** on p. 283 as a guide. These are the **olfactory bulbs** (synapse point of cranial nerve I) and **tracts, optic nerves** (cranial nerve II), **optic chiasma** (where the medial fibers of the optic nerves cross over), **optic tracts, pituitary gland**, and **mammillary bodies**.

Brain Stem

Continue to identify the **brain stem** structures—the **cerebral peduncles** (fiber tracts in the **midbrain** connecting the pons below with cerebrum above), the pons, and the medulla oblongata. *Pons* means "bridge," and the **pons** consists primarily of motor and sensory fiber tracts connecting the brain with lower CNS centers. The lowest brain stem region, the **medulla oblongata** (often shortened to just *medulla*), is also composed primarily of fiber tracts. You can see the **decussation of pyramids**, a crossover point for the major motor tracts (pyramidal tracts) descending

Table 17.1	Important Functional Areas of the Cerebral Cortex	
Functional sensory areas	**Location**	**Functions**
Primary somatosensory cortex	Postcentral gyrus of the parietal lobe	Receives information from the body's sensory receptors in the skin and from proprioceptors in the skeletal muscles, joints, and tendons
Primary visual cortex	Occipital lobe	Receives visual information that originates in the retina of the eye
Primary auditory cortex	Temporal lobe in the gyrus bordering the lateral sulcus	Receives sound information from the receptors for hearing in the internal ear
Olfactory cortex	Medial surface of the temporal lobe, in a region called the uncus	Receives information from olfactory (smell) receptors in the superior nasal cavity
Functional motor areas	**Location**	**Functions**
Primary motor cortex	Precentral gyrus of the frontal lobe	Conscious control of voluntary movement of skeletal muscles
Broca's area	At the base of the precentral gyrus of the frontal lobe just above the superior sulcus, in only one hemisphere	Controls the muscles involved in speech production and also plays a role in the planning of nonspeech motor functions

Text continues on page 283. ➡

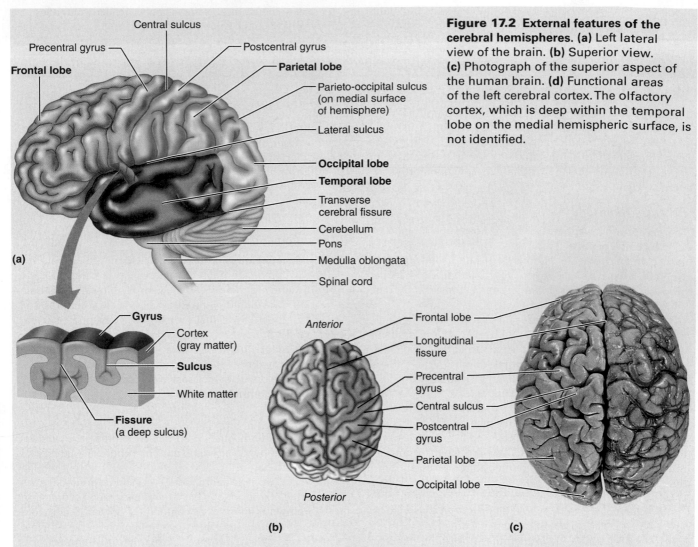

Figure 17.2 External features of the cerebral hemispheres. (a) Left lateral view of the brain. **(b)** Superior view. **(c)** Photograph of the superior aspect of the human brain. **(d)** Functional areas of the left cerebral cortex. The olfactory cortex, which is deep within the temporal lobe on the medial hemispheric surface, is not identified.

(a)

Central sulcus
Precentral gyrus
Postcentral gyrus
Frontal lobe
Parietal lobe
Parieto-occipital sulcus (on medial surface of hemisphere)
Lateral sulcus
Occipital lobe
Temporal lobe
Transverse cerebral fissure
Cerebellum
Pons
Medulla oblongata
Spinal cord

Gyrus
Cortex (gray matter)
Sulcus
White matter
Fissure (a deep sulcus)

Anterior
Frontal lobe
Longitudinal fissure
Precentral gyrus
Central sulcus
Postcentral gyrus
Parietal lobe
Occipital lobe
Posterior

(b)

(c)

17

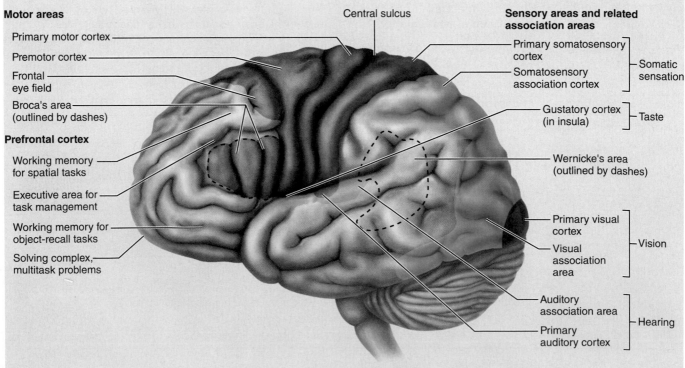

(d)

Motor areas
Primary motor cortex
Premotor cortex
Frontal eye field
Broca's area (outlined by dashes)
Prefrontal cortex
Working memory for spatial tasks
Executive area for task management
Working memory for object-recall tasks
Solving complex, multitask problems

Central sulcus

Sensory areas and related association areas
Primary somatosensory cortex
Somatosensory association cortex
— Somatic sensation
Gustatory cortex (in insula) — Taste
Wernicke's area (outlined by dashes)
Primary visual cortex
Visual association area — Vision
Auditory association area
Primary auditory cortex — Hearing

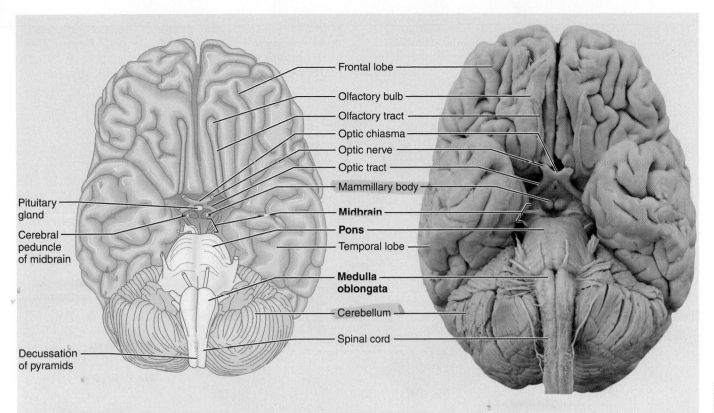

Figure 17.3 Ventral (inferior) aspect of the human brain, showing the three regions of the brain stem. Only a small portion of the midbrain can be seen; the rest is surrounded by other brain regions.

17

from the motor areas of the cerebrum to the spinal cord, on the medulla's surface. The medulla also houses many vital autonomic centers involved in the control of heart rate, respiratory rhythm, and blood pressure as well as involuntary centers involved in vomiting and swallowing.

Cerebellum

1. Turn the brain model so you can see the dorsal aspect. Identify the large cauliflower-shaped **cerebellum**, which projects dorsally from under the occipital lobe of the cerebrum. Notice that, like the cerebrum, the cerebellum has two major hemispheres and a convoluted surface (see Figure 17.6). It also has an outer cortex made up of gray matter with an inner region of white matter.

2. Remove the cerebellum to view the **corpora quadrigemina** (Figure 17.4), located on the posterior aspect of the midbrain, a brain stem structure. The two superior prominences are the **superior colliculi** (visual reflex centers); the two smaller inferior prominences are the **inferior colliculi** (auditory reflex centers).

Activity 2

Identifying Internal Brain Structures

The deeper structures of the brain have also been well mapped. As the internal brain areas are described, identify them on the figures cited. Also, use the brain model as indicated to help you in this study.

Cerebral Hemispheres

1. Take the brain model apart so you can see a medial view of the internal brain structures (**Figure 17.4**, p. 284). Observe the model closely to see the extent of the outer cortex (gray matter), which contains the cell bodies of cerebral neurons. The pyramidal cells of the cerebral motor cortex (studied in Exercise 15 and Figure 15.5) are representative of the neurons seen in the precentral gyrus.

2. Observe the deeper area of white matter, which is composed of fiber tracts. The fiber tracts found in the cerebral hemisphere white matter are called *association tracts* if they connect two portions of the same hemisphere, *projection tracts* if they run between the cerebral cortex and lower brain structures or spinal cord, and *commissures* if they run from one hemisphere to another. Observe the large **corpus callosum**, the major commissure connecting the cerebral hemispheres. The corpus callosum arches above the structures of the diencephalon and roofs over the lateral ventricles. Notice also the **fornix,** a bandlike

Text continues on next page. →

fiber tract concerned with olfaction as well as limbic system functions, and the membranous **septum pellucidum,** which separates the lateral ventricles of the cerebral hemispheres.

3. In addition to the gray matter of the cerebral cortex, there are several clusters of neuron cell bodies called **nuclei** buried deep within the white matter of the cerebral hemispheres. One important group of cerebral

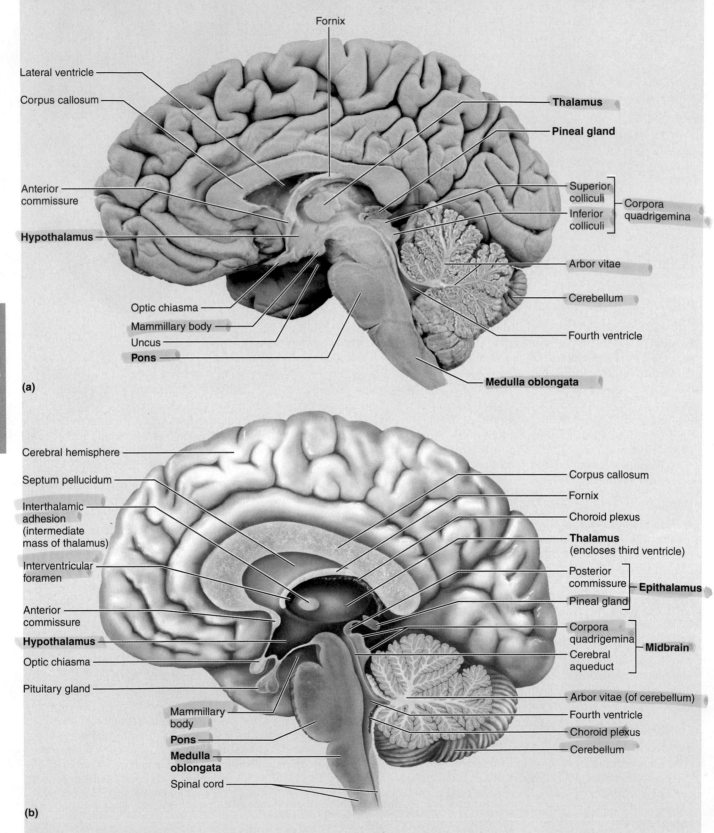

(a)

(b)

Figure 17.4 Diencephalon and brain stem structures as seen in a medial section of the brain. (a) Photograph. **(b)** Diagram.

nuclei, called the **basal nuclei** or **basal ganglia,*** flank the lateral and third ventricles. You can see these nuclei if you have a dissectible model or a coronally or cross-sectioned human brain slice. (Otherwise, **Figure 17.5** will suffice.)

The basal nuclei are involved in regulating voluntary motor activities. The most important of them are the arching, comma-shaped **caudate nucleus,** the **putamen,** and the **globus pallidus.** The closely associated *amygdaloid body* (located at the tip of the tail of the caudate nucleus) is part of the *limbic system.*

*The historical term for these nuclei, *basal ganglia,* is misleading because ganglia are PNS structures. Although technically not the correct anatomical term, "basal ganglia" is included here because it is widely used in clinical settings.

Text continues on next page. →

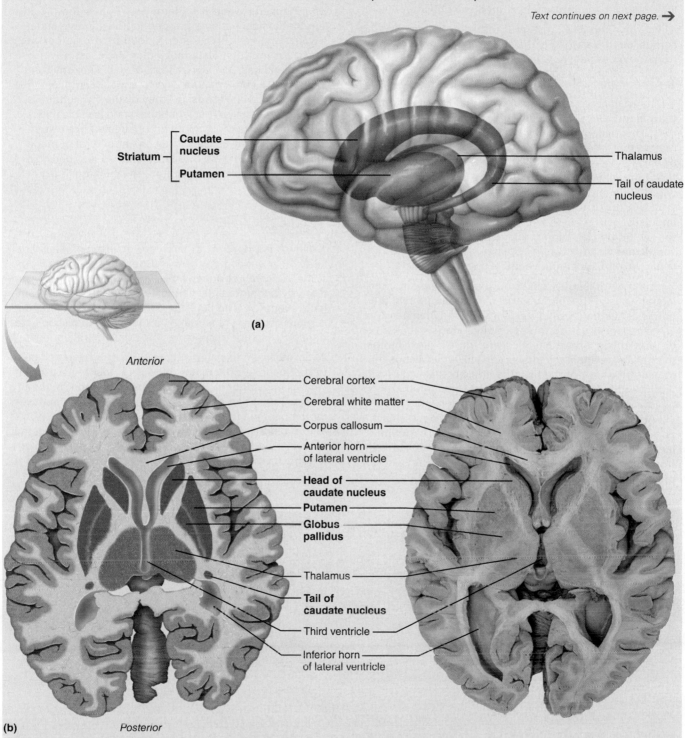

(a)

Striatum — {
Caudate nucleus
Putamen
}

Thalamus

Tail of caudate nucleus

Anterior

Posterior

Cerebral cortex
Cerebral white matter
Corpus callosum
Anterior horn of lateral ventricle
Head of caudate nucleus
Putamen
Globus pallidus
Thalamus
Tail of caudate nucleus
Third ventricle
Inferior horn of lateral ventricle

(b)

Figure 17.5 Basal nuclei. (a) Three-dimensional view of the basal nuclei showing their positions within the cerebrum. **(b)** A transverse section of the cerebrum and diencephalon showing the relationship of the basal nuclei to the thalamus and the lateral and third ventricles.

17

The **corona radiata,** a spray of projection fibers coursing down from the precentral (motor) gyrus, combines with sensory fibers traveling to the sensory cortex to form a broad band of fibrous material called the **internal capsule.** The internal capsule passes between the thalamus and the basal nuclei and through parts of the basal nuclei, giving them a striped appearance. This is why the caudate nucleus and the putamen are sometimes referred to collectively as the **striatum,** or "striped body" (Figure 17.5a).

4. Examine the relationship of the lateral ventricles and corpus callosum to the thalamus and third ventricle—from the cross-sectional viewpoint (see Figure 17.5b).

Diencephalon

1. The major internal structures of the diencephalon are the thalamus, hypothalamus, and epithalamus (see Figure 17.4). The **thalamus** consists of two large lobes of gray matter that laterally enclose the narrow third ventricle of the brain. A slender stalk of thalamic tissue, the **interthalamic adhesion,** or **intermediate mass,** connects the two thalamic lobes and bridges the ventricle. The thalamus is a major integrating and relay station for sensory impulses passing upward to the cortical sensory areas for localization and interpretation. Locate also the **interventricular foramen,** a tiny opening connecting the third ventricle with the lateral ventricle on the same side.

2. The **hypothalamus** makes up the floor and the inferolateral walls of the third ventricle. It is an important autonomic center involved in regulation of body temperature, water balance, and fat and carbohydrate metabolism as well as many other activities and drives (sex, hunger, thirst). Locate again the pituitary gland, which hangs from the anterior floor of the hypothalamus by a slender stalk, the **infundibulum.** In life, the pituitary rests in the hypophyseal fossa of the sella turcica of the sphenoid bone.

Anterior to the pituitary, identify the optic chiasma portion of the optic pathway to the brain. The **mammillary bodies,** relay stations for olfaction, bulge exteriorly from the floor of the hypothalamus just posterior to the pituitary gland.

3. The **epithalamus** forms the roof of the third ventricle and is the most dorsal portion of the diencephalon. Important structures in the epithalamus are the **pineal gland,** and the **choroid plexus** of the third ventricle.

WHY THIS MATTERS | **Vegetative State (VS)**

We appreciate the thalamus more when we look at what happens when it is damaged. This is the problem that afflicts patients trapped in a vegetative state. A vegetative state (VS) is described as a disorder of consciousness that lasts more than 4 weeks in which the patient has lost awareness of self and environment. Unlike coma patients, who cannot be awakened, VS patients can be aroused—they display wakefulness without awareness. The vegetative state is clinical loss of cerebral cortex function with unaffected brain stem function. In many cases, there is no structural damage to the cerebral cortex, but its function is lost because the relay of signals to the cortex—the thalamus—is not working. ■

Brain Stem

1. Now trace the short midbrain from the mammillary bodies to the rounded pons below. (Continue to refer to Figure 17.4). The **cerebral aqueduct** is a slender canal traveling through the midbrain; it connects the third ventricle to the fourth ventricle. The cerebral peduncles and the rounded corpora quadrigemina make up the midbrain tissue anterior and posterior (respectively) to the cerebral aqueduct.

2. Locate the hindbrain structures. Trace the rounded pons to the medulla oblongata below, and identify the fourth ventricle posterior to these structures. Attempt to identify the single median aperture and the two lateral apertures, three openings found in the walls of the fourth ventricle. These apertures serve as passageways for cerebrospinal fluid to circulate into the subarachnoid space from the fourth ventricle.

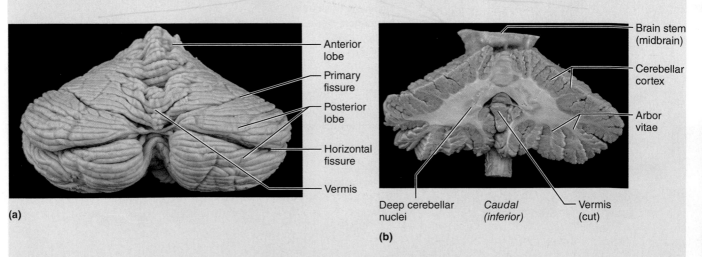

(a)

Anterior lobe

Primary fissure

Posterior lobe

Horizontal fissure

Vermis

Brain stem (midbrain)

Cerebellar cortex

Arbor vitae

Deep cerebellar nuclei

Caudal (inferior)

Vermis (cut)

(b)

Figure 17.6 Cerebellum. (a) Posterior (dorsal) view. **(b)** Sectioned to reveal the cerebellar cortex. (The cerebellum is sectioned frontally, and the brain stem is sectioned transversely, in this posterior view.)

Cerebellum

Examine the cerebellum. Notice that it is composed of two lateral hemispheres, each with three lobes (*anterior, posterior,* and a deep *flocculonodular*) connected by a midline lobe called the **vermis** (**Figure 17.6**). As in the cerebral hemispheres, the cerebellum has an outer cortical area of gray matter and an inner area of white matter. The treelike branching of the cerebellar white matter is referred to as the **arbor vitae,** or "tree of life." The cerebellum controls the unconscious coordination of skeletal muscle activity along with balance and equilibrium.

Meninges of the Brain

The brain and spinal cord are covered and protected by three connective tissue membranes called **meninges** (**Figure 17.7**). The outermost meninx is the leathery **dura mater,** a double-layered membrane. One of its layers (the *periosteal layer*) is attached to the inner surface of the skull, forming the periosteum. The other (the *meningeal layer*) forms the outermost brain covering and is continuous with the dura mater of the spinal cord.

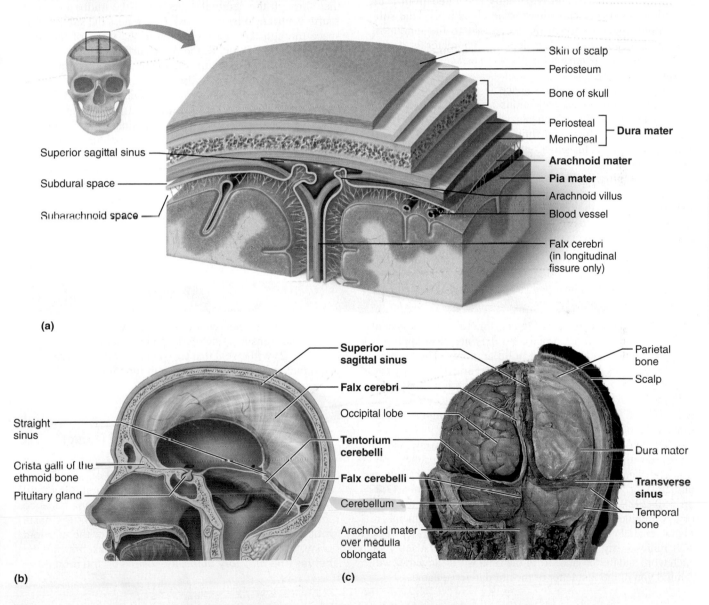

(a)

(b)

(c)

Figure 17.7 Meninges of the brain. (a) Three-dimensional frontal section showing the relationship of the dura mater, arachnoid mater, and pia mater. The meningeal dura forms the falx cerebri fold, which extends into the longitudinal fissure and attaches the brain to the ethmoid bone of the skull. The superior sagittal sinus is enclosed by the dural membranes superiorly. Arachnoid villi, which return cerebrospinal fluid to the dural venous sinus, are also shown. **(b)** Medial view showing the position of the dural folds: the falx cerebri, tentorium cerebelli, and falx cerebelli. **(c)** Posterior view of the brain in place, surrounded by the dura mater. Sinuses between periosteal and meningeal dura contain venous blood.

17

The dural layers are fused together except in three places where the inner membrane extends inward to form a septum that secures the brain to structures inside the cranial cavity. One such extension, the **falx cerebri,** dips into the longitudinal fissure between the cerebral hemispheres to attach to the crista galli of the ethmoid bone of the skull (Figure 17.7a). The cavity created at this point is the large **superior sagittal sinus,** which collects blood draining from the brain tissue. The **falx cerebelli,** separating the two cerebellar hemispheres, and the **tentorium cerebelli,** separating the cerebrum from the cerebellum below, are two other important inward folds of the inner dural membrane.

The middle meninx, the weblike **arachnoid mater,** underlies the dura mater and is partially separated from it by the **subdural space.** Threadlike projections bridge the **subarachnoid space** to attach the arachnoid to the innermost meninx, the **pia mater.** The delicate pia mater is highly vascular and clings tenaciously to the surface of the brain, following its gyri.

In life, the subarachnoid space is filled with cerebrospinal fluid. Specialized projections of the arachnoid tissue called **arachnoid villi** protrude through the dura mater. These villi allow the cerebrospinal fluid to drain back into the venous circulation via the superior sagittal sinus and other dural venous sinuses.

Meningitis, inflammation of the meninges, is a serious threat to the brain because of the intimate association between the brain and meninges. Should infection spread to the neural tissue of the brain itself, life-threatening **encephalitis** may occur. Meningitis is often diagnosed by taking a sample of cerebrospinal fluid (via a spinal tap) from the subarachnoid space. ✚

Cerebrospinal Fluid

The cerebrospinal fluid (CSF), much like plasma in composition, is continually formed by the **choroid plexuses,** small capillary knots hanging from the roof of the ventricles of the brain. The cerebrospinal fluid in and around the brain forms a watery cushion that protects the delicate brain tissue against blows to the head.

Within the brain, the cerebrospinal fluid circulates from the two lateral ventricles (in the cerebral hemispheres) into the third ventricle via the **interventricular foramina,** and then through the cerebral aqueduct of the midbrain into the fourth ventricle (**Figure 17.8**). CSF enters the subarachnoid space through the three foramina in the walls of the fourth ventricle. There it bathes the outer surfaces of the brain and spinal cord. The fluid returns to the blood in the dural venous sinuses via the arachnoid villi.

Ordinarily, cerebrospinal fluid forms and drains at a constant rate. However, under certain conditions—for example, obstructed drainage or circulation resulting from tumors or anatomical deviations—cerebrospinal fluid accumulates and exerts increasing pressure on the brain which, uncorrected, causes neurological damage in adults. In infants, **hydrocephalus** (literally, "water on the brain") is indicated by a gradually enlarging head. The infant's skull is still flexible and contains fontanelles, so it can expand to accommodate the increasing size of the brain. ✚

Cranial Nerves

The **cranial nerves** are part of the peripheral nervous system and not part of the brain proper, but they are most appropriately identified while studying brain anatomy. The 12 pairs of cranial nerves primarily serve the head and neck. Only one pair, the vagus nerves, extends into the thoracic and abdominal cavities. All but the first two pairs (olfactory and optic nerves) arise from the brain stem and pass through foramina in the base of the skull to reach their destination.

The cranial nerves are numbered consecutively, and in most cases their names reflect the major structures they control. The cranial nerves are described by name, number (Roman numeral), origin, course, and function in the list (**Table 17.2,** pp. 290–292). This information should be committed to memory. A mnemonic device that might be helpful for remembering the cranial nerves in order is "*O*n *o*ccasion, *o*ur *t*rusty *t*ruck *a*cts *f*unny—*v*ery *g*ood *v*ehicle *a*ny*h*ow." The first letter of each word and the "a" and "h" of the final word "anyhow" will remind you of the first letter of the cranial nerve name.

Most cranial nerves are mixed nerves (containing both motor and sensory fibers). But close scrutiny of the list (Table 17.2) will reveal that two pairs of cranial nerves (optic and olfactory) are purely sensory in function.

Activity 3

Identifying and Testing the Cranial Nerves

1. Observe the ventral surface of the brain model to identify the cranial nerves. (**Figure 17.9** on p. 290 may also aid you in this study.) Notice that the first (olfactory) cranial nerves are not visible on the model because they consist only of short axons that run from the nasal mucosa through the cribriform foramina of the ethmoid bone. (However, the synapse points of the first cranial nerves, the *olfactory bulbs,* are visible on the model.)

Text continues on page 290. →

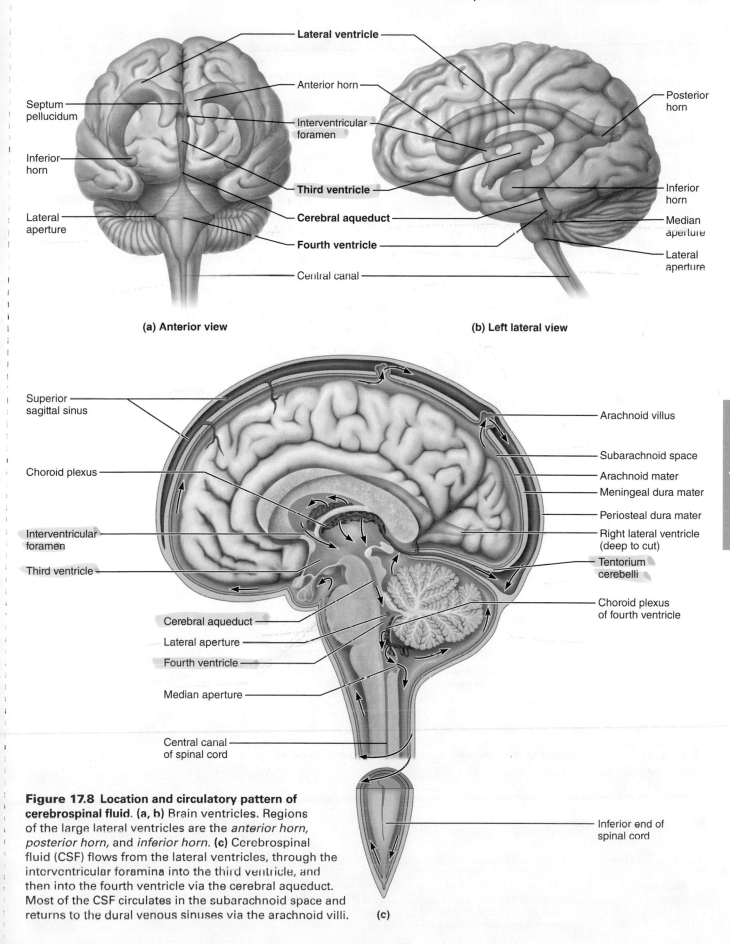

(a) Anterior view

(b) Left lateral view

Figure 17.8 Location and circulatory pattern of cerebrospinal fluid. (a, b) Brain ventricles. Regions of the large lateral ventricles are the *anterior horn*, *posterior horn*, and *inferior horn*. **(c)** Cerebrospinal fluid (CSF) flows from the lateral ventricles, through the interventricular foramina into the third ventricle, and then into the fourth ventricle via the cerebral aqueduct. Most of the CSF circulates in the subarachnoid space and returns to the dural venous sinuses via the arachnoid villi.

(c)

17

Activity 3: Cranial Nerve Ganglia

Cranial nerve ganglion	Cranial nerve	Site of ganglion
Trigeminal		
Geniculate		
Inferior		
Superior		
Spiral		
Vestibular		

2. Testing cranial nerves is an important part of any neurological examination. See the last column of Table 17.2 for techniques you can use for such tests. Conduct tests of cranial nerve function following directions given in the "testing" column of the table. The results may help you understand cranial nerve function, especially as it pertains to some aspects of brain function.

3. Several cranial nerve ganglia are named in the **Activity 3 chart**. *Using your textbook or an appropriate reference,* fill in the chart by naming the cranial nerve the ganglion is associated with and stating the ganglion location.

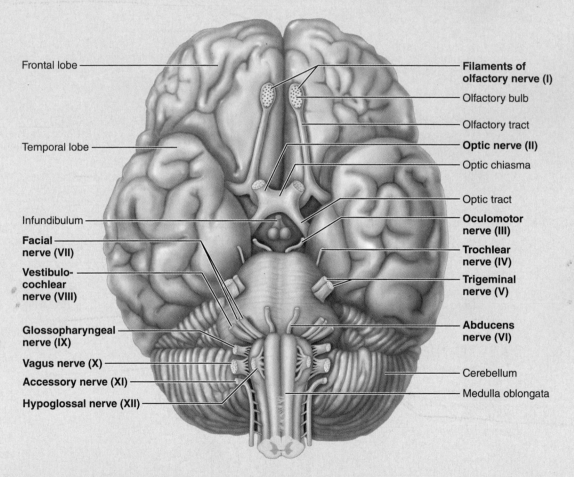

Figure 17.9 Ventral aspect of the human brain, showing the cranial nerves. (See also Figure 17.3.)

Table 17.2 — The Cranial Nerves (Figure 17.9)

Number and name	Origin and course	Function*	Testing
I. Olfactory	Fibers arise from olfactory epithelium and run through cribriform foramina of ethmoid bone to synapse in olfactory bulb.	Purely sensory—carries afferent impulses associated with sense of smell.	Person is asked to sniff aromatic substances, such as oil of cloves and vanilla, and to identify each.

Table 17.2	(continued)		
Number and name	**Origin and course**	**Function***	**Testing**
II. Optic	Fibers arise from retina of eye and pass through optic canal of sphenoid bone. Fibers partially cross over at the optic chiasma and continue on to the thalamus as the optic tracts. Final fibers of this pathway travel from the thalamus to the primary visual cortex as the optic radiation.	Purely sensory—carries afferent impulses associated with vision.	Vision and visual field are determined with eye chart and by testing the point at which the person first sees an object (finger) moving into the visual field. Fundus of eye viewed with ophthalmoscope to detect papilledema (swelling of optic disc, or point at which optic nerve leaves the eye) and to observe blood vessels.
III. Oculomotor	Fibers emerge from ventral midbrain and course ventrally to enter the orbit. They exit from skull via superior orbital fissure.	Primarily motor—somatic motor fibers to inferior oblique and superior, inferior, and medial rectus muscles, which direct eyeball, and to levator palpebrae muscles of the superior eyelid; parasympathetic fibers to smooth muscle controlling lens shape and pupil size.	Pupils are examined for size, shape, and equality. Pupillary reflex is tested with penlight (pupils should constrict when illuminated). Convergence for near vision is tested, as is subject's ability to follow objects with the eyes.
IV. Trochlear	Fibers emerge from midbrain and exit from skull via superior orbital fissure.	Primarily motor—provides somatic motor fibers to superior oblique muscle that moves the eyeball.	Tested in common with cranial nerve III.
V. Trigeminal	Fibers run from face to pons and form three divisions: mandibular division fibers pass through foramen ovale in sphenoid bone, maxillary division fibers pass via foramen rotundum in sphenoid bone, and ophthalmic division fibers pass through superior orbital fissure of sphenoid bone.	Mixed—major sensory nerve of face; conducts sensory impulses from skin of face and anterior scalp, from mucosae of mouth and nose, and from surface of eyes; mandibular division also contains motor fibers that innervate muscles of mastication and muscles of floor of mouth.	Sensations of pain, touch, and temperature are tested with safety pin and hot and cold probes. Corneal reflex tested with wisp of cotton. Motor branch assessed by asking person to clench the teeth, open mouth against resistance, and move jaw side to side.
VI. Abducens	Fibers leave inferior pons and exit from skull via superior orbital fissure.	Primarily motor—carries somatic motor fibers to lateral rectus muscle that abducts the eyeball.	Tested in common with cranial nerve III.
VII. Facial	Fibers leave pons and travel through temporal bone via internal acoustic meatus, exiting via stylomastoid foramen to reach the face.	Mixed—supplies somatic motor fibers to muscles of facial expression and the posterior belly of the digastric muscle; parasympathetic motor fibers to lacrimal and salivary glands; carries sensory fibers from taste receptors of anterior tongue.	Anterior two-thirds of tongue is tested for ability to taste sweet (sugar), salty, sour (vinegar), and bitter (quinine) substances. Symmetry of face is checked. Subject is asked to close eyes, smile, whistle, and so on. Tearing is assessed with ammonia fumes.
VIII. Vestibulocochlear	Fibers run from inner ear equilibrium and hearing apparatus, housed in temporal bone, through internal acoustic meatus to enter pons.	Mostly sensory—vestibular branch transmits impulses associated with sense of equilibrium from vestibular apparatus and semicircular canals; cochlear branch transmits impulses associated with hearing from cochlea. Small motor component adjusts the sensitivity of the sensory receptors.	Hearing is checked by air and bone conduction using tuning fork.
IX. Glossopharyngeal	Fibers emerge from medulla oblongata and leave skull via jugular foramen to run to throat.	Mixed—somatic motor fibers serve pharyngeal muscles, and parasympathetic motor fibers serve salivary glands; sensory fibers carry impulses from pharynx, tonsils, posterior tongue (taste buds), and from chemoreceptors and pressure receptors of carotid artery.	A tongue depressor is used to check the position of the uvula. Gag and swallowing reflexes are checked. Subject is asked to speak and cough. Posterior third of tongue may be tested for taste.

17

(Table continues on next page.)

Table 17.2	The Cranial Nerves *(continued)*		
Number and name	**Origin and course**	**Function***	**Testing**
X. Vagus	Fibers emerge from medulla oblongata and pass through jugular foramen and descend through neck region into thorax and abdomen.	Mixed—fibers carry somatic motor impulses to pharynx and larynx and sensory fibers from same structures; very large portion is composed of parasympathetic motor fibers, which supply heart and smooth muscles of abdominal visceral organs; transmits sensory impulses from viscera.	As for cranial nerve IX (IX and X are tested in common, since they both innervate muscles of throat and mouth).
XI. Accessory	Fibers arise from the superior aspect of spinal cord, enter the skull, and then travel through jugular foramen to reach muscles of neck and back.	Mixed (but primarily motor in function)—provides somatic motor fibers to sternocleidomastoid and trapezius muscles.	Sternocleidomastoid and trapezius muscles are checked for strength by asking person to rotate head and shrug shoulders against resistance.
XII. Hypoglossal	Fibers arise from medulla oblongata and exit from skull via hypoglossal canal to travel to tongue.	Mixed (but primarily motor in function)—carries somatic motor fibers to muscles of tongue.	Person is asked to protrude and retract tongue. Any deviations in position are noted.

*Does not include sensory impulses from proprioceptors.

 DISSECTION

The Sheep Brain

The sheep brain is enough like the human brain to warrant comparison. Obtain a sheep brain, disposable gloves, dissecting tray, and instruments, and bring them to your laboratory bench.

1. Don disposable gloves. If the dura mater is present, remove it as described here. Place the intact sheep brain ventral surface down on the dissecting pan, and observe the dura mater. Feel its consistency and note its toughness. Cut through the dura mater along the line of the longitudinal fissure (which separates the cerebral hemispheres) to enter the superior sagittal sinus. Gently force the cerebral hemispheres apart laterally to expose the corpus callosum deep to the longitudinal fissure.

2. Carefully remove the dura mater and examine the superior surface of the brain. Notice that its surface, like that of the human brain, is thrown into convolutions (fissures and gyri). Locate the arachnoid mater, which appears on the brain surface as a delicate "cottony" material spanning the fissures. In contrast, the innermost meninx, the pia mater, closely follows the cerebral contours.

3. Before beginning the dissection, turn your sheep brain so that you are viewing its left lateral aspect. Compare the various areas of the sheep brain (cerebrum, brain stem, cerebellum) to the photo of the human brain (**Figure 17.10**). Relatively speaking, which of these structures is obviously much larger in the human brain?

Ventral Structures

Turn the brain so that its ventral surface is uppermost. (**Figure 17.11a** and **b** show the important features of the ventral surface of the brain.)

1. Look for the clublike olfactory bulbs anteriorly, on the inferior surface of the frontal lobes of the cerebral hemispheres. Axons of olfactory neurons run from the nasal mucosa through the cribriform foramina of the ethmoid bone to synapse with the olfactory bulbs.

How does the size of these olfactory bulbs compare with those of humans?

_____

Is the sense of smell more important for protection and foraging in sheep or in humans?

_____

2. The optic nerve (II) carries sensory impulses from the retina of the eye. Thus this cranial nerve is involved in the

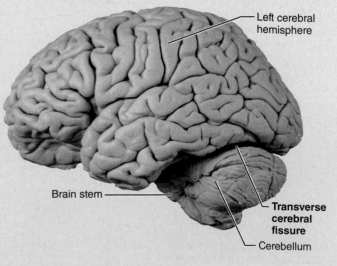

Figure 17.10 Photograph of lateral aspect of the human brain.

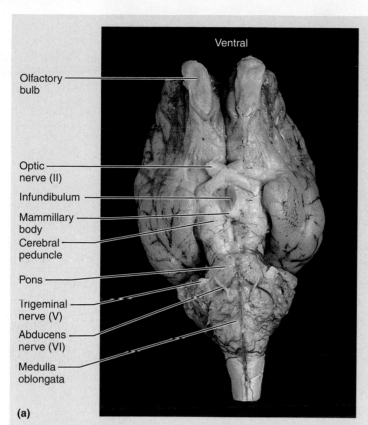

Ventral

Olfactory bulb

Optic nerve (II)

Infundibulum

Mammillary body

Cerebral peduncle

Pons

Trigeminal nerve (V)

Abducens nerve (VI)

Medulla oblongata

(a)

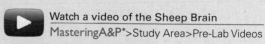
Watch a video of the Sheep Brain
MasteringA&P®>Study Area>Pre-Lab Videos

Figure 17.11 Intact sheep brain. (a) Photograph of ventral view. **(b)** Diagram of ventral view.

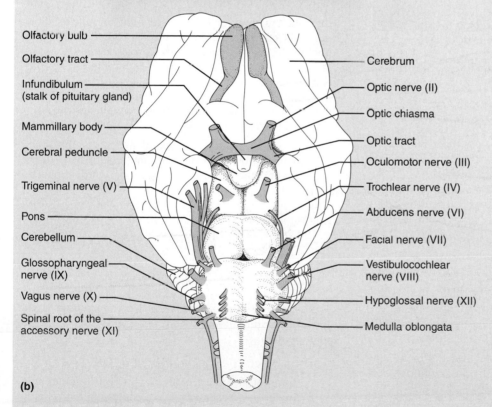

Olfactory bulb

Olfactory tract

Infundibulum (stalk of pituitary gland)

Mammillary body

Cerebral peduncle

Trigeminal nerve (V)

Pons

Cerebellum

Glossopharyngeal nerve (IX)

Vagus nerve (X)

Spinal root of the accessory nerve (XI)

Cerebrum

Optic nerve (II)

Optic chiasma

Optic tract

Oculomotor nerve (III)

Trochlear nerve (IV)

Abducens nerve (VI)

Facial nerve (VII)

Vestibulocochlear nerve (VIII)

Hypoglossal nerve (XII)

Medulla oblongata

(b)

17

sense of vision. Identify the optic nerves, optic chiasma, and optic tracts.

3. Posterior to the optic chiasma, two structures protrude from the ventral aspect of the hypothalamus—the infundibulum (stalk of the pituitary gland) immediately posterior to the optic chiasma and the mammillary body. Notice that the sheep's mammillary body is a single rounded eminence. In humans it is a double structure.

4. Identify the cerebral peduncles on the ventral aspect of the midbrain, just posterior to the mammillary body of the hypothalamus. The cerebral peduncles are fiber tracts connecting the cerebrum and medulla oblongata. Identify the large oculomotor nerves (III), which arise from the ventral midbrain surface, and the tiny trochlear nerves (IV), which can be seen at the junction of the

Text continues on next page. ➡

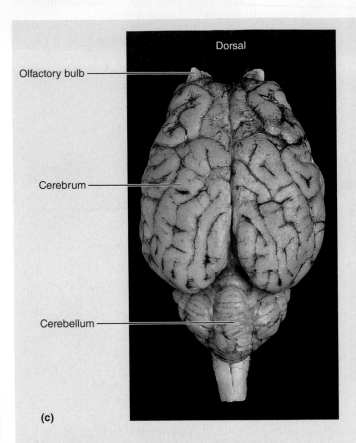

Dorsal

Olfactory bulb

Cerebrum

Cerebellum

(c)

Figure 17.11 *(continued)* **Intact sheep brain.**
(c) Photograph of the dorsal view.

 Watch a video of the Sheep Brain
MasteringA&P®>Study Area>Pre-Lab Videos

midbrain and pons. Both of these cranial nerves provide motor fibers to extrinsic muscles of the eyeball.

5. Move posteriorly from the midbrain to identify first the pons and then the medulla oblongata, structures composed primarily of ascending and descending fiber tracts.

6. Return to the junction of the pons and midbrain, and proceed posteriorly to identify the following cranial nerves, all arising from the pons. Check them off as you locate them.

☐ Trigeminal nerves (V), which are involved in chewing and sensations of the head and face.

☐ Abducens nerves (VI), which abduct the eye (and thus work in conjunction with cranial nerves III and IV).

☐ Facial nerves (VII), large nerves involved in taste sensation, gland function (salivary and lacrimal glands), and facial expression.

7. Continue posteriorly to identify and check off:

☐ Vestibulocochlear nerves (VIII), mostly sensory nerves that are involved with hearing and equilibrium.

☐ Glossopharyngeal nerves (IX), which contain motor fibers innervating throat structures and sensory fibers transmitting taste stimuli (in conjunction with cranial nerve VII).

☐ Vagus nerves (X), often called "wanderers," which serve many organs of the head, thorax, and abdominal cavity.

☐ Accessory nerves (XI), which serve muscles of the neck, larynx, and shoulder; actually arise from the spinal

cord (C_1 through C_5) and travel superiorly to enter the skull before running to the muscles that they serve.

☐ Hypoglossal nerves (XII), which stimulate tongue and neck muscles.

It is likely that some of the cranial nerves will have been broken off during brain removal. If so, observe sheep brains of other students to identify those missing from your specimen, using your check marks as a guide.

Dorsal Structures

1. Refer to the dorsal view photograph (Figure 17.11c) as a guide in identifying the following structures. Reidentify the now exposed cerebral hemispheres. How does the depth of the fissures in the sheep's cerebral hemispheres compare to that of the fissures in the human brain?

_____

2. Examine the cerebellum. Notice that, in contrast to the human cerebellum, it is not divided longitudinally, and that its fissures are oriented differently. What dural fold (falx cerebri or falx cerebelli) is missing that is present in humans?

_____

3. Locate the three pairs of cerebellar peduncles, fiber tracts that connect the cerebellum to other brain structures, by lifting the cerebellum dorsally away from the brain stem. The most posterior pair, the inferior cerebellar peduncles, connect the cerebellum to the medulla. The middle cerebellar peduncles attach the cerebellum to the pons, and the superior cerebellar peduncles run from the cerebellum to the midbrain.

4. To expose the dorsal surface of the midbrain, gently separate the cerebrum and cerebellum (as shown in **Figure 17.12**.) Identify the corpora quadrigemina, which appear as four rounded prominences on the dorsal midbrain surface.

What is the function of the corpora quadrigemina?

_____

_____

Also locate the pineal gland, which appears as a small oval protrusion in the midline just anterior to the corpora quadrigemina.

Internal Structures

1. The internal structure of the brain can be examined only after further dissection. Place the brain ventral side down on the dissecting tray and make a cut completely through it in a superior to inferior direction. Cut through the longitudinal fissure, corpus callosum, and midline of the cerebellum. Refer to **Figure 17.13** as you work.

2. A thin nervous tissue membrane immediately ventral to the corpus callosum that separates the lateral ventricles is the septum pellucidum. If it is still intact, pierce this membrane and probe the lateral ventricle cavity. The fiber tract ventral to the septum pellucidum and anterior to the third ventricle is the fornix.

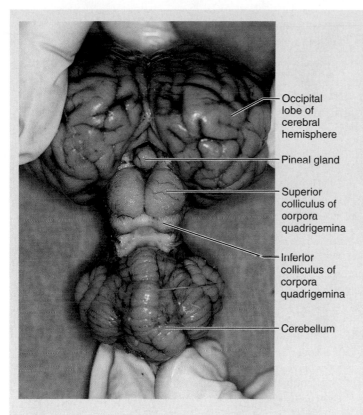

Occipital lobe of cerebral hemisphere

Pineal gland

Superior colliculus of corpora quadrigemina

Inferior colliculus of corpora quadrigemina

Cerebellum

Figure 17.12 Means of exposing the dorsal midbrain structures of the sheep brain.

 Watch a video of the Sheep Brain
MasteringA&P®>Study Area>Pre-Lab Videos

How does the size of the fornix in this brain compare with the size of the human fornix?

_____

_____

Why do you suppose this is so? (Hint: What is the function of this band of fibers?)

_____

3. Identify the thalamus, which forms the walls of the third ventricle and is located posterior and ventral to the fornix. The interthalamic adhesion appears as an oval protrusion of the thalamic wall.

4. The hypothalamus forms the floor of the third ventricle. Identify the optic chiasma, infundibulum, and mammillary body on its exterior surface. The pineal gland is just beneath the junction of the corpus callosum and fornix.

5. Locate the midbrain by identifying the corpora quadrigemina that form its dorsal roof. Follow the cerebral aqueduct through the midbrain tissue to the fourth ventricle. Identify the cerebral peduncles, which form its anterior walls.

6. Identify the pons and medulla oblongata, which lie anterior to the fourth ventricle. The medulla continues into the spinal cord without any obvious anatomical change, but the point at which the fourth ventricle narrows to a small canal is generally accepted as the beginning of the spinal cord.

7. Identify the cerebellum posterior to the fourth ventricle. Notice its internal treelike arrangement of white matter, the arbor vitae.

Text continues on next page. →

17

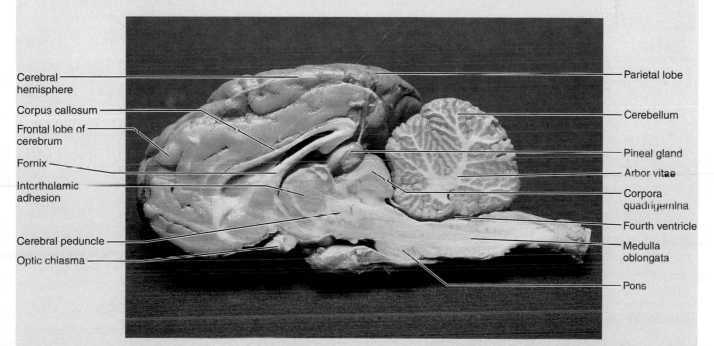

Cerebral hemisphere

Corpus callosum

Frontal lobe of cerebrum

Fornix

Interthalamic adhesion

Cerebral peduncle

Optic chiasma

Parietal lobe

Cerebellum

Pineal gland

Arbor vitae

Corpora quadrigemina

Fourth ventricle

Medulla oblongata

Pons

Figure 17.13 Photograph of median section of the sheep brain showing internal structures.

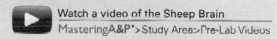 Watch a video of the Sheep Brain
MasteringA&P®>Study Area>Pre-Lab Videos

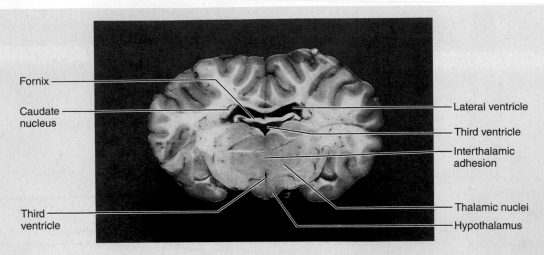

Fornix

Caudate nucleus

Third ventricle

Lateral ventricle

Third ventricle

Interthalamic adhesion

Thalamic nuclei

Hypothalamus

Figure 17.14 Frontal section of a sheep brain. Major structures include the thalamus, hypothalamus, and lateral and third ventricles.

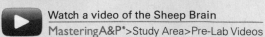

Watch a video of the Sheep Brain
MasteringA&P*>Study Area>Pre-Lab Videos

8. If time allows, obtain another sheep brain and section it along the frontal plane so that the cut passes through the infundibulum. Compare your specimen with the photograph of a frontal section (**Figure 17.14**), and attempt to identify all the structures shown in the figure.

9. Check with your instructor to determine if a small portion of the spinal cord from your brain specimen should be saved for spinal cord studies (Exercise 19). Otherwise, dispose of all the organic debris in the appropriate laboratory containers and clean the laboratory bench, the dissection instruments, and the tray before leaving the laboratory.

Group Challenge

Odd (Cranial) Nerve Out

The following boxes each contain four cranial nerves. One of the listed nerves does not share a characteristic with the other three. Working in groups of three, discuss the characteristics of the four cranial nerves in each set. On a separate piece of paper, one student will record the characteristics for each nerve for the group. For each set of four nerves, discuss the possible candidates for the "odd nerve" and which characteristic it lacks based upon your notes. Once you have come to a consensus within your group, circle the cranial nerve that doesn't belong with the others, and explain why it is singled out. What characteristic is it missing? Sometimes there may be multiple reasons why the cranial nerve doesn't belong with the others.

1. Which is the "odd" nerve?	Why is it the odd one out?
Optic nerve (II) Oculomotor nerve (III) Olfactory nerve (I) Vestibulocochlear nerve (VIII)	
2. Which is the "odd" nerve?	**Why is it the odd one out?**
Oculomotor nerve (III) Trochlear nerve (IV) Abducens nerve (VI) Hypoglossal nerve (XII)	
3. Which is the "odd" nerve?	**Why is it the odd one out?**
Facial nerve (VII) Hypoglossal nerve (XII) Trigeminal nerve (V) Glossopharyngeal nerve (IX)	

Electroencephalography

Objectives

☐ Define *electroencephalogram (EEG)*, and discuss its clinical significance.

☐ Describe or recognize typical tracings of alpha, beta, theta, and delta brain waves, and indicate the conditions when each is most likely to occur.

☐ Indicate the source of brain waves.

☐ Define *alpha block*.

☐ Monitor the EEG in a human subject.

☐ Describe the effect of a sudden sound, mental concentration, and respiratory alkalosis on the EEG.

Materials

- Oscilloscope and EEG lead-selector box or physiograph and high-gain preamplifier
- Cot (if available) or pillow
- Electrode gel
- EEG electrodes and leads
- Collodion gel or long elastic EEG straps
- Abrasive pad

BIOPAC® BIOPAC® BSL System with BSL software version 3.7.5 to 3.7.7 (for Windows 7/Vista/XP or Mac OS X 10.4–10.6), data acquisition unit MP36/35 or MP45, PC or Mac computer, Biopac Student Lab electrode lead set, disposable vinyl electrodes, Lycra® swim cap (such as Speedo® brand) or supportive wrap (such as 3M Coban™ Self-adhering Support Wrap) to press electrodes against head for improved contact, and a cot or lab bench and pillow.

Instructors using the MP36/35/30 data acquisition unit with BSL software versions earlier than 3.7.5 (for Windows or Mac) will need slightly different channel settings and collection strategies. Instructions for using the older data acquisition unit can be found on MasteringA&P.

MasteringA&P®

For related exercise study tools, go to the Study Area of **MasteringA&P**. There you will find:

- Practice Anatomy Lab PAL
- PhysioEx **PEx**
- A&PFlix **A&PFlix**
- Practice quizzes, Histology Atlas, eText, Videos, and more!

Pre-Lab Quiz

1. What does an electroencephalogram (EEG) measure?
   a. electrical activity of the brain
   b. electrical activity of the heart
   c. emotions
   d. physical activity of the subject
2. Circle the correct underlined term. <u>Alpha waves</u> / <u>Beta waves</u> are typical of the attentive or awake state.
3. Circle True or False. Brain waves can change with age, sensory stimuli, and the chemical state of the body.
4. Where will you place the indifferent (ground) electrode on your subject?
   a. the earlobe
   b. the forehead
   c. over the occipital lobe
   d. over the temporal bone
5. During today's activity, students will instruct subjects to *hyperventilate*. What should the subjects do?
   a. breathe in a normal manner
   b. breathe rapidly
   c. breathe very slowly
   d. hold their breath until they almost pass out

As curious humans we are particularly interested in how the brain thinks, reasons, learns, remembers, and controls consciousness. As students we have learned that the brain accomplishes its tasks through electrical activities of neurons. The remarkable noninvasive technologies of twenty-first-century neuroscience have advanced our understanding of brain functions, as has the long-used technique of electroencephalography—the recording of electrical activity from the surface portions of the brain. It is incredible that the sophisticated equipment used to record an electroencephalogram (EEG) is commonly available in undergraduate laboratories. As you use it, you begin to explore the complex higher functions of the human brain.

Brain Wave Patterns and the Electroencephalogram

The **electroencephalogram (EEG),** a record of the electrical activity of the brain, can be obtained through electrodes placed at various points on the skin or scalp of the head. This electrical activity, which is recorded as waves (**Figure 18.1**), represents the summed synaptic activity of many neurons at the surface of the cerebral cortex.

Certain characteristics of brain waves are known. They have a frequency of 1 to 30 hertz (Hz) or cycles per second, a dominant rhythm of 10 Hz, and an average amplitude (voltage) of 20 to 100 microvolts (µV). They vary in frequency in different brain areas, occipital waves having a lower frequency than those associated with the frontal and parietal lobes.

The first of the brain waves to be described by scientists were the alpha waves (or alpha rhythm). **Alpha waves** have an average frequency range of 8 to 13 Hz and are produced when the individual is in a relaxed state with the eyes closed. **Alpha block,** suppression of the alpha rhythm, occurs if the eyes are opened or if the individual begins to concentrate on some mental problem or visual stimulus. Under these conditions, the waves decrease in amplitude but increase in frequency. Under conditions of fright or excitement, the frequency increases even more.

Beta waves, closely related to alpha waves, are faster (14 to 30 Hz) and have a lower amplitude. They are typical of the attentive or alert state.

Very large (high-amplitude) waves with a frequency of 4 Hz or less that are seen in deep sleep are **delta waves. Theta waves** are large, irregular waves with a frequency of 4 to 7 Hz. Although theta waves are normal in children, they are uncommon in awake adults.

Brain waves change with age, sensory stimuli, brain pathology, and the chemical state of the body. Glucose deprivation, oxygen poisoning, and sedatives all interfere with the rhythmic activity of brain output by disturbing the metabolism of neurons. Sleeping individuals and patients in a coma have EEGs that are slower (lower frequency) than the alpha rhythm of normal adults. Fright, epileptic seizures, and various types of drug intoxication can be associated with comparatively faster cortical activity. As these examples show, impairment of cortical function is indicated by neuronal activity that is either too fast or too slow; unconsciousness occurs at both extremes of the frequency range.

Because spontaneous brain waves are always present, even during unconsciousness and coma, the absence of brain waves (a "flat" EEG) is taken as clinical evidence of brain death. The EEG is used clinically to diagnose and localize many types of brain lesions, including epileptic foci (cortical area responsible for causing the seizure), infections, abscesses, and tumors. ✚

Activity 1

Observing Brain Wave Patterns Using an Oscilloscope or Physiograph

If one electrode (the *active electrode*) is placed over a particular cortical area and another (the *indifferent, or ground, electrode*) is placed over an inactive part of the head, such as the earlobe, all of the activity of the cortex underlying the active electrode will, theoretically, be recorded. The inactive area provides a zero reference point, or a baseline, and the EEG represents the difference between "activities" occurring under the two electrodes.

1. Connect the EEG lead-selector box to the oscilloscope preamplifier, or connect the high-gain preamplifier to the physiograph channel amplifier. Adjust the horizontal sweep and sensitivity according to the directions given in the instrument manual or by your instructor.

2. Prepare the subject. The subject should lie undisturbed on a cot or on the lab bench with eyes closed in a quiet,

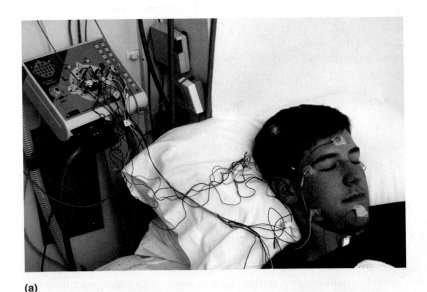

(a)

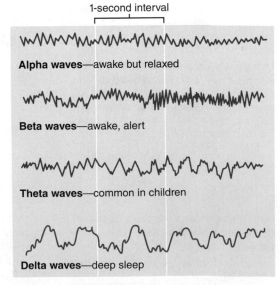

(b)

Figure 18.1 Electroencephalography and brain waves. (a) Scalp electrodes are positioned on the patient to record brain waves. **(b)** Typical EEGs.

dimly lit area. (Someone who is able to relax easily makes a good subject.) Apply a small amount of electrode gel to the subject's forehead above the left eye and on the left earlobe. Press an electrode to each prepared area and secure each by (1) applying a film of collodion gel to the electrode surface and the adjacent skin or (2) using a long elastic EEG strap (knot tied at the back of the head). If collodion gel is used, allow it to dry before you continue.

3. Connect the active frontal lead (forehead) to the EEG lead-selector box outlet marked "L Frontal." Connect the lead from the indifferent electrode (earlobe) to the ground outlet (or to the appropriate input terminal on the high-gain preamplifier).

4. Turn the oscilloscope or physiograph on, and observe the EEG pattern of the relaxed subject for a period of 5 minutes. If the subject is truly relaxed, you should see a typical alpha-wave pattern. (If the subject is unable to relax and the alpha-wave pattern does not appear in this time interval, test another subject.) Since the electrical activity of muscles interferes with EEG recordings, discourage all muscle movement during the monitoring period. If 60-cycle "noise" (appearing as fast, regular, low-amplitude waves superimposed on the more irregular brain waves) is present in your record because of the presence of other electronic equipment, consult your instructor to eliminate it.

5. Abruptly and loudly clap your hands. The subject's eyes should open, and alpha block should occur. Observe the immediate brain wave pattern. How do the frequency and amplitude (rhythm) of the brain waves change?

_____

_____

Would you characterize this as beta rhythm? _____

Why? _____

_____

6. Allow the subject about 5 minutes to achieve relaxation once again, then ask him or her to compute a number problem that requires concentration (for example, add 3 and 36, subtract 7, multiply by 2, add 50, etc.). Observe the brain wave pattern during the period of mental computation.

Observations: _____

_____

_____

7. Once again allow the subject to relax until alpha rhythm resumes. Then, instruct him or her to hyperventilate (breathe rapidly) for 3 minutes. *Be sure to tell the subject when to stop hyperventilating.* Hyperventilation rapidly flushes carbon dioxide out of the lungs, decreasing carbon dioxide levels in the blood and producing respiratory alkalosis.

Observe the changes in the rhythm and amplitude of the brain waves occurring during the period of hyperventilation.

Observations: _____

_____

_____

8. Think of other stimuli that might affect brain wave patterns. Test your hypotheses. Describe what stimuli you tested and what responses you observed.

_____

_____

_____

_____

Activity 2

Electroencephalography Using BIOPAC®

In this activity, the EEG of the subject will be recorded during a relaxed state, first with the eyes closed, then with the eyes open while silently counting to ten, and finally with the eyes closed again.

Setting Up the Equipment

1. Connect the BIOPAC® unit to the computer and turn the computer **ON**.

2. Make sure the BIOPAC® unit is **OFF**.

3. Plug in the equipment as shown in **Figure 18.2** on p. 306.

- Electrode lead set—CH 1

4. Turn the BIOPAC® unit **ON**.

5. Attach three electrodes to the subject's scalp and earlobe as shown in **Figure 18.3** on p. 306. Follow these important guidelines to assist in effective electrode placement:

- Select subjects with the easiest access to the scalp.

- Move hair apart at electrode site and gently abrade skin.

- Apply some gel to the electrode. A fair amount of gel must be used to obtain a good electrode-to-scalp connection.

- Apply pressure to the electrodes for 1 minute to ensure attachment.

- Use a swimcap or supportive wrap to maintain attachment. Subject should not press electrodes against scalp.

- Do not touch the electrodes while recording.

- The earlobe electrode may be folded under the lobe itself.

6. When the electrodes are attached, the subject should lie down and relax with eyes closed for 5 minutes before recording.

Text continues on next page. →

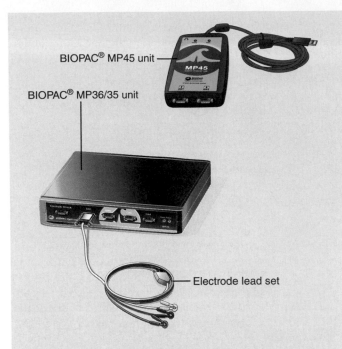

Figure 18.2 Setting up the BIOPAC® equipment. Plug the electrode set into Channel 1. Electrode leads are shown connected to the MP36/35 unit.

7. Start the Biopac Student Lab program on the computer by double-clicking the icon on the desktop or by following your instructor's guidance.

8. Select lesson **L03-EEG-1** from the menu, and click **OK**.

9. Type in a filename that will save this subject's data on the computer hard drive. You may want to use the subject's last name followed by EEG-1 (for example, SmithEEG-1). Then click **OK**.

10. During this preparation, the subject should be very still and in a relaxed state with eyes closed. Allow the subject to relax with minimal stimuli.

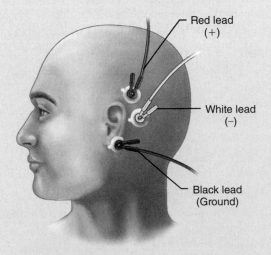

Figure 18.3 Placement of electrodes and the appropriate attachment of electrode leads by color.

Calibrating the Equipment

1. Make sure that the electrodes remain firmly attached to the surface of the scalp and earlobe. The subject should remain absolutely still and try to avoid movement of the body or face.

2. With the subject in a relaxed position, click **Calibrate**.

3. You will be prompted to check electrode attachment one final time. When ready, click **OK**; the computer will record for 8 seconds and stop automatically.

4. Observe the recording of the calibration data (it should look like **Figure 18.4** with baseline at zero).

- If the data look very different, click **Redo Calibration** and repeat the steps above.

- Once the data look similar to Figure 18.4, proceed to the next section. Note that despite your best efforts, electrode adhesion may not be strong enough to record data; try another subject or different electrode placement.

Recording the Data

1. The subject should remain relaxed with eyes closed.

2. After clicking **Record,** the "director" will instruct the subject to keep his or her eyes closed for the first 20 seconds of recording, then open the eyes and *mentally* (not verbally) count to 20, then close the eyes again and relax for 20 seconds. The director will insert a marker by pressing the **F4** key (PC or Mac) when the command to open eyes is given, and another marker by pressing the **F5** key (PC or Mac) when the subject reaches the count of twenty and closes the eyes. Click **Suspend** 20 seconds after the subject recloses the eyes.

3. Observe the recording of the data (it should look similar to the data in **Figure 18.5**).

- If the subject moved too much during the recording, it is likely that artifact spikes will appear in the data. Remind the subject to be very still.

- Look carefully at the **alpha rhythm** band of data. The intensity of the alpha signal should decrease during the "eyes open" phase of the recording. If the data do not demonstrate this change, make sure that the electrodes are firmly attached.

- If the data show artifact spikes or the alpha signal fails to decrease when the eyes are open, click **Redo.**

- If the data look similar to the example (Figure 18.5), proceed to the next step.

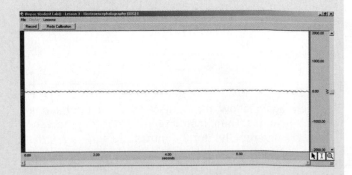

Figure 18.4 Example of calibration data.

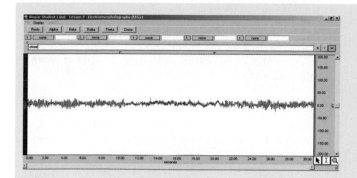

Figure 18.5 Example of EEG data.

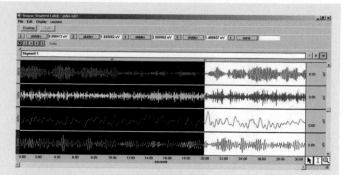

Figure 18.6 Highlighting the first data segment.

4. When finished, click **Done**. If you are certain you want to stop recording, click **YES**. Remove the electrodes from the subject's scalp.

5. A pop-up window will appear. To record from another subject, select **Record from another subject** and return to step 5 under Setting Up the Equipment. If continuing to the Data Analysis section, select **Analyze current data file** and proceed to step 2.

Data Analysis

1. If just starting the BIOPAC® program to perform data analysis, enter **Review Saved Data** mode, and choose the file with the subject's EEG data (for example, SmithEEG-1).

2. Observe the way the channel numbers are designated: CH 40–**alpha**; CH 41–**beta**; CH 42–**delta**; and CH 43– **theta**. CH 1 (raw EEG) is hidden. The software used it to extract and display each frequency band. If you want to see CH 1, hold down the **Ctrl** key (PC) or **Option** key (Mac) while using the cursor to click channel box 1 (the small box with a 1 at the upper left of the screen).

3. To analyze the data, set up the first four pairs of channel/measurement boxes at the top of the screen by selecting the following channels and measurement types from the drop-down menus:

Channel	Measurement	Data
CH 40	stddev	alpha
CH 41	stddev	beta
CH 42	stddev	delta
CH 43	stddev	theta

stddev (standard deviation): This is a statistical calculation that estimates the variability of the data in the area high-

lighted by the I-beam cursor. This function minimizes the effects of extreme values and electrical artifacts that may unduly influence interpretation of the data.

4. Use the arrow cursor and click the I-beam cursor box at the lower right of the screen to activate the "area selection" function. Using the activated I-beam cursor, highlight the first 20-second segment of EEG data, which represents the subject at rest with eyes closed (**Figure 18.6**).

5. Observe that the computer automatically calculates the stddev for each of the channels of data (alpha, beta, delta, and theta).

6. Record the data for each rhythm in the **Standard Deviations chart,** rounding to the nearest 0.01 μV.

7. Repeat steps 4–6 to analyze and record the data for the next two segments of data, with eyes open, and with eyes reclosed. The triangular markers inserted at the top of the data should provide guidance for highlighting.

8. To continue the analysis, change the settings in the first four pairs of channel/measurement boxes. Select the following channels and measurement types:

Channel	Measurement	Data
CH 40	Freq	alpha
CH 41	Freq	beta
CH 42	Freq	delta
CH 43	Freq	theta

Freq (frequency): This gives the frequency in hertz (Hz) of an individual wave that is highlighted by the I-beam cursor.

9. To view an individual wave from among the high-frequency waveforms, you must use the zoom function.

Text continues on next page. →

18

Standard Deviations (stddev) of Signals in Each Segment				
Rhythm	**Channel**	**Eyes closed Segment 1 Seconds 0–20**	**Eyes open Segment 2 Seconds 21–40**	**Eyes reclosed Segment 3 Seconds 41–60**
Alpha	CH 40			
Beta	CH 41			
Delta	CH 42			
Theta	CH 43			

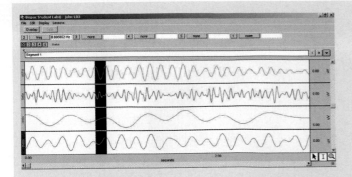

Figure 18.7 Highlighting a single alpha wave.

To activate the zoom function, use the cursor to click the magnifying glass at the lower-right corner of the screen (near the I-beam cursor box). The cursor will become a magnifying glass.

10. As the analysis begins, CH 40—the alpha data—will be automatically activated. To examine individual waves within the **alpha** data, click that band with the magnifying glass until it is possible to observe the peaks and troughs of individual waves within **Segment 1.**

• To properly view each of the waveforms, you may have to click the **Display** menu and select **Autoscale Waveforms.** This function rescales the data for the rhythm band that is selected.

11. At this time, focus on alpha waves only. Reactivate the I-beam cursor by clicking its box in the lower-right corner. Highlight a *single* alpha wave from peak to peak (as shown in **Figure 18.7**).

12. Read the calculated frequency (in Hz) in the measurement box for CH 40, and record this as the frequency of Wave 1 for alpha rhythm in the **Frequencies of Waves chart.**

13. Use the I-beam cursor to select two more individual **alpha** waves, and record their frequencies in the chart.

14. You will now perform the same frequency measurements for three waves in each of the **beta** (CH 41), **delta** (CH 42), and **theta** (CH 43) data sets. Record these measurements in the chart.

15. Calculate the average of the three waves measured for each of the brain rhythms, and record the average in the chart.

16. When finished, answer the following questions, and then exit the program by going to the **File** menu at the top of the page and clicking **Quit.**

Look at the waveforms you recorded, and carefully examine all three segments of the alpha rhythm record. Is there a difference in electrical activity in this frequency range when the eyes are open versus closed? Describe your observations.

_____

_____

_____

Carefully examine all three segments of the beta rhythm record. Is there a difference in electrical activity in this frequency range when the eyes are open versus closed? Describe what you observe.

_____

_____

_____

This time, compare the intensity (height) of the alpha and beta waveforms throughout all three segments. Does the intensity of one signal appear more varied than the other in the record? Describe your observations.

_____

_____

_____

Examine the data for the delta and theta rhythms. Is there any change in the waveform as the subject changes states? If so, describe the change observed.

_____

_____

_____

The degree of variation in the intensity of the signal was estimated by calculating the standard deviation of the waves in each segment of data. In which time segment (eyes open, eyes closed, or eyes reclosed) is the difference in the standard deviations the greatest?

_____

Frequencies of Waves for Each Rhythm (Hz)					
Rhythm	**Channel**	**Wave 1**	**Wave 2**	**Wave 3**	**Average**
Alpha	CH 40				
Beta	CH 41				
Delta	CH 42				
Theta	CH 43				

REVIEW SHEET
Electroencephalography

Name _____ Lab Time/Date _____

Brain Wave Patterns and the Electroencephalogram

1. Define *EEG*. _____

2. Identify the type of brain wave pattern described in each statement below.

   _____ below 4 Hz; slow, large waves; normally seen during deep sleep

   _____ rhythm generally apparent when an individual is in a relaxed, nonattentive state with the eyes closed

   _____ correlated to the alert state; usually about 14 to 30 Hz

3. What is meant by the term *alpha block*? _____

   _____

4. List at least four types of brain lesions that may be determined by EEG studies. _____

   _____

5. What is the common result of hypoactivity or hyperactivity of the brain neurons? _____

Observing Brain Wave Patterns

6. How was alpha block demonstrated in the laboratory experiment? _____

   _____

7. What was the effect of mental concentration on the brain wave pattern? ___ _____

   _____

   _____

8. What effect on the brain wave pattern did hyperventilation have? _____

   _____

   _____

Electroencephalography Using BIOPAC®

9. Observe the average frequency of the waves you measured for each rhythm. Did the calculated average for each fall within the specified range indicated in the introduction to encephalograms?

_____

10. Suggest the possible advantages and disadvantages of using electroencephalography in a clinical setting.

_____

_____

_____

_____

The Spinal Cord and Spinal Nerves

Objectives

☐ List two major functions of the spinal cord.

☐ Define *conus medullaris, cauda equina,* and *filum terminale.*

☐ Name the meninges of the spinal cord, and state their function.

☐ Indicate two major areas where the spinal cord is enlarged, and explain the reasons for the enlargement.

☐ Identify important anatomical areas on a model or image of a cross section of the spinal cord, and where applicable, name the neuron type found in these areas.

☐ Locate on a diagram the fiber tracts in the spinal cord, and state their functions.

☐ Note the number of pairs of spinal nerves that arise from the spinal cord, describe their division into groups, and identify the number of pairs in each group.

☐ Describe the origin and fiber composition of the spinal nerves, differentiating between roots, the spinal nerve proper, and rami, and discuss the result of transecting these structures.

☐ Discuss the distribution of the dorsal and ventral rami of the spinal nerves.

☐ Identify the four major nerve plexuses on a model or image, name the major nerves of each plexus, and describe the destination and function of each.

Materials

- Spinal cord model (cross section)
- Three-dimensional models or laboratory charts of the spinal cord and spinal nerves
- Red and blue pencils
- Preserved cow spinal cord sections with meninges and nerve roots intact (or spinal cord segment saved from the brain dissection in Exercise 17)
- Dissecting instruments and tray
- Disposable gloves

Text continues on next page. →

MasteringA&P®

For related exercise study tools, go to the Study Area of **MasteringA&P** There you will find:

- Practice Anatomy Lab **PAL**
- PhysioEx **PEx**
- A&PFlix **A&PFlix**
- Practice quizzes, Histology Atlas, eText, Videos, and more!

Pre-Lab Quiz

1. The spinal cord extends from the foramen magnum of the skull to the first or second lumbar vertebra, where it terminates in the:
   a. conus medullaris
   b. denticulate ligament
   c. filum terminale
   d. gray matter

2. How many pairs of spinal nerves do humans have?
   a. 10
   b. 12
   c. 31
   d. 47

3. Circle the correct underlined term. In cross section, the gray / white matter of the spinal cord looks like a butterfly or the letter H.

4. Circle True or False. The cell bodies of sensory neurons are found in an enlarged area of the dorsal root called the gray commissure.

5. Circle the correct underlined term. Fiber tracts conducting impulses to the brain are called ascending or sensory / motor tracts.

6. Circle True or False. Because the spinal nerves arise from fusion of the ventral and dorsal roots of the spinal cord, and contain motor and sensory fibers, all spinal nerves are considered mixed nerves.

7. The ventral rami of all spinal nerves except T_2 through T_{12} form complex networks of nerves known as:
   a. fissures
   b. ganglia
   c. plexuses
   d. sulci

Text continues on next page. →

- Dissecting microscope
- Prepared slide of spinal cord (x.s.)
- Compound microscope
- Post-it® Notes

For instructions on animal dissections, see the dissection exercises (starting on p. 705) in the cat and fetal pig editions of this manual.

For instructions on animal dissections, see the dissection exercises (starting on p. 705) in the cat and fetal pig editions of this manual.

8. Severe injuries to the _____ plexus cause weakness or paralysis of the entire upper limb.
   a. brachial   c. lumbar
   b. cervical   d. sacral
9. Circle True or False. The femoral nerve is the largest nerve from the sacral plexus.
10. Circle the correct underlined term. The sciatic nerve divides into the tibial and posterior femoral cutaneous / common fibular nerves.

The cylindrical **spinal cord,** a continuation of the brain stem, is a communication center. It plays a major role in spinal reflex activity and provides neural pathways to and from the brain.

Anatomy of the Spinal Cord

Enclosed within the vertebral canal of the spinal column, the spinal cord extends from the foramen magnum of the skull to the first or second lumbar vertebra, where it terminates in the cone-shaped **conus medullaris.** Like the brain, the cord is cushioned and protected by meninges (**Figure 19.1**). The dura mater and arachnoid mater extend beyond the conus medullaris, approximately to the level of S₂, and the **filum terminale,** a fibrous extension of the conus medullaris covered by pia mater, extends even farther to attach to the posterior coccyx (**Figure 19.2**). **Denticulate ligaments,** saw-toothed shelves of pia mater, secure the spinal cord to the dura mater (Figure 19.2c).

The cerebrospinal fluid–filled meninges extend beyond the end of the spinal cord, providing an excellent site for removing cerebrospinal fluid without endangering the spinal cord. Analysis of the fluid can provide information about suspected bacterial or viral infections of the meninges. This procedure, called a *lumbar puncture,* is usually performed below L₃.

In humans, 31 pairs of spinal nerves arise from the spinal cord and pass through intervertebral foramina to serve the body area near their level of emergence. The cord is about the circumference of a thumb for most of its length, but there are enlargements in the *cervical* and *lumbar* areas where the nerves serving the upper and lower limbs arise.

Because the spinal cord does not extend to the end of the vertebral column, the lumbar and sacral spinal nerve roots must travel through the vertebral canal before reaching their intervertebral foramina. This collection of spinal nerve roots is called the **cauda equina** (Figure 19.2a and d) because of its similarity to a horse's tail.

Figure 19.3 (p. 314) illustrates the spinal cord in cross section. Notice that the gray matter looks like a butterfly.

Activity 1

Identifying Structures of the Spinal Cord

Obtain a three-dimensional model or laboratory chart of a cross section of a spinal cord, and identify its structures as they are described in **Table 19.1** on p. 314 and shown in Figure 19.3.

Text continues on page 315. →

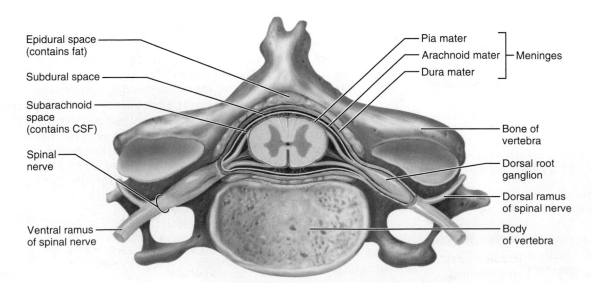

Epidural space (contains fat)

Subdural space

Subarachnoid space (contains CSF)

Spinal nerve

Ventral ramus of spinal nerve

Pia mater
Arachnoid mater — Meninges
Dura mater

Bone of vertebra

Dorsal root ganglion

Dorsal ramus of spinal nerve

Body of vertebra

Figure 19.1 Cross section through the spinal cord illustrating its relationship to the surrounding vertebra.

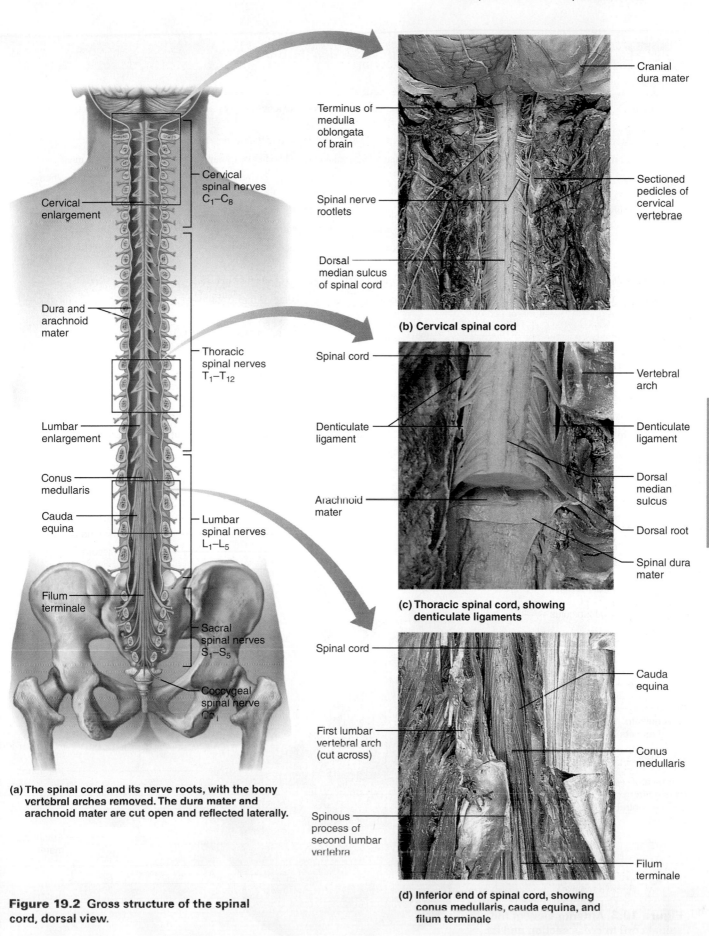

Cranial dura mater

Terminus of medulla oblongata of brain

Spinal nerve rootlets

Dorsal median sulcus of spinal cord

Sectioned pedicles of cervical vertebrae

(b) Cervical spinal cord

Spinal cord

Denticulate ligament

Arachnoid mater

Vertebral arch

Denticulate ligament

Dorsal median sulcus

Dorsal root

Spinal dura mater

(c) Thoracic spinal cord, showing denticulate ligaments

Cervical spinal nerves C_1–C_8

Cervical enlargement

Dura and arachnoid mater

Thoracic spinal nerves T_1–T_{12}

Lumbar enlargement

Conus medullaris

Cauda equina

Filum terminale

Lumbar spinal nerves L_1–L_5

Sacral spinal nerves S_1–S_5

Coccygeal spinal nerve Co_1

(a) The spinal cord and its nerve roots, with the bony vertebral arches removed. The dura mater and arachnoid mater are cut open and reflected laterally.

Spinal cord

First lumbar vertebral arch (cut across)

Spinous process of second lumbar vertebra

Cauda equina

Conus medullaris

Filum terminale

(d) Inferior end of spinal cord, showing conus medullaris, cauda equina, and filum terminale

Figure 19.2 Gross structure of the spinal cord, dorsal view.

19

Table 19.1	Anatomy of the Spinal Cord in Cross Section (Figure 19.3)
Structure	**Description**
Gray matter	Located in the center of the spinal cord and shaped like a butterfly or the letter H. Areas of the gray matter are individually named and described below.
Dorsal (posterior) horns	Posterior projections of the gray matter that contain primarily interneurons and the axons of sensory neurons.
Ventral (anterior) horns	Anterior projections of the gray matter that contain the cell bodies of somatic motor neurons and some interneurons.
Lateral horns	Small lateral projections that are present only in the thoracic and lumbar regions of the gray matter. When present, they contain the cell bodies of motor neurons of the autonomic nervous system.
Gray commissure	The cross bar of the H that surrounds the central canal.
Central canal	A narrow central cavity that is continuous with the ventricles of the brain.
Dorsal root	A nerve root that fans out into dorsal rootlets to connect to the posterior spinal cord. It contains the axons of sensory neurons.
Dorsal root ganglion	A bulge on the dorsal root that contains the cell bodies of sensory neurons.
Ventral root	A nerve root that is formed by the ventral rootlets connected to the anterior spinal cord. It contains the axons of motor neurons.
Spinal nerve	Formed by the fusion of dorsal and ventral roots. They are mixed nerves because they contain both sensory and motor fibers.
White matter	Forms the outer region of the spinal cord and is composed of myelinated and nonmyelinated axons organized into tracts.
Ventral median fissure	The anterior, more open of the two grooves that partially divide the spinal cord into left and right halves.
Dorsal median sulcus	The posterior, shallower of the two grooves that partially divide the spinal cord into left and right halves.
White columns	Each side of the spinal cord has three funiculi (columns): dorsal (posterior) funiculus, lateral funiculus, and ventral (anterior) funiculus, which are further divided into tracts.
Ascending (sensory) tracts	Carry sensory information from the sensory neurons to the brain.
Descending (motor) tracts	Carry motor instructions from the brain to the body's muscles and glands.

19

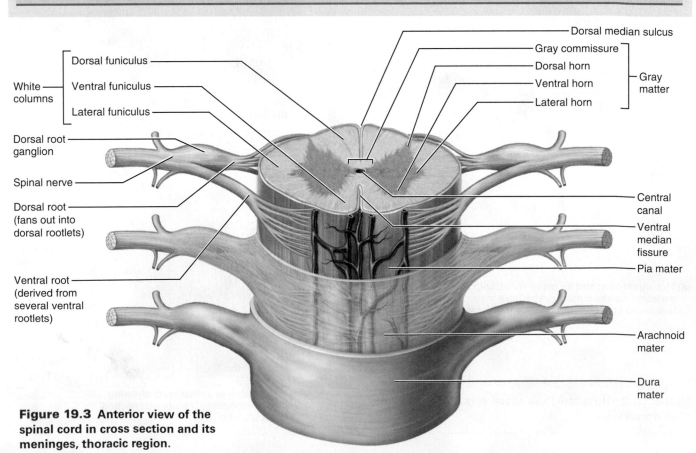

Figure 19.3 Anterior view of the spinal cord in cross section and its meninges, thoracic region.

Did you have chickenpox as a child or perhaps were immunized with the vaccine? If you had chickenpox, you have a one in three chance of experiencing shingles in your lifetime. Receiving the vaccine appears to prevent both chickenpox and shingles. The virus that causes both diseases is the varicella-zoster virus. After a person recovers from chickenpox, the virus lies dormant (latent) in the cell body of sensory neurons in the dorsal root ganglion. The virus can lie dormant in multiple levels of the spinal cord and in cranial nerves, as well. Later in life, the virus can be reactivated and travel within a sensory neuron to the surface of the skin, usually in response to stress or a weakened immune response. The later activation of the virus, called *shingles*, results in a rash and extreme pain. ■

Even though the spinal cord is protected by meninges and cerebrospinal fluid in the vertebral canal, it is highly vulnerable to traumatic injuries, such as might occur in an automobile accident.

When the cord is damaged, both motor and sensory functions are lost in body areas normally served by that region and lower regions of the spinal cord. Injury to certain spinal cord areas may even result in a permanent flaccid paralysis of both legs, called **paraplegia,** or of all four limbs, called **quadriplegia. ✚**

Activity 2

Identifying Spinal Cord Tracts

With the help of your textbook, label the spinal cord diagram **(Figure 19.4)** with the tract names that follow.

Each tract is represented on both sides of the cord, but for clarity, label the motor tracts on the right side of the diagram and the sensory tracts on the left side of the diagram. *Color ascending (sensory) tracts blue, and descending (motor) tracts red.* Then fill in the functional importance of each tract beside its name below. As you work, try to be aware of how the naming of the tracts is related to their anatomical distribution.

Dorsal columns

    Fasciculus gracilis _____

    Fasciculus cuneatus _____

Dorsal spinocerebellar _____

Ventral spinocerebellar _____

Lateral spinothalamic _____

Ventral spinothalamic _____

Lateral corticospinal _____

Ventral corticospinal _____

Rubrospinal _____

Tectospinal _____

Vestibulospinal _____

Medial reticulospinal _____

Lateral reticulospinal _____

19

Ascending (sensory) tracts　　　　　　　　**Descending (motor) tracts**

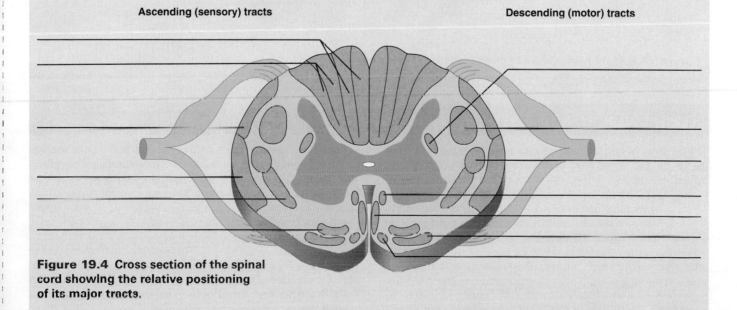

Figure 19.4 Cross section of the spinal cord showing the relative positioning of its major tracts.

DISSECTION

Spinal Cord

1. Obtain a dissecting tray and instruments, disposable gloves, and a segment of preserved spinal cord (from a cow or saved from the brain specimen used in Exercise 17). Identify the dura mater and the weblike arachnoid mater.

What name is given to the third meninx, and where is it found?

_____

_____

Peel back the dura mater, and observe the fibers making up the dorsal and ventral roots. If possible, identify a dorsal root ganglion.

2. Cut a thin cross section of the cord, and identify the ventral and dorsal horns of the gray matter with the naked eye or with the aid of a dissecting microscope.

Notice that the dorsal horns are more tapered than the ventral horns and that they extend closer to the edge of the spinal cord.

Also identify the central canal, white matter, ventral median fissure, dorsal median sulcus, and dorsal, ventral, and lateral funiculi.

3. Obtain a prepared slide of the spinal cord (cross section) and a compound microscope. Examine the slide carefully under low power (refer to **Figure 19.5** to identify spinal cord features). Observe the shape of the central canal.

Is it basically circular or oval? _____

Name the type of neuroglia that lines this canal. _____

_____

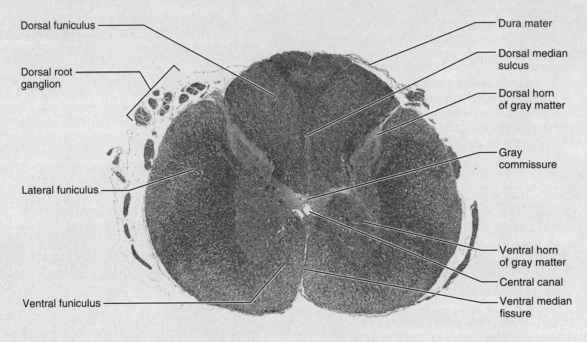

Figure 19.5 Light micrograph of cross section of the spinal cord (8×).

Labels: Dorsal funiculus; Dorsal root ganglion; Lateral funiculus; Ventral funiculus; Dura mater; Dorsal median sulcus; Dorsal horn of gray matter; Gray commissure; Ventral horn of gray matter; Central canal; Ventral median fissure

Spinal Nerves and Nerve Plexuses

The 31 pairs of human spinal nerves arise from the fusions of the ventral and dorsal roots of the spinal cord and are therefore **mixed nerves** containing both sensory and motor fibers (see Figure 19.1). There are 8 pairs of cervical nerves (C_1–C_8), 12 pairs of thoracic nerves (T_1–T_{12}), 5 pairs of lumbar nerves (L_1–L_5), 5 pairs of sacral nerves (S_1–S_5), and 1 pair of coccygeal nerves (Co_1) (**Figure 19.6a**). The first pair of spinal nerves leaves the vertebral canal between the base of the skull and the atlas, but all the rest exit via the intervertebral foramina. The first through seventh pairs of cervical nerves emerge *above* the vertebra for which they are named; C_8 emerges between C_7 and T_1. (Notice that there are 7 cervical vertebrae, but 8 pairs of cervical nerves.) The remaining spinal nerve pairs emerge from the spinal cord area *below* the same-numbered vertebra.

Almost immediately after emerging, each nerve divides into **dorsal** and **ventral rami.** Thus each spinal nerve is only about 1 or 2 cm long. The rami, like the spinal nerves, contain both motor and sensory fibers. The smaller dorsal rami serve the skin and musculature of the posterior body trunk at their approximate level of emergence. The ventral rami of spinal nerves T_2 through T_{12} pass anteriorly as the **intercostal nerves** to supply the muscles of intercostal spaces, and the skin and muscles of the anterior and lateral trunk. The ventral rami of all other spinal nerves form complex networks of nerves called **nerve**

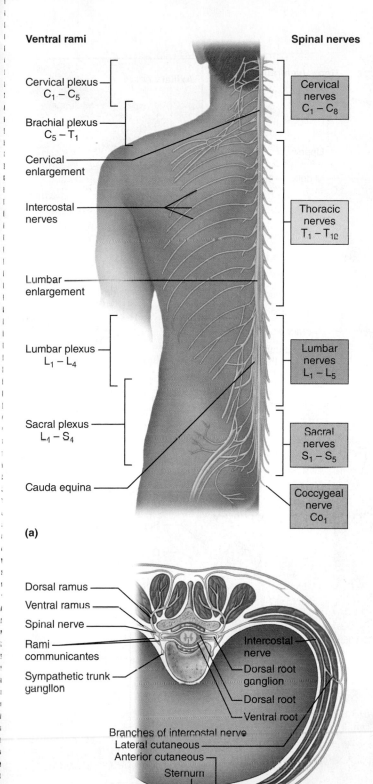

Figure 19.6 Human spinal nerves. (a) Spinal nerves are shown at right; ventral rami and the major nerve plexuses are shown at left. **(b)** Relative distribution of the ventral and dorsal rami of a spinal nerve (cross section of thorax).

plexuses. These plexuses primarily serve the muscles and skin of the limbs. The fibers of the ventral rami unite in the plexuses (with a few rami supplying fibers to more than one plexus). From the plexuses the fibers diverge again to form peripheral nerves, each of which contains fibers from more than one spinal nerve. (The four major nerve plexuses and their chief peripheral nerves are described in Tables 19.2–19.5 and illustrated in Figures 19.7–19.10. Their names and site of origin should be committed to memory). The tiny S_5 and Co_1 spinal nerves contribute to a small plexus that serves part of the pelvic floor.

Cervical Plexus and the Neck

The **cervical plexus** (**Figure 19.7** and **Table 19.2**, p. 319) arises from the ventral rami of C_1 through C_5 to supply muscles of the shoulder and neck. The major motor branch of this plexus is the **phrenic nerve,** which arises from C_3 through C_4 (plus some fibers from C_5) and passes into the thoracic cavity in front of the first rib to innervate the diaphragm. The primary danger of a broken neck is that the phrenic nerve may be severed, leading to paralysis of the diaphragm and cessation of breathing. A rhyme to help you remember the rami (roots) forming the phrenic nerves is "C_3, C_4, C_5 keep the diaphragm alive."

Brachial Plexus and the Upper Limb

The **brachial plexus** is large and complex, arising from the ventral rami of C_5 through C_8 and T_1 (**Table 19.3**, p. 319). The plexus, after being rearranged consecutively into *trunks, divisions,* and *cords,* finally becomes subdivided into five major *peripheral nerves* (**Figure 19.8**, p. 318).

19

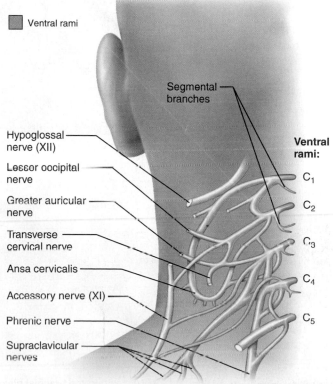

Figure 19.7 The cervical plexus. The nerves colored gray connect to the plexus but do not belong to it. (See Table 19.2.)

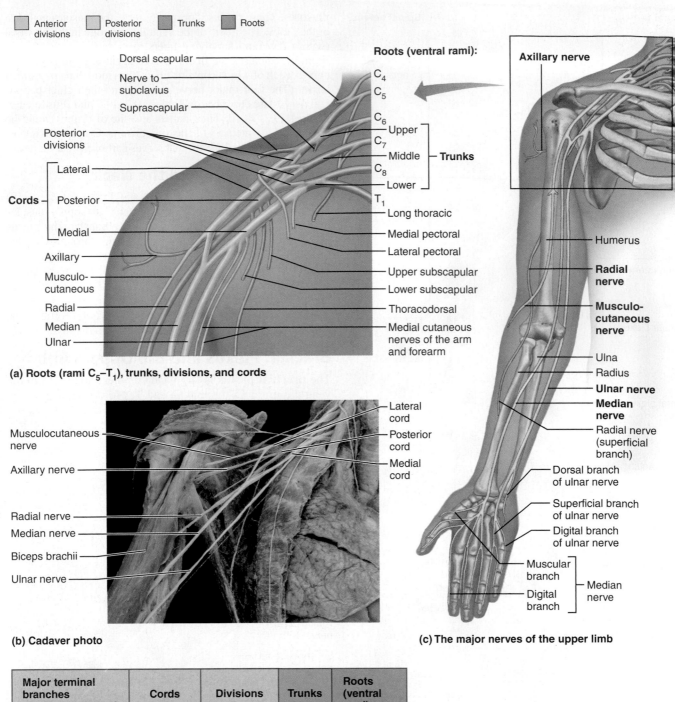

Anterior divisions | Posterior divisions | Trunks | Roots

Dorsal scapular
Nerve to subclavius
Suprascapular

Roots (ventral rami):

C_4
C_5
C_6 — Upper
C_7 — Middle — **Trunks**
C_8 — Lower
T_1

Posterior divisions

Cords — Lateral / Posterior / Medial

Long thoracic
Medial pectoral
Lateral pectoral
Upper subscapular
Lower subscapular
Thoracodorsal
Medial cutaneous nerves of the arm and forearm

Axillary
Musculo-cutaneous
Radial
Median
Ulnar

(a) Roots (rami C_5–T_1), trunks, divisions, and cords

Axillary nerve

Humerus

Radial nerve

Musculo-cutaneous nerve

Ulna
Radius
Ulnar nerve
Median nerve
Radial nerve (superficial branch)

Dorsal branch of ulnar nerve
Superficial branch of ulnar nerve
Digital branch of ulnar nerve
Muscular branch — Median nerve
Digital branch

(c) The major nerves of the upper limb

Musculocutaneous nerve
Axillary nerve
Radial nerve
Median nerve
Biceps brachii
Ulnar nerve

Lateral cord
Posterior cord
Medial cord

(b) Cadaver photo

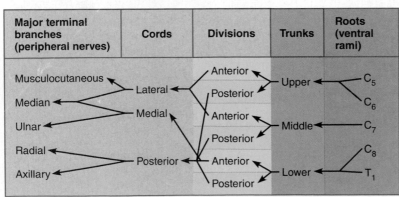

Major terminal branches (peripheral nerves)	Cords	Divisions	Trunks	Roots (ventral rami)
Musculocutaneous	Lateral	Anterior	Upper	C_5
Median		Posterior		C_6
Ulnar	Medial	Anterior	Middle	C_7
Radial		Posterior		C_8
Axillary	Posterior	Anterior	Lower	T_1
		Posterior		

(d) Flowchart summarizing relationships within the brachial plexus

Figure 19.8 The brachial plexus. (See Table 19.3.)

Table 19.2	Branches of the Cervical Plexus (Figure 19.7)	
Nerves	**Ventral rami**	**Structures served**
Cutaneous Branches (Superficial)		
Lesser occipital	C_2 (C_3)	Skin on posterolateral aspect of neck
Greater auricular	C_2, C_3	Skin of ear, skin over parotid gland
Transverse cervical	C_2, C_3	Skin on anterior and lateral aspect of neck
Supraclavicular (medial, intermediate, and lateral)	C_3, C_4	Skin of shoulder and clavicular region
Motor Branches (Deep)		
Ansa cervicalis (superior and inferior roots)	C_1–C_3	Infrahyoid muscles of neck (omohyoid, sternohyoid, and sternothyroid)
Segmental and other muscular branches	C_1–C_5	Deep muscles of neck (geniohyoid and thyrohyoid) and portions of scalenes, levator scapulae, trapezius, and sternocleidomastoid muscles
Phrenic	C_3–C_5	Diaphragm (sole motor nerve supply)

Table 19.3	Branches of the Brachial Plexus (Figure 19.8)	
Nerves	**Cord and ventral rami**	**Structures served**
Axillary	Posterior cord (C_5, C_6)	Muscular branches: deltoid and teres minor muscles Cutaneous branches: some skin of shoulder region
Musculocutaneous	Lateral cord (C_5–C_7)	Muscular branches: flexor muscles in anterior arm (biceps brachii, brachialis, coracobrachialis) Cutaneous branches: skin on anterolateral forearm (extremely variable)
Median	By two branches, one from medial cord (C_8, T_1) and one from the lateral cord (C_5–C_7)	Muscular branches to flexor group of anterior forearm (palmaris longus, flexor carpi radialis, flexor digitorum superficialis, flexor pollicis longus, lateral half of flexor digitorum profundus, and pronator muscles); intrinsic muscles of lateral palm and digital branches to the fingers Cutaneous branches: skin of lateral two-thirds of hand on ventral side and dorsum of fingers 2 and 3
Ulnar	Medial cord (C_8, T_1)	Muscular branches: flexor muscles in anterior forearm (flexor carpi ulnaris and medial half of flexor digitorum profundus); most intrinsic muscles of hand Cutaneous branches: skin of medial third of hand, both anterior and posterior aspects
Radial	Posterior cord (C_5–C_8, T_1)	Muscular branches: posterior muscles of arm and forearm (triceps brachii, anconeus, supinator, brachioradialis, extensors carpi radialis longus and brevis, extensor carpi ulnaris, and several muscles that extend the fingers) Cutaneous branches: skin of posterolateral surface of entire limb (except dorsum of fingers 2 and 3)
Dorsal scapular	Branches of C_5 rami	Rhomboid muscles and levator scapulae
Long thoracic	Branches of C_5–C_7 rami	Serratus anterior muscle
Subscapular	Posterior cord; branches of C_5 and C_6 rami	Teres major and subscapularis muscles
Suprascapular	Upper trunk (C_5, C_6)	Shoulder joint; supraspinatus and infraspinatus muscles
Pectoral (lateral and medial)	Branches of lateral and medial cords (C_5–T_1)	Pectoralis major and minor muscles

The **axillary nerve,** which serves the muscles and skin of the shoulder, has the most limited distribution. The large **radial nerve** passes down the posterolateral surface of the arm and forearm, supplying all the extensor muscles of the arm, forearm, and hand and the skin along its course. The radial nerve is often injured in the axillary region by the pressure of a crutch or by hanging one's arm over the back of a chair. The **median nerve** passes down the anteromedial surface of the arm to supply most of the flexor muscles in the forearm and several muscles in the hand (plus the skin of the lateral surface of the palm of the hand).

☐ Hyperextend at the wrist to identify the long, obvious tendon of your palmaris longus muscle, which crosses the exact midline of the anterior wrist. Your median nerve lies immediately deep to that tendon, and the radial nerve lies just *lateral* to it.

The **musculocutaneous nerve** supplies the arm muscles that flex the forearm and the skin of the lateral surface of the forearm. The **ulnar nerve** travels down the posteromedial surface of the arm. It courses around the medial epicondyle of the humerus to supply the flexor carpi ulnaris, the ulnar head of the flexor digitorum profundus of the forearm, and all intrinsic muscles of the hand not served by the median nerve. It supplies the skin of the medial third of the hand, both the anterior and posterior surfaces. Trauma to the ulnar nerve, which often occurs when the elbow is hit, is commonly referred to as "hitting the funny bone" because of the odd, painful, tingling sensation it causes.

Severe injuries to the brachial plexus cause weakness or paralysis of the entire upper limb. Such injuries may occur when the upper limb is pulled hard and the plexus is stretched (as when a football tackler yanks the arm of the halfback), and by blows to the shoulder that force the humerus inferiorly (as when a cyclist is pitched headfirst off his motorcycle and grinds his shoulder into the pavement). ✚

Lumbosacral Plexus and the Lower Limb

The **lumbosacral plexus,** which serves the pelvic region of the trunk and the lower limbs, is actually a complex of two plexuses, the lumbar plexus and the sacral plexus (Figures 19.9 and 19.10). These plexuses interweave considerably and many fibers of the lumbar plexus contribute to the sacral plexus.

The Lumbar Plexus

The **lumbar plexus** arises from ventral rami of L_1 through L_4 (and sometimes T_{12}). Its nerves serve the lower abdominopelvic region and the anterior thigh (**Table 19.4** and **Figure 19.9**). The largest nerve of this plexus is the **femoral nerve,** which passes beneath the inguinal ligament to innervate the anterior thigh muscles. The cutaneous branches of the femoral nerve (median and anterior femoral cutaneous and the saphenous nerves) supply the skin of the anteromedial surface of the entire lower limb.

The Sacral Plexus

Arising from L_4 through S_4, the nerves of the **sacral plexus** supply the buttock, the posterior surface of the thigh, and virtually all sensory and motor fibers of the leg and foot (**Table 19.5**, p. 322, and **Figure 19.10**). The major peripheral nerve of this plexus is the **sciatic nerve,** the largest nerve in the body. The sciatic nerve leaves the pelvis through the greater sciatic notch and travels down the posterior thigh, serving its flexor muscles and skin. In the popliteal region, the sciatic

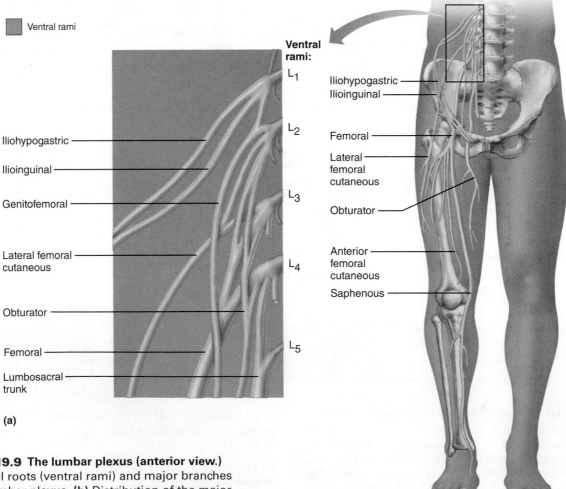

Ventral rami

Ventral rami:
L_1

Iliohypogastric
Ilioinguinal

L_2

Femoral

Lateral femoral cutaneous

L_3

Obturator

L_4

Anterior femoral cutaneous

Saphenous

L_5

Iliohypogastric

Ilioinguinal

Genitofemoral

Lateral femoral cutaneous

Obturator

Femoral

Lumbosacral trunk

(a)

(b)

Figure 19.9 The lumbar plexus (anterior view.)
(a) Spinal roots (ventral rami) and major branches of the lumbar plexus. **(b)** Distribution of the major peripheral nerves of the lumbar plexus in the lower limb. (See Table 19.4.)

19

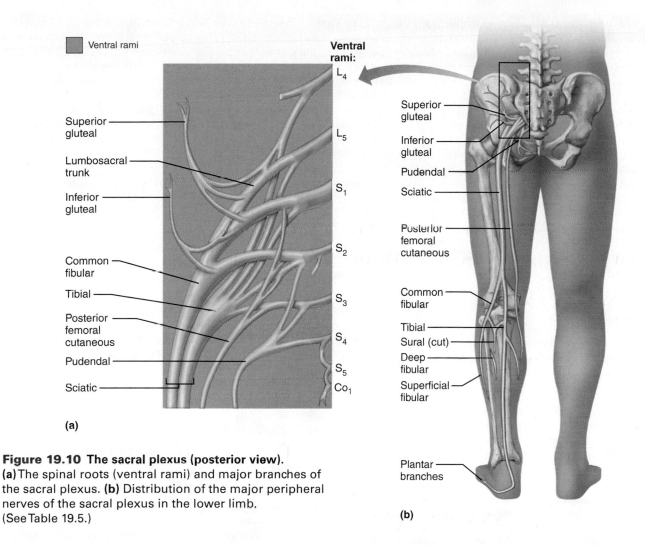

Figure 19.10 The sacral plexus (posterior view).
(a) The spinal roots (ventral rami) and major branches of the sacral plexus. **(b)** Distribution of the major peripheral nerves of the sacral plexus in the lower limb. (See Table 19.5.)

19

Table 19.4	Branches of the Lumbar Plexus (Figure 19.9)	
Nerves	**Ventral rami**	**Structures served**
Femoral	L_2–L_4	Skin of anterior and medial thigh via *anterior femoral cutaneous* branch; skin of medial leg and foot, hip and knee joints via *saphenous* branch; motor to anterior muscles (quadriceps and sartorius) of thigh and to pectineus, iliacus
Obturator	L_2–L_4	Motor to adductor magnus (part), longus, and brevis muscles, gracilis muscle of medial thigh, obturator externus; sensory for skin of medial thigh and for hip and knee joints
Lateral femoral cutaneous	L_2, L_3	Skin of lateral thigh; some sensory branches to peritoneum
Iliohypogastric	L_1	Skin of lower abdomen and hip; muscles of anterolateral abdominal wall (internal obliques and transversus abdominis)
Ilioinguinal	L_1	Skin of external genitalia and proximal medial aspect of the thigh; inferior abdominal muscles
Genitofemoral	L_1, L_2	Skin of scrotum in males, of labia majora in females, and of anterior thigh inferior to middle portion of inguinal region; cremaster muscle in males

Table 19.5	Branches of the Sacral Plexus (Figure 19.10)	
Nerves	**Ventral rami**	**Structures served**
Sciatic nerve	L₄–S₃	Composed of two nerves (tibial and common fibular) in a common sheath; they diverge just proximal to the knee
• Tibial (including sural, medial and lateral plantar, and medial calcaneal branches)	L₄–S₃	Cutaneous branches: to skin of posterior surface of leg and sole of foot Motor branches: to muscles of back of thigh, leg, and foot (hamstrings [except short head of biceps femoris], posterior part of adductor magnus, triceps surae, tibialis posterior, popliteus, flexor digitorum longus, flexor hallucis longus, and intrinsic muscles of foot)
• Common fibular (superficial and deep branches)	L₄–S₂	Cutaneous branches: to skin of anterior and lateral surface of leg and dorsum of foot Motor branches: to short head of biceps femoris of thigh, fibularis muscles of lateral leg, tibialis anterior, and extensor muscles of toes (extensor hallucis longus, extensors digitorum longus and brevis)
Superior gluteal	L₄–S₁	Motor branches to gluteus medius and minimus and tensor fasciae latae
Inferior gluteal	L₅–S₂	Motor branches to gluteus maximus
Posterior femoral cutaneous	S₁–S₃	Skin of buttock, posterior thigh, and popliteal region; length variable; may also innervate part of skin of calf and heel
Pudendal	S₂–S₄	Supplies most of skin and muscles of perineum (region encompassing external genitalia and anus and including clitoris, labia, and vaginal mucosa in females, and scrotum and penis in males); external anal sphincter

19

nerve divides into the **common fibular nerve** and the **tibial nerve,** which together supply the balance of the leg muscles and skin, both directly and via several branches.

 Injury to the proximal part of the sciatic nerve, as might follow a fall or disc herniation, results in a number of lower limb impairments. **Sciatica,** characterized by stabbing pain radiating over the course of the sciatic nerve, is common. When the sciatic nerve is completely severed, the leg is nearly useless. The leg cannot be flexed and the foot drops into plantar flexion (it dangles), a condition called **footdrop. +**

Activity 3

Identifying the Major Nerve Plexuses and Peripheral Nerves

Identify each of the four major nerve plexuses and their major nerves (Figures 19.7–19.10) on a large laboratory chart or model. Trace the courses of the nerves and relate those observations to the information provided in Tables 19.2–19.5.

👥 Group Challenge

Fix the Sequence

Listed below are sets consisting of a plexus, a nerve, and a muscle possibly innervated by the listed nerve. Working in a small group, one student will act as the recorder. For each set, the recorder will write each structure on a separate Post-it® Note. If the sequence is correct, simply write "all correct." If incorrect, identify the structure that doesn't match, cross it out on the Post-it® Note, and write the correct structure. Note that there may be more than one way to correct the sequence. Explain how you corrected the sequence for each numbered set.

1. Cervical plexus, phrenic nerve, diaphragm _____

_____

2. Brachial plexus, ulnar nerve, palmaris longus _____

_____

3. Brachial plexus, radial nerve, triceps brachii

_____

4. Cervical plexus, axillary nerve, deltoid _____

_____

5. Lumbar plexus, femoral nerve, gracilis _____

_____

6. Lumbar plexus, sciatic nerve, common fibular nerve, tibialis anterior _____

_____

7. Sacral plexus, superior gluteal nerve, gluteus maximus

_____

_____

Objectives

☐ State how the autonomic nervous system differs from the somatic nervous system.

☐ Identify the site of origin and the function of the sympathetic and parasympathetic divisions of the autonomic nervous system.

☐ Identify the neurotransmitters associated with the sympathetic and parasympathetic fibers.

☐ Record and analyze data associated with the galvanic skin response.

Materials

- Laboratory chart or three-dimensional model of the sympathetic trunk (chain)

BIOPAC BIOPAC® BSL System with BSL software version 3.7.6 to 3.7.7 (for Windows 7/Vista/XP or Mac OS X 10.4–10.6), data acquisition unit MP36/35 or MP45, PC or Mac computer, respiratory transducer belt, EDA/GSR finger leads or disposable finger electrodes with EDA pinch leads, Biopac Student Lab electrode lead set, disposable vinyl electrodes, conduction gel, and nine 8½ × 11 inch sheets of paper of different colors (white, black, red, blue, green, yellow, orange, brown, and purple) to be viewed in this sequence.

Instructors using the MP36/35/30 data acquisition unit with BSL software versions earlier than 3.7.6 (for Windows or Mac) will need slightly different channel settings and collection strategies. Instructions for using the older data acquisition unit can be found on MasteringA&P.

✂ For instructions on animal dissections, see the dissection exercises (starting on p. 705) in the cat and fetal pig editions of this manual.

MasteringA&P®

For related exercise study tools, go to the Study Area of **MasteringA&P**. There you will find:

- Practice Anatomy Lab **PAL**
- **PhysioEx** **PEx**
- A&PFlix **A&PFlix**
- Practice quizzes, Histology Atlas, eText, Videos, and more!

Pre-Lab Quiz

1. The _____ nervous system is the subdivision of the peripheral nervous system which regulates body activities that are generally not under conscious control.
   - **a.** autonomic
   - **b.** cephalic
   - **c.** somatic
   - **d.** vascular

2. Circle the correct underlined term. The parasympathetic division of the autonomic nervous system is also known as the <u>craniosacral</u> / <u>thoracolumbar</u> division.

3. Circle True or False. Cholinergic fibers release epinephrine.

4. The _____ division of the autonomic nervous system is responsible for the "fight-or-flight" response because it adapts the body for extreme conditions such as exercise.

5. Circle True or False. The galvanic skin response measures an increase in water and electrolytes at the skin surface.

The **autonomic nervous system (ANS)** is the subdivision of the peripheral nervous system (PNS) that regulates body activities that are generally not under conscious control. For this reason, the ANS is also called the *involuntary nervous system.*

The motor pathways of the **somatic** (voluntary) **nervous system,** innervate the skeletal muscles, and the motor pathways of the autonomic nervous system innervate smooth muscle, cardiac muscle, and glands. In the somatic division, the cell bodies of the motor neurons reside in the brain stem or ventral horns of the spinal cord, and their axons extend directly to the skeletal muscles they serve. However, the autonomic nervous system consists of chains of two motor neurons. The first motor neuron of each pair, called the *preganglionic neuron,* resides in the brain stem or the spinal cord. Its axon leaves the central nervous system (CNS) to synapse with the second motor neuron, the *postganglionic neuron,* whose cell body is located in an autonomic ganglion outside the CNS. The axon of the postganglionic neuron then extends to the organ it serves.

The ANS has two major functional subdivisions: the sympathetic and parasympathetic divisions. Both serve most of the same organs but generally cause opposing, or antagonistic, effects. **Table 20.1** on p. 328 compares the parasympathetic and sympathetic divisions. Refer also to **Figure 20.1,** p. 328, and **Figure 20.2,** p. 329.

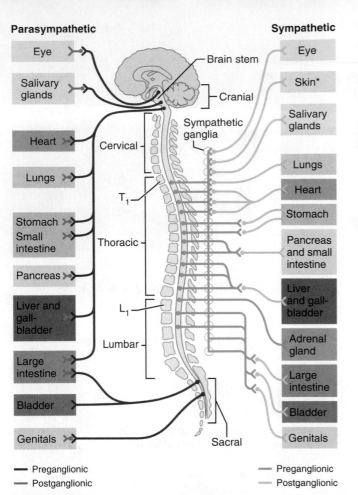

Parasympathetic

Eye
Salivary glands
Heart
Lungs
Stomach
Small intestine
Pancreas
Liver and gall-bladder
Large intestine
Bladder
Genitals

Brain stem
Cranial
Sympathetic ganglia
Cervical
T₁
Thoracic
L₁
Lumbar
Sacral

Sympathetic

Eye
Skin*
Salivary glands
Lungs
Heart
Stomach
Pancreas and small intestine
Liver and gall-bladder
Adrenal gland
Large intestine
Bladder
Genitals

— Preganglionic
— Postganglionic

— Preganglionic
— Postganglionic

Figure 20.1 Overview of the subdivisions of the autonomic nervous system. Although sympathetic innervation to the skin(*) is shown here mapped to the cervical area, all nerves to the periphery carry postganglionic sympathetic fibers.

Activity 1

Locating the Sympathetic Trunk

Locate the sympathetic trunk (chain) on the spinal nerve chart or three-dimensional model. Notice that the trunk resembles a string of beads. There is a trunk on either side of the vertebral column.

As we grow older, our sympathetic nervous system gradually becomes less and less efficient, particularly in causing vasoconstriction of blood vessels. When elderly people stand up quickly after sitting or lying down, they often become light-headed or faint. The sympathetic nervous system is not able to react quickly enough to counteract the pull of gravity by activating the vasoconstrictor fibers. So, blood pools in the feet. This condition, **orthostatic hypotension,** is a type of low blood pressure resulting from changes in body position as described. Orthostatic hypotension can be prevented to some degree if the person changes position *slowly*. This gives the sympathetic nervous system a little more time to react and adjust. ✚

20

Table 20.1	Anatomical and Physiological Comparison of the Parasympathetic and Sympathetic Divisions (Figure 20.1)	
Characteristic	**Parasympathetic (craniosacral)**	**Sympathetic (thoracolumbar)**
Origin in the CNS (contains the preganglionic cell body)	Brain stem nuclei of cranial nerves III (oculomotor), VII (facial), IX (glossopharyngeal), and X (vagus); spinal cord segments S_2 through S_4	Lateral horns of the gray matter of the spinal cord T_1 through L_2
Location of ganglia (contains the postganglionic cell body)	Located close to the target organ (**terminal ganglia**) or within the wall of the target organ (**intramural ganglia**)	Located close to the CNS: alongside the vertebral column (**sympathetic trunk ganglia**) or anterior to the vertebral column (**collateral ganglia**)
Axon lengths	Long preganglionic axons	Short preganglionic axons
	Short postganglionic axons	Long postganglionic axons
White and gray rami communicantes (see Figure 20.2)	None	Each white ramus communicans contains myelinated preganglionic axons. Each gray ramus communicans contains nonmyelinated postganglionic axons.
Degree of branching of preganglionic axons	Minimal	Extensive
Functional role	Performs maintenance functions; conserves and stores energy; rest-and-digest response	Prepares the body for emergency situations and vigorous physical activity; fight-or-flight response
Neurotransmitters	Preganglionic axons release acetylcholine (cholinergic fibers)	Preganglionic axons release acetylcholine
	Postganglionic axons release acetylcholine	Most postganglionic axons release norepinephrine (adrenergic fibers); postganglionic axons serving sweat glands and the blood vessels of skeletal muscle release acetylcholine; neurotransmitter activity is supplemented by the release of epinephrine and norepinephrine by the adrenal medulla
Effects of the division	More specific and local	More general and widespread

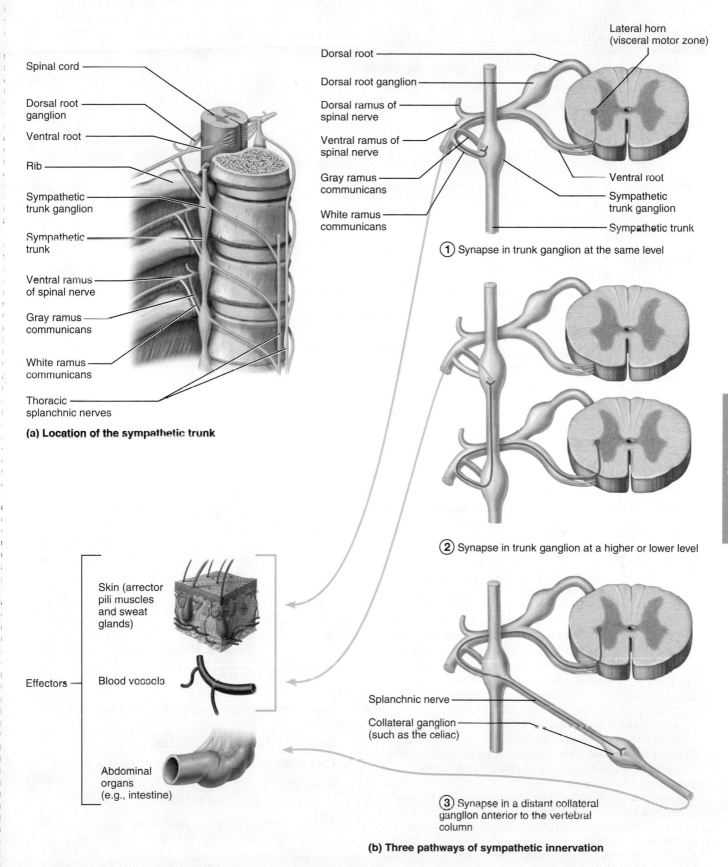

(a) **Location of the sympathetic trunk**

Spinal cord

Dorsal root ganglion

Ventral root

Rib

Sympathetic trunk ganglion

Sympathetic trunk

Ventral ramus of spinal nerve

Gray ramus communicans

White ramus communicans

Thoracic splanchnic nerves

Dorsal root

Dorsal root ganglion

Dorsal ramus of spinal nerve

Ventral ramus of spinal nerve

Gray ramus communicans

White ramus communicans

Lateral horn (visceral motor zone)

Ventral root

Sympathetic trunk ganglion

Sympathetic trunk

① Synapse in trunk ganglion at the same level

② Synapse in trunk ganglion at a higher or lower level

Effectors

Skin (arrector pili muscles and sweat glands)

Blood vessels

Abdominal organs (e.g., intestine)

Splanchnic nerve

Collateral ganglion (such as the celiac)

③ Synapse in a distant collateral ganglion anterior to the vertebral column

(b) **Three pathways of sympathetic innervation**

Figure 20.2 Sympathetic trunks and pathways. (a) Diagram of the right sympathetic trunk in the posterior thorax. **(b)** Synapses between preganglionic and postganglionic sympathetic neurons can occur at three different locations.

20

Activity 2

Comparing Sympathetic and Parasympathetic Effects

Several body organs are listed in the **Activity 2 chart**. Using your textbook as a reference, list the effect of the sympathetic and parasympathetic divisions on each.

Activity 2: Parasympathetic and Sympathetic Effects		
Organ	**Parasympathetic effect**	**Sympathetic effect**
Heart		
Bronchioles of lungs		
Digestive tract		
Urinary bladder		
Iris of the eye		
Blood vessels (most)		
Penis/clitoris		
Sweat glands		
Adrenal medulla		
Pancreas		

Activity 3

Exploring the Galvanic Skin Response (Electrodermal Activity) Within a Polygraph Using BIOPAC®

The autonomic nervous system is closely integrated with the emotions of an individual. A sad event, sharp pain, or simple stress can bring about measurable changes in autonomic regulation of heart rate, respiration, and blood pressure. In addition to these obvious physiological signs, more subtle autonomic changes can occur in the skin. Specifically, changes in autonomic tone in response to external circumstances can influence the rate of sweat gland secretion and blood flow to the skin that may not be readily seen but can be measured. The **galvanic skin response** is an electrophysiological measurement of changes that occur in the skin due to changes in autonomic stimulation.

The galvanic skin response, also referred to as electrodermal activity (EDA), is measured by recording the changes in **galvanic skin resistance (GSR)** and **galvanic skin potential (GSP)**. Resistance, recorded in *ohms* (Ω), is a measure of the opposition to the flow of current from one electrode to another. Increasing resistance results in decreased current. Potential, measured in *volts* (V), is a measure of the amount of charge separation between two points. Increased sympathetic stimulation of sweat glands decreases resistance on the skin because of increased water and electrolytes on the skin surface.

In this experiment you will record heart rate, respiration, and EDA/GSR while the subject is exposed to various conditions. Because "many" variables will be "recorded," this process is often referred to as a **polygraph**. The goal of this exercise is to record and analyze data to observe how this process works. This is not a "lie detector test," as its failure rate is far too high to provide true scientific or legal certainty. However, the polygraph can be used as an investigative tool.

Setting Up the Equipment

1. Connect the BIOPAC® unit to the computer, and turn the computer **ON**.

2. Make sure the BIOPAC® unit is **OFF**.

3. Plug in the equipment (as shown in **Figure 20.3**).

- Respiratory transducer belt—CH 1

- Electrode lead set—CH 2

- EDA/GSR finger leads or disposable finger electrodes and EDA pinch leads—CH 3

4. Turn the BIOPAC® unit **ON**.

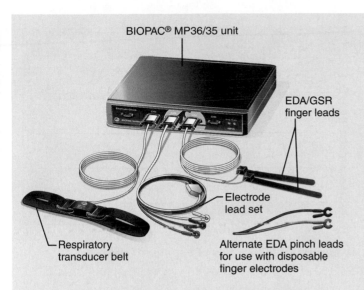

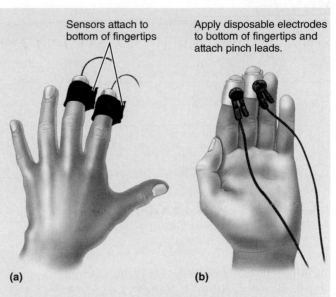

Sensors attach to bottom of fingertips

Apply disposable electrodes to bottom of fingertips and attach pinch leads.

(a)   (b)

Figure 20.5 Placement of the EDA/GSR finger lead sensors or disposable electrodes on the fingers.

EDA/GSR finger leads

Electrode lead set

Alternate EDA pinch leads for use with disposable finger electrodes

Respiratory transducer belt

BIOPAC® MP36/35 unit

Figure 20.3 Setting up the BIOPAC® equipment. Plug the respiratory transducer belt into Channel 1, the electrode lead set into Channel 2, and the EDA/GSR finger leads into Channel 3.

5. Attach the respiratory transducer belt to the subject (as shown in **Figure 20.4**). It should be fastened so that it is slightly tight even at the point of maximal expiration.

6. To pick up a good EDA/GSR signal, it is important that the subject's hand have enough sweat (as it normally would). *The subject should not have freshly washed or cold hands.* Place the electrodes on the middle and index fingers with the sensors on the skin, not the fingernail. They should fit snugly but not be so tight as to cut off circulation. If using EDA/GSR finger lead sensors, fill both cavities of the leads with conduction gel, and attach the sensors to the subject's fingers (as shown in **Figure 20.5a**). If using disposable finger electrodes, apply to subject's fingers and attach pinch leads (as shown in Figure 20.5b). Attach the electrodes at least 5 minutes before recording.

7. In order to record the heart rate, place the electrodes on the subject (as shown in **Figure 20.6**). Place an electrode on the medial surface of each leg, just above the ankle. Place another electrode on the right anterior forearm just above the wrist.

Text continues on next page. →

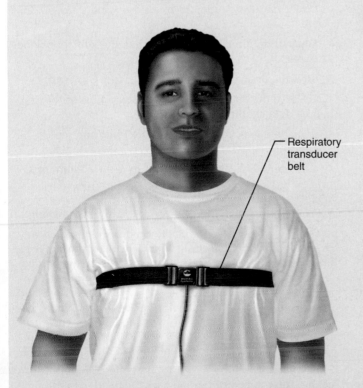

Respiratory transducer belt

Figure 20.4 Proper placement of the respiratory transducer belt around the subject's thorax.

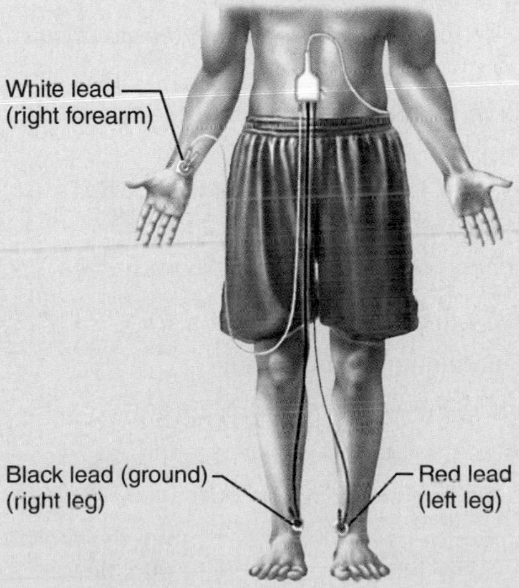

White lead (right forearm)

Black lead (ground) (right leg)

Red lead (left leg)

Figure 20.6 Placement of electrodes and the appropriate attachment of electrode leads by color.

20

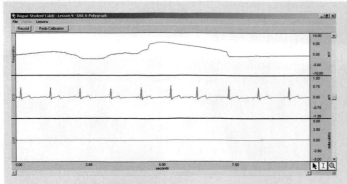

Figure 20.7 Example of waveforms during the calibration procedure.

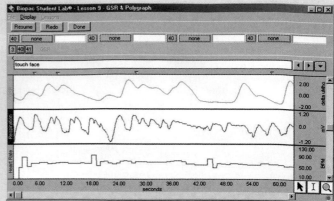

Figure 20.8 Example of Segment 1 data.

8. Attach the electrode lead set to the electrodes according to the colors shown in the example (Figure 20.6). Wait 5 minutes before starting the calibration procedure.

9. Start the Biopac Student Lab program on the computer by double-clicking the icon on the desktop or by following your instructor's guidance.

10. Select lesson **L09-Poly-1** from the menu, and click **OK**.

11. Type in a filename that will save this subject's data on the computer hard drive. You may want to use the subject's last name followed by Poly-1 (for example, SmithPoly-1), then click **OK.**

Calibrating the Equipment

1. Have the subject sit facing the director, but do not allow the subject to see the computer screen. The subject should remain immobile but be relaxed with legs and arms in a comfortable position.

2. When the subject is ready, click **Calibrate** and then click **Yes** and **OK** if prompted. After 3 seconds, the subject will hear a beep and should inhale and exhale deeply for one breath.

3. Wait for the calibration to stop automatically after 10 seconds.

4. Observe the data, which should look similar to that in **(Figure 20.7)**.

- If the data look very different, click **Redo Calibration** and repeat the steps above.
- If the data look similar, proceed to the next section.

Recording the Data

Hints to obtaining the best data:

- Do not let the subject see the data as it is being recorded.
- Conduct the exam in a quiet setting.
- Keep the subject as still as possible.
- Take care to have the subject move the mouth as little as possible when responding to questions.
- Make sure the subject is relaxed at resting heart rate before the exam begins.

The data will be recorded in three segments. The director must read through the directions for the entire segment before proceeding so that the subject can be prompted and questioned appropriately.

Segment 1: Baseline Data

1. When the subject and director are ready, click **Record**.

2. After waiting 5 seconds, the director will ask the subject to respond to the following questions and should remind the subject to minimize mouth movements when answering. Use the **F9** key (PC) or **ESC** key (Mac) to insert a marker after each response. Wait about 5 seconds after each answer.

- Quietly state your name.
- Slowly count down from 10 to zero.
- Count backward from 30 by odd numbers (29, 27, 25, etc.).
- Finally, the director lightly touches the subject on the cheek.

3. After the final, cheek-touching test, click **Suspend**.

4. Observe the data, which should look similar to the Segment 1 data example **(Figure 20.8)**.

- If the data look very different, click **Redo** and repeat the steps above.
- If the data look similar, proceed to record Segment 2.

Segment 2: Response to Different Colors

1. When the subject and director are ready, click **Resume**.

2. The director will sequentially hold up nine differently colored paper squares about 2 feet in front of the subject's face. He or she will ask the subject to focus on the particular color for 10 seconds before moving to the next color in the sequence. The director will display the colors and insert a marker in the following order: white, black, red, blue, green, yellow, orange, brown, and purple. The director or assistant will use the **F9** key (PC) or **ESC** key (Mac) to insert a marker at the start of each color.

3. The subject will be asked to view the complete set of colors. After the color purple, click **Suspend**.

4. Observe the data, which should look similar to the Segment 2 data example **(Figure 20.9)**.

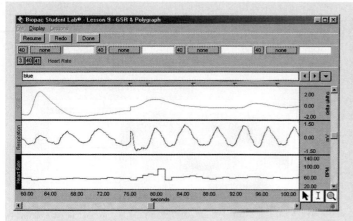

Figure 20.9 Example of Segment 2 data.

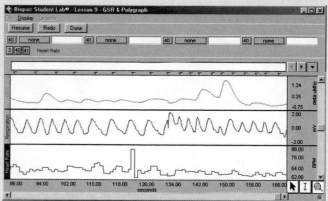

Figure 20.10 Example of Segment 3 data.

- If the data look very different, click **Redo** and repeat the steps above.
- If the data look similar, proceed to record Segment 3.

Segment 3: Response to Different Questions

1. When the subject and director are ready, click **Resume.**

2. The director will ask the subject the 10 questions in step 3 and note if the answer is Yes or No. In this segment, the recorder will use the **F9** key (PC) or **ESC** key (Mac) to insert a marker at the end of each question and the end of each answer. The director will circle the Yes or No response of the subject in the "Response" column of the Segment 3 Measurements chart (p. 335).

3. The following questions are to be asked and answered either Yes or No:

- Are you currently a student?
- Are your eyes blue?
- Do you have any brothers?
- Did you earn an "A" on the last exam?
- Do you drive a motorcycle?
- Are you less than 25 years old?
- Have you ever traveled to another planet?
- Have aliens from another planet ever visited you?
- Do you watch *Sesame Street* ?
- Have you answered all of the preceding questions truthfully?

4. After the last question is answered, click **Suspend.**

5. Observe the data, which should look similar to the Segment 3 data example **(Figure 20.10)**.

- If the data look very different, click **Redo** and repeat the steps above.
- If the data look similar, click **Done.** Click **Yes** if you are finished recording.

6. Without recording, simply ask the subject to respond once again to all of the questions as honestly as possible. The director circles the Yes or No response of the subject in the "Truth" column of the Segment 3 Measurements chart (p. 335).

7. Remove all of the sensors and equipment from the subject, and continue to Data Analysis.

Data Analysis

1. If you are just starting the BIOPAC® program to perform data analysis, enter **Review Saved Data** mode and choose the file with the subject's EDA/GSR data (for example, SmithPoly-1). If **Analyze Current Data File** was previously chosen, proceed to analysis.

2. Observe how the channel numbers are designated (as shown in **Figure 20.11**): CH 3—**EDA/GSR;** CH 40—**Respiration;** CH 41—**Heart Rate.**

3. You may need to use the following tools to adjust the data in order to clearly view and analyze the first 5 seconds of the recording.

- Click the magnifying glass in the lower right corner of the screen (near the I-beam box) to activate the **zoom** function. Use the magnifying glass cursor to click on the very first waveforms until the first 5 seconds of data are represented (see horizontal time scale at the bottom of the screen).

- Select the **Display** menu at the top of the screen, and click **Autoscale Waveforms** in the drop-down menu. This function will adjust the data for better viewing.

Text continues on next page. →

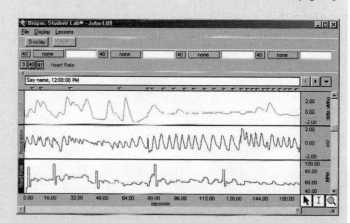

Figure 20.11 Example of polygraph recording with EDA/GSR, respiration, and heart rate.

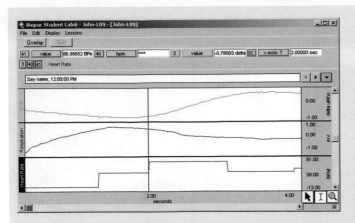

Figure 20.12 Selecting the 2-second point for data analysis.

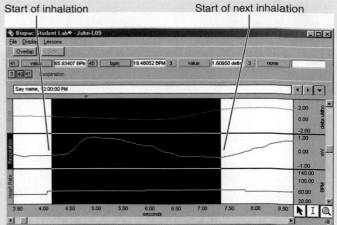

Figure 20.13 Highlighting the waveforms from the start of one inhalation to the start of the next.

4. To analyze the data, note the first three pairs of channel/measurement boxes at the top of the screen. (Each box activates a drop-down menu when you click it.) The following channels and measurement types should already be set:

Channel	Measurement	Data
CH 41	value	heart rate
CH 40	value	respiration
CH 3	value	EDA/GSR

Value: Displays the value of the measurement (for example, heart rate or EDA/GSR) at the point in time that is selected.

BPM: In this analysis, the BPM calculates breaths per minute when the area that is highlighted starts at the beginning of one inhalation and ends at the beginning of the next inhalation.

5. Use the arrow cursor and click the I-beam cursor box at the lower right side of the screen to activate the "area selection" function. Using the activated I-beam cursor, select the 2-second point on the data (as shown in **Figure 20.12**). Record the heart rate and EDA/GSR values for Segment 1 data in the Segment 1 Measurements chart. This point represents the resting or baseline data.

6. Using data from the first 5 seconds, use the I-beam cursor tool to highlight an area from the start of one inhalation to the start of the next inhalation (as shown in **Figure 20.13**). The start of an inhalation is indicated by the beginning of the ascension of the waveform. Record this as the baseline respiratory rate in the Segment 1 Measurements chart.

7. Using the markers as guides, scroll along the bottom scroll bar until the data from Segment 1 appears.

8. Analyze all parts of Segment 1. Using the tools described in steps 5 and 6, acquire the measurements for the heart rate, EDA/GSR, and respiration rate soon after each subject response. Use the maximum EDA/GSR value in that time frame as the point of measurement for EDA/GSR and heart rate. Use the beginning of two consecutive inhalations in that same time frame to measure respiration rate. Record these data in the Segment 1 Measurements chart.

9. Repeat these same procedures to measure EDA/GSR, heart rate, and respiration rate for each color in Segment 2. Record these data in the Segment 2 Measurements chart.

Segment 1 Measurements			
Procedure	**Heart rate [CH 41 value]**	**Respiratory rate [CH 40 BPM]**	**EDA/GSR [CH 3 value]**
Baseline			
Quietly say name			
Count from 10			
Count from 30			
Face is touched			

Segment 2 Measurements

Color	Heart rate [CH 41 value]	Respiratory rate [CH 40 BPM]	EDA/GSR [CH 3 value]
White			
Black			
Red			
Blue			
Green			
Yellow			
Orange			
Brown			
Purple			

Segment 3 Measurements

Question	Response		Truth		Heart rate [CH 41 value]	Resp. rate [CH 40 BPM]	EDA/GSR [CH 3 value]
Student?	Y	N	Y	N			
Blue eyes?	Y	N	Y	N			
Brothers?	Y	N	Y	N			
Earn "A"?	Y	N	Y	N			
Motorcycle?	Y	N	Y	N			
Under 25?	Y	N	Y	N			
Planet?	Y	N	Y	N			
Aliens?	Y	N	Y	N			
Sesame?	Y	N	Y	N			
Truthful?	Y	N	Y	N			

10. Repeat these same procedures to measure EDA/GSR, heart rate, and respiration rate for responses to each question in Segment 3. Record these data in the Segment 3 Measurements chart.

11. Examine EDA/GSR, heart rate, and respiration rate of the baseline data in the Segment 1 Measurements chart.

12. For every condition to which the subject was exposed, write **H** if that value is higher than baseline, write **L** if the value is lower, and write **NC** if there is no significant change. Repeat this analysis for Segments 2 and 3.

Examine the data in the Segment 1 Measurements chart. Is there any noticeable difference between the baseline EDA/GSR, heart rate, and respiration rate after each prompt? Under which prompts is the most significant change noted?

Examine the data in the Segment 2 Measurements chart. Is there any noticeable difference between the baseline EDA/GSR, heart rate, and respiration rate after each color presentation? Under which colors is the most significant change noted?

Examine the data in the Segment 3 Measurements chart. Is there any noticeable difference between the baseline EDA/GSR, heart rate, and respiration rate after each question? After which is the most significant change noted?

Text continues on next page. →

20

Speculate as to the reasons why a subject may demonstrate a change in EDA/GSR from baseline under different color conditions.

_____

_____

_____

Speculate as to the reasons why a subject may demonstrate a change in EDA/GSR from baseline when a particular question is asked.

_____

_____

_____

Which branch of the autonomic nervous system is dominant during a galvanic skin response?

_____

20

REVIEW SHEET
The Autonomic Nervous System

Name _____ LabTime/Date _____

Parasympathetic and Sympathetic Divisions

1. List the names of the two motor neurons of the autonomic nervous system.

_____

2. List the names and numbers of the four cranial nerves that the parasympathetic division of the ANS arises from.

_____

3. List two types of sympathetic ganglia that contain postganglionic cell bodies.

_____

4. List two types of parasympathetic ganglia that contain postganglionic cell bodies.

_____

5. Which part of the rami communicantes contains nonmyelinated fibers? _____

_____

6. The following chart states a number of characteristics. Use a check mark to show which division of the autonomic nervous system is involved in each.

Sympathetic division	Characteristic	Parasympathetic division
	Postganglionic axons secrete norepinephrine; adrenergic fibers	
	Postganglionic axons secrete acetylcholine; cholinergic fibers	
	Long preganglionic axon; short postganglionic axon	
	Short preganglionic axon; long postganglionic axon	
	Arises from cranial and sacral nerves	
	Arises from spinal nerves T_1 through L_3	
	Normally in control	
	"Fight-or-flight" system	
	Has more specific effects	
	Has rami communicantes	
	Has extensive branching of preganglionic axons	

Galvanic Skin Response (Electrodermal Activity) Within a Polygraph Using BIOPAC®

7. Describe exactly how, from a physiological standpoint, EDA/GSR can be correlated with activity of the autonomic nervous system.

_____

_____

_____

8. Based on this brief exposure to a polygraph, explain why this might not be an exact tool for testing the sincerity and honesty of a subject. Refer to your data to support your conclusions.

_____

_____

_____

Human Reflex Physiology

Objectives

☐ Define *reflex* and *reflex arc*.

☐ Describe the differences between autonomic and somatic reflexes.

☐ Explain why reflex testing is an important part of every physical examination.

☐ Name, identify, and describe the function of each element of a reflex arc.

☐ Describe and discuss several types of reflex activities as observed in the laboratory; indicate the functional or clinical importance of each; and categorize each as a somatic or autonomic reflex action.

☐ Explain why cord-mediated reflexes are generally much faster than those involving input from the higher brain centers.

☐ Investigate differences in reaction time between intrinsic and learned reflexes.

Materials

- Reflex hammer
- Sharp pencils
- Cot (if available)
- Absorbent cotton (sterile)
- Tongue depressor
- Metric ruler
- Flashlight
- 100- or 250-ml beaker
- 10- or 25-ml graduated cylinder
- Lemon juice in dropper bottle
- Wide-range pH paper
- Large laboratory bucket containing freshly prepared 10% household bleach solution for saliva-containing glassware
- Disposable autoclave bag

Text continues on next page. →

MasteringA&P®

For related exercise study tools, go to the Study Area of **MasteringA&P**. There you will find:

- Practice Anatomy Lab **PAL**
- PhysioEx **PEx**
- A&PFlix **A&PFlix**
- Practice quizzes, Histology Atlas, eText, Videos, and more!

Pre-Lab Quiz

1. Define *reflex*. _____

2. Circle the correct underlined term. <u>Autonomic</u> / <u>Somatic</u> reflexes include all those reflexes that involve stimulation of skeletal muscles.

3. In a reflex arc, the _____ transmits afferent impulses to the central nervous system.
   a. integration center
   b. motor neuron
   c. receptor
   d. sensory neuron

4. Circle True or False. Most reflexes are simple, two-neuron, monosynaptic reflex arcs.

5. Stretch reflexes are initiated by tapping a _____, which stretches the associated muscle.
   a. bone
   b. muscle
   c. tendon or ligament

6. An example of an autonomic reflex that you will be studying in today's lab is the _____ reflex.
   a. crossed-extensor
   b. gag
   c. plantar
   d. salivary

7. Circle True or False. A reflex that occurs on the same side of the body that was stimulated is an ipsilateral response.

8. Name one of the pupillary reflexes you will be examining today. _____

9. Circle the correct underlined term. The effectors of the salivary reflex are <u>muscles</u> / <u>glands</u>.

10. Circle True or False. Learned reflexes involve far fewer neural pathways and fewer types of higher intellectual activities than intrinsic reflexes, which shortens their response time.

- Wash bottle containing 10% bleach solution
- Reaction time ruler (if available)

BIOPAC® BIOPAC® BSL System with BSL software version 3.7.6 to 3.7.7 (for Windows 7/Vista/XP or Mac OS X 10.4–10.6), data acquisition unit MP36/35 or MP45, PC or Mac computer, hand switch and headphones.

Instructors using the MP36/35/30 data acquisition unit with BSL software versions earlier than 3.7.6 (for Windows or Mac) will need slightly different channel settings and collection strategies. Instructions for using the older data acquisition unit can be found on MasteringA&P.

Note: *Instructions for using PowerLab® equipment can be found on MasteringA&P.*

Reflexes are rapid, predictable, involuntary motor responses to stimuli; they are mediated over neural pathways called reflex arcs. Many of the body's control systems are reflexes, which can be either inborn (intrinsic) or learned (acquired).

Another way to categorize reflexes is into one of two large groups: autonomic reflexes and somatic reflexes. **Autonomic** (or visceral) **reflexes** are mediated through the autonomic nervous system, and we are not usually aware of them. These reflexes activate smooth muscles, cardiac muscle, and the glands of the body, and they regulate body functions such as digestion, elimination, blood pressure, salivation, and sweating. **Somatic reflexes** include all those reflexes that involve stimulation of skeletal muscles by the somatic division of the nervous system.

Reflex testing is an important diagnostic tool for assessing the condition of the nervous system. If the spinal cord is damaged, the easily performed reflex tests can help pinpoint the area (level) of spinal cord injury. Motor nerves above the injured area may be unaffected, whereas those at or below the lesion site may be unable to participate in normal reflex activity. ✚

Components of a Reflex Arc

Reflex arcs have five basic components (**Figure 21.1**):

1. The *receptor* is the site of stimulus action.

2. The *sensory neuron* transmits afferent impulses to the CNS.

3. The *integration center* consists of one or more neurons in the CNS.

4. The *motor neuron* conducts efferent impulses from the integration center to an effector organ.

5. The *effector,* a muscle fiber or a gland cell, responds to efferent impulses by contracting or secreting, respectively.

The simple patellar, or knee-jerk, reflex (**Figure 21.2a**) is an example of a simple, two-neuron, *monosynaptic* (literally, "one synapse") reflex arc. It will be demonstrated in the laboratory. However, most reflexes are more complex and *polysynaptic,* involving the participation of one or more interneurons in the reflex arc pathway. An example of a polysynaptic reflex is the flexor reflex (Figure 21.2b). Because the reflex may be delayed or inhibited at the synapses, the more synapses encountered in a reflex pathway, the more time is required for the response.

Reflexes of many types may be considered programmed into the neural anatomy. Many *spinal reflexes,* reflexes that are initiated and completed at the spinal cord level, occur without the direct involvement of higher brain centers. Generally, these reflexes are present in animals whose brains have been destroyed, provided that the spinal cord is functional. Although many spinal reflexes do not require the involvement of higher centers, the brain is "advised" of spinal cord reflex activity and may alter it by facilitating or inhibiting the reflexes.

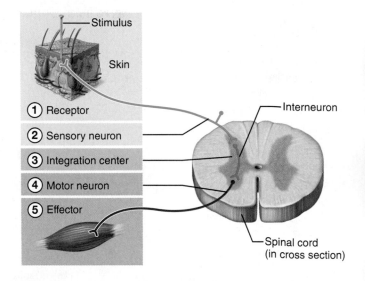

Figure 21.1 The five basic components of reflex arcs. The reflex illustrated is polysynaptic.

Somatic Reflexes

There are several types of somatic reflexes, including several that you will be eliciting during this laboratory session—the stretch, crossed-extensor, superficial, corneal, and gag reflexes. Some require only spinal cord activity; others require brain involvement as well. Some somatic reflexes are mediated by cranial nerves.

Spinal Reflexes

Stretch Reflexes

Stretch reflexes are important for maintaining and adjusting muscle tone for posture, balance, and locomotion.

Stretch reflexes are initiated by tapping a tendon or ligament, which stretches the muscle to which the tendon is attached (**Figure 21.3**, p. 342). This stimulates the muscle spindles and causes reflex contraction of the stretched muscle or muscles. Branches of the afferent fibers from the muscle spindles also synapse with interneurons controlling the antagonist muscles. The inhibition of those interneurons and the antagonist muscles, called *reciprocal inhibition,* causes them to relax and prevents them from resisting (or reversing) the contraction of the stretched muscle. Additionally, impulses are relayed to higher brain centers (largely via the dorsal white

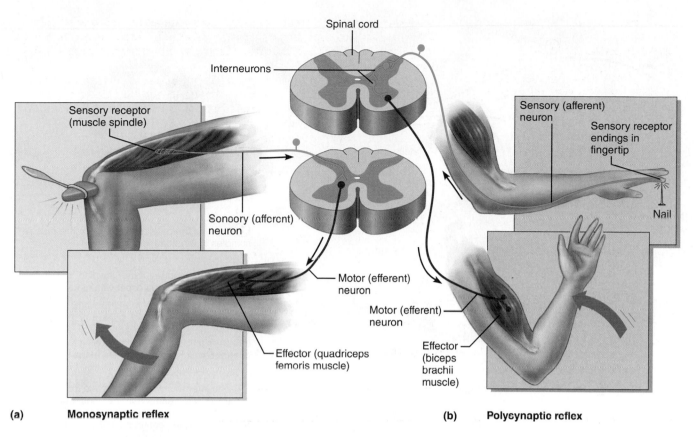

(a)   Monosynaptic reflex

(b)   Polysynaptic reflex

Figure 21.2 Monosynaptic and polysynaptic reflex arcs. The integration center is in the spinal cord, and in each example the receptor and effector are in the same limb. **(a)** The patellar reflex, a two-neuron monosynaptic reflex. **(b)** A flexor reflex, an example of a polysynaptic reflex.

columns) to advise of muscle length, speed of shortening, and the like—information needed to maintain muscle tone and posture. Stretch reflexes tend to be hypoactive or absent in cases of peripheral nerve damage or ventral horn disease and hyperactive in corticospinal tract lesions. They are absent in deep sedation and coma.

 Prepare for lab: Watch the Pre-Lab Video
MasteringA&P®>Study Area>Pre-Lab Videos

Activity 1

Initiating Stretch Reflexes

1. Test the **patellar**, or **knee-jerk, reflex** by seating a subject on the laboratory bench with legs hanging free (or with knees crossed). Tap the patellar ligament sharply with the broad side of the reflex hammer just below the knee between the patella and the tibial tuberosity, as shown in **Figure 21.4** on p. 342. The knee-jerk response assesses the L_2–L_4 level of the spinal cord. Test both knees and record your observations. (Sometimes a reflex can be altered by your actions. If you encounter difficulty, consult your instructor for helpful hints.)

_____

_____

Which muscles contracted? _____

_____

What nerve is carrying the afferent and efferent impulses?

_____

2. Test the effect of mental distraction on the patellar reflex by having the subject add a column of three-digit numbers while you test the reflex again. Is the response more *or* less vigorous than the first response?

_____

What are your conclusions about the effect of mental distraction on reflex activity?

_____

3. Now test the effect of muscular activity occurring simultaneously in other areas of the body. Have the subject

Text continues on next page. →

21

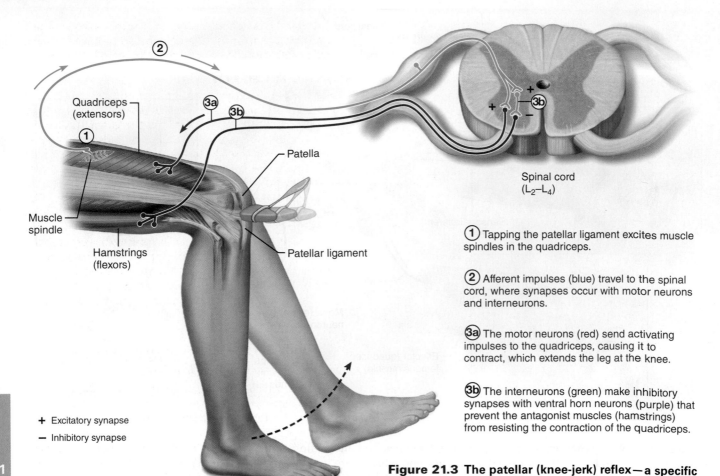

Figure 21.3 The patellar (knee-jerk) reflex—a specific example of a stretch reflex.

Quadriceps (extensors)

Muscle spindle

Hamstrings (flexors)

Patella

Patellar ligament

Spinal cord (L₂–L₄)

① Tapping the patellar ligament excites muscle spindles in the quadriceps.

② Afferent impulses (blue) travel to the spinal cord, where synapses occur with motor neurons and interneurons.

③ₐ The motor neurons (red) send activating impulses to the quadriceps, causing it to contract, which extends the leg at the knee.

③ᵦ The interneurons (green) make inhibitory synapses with ventral horn neurons (purple) that prevent the antagonist muscles (hamstrings) from resisting the contraction of the quadriceps.

+ Excitatory synapse
− Inhibitory synapse

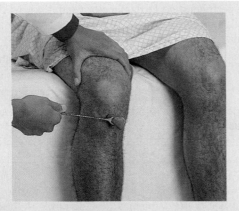

Figure 21.4 Testing the patellar reflex. The examiner supports the subject's knee so that the subject's muscles are relaxed, and then strikes the patellar ligament with the reflex hammer. The proper location may be ascertained by palpation of the patella.

clasp the edge of the laboratory bench and vigorously attempt to pull it upward with both hands. At the same time, test the patellar reflex again. Is the response more or less vigorous than the first response?

_____

4. Fatigue also influences the reflex response. The subject should jog in position until she or he is very fatigued

(*really fatigued*—no slackers). Test the patellar reflex again, and record whether it is more or less vigorous than the first response.

_____

Would you say that nervous system activity *or* muscle function is responsible for the changes you have just observed?

Explain your reasoning. _____

_____

_____

_____

5. The **calcaneal tendon,** or **ankle-jerk, reflex** assesses the first two sacral segments of the spinal cord. With your shoe removed and your foot dorsiflexed slightly to increase the tension of the gastrocnemius muscle, have your partner sharply tap your calcaneal tendon with the broad side of the reflex hammer (**Figure 21.5**).

What is the result? _____

_____

During walking, what is the action of the gastrocnemius?

_____

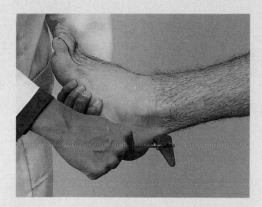

Figure 21.5 Testing the calcaneal tendon reflex. The examiner slightly dorsiflexes the subject's ankle by supporting the foot lightly in the hand, and then taps the calcaneal tendon just above the ankle.

Crossed-Extensor Reflex

The **crossed-extensor reflex** is more complex than the stretch reflex. It consists of a flexor, or withdrawal, reflex followed by extension of the opposite limb.

This reflex is quite obvious when, for example, a stranger suddenly and strongly grips one's arm. The immediate response is to withdraw the clutched arm and push the intruder away with the other arm. The reflex is more difficult to demonstrate in a laboratory because it is anticipated, and under these conditions the extensor part of the reflex may be inhibited.

Activity 2

Initiating the Crossed-Extensor Reflex

The subject should sit with eyes closed and with the back of one hand resting on the laboratory bench. Obtain a sharp pencil, and suddenly prick the subject's index finger. What are the results?

_____

_____

_____

Did the extensor part of this reflex occur simultaneously or more slowly than the other reflexes you have observed?

_____

What are the reasons for this? _____

_____

_____

The reflexes that have been demonstrated so far—the stretch and crossed-extensor reflexes—are examples of reflexes in which the reflex pathway is mediated at the spinal cord level only.

Superficial Reflexes

The **superficial reflexes** (abdominal, cremaster, and plantar reflexes) result from pain and temperature changes. They are initiated by stimulation of receptors in the skin and mucosae. The superficial reflexes depend _both_ on functional upper-motor pathways and on the spinal cord–level reflex arc. Since only the plantar reflex can be tested conveniently in a laboratory setting, we will use this as our example.

The **plantar reflex,** an important neurological test, is elicited by stimulating the cutaneous receptors in the sole of the foot. In adults, stimulation of these receptors causes the toes to flex and move closer together. Damage to the corticospinal tract, however, produces _Babinski's sign,_ an abnormal response in which the toes flare and the great toe moves in an upward direction. In a newborn infant, it is normal to see Babinski's sign because myelination of the nervous system is incomplete.

Activity 3

Initiating the Plantar Reflex

Have the subject remove a shoe and sock and lie on the cot or laboratory bench with knees slightly bent and thighs rotated so that the posterolateral side of the foot rests on the cot. Alternatively, the subject may sit up and rest the lateral surface of the foot on a chair. Draw the handle of the reflex hammer firmly along the lateral side of the exposed sole from the heel to the base of the great toe (**Figure 21.6**).

What is the response? _____

_____

Is this a normal plantar reflex or a Babinski's sign?

_____

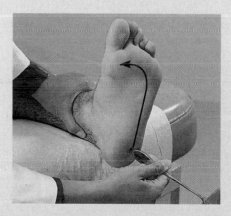

Figure 21.6 Testing the plantar reflex. Using a moderately sharp object, the examiner strokes the lateral border of the subject's sole, starting at the heel and continuing toward the great toe across the ball of the foot.

21

Cranial Nerve Reflex Tests

In these experiments, you will be working with your lab partner to illustrate two somatic reflexes mediated by cranial nerves.

Corneal Reflex

The **corneal reflex** is mediated through the trigeminal nerve. The absence of this reflex is an ominous sign because it often indicates damage to the brain stem resulting from compression of the brain or other trauma.

Activity 4

Initiating the Corneal Reflex

Stand to one side of the subject; the subject should look away from you toward the opposite wall. Wait a few seconds and then quickly, *but gently,* touch the subject's cornea (on the side toward you) with a wisp of absorbent cotton. What reflexive reaction occurs when something touches the cornea?

_____

What is the function of this reflex?

_____

_____

Gag Reflex

The **gag reflex** tests the somatic motor responses of cranial nerves IX and X. When the oral mucosa on the side of the uvula is stroked, each side of the mucosa should rise, and the amount of elevation should be equal. The uvula is the fleshy tab hanging from the roof of the mouth.

Activity 5

Initiating the Gag Reflex

For this experiment, select a subject who does not have a queasy stomach, because regurgitation is a possibility. Gently stroke the oral mucosa on each side of the subject's uvula with a tongue depressor. What happens?

_____

_____

⚠️ Discard the used tongue depressor in the disposable autoclave bag before continuing. *Do not* lay it on the laboratory bench at any time.

Autonomic Reflexes

The autonomic reflexes include the pupillary, ciliospinal, and salivary reflexes, as well as many other reflexes. Work with a partner to demonstrate these four autonomic reflexes.

Pupillary Reflexes

There are several types of pupillary reflexes. The **pupillary light reflex** and the **consensual reflex** will be examined here. In both of these pupillary reflexes, the retina of the eye is the receptor, the optic nerve contains the afferent fibers, the oculomotor nerve contains the efferent fibers, and the smooth muscle of the iris is the effector. Absence of normal pupillary reflexes is generally a late indication of severe trauma or deterioration of the vital brain stem tissue due to metabolic imbalance.

Activity 6

Initiating Pupillary Reflexes

1. Conduct the reflex testing in an area where the lighting is relatively dim. Before beginning, obtain a metric ruler and a flashlight. Measure and record the size of the subject's pupils as best you can.

Right pupil: _____ mm   Left pupil: _____ mm

2. Stand to the left of the subject to conduct the testing. The subject should shield his or her right eye by holding a hand vertically between the eye and the right side of the nose.

3. Shine a flashlight into the subject's left eye. What is the pupillary response?

_____

Measure the size of the left pupil: _____ mm

4. Without moving the flashlight, observe the right pupil. Has the same type of change (called a *consensual response*) occurred in the right eye?

_____

Measure the size of the right pupil: _____ mm

The consensual response, or any reflex observed on one side of the body when the other side has been stimulated, is called a **contralateral response**. The pupillary light response, or any reflex occurring on the same side stimulated, is referred to as an **ipsilateral response**.

What does the occurrence of a contralateral response indicate about the pathways involved?

_____

_____

What is the function of these pupillary responses?

_____

_____

_____

Ciliospinal Reflex

The **ciliospinal reflex** is another example of reflex activity in which pupillary responses can be observed. This response may initially seem a little bizarre, especially in view of the consensual reflex just demonstrated.

Activity 7

Initiating the Ciliospinal Reflex

1. While observing the subject's eyes, gently stroke the skin (or just the hairs) on the left side of the back of the subject's neck, close to the hairline.

What is the reaction of the left pupil? _____

The reaction of the right pupil? _____

2. If you see no reaction, repeat the test using a gentle pinch in the same area.

The response you should have noted—pupillary dilation—is consistent with the pupillary changes occurring when the sympathetic nervous system is stimulated. Such a response may also be elicited in a single pupil when more impulses from the sympathetic nervous system reach it for any reason. For example, when the left side of the subject's neck was stimulated, sympathetic impulses to the left iris increased, resulting in the ipsilateral reaction of the left pupil.

On the basis of your observations, would you say that the sympathetic innervation of the two irises is closely integrated?

_____ Why or why not? _____

_____

_____

Salivary Reflex

Unlike the other reflexes, in which the effectors were smooth or skeletal muscles, the effectors of the **salivary reflex** are glands. The salivary glands secrete varying amounts of saliva in response to reflex activation.

Activity 8

Initiating the Salivary Reflex

1. Obtain a small beaker, a graduated cylinder, lemon juice, and wide-range pH paper. After refraining from swallowing for 2 minutes, the subject is to expectorate (spit) the accumulated saliva into a small beaker. Using the graduated cylinder, measure the volume of the expectorated saliva and determine its pH.

Volume: _____ cc   pH: _____

2. Now place 2 or 3 drops of lemon juice on the subject's tongue. Allow the lemon juice to mix with the saliva for 5 to 10 seconds, and then determine the pH of the subject's saliva by touching a piece of pH paper to the tip of the tongue.

pH: _____

As before, the subject is to refrain from swallowing for 2 minutes. After the 2 minutes is up, again collect and measure the volume of the saliva and determine its pH.

Volume: _____ cc   pH: _____

3. How does the volume of saliva collected after the application of the lemon juice compare with the volume of the first saliva sample?

_____

How does the final saliva pH reading compare to the initial reading?

_____

How does the final saliva pH reading compare to that obtained 10 seconds after the application of lemon juice?

_____

⚠ Dispose of the saliva-containing beakers and the graduated cylinders in the laboratory bucket that contains bleach, and put the used pH paper into the disposable autoclave bag. Wash the bench down with 10% bleach solution before continuing.

21

Reaction Time of Intrinsic and Learned Reflexes

The time required for reaction to a stimulus depends on many factors—sensitivity of the receptors, velocity of nerve conduction, the number of neurons and synapses involved, and the speed of effector activation, to name just a few. There is no clear-cut distinction between intrinsic and learned reflexes, as most reflex actions are subject to modification by learning or conscious effort. In general, however, if the response involves a simple reflex arc, the response time is short. Learned reflexes involve a far larger number of neural pathways and many types of higher intellectual activities, including choice and decision making, which lengthens the response time.

There are various ways of testing reaction time of reflexes. The following activities provide an opportunity to demonstrate the major time difference between a simple reflex arc and learned reflexes and to measure response time under various conditions.

Activity 9

Testing Reaction Time for Intrinsic and Learned Reflexes

1. Using a reflex hammer, elicit the patellar reflex in your partner. Note the relative reaction time needed for this intrinsic reflex to occur.

2. Now test the reaction time for learned reflexes. The subject should hold a hand out, with the thumb and index finger extended. Hold a metric ruler so that its end is exactly 3 cm above the subject's outstretched hand. The ruler should be in the vertical position with the numbers reading from the bottom up. When the ruler is dropped, the subject should be able to grasp it between thumb and index finger as it passes, without having to change position. Have the subject catch the ruler five times, varying the time between trials. The relative speed of reaction can be determined by reading the number on the ruler at the point of the subject's fingertips.* (Thus if the number at the fingertips is 15 cm, the subject was unable to catch the ruler until 18 cm of length had passed through the fingers; 15 cm of ruler length plus 3 cm to account for the distance of the ruler above the hand.)† Record the number of centimeters that pass through the subject's fingertips (or the number of seconds required for reaction) for each trial:

Trial 1: _____ cm   Trial 4: _____ cm

_____ sec   _____ sec

Trial 2: _____ cm   Trial 5: _____ cm

_____ sec   _____ sec

Trial 3: _____ cm

_____ sec

3. Perform the test again, but this time say a simple word each time you release the ruler. Designate a specific word as a signal for the subject to catch the ruler. On all other words, the subject is to allow the ruler to pass through the fingers. Trials in which the subject

erroneously catches the ruler are to be disregarded. Record the distance the ruler travels (or the number of seconds required for reaction) in five *successful* trials:

Trial 1: _____ cm   Trial 4: _____ cm

_____ sec   _____ sec

Trial 2: _____ cm   Trial 5: _____ cm

_____ sec   _____ sec

Trial 3: _____ cm

_____ sec

Did the addition of a specific word to the stimulus increase or decrease the reaction time?

_____

4. Perform the testing once again to investigate the subject's reaction to word association. As you drop the ruler, say a word—for example, *hot*. The subject is to respond with a word he or she associates with the stimulus word— for example, *cold*—catching the ruler while responding. If unable to make a word association, the subject must allow the ruler to pass through the fingers. Record the distance the ruler travels (or the number of seconds required for reaction) in five successful trials, as well as the number of times the subject does not catch the ruler.

Trial 1: _____ cm   Trial 4: _____ cm

_____ sec   _____ sec

Trial 2: _____ cm   Trial 5: _____ cm

_____ sec   _____ sec

Trial 3: _____ cm

_____ sec

Number of times the subject did *not* catch the ruler:

_____

*Distance (d) can be converted to time (t) using the simple formula:

$$d \text{ (in cm)} = (1/2)(980 \text{ cm/sec}^2)t^2$$

$$t^2 = (d/490 \text{ cm/sec}^2)$$

$$t = \sqrt{(d/(490 \text{ cm/sec}^2)}$$

†An alternative would be to use a reaction time ruler, which converts distance to time (seconds).

Activity 10

Measuring Reaction Time Using BIOPAC®

Setting Up the Equipment

1. Connect the BIOPAC® unit to the computer, and turn the computer **ON.**

2. Make sure the BIOPAC® unit is **OFF.**

3. Plug in the equipment (as shown in **Figure 21.7**).

- Hand switch—CH1

- Headphones—back of MP36/35 unit

4. Turn the BIOPAC® unit **ON.**

5. Start the Biopac Student Lab program on the computer by double-clicking the icon on the desktop or by following your instructor's guidance.

6. Select lesson **L11-React-1** from the menu and click **OK.**

7. Type in a filename that will save this subject's data on the computer hard drive. You may want to use the subject's last name followed by React-1 (for example, SmithReact-1), then click **OK.**

Calibrating the Equipment

1. Seat the subject comfortably so that he or she cannot see the computer screen and keyboard.

2. Put the headphones on the subject, and give the subject the hand switch to hold.

3. Tell the subject that he or she is to push the hand switch button when a "click" is heard.

4. Click **Calibrate,** and then click **OK** when the subject is ready.

5. Observe the recording of the calibration data, which should look like the waveforms in the calibration example (**Figure 21.8**).

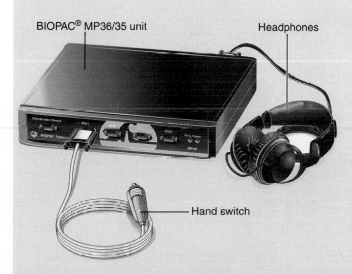

BIOPAC® MP36/35 unit   Headphones

Hand switch

Figure 21.7 Setting up the BIOPAC® equipment. Plug the headphones into the back of the MP36/35 data acquisition unit and the hand switch into Channel 1. Hand switch and headphones are shown connected to the MP36/35 unit.

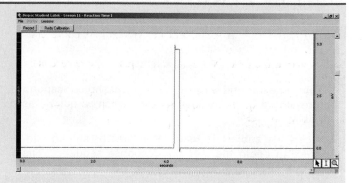

Figure 21.8 Example of waveforms during the calibration procedure.

- If the data look very different, click **Redo Calibration** and repeat the steps above.

- If the data look similar, proceed to the next section.

Recording the Data

In this experiment, you will record four different segments of data. In Segments 1 and 2, the subject will respond to random click stimuli. In Segments 3 and 4, the subject will respond to click stimuli at fixed intervals (about 4 seconds). The director will click **Record** to initiate the Segment 1 recording, and **Resume** to initiate Segments 2, 3, and 4. The subject should focus only on responding to the sound.

Segment 1: Random Trial 1

1. Each time a sound is heard, the subject should respond by pressing the button on the hand switch as quickly as possible.

2. When the subject is ready, the director should click **Record** to begin the stimulus-response sequence. The recording will stop automatically after 10 clicks.

- A triangular marker will be inserted above the data each time a "click" stimulus occurs.

- An upward-pointing "pulse" will be inserted each time the subject responds to the stimulus.

3. Observe the recording of the data, which should look similar to the data-recording example (**Figure 21.9**).

Text continues on next page. →

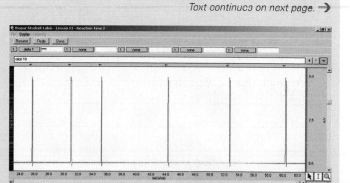

Figure 21.9 Example of waveforms during the recording of data.

- If the data look very different, click **Redo** and repeat the steps above.
- If the data look similar, move on to recording the next segment.

Segment 2: Random Trial 2

1. Each time a sound is heard, the subject should respond by pressing the button on the hand switch as quickly as possible.

2. When the subject is ready, the director should click **Resume** to begin the stimulus-response sequence. The recording will stop automatically after 10 clicks.

3. Observe the recording of the data, which should again look similar to the data-recording example (Figure 21.9).

- If the data look very different, click **Redo** and repeat the steps above.
- If the data look similar, move on to recording the next segment.

Segment 3: Fixed Interval Trial 3

1. Repeat the steps for Segment 2 above.

Segment 4: Fixed Interval Trial 4

1. Repeat the steps for Segment 2 above.

2. If the data after this final segment are fine, click **Done**. A pop-up window will appear; to record from another subject select **Record from another subject,** and return to step 7 under Setting Up the Equipment. If continuing to the Data Analysis section, select **Analyze current data file** and proceed to step 2 in the Data Analysis section.

Data Analysis

1. If just starting the BIOPAC® program to perform data analysis, enter **Review Saved Data** mode and choose the file with the subject's reaction data (for example, SmithReact-1).

2. Observe that all 10 reaction times are automatically calculated for each segment and are placed in the journal at the bottom of the computer screen.

3. Write the 10 reaction times for each segment in the **chart Reaction Times.**

4. Delete the highest and lowest values of each segment, then calculate and record the average for the remaining eight data points.

5. When finished, exit the program by going to the **File** menu at the top of the page and clicking **Quit**.

Do you observe a significant difference between the average response times of Segment 1 and Segment 2? If so, what might account for the difference, even though they are both random trials?

_____

_____

Likewise, do you observe a significant difference between the average response times of Segment 3 and Segment 4? If so, what might account for the difference, even though they are both fixed interval trials?

_____

_____

Reaction Times (seconds)				
	Random		**Fixed interval**	
Stimulus #	**Segment 1**	**Segment 2**	**Segment 3**	**Segment 4**
1				
2				
3				
4				
5				
6				
7				
8				
9				
10				
Average				

22

General Sensation

Objectives

☐ List the stimuli that activate general sensory receptors.

☐ Define *exteroceptor, interoceptor,* and *proprioceptor.*

☐ Recognize and describe the various types of general sensory receptors as studied in the laboratory, and list the function and locations of each.

☐ Explain the tactile two-point discrimination test, and state its anatomical basis.

☐ Define *tactile localization,* and describe how this ability varies in different areas of the body.

☐ Define *adaptation,* and describe how this phenomenon can be demonstrated.

☐ Discuss *negative afterimages* as they are related to temperature receptors.

☐ Define *referred pain,* and give an example of it.

Materials

- Compound microscope
- Immersion oil
- Prepared slides (longitudinal sections) of lamellar corpuscles, tactile corpuscles, tendon organs, and muscle spindles
- Calipers or esthesiometer
- Small metric rulers
- Fine-point, felt-tipped markers (black, red, and blue)
- Large beaker of ice water; chipped ice
- Hot water bath set at 45°C; laboratory thermometer
- Towel
- Four coins (nickels or quarters)
- Three large finger bowls or 1000-ml beakers

MasteringA&P®

For related exercise study tools, go to the Study Area of **MasteringA&P**. There you will find:

- Practice Anatomy Lab PAL
- A&PFlix *A&PFlix*
- PhysioEx PEx
- Practice quizzes, Histology Atlas, eText, videos, and more!

Pre-Lab Quiz

1. Name one of the special senses. _____

2. Sensory receptors can be classified according to their source of stimulus. _____ are found close to the body surface and react to stimuli in the external environment.
   a. Exteroceptors
   b. Interoceptors
   c. Proprioceptors
   d. Visceroceptors

3. Circle True or False. General sensory receptors are widely distributed throughout the body and respond to, among other things, touch, pain, stretch, and changes in body position.

4. Tactile corpuscles respond to light touch. Where would you expect to find tactile corpuscles?
   a. deep within the dermal layer of hairy skin
   b. in the dermal papillae of hairless skin
   c. in the hypodermis of hairless skin
   d. in the uppermost portion of the epidermis

5. Lamellar corpuscles respond to:
   a. deep pressure and vibrations
   b. light touch
   c. pain and temperature

6. Circle True or False. A map of the sensory receptors for touch, heat, cold, and pain shows that they are not evenly distributed throughout the body.

7. Circle the correct underlined term. Two-point threshold / Tactile localization is the ability to determine where on the body the skin has been touched.

8. When a stimulus is applied for a prolonged period, the rate of receptor discharge slows, and conscious awareness of the stimulus declines. This phenomenon is known as:
   a. accommodation
   b. adaptation
   c. adjustment
   d. discernment

9. Circle True or False. Pain is always perceived in the same area of the body that is receiving the stimulus.

Text continues on next page. →

10. You will test referred pain in this activity by immersing the subject's:
   a. face in ice water to test the cranial nerve response
   b. elbow in ice water to test the ulnar nerve response
   c. hand in ice water to test the axillary nerve response
   d. leg in ice water to test the sciatic nerve response

People are very responsive to *stimuli*, which are changes within a person's environment. Hold a freshly baked apple pie before them, and their mouths water. Tickle them, and they giggle. These and many other stimuli continually surround us.

The body's **sensory receptors** react to stimuli. The tiny sensory receptors of the **general senses** react to touch, pressure, pain, heat, cold, stretch, vibration, and changes in body position and are distributed throughout the body. In contrast to these widely distributed *general sensory receptors*, the receptors of the special senses are large, complex *sense organs* or small, localized groups of receptors. The **special senses** include vision, hearing, equilibrium, smell, and taste. Only the anatomically simpler **general sensory receptors** will be studied in this exercise. Sensory receptors may be classified by the location of the stimulus.

- **Exteroceptors** react to stimuli in the external environment, and typically they are found close to the body surface. Exteroceptors include the simple cutaneous receptors in the skin (**Figure 22.1**) and the highly specialized

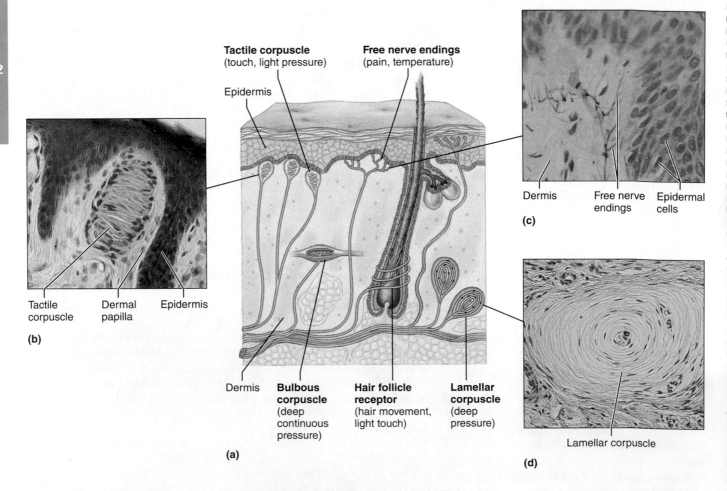

Figure 22.1 Examples of cutaneous receptors. Drawing **(a)** and photomicrographs **(b–d)**. **(a)** Free nerve endings, hair follicle receptor, tactile corpuscles, lamellar corpuscles, and bulbous corpuscle. Tactile (Merkel) discs are not illustrated. **(b)** Tactile corpuscle in a dermal papilla (300×). **(c)** Free nerve endings at dermal-epidermal junction (330×). **(d)** Cross section of a lamellar corpuscle in the dermis (220×).

receptor structures of the special senses (the vision apparatus of the eye, for example).

- **Interoceptors,** or *visceroceptors,* respond to stimuli arising within the body. Interoceptors are found in the internal visceral organs and include stretch receptors (in walls of hollow organs), chemoreceptors, and others.

- **Proprioceptors,** like interoceptors, respond to internal stimuli but are restricted to skeletal muscles, tendons, joints, ligaments, and connective tissue coverings of bones and muscles. They provide information about body movements and position by monitoring the degree of stretch of those structures.

Structure of General Sensory Receptors

Anatomically, general sensory receptors are nerve endings that are either nonencapsulated or encapsulated. General sensory receptors are summarized in **Table 22.1** on p. 356.

Activity 1

Studying the Structure of Selected Sensory Receptors

1. Obtain a compound microscope and histologic slides of lamellar and tactile corpuscles. Locate, under low power, a tactile corpuscle in the dermal layer of the skin. As mentioned above, these are usually found in the dermal papillae. Then switch to the oil immersion lens for a detailed study. Notice that the free nerve fibers within the corpuscle are aligned parallel to the skin surface. Compare your observations to the photomicrograph of a tactile corpuscle (Figure 22.1b).

2. Next observe a lamellar corpuscle located much deeper in the dermis. Try to identify the slender naked nerve ending in the center of the receptor and the many layers of connective tissue surrounding it (which looks rather like an onion cut lengthwise). Also, notice how much larger the lamellar corpuscles are than the tactile corpuscles. Compare your observations to the photomicrograph of a lamellar corpuscle (Figure 22.1d).

3. Obtain slides of muscle spindles and tendon organs, the two major types of proprioceptors (**Figure 22.2**). In the slide of **muscle spindles,** note that minute extensions of the nerve endings of the sensory neurons coil around specialized slender skeletal muscle cells called **intrafusal fibers.** The **tendon organs** are composed of nerve endings that ramify through the tendon tissue close to the attachment between muscle and tendon. Stretching of muscles or tendons excites these receptors, which then transmit impulses that ultimately reach the cerebellum for interpretation. Compare your observations to Figure 22.2.

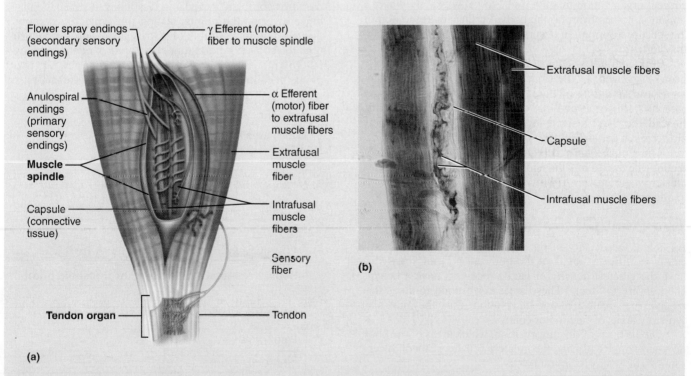

(a)

(b)

Figure 22.2 Proprioceptors. (a) Diagram of a muscle spindle and tendon organ. Myelin has been omitted from all nerve fibers for clarity. **(b)** Photomicrograph of a muscle spindle (80×).

Table 22.1 Receptors of the General Senses (Figures 22.1 and 22.2)

Structural class	Body location(s)	Stimulus type
Nonencapsulated		
Free nerve endings	Most body tissues; especially the epithelia and connective tissues	Pain, heat and cold
Tactile (Merkel) discs	Stratum basale of the epidermis	Light touch
Hair follicle receptors	Wrapped basket-like around hair follicles	Light touch and the bending of hairs
Encapsulated		
Tactile (Meissner's) corpuscles	Dermal papillae of hairless skin	Light pressure, discriminative touch
Bulbous corpuscles	Deep in the dermis, subcutaneous tissue, and joint capsules	Deep pressure and stretch
Lamellar corpuscles	Dermis, subcutaneous tissue, periosteum, tendons, and joint capsules	Deep pressure, stretch and vibration
Muscle spindles	Skeletal muscles	Muscle stretch (proprioception)
Tendon organs	Tendons	Tendon stretch, tension (proprioception)

Receptor Physiology

Sensory receptors act as **transducers,** changing environmental *stimuli* into nerve impulses that are relayed to the CNS. *Sensation* (awareness of the stimulus) and *perception* (interpretation of the meaning of the stimulus) occur in the brain. Nerve impulses from cutaneous receptors are relayed to the primary somatosensory cortex, where stimuli from different body regions form a body map. Therefore, each location on the body is represented by a specific cortical area. It is this cortical organization that allows us to know exactly where a sensation comes from on the body. Further interpretation of the sensory information occurs in the somatosensory association cortex.

Four qualities of cutaneous sensations have traditionally been recognized: tactile (touch), heat, cold, and pain. Mapping these sensations on the skin has revealed that the sensory receptors for these qualities are not distributed uniformly. Instead, they have discrete locations and are characterized by clustering at certain points—**punctate distribution.**

The simple pain receptors, extremely important in protecting the body, are the most numerous. Touch receptors cluster where greater sensitivity is desirable, as on the hands and face.

Two-Point Discrimination Test

As noted, the density of the touch receptors varies significantly in different areas of the body. In general, areas that have the greatest density of tactile receptors have a heightened ability to "feel." These areas correspond to areas that receive the greatest motor innervation; thus they are also typically areas of fine motor control.

On the basis of this information, which areas of the body do you *predict* will have the greatest density of touch receptors? Write your prediction below.

_____

_____

Activity 2

Determining the Two-Point Threshold

1. Using calipers or an esthesiometer and a metric ruler, test the ability of the subject to differentiate two distinct sensations when the skin is touched simultaneously at two points. Beginning with the face, start with the caliper arms completely together. Gradually increase the distance between the points, testing the subject's skin after each adjustment. Continue with this testing procedure until the subject reports that *two points* of contact can be felt. This measurement, the smallest distance at which two points of contact can be felt, is the **two-point threshold.**

2. Repeat this procedure on the body areas listed and record your results in the Activity 2 chart.

3. Which area has the smallest two-point threshold?

_____

4. How well did this compare with your prediction?

_____

Activity 2: Determining Two-Point Threshold	
Body area tested	**Two-point threshold (mm)**
Face	
Back of hand	
Palm of hand	
Fingertip	
Lips	
Back of neck	
Anterior forearm	

Tactile Localization

Tactile localization is the ability to determine which portion of the skin has been touched. The tactile receptor field of the body periphery has a corresponding "touch" field in the brain's primary somatosensory cortex. Some body areas are well represented with touch receptors, allowing tactile stimuli to be localized with great accuracy, but in other body areas, touch-receptor density allows only crude discrimination.

Activity 3

Testing Tactile Localization

1. The subject's eyes should be closed during the testing. The experimenter touches the palm of the subject's hand with a pointed black felt-tipped marker. The subject should then try to touch the exact point with his or her own marker, which should be of a different color. Measure the error of localization (the distance between the two marks) in millimeters.

2. Repeat the test in the same spot twice more, recording the error of localization for each test. Average the results of the three determinations, and record it in the **Activity 3 chart**.

| Activity 3: Testing Tactile Localization ||
Body area tested	**Average error (mm)**
Palm of hand	
Fingertip	
Anterior forearm	
Back of hand	
Back of neck	

Does the ability to localize the stimulus improve the second

time? _____ The third time? _____

Explain. _____

_____

3. Repeat the procedure on the body areas listed and record the averaged results in the chart above.

4. Which area has the smallest error of localization?

_____

Adaptation of Sensory Receptors

The number of impulses transmitted by sensory receptors often changes both with the intensity of the stimulus and with the length of time the stimulus is applied. In many cases, when a stimulus is applied for a prolonged period without movement, the rate of receptor discharge slows and conscious awareness of the stimulus declines or is lost until some type of stimulus change occurs. This phenomenon is referred to as **adaptation.** The touch receptors adapt particularly rapidly,

which is highly desirable. Who, for instance, would want to be continually aware of the pressure of clothing on their skin?

Activity 4

Demonstrating Adaptation of Touch Receptors

1. The subject's eyes should be closed. Obtain four coins. Place one coin on the anterior surface of the subject's forearm, and determine how long the sensation persists for the subject. Duration of the sensation:

_____ sec

2. Repeat the test, placing the coin at a different forearm location. How long does the sensation persist at the second location?

_____ sec

3. After awareness of the sensation has been lost at the second site, stack three more coins atop the first one.

Does the pressure sensation return? _____

If so, for how long is the subject aware of the pressure in this instance?

_____ sec

Are the same receptors being stimulated when the four

coins, rather than the one coin, are used? _____

Explain how perception of the stimulus intensity has

changed. _____

_____

Activity 5

Demonstrating Adaptation of Temperature Receptors

Adaptation of temperature receptors also can be tested.

1. Obtain three large finger bowls or 1000-ml beakers and fill the first with 45°C water. Have the subject immerse her or his left hand in the water and report the sensation. Keep the left hand immersed for 1 minute, and then also immerse the right hand in the same bowl.

What is the sensation of the left hand when it is first immersed?

_____

What is the sensation of the left hand after 1 minute as compared to the sensation in the right hand just immersed?

_____

22

Had adaptation occurred in the left hand? _____

Which hand seemed to adapt more quickly?

_____

2. Rinse both hands in tap water, dry them, and wait 5 minutes before conducting the next test. Just before beginning the test, refill the finger bowl with fresh 45°C water, fill a second with ice water, and fill a third with water at room temperature.

3. Place the *left* hand in the ice water and the *right* hand in the 45°C water. What is the sensation in each hand after 2 minutes as compared to the sensation perceived when the hands were first immersed?

4. After reporting these observations, the subject should then place both hands simultaneously into the finger bowl containing the water at room temperature. Record the

sensation in the left hand: _____

The right hand: _____

_____

_____

Referred Pain

Pain receptors are densely distributed in the skin, and they adapt very little, if at all. This lack of adaptability is due to the protective function of the receptors. The sensation of pain often indicates tissue damage or trauma to body structures. Thus no attempt will be made in this exercise to localize the pain receptors or to prove their nonadaptability, since both would cause needless discomfort.

However, the phenomenon of referred pain is easily demonstrated in the laboratory, and such experiments provide information that may be useful in explaining common examples of this phenomenon. **Referred pain** is a sensory experience in which pain is perceived as arising in one area of the body when in fact another, often quite remote area, is receiving the painful stimulus. Thus the pain is said to be "referred" to a different area.

Activity 6

Demonstrating the Phenomenon of Referred Pain

Immerse the subject's elbow in a finger bowl containing ice water. In the **Activity 6 chart**, record the quality of the sensation (such as discomfort, tingling, or pain) and the localization of the sensations he or she reports for the intervals indicated. The elbow should be removed from ice water after the 2-minute reading. The last recording is to occur 3 minutes after removal of the subject's elbow from the ice water.

The ulnar nerve, which serves the medial third of the hand, is involved in the phenomenon of referred pain experienced during this test. How does the localization of this referred pain correspond to the areas served by the ulnar nerve?

_____

_____

WHY THIS **MATTERS** | "Brain Freeze"

Slurp a smoothie too quickly, and you will likely experience a common type of referred pain often called "brain freeze." One theory for this phenomenon is that the cold stimulus on the palate and the throat causes the capillaries in the frontal sinuses to constrict and then quickly dilate. The dilation activates pain receptors that send sensory information to the brain via the trigeminal nerve. The brain interprets the pain as coming from the forehead. Another explanation proposes that the decrease in temperature at the back of the throat increases blood flow to the brain, which increases the pressure in the anterior cerebral artery serving the brain. The brain interprets the pressure change as pain in the forehead. ■

Activity 6: Demonstrating Referred Pain		
Time of observation	**Quality of sensation**	**Localization of sensation**
On immersion		
After 1 min		
After 2 min		
3 min after removal		

REVIEW SHEET
General Sensation

Name_____ Lab Time/Date _____

Structure of General Sensory Receptors

1. Differentiate between interoceptors and exteroceptors relative to location and stimulus source.

   interoceptor: _____

   exteroceptor: _____

2. A number of activities and sensations are listed in the chart below. For each, check whether the receptors would be exteroceptors or interoceptors; and then name the specific receptor types. (Because specific visceral receptors were not described in detail in this exercise, you need only indicate that the receptor is a visceral receptor if it falls into that category.)

Activity or sensation	Exteroceptor	Interoceptor	Specific receptor type
Backing into a sun-heated iron railing			
Someone steps on your foot			
Reading a street sign			
Leaning on your elbows			
Doing sit-ups			
The "too full" sensation			

Receptor Physiology

3. Explain how the sensory receptors act as transducers. _____

_____

4. Define *stimulus.* _____

5. What was demonstrated by the two-point discrimination test? _____

_____

   How well did your results correspond to your predictions? _____

_____

   What is the relationship between the accuracy of the subject's tactile localization and the results of the two-point

   discrimination test? _____

6. Define *punctate distribution.* _____

_____

7. Several questions regarding general sensation are posed below. Answer each by placing your response in the appropriately numbered blanks to the right.

    1. Which cutaneous receptors are the most numerous?

    2–3. Which two body areas tested were most sensitive to touch?

    4–5. Which two body areas tested were least sensitive to touch?

    6–8. Where would referred pain appear if the following organs were receiving painful stimuli: (6) gallbladder, (7) kidneys, and (8) appendix? (Use your textbook if necessary.)

    9. Where was referred pain felt when the elbow was immersed in ice water during the laboratory experiment?

    10. What region of the cerebrum interprets the kind and intensity of stimuli that cause cutaneous sensations?

1. _____

2–3. _____

4–5. _____

6. _____

7. _____

8. _____

9. _____

10. _____

8. Define *adaptation of sensory receptors.* _____

_____

9. Why is it advantageous to have pain receptors that are sensitive to all vigorous stimuli, whether heat, cold, or pressure?

_____

    Why is the nonadaptability of pain receptors important?_____

_____

10. Imagine yourself without any cutaneous sense organs. Why might this be very dangerous?_____

_____

_____

WHY THIS MATTERS | **11.** Explain why "brain freeze" is a type of referred pain. _____

_____

    12. Describe the location of the stimulus for "brain freeze" and the location of the pain (as interpreted by

    the brain). _____

_____

23

Special Senses: Anatomy of the Visual System

Objectives

☐ Identify the anatomy of the eye and the accessory anatomical structures of the eye on a model or appropriate image, and list the function(s) of each; identify the structural components that are present in a preserved sheep or cow eye (if available).

☐ Define *conjunctivitis*, *cataract*, and *glaucoma*.

☐ Describe the cellular makeup of the retina.

☐ Explain the difference between rods and cones with respect to visual perception and retinal localization.

☐ Trace the visual pathway to the primary visual cortex, and indicate the effects of damage to various parts of this pathway.

Materials

- Chart of eye anatomy
- Dissectible eye model
- Prepared slide of longitudinal section of an eye showing retinal layers
- Compound microscope
- Preserved cow or sheep eye
- Dissecting instruments and tray
- Disposable gloves

MasteringA&P®

For related exercise study tools, go to the Study Area of **MasteringA&P**. There you will find:

- Practice Anatomy Lab PAL
- A&PFlix *A&PFlix*
- PhysioEx PEx
- Practice quizzes, Histology Atlas, eText, Videos, and more!

Pre-Lab Quiz

1. Name the mucous membrane that lines the internal surface of the eyelids and continues over the anterior surface of the eyeball.

   _____

2. How many extrinsic eye muscles are attached to the exterior surface of each eyeball?
   a. three   c. five
   b. four   d. six

3. The wall of the eye has three layers. The outermost fibrous layer is made up of the opaque white sclera and the transparent:
   a. choroid   c. cornea
   b. ciliary gland   d. lacrima

4. Circle the correct underlined term. The <u>aqueous humor</u> / <u>vitreous humor</u> is a clear, watery fluid that helps to maintain the intraocular pressure of the eye and provides nutrients for the avascular lens and cornea.

5. Circle True or False. At the optic chiasma, the fibers from the medial side of each eye cross over to the opposite side.

Anatomy of the Eye

Accessory Structures

The adult human eye is a sphere measuring about 2.5 cm (1 inch) in diameter. Only about one-sixth of the eye's anterior surface is observable (**Figure 23.1**, p. 362); the remainder is enclosed and protected by a cushion of fat and the walls of the bony orbit.

The accessory structures of the eye include the eyebrows, eyelids, conjunctivae, lacrimal apparatus, and extrinsic eye muscles (**Table 23.1**, p. 362, and **Figure 23.2**, p. 363).

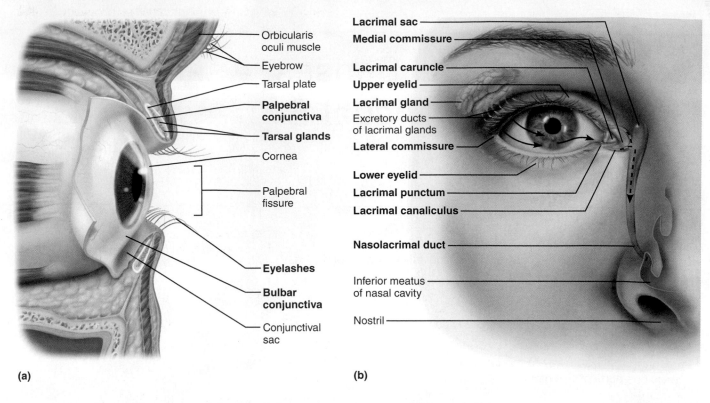

Figure 23.1 External anatomy of the eye and accessory structures. **(a)** Lateral view; some structures shown in sagittal section. **(b)** Anterior view with lacrimal apparatus.

Table 23.1	Accessory Structures of the Eye (Figures 23.1 and 23.2)	
Structure	**Description**	**Function**
Eyebrows	Short hairs located on the supraorbital margins	Shade and prevent sweat from entering the eyes.
Eyelids (palpebrae)	Skin-covered upper and lower lids, with eyelashes projecting from their free margin	Protect the eyes and spread lacrimal fluid (tears) with blinking.
Tarsal glands	Modified sebaceous glands embedded in the tarsal plate of the eyelid	Secrete an oily secretion that lubricates the surface of the eye.
Ciliary glands	Typical sebaceous and modified sweat glands that lie between the eyelash follicles	Secrete an oily secretion that lubricates the surface of the eye and the eyelashes. An infection of a ciliary gland is called a **sty.**
Conjunctivae	A clear mucous membrane that lines the eyelids (palpebral conjunctivae) and the anterior white of the eye (bulbar conjunctiva)	Secrete mucus to lubricate the eye. Inflammation of the conjunctiva results in conjunctivitis, (commonly called "pinkeye").
Medial and lateral commissures	Junctions where the eyelids meet medially and laterally	Form the corners of the eyes. The medial commissure contains the lacrimal caruncle.
Lacrimal caruncle	Fleshy reddish elevation that contains sebaceous and sweat glands	Secretes a whitish oily secretion for lubrication of the eye (can dry and form "eye sand").
Lacrimal apparatus	Includes the lacrimal gland and a series of ducts that drain the lacrimal fluid into the nasal cavity	Protects the eye by keeping it moist. Blinking spreads the lacrimal fluid.
Lacrimal gland	Located in the superior and lateral aspect of the orbit of the eye	Secretes lacrimal fluid, which contains mucus, antibodies, and lysozyme.
Lacrimal puncta	Two tiny openings on the medial margin of each eyelid	Allow lacrimal fluid to drain into the superior and inferiorly located lacrimal canaliculi.
Lacrimal canaliculi	Two tiny canals that are located in the eyelids	Allow lacrimal fluid to drain into the lacrimal sac.
Lacrimal sac	A single pouch located in the medial orbital wall	Allows lacrimal fluid to drain into the nasolacrimal duct.
Nasolacrimal duct	A single tube that empties into the nasal cavity	Allows lacrimal fluid to flow into the nasal cavity.
Extrinsic eye muscles	Six muscles for each eye; four recti and two oblique muscles (see Figure 23.2)	Control the movement of each eyeball and hold the eyes in the orbits.

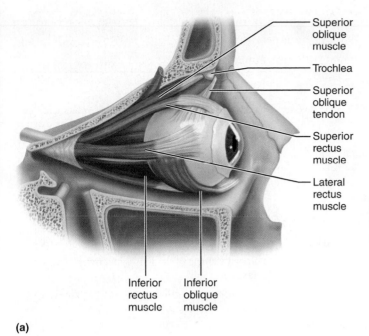

(a)

Muscle	Action
Lateral rectus	Moves eye laterally
Medial rectus	Moves eye medially
Superior rectus	Elevates eye and turns it medially
Inferior rectus	Depresses eye and turns it medially
Inferior oblique	Elevates eye and turns it laterally
Superior oblique	Depresses eye and turns it laterally

(b)

Figure 23.2 Extrinsic muscles of the eye. (a) Lateral view of the right eye. **(b)** Summary of actions of the extrinsic eye muscles.

Identifying Accessory Eye Structures

Using a chart of eye anatomy or Figure 23.1, observe the eyes of another student, and identify as many of the accessory structures as possible. Ask the student to look to the left. Which extrinsic eye muscles are responsible for this action?

Right eye: _____

Left eye: _____

Internal Anatomy of the Eye

Anatomically, the wall of the eye is constructed of three layers: the **fibrous layer,** the **vascular layer,** and the **inner layer** (**Table 23.2,** p. 364, and **Figure 23.3,** p. 365).

Distribution of Photoreceptors

The photoreceptor cells are distributed over the entire neural retina, except where the optic nerve leaves the eyeball. This site is called the **optic disc,** or *blind spot,* and is located in a weak spot in the **fundus** (posterior wall). Lateral to each blind spot, and directly posterior to the lens, is an area called the **macula lutea** ("yellow spot"), an area of high cone density. In its center is the **fovea centralis,** a tiny pit, which contains only cones and is the area of greatest visual acuity. Focusing for detailed color vision occurs in the fovea centralis.

Internal Chambers and Fluids

The lens divides the eye into two segments: the **anterior segment** anterior to the lens, which contains a clear watery fluid called the **aqueous humor,** and the **posterior segment** behind the lens, filled with a gel-like substance, the **vitreous humor.** The anterior segment is further divided into **anterior** and **posterior chambers,** located before and after the iris, respectively. The aqueous humor is continually formed by the capillaries of the **ciliary processes** of the ciliary body. It helps to maintain the intraocular pressure of the eye and provides nutrients for the avascular lens and cornea. The aqueous humor is drained into the **scleral venous sinus.** The vitreous humor provides the major internal reinforcement of the posterior part of the eyeball, and helps to keep the retina pressed firmly against the wall of the eyeball. It is formed *only* before birth and is not renewed. In addition to the cornea, the light-bending media of the eye include the lens, the aqueous humor, and the vitreous humor.

Anything that interferes with drainage of the aqueous fluid increases intraocular pressure. When intraocular pressure reaches dangerously high levels, the retina and optic nerve are compressed, resulting in pain and possible blindness, a condition called **glaucoma. ✚**

Identifying Internal Structures of the Eye

Obtain a dissectible eye model and identify its internal structures described above. (As you work, also refer to Figure 23.3.)

23

Table 23.2	Layers of the Eye (Figure 23.3)	
Structure	**Description**	**Function**
Fibrous Layer (External Layer)		
Sclera	Opaque white connective tissue that forms the "white of the eye."	Helps to maintain the shape of the eyeball and provides an attachment point for the extrinsic eye muscles.
Cornea	Structurally continuous with the sclera; modified to form a transparent layer that bulges anteriorly.	Forms a clear window that is the major light bending (refracting) medium of the eye.
Vascular Layer (Middle Layer)		
Choroid	A blood vessel–rich, dark membrane.	The blood vessels nourish the other layers of the eye, and the melanin helps to absorb excess light.
Cilary body	Modification of the choroid that encircles the lens.	Contains the ciliary muscle and the ciliary process.
Ciliary muscle	Smooth muscle found within the ciliary body.	Alters the shape of the lens with contraction and relaxation.
Ciliary process	Radiating folds of the ciliary muscle.	Capillaries of the ciliary process form the aqueous humor by filtering plasma.
Ciliary zonule (Suspensory ligament)	A halo of fine fibers that extends from the ciliary process around the lens.	Attaches the lens to the ciliary process.
Iris	The anterior portion of the vascular layer that is pigmented. It contains two layers of smooth muscle (sphincter pupillae and dilator pupillae).	Controls the amount of light entering the eye by changing the size of the pupil diameter. The sphincter pupillae contract to constrict the pupil. The dilator pupillae contract to dilate the pupil.
Pupil	The round central opening of the iris.	Allows light to enter the eye.
Inner Layer (Retina)		
Pigmented layer of the retina	The outer layer that is composed of only a single layer of pigment cells (melanocytes).	Absorbs light and prevents it from scattering in the eye. Pigment cells act as phagocytes for cleaning up cell debris and also store vitamin A needed for photoreceptor renewal.
Neural layer of the retina	The thicker inner layer composed of three main types of neurons: photoreceptors (rods and cones), bipolar cells, and ganglion cells.	Photoreceptors respond to light and convert the light energy into action potentials that travel to the primary visual cortex of the brain.

Microscopic Anatomy of the Retina

Cells of the retina include the pigment cells of the outer pigmented layer and the neurons of the neural layer (**Figure 23.4**, p. 366). The inner neural layer is composed of three major populations of neurons. These are, from outer to inner aspect, the **photoreceptors,** the **bipolar cells,** and the **ganglion cells.**

The **rods** are the specialized photoreceptors for dim light. Visual interpretation of their activity is in gray tones. The **cones** are color photoreceptors that permit high levels of visual acuity, but they function only under conditions of high light intensity; thus, for example, no color vision is possible in moonlight. The fovea centralis contains only cones, the macula lutea contains mostly cones, and from the edge of the macula to the retina periphery, cone density declines gradually. By contrast, rods are most numerous in the periphery, and their density decreases as the macula is approached.

Light must pass through the ganglion cell layer and the bipolar cell layer to reach and excite the rods and cones. As a result of a light stimulus, the photoreceptors undergo changes in their membrane potential that influence the bipolar cells. These in turn stimulate the ganglion cells, whose axons leave the retina in the tight bundle of fibers known as the **optic nerve** (Figure 23.3). In addition to these three major types of neurons, the retina also contains horizontal cells and amacrine cells (both are interneurons), which play a role in visual processing.

Activity 3

Studying the Microscopic Anatomy of the Retina

Use a compound microscope to examine a histologic slide of a longitudinal section of the eye. Identify the retinal layers by comparing your view to the photomicrograph (Figure 23.4b).

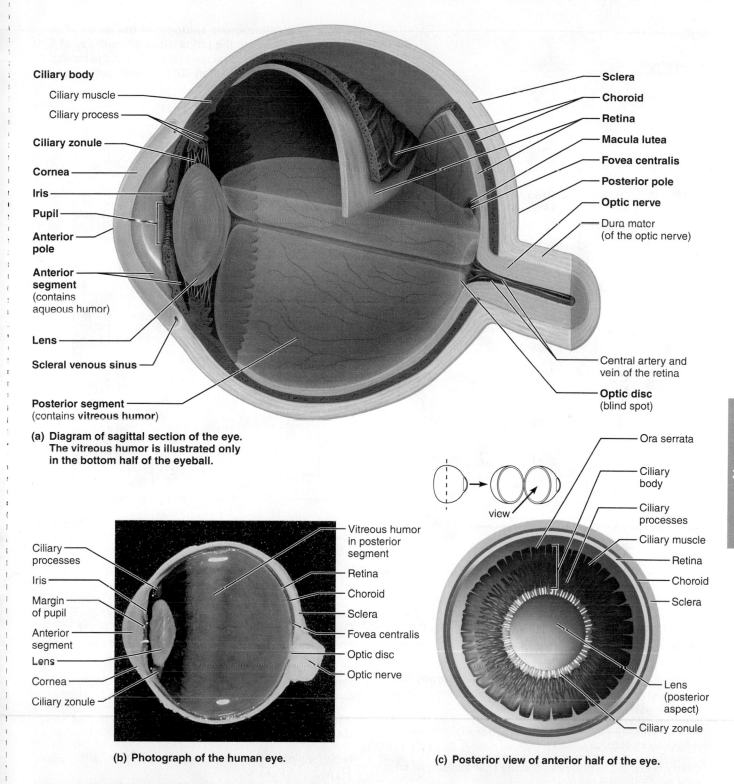

(a) **Diagram of sagittal section of the eye. The vitreous humor is illustrated only in the bottom half of the eyeball.**

(b) **Photograph of the human eye.**

(c) **Posterior view of anterior half of the eye.**

Figure 23.3 Internal anatomy of the eye.

 DISSECTION

The Cow (Sheep) Eye

1. Obtain a preserved cow or sheep eye, dissecting instruments, and a dissecting tray. Don disposable gloves.

2. Examine the external surface of the eye, noting the thick cushion of adipose tissue. Identify the optic nerve as it leaves the eyeball, the remnants of the extrinsic eye

muscles, the conjunctiva, the sclera, and the cornea. (Refer to **Figure 23.5** on p. 367 as you work.)

3. Trim away most of the fat and connective tissue, but leave the optic nerve intact. Holding the eye with the cornea facing downward, carefully make an incision with a sharp scalpel

Text continues on next page. →

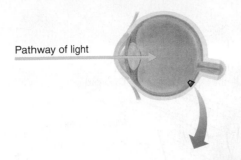

Pathway of light

Figure 23.4 Microscopic anatomy of the retina.
(a) Diagram of cells of the retina. Note the pathway of light through the retina. Neural signals (output of the retina) flow in the opposite direction. **(b)** Photomicrograph of the retina (140×).

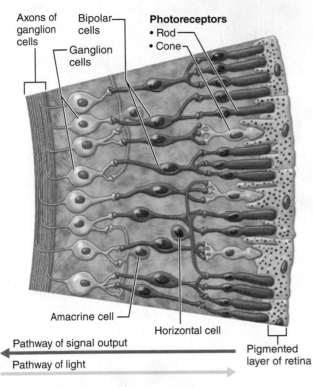

Axons of ganglion cells

Bipolar cells

Ganglion cells

Photoreceptors
• Rod
• Cone

Amacrine cell

Horizontal cell

Pathway of signal output

Pathway of light

Pigmented layer of retina

(a)

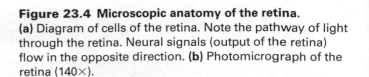

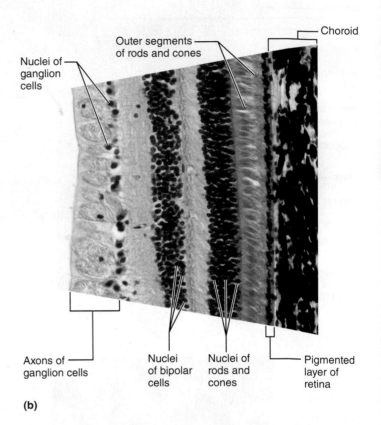

Nuclei of ganglion cells

Outer segments of rods and cones

Choroid

Axons of ganglion cells

Nuclei of bipolar cells

Nuclei of rods and cones

Pigmented layer of retina

(b)

23

into the sclera about 6 mm (¼ inch) above the cornea. Using scissors, complete the incision around the circumference of the eyeball paralleling the corneal edge.

4. Carefully lift the anterior part of the eyeball away from the posterior portion. The vitreous humor should remain with the posterior part of the eyeball.

5. Examine the anterior part of the eye, and identify the following structures:

Ciliary body: Black pigmented body that appears to be a halo encircling the lens.

Lens: Biconvex structure that is opaque in preserved specimens.

Carefully remove the lens and identify the adjacent structures:

Iris: Anterior continuation of the ciliary body penetrated by the pupil.

Cornea: More convex anteriormost portion of the sclera; normally transparent but cloudy in preserved specimens.

6. Examine the posterior portion of the eyeball. Carefully remove the vitreous humor, and identify the following structures:

Retina: Appears as a delicate tan membrane that separates easily from the choroid.

Note its posterior point of attachment. What is this point called?

_____

Pigmented choroid coat: Appears iridescent in the cow or sheep eye owing to a modification called the **tapetum lucidum.** This specialized surface reflects the light within the eye and is found in the eyes of animals that live under conditions of dim light. It is not found in humans.

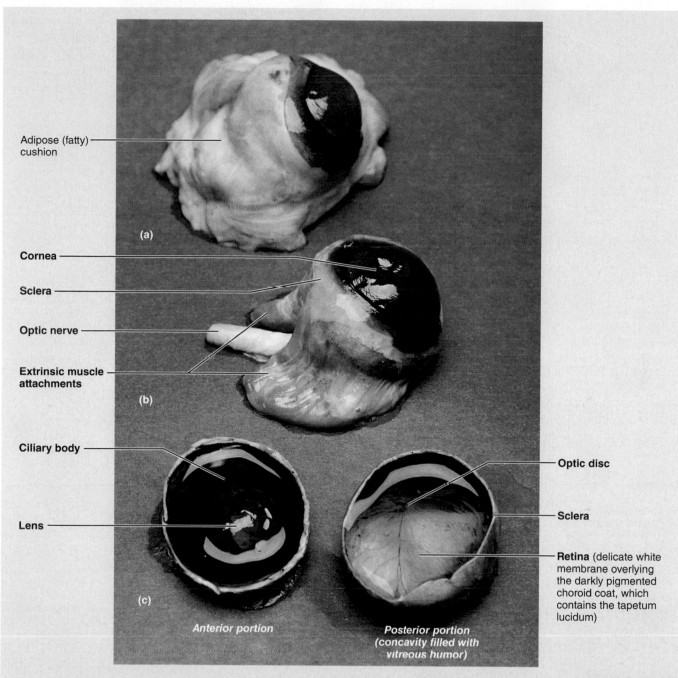

Adipose (fatty) cushion

(a)

Cornea

Sclera

Optic nerve

Extrinsic muscle attachments

(b)

Ciliary body

Lens

Optic disc

Sclera

Retina (delicate white membrane overlying the darkly pigmented choroid coat, which contains the tapetum lucidum)

(c)

Anterior portion

Posterior portion (concavity filled with vitreous humor)

Figure 23.5 Anatomy of the cow eye. (a) Cow eye (entire) removed from the bony orbit (notice the large amount of fat cushioning the eyeball). **(b)** Cow eye (entire) with fat removed to show the extrinsic muscle attachments and optic nerve. **(c)** Cow eye cut along the frontal plane to reveal internal structures.

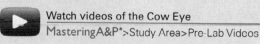 Watch videos of the Cow Eye
MasteringA&P®>Study Area>Pre-Lab Videos

Visual Pathways to the Brain

The axons of the ganglion cells of the retina converge at the posterior aspect of the eyeball and exit from the eye as the optic nerve. At the **optic chiasma,** the fibers from the medial side of each eye cross over to the opposite side (**Figure 23.6,** p. 368). The fiber tracts thus formed are called the **optic tracts.** Each optic tract contains fibers from the lateral side of the eye on the same side and from the medial side of the opposite eye.

The optic tract fibers synapse with neurons in the **lateral geniculate nucleus** of the thalamus, whose axons form the **optic radiation,** terminating in the **primary visual cortex** in the occipital lobe of the brain. Here they synapse with the cortical neurons, and visual interpretation occurs.

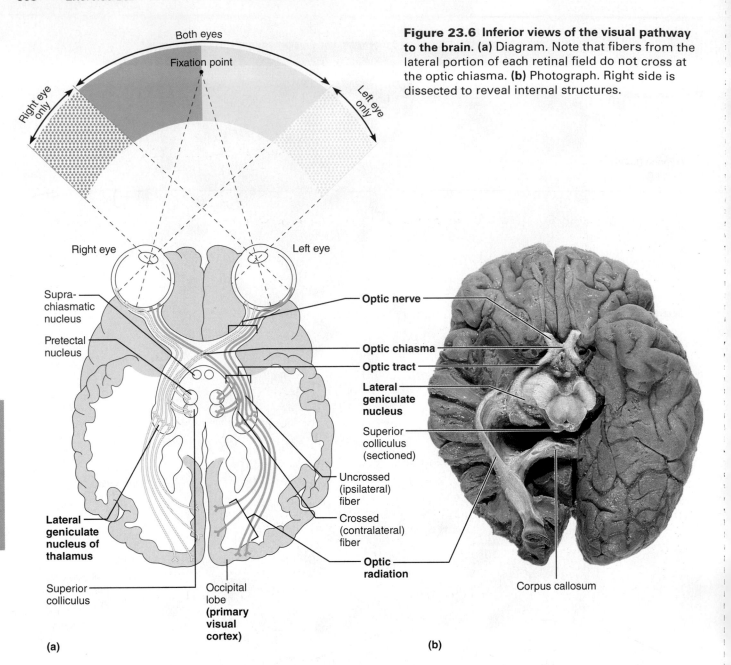

Figure 23.6 Inferior views of the visual pathway to the brain. (a) Diagram. Note that fibers from the lateral portion of each retinal field do not cross at the optic chiasma. **(b)** Photograph. Right side is dissected to reveal internal structures.

(a)

(b)

Activity 4

Predicting the Effects of Visual Pathway Lesions

After examining the visual pathway diagram (Figure 23.6a), determine what effects lesions in the following areas would have on vision:

In the right optic nerve: _____

_____

Through the optic chiasma: _____

_____

In the left optic tract: _____

_____

In the right cerebral cortex (primary visual cortex): _____

_____

Special Senses: Visual Tests and Experiments

Objectives

☐ Discuss the mechanism of image formation on the retina.

☐ Define the following terms: *accommodation, astigmatism, emmetropic, hyperopia, myopia, refraction,* and *presbyopia,* and describe several simple visual tests to which the terms apply.

☐ Discuss the benefits of binocular vision.

☐ Define *convergence,* and discuss the importance of the pupillary and convergence reflexes.

☐ State the importance of an ophthalmoscopic examination.

Materials

- Metric ruler; meter stick
- Common straight pins
- Snellen eye chart, floor marked with chalk or masking tape to indicate 20-ft distance from posted Snellen chart
- Ishihara's color plates
- Two pencils
- Test tubes large enough to accommodate a pencil
- Laboratory lamp or penlight
- Ophthalmoscope (if available)

Pre-Lab Quiz

1. Circle the correct underlined term. Photoreceptors are distributed over the entire neural retina, except where the optic nerve leaves the eyeball. This site is called the <u>macula lutea</u> / <u>optic disc</u>.

2. Circle True or False. People with difficulty seeing objects at a distance are said to have myopia.

3. A condition that results in the loss of elasticity of the lens and difficulty focusing on a close object is called:
   - **a.** myopia
   - **c.** hyperopia
   - **b.** presbyopia
   - **d.** astigmatism

4. Photoreceptors of the eye include rods and cones. Which one is responsible for interpreting color; which can function only under conditions of high light intensity?

   _____

5. Circle the correct underlined term. <u>Extrinsic</u> / <u>Intrinsic</u> eye muscles are controlled by the autonomic nervous system.

MasteringA&P®

For related exercise study tools, go to the Study Area of **MasteringA&P**. There you will find:

- Practice Anatomy Lab **PAL**
- A&PFlix **A&PFlix**
- PhysioEx **PEx**
- Practice quizzes, Histology Atlas, eText, Videos, and more!

The Optic Disc

In this exercise, you will perform several visual tests and experiments focusing on the physiology of vision. The first test involves demonstrating the blind spot (optic disc), the site where the optic nerve exits the eyeball.

Activity 1

Demonstrating the Blind Spot

1. Hold the figure for the blind spot test (**Figure 24.1**, p. 374) about 46 cm (18 inches) from your eyes. Close your left eye, and focus your right eye on the X, which should be positioned so that it is directly in line with your right eye. Move the figure slowly toward your face, keeping your right eye focused on the X. When the dot focuses on the blind spot, which lacks photoreceptors, it will disappear.

text continues on next page →

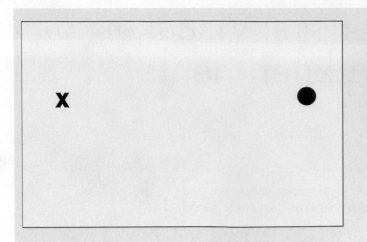

2. Have your laboratory partner record in metric units the distance at which this occurs. The dot will reappear as the figure is moved closer. Distance at which the dot disappears:

Right eye _____

Repeat the test for the left eye, this time closing the right eye and focusing the left eye on the dot. Record the distance at which the X disappears:

Left eye _____

Figure 24.1 Blind spot test figure.

Refraction, Visual Acuity, and Astigmatism

When light rays pass from one medium to another, their speed changes, and the rays are bent, or **refracted.** Thus the light rays in the visual field are refracted as they encounter the cornea, lens, and vitreous humor of the eye.

The refractive index (bending power) of the cornea and vitreous humor are constant. But the lens's refractive index can be varied by changing the lens's shape. The greater the lens convexity, or bulge, the more the light will be bent. Conversely, the less convex the lens (the flatter it is), the less it bends the light.

In general, light from a distant source (over 6 m, or 20 feet) approaches the eye as parallel rays, and no change in lens convexity is necessary for it to focus properly on the retina. However, light from a close source tends to diverge, and the convexity of the lens must increase to make close vision possible. To achieve this, the ciliary muscle contracts, decreasing the tension on the ciliary zonule attached to the lens and allowing the elastic lens to bulge. Thus, a lens capable of bringing a *close* object into sharp focus is more convex than a lens focusing on a more distant object. The ability of the eye to focus differentially for objects of close vision (less than 6 m, or 20 feet) is called **accommodation.** It should be noted that the image formed on the retina as a result of the refractory activity of the lens (**Figure 24.2**) is a **real image** (reversed from left to right, inverted, and smaller than the object).

The normal, or **emmetropic,** eye is able to accommodate properly. However, visual problems may result (1) from lenses that are too strong or too "lazy" (overconverging and underconverging, respectively); (2) from structural problems, such as an eyeball that is too long or too short to provide for proper focusing by the lens; or (3) from a cornea or lens with improper curvatures. Problems of refraction are summarized in **Figure 24.3**.

Irregularities in the curvatures of the lens and/or the cornea lead to a blurred vision problem called **astigmatism.** Cylindrically ground lenses are prescribed to correct the condition. ✚

Near-Point Accommodation

The elasticity of the lens decreases dramatically with age, resulting in difficulty in focusing for near or close vision, especially when reading. This condition is called **presbyopia**—literally, old vision. Lens elasticity can be tested by measuring the **near point of accommodation.** The near point of accommodation is about 10 cm from the eye in young adults. It is closer in children and farther in old age.

Activity 2

Determining Near Point of Accommodation

To determine your near point of accommodation, hold a common straight pin (or other object) at arm's length in front of one eye. Slowly move the pin toward that eye until the pin image becomes distorted. Have your lab partner use a metric ruler to measure the distance in centimeters from your eye to the pin at this point, and record the distance. Repeat the procedure for the other eye.

Near point for right eye: _____

Near point for left eye: _____

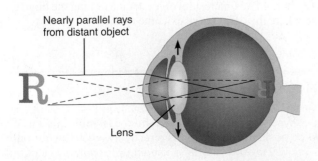

Figure 24.2 Refraction and real images. The refraction of light in the eye produces a real image (reversed, inverted, and reduced) on the retina.

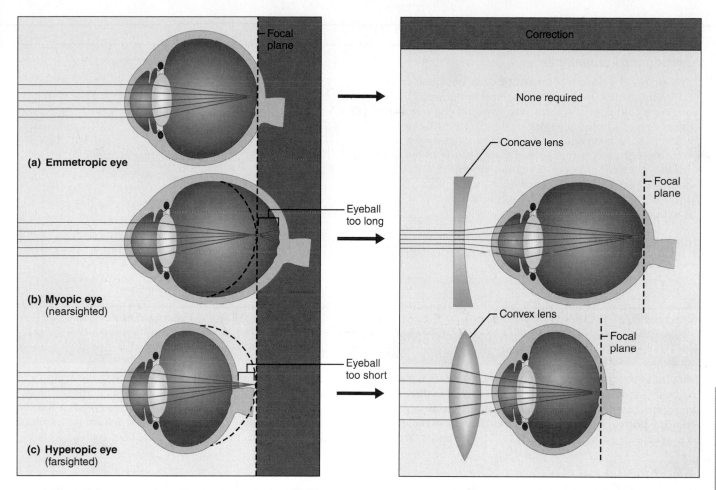

Figure 24.3 Problems of refraction. (a) In the emmetropic eye, light from near and far objects is focused properly on the retina. **(b)** In a myopic eye, light from distant objects is brought to a focal point before reaching the retina. Applying a concave lens focuses objects properly on the retina. **(c)** In the hyperopic eye, light from a near object is brought to a focal point behind the retina. Applying a convex lens focuses objects properly on the retina. The refractory effect of the cornea is ignored here.

Visual Acuity

Visual acuity, or sharpness of vision, is generally tested with a Snellen eye chart. This test is based on the fact that letters of a certain size can be seen clearly with normal vision at a specific distance. The distance at which the emmetropic eye can read a line of letters is printed at the end of that line.

Activity 3

Testing Visual Acuity

1. Have your partner stand 6 m (20 feet) from the posted Snellen eye chart and cover one eye with a card or hand. As your partner reads each consecutive line aloud, check for accuracy. If this individual wears glasses, give the test twice—first with glasses off and then with glasses on. *Do not remove contact lenses, but note that they were in place during the test.*

2. Record the number of the line with the smallest-sized letters read. If it is 20/20, the person's vision for that eye

is normal. If it is 20/40, or any ratio with a value less than one, he or she has less than the normal visual acuity. (Such an individual is myopic.) If the visual acuity is 20/15, vision is better than normal, because this person can stand at 6 m (20 feet) from the chart and read letters that are discernible by the normal eye only at 4.5 m (15 feet). Give your partner the number of the line corresponding to the smallest letters read, to record in step 4.

3. Repeat the process for the other eye.

4. Have your partner test and record your visual acuity.

Visual acuity, right eye without glasses: _____

Visual acuity, right eye with glasses: _____

Visual acuity, left eye without glasses: _____

Visual acuity, left eye with glasses: _____

Activity 4

Testing for Astigmatism

The astigmatism chart (**Figure 24.4**) is designed to test for unequal curvatures of the lens and/or cornea.

View the chart first with one eye and then with the other, focusing on the center of the chart. If all the radiating lines appear equally dark and distinct, there is no distortion of your refracting surfaces. If some of the lines are blurred or appear less dark than others, at least some degree of astigmatism is present.

Is astigmatism present in your left eye? _____

Right eye? _____

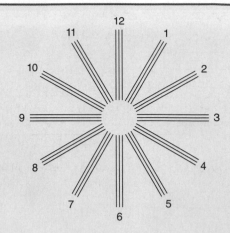

Figure 24.4 Astigmatism testing chart.

Color Blindness

Ishihara's color plates are designed to test for deficiencies in the cones or color photoreceptor cells. There are three cone types, each containing a different light-absorbing pigment. One type primarily absorbs the red wavelengths of the visible light spectrum, another the blue wavelengths, and a third the green wavelengths. Nerve impulses reaching the brain from these different photoreceptor types are then interpreted (seen) as red, blue, and green, respectively. Interpretation of the intermediate colors of the visible light spectrum is a result of overlapping input from more than one cone type.

Activity 5

Testing for Color Blindness

1. Find the interpretation table that accompanies the Ishihara color plates, and prepare a sheet to record data for the test. Note which plates are patterns rather than numbers.

2. View the color plates in bright light or sunlight while holding them about 0.8 m (30 inches) away and at right angles to your line of vision. Report to your laboratory partner what you see in each plate. Take no more than 3 seconds for each decision.

3. Your partner should record your responses and then check their accuracy with the correct answers provided in the color plate book. Is there any indication that you have some degree of color blindness? _____ If so, what type?

_____.

Binocular Vision

Humans, cats, predatory birds, and most primates are endowed with *binocular vision*. Their visual fields, each about 170 degrees, overlap to a considerable extent, and each eye sees a slightly different view (**Figure 24.5**). The primary visual cortex fuses the slightly different images, providing **depth perception** (or **three-dimensional vision**).

In contrast, the eyes of rabbits, pigeons, and many other animals are on the sides of their head. Such animals see in two different directions and thus have a panoramic field of view and *panoramic vision*. A mnemonic device to keep these straight is "Eyes in the front—likes to hunt; eyes to the side—likes to hide."

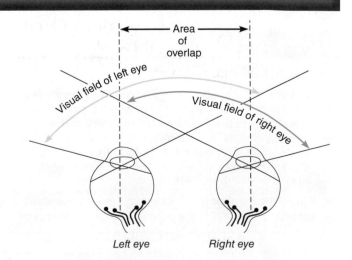

Figure 24.5 Overlapping of the visual fields.

Activity 6

Testing for Depth Perception

1. To demonstrate that a slightly different view is seen by each eye, perform the following simple experiment.

Close your left eye. Hold a pencil at arm's length directly in front of your right eye. Position another pencil directly beneath it and then move the lower pencil about half the distance toward you. As you move the lower pencil, make sure it remains in the *same plane* as the stationary pencil, so that the two pencils continually form a straight line. Then, without moving the pencils, close your right eye and open your left eye. Notice that with only the right eye open, the moving pencil stays in the same plane as the fixed pencil, but that when viewed with the left eye, the moving pencil is displaced laterally away from the plane of the fixed pencil.

2. To demonstrate the importance of two-eyed binocular vision for depth perception, perform this second experiment. Have your laboratory partner hold a test tube vertical about arm's length in front of you. With both eyes open, quickly insert a pencil into the test tube. Remove the pencil, bring it back close to your body, close one eye, and quickly insert the pencil into the test tube. *(Do not feel for the test tube with the pencil!)* Repeat with the other eye closed.

Was it as easy to dunk the pencil with one eye closed as with both eyes open?

_____

Eye Reflexes

Both intrinsic (internal) and extrinsic (external) muscles are necessary for proper eye functioning. The *intrinsic muscles,* controlled by the autonomic nervous system, are those of the ciliary body (which alters the lens curvature in focusing) and the sphincter pupillae and dilator pupillae muscles of the iris (which control pupil size and thus regulate the amount of light entering the eye). The *extrinsic muscles* are the rectus and oblique muscles, which are attached to the eyeball exterior (see Figure 23.2). These muscles control eye movement and make it possible to keep moving objects focused on the fovea centralis. They are also responsible for **convergence,** or medial eye movements, which is essential for near vision. When convergence occurs, both eyes are directed toward the near object viewed. The extrinsic eye muscles are controlled by the somatic nervous system.

Activity 7

Demonstrating Reflex Activity of Intrinsic and Extrinsic Eye Muscles

Involuntary activity of both the intrinsic and extrinsic muscle types is brought about by reflex actions that can be observed in the following experiments.

Photopupillary Reflex

Sudden illumination of the retina by a bright light causes the pupil to constrict reflexively in direct proportion to the light intensity. This protective response prevents damage to the delicate photoreceptor cells.

Obtain a laboratory lamp or penlight. Have your laboratory partner sit with eyes closed and hands over the eyes. Turn on the light and position it so that it shines on the subject's right hand. After 1 minute, ask your partner to uncover and open the right eye. Quickly observe the pupil of that eye. What happens to the pupil?

_____

Shut off the light, and ask your partner to uncover and open the opposite eye. What are your observations of the pupil?

_____

_____

Accommodation Pupillary Reflex

Have your partner gaze for approximately 1 minute at a distant object in the lab—*not* toward the windows or another light source. Observe your partner's pupils. Then hold some printed material 15 to 25 cm (6 to 10 inches) from his or her face, and direct him or her to focus on it.

How does pupil size change as your partner focuses on the printed material?

_____

Explain the value of this reflex. _____

_____

_____

_____

Convergence Reflex

Repeat the previous experiment, this time using a pen or pencil as the close object to be focused on. Note the position of your partner's eyeballs while he or she gazes at the distant object, and then at the close object. Do they change position as the object of focus is changed?

_____ In what way? _____

_____

_____

24

Ophthalmoscopic Examination of the Eye (Optional)

The ophthalmoscope is an instrument used to examine the *fundus,* or eyeball interior, to determine visually the condition of the retina, optic disc, and internal blood vessels. Certain pathological conditions such as diabetes mellitus, arteriosclerosis, and degenerative changes of the optic nerve and retina can be detected by such an examination. The ophthalmoscope consists of a set of lenses mounted on a rotating disc (the **lens selection disc**), a light source regulated by a **rheostat control,** and a mirror that reflects the light so that the eye interior can be illuminated.

The lens selection disc is positioned in a small slit in the mirror, and the examiner views the eye interior through this slit, appropriately called the **viewing window.** The focal length of each lens is indicated in diopters preceded by a plus (+) sign if the lens is convex and by a negative (−) sign if

the lens is concave. When the zero (0) is seen in the **diopter window,** on the examiner side of the instrument, there is no lens positioned in the slit. The depth of focus for viewing the eye interior is changed by changing the lens.

The light is turned on by depressing the red **rheostat lock button** and then rotating the rheostat control in the clockwise direction. The **aperture selection dial** on the front of the instrument allows the nature of the light beam to be altered. The **filter switch,** also on the front, allows the choice of a green, unfiltered, or polarized light beam. Generally, green light allows for clearest viewing of the blood vessels in the eye interior and is most comfortable for the subject.

Once you have examined the ophthalmoscope and have become familiar with it, you are ready to conduct an eye examination.

Activity 8

Conducting an Ophthalmoscopic Examination

1. Conduct the examination in a dimly lit or darkened room with the subject comfortably seated and gazing straight ahead. To examine the right eye, sit face-to-face with the subject, hold the instrument in your right hand, and use your right eye to view the interior of the eye. You may want to steady yourself by resting your left hand on the subject's shoulder. To view the left eye, use your left eye, hold the instrument in your left hand, and steady yourself with your right hand.

2. Begin the examination with the 0 (no lens) in position. Grasp the instrument so that the lens disc may be rotated with the index finger. Holding the ophthalmoscope about 15 cm (6 inches) from the subject's eye, direct the light into the pupil at a slight angle—through the pupil edge rather than directly through its center. You will see a red circular area that is the illuminated eye interior.

3. Move in as close as possible to the subject's cornea (to within 5 cm, or 2 inches) as you continue to observe the area. Steady your instrument-holding hand on the subject's cheek if necessary. If both your eye and that of the subject are normal, the fundus can be viewed clearly without further adjustment of the ophthalmoscope. If the fundus cannot be focused, slowly rotate the lens disc counterclockwise until the fundus can be clearly seen. When the ophthalmoscope is correctly set, the fundus of the right eye should appear as in the photograph (**Figure 24.6**). (**Note:** If a positive [convex] lens is required and your eyes are normal, the subject has hyperopia. If a negative [concave] lens is necessary to view the fundus and your eyes are normal, the subject is myopic.)

When the examination is proceeding correctly, the subject can often see images of retinal vessels in his own eye that appear rather like cracked glass. If you are unable

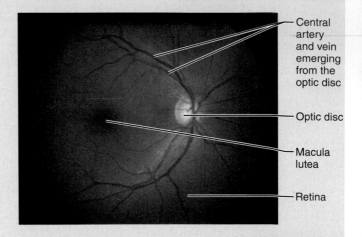

Figure 24.6 Fundus (posterior wall) of right retina.

- Central artery and vein emerging from the optic disc
- Optic disc
- Macula lutea
- Retina

to achieve a sharp focus or to see the optic disc, move medially or laterally and begin again.

4. Examine the optic disc for color, elevation, and sharpness of outline, and observe the blood vessels radiating from near its center. Locate the macula lutea, lateral to the optic disc. It is a darker area in which blood vessels are absent, and the fovea centralis appears to be a slightly lighter area in its center. The macula lutea is most easily seen when the subject looks directly into the light of the ophthalmoscope.

 Do not examine the macula lutea for longer than 1 second at a time.

5. When you have finished examining your partner's retina, shut off the ophthalmoscope. Change places with your partner (become the subject), and repeat steps 1–4.

Special Senses: Hearing and Equilibrium

Objectives

- [ ] Identify the anatomical structures of the external, middle, and internal ear on a model or appropriate diagram, and explain their functions.
- [ ] Describe the anatomy of the organ of hearing (spiral organ in the cochlea), and explain its function in sound reception.
- [ ] Discuss how one is able to localize the source of sounds.
- [ ] Define *sensorineural deafness* and *conduction deafness,* and relate these conditions to the Weber and Rinne tests.
- [ ] Describe the anatomy of the organs of equilibrium in the internal ear, and explain their relative function in maintaining equilibrium.
- [ ] State the locations and functions of endolymph and perilymph.
- [ ] Discuss the effects of acceleration on the semicircular canals.
- [ ] Define *nystagmus,* and relate this event to the balance and Barany tests.
- [ ] State the purpose of the Romberg test.
- [ ] Explain the role of vision in maintaining equilibrium.

Materials

- Three-dimensional dissectible ear model and/or chart of ear anatomy
- Otoscope (if available)
- Disposable otoscope tips (if available) and autoclave bag
- Alcohol swabs
- Compound microscope
- Prepared slides of the cochlea of the ear
- Absorbent cotton
- Pocket watch or clock that ticks
- Metric ruler
- Tuning forks (range of frequencies)

Text continues on next page. →

MasteringA&P®

For related exercise study tools, go to the Study Area of **MasteringA&P**. There you will find:

- Practice Anatomy Lab PAL
- A&PFlix **A&PFlix**
- PhysioEx **PEx**
- Practice quizzes, Histology Atlas, eText, Videos, and more!

Pre-Lab Quiz

1. Circle the correct underlined term. The ear is divided into <u>three</u> / <u>four</u> major areas.
2. The external ear is composed primarily of the _____ and the external acoustic meatus.
   - **a.** auricle
   - **b.** cochlea
   - **c.** eardrum
   - **d.** stapes
3. Circle the correct underlined term. Sound waves that enter the external acoustic meatus eventually encounter the <u>tympanic membrane</u> / <u>oval window</u>, which then vibrates at the same frequency as the sound waves hitting it.
4. Three small bones found within the middle ear are the malleus, incus, and _____.
5. The snail-like _____, found in the internal ear, contains sensory receptors for hearing.
6. Circle the correct underlined term. Today you will use an <u>ophthalmoscope</u> / <u>otoscope</u> to examine the ear.
7. The _____ test is used for comparing bone and air-conduction hearing.
   - **a.** Barany
   - **b.** Rinne
   - **c.** Weber
8. The equilibrium apparatus of the ear, the vestibule, is found in the:
   - **a.** external ear
   - **b.** internal ear
   - **c.** middle ear
9. Circle the correct underlined terms. The <u>crista ampullaris</u> / <u>macula</u> located in the semicircular <u>duct</u> / <u>vestibule</u> is essential for detecting static equilibrium.

Text continues on next page →

- Rubber mallet
- Audiometer and earphones
- Red and blue pencils
- Demonstration: Microscope focused on a slide of a crista ampullaris receptor of a semicircular canal
- Three coins of different sizes
- Rotating chair or stool
- Blackboard and chalk or whiteboard and markers

10. Nystagmus is:
    a. ability to hear only high-frequency tones
    b. ability to hear only low-frequency tones
    c. involuntary trailing of eyes in one direction, then rapid movement in the other
    d. sensation of dizziness

The ear is a complex structure containing sensory receptors for hearing and equilibrium. The ear is divided into three major areas: the *external ear,* the *middle ear,* and the *internal ear* (**Figure 25.1**). The external and middle ear structures serve the needs of the sense of hearing *only,* whereas internal ear structures function both in equilibrium and hearing.

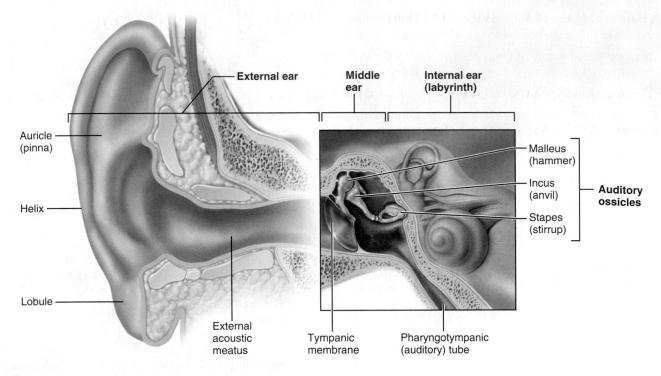

Figure 25.1 Anatomy of the ear.

Anatomy of the Ear

Gross Anatomy

Activity 1

Identifying Structures of the Ear

Obtain a dissectible ear model or chart of ear anatomy, and identify the structures summarized in (**Table 25.1**).

Because the mucosal membranes of the middle ear cavity and nasopharynx are continuous through the pharyngotympanic tube, **otitis media,** or inflammation of the middle ear, is a fairly common condition, especially among youngsters prone to sore throats. In cases where large amounts of fluid or pus accumulate in the middle ear cavity, an emergency myringotomy (lancing of the eardrum) may be necessary to relieve the pressure. Frequently, tiny ventilating tubes are put in during the procedure. ✚

The **internal ear** consists of a system of mazelike chambers and is therefore also referred to as the **labyrinth** (**Figure 25.2**). The internal ear (labyrinth) has two main divisions: the bony labyrinth and the membranous labyrinth. The **bony labyrinth** is a cavity in the temporal bone that includes the semicircular canals, the vestibule, and the cochlea. The **membranous labyrinth** is a collection of ducts and sacs including the semicircular ducts, the utricle, the saccule, and the cochlear duct. The bony labyrinth is filled with an aqueous fluid called **perilymph.** The membranous labyrinth is filled with a more viscous fluid called **endolymph.** The ducts and sacs of the

Text continues on page 386. ➡

Table 25.1   Structures of the External, Middle, and Internal Ear (Figure 25.1)

External Ear

Structure	Description	Function
Auricle (pinna)	Elastic cartilage covered with skin	Collects and directs sound waves into the external acoustic meatus
Lobule ("earlobe")	Portion of the auricle that is inferior to the external acoustic meatus	Completes the formation of the auricle
External acoustic meatus	Short, narrow canal carved into the temporal bone; lined with ceruminous glands	Transmits sound waves from the auricle to the tympanic membrane
Tympanic membrane (eardrum)	Thin membrane that separates the external ear from the middle ear	Vibrates at exactly the same frequency as the sound wave(s) hitting it and transmits vibrations to the auditory ossicles

Middle Ear	A small air-filled chamber — the tympanic cavity	Contains the auditory ossicles (malleus, incus and stapes)

Structure	Description	Function
Malleus (hammer)	Tiny bone shaped like a hammer; its "handle" is attached to the eardrum	Transmits and amplifies vibrations from the tympanic membrane to the incus
Incus (anvil)	Tiny bone shaped like an anvil that articulates with the malleus and the stapes	Transmits and amplifies vibrations from the malleus to the stapes
Stapes (stirrup)	Tiny bone shaped like a stirrup; its "base" fits into the oval window	Transmits and amplifies vibrations from the incus to the oval window
Oval window	Oval-shaped membrane located deep to the stapes.	Transmits vibrations from the stapes to the perilymph of the scala vestibuli
Pharyngotympanic (auditory) tube	A tube that connects the middle ear to the superior portion of the pharynx (throat).	Equalizes the pressure in the middle ear cavity with the external air pressure so that the tympanic membrane can vibrate properly

Internal Ear

Bony labyrinth	Membranous labyrinth (within the bony labyrinth)	Structure that contains the receptors	Function of the receptors
Cochlea	Cochlea duct	Spiral organ	Hearing
Vestibule	Utricle and saccule	Maculae	Equilibrium: static equilibrium and linear acceleration of the head
Semicircular canals	Semicircular ducts	Ampullae	Equilibrium: rotational acceleration of the head

25

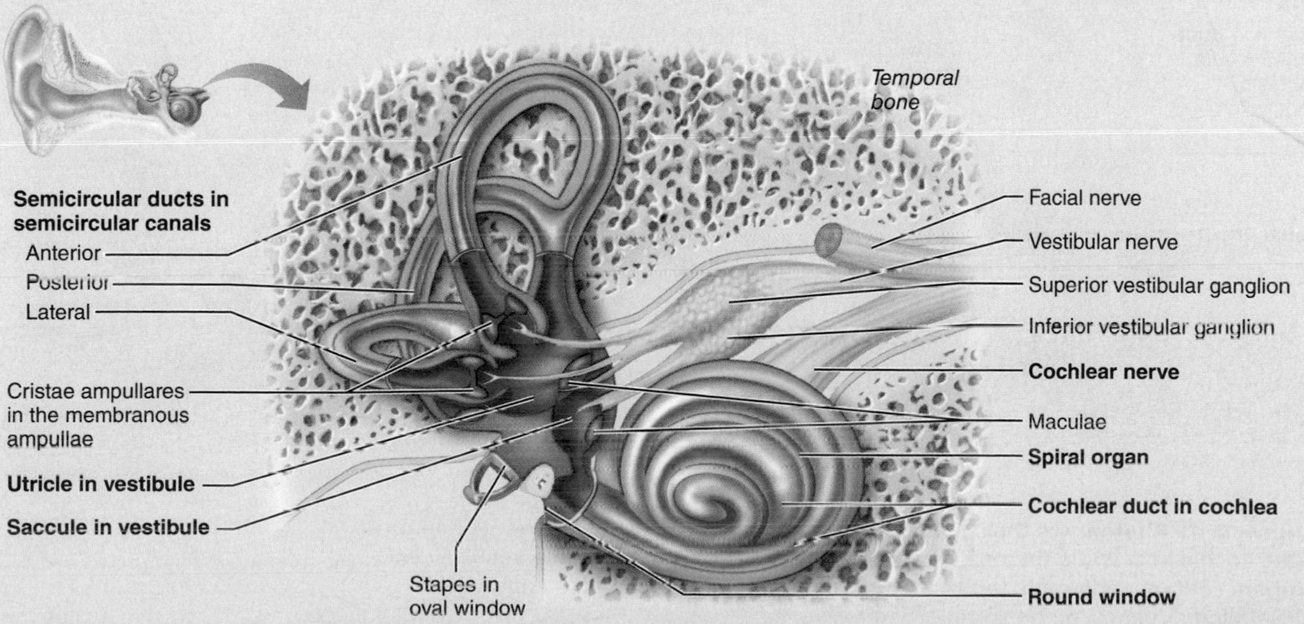

Figure 25.2 Internal ear. Right membranous labyrinth (blue) shown within the bony labyrinth (tan). The locations of sensory organs for hearing and equilibrium are shown in purple.

membranous labyrinth are suspended in the perilymph. The snail-shaped **cochlea** contains the sensory receptors for hearing. The **vestibule** and **semicircular canals** are involved with equilibrium.

WHY THIS MATTERS | **Acute Labyrinthitis**

Imagine climbing out of bed in the morning only to struggle to catch your balance because the room appears to be spinning. A visit to the doctor reveals a diagnosis of acute labyrinthitis, a sudden and severe inflammation of the membranous labyrinth of the internal ear. The structures affected include any or all of the following: the semicircular ducts, utricle, saccule, and the cochlea duct. In addition to the loss of balance and vertigo, temporary hearing loss is possible. The exact cause of labyrinthitis is difficult to determine, but a viral or bacterial respiratory infection often precedes the onset of the inflammation. ■

Activity 2

Examining the Ear with an Otoscope (Optional)

1. Obtain an otoscope and two alcohol swabs. Inspect your partner's external acoustic meatus, and then select the speculum with the largest *diameter* that will fit comfortably into his or her ear to permit full visibility. Clean the speculum thoroughly with an alcohol swab, and then attach the speculum to the battery-containing otoscope handle. Before beginning, check that the otoscope light beam is strong. If not, obtain another otoscope or new batteries. Some otoscopes come with disposable tips. Be sure to use a new tip for each ear examined. Dispose of these tips in an autoclave bag after use.

2. When you are ready to begin the examination, hold the lighted otoscope securely between your thumb and forefinger (like a pencil), and rest the little finger of the otoscope-holding hand against your partner's head. This maneuver forms a brace that allows the speculum to move as your partner moves and prevents the speculum from penetrating too deeply into the external acoustic meatus during unexpected movements.

3. Grasp the ear auricle firmly and pull it up, back, and slightly laterally. If your partner experiences pain or discomfort when the auricle is manipulated, an inflammation or infection of the external ear may be present. If this occurs, do not attempt to examine the ear canal.

4. Carefully insert the speculum of the otoscope into the external acoustic meatus in a downward and forward direction only far enough to permit examination of the tympanic membrane, or eardrum. Note its shape, color, and vascular network. The healthy tympanic membrane is pearly white. During the examination, notice whether there is any discharge or redness in the external acoustic meatus, and identify earwax.

5. After the examination, thoroughly clean the speculum with the second alcohol swab before returning the otoscope to the supply area.

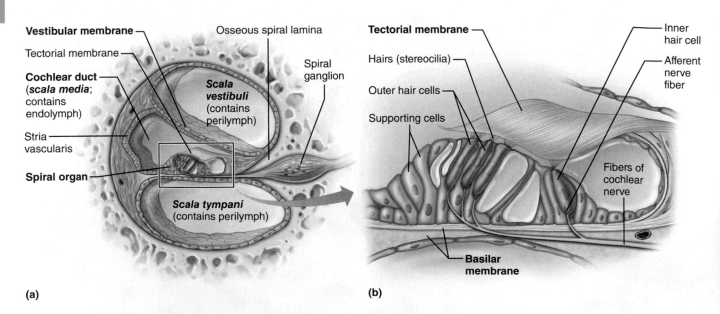

(a)

(b)

Figure 25.3 Anatomy of the cochlea. (a) Magnified cross-sectional view of one turn of the cochlea, showing the relationship of the three scalae. The scalae vestibuli and tympani contain perilymph; the cochlear duct (scala media) contains endolymph. **(b)** Detailed structure of the spiral organ.

Microscopic Anatomy of the Spiral Organ

The membranous **cochlear duct** is a soft wormlike tube about 3.8 cm long. It winds through the full two and three-quarter turns of the cochlea and separates the perilymph-containing cochlear cavity into upper and lower chambers, the **scala vestibuli** and **scala tympani** (**Figure 25.3**). The scala vestibuli terminates at the oval window, which "seats" the foot plate of the stapes located laterally in the tympanic cavity. The scala tympani is bounded by a membranous area called the **round window.** The cochlear duct is the middle **scala media.** It is filled with endolymph and supports the **spiral organ,** which contains the receptors for hearing—the sensory hair cells and nerve endings of the **cochlear nerve,** a division of the vestibulocochlear nerve (VIII).

In the spiral organ, the auditory receptors are hair cells that rest on the **basilar membrane,** which forms the floor of the cochlear duct (Figure 25.3). Their hairs are stereocilia that project into a gelatinous membrane, the **tectorial membrane,** that overlies them. The roof of the cochlear duct is called the **vestibular membrane.**

Activity 3

Examining the Microscopic Structure of the Cochlea

Obtain a compound microscope and a prepared microscope slide of the cochlea, and identify the areas shown in **Figure 25.4**.

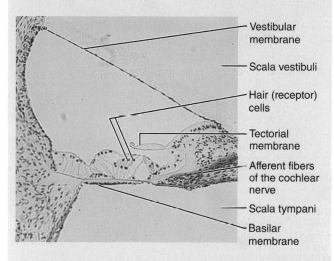

Figure 25.4 Histological image of the spiral organ (100×).

Labels: Vestibular membrane; Scala vestibuli; Hair (receptor) cells; Tectorial membrane; Afferent fibers of the cochlear nerve; Scala tympani; Basilar membrane

25

The Mechanism of Hearing

The mechanism of hearing begins as sound waves pass through the external acoustic meatus and through the middle ear into the internal ear, where the vibration eventually reaches the spiral organ, which contains the receptors for hearing.

Vibration of the stapes at the oval window creates pressure waves in the perilymph of the scala vestibuli, which are transferred to the endolymph in the cochlear duct (scala media). As the waves travel through the cochlear duct, they displace the basilar membrane and bend the hairs (stereocilia) of the hair cells. Hair cells at any given location on the basilar membrane are stimulated by sounds of a specific frequency (pitch). High-frequency waves (high-pitched sounds) displace the basilar membrane near the base where the fibers are short and stiff. Low-frequency waves (low-pitched sounds) displace the basilar membrane near the apex, where the fibers are long and flexible.

Once the hair cells are stimulated, they depolarize and begin the chain of nervous impulses that travel along the cochlear nerve to the auditory centers of the temporal lobe cortex. This series of events results in the phenomenon we call hearing (**Figure 25.5**, p. 388).

Sensorineural deafness results from damage to neural structures anywhere from the cochlear hair cells through neurons of the auditory cortex. **Presbycusis** is a type of sensorineural deafness that occurs commonly in people by the time they are in their sixties. It results from a gradual deterioration and atrophy of the spiral organ and leads to a loss in the ability to hear high tones and speech sounds.

Although presbycusis is considered to be a disability of old age, it is becoming much more common in younger people as our world grows noisier. Prolonged or excessive noise tears the cilia from hair cells, and the damage is progressive and cumulative. Each assault causes a bit more damage. Music played and listened to at deafening levels definitely contributes to the deterioration of hearing receptors. ✚

Activity 4

Conducting Laboratory Tests of Hearing

Perform the following hearing tests in a quiet area. Test both the right and left ears.

Acuity Test

Have your lab partner pack one ear with cotton and sit quietly with eyes closed. Obtain a ticking clock or pocket watch, and hold it very close to his or her *unpacked* ear. Then slowly move it away from the ear until your partner signals that the ticking is no longer audible. Record the distance in centimeters at which ticking is inaudible, and then remove the cotton from the packed ear.

Right ear: _____ Left ear: _____

Is the threshold of audibility sharp or indefinite?

_____

Text continues on next page. →

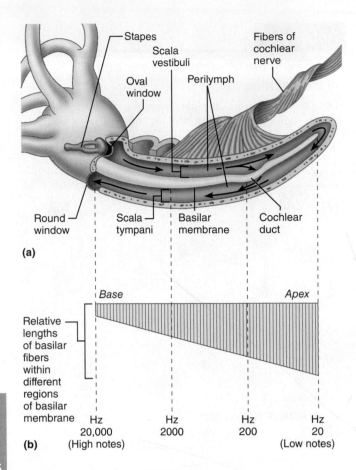

(a)

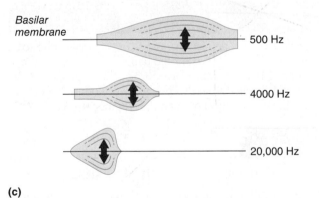

(b)

(c)

Figure 25.5 Resonance of the basilar membrane.
The cochlea is depicted as if it has been uncoiled.
(a) Perilymph movement in the cochlea following the stapes pushing the oval window. The compressional wave thus created causes the round window to bulge into the middle ear. Pressure waves set up vibrations in the basilar membrane. **(b)** Fibers span the basilar membrane. The stiffness of the fibers "tunes" specific regions to vibrate at specific frequencies. **(c)** Different frequencies of pressure waves in the spiral organ stimulate particular hair cells located in the basilar membrane.

Sound Localization

Ask your partner to close both eyes. Hold the pocket watch at an audible distance (about 15 cm) from his or her ear, and move it to various locations (front, back, sides, and above his or her head). Have your partner locate the position by pointing in each instance. Can the sound be localized equally well at all positions?

_____

If not, at what position(s) was the sound _less_ easily located?

_____

_____

The ability to localize the source of a sound depends on two factors—the difference in the loudness of the sound reaching each ear and the time of arrival of the sound at each ear. How does this information help to explain your findings?

_____

_____

Frequency Range of Hearing

Obtain three tuning forks: one with a low frequency (75 to 100 Hz [cps]), one with a frequency of approximately 1000 Hz, and one with a frequency of 4000 to 5000 Hz. Strike the lowest-frequency fork on the heel of your hand or with a rubber mallet, and hold it close to your partner's ear. Repeat with the other two forks.

Which fork was heard _most_ clearly and comfortably?

_____ Hz

Which was heard _least_ well? _____ Hz

Weber Test to Determine Conduction and Sensorineural Deafness

Strike a tuning fork and place the handle of the tuning fork medially on your partner's head (**Figure 25.6a**). Is the tone equally loud in both ears, or is it louder in one ear?

_____

If it is equally loud in both ears, your partner has equal hearing or equal loss of hearing in both ears. If sensorineural deafness is present in one ear, the tone will be heard in the unaffected ear but not in the ear with sensorineural deafness.

Conduction deafness occurs when something prevents sound waves from reaching the fluids of the internal ear. Compacted earwax, a perforated eardrum, inflammation of the middle ear (otitis media), and damage to the

25

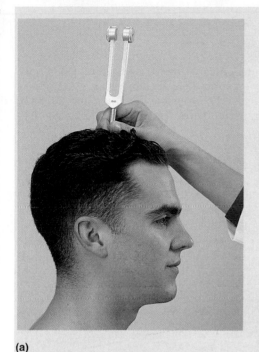
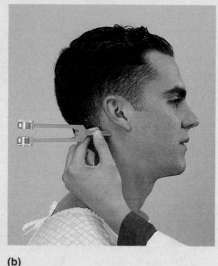
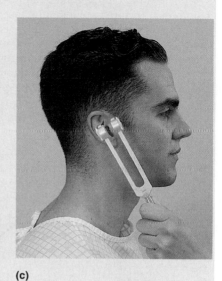

(a)   (b)   (c)

Figure 25.6 The Weber and Rinne tuning fork tests. (a) The Weber test to evaluate whether the sound remains centralized (normal) or lateralizes to one side or the other (indicative of some degree of conduction or sensorineural deafness). **(b, c)** The Rinne test to compare bone conduction and air conduction.

ossicles are all causes of conduction deafness. If conduction deafness is present, the sound will be heard more strongly in the ear in which there is a hearing loss due to sound conduction by the bone of the skull. Conduction deafness can be simulated by plugging one ear with cotton.

Rinne Test for Comparing Bone- and Air-Conduction Hearing

1. Strike the tuning fork, and place its handle on your partner's mastoid process (Figure 25.6b).

2. When your partner indicates that the sound is no longer audible, hold the still-vibrating prongs close to his or her external acoustic meatus (Figure 25.6c). If your partner hears the fork again (by air conduction) when it is moved to that position, hearing is not impaired, and you record the test result as positive (+). (Record below step 5.)

3. Repeat the test on the same ear, but this time test air-conduction hearing first.

4. After the tone is no longer heard by air conduction, hold the handle of the tuning fork on the bony mastoid process. If the subject hears the tone again by bone conduction after hearing by air conduction is lost, there is some conduction deafness and the result is recorded as negative (–).

5. Repeat the sequence for the opposite ear.

Right ear: _____ Left ear: _____

Does the subject hear better by bone or by air conduction?

_____

Audiometry

When the simple tuning fork tests reveal a problem in hearing, audiometer testing is usually prescribed to determine the precise nature of the hearing deficit. An *audiometer* is an instrument (specifically, an electronic oscillator with earphones) used to determine hearing acuity by exposing each ear to sound stimuli of differing *frequencies* and *intensities*. The hearing range of human beings during youth is from 20 to 20,000 Hz, but hearing acuity declines with age, with reception for the high-frequency sounds lost first. Most of us tend to be fairly unconcerned until we begin to have problems hearing sounds in the range of 125 to 8000 Hz, the normal frequency range of speech.

The basic procedure of audiometry is to initially deliver tones of different frequencies to one ear of the subject at an intensity (loudness) of 0 decibels (dB). Zero decibels is not the complete absence of sound, but rather the softest sound intensity that can be heard by a person of normal hearing at each frequency. If the subject cannot hear a particular frequency stimulus of 0 dB, the hearing threshold level control is adjusted until the subject reports that he or she can hear the tone. The number of decibels of intensity required above 0 dB is recorded as the hearing loss. For example, if the subject cannot hear a particular frequency tone until it is delivered at 30 dB intensity, then he or she has a hearing loss of 30 dB for that frequency.

25

Activity 5

Audiometry Testing

1. Obtain an audiometer and earphones, and a red and a blue pencil. Before beginning the tests, examine the audiometer to identify the two tone controls: one to regulate frequency and a second to regulate the intensity (loudness) of the sound stimulus. Identify the two output control switches that regulate the delivery of sound to one ear or the other (*red* to the right ear, *blue* to the left ear). Also find the *hearing threshold level control,* which is calibrated to deliver a basal tone of 0 dB to the subject's ears.

2. Place the earphones on the subject's head so that the red cord or ear-cushion is over the right ear and the blue cord or ear-cushion is over the left ear. Instruct the subject to raise one hand when he or she hears a tone.

3. Set the frequency control at 125 Hz and the intensity control at 0 dB. Press the red output switch to deliver

a tone to the subject's right ear. If the subject does not respond, raise the sound intensity slowly by rotating the hearing level control counterclockwise until the subject reports (by raising a hand) that a tone is heard. Repeat this procedure for frequencies of 250, 500, 1000, 2000, 4000, and 8000.

4. Record the results in the grid (above) for frequency versus hearing loss by marking a small red circle on the grid at each frequency-dB junction at which a tone was heard. Then connect the circles with a red line to produce a hearing acuity graph for the right ear.

5. Repeat steps 3 and 4 for the left (blue) ear, and record the results with blue circles and connecting lines on the grid.

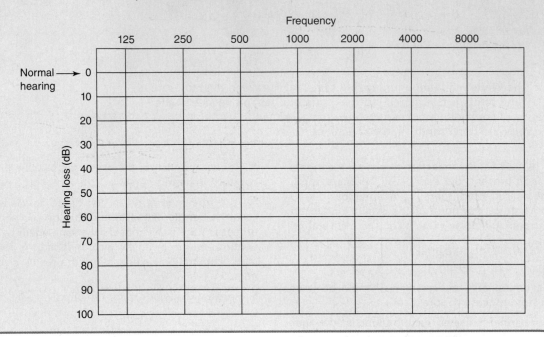

Microscopic Anatomy of the Equilibrium Apparatus and Mechanisms of Equilibrium

The equilibrium receptors of the internal ear are collectively called the **vestibular apparatus** and are found in the vestibule and semicircular canals of the bony labyrinth. The vestibule contains the saclike **utricle** and **saccule,** and the semicircular chambers contain membranous **semicircular ducts.** These membranes are filled with endolymph and contain receptor cells that are activated by the bending of the hairs on their hair cells.

Semicircular Canals

The semicircular canals monitor rotational acceleration of the head. This process is called **dynamic equilibrium.** The canals are 1.2 cm in circumference and are oriented in three planes—horizontal, frontal, and sagittal. At the base of each semicircular duct is an enlarged region, the **ampulla.** Within

each ampulla is a receptor region called a **crista ampullaris,** which consists of a tuft of hair cells covered with a gelatinous cap, or **ampullary cupula** (**Figure 25.7**).

The cristae respond to changes in the velocity of rotational head movements. During acceleration, as when you begin to twirl around, the endolymph in the canal lags behind the head movement because inertia causes the ampullary cupula to push—like a swinging door—in the opposite direction. The head movement depolarizes the hair cells, and as a result, impulse transmission is enhanced in the vestibular division of the eighth cranial nerve to the brain (Figure 25.7c). If the body continues to rotate at a constant rate, the endolymph eventually comes to rest and moves at the same speed as the body. The ampullary cupula returns to its upright position, hair cells are no longer stimulated, and you lose the sensation of spinning. When rotational movement stops suddenly, the endolymph keeps on going in the direction of head movement. This pushes the ampullary cupula in the *same* direction as the previous head movement and hyperpolarizes

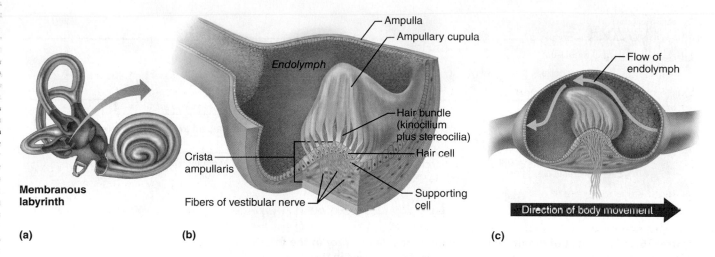

Membranous labyrinth

(a)

Endolymph

Crista ampullaris

Fibers of vestibular nerve

(b)

Ampulla

Ampullary cupula

Hair bundle (kinocilium plus stereocilia)

Hair cell

Supporting cell

Flow of endolymph

Direction of body movement

(c)

Figure 25.7 Structure and function of the crista ampullaris. (a) Arranged in the three spatial planes, the semicircular ducts in the semicircular canals each have a swelling called an ampulla at their base. **(b)** Each ampulla contains a crista ampullaris, a receptor that is essentially a cluster of hair cells with hairs projecting into a gelatinous cap called the ampullary cupula. **(c)** Movement of the cupula during rotational acceleration of the head.

the hair cells; as a result, fewer impulses are transmitted to the brain. This tells the brain that you have stopped moving and accounts for the reversed motion sensation you feel when you stop twirling suddenly.

Activity 6

Examining the Microscopic Structure of the Crista Ampullaris

Go to the demonstration area and examine the slide of a crista ampullaris. Identify the areas depicted in the photomicrograph (**Figure 25.8**) and labeled diagram (Figure 25.7b).

Ampullary cupula

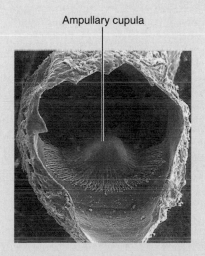

Figure 25.8 Scanning electron micrograph of a crista ampullaris (14×).

Maculae

Maculae in the membranous utricle and saccule contain another set of **hair cells,** receptors that in this case monitor head position and acceleration in a straight line. This monitoring process is called **static equilibrium.** The maculae respond to gravitational pull, thus providing information on which way is up or down as well as changes in linear speed. The hair cells in each macula have **stereocilia** plus one **kinocilium** that are embedded in the **otolith membrane,** a gelatinous material containing small grains of calcium carbonate called **otolith.** When the head moves, the otoliths move in response to variations in gravitational pull. As they deflect different hair cells, they trigger hyperpolarization or depolarization of the hair cells and modify the rate of impulse transmission along the vestibular nerve (**Figure 25.9**, p. 392).

Although the receptors of the semicircular canals and the vestibule are responsible for dynamic and static equilibrium respectively, they rarely act independently. Complex interaction of many of the receptors is the rule. Processing is also complex and involves the brain stem and cerebellum as well as input from proprioceptors and the eyes.

Activity 7

Conducting Laboratory Tests on Equilibrium

The function of the semicircular canals and vestibule are not routinely tested in the laboratory, but the following simple tests illustrate normal equilibrium apparatus function as well as some of the complex processing interactions.

In the balance test and the Barany test, you will look for **nystagmus,** which is the involuntary rolling of the eyes in any direction or the trailing of the eyes slowly in

Text continues on next page. →

25

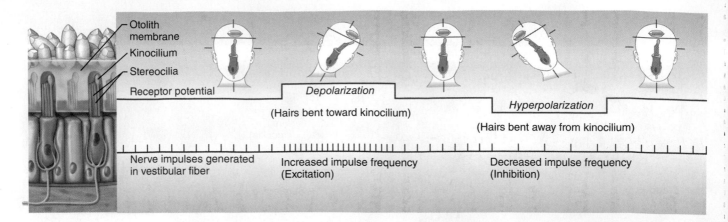

Figure 25.9 The effect of gravitational pull on a macula receptor in the utricle.
When movement of the otolith membrane bends the hair cells in the direction of
the kinocilium, the hair cells depolarize, exciting the nerve fibers, which generates
action potentials more rapidly. When the hairs are bent in the direction away from
the kinocilium, the hair cells become hyperpolarized, inhibiting the nerve fibers and
decreasing the action potential rate (i.e., below the resting rate of discharge).

one direction (slow phase), followed by their rapid movement in the opposite direction (rapid phase). During rotation, the slow drift of the eyes is related to the backflow
of endolymph in the semicircular ducts. Nystagmus is
normal during and after rotation; abnormal otherwise.

Nystagmus is often accompanied by **vertigo**—a sensation of dizziness and rotational movement when such
movement is not occurring or has ceased.

Balance Tests

1. Have your partner walk a straight line, placing one foot
directly in front of the other.

Is he or she able to walk without significant wobbling

from side to side? _____

Did he or she experience any dizziness? _____

The ability to walk with balance and without dizziness, unless subject to rotational forces, indicates normal function
of the equilibrium apparatus.

Was nystagmus present? _____

2. Place three coins of different sizes on the floor. Ask your
lab partner to pick up the coins, and carefully observe his
or her muscle activity and coordination.

Did your lab partner have any difficulty locating and pick-

ing up the coins? _____

Describe your observations and your lab partner's observations during the test.

_____

_____

What kinds of complex interactions involving balance and
coordination must occur for a person to move fluidly during this test?

_____

_____

3. If a person has a depressed nervous system, mental
concentration may result in a loss of balance. Ask your
lab partner to stand up and count backward from ten as
rapidly as possible.

Did your lab partner lose balance? _____

*Barany Test (Induction
of Nystagmus and Vertigo)*

This experiment evaluates the semicircular canals and
should be conducted as a group effort to protect the test
subject(s) from possible injury.

⚠ Read the following precautionary notes before
beginning:

• The subject(s) chosen should not be easily inclined to
dizziness during rotational or turning movements.

• Rotation should be stopped immediately if the subject
feels nauseated.

• Because the subject(s) will experience vertigo and
loss of balance as a result of the rotation, several classmates should be prepared to catch, hold, or support the
subject(s) as necessary until the symptoms pass.

1. Instruct the subject to sit on a rotating chair or stool,
and to hold on to the arms or seat of the chair, feet on
stool rungs. The subject's head should be tilted forward approximately 30 degrees (chin almost touching
the chest). The horizontal (lateral) semicircular canal is
stimulated when the head is in this position. The subject's eyes are to remain *open* during the test.

2. Four classmates should position themselves so that the subject is surrounded on all sides. The classmate posterior to the subject will rotate the chair.

3. Rotate the chair to the subject's right approximately 10 revolutions in 10 seconds, then suddenly stop the rotation.

4. Immediately note the direction of the subject's resultant nystagmus; and ask him or her to describe the feelings of movement, indicating speed and direction sensation. Record this information below.

_____

_____

_____

If the semicircular canals are operating normally, the subject will experience a sensation that the stool is still rotating immediately after it has stopped and *will* demonstrate nystagmus.

When the subject is rotated to the right, the ampullary cupula will be bent to the left, causing nystagmus during rotation in which the eyes initially move slowly to the left and then quickly to the right. Nystagmus will continue until the ampullary cupula has returned to its initial position. Then, when rotation is stopped abruptly, the ampullary cupula will be bent to the right, producing nystagmus with its slow phase to the right and its rapid phase to the left. In many subjects, this will be accompanied by a feeling of vertigo and a tendency to fall to the right.

Romberg Test

The Romberg test determines the integrity of the dorsal white column of the spinal cord, which transmits impulses to the brain from the proprioceptors involved with posture.

1. Have your partner stand with his or her back to the blackboard or whiteboard.

2. Draw one line parallel to each side of your partner's body. He or she should stand erect, with eyes open and staring straight ahead for 2 minutes while you observe any movements. Did you see any gross swaying movements?

_____

3. Repeat the test. This time the subject's eyes should be closed. Note and record the degree of side-to-side movement.

_____

4. Repeat the test with the subject's eyes first open and then closed. This time, however, the subject should be positioned with his or her left shoulder toward, but not touching, the board so that you may observe and record the degree of front-to-back swaying.

_____

Do you think the equilibrium apparatus of the internal ear was operating equally well in all these tests?

_____

The proprioceptors? _____

Why was the observed degree of swaying greater when the eyes were closed?

_____

_____

What conclusions can you draw regarding the factors necessary for maintaining body equilibrium and balance?

_____

_____

Role of Vision in Maintaining Equilibrium

To further demonstrate the role of vision in maintaining equilibrium, perform the following experiment. (Ask your lab partner to record observations and act as a "spotter.") Stand erect, with your eyes open. Raise your left foot approximately 30 cm off the floor, and hold it there for 1 minute.

Record the observations: _____

_____

_____

Rest for 1 or 2 minutes; and then repeat the experiment with the same foot raised but with your eyes closed. Record the observations:

_____

_____

25

26

Special Senses: Olfaction and Taste

Objectives

☐ State the location and cellular composition of the olfactory epithelium.

☐ Describe the structure of olfactory sensory neurons, and state their function.

☐ Discuss the locations and cellular composition of taste buds.

☐ Describe the structure of gustatory epithelial cells, and state their function.

☐ Identify the cranial nerves that carry the sensations of olfaction and taste.

☐ Name five basic qualities of taste sensation, and list the chemical substances that elicit them.

☐ Explain the interdependence between the senses of smell and taste.

☐ Name two factors other than olfaction that influence taste appreciation of foods.

☐ Define *olfactory adaptation*.

Materials

- Prepared slides: nasal olfactory epithelium (l.s.); the tongue showing taste buds (x.s.)
- Compound microscope
- Paper towels
- Packets of granulated sugar
- Disposable autoclave bag
- Paper plates
- Equal-sized food cubes of apple, raw potato, dried prunes, banana, and raw carrot (These prepared foods should be in an opaque container; a foil-lined egg carton would work well.)
- Toothpicks
- Disposable gloves
- Cotton-tipped swabs
- Paper cups
- Flask of distilled or tap water

Text continues on next page. →

MasteringA&P®

For related exercise study tools, go to the Study Area of **MasteringA&P**. There you will find:

- Practice Anatomy Lab PAL
- A&PFlix *A&PFlix*
- PhysioEx PEx
- Practice quizzes, Histology Atlas, eText, Videos, and more!

Pre-Lab Quiz

1. Circle True or False. Receptors for olfaction and taste are classified as chemoreceptors because they respond to dissolved chemicals.

2. The organ of smell is the _____, located in the roof of the nasal cavity.
   - **a.** nares
   - **b.** nostrils
   - **c.** olfactory epithelium
   - **d.** olfactory nerve

3. Circle the correct underlined term. Olfactory receptors are <u>bipolar</u> / <u>unipolar</u> sensory neurons whose olfactory cilia extend outward from the epithelium.

4. Most taste buds are located in _____, peglike projections of the tongue mucosa.
   - **a.** cilia
   - **b.** concha
   - **c.** papillae
   - **d.** supporting cells

5. Circle the correct underlined term. Vallate papillae are arranged in a V formation on the <u>anterior</u> / <u>posterior</u> surface of the tongue.

6. Circle the correct underlined term. Most taste buds are made of <u>two</u> / <u>three</u> types of modified epithelial cells.

7. There are five basic taste sensations. Name one. _____

8. Circle True or False. Taste buds typically respond optimally to one of the five basic taste sensations.

9. Circle True or False. Texture, temperature, and smell have little or no effect on the sensation of taste.

10. You will use absorbent cotton and oil of wintergreen, peppermint, or cloves to test for olfactory:
    - **a.** accommodation
    - **b.** adaptation
    - **c.** identification
    - **d.** recognition

- Prepared vials of oil of cloves, oil of peppermint, and oil of wintergreen or corresponding flavorings found in the condiment section of a supermarket
- Chipped ice
- Five numbered vials containing common household substances with strong odors (herbs, spices, etc.)
- Nose clips
- Absorbent cotton

The receptors for olfaction and taste are classified as **chemoreceptors** because they respond to chemicals in solution. Although five relatively specific types of taste receptors have been identified, the olfactory receptors are considered sensitive to a much wider range of chemical sensations.

Olfactory Epithelium and Olfaction

A pseudostratified epithelium called the **olfactory epithelium** is the organ of smell. It occupies an area lining the roof of the nasal cavity (**Figure 26.1a**). Since most of the air entering the nasal cavity enters the respiratory passages below, the superiorly located nasal epithelium is in a rather poor position for performing its function. This is why sniffing, which brings more air into contact with the receptors, increases your ability to detect odors.

Three cell types are found within the olfactory epithelium:

- **Olfactory sensory neurons:** Specialized receptor cells that are bipolar neurons with nonmotile olfactory cilia.

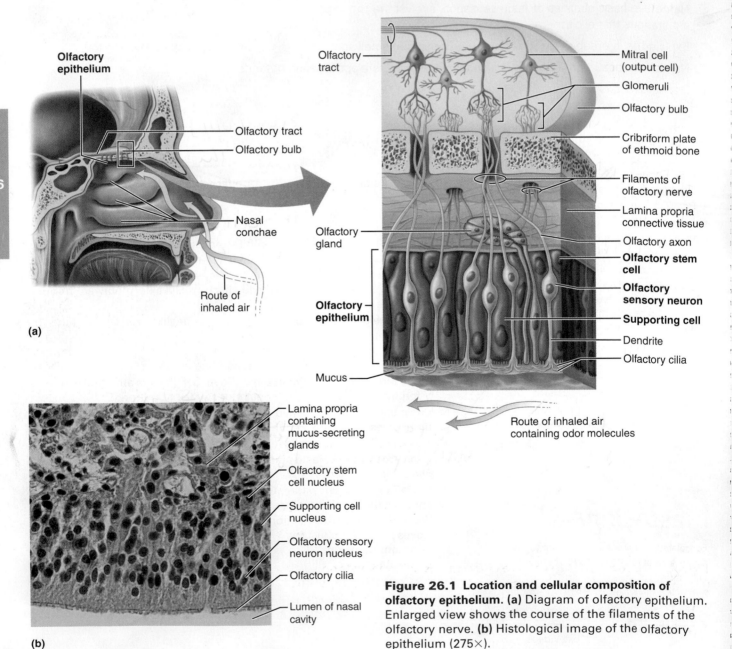

Figure 26.1 Location and cellular composition of olfactory epithelium. (a) Diagram of olfactory epithelium. Enlarged view shows the course of the filaments of the olfactory nerve. **(b)** Histological image of the olfactory epithelium (275×).

(a)

(b)

- **Supporting cells:** Columnar cells that surround and support the olfactory sensory neurons.
- **Olfactory stem cells:** Located near the basal surface of the epithelium, they divide to form new olfactory sensory neurons.

The axons of the olfactory sensory neurons form small fascicles called the *filaments of the olfactory nerve* (cranial nerve I), which penetrate the cribriform foramina and synapse in the olfactory bulbs.

Taste Buds and Taste

The **taste buds,** containing specific receptors for the sense of taste, are widely but not uniformly distributed in the oral cavity. Most are located in **papillae,** peglike projections of the mucosa, on the dorsal surface of the tongue. A few are found on the soft palate, epiglottis, pharynx, and inner surface of the cheeks.

Taste buds are located in the side walls of the large **vallate papillae** (arranged in a V formation on the posterior surface of the tongue); in the side walls of the **foliate papillae;** and on the tops of the more numerous, mushroom-shaped **fungiform papillae** (**Figure 26.2**).

Each taste bud consists largely of an arrangement of two types of modified epithelial cells:

- **Gustatory epithelial cells:** The receptors for taste; they have long microvilli called **gustatory hairs** that project through the epithelial surface through a **taste pore.**

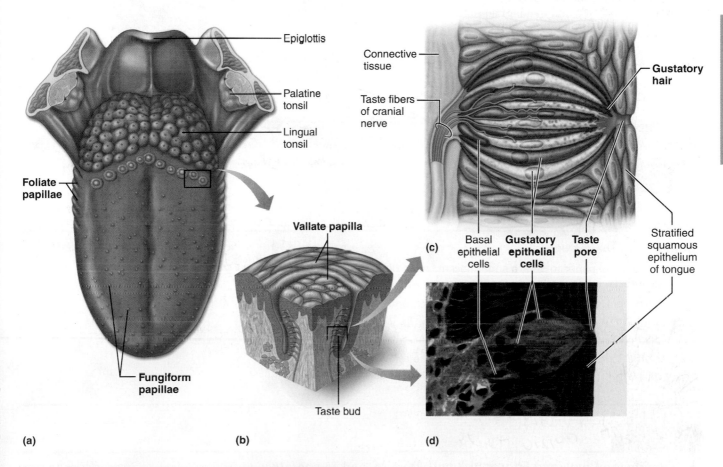

Figure 26.2 Location and structure of taste buds. (a) Taste buds on the tongue are associated with papillae, projections of the tongue mucosa. **(b)** A sectioned vallate papilla shows the position of the taste buds in its lateral walls. **(c)** An enlarged view of a taste bud. **(d)** Photomicrograph of a taste bud (445×).

• **Basal epithelial cells:** Precursor cells that divide to replace the gustatory epithelial cells.

When the gustatory hairs contact food molecules dissolved in saliva, the gustatory epithelial cells depolarize. The sensory (afferent) neurons that innervate the taste buds are located in three cranial nerves: the *facial nerve (VII)* serves the anterior two-thirds of the tongue; the *glossopharyngeal nerve (IX)* serves the posterior third of the tongue; and the *vagus nerve (X)* carries a few fibers from the pharyngeal region.

When taste is tested with pure chemical compounds, most taste sensations can be grouped into one of five basic qualities—sweet, sour, bitter, salty, or umami (oo-mom'ē; "delicious"). Although all taste buds are believed to respond in some degree to all five classes of chemical stimuli, each type responds optimally to only one.

The *sweet* receptors respond to a number of seemingly unrelated compounds such as sugars (fructose, sucrose, glucose), saccharine, some lead salts, and some amino acids. *Sour* receptors are activated by acids, specifically their hydrogen ions (H^+) in solution. *Salty* taste seems to be due to an influx of metal ions, particularly Na^+, while *umami* is elicited by the amino acids glutamate and aspartate, which are responsible for the "meat taste" of beef and the flavor of monosodium glutamate (MSG). *Bitter* taste is elicited by alkaloids (e.g., caffeine and quinine) and other nonalkaloid substances such as aspirin.

Activity 2

Microscopic Examination of Taste Buds

Obtain a microscope and a prepared slide of a tongue cross section. Locate the taste buds on the tongue papillae (use Figure 26.2b as a guide). Make a detailed study of one taste bud. Identify the taste pore and gustatory hairs if observed. Compare your observations to the photomicrograph (**Figure 26.3**).

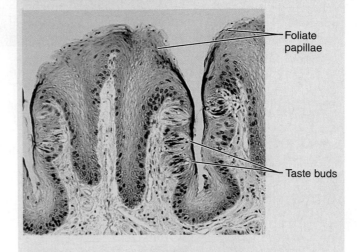

Foliate papillae

Taste buds

Figure 26.3 Taste buds on the lateral aspects of foliate papillae of the tongue (140×).

Laboratory Experiments

 Notify instructor of any food or scent allergies or restrictions before beginning experiments.

Activity 3

Stimulating Taste Buds

1. Obtain several paper towels, a sugar packet, and a disposable autoclave bag, and bring them to your bench.

2. With a paper towel, dry the dorsal surface of your tongue.

 Immediately dispose of the paper towel in the autoclave bag.

3. Tear off a corner of the sugar packet, and shake a few sugar crystals on your dried tongue. Do *not* close your mouth.

How long does it take to taste the sugar? _____ sec

Why couldn't you taste the sugar immediately?

_____

Activity 4

Examining the Combined Effects of Smell, Texture, and Temperature on Taste

Effects of Smell and Texture

1. Ask the subject to sit with eyes closed and to pinch his or her nostrils shut.

2. Using a paper plate, obtain samples of the food items provided by your laboratory instructor. At no time should the subject be allowed to see the foods being tested. Wear disposable gloves, and use toothpicks to handle food.

3. For each test, place a cube of food in the subject's mouth, and ask him or her to identify the food by using the following sequence of activities:

- First, manipulate the food with the tongue.
- Second, chew the food.
- Third, if the subject does not make a positive identification with the first two techniques and the taste sense, ask the subject to release the pinched nostrils and to continue chewing with the nostrils open to determine whether he or she can make a positive identification.

In the **Activity 4 chart**, record the type of food, and then put a check mark in the appropriate column for the result.

Was the sense of smell equally important in all cases?

_____

_____

Where did it seem to be important, and why?

_____

_____

Discard gloves in autoclave bag.

Effect of Olfactory Stimulation

What is commonly referred to as taste depends heavily on stimulation of the olfactory receptors, particularly in the case of strongly odoriferous substances. The following experiment should illustrate this fact.

1. Obtain vials of oil of wintergreen, peppermint, and cloves, paper cup, flask of water, paper towels, and some fresh cotton-tipped swabs. Ask the subject to sit so that he or she cannot see which vial is being used, and to dry the tongue and pinch the nostrils shut.

2. Use a cotton swab to apply a drop of one of the oils to the subject's tongue. Can he or she distinguish the flavor?

_____

 Put the used swab in the autoclave bag. *Do not redip the swab into the oil.*

3. Have the subject open the nostrils, and record the change in sensation he or she reports.

_____

4. Have the subject rinse the mouth well and dry the tongue.

5. Prepare two swabs, each with one of the two remaining oils.

6. Hold one swab under the subject's open nostrils, while touching the second swab to the tongue.

Record the reported sensations. _____

_____

_____

! **7.** Dispose of the used swabs and paper towels in the autoclave bag before continuing.

Which sense, taste or smell, appears to be more important in the proper identification of a strongly flavored volatile substance?

_____

Effect of Temperature

In addition to the effect that olfaction and food texture have in determining our taste sensations, the temperature of foods also helps determine if the food is appreciated or even tasted. To illustrate this, have your partner hold some chipped ice on the tongue for approximately a minute and then close the eyes. Immediately place any of the foods previously identified in the mouth, and ask for an identification.

Results? _____

_____

_____

26

Activity 4: Identification by Texture and Smell				
Food tested	Texture only	Chewing with nostrils pinched	Chewing with nostrils open	Identification not made

Activity 5

Assessing the Importance of Taste and Olfaction in Odor Identification

1. Go to the designated testing area. Close your nostrils with a nose clip, and breathe through your mouth. Breathing through your mouth only, attempt to identify the odors of common substances in the numbered vials at the testing area. Do not look at the substance in the container. Record your responses on the chart above.

2. Remove the nose clips, and repeat the tests using your nose to sniff the odors. Record your responses in the **Activity 5 chart.**

3. Record any other observations you make as you conduct the tests.

4. Which method gave the best identification results?

_____

What can you conclude about the effectiveness of the senses of taste and olfaction in identifying odors?

_____

Activity 5: Identification by Mouth and Nasal Inhalation			
Vial number	**Identification with nose clips**	**Identification without nose clips**	**Other observations**
1			
2			
3			
4			
5			

Activity 6

Demonstrating Olfactory Adaptation

Obtain some absorbent cotton and two of the following oils (oil of wintergreen, peppermint, or cloves). Place several drops of oil on the absorbent cotton. Press one nostril shut.

Hold the cotton under the open nostril, and exhale through the mouth. Record the time required for the odor to disappear (for olfactory adaptation to occur).

_____ sec

Repeat the procedure with the other nostril.

_____ sec

Immediately test another oil with the nostril that has just experienced olfactory adaptation. What are the results?

_____

What conclusions can you draw? _____

_____

REVIEW SHEET
Special Senses: Olfaction and Taste

Name _____ Lab Time/Date _____

Olfactory Epithelium and Olfaction

1. Describe the location and cellular composition of the olfactory epithelium. _____

   _____

   _____

2. How and why does sniffing increase your ability to detect an odor? _____

   _____

Taste Buds and Taste

3. Name five sites where receptors for taste are found, and circle the predominant site.

   _____, _____, _____,

   _____, and _____

4. Describe the cellular makeup and arrangement of a taste bud. (Use a diagram, if helpful.) _____

   _____

5. Taste and smell receptors are both classified as _____, because they both

   respond to _____

6. Why is it impossible to taste substances if your tongue is dry? _____

   _____

7. The basic taste sensations are mediated by specific chemical substances or groups. Name them for the following taste modalities.

   salt: _____   sour: _____   umami: _____

   bitter: _____   sweet: _____

Laboratory Experiments

8. Name three factors that influence our enjoyment of foods. Substantiate each choice with an example from the laboratory experience.

1. _____ Substantiation: _____

_____

2. _____ Substantiation: _____

_____

3. _____ Substantiation: _____

_____

Which of the factors chosen is most important? _____ Substantiate your choice with an example from

everyday life. _____

_____

Expand on your explanation and choices by explaining why a cold, greasy hamburger is unappetizing to most people.

_____

_____

9. How palatable is food when you have a cold? _____ Explain your answer. _____

_____

10. In your opinion, is olfactory adaptation desirable? _____ Explain your answer.

_____

_____

27 Functional Anatomy of the Endocrine Glands

Objectives

☐ Identify the major endocrine glands of the body using an appropriate image.

☐ List the major hormones and discuss the target and general function of each.

☐ Explain how hormones contribute to body homeostasis using appropriate examples.

☐ Discuss some mechanisms that stimulate release of hormones from endocrine glands.

☐ Describe the structural and functional relationship between the hypothalamus and the pituitary gland.

☐ Correctly identify the histology of the thyroid, parathyroid, pancreas, anterior and posterior pituitary, adrenal cortex, and adrenal medulla by microscopic inspection or in an image.

☐ Name and identify the specialized hormone-secreting cells in the above tissues.

☐ Describe the pathology of hypersecretion and hyposecretion of several of the hormones studied.

Materials

- Human torso model
- Anatomical chart of the human endocrine system
- Compound microscope
- Prepared slides of the anterior pituitary and pancreas (with differential staining), posterior pituitary, thyroid gland, parathyroid glands, and adrenal gland

 For instructions on animal dissections, see the dissection exercises (starting on page 705) in the cat and fetal pig editions of this manual.

MasteringA&P®

For related exercise study tools, go to the Study Area of **MasteringA&P**. There you will find:

- Practice Anatomy Lab PAL
- PhysioEx PEx
- A&PFlix **A&PFlix**
- Practice quizzes, Histology Atlas, eText, Videos, and more!

Pre-Lab Quiz

1. Define *hormone*. _____
2. Circle the correct underlined term. An <u>endocrine</u> / <u>exocrine</u> gland is a ductless gland that empties its hormone into the extracellular fluid.
3. The pituitary gland, also known as the _____, is located in the sella turcica of the sphenoid bone.
   a. hypophysis   b. hypothalamus   c. thalamus
4. Circle True or False. The anterior pituitary gland is sometimes referred to as the master endocrine gland because it controls the activity of many other endocrine glands.
5. The _____ gland is composed of two lobes and located in the throat, just inferior to the larynx.
   a. pancreas   c. thymus
   b. posterior pituitary   d. thyroid
6. The pancreas produces two hormones that are responsible for regulating blood sugar levels. Name the hormone that increases blood glucose levels. _____
7. Circle True or False. The gonads are considered to be both endocrine and exocrine glands.
8. This gland is rather large in an infant, begins to atrophy at puberty, and is relatively inconspicuous by old age. It produces hormones that direct the maturation of T cells. It is the _____ gland.
   a. pineal   c. thymus
   b. testes   d. thyroid
9. Circle the correct underlined term. <u>Pancreatic islets</u> / <u>Acinar cells</u> form the endocrine portion of the pancreas.
10. The outer cortex of the adrenal gland is divided into three areas. Which one produces aldosterone?
   a. zona fasciculata
   b. zona glomerulosa
   c. zona reticularis

The **endocrine system** is the second major control system of the body. Acting with the nervous system, it helps coordinate and integrate the activity of the body. The nervous system uses electrochemical impulses to bring about rapid control, whereas the more slowly acting endocrine system uses chemical messengers, or **hormones.**

The term *hormone* comes from a Greek word meaning "to arouse." The body's hormones, which are steroids or amino acid–based molecules, arouse the body's tissues and cells by stimulating changes in their metabolic activity. These changes lead to growth and development and to the physiological homeostasis of many body systems. Although hormones travel through the blood, a given hormone affects only a specific organ or organs. Cells within an organ that respond to a particular hormone are referred to as the **target cells** (also **target**) of that hormone. The ability of the target to respond depends on the ability of the hormone to bind with specific cellular receptors.

Although the function of most hormone-producing glands is purely endocrine, the function of others (the pancreas and gonads) is mixed—both endocrine and exocrine. The endocrine glands release their hormones directly into the extracellular fluid, from which the hormones enter blood or lymph.

Activity 1

Identifying the Endocrine Organs

Locate the endocrine organs in **Figure 27.1**. Also locate these organs on the anatomical charts or torso model. As you locate the organs, read through Tables 27.1–27.4.

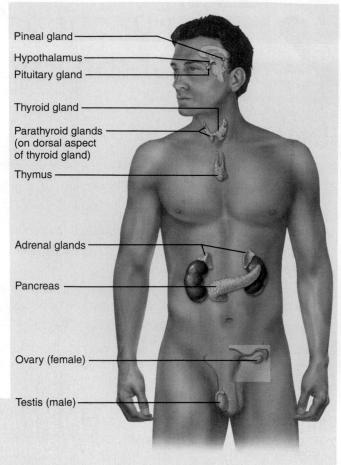

Figure 27.1 Human endocrine organs.

Labels: Pineal gland, Hypothalamus, Pituitary gland, Thyroid gland, Parathyroid glands (on dorsal aspect of thyroid gland), Thymus, Adrenal glands, Pancreas, Ovary (female), Testis (male)

Endocrine Glands

Pituitary Gland (Hypophysis)

The **pituitary gland,** or **hypophysis,** is located in the sella turcica of the sphenoid bone. It consists largely of two functional *lobes,* the **adenohypophysis,** or **anterior pituitary,** and the **neurohypophysis,** consisting of the **posterior pituitary** and the **infundibulum**—the stalk that attaches the pituitary gland to the hypothalamus (**Figure 27.2**).

The anterior pituitary produces and secretes a number of hormones, four of which are **tropic hormones.** The target organ of a tropic hormone is another endocrine gland.

Because the anterior pituitary controls the activity of many other endocrine glands, it is sometimes called the *master endocrine gland.* However, because *releasing* or *inhibiting hormones* from neurons of the ventral hypothalamus control anterior pituitary cells, the hypothalamus supersedes the anterior pituitary as the major controller of endocrine glands.

The ventral hypothalamic hormones control production and secretion of the anterior pituitary hormones. The hypothalamic hormones reach the cells of the anterior pituitary through the **hypophyseal portal system** (Figure 27.2), a complex vascular arrangement of two capillary beds that are connected by the hypophyseal portal veins.

The posterior pituitary is not an endocrine gland because it does not synthesize the hormones it releases. Instead, it acts as a storage area for two *neurohormones* transported to it via the axons of neurons in the paraventricular and supraoptic nuclei of the hypothalamus. Refer to **Table 27.1,** on pp. 409–410, which summarizes the hormones released by the pituitary gland.

Pineal Gland

The *pineal gland* is a small cone-shaped gland located in the roof of the third ventricle of the brain. Its major endocrine product is **melatonin,** which exhibits a diurnal (daily) cycle. It peaks at night, making us drowsy, and is lowest around noon. Recent evidence suggests that melatonin has anti-aging properties. Melatonin appears to play a role in the production of antioxidants.

Thyroid Gland

The *thyroid gland* is composed of two lobes joined by a central mass, or isthmus. It is located in the throat, just inferior to the larynx.

Parathyroid Glands

The *parathyroid glands* are found embedded in the posterior surface of the thyroid gland. Typically, there are two small

27

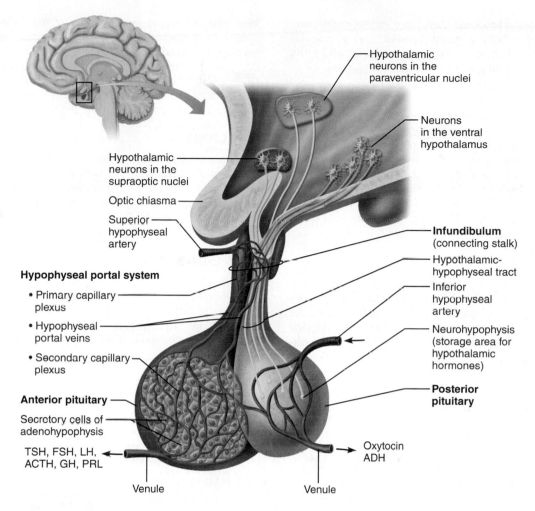

Figure 27.2 Hypothalamus and pituitary gland. Neural and vascular relationships between the hypothalamus and the anterior and posterior lobes of the pituitary are depicted.

Table 27.1A	Pituitary Gland Hormones (Figure 27.2)		
Hormone	**Stimulus for release**	**Target**	**Effects**
Anterior Pituitary Gland: Tropic Hormones			
Thyroid-stimulating hormone (TSH)	Thyrotropin-releasing hormone (TRH)*	Thyroid gland	Stimulates the secretion of thyroid hormones (T_3 and T_4)
Follicle-stimulating hormone (FSH)	Gonadotropin-releasing hormone (GnRH)*	Ovaries and testes (gonads)	**Females**—stimulates ovarian follicle maturation and estrogen production **Males**—stimulates sperm production
Luteinizing hormone (LH)	Gonadotropin-releasing hormone (GnRH)*	Ovaries and testes (gonads)	**Females**—triggers ovulation and stimulates ovarian production of estrogen and progesterone **Males**—stimulates testosterone production
Adrenocorticotropic hormone (ACTH)	Corticotropin-releasing hormone (CRH)*	Adrenal cortex	Stimulates the release of glucocorticoids and androgens (mineralocorticoids to a lesser extent)
Anterior Pituitary Gland: Other Hormones (Not Tropic)			
Growth hormone (GH)	Growth hormone–releasing hormone (GHRH)*	Liver, muscle, bone, and cartilage, mostly	Stimulates body growth and protein synthesis, mobilizes fat and conserves glucose
Prolactin (PRL)	A decrease in the amount of prolactin-inhibiting hormone (PIH)*	Mammary glands in the breasts	Stimulates milk production (lactation)

* Indicates hormones produced by the hypothalamus.

oval glands on each lobe, but there may be more and some may be located in other regions of the neck.

Table 27.2 summarizes the hormones secreted by the thyroid and parathyroid glands.

Thymus

The *thymus* is a bilobed gland situated in the superior thorax, posterior to the sternum and overlying the heart. Conspicuous in the infant, it begins to atrophy at puberty, and by old age it is relatively inconspicuous. The thymus produces several different families of peptide hormones, including **thymulin, thymosins,** and **thymopoietins.** These hormones are thought to be involved in the development of T lymphocytes and the immune response, but their roles are poorly understood. They appear to act locally as paracrines.

Adrenal Glands

The two *adrenal,* or *suprarenal, glands* are located atop the kidneys. Anatomically, the **adrenal medulla** develops from neural crest tissue, and it is directly controlled by the sympathetic nervous system. The medullary cells respond to this stimulation by releasing a hormone mix of **epinephrine** (80%) and **norepinephrine** (20%), which act with the sympathetic nervous system to elicit the fight-or-flight response to stressors. **Table 27.3** summarizes the hormones secreted by the adrenal glands.

The gonadocorticoids are produced throughout life in relatively insignificant amounts; however, hypersecretion of these hormones produces abnormal hairiness (**hirsutism**) and masculinization. ✚

Pancreas

The *pancreas,* located behind the stomach and close to the small intestine, functions as both an endocrine and exocrine gland. It produces digestive enzymes as well as insulin and glucagon, important hormones concerned with the regulation of blood sugar levels. **Table 27.4** summarizes two of the hormones produced by the pancreas.

The Gonads

The *female gonads,* or *ovaries,* are paired, almond-sized organs located in the pelvic cavity. In addition to producing the female sex cells (ova), the ovaries produce two steroid hormone groups, the estrogens and progesterone. The endocrine and exocrine functions of the ovaries do not begin until the onset of puberty.

Table 27.1B	Pituitary Gland Hormones (Figure 27.2)		
Hormone	**Stimulus for release**	**Target**	**Effects**
Posterior Pituitary Gland (Hormones That Are Synthesized by the Hypothalamus and Stored in the Posterior Pituitary)			
Oxytocin*	Nerve impulses from hypothalamic neurons in response to cervical/uterine stretch or suckling of an infant	Uterus and mammary glands	Stimulates powerful uterine contractions during birth and stimulates milk ejection (let-down) in lactating mothers
Antidiuretic hormone (ADH)*	Nerve impulses from hypothalamic neurons in response to increased blood solute concentration or decreased blood volume	Kidneys	Stimulates the kidneys to reabsorb more water, reducing urine output and conserving body water

* Indicates hormones produced by the hypothalamus.

Table 27.2	Thyroid and Parathyroid Gland Hormones		
Hormone(s)	**Stimulus for release**	**Target**	**Effects**
Thyroid Gland			
Thyroxine (T_4) and Triiodothyronine (T_3), collectively referred to as thyroid hormone (TH)	Thyroid-stimulating hormone (TSH)	Most cells of the body	Increases basal metabolic rate (BMR); regulates tissue growth and development.
Calcitonin	High levels of calcium in the blood	Bones	No known physiological role in humans. When the hormone is supplemented at doses higher than normally found in humans, it does have some pharmaceutical applications.
Parathyroid Gland (Located on the Posterior Aspect of the Thyroid Gland)			
Parathyroid hormone (PTH)	Low levels of calcium in the blood	Bones and kidneys	Increases blood calcium by stimulating osteoclasts and by stimulating the kidneys to reabsorb more calcium. PTH also stimulates the kidneys to convert vitamin D to calcitriol, which is required for the absorption of calcium in the intestines.

Table 27.3   Adrenal Gland Hormones

Cortical area	Hormone(s)	Stimulus for release	Target	Effects
Adrenal Cortex				
Zona glomerulosa	Mineralcorticoids: mostly aldosterone	Angiotensin II release and increased potassium in the blood (ACTH only in times of severe stress)	Kidneys	Increases the reabsorption of sodium and water by the kidney tubules. Increases the secretion of potassium in the urine.
Zona fasciculata	Glucocorticoids: mostly cortisol	ACTH	Most body cells	Promotes the breakdown of fat and protein, promotes stress resistance, and inhibits the immune response.
Zona reticularis	Gonadocorticoids: androgens (most are converted to testosterone and some to estrogen)	ACTH	Bone, muscle, integument, and other tissues	In females, androgens contribute to body growth, contribute to the development of pubic and axillary hair, and enhance sex drive. They have insignificant effects in males.

Cells	Hormone(s)	Stimulus for release	Target	Effects
Adrenal Medulla				
Chromaffin cells	Catecholamines: epinephrine and norepinephrine	Nerve impulses from preganglionic sympathetic fibers	Most body cells	Mimics sympathetic nervous system activation, "fight-or-flight response."

Table 27.4   Pancreas and Gonad Hormones

Hormone	Stimulus for release	Target(s)	Effects
Pancreas			
Insulin	Increased blood glucose levels, parasympathetic nervous system stimulation	Most cells of the body	Accelerates the transport of glucose into body cells; promotes glycogen, fat, and protein synthesis
Glucagon	Decreased blood glucose levels, sympathetic nervous system stimulation	Primarily the liver and adipose	Accelerates the breakdown of glycogen to glucose, stimulates the conversion of lactic acid into glucose, releases glucose into the blood from the liver
Ovaries (Female Gonads)			
Estrogens	Luteinizing hormone (LH) and follicle-stimulating hormone (FSH)	Most cells of the body	Promote the maturation of the female reproductive organs and the development of secondary sex characteristics
Estrogens and progesterone (together)	LH and FSH	Uterus and mammary glands	Regulate the menstrual cycle and promote breast development
Testes (Male Gonads)			
Testosterone	LH and FSH	Most cells of the body	Promotes the maturation of the male reproductive organs, the development of secondary sex characteristics, sperm production, and sex drive

The paired oval *testes* of the male are suspended in a pouchlike sac, the scrotum, outside the pelvic cavity. In addition to the male sex cells (sperm), the testes produce the male sex hormone, testosterone. Both the endocrine and exocrine functions of the testes begin at puberty. (For a more detailed discussion of the function and histology of the ovaries and testes, see Exercises 42 and 43.) Table 27.4 summarizes the hormones produced by the gonads.

Microscopic Anatomy of Selected Endocrine Glands

Activity 2

Examining the Microscopic Structure of Endocrine Glands

Obtain a microscope and one of each assigned slide. Compare your observations with the images (**Figure 27.3a–f**).

Thyroid Gland

1. Scan the thyroid under low power, noting the **follicles,** generally spherical sacs containing a pink-stained material (colloid). Stored T_3 and T_4 are attached to the protein colloidal material stored in the follicles as **thyroglobulin** and are released gradually to the blood. Compare the tissue viewed to the photomicrograph of thyroid tissue (Figure 27.3a).

2. Observe the tissue under high power. Notice that the walls of the follicles are formed by simple cuboidal or squamous epithelial cells. The **parafollicular,** or **C, cells** you see between the follicles produce calcitonin.

When the thyroid gland is actively secreting, the follicles appear small. When the thyroid is hypoactive or inactive, the follicles are large and plump, and the follicular epithelium appears to be squamous.

Parathyroid Glands

Observe the parathyroid tissue under low power to view its two major cell types, the parathyroid cells and the oxyphil cells. Compare your observations to the photomicrograph of parathyroid tissue (Figure 27.3b). The **parathyroid cells,** which synthesize parathyroid hormone (PTH), are small and abundant. The function of the scattered, much larger **oxyphil cells** is unknown.

Pancreas

1. Observe pancreas tissue under low power to identify the roughly circular **pancreatic islets** (also called islets of Langerhans), the endocrine portions of the pancreas. The islets are scattered amid the more numerous **acinar cells** and stain differently (usually lighter), which makes it possible to identify them. The deeper-staining acinar cells form the major portion of the pancreatic tissue. Acinar cells produce the exocrine secretion of digestive enzymes. Alkaline fluid produced by duct cells accompanies the hydrolytic enzymes. (See Figure 27.3c.)

2. Focus on islet cells under high power. Notice that they are densely packed and have no definite arrangement (Figure 27.3c). In contrast, the cuboidal acinar cells are arranged around secretory ducts. In Figure 27.3c, it is possible to distinguish the **alpha (α) cells,** which stain darker and produce glucagon, from the **beta (β) cells,** which stain lighter and synthesize insulin. If differential staining is used, the beta cells are larger and stain gray-blue, and the alpha cells are smaller and appear bright pink.

Pituitary Gland

1. Observe the general structure of the pituitary gland under low power to differentiate the glandular anterior pituitary from the neural posterior pituitary.

2. Using the high-power lens, focus on the nests of cells of the anterior pituitary. When differential stains are used, it is possible to identify the specialized cell types that secrete the specific hormones. Using Figure 27.3d as a guide, locate the reddish pink–stained **acidophil cells,** which produce growth hormone and prolactin, and the **basophil cells,** stained blue to purple in color, which produce the tropic hormones (TSH, ACTH, FSH, and LH). **Chromophobe cells,** the third cellular population, do not take up the stain and appear colorless. The role of the chromophobe cells is controversial, but they apparently are not directly involved in hormone production.

3. Now focus on the posterior pituitary, where two hormones (oxytocin and ADH) synthesized by hypothalamic neurons are stored. Observe the nerve fibers of hypothalamic neurons. Also note the **pituicytes** (Figure 27.3e).

WHY THIS MATTERS | **Growth Hormone Uses and Abuses**

Growth hormone is an anabolic hormone that contributes to tissue building while mobilizing fat stores. Its primary targets include cartilage, bone, and muscle. Growth hormone was first approved by the U.S. Food and Drug Administration (FDA) for the treatment of children with growth disorders. Since then it has been approved to treat muscle-wasting disease associated with HIV/AIDS and a small number of other disorders. Ironically, growth hormone is more famous for its abuse than for its therapeutic applications. For example, athletes inject growth hormone to improve strength, speed, and endurance; actors use growth hormone as a chemical "fountain of youth" to smooth wrinkles and decrease body fat. ∎

Adrenal Gland

1. Hold the slide of the adrenal gland up to the light to distinguish the cortex and medulla areas. Then scan the cortex under low power to distinguish the differences in cell appearance and arrangement in the three cortical areas. Refer to Figure 27.3f as you work. In the outermost **zona glomerulosa,** where most mineralocorticoid production occurs, the cells are arranged in spherical clusters. The deeper intermediate **zona fasciculata** produces glucocorticoids. Its cells are arranged in parallel cords. The innermost cortical zone, the **zona reticularis** produces sex hormones and some glucocorticoids. The cells here stain intensely and form a branching network.

2. Switch to higher power to view the large, lightly stained cells of the adrenal medulla, which produce epinephrine and norepinephrine. Notice their clumped arrangement.

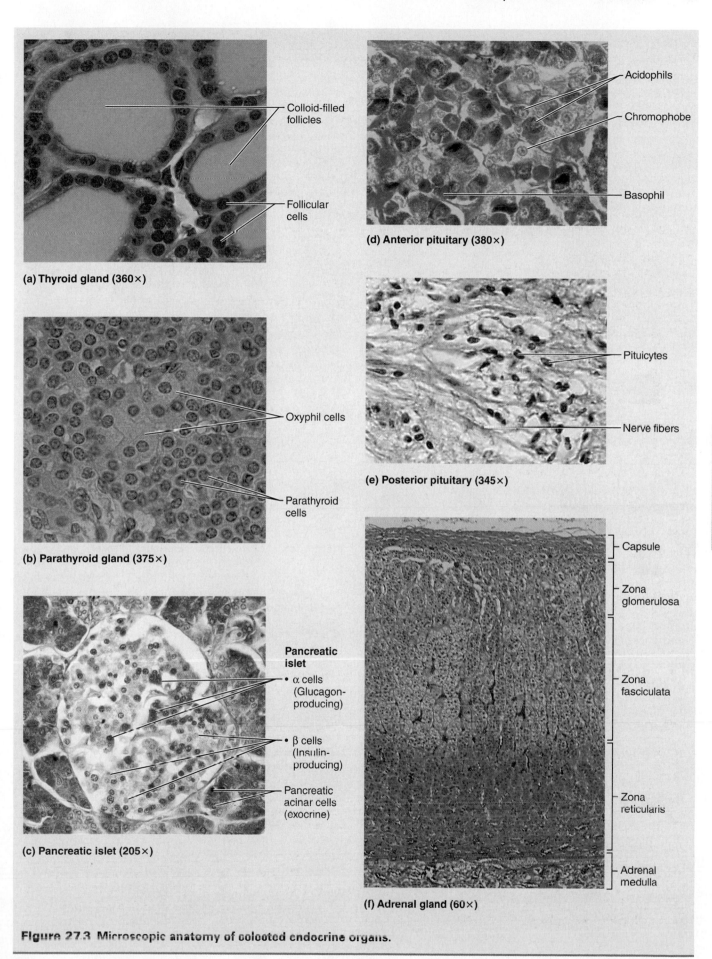

(a) Thyroid gland (360×)

Colloid-filled follicles

Follicular cells

(b) Parathyroid gland (375×)

Oxyphil cells

Parathyroid cells

(c) Pancreatic islet (205×)

Pancreatic islet
• α cells (Glucagon-producing)

• β cells (Insulin-producing)

Pancreatic acinar cells (exocrine)

(d) Anterior pituitary (380×)

Acidophils

Chromophobe

Basophil

(e) Posterior pituitary (345×)

Pituicytes

Nerve fibers

(f) Adrenal gland (60×)

Capsule

Zona glomerulosa

Zona fasciculata

Zona reticularis

Adrenal medulla

27

Figure 27.3 Microscopic anatomy of selected endocrine organs.

Endocrine Disorders

Many endocrine disorders are a result of either hyposecretion (underproduction) or hypersecretion (overproduction) of a given hormone. The characteristics of select endocrine disorders are summarized in **Table 27.5**. As you read through the table, recall the targets for the hormones and the effects of normal secretion levels.

Table 27.5	Summary of Select Endocrine Homeostatic Imbalances	
Hormone	**Effects of hyposecretion**	**Effects of hypersecretion**
Growth hormone	*In children:* **pituitary dwarfism,** which results in short stature with normal proportions	*In children:* **gigantism,** abnormally tall *In adults:* **acromegaly,** abnormally large bones of the face, feet, and hands
Antidiuretic hormone	**Diabetes insipidus,** a condition characterized by thirst and excessive urine output	**Syndrome of inappropriate ADH secretion,** a condition characterized by fluid retention, headache, and disorientation
Thyroid hormone	*In children:* **cretinism,** mental retardation with a disproportionately short-sized body *In adults:* **myxedema,** low metabolic rate, edema, physical and mental sluggishness	**Graves' disease,** elevated metabolic rate, sweating, irregular heart rate, weight loss, protrusion of the eyeballs, and nervousness
Parathyroid hormone	**Hypoparathyroidism,** neural excitability with tetany (muscle spasms) and convulsions	**Hyperparathyroidism,** loss of calcium from bones, causing deformation, and spontaneous fractures
Insulin	**Diabetes mellitus,** which results in an inability of cells to take up and utilize glucose and in loss of glucose in the urine (may be due to hyposecretion or hypoactivity of insulin)	**Hypoglycemia,** which results in low blood sugar and is characterized by anxiety, nervousness, tremors, and weakness

Group Challenge

Odd Hormone Out

Each box below contains four hormones. One of the listed hormones does not share a characteristic that the other three do. Work in groups of three, and discuss the characteristics of the four hormones in each group. On a separate piece of paper, one student will record the characteristics for each hormone for the group. For each set of four hormones, discuss the possible candidates for the "odd hormone" and which characteristic it lacks based upon your recorded notes. Once you have come to a consensus among your group, circle the hormone that doesn't belong with the others and explain why it is singled out. Sometimes there may be multiple reasons why the hormone doesn't belong with the others.

1. Which is the "odd hormone"?	Why is it the odd one out?
ACTH oxytocin LH FSH	
2. Which is the "odd hormone"?	**Why is it the odd one out?**
aldosterone cortisol epinephrine ADH	
3. Which is the "odd hormone"?	**Why is it the odd one out?**
PTH testosterone LH FSH	
4. Which is the "odd hormone"?	**Why is it the odd one out?**
insulin cortisol calcitonin glucagon	

27

REVIEW SHEET
Functional Anatomy
of the Endocrine Glands

Name _____   Lab Time/Date _____

Gross Anatomy and Basic Function of the Endocrine Glands

1. Both the endocrine and nervous systems are major regulating systems of the body; however, the nervous system has been compared to an airmail delivery system, and the endocrine system to the Pony Express. Briefly explain this comparison.

_____

_____

_____

2. Define *hormone*. _____

_____

3. Chemically, hormones belong chiefly to two molecular groups, the _____

and the _____.

4. Define *target cell*. _____

_____

5. If hormones travel in the bloodstream, why don't all tissues respond to all hormones? _____

_____

6. Identify the endocrine organ described by each of the following statements.

_____ 1. located in the throat; bilobed gland connected by an isthmus

_____ 2. found atop the kidney

_____ 3. a mixed gland, located behind the stomach and close to the small intestine

_____ 4. paired glands suspended in the scrotum

_____ 5. ride "horseback" on the thyroid gland

_____ 6. found in the pelvic cavity of the female, concerned with ova and female hormone production

_____ 7. found in the upper thorax overlying the heart; large during youth

_____ 8. found in the roof of the third ventricle of the brain

7. The table below lists the functions of many of the hormones you have studied. From the keys below, fill in the hormones responsible for each function, and the endocrine glands that produce each hormone. Glands may be used more than once.

Hormones Key:

ACTH	FSH	prolactin
ADH	glucagon	PTH
aldosterone	insulin	T_3/T_4
cortisol	LH	testosterone
epinephrine	oxytocin	TSH
estrogens	progesterone	

Glands Key:

adrenal cortex	parathyroid glands
adrenal medulla	posterior pituitary
anterior pituitary	testes
hypothalamus	thyroid gland
ovaries	
pancreas	

Function	Hormone(s)	Synthesizing Gland(s)
Regulate the function of another endocrine gland (tropic)	1.	
	2.	
	3.	
	4.	
Maintain salt and water balance in the extracellular fluid	1.	
	2.	
Directly involved in milk production and ejection	1.	
	2.	
Controls the rate of body metabolism and cellular oxidation	1.	
Regulates blood calcium levels	1.	
Regulate blood glucose levels; produced by the same "mixed" gland	1.	
	2.	
Released in response to stressors	1.	
	2.	
Drive development of secondary sex characteristics in males	1.	
Directly responsible for regulation of the menstrual cycle	1.	
	2.	

8. Although the pituitary gland is sometimes referred to as the master gland of the body, the hypothalamus exerts control over the pituitary gland. How does the hypothalamus control both anterior and posterior pituitary functioning?

_____

_____

9. Indicate whether the release of the hormones listed below is stimulated by (A) another hormone; (B) the nervous system (neurotransmitters, or neurosecretions); or (C) humoral factors (the concentration of specific nonhormonal substances in the blood or extracellular fluid).

_____ 1. ACTH   _____ 4. insulin   _____ 7. T₄/T₃

_____ 2. calcitonin   _____ 5. norepinephrine   _____ 8. testosterone

_____ 3. estrogens   _____ 6. parathyroid hormone _____ 9. TSH, FSH

10. Name the hormone(s) produced in *inadequate* amounts that directly result in the following conditions.

_____ 1. tetany

_____ 2. excessive urine output without high blood glucose levels

_____ 3. loss of glucose in the urine

_____ 4. abnormally small stature, normal proportions

_____ 5. low BMR, mental and physical sluggishness

11. Name the hormone(s) produced in *excessive* amounts that directly result in the following conditions.

_____ 1. large hands and feet in the adult, large facial bones

_____ 2. nervousness, irregular pulse rate, sweating

_____ 3. demineralization of bones, spontaneous fractures

Microscopic Anatomy of Selected Endocrine Glands

12. Choose a response from the key below to name the hormone(s) produced by the cell types listed.

Key: a. calcitonin   d. glucocorticoids   g. PTH
      b. GH, prolactin   e. insulin   h. T₄/T₃
      c. glucagon   f. mineralocorticoids   i. TSH, ACTH, FSH, LH

_____ 1. parafollicular cells of the thyroid   _____ 6. zona fasciculata cells

_____ 2. follicular cells of the thyroid   _____ 7. zona glomerulosa cells

_____ 3. beta cells of the pancreatic islets   _____ 8. parathyroid cells

_____ 4. alpha cells of the pancreatic islets   _____ 9. acidophil cells of the anterior pituitary

_____ 5. basophil cells of the anterior pituitary

WHY THIS MATTERS | 13. Explain why growth hormone is an anabolic hormone. _____

_____

14. Considering the primary target organs of growth hormone, explain why growth hormone is not a tropic

hormone. _____

_____

15. Six diagrams of the microscopic structures of the endocrine glands are presented here. Identify each and *name all structures indicated by a leader line or bracket*.

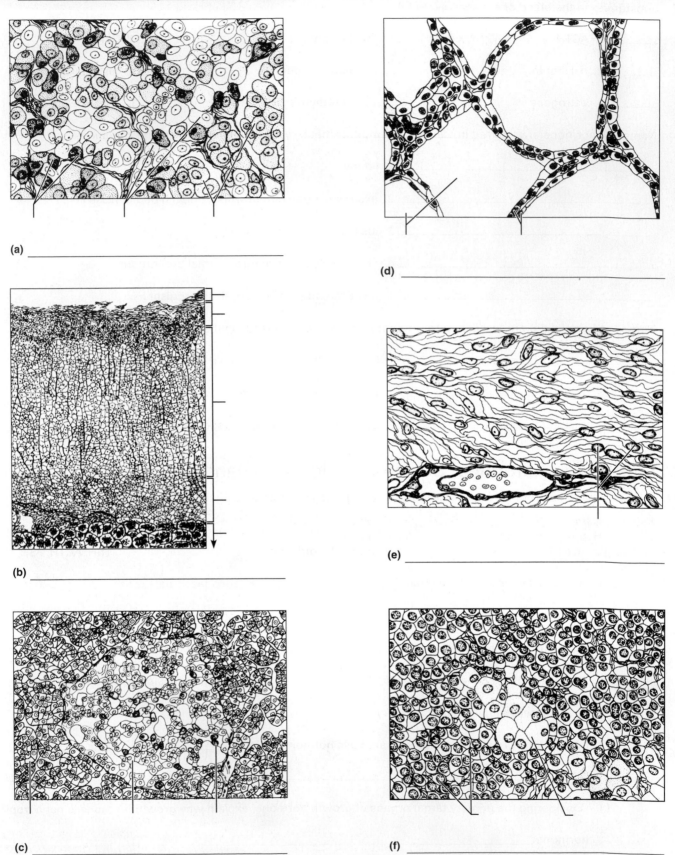

(a) _____

(b) _____

(c) _____

(d) _____

(e) _____

(f) _____

Endocrine Wet Labs and Human Metabolism

Objectives

☐ Describe the effects of pituitary extract in the frog, and indicate which hormone(s) is/are responsible for these effects.

☐ Describe the symptoms of hyperinsulinism in the fish, and explain how these symptoms were reversed.

☐ Define *metabolism*.

☐ State the functions of thyroid hormone in the body.

☐ Explain how negative feedback mechanisms regulate thyroid hormone secretion.

☐ Describe and explain the various pathologies associated with hypothyroidism and hyperthyroidism.

Materials

Activity 1: Pituitary hormone and ovary*

* Female frogs *(Rana pipiens)*
* Disposable gloves
* Battery jars
* Syringe (2-ml capacity)
* 20- to 25-gauge needle
* Frog pituitary extract
* Physiological saline
* Spring or pond water
* Wax marking pencils

Activity 2: Hyperinsulinism*

* 500- or 600-ml beakers
* 20% glucose solution
* Commercial insulin solution (400 international units [IU] per 100 ml of H_2O)
* Finger bowls
* Small (4–5 cm, or 1½–2 in.) freshwater fish (guppy, bluegill, or sunfish—listed in order of preference)
* Wax marking pencils

Text continues on next page. →

MasteringA&P®

For related exercise study tools, go to the Study Area of **MasteringA&P**. There you will find:

* Practice Anatomy Lab **PAL**
* **PhysioEx** **PEx**
* A&PFlix *A&PFlix*
* Practice quizzes, Histology Atlas, eText, Videos, and more!

Pre-Lab Quiz

1. Circle True or False. Gonadotropins are produced by the anterior pituitary gland.
2. Circle the correct underlined term. Many people with diabetes mellitus need injections of <u>insulin</u> / <u>glucagon</u> to maintain glucose homeostasis.
3. Circle the correct underlined term. <u>Catabolism</u> / <u>Anabolism</u> is the process by which substances are broken down into simpler compounds.
4. _____ is the single most important hormone responsible for influencing the rate of cellular metabolism and body heat production.
   a. Calcitonin
   b. Estrogen
   c. Insulin
   d. Thyroid hormone
5. Basal metabolic rate (BMR) is:
   a. decreased in individuals with hyperthyroidism
   b. increased in individuals with hyperthyroidism
   c. increased in obese individuals

The endocrine system performs many complex and interrelated effects on the body as a whole, as well as on specific organs and tissues. Most scientific knowledge about this system is recent, and new information is constantly being reported. Many experiments on the endocrine system require relatively large laboratory animals; are time-consuming (requiring days to weeks of observation); and often involve technically difficult surgical procedures to remove the glands or parts of them, all of which makes it difficult to conduct more general types of laboratory experiments. Nevertheless, the two technically unsophisticated experiments presented here should illustrate how dramatically hormones affect body functioning. (Also, students may perform simulated endocrine wet labs in PhysioEx Exercise 4.)

The Selected Actions of Hormones and Other Chemical Messengers video (available to qualified adopters from Pearson Education) may be used in lieu of student participation in Activities 1 and 2.

PEx PhysioEx™ 9.1 Computer Simulation Ex. 4 on p. PEx-59.

Endocrine Experiments: Gonadotropins and Insulin

Activity 1

Determining the Effect of Pituitary Hormones on the Ovary

The anterior pituitary gonadotropic hormones—follicle-stimulating hormone (FSH) and luteinizing hormone (LH)—regulate the ovarian cycles of the female (see Exercise 43). Although amphibians normally ovulate seasonally, many can be stimulated to ovulate "on demand" by injecting an extract of pituitary hormones. In the following experiment, you will need to inject the frog the day before the lab session or return to check results the day after the scheduled lab session.

⚠ 1. Don disposable gloves, and obtain two frogs. Place them in separate battery jars to bring them to your laboratory bench. Also bring back a syringe and needle, a wax marking pencil, pond or spring water, and containers of pituitary extract and physiological saline.

2. Before beginning, examine each frog for the presence of eggs. Hold the frog firmly with one hand and exert pressure on its abdomen toward the cloaca (in the direction of the legs). If ovulation has occurred, any eggs present in the uterine tube will be forced out and will appear at the cloacal opening. If no eggs are present, continue with step 3.

If eggs are expressed, return the animal to your instructor and obtain another frog for experimentation. Repeat the procedure for determining if eggs are present until two frogs that lack eggs have been obtained.

3. Draw 1 to 2 ml of the pituitary extract into a syringe. Inject the extract subcutaneously into the anterior abdominal (peritoneal) cavity of the frog you have selected to be the experimental animal. To inject into the peritoneal cavity, hold the frog with its ventral surface superiorly. Insert the needle through the skin and muscles of the abdominal wall in the lower quarter of the abdomen. Do not insert the needle far enough to damage any of the vital organs. With a wax marker, label its large battery jar "experimental," and place the frog in it. Add a small amount of pond or spring water to the battery jar before continuing.

4. Draw 1 to 2 ml of physiological saline into a syringe and inject it into the peritoneal cavity of the second frog—this will be the control animal. (Make sure you inject the same volume of fluid into both frogs.) Place this frog into the second battery jar, marked "control." Add a small amount of pond or spring water to the battery jar. Allow the animals to remain undisturbed for 24 hours.

5. After 24 hours, again check each frog for the presence of eggs in the cloacal opening. (See step 2.) If no eggs are present, make arrangements with your laboratory instructor to return to the lab on the next day (at 48 hours after injection) to check your frogs for the presence of eggs.

6. Return the frogs to the terrarium before leaving or continuing with the lab.

In which of the prepared frogs was ovulation induced?

_____

Specifically, what hormone in the pituitary extract causes ovulation to occur?

_____

Activity 2

Observing the Effects of Hyperinsulinism

Many people with diabetes mellitus need injections of insulin to maintain normal blood glucose levels. Adequate amounts of blood glucose are essential for proper functioning of the nervous system; thus, the administration of insulin must be carefully controlled. If blood glucose levels fall sharply, the patient will go into insulin shock.

A small fish will be used to demonstrate the effects of hyperinsulinism. Since the action of insulin on the fish parallels that in the human, this experiment should provide valid information concerning its administration to humans.

1. Prepare two finger bowls. Using a wax marking pencil, mark one A and the other B. To finger bowl A, add 100 ml of the commercial insulin solution. To finger bowl B, add 200 ml of 20% glucose solution.

2. Place a small fish in finger bowl A and observe its actions carefully as the insulin diffuses into its bloodstream through the capillary circulation of its gills.

Approximately how long did it take for the fish to become comatose?

_____

What types of activity did you observe in the fish before it became comatose?

_____

_____

_____

_____

3. When the fish is comatose, carefully transfer it to finger bowl B and observe its actions. What happens to the fish after it is transferred?

_____

_____

_____

Approximately how long did it take for this recovery?

_____

4. After all observations have been made and recorded, carefully return the fish to the aquarium.

Human Metabolism and Thyroid Hormones

Metabolism is a broad term referring to all chemical reactions that are necessary to maintain life. It involves both *catabolism,* enzymatically controlled processes in which substances are broken down to simpler substances, and *anabolism,* processes in which larger molecules or structures are built from smaller ones. Most catabolic reactions in the body are accompanied by a net release of energy. Some of the liberated energy is captured to make ATP, the energy rich molecule used by body cells to energize all their activities; the balance is lost in the form of thermal energy or heat. Maintaining body temperature is linked to the heat-liberating aspects of metabolism.

Various foods make different contributions to the process of metabolism. For example, carbohydrates, particularly glucose, are generally broken down or oxidized to make ATP, whereas fats are utilized to form cell membranes and myelin sheaths and to insulate the body with a fatty cushion. Fats are used secondarily for producing ATP, particularly when the diet is inadequate in carbohydrates. Proteins and amino acids tend to be conserved by body cells, and understandably so, since most structural elements of the body are built with proteins.

Thyroid hormone (TH, collectively T_3 and T_4), produced by the thyroid gland, is the single most important hormone influencing an individual's basal metabolic rate (BMR) and body heat production. Basal metabolic rate, often called the "energy cost of living," is the energy needed to perform essential activity such as breathing and maintaining organ function. The level of thyroid hormone produced directly affects BMR; the more thyroid hormone produced, the higher the BMR. In addition, thyroid hormone regulates growth and development.

The tropic hormone thyroid-stimulating hormone (TSH), produced by the anterior pituitary, controls the secretory activity of the thyroid gland. The hypothalamic hormone thyrotropin-releasing hormone (TRH) stimulates the release of TSH from cells of the anterior pituitary gland. Rising levels of thyroid hormone act on both the anterior pituitary and the hypothalamus to inhibit secretion of TSH. (**Figure 28.1**

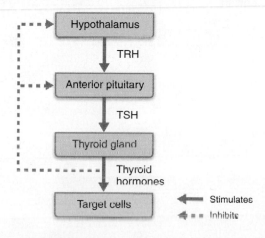

Figure 28.1 Regulation of thyroid hormone secretion.

28

illustrates the feedback loop that regulates thyroid hormone secretion.)

 A **goiter** is an enlargement of the thyroid gland. Both *hypothyroidism* and *hyperthyroidism* can result in production of a goiter. In either case, the goiter is a result of excessive stimulation of the thyroid gland.

Hypothyroidism, also called **myxedema**, produces symptoms including low metabolic rate; feeling chilled; constipation; thick, dry skin and puffy skin ("bags") beneath the eyes; edema; lethargy; and mental sluggishness. A goiter occurs when hypothyroidism is caused by (1) primary failure of the thyroid gland or (2) an iodine-deficient diet that prevents the thyroid gland from producing TH. In both cases, the low levels of TH remove the inhibition for secretion of TSH, and its levels rise. When hypothyroidism is secondary to hypothalamic or anterior pituitary failure, TRH and/or TSH levels fall, and no goiter is produced.

Symptoms of *hyperthyroidism* include elevated metabolism; sweating; a rapid, more forceful heartbeat; nervousness; weight loss; difficulty concentrating; and changes in skin texture. The most common cause of hyperthyroidism is Graves' disease. Protrusion of the eyeballs sometimes occurs in patients with Graves' disease and is a unique symptom of this type of hyperthyroidism. Graves' disease is an autoimmune disorder in which the body makes abnormal antibodies that mimic the action of TSH on follicular cells of the thyroid. Despite low levels of TSH, the thyroid is being powerfully stimulated and produces a large goiter. Hyperthyroidism can also arise secondary to excess hypothalamic or anterior pituitary secretion. In this case, TSH levels are high and a goiter also occurs. A hypersecreting thyroid tumor also causes hyperthyroidism. TSH levels are low when such a tumor is present, and there is no goiter. ✚

Use the information above to answer the questions associated with the case studies in the following Group Challenge.

👥 Group Challenge

Thyroid Hormone Case Studies

Work in groups of three, and discuss the two cases presented. Record your group's answers to the questions in the space provided.

Case 1: Marty is a 24-year-old male. He has noticed a bulge on his neck that has been increasing in size over the past few months. His physician orders a blood test with the following results:

Component	Results	Normal range	Units
TSH	<0.1	0.1–5.5	µIU/ml
Free T$_4$	5.3	0.8–1.7	ng/dL

Does Marty have hypothyroidism or hyperthyroidism?

_____

Name and briefly describe the most likely cause of his

thyroid disorder. _____

_____

What other signs and symptoms might Marty be

experiencing? _____

_____

Case 2: Heather is a 60-year-old female. She complains of swelling in her limbs and fatigue. Her physician orders a blood test with the following results:

Component	Results	Normal range	Units
TSH	5.7	0.1–5.5	µIU/ml
Free T$_4$	0.5	0.8–1.7	ng/dL

Does Heather have hypothyroidism or hyperthyroidism?

_____

Name and briefly describe the most likely cause(s) of her

thyroid disorder. _____

_____

Does Heather have a goiter? _____

What other signs and symptoms might Heather be

experiencing? _____

_____

REVIEW SHEET
Endocrine Wet Labs and Human Metabolism

Name _____ Lab Time/Date _____

Determining the Effect of Pituitary Hormones on the Ovary

1. In the experiment on the effects of pituitary hormones, two anterior pituitary hormones caused ovulation to occur in the experimental animal. Which of these actually triggered ovulation?

   _____ _____The normal function of the second hormone involved, _____,

   is to _____.

2. Why was a second frog injected with saline? _____

Observing the Effects of Hyperinsulinism

3. Briefly explain what was happening within the fish's system when the fish was immersed in the insulin solution.

   _____

   _____

4. What is the mechanism of the recovery process observed? _____

   _____

5. What would you do to help a friend who had inadvertently taken an overdose of insulin? _____

   _____ Why? _____

   _____

Human Metabolism and Thyroid Hormones

6. Use an appropriate reference to indicate which of the following would be associated with increased or decreased BMR. Indicate increase by ↑ and decrease by ↓.

   increased exercise _____   aging _____   infection/fever _____

   increased stress _____   obesity _____   sex (♂ or ♀) _____

7. What are some possible treatments for myxedema? (Use your textbook or another appropriate reference.)

_____

_____

8. What are some possible treatments for Graves' disease? (Use your textbook or another appropriate reference.)

_____

_____

EXERCISE
29 Blood

Objectives

☐ Name the two major components of blood, and state their average percentages in whole blood.

☐ Describe the composition and functional importance of plasma.

☐ Define *formed elements,* and list the cell types composing them, state their relative percentages, and describe their major functions.

☐ Identify erythrocytes, basophils, eosinophils, monocytes, lymphocytes, and neutrophils when provided with a microscopic preparation or appropriate image.

☐ Provide the normal values for a total white blood cell count and a total red blood cell count, and state the importance of these tests.

☐ Conduct the following blood tests in the laboratory, and state their norms and the importance of each: differential white blood cell count, hematocrit, hemoglobin determination, clotting time, and plasma cholesterol concentration.

☐ Define *leukocytosis, leukopenia, leukemia, polycythemia,* and *anemia;* cite a possible cause for each condition.

☐ Perform an ABO and Rh blood typing test in the laboratory, and discuss the reason for transfusion reactions resulting from the administration of mismatched blood.

Materials

General supply area:
- Disposable gloves
- Safety glasses (student-provided)
- Bucket or large beaker containing 10% household bleach solution for slide and glassware disposal
- Spray bottles containing 10% bleach solution
- Autoclave bag
- Designated lancet (sharps) disposal container

Text continues on next page. →

MasteringA&P®

For related exercise study tools, go to the Study Area of **MasteringA&P**. There you will find:

- Practice Anatomy Lab **PAL**
- A&PFlix **A&PFlix**
- PhysioEx **PEx**
- Practice quizzes, Histology Atlas, eText, Videos, and more!

Pre-Lab Quiz

1. Circle True or False. There are no special precautions that I need to observe when performing today's lab.

2. Three types of formed elements found in blood include erythrocytes, leukocytes, and:
   a. electrolytes   b. fibers   c. platelets   d. sodium salts

3. Circle the correct underlined term. Mature <u>erythrocytes</u> / <u>leukocytes</u> are the most numerous blood cells and do not have a nucleus.

4. The least numerous but largest of all agranulocytes is the:
   a. basophil   b. lymphocyte   c. monocyte   d. neutrophil

5. _____ are the leukocytes responsible for releasing histamine and other mediators of inflammation.
   a. Basophils   b. Eosinophils   c. Monocytes   d. Neutrophils

6. _____ are essential for blood clotting.

7. Circle the correct underlined term. When determining the <u>hematocrit</u> / <u>hemoglobin</u>, you will centrifuge whole blood in order to allow the formed elements to sink to the bottom of the sample.

8. Circle the correct underlined term. The normal hematocrit value for <u>females</u> / <u>males</u> is generally higher than that of the opposite sex.

9. Circle the correct underlined term. Blood typing is based on the presence of proteins known as <u>antigens</u> / <u>antibodies</u> on the outer surface of the red blood cell plasma membrane.

10. Circle True or False. If an individual is transfused with the wrong blood type, the recipient's antibodies react with the donor's blood antigens, eventually clumping and hemolyzing the donated RBCs.

- Plasma (obtained from an animal hospital or prepared by centrifuging animal [for example, cattle or sheep] blood obtained from a biological supply house)
- Test tubes and test tube racks
- Wide-range pH paper
- Stained smears of human blood from a biological supply house or, if desired by the instructor, heparinized animal blood obtained from a biological supply house or an animal hospital (for example, dog blood), or EDTA-treated red cells (reference cells*) with blood type labels obscured (available from Immucor, Inc.)

Note to the Instructor: *See directions below for handling of soiled glassware and disposable items.*

*The blood in these kits (each containing four blood cell types—A1, A2, B, and O—individually supplied in 10-ml vials) is used to calibrate cell counters and other automated clinical laboratory equipment. This blood has been carefully screened and can be safely used by students for blood typing and deter-mining hematocrits. It is not usable for hemo-globin determinations or coagulation studies.

- Clean microscope slides
- Glass stirring rods
- Wright's stain in a dropper bottle

- Distilled water in a dropper bottle
- Sterile lancets
- Absorbent cotton balls
- Alcohol swabs (wipes)
- Paper towels
- Compound microscope
- Immersion oil
- Assorted slides of white blood count pathologies labeled "Unknown Sample _____"
- Timer

Because many blood tests are to be conducted in this exercise, it is advisable to set up a number of appropriately labeled supply areas for the various tests, as designated below. Some needed supplies are located in the general supply area.

Note: *Artificial blood prepared by Ward's Natural Science can be used for differential counts, hematocrit, and blood typing.*

Activity 4: Hematocrit

- Heparinized capillary tubes
- Microhematocrit centrifuge and reading gauge (if the reading gauge is not available, a millimeter ruler may be used)
- Capillary tube sealer or modeling clay

Activity 5: Hemoglobin determination

- Hemoglobinometer, hemolysis applicator, and lens paper; or Tallquist hemoglobin scale and test paper

Activity 6: Coagulation time

- Capillary tubes (nonheparinized)
- Fine triangular file

Activity 7: Blood typing

- Blood typing sera (anti-A, anti-B, and anti-Rh [anti-D])
- Rh typing box
- Wax marking pencil
- Toothpicks
- Medicine dropper
- Blood test cards or microscope slides

Activity 8: Demonstration

- Microscopes set up with prepared slides demonstrating the following bone (or bone marrow) conditions: macrocytic hypochromic anemia, microcytic hypochromic anemia, sickle cell anemia, lymphocytic leukemia (chronic), and eosinophilia

Activity 9: Cholesterol measurement

- Cholesterol test cards and color scale

PEx PhysioEx™ 9.1 Computer Simulation Ex. 11 on p. PEx-161.

I n this exercise, you will study plasma and formed elements of blood and conduct various hematologic tests. These tests are useful diagnostic tools for the physician because blood composition (number and types of blood cells, and chemical composition) reflects the status of many body functions and malfunctions.

! **ALERT: Special precautions when handling blood.** This exercise provides information on blood from several sources: human, animal, human treated, and artificial blood. The instructor will decide whether to use animal blood for testing or to have students test their own blood in accordance with the educational goals of the student group. For example, for students in the nursing or laboratory technician curricula, learning how to safely handle human blood or other human wastes is essential. Whenever blood is being handled, special attention must be paid to safety precautions. Instructors who opt to use human blood are responsible for its safe handling. Precautions should be used regardless of the source of the blood. This will both teach good technique and ensure the safety of the students.

Follow exactly the safety precautions listed below.

1. Wear safety gloves at all times. Discard appropriately.

2. Wear safety glasses throughout the exercise.

3. Handle only your own, freshly drawn (human) blood.

4. Be sure you understand the instructions and have all supplies on hand before you begin any part of the exercise.

5. Do not reuse supplies and equipment once they have been exposed to blood.

6. Keep the lab area clean. Do not let anything that has come in contact with blood touch surfaces or other individuals in the lab. Keep track of the location of any supplies and equipment that come into contact with blood.

7. Immediately after use dispose of lancets in a designated disposal container. Do not put them down on the lab bench, even temporarily.

8. Dispose of all used cotton balls, alcohol swabs, blotting paper, and so forth, in autoclave bags; place all soiled glassware in containers of 10% bleach solution.

9. Wipe down the lab bench with 10% bleach solution when you finish.

Composition of Blood

Circulating blood is a rather viscous substance that varies from bright red to a dull brick red, depending on the amount of oxygen it is carrying. Oxygen-rich blood is bright red. The average volume of blood in the body is about 5–6 L in adult males and 4–5 L in adult females.

Blood is classified as a type of connective tissue because it consists of a nonliving fluid matrix (the **plasma**) in which living cells **(formed elements)** are suspended. The fibers typical of a connective tissue matrix become visible in blood only when clotting occurs. They then appear as fibrin threads, which form the structural basis for clot formation.

More than 100 different substances are dissolved or suspended in plasma (**Figure 29.1**), which is over 90% water. These include nutrients, gases, hormones, various wastes and metabolites, many types of proteins, and electrolytes. The composition of plasma varies continuously as cells remove or add substances to the blood.

Three types of formed elements are present in blood (**Table 29.1**, p. 428). Most numerous are **erythrocytes,** or **red blood cells (RBCs),** which are literally sacs of hemoglobin molecules that transport the bulk of the oxygen carried in the blood (and a small percentage of the carbon dioxide). **Leukocytes,** or **white blood cells (WBCs),** are part of the body's nonspecific defenses and the immune system, and **platelets** function in hemostasis (blood clot formation); together they make up <1% of whole blood. Formed elements normally constitute about 45% of whole blood; plasma accounts for the remaining 55%.

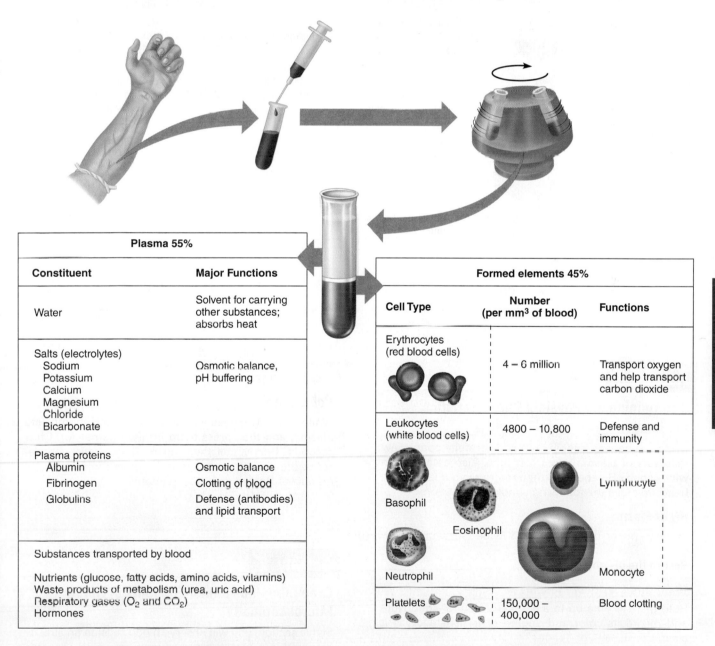

Plasma 55%	
Constituent	**Major Functions**
Water	Solvent for carrying other substances; absorbs heat
Salts (electrolytes) 　Sodium 　Potassium 　Calcium 　Magnesium 　Chloride 　Bicarbonate	Osmotic balance, pH buffering
Plasma proteins 　Albumin 　Fibrinogen 　Globulins	Osmotic balance Clotting of blood Defense (antibodies) and lipid transport
Substances transported by blood Nutrients (glucose, fatty acids, amino acids, vitamins) Waste products of metabolism (urea, uric acid) Respiratory gases (O_2 and CO_2) Hormones	

Formed elements 45%		
Cell Type	**Number (per mm³ of blood)**	**Functions**
Erythrocytes (red blood cells)	4 – 6 million	Transport oxygen and help transport carbon dioxide
Leukocytes (white blood cells)	4800 – 10,800	Defense and immunity
Basophil　Eosinophil Neutrophil		Lymphocyte Monocyte
Platelets	150,000 – 400,000	Blood clotting

Figure 29.1 The composition of blood. Note that leukocytes and platelets are found in the band between plasma (above) and erythrocytes (below).

29

Table 29.1   Summary of Formed Elements of Blood

Cell type	Illustration	Description*	Cells/mm³ (µl) of blood	Function
Erythrocytes (red blood cells, RBCs)		Biconcave, anucleate disc; orange-pink color; diameter 7–8 µm	4–6 million	Transport oxygen and carbon dioxide
Leukocytes (white blood cells, WBCs)		Spherical, nucleated cells	4800–10,800	
Granulocytes Neutrophil		Nucleus multilobed; pale red and blue cytoplasmic granules; diameter 10–12 µm	3000–7000 Differential count: 50–70%	Phagocytize pathogens or debris
Eosinophil		Nucleus bilobed; red cytoplasmic granules; diameter 10–14 µm	100–400 Differential count: 2–4%	Kill parasitic worms; slightly phagocytic; complex role in allergy and asthma
Basophil		Nucleus lobed; large blue-purple cytoplasmic granules; diameter 10–14 µm	20–50 Differential count: <1%	Release histamine and other mediators of inflammation; contain heparin, an anticoagulant
Agranulocytes Lymphocyte		Nucleus spherical or indented; pale blue cytoplasm; diameter 5–17 µm	1500–3000 Differential count: 20–40%	Mount immune response by direct cell attack or via antibody production
Monocyte		Nucleus U- or kidney-shaped; gray-blue cytoplasm; diameter 14–24 µm	100–700 Differential count: 3–8%	Develop into macrophages in tissues and phagocytize pathogens or debris
Platelets		Cytoplasmic fragments containing granules; stain deep purple; diameter 2–4 µm	150,000–400,000	Seal small tears in blood vessels; instrumental in blood clotting

*Appearance when stained with Wright's stain.

Activity 1

Determining the Physical Characteristics of Plasma

Go to the general supply area and carefully pour a few milliliters of plasma into a test tube. Also obtain some wide-range pH paper, and then return to your laboratory bench to make the following simple observations.

pH of Plasma

Test the pH of the plasma with wide-range pH paper.

Record the pH observed. _____

Color and Clarity of Plasma

Hold the test tube up to a source of natural light. Note and record its color and degree of transparency. Is it clear, translucent, or opaque?

Color _____

Degree of transparency _____

Consistency

While wearing gloves, dip your finger and thumb into plasma, and then press them firmly together for a few seconds. Gently pull them apart. How would you describe the consistency of plasma (slippery, watery, sticky, granular)? Record your observations.

_____

Activity 2

Examining the Formed Elements of Blood Microscopically

In this section, you will observe blood cells on an already prepared (purchased) blood slide or on a slide prepared from your own blood or blood provided by your instructor.

• If you are using the purchased blood slide, obtain a slide and begin your observations at step 6.

• If you are testing blood provided by a biological supply source or an animal hospital, obtain a tube of the supplied blood, disposable gloves, and the supplies listed in step 1, except for the lancets and alcohol swabs. After donning gloves, go to step 3b to begin your observations.

• If you are examining your own blood, you will perform all the steps described below *except* step 3b.

1. Obtain two glass slides, a glass stirring rod, dropper bottles of Wright's stain and distilled water, two or three lancets, cotton balls, and alcohol swabs. Bring this equipment to the laboratory bench. Clean the slides thoroughly and dry them.

2. Open the alcohol swab packet, and scrub your third or fourth finger with the swab. (Because the pricked finger may be a little sore later, it is better to prepare a finger on the nondominant hand.) Swing your hand in a cone-shaped path for 10 to 15 seconds. This will dry the alcohol and cause your fingers to become filled with blood. Then, open the lancet packet and grasp the lancet by its blunt end. Quickly jab the pointed end into the prepared finger to produce a free flow of blood. It is *not* a good idea to squeeze or "milk" the finger, because this forces out tissue fluid as well as blood. If the blood is not flowing freely, make another puncture.

⚠ <u>*Under no circumstances is a lancet to be used for more than one puncture.*</u> Dispose of the lancets in the designated disposal container immediately after use.

3a. With a cotton ball, wipe away the first drop of blood; then allow another large drop of blood to form. Touch the blood to one of the cleaned slides approximately 1.3 cm, or ½ inch, from the end. Then quickly (to prevent clotting) use the second slide to form a blood smear (**Figure 29.2**). When properly prepared, the blood smear is uniformly thin. If the blood smear appears streaked, the blood probably began to clot or coagulate before the smear was made, and another slide should be prepared. Continue at step 4.

3b. Dip a glass rod in the blood provided, and transfer a generous drop of blood to the end of a cleaned microscope slide. For the time being, lay the glass rod on a paper towel on the bench. Then, as described in step 3a (Figure 29.2), use the second slide to make your blood smear.

4. Allow the blood smear slide to air dry. When it is completely dry, it will look dull. Place it on a paper towel, and add 5 to 10 drops of Wright's stain. Count the number of drops of stain used. Allow the stain to remain on the slide for 3 to 4 minutes, and then add an equal number of drops of distilled water. Allow the water and Wright's stain mixture to remain on the slide for 4 or 5 minutes or until a metallic green film or scum is apparent on the fluid surface.

5. Rinse the slide with a stream of distilled water. Then flood it with distilled water, and allow it to lie flat until the slide becomes translucent and takes on a pink cast. Then stand the slide on its long edge on the paper towel, and

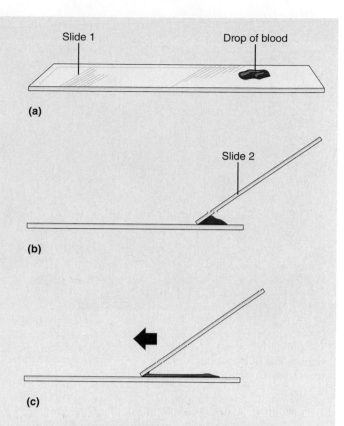

Figure 29.2 Procedure for making a blood smear.
(a) Place a drop of blood on slide 1 approximately ½ inch from one end. (b) Hold slide 2 at a 30° to 40° angle to slide 1 (it should touch the drop of blood) and allow blood to spread along entire bottom edge of angled slide. (c) Smoothly advance slide 2 to end of slide 1 (blood should run out before reaching the end of slide 1). Then lift slide 2 away from slide 1 and place slide 1 on a paper towel. Dispose of slide 2 in the appropriate container.

allow it to dry completely. Once the slide is dry, you can begin your observations.

6. Obtain a microscope and scan the slide under low power to find the area where the blood smear is the thinnest. After scanning the slide in low power to find the areas with the largest numbers of WBCs, read the following descriptions of cell types, and find each one in the art illustrating blood cell types (in Figure 29.1 and Table 29.1). (The formed elements are also shown in **Figure 29.3**, p. 430, and Figure 29.4.) Then, switch to the oil immersion lens, and observe the slide carefully to identify each cell type.

7. Set your prepared slide aside for use in Activity 3.

Erythrocytes

Erythrocytes, or red blood cells, which average 7.5 μm in diameter, vary in color from an orange-pink color to pale pink, depending on the effectiveness of the stain. They have a distinctive biconcave disc shape and appear paler in the center than at the edge (see Figure 29.3).

As you observe the slide, notice that the red blood cells are by far the most numerous blood cells seen in the

Text continues on next page. →

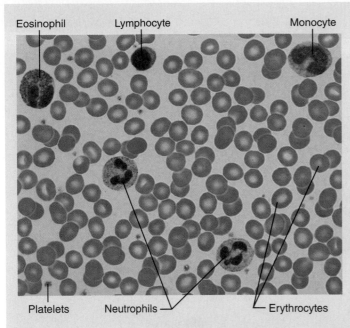

Figure 29.3 **Photomicrograph of a human blood smear stained with Wright's stain (765×).**

field. Their number averages 4.5 million to 5.5 million cells per cubic millimeter of blood (for women and men, respectively).

Red blood cells differ from the other blood cells because they are anucleate (lacking a nucleus) when mature and circulating in the blood. As a result, they are unable to reproduce or repair damage and have a limited life span of 100 to 120 days, after which they begin to fragment and are destroyed, mainly in the spleen.

In various anemias, the red blood cells may appear pale (an indication of decreased hemoglobin content) or may be nucleated (an indication that the bone marrow is turning out cells prematurely). ✚

Leukocytes

Leukocytes, or white blood cells, are nucleated cells that are formed in the bone marrow from the same stem cells *(hemocytoblast)* as red blood cells. They are much less numerous than the red blood cells, averaging from 4800 to 10,800 cells per cubic millimeter. The life span of leukocytes varies. They can survive for minutes or decades, depending on the type of leukocyte and tissue activity. Basically, white blood cells are protective, pathogen-destroying cells that are transported to all parts of the body in the blood or lymph. Important to their protective function is their ability to move in and out of blood vessels, a process called **diapedesis,** and to wander through body tissues by **amoeboid motion** to reach sites of inflammation or tissue destruction. They are classified into two major groups, depending on whether or not they contain conspicuous granules in their cytoplasm.

Granulocytes make up the first group. The granules in their cytoplasm stain differentially with Wright's stain, and they have peculiarly lobed nuclei, which often consist of expanded nuclear regions connected by thin strands of nucleoplasm. There are three types of granulocytes: **neutrophils, eosinophils,** and **basophils.**

The second group, **agranulocytes,** contains no *visible* cytoplasmic granules. Although found in the bloodstream, they are much more abundant in lymphoid tissues. There are two types of agranulocytes: **lymphocytes** and **monocytes.** The specific characteristics of leukocytes are described in Table 29.1. Photomicrographs of the leukocytes illustrate their different appearances (**Figure 29.4**).

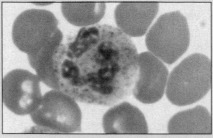

(a) Neutrophil; multilobed nucleus, pale red and blue cytoplasmic granules

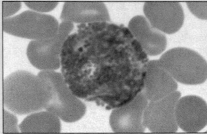

(b) Eosinophil; bilobed nucleus, red cytoplasmic granules

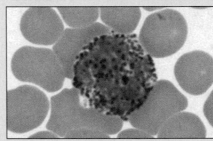

(c) Basophil; bilobed nucleus, purplish black cytoplasmic granules

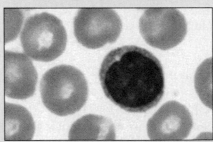

(d) Small lymphocyte; large spherical nucleus

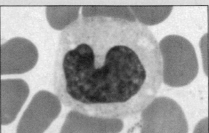

(e) Monocyte; kidney-shaped nucleus

Figure 29.4 **Leukocytes.** In each case, the leukocytes are surrounded by erythrocytes (1330×, Wright's stain).

Students are often asked to list the leukocytes in order from the most abundant to the least abundant. The following silly phrase may help you with this task: *Never let monkeys eat bananas* (neutrophils, lymphocytes, monocytes, eosinophils, basophils).

Platelets

Platelets are cell fragments of large multinucleate cells (**megakaryocytes**) formed in the bone marrow. They appear as darkly staining, irregularly shaped bodies interspersed among the blood cells (see Figure 29.3). The normal platelet count in blood ranges from 150,000 to 400,000 per cubic millimeter. Platelets are instrumental in the clotting process that occurs in plasma when blood vessels are ruptured.

After you have identified these cell types on your slide, observe charts and three-dimensional models of blood cells if these are available. Do not dispose of your slide, as you will use it later for the differential white blood cell count.

Hematologic Tests

When someone enters a hospital as a patient, several hematologic tests are routinely done to determine general level of health as well as the presence of pathologic conditions. You will be conducting the most common of these tests in this exercise.

⚠ Materials such as cotton balls, lancets, and alcohol swabs are used in nearly all of the following diagnostic tests. These supplies are at the general supply area and should be properly disposed of (glassware to the bleach bucket, lancets in a designated disposal container, and disposable items to the autoclave bag) immediately after use.

Other necessary supplies and equipment are at specific supply areas marked according to the test with which they are used. Since nearly all of the tests require a finger stick, if you will be using your own blood it might be wise to quickly read through the tests to determine in which instances more than one preparation can be done from the same finger stick. A little planning will save you the discomfort of multiple finger sticks.

An alternative to using blood obtained from the finger stick technique is using heparinized blood samples supplied by your instructor. The purpose of using heparinized tubes is to prevent the blood from clotting. Thus blood collected and stored in such tubes will be suitable for all tests except coagulation time testing.

Total White and Red Blood Cell Counts

A **total WBC count** or **total RBC count** determines the total number of that cell type per unit volume of blood. Total WBC and RBC counts are a routine part of any physical exam. Most clinical agencies use computers to conduct these counts. Total WBC and RBC counts will not be done here, but the importance of such counts (both normal and abnormal values) is briefly described below.

Total White Blood Cell Count

Since white blood cells are an important part of the body's defense system, it is essential to note any abnormalities in them.

Leukocytosis, an abnormally high WBC count, may indicate bacterial or viral infection, metabolic disease, hemorrhage, or poisoning by drugs or chemicals. A decrease in the white cell number below 4000/mm³ (**leukopenia**) may indicate infectious hepatitis or cirrhosis, tuberculosis, or excessive antibiotic or X-ray therapy. A person with leukopenia lacks the usual protective mechanisms. **Leukemia,** a malignant disorder of the lymphoid tissues characterized by uncontrolled proliferation of abnormal WBCs accompanied by a reduction in the number of RBCs and platelets, is detectable not only by a total WBC count but also by a differential WBC count. ✚

Total Red Blood Cell Count

Since RBCs are absolutely necessary for oxygen transport, a doctor typically investigates any excessive change in their number immediately.

An increase in the number of RBCs (**polycythemia**) may result from bone marrow cancer or from living at high altitudes where less oxygen is available. A decrease in the number of RBCs results in anemia. The term **anemia** simply indicates a decreased oxygen-carrying capacity of blood that may result from a decrease in RBC number or size or a decreased hemoglobin content of the RBCs. A decrease in RBCs may result suddenly from hemorrhage or more gradually from conditions that destroy RBCs or hinder RBC production. ✚

Differential White Blood Cell Count

To make a **differential white blood cell count,** 100 WBCs are counted and classified according to type. Such a count is routine in a physical examination and in diagnosing illness, since any abnormality in percentages of WBC types may indicate a problem and the source of pathology.

Activity 3

Conducting a Differential WBC Count

1. Use the slide prepared for the identification of the blood cells in Activity 2 or a prepared slide provided by your instructor. Begin at the edge of the smear and move the slide in a systematic manner on the microscope stage—either up and down or from side to side (as indicated in **Figure 29.5** on p. 432).

2. Record each type of white blood cell you observe by making a count in the first blank column of the **Activity 3 chart** on p. 432 (for example, ⊮⊮ ‖ = 7 cells) until you have observed and recorded a total of 100 WBCs. Using the following equation, compute the percentage of each WBC type counted, and record the percentages on

Text continues on next page. →

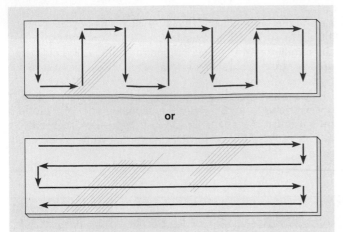

Figure 29.5 Alternative methods of moving the slide for a differential WBC count.

the Hematologic Test Data Sheet on the last page of the exercise, preceding the Review Sheet.

$$\text{Percent (\%)} = \frac{\text{\# observed}}{\text{Total \# counted}} \times 100$$

3. Select a slide marked "Unknown sample," record the slide number, and use the count chart below to conduct a differential count. Record the percentages on the data sheet (p. 438).

How does the differential count from the unknown sample slide compare to the normal percentages given for each type in Table 29.1?

_____

_____

Activity 3: Count of 100 WBCs

| Cell type | Number observed | |
	Student blood smear	Unknown sample # ____
Neutrophils		
Eosinophils		
Basophils		
Lymphocytes		
Monocytes		

Using the text and other references, try to determine the blood pathology on the unknown slide. Defend your answer.

_____

_____

4. How does your differential white blood cell count compare to the percentages given in Table 29.1?

_____

Hematocrit

The **hematocrit** is routinely determined when anemia is suspected. Centrifuging whole blood spins the formed elements to the bottom of the tube, with plasma forming the top layer (see Figure 29.1). Since the blood cell population is primarily RBCs, the hematocrit is generally considered equivalent to the RBC volume, and this is the only value reported. However, the relative percentage of WBCs can be differentiated, and both WBC and plasma volume will be reported here. Normal hematocrit values for the male and female, respectively, are 47.0 ± 5 and 42.0 ± 5.

 Prepare for lab: Watch the Pre-Lab Video
MasteringA&P®>Study Area > Pre-Lab Videos

Activity 4

Determining the Hematocrit

The hematocrit is determined by the micromethod, so only a drop of blood is needed. If possible (and the centrifuge allows), all members of the class should prepare their capillary tubes at the same time so the centrifuge can be run only once.

1. Obtain two heparinized capillary tubes, capillary tube sealer or modeling clay, a lancet, alcohol swabs, and some cotton balls.

2. If you are using your own blood, use an alcohol swab to cleanse a finger, prick the finger with a lancet, and allow the blood to flow freely. Wipe away the first few drops and, holding the red-line-marked end of the capillary tube to the blood drop, allow the tube to fill at least three-fourths full by capillary action (**Figure 29.6a**). If the blood is not flowing freely, the end of the capillary tube will not be completely submerged in the blood during filling, air will enter, and you will have to prepare another sample.

If you are using instructor-provided blood, simply immerse the red-marked end of the capillary tube in the blood sample and fill it three-quarters full as just described.

3. Plug the blood-containing end by pressing it into the capillary tube sealer or clay (Figure 29.6b). Prepare a second tube in the same manner.

4. Place the prepared tubes opposite one another in the radial grooves of the microhematocrit centrifuge with the sealed ends abutting the rubber gasket at the centrifuge periphery (Figure 29.6c). This loading procedure balances the centrifuge and prevents blood from spraying everywhere by centrifugal force. *Make a note of the numbers of the grooves your tubes are in.* When all the tubes have been loaded, make sure the centrifuge is properly balanced, and secure the centrifuge cover. Turn the centrifuge on, and set the timer for 4 or 5 minutes.

29

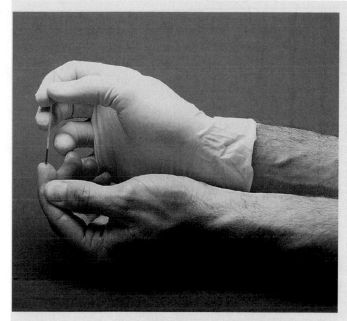

(a)

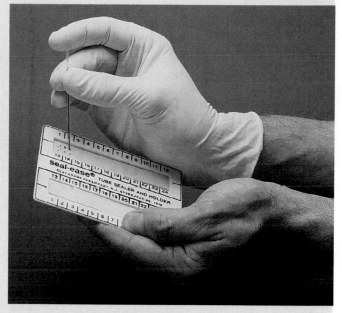

(b)

(c)

Figure 29.6 Steps in a hematocrit determination.
(a) Fill a heparinized capillary tube with blood.
(b) Plug the blood-containing end of the tube with clay.
(c) Place the tube in a microhematocrit centrifuge.
(Centrifuge must be balanced.)

5. Determine the percentage of RBCs, WBCs, and plasma by using the microhematocrit reader. The RBCs are the bottom layer, the plasma is the top layer, and the WBCs are the buff-colored layer between the two. If the reader is not available, use a millimeter ruler to measure the length of the filled capillary tube occupied by each element, and compute its percentage by using the following formula:

$$\frac{\text{Height of the column composed of the element (mm)}}{\text{Height of the original column of whole blood (mm)}} \times 100$$

Record your calculations below and on the data sheet (p. 438).

% RBC _____   % WBC _____   % plasma _____

Usually WBCs constitute 1% of the total blood volume. How do your blood values compare to this figure and to the normal percentages for RBCs and plasma? (See Figure 29.1.)

As a rule, a hematocrit is considered a more accurate test than the total RBC count for determining the RBC composition of the blood. A hematocrit within the normal range generally indicates a normal RBC number, whereas an abnormally high or low hematocrit is cause for concern.

Hemoglobin Concentration

As noted earlier, a person can be anemic even with a normal RBC count. Since hemoglobin (Hb) is the RBC protein responsible for oxygen transport, perhaps the most accurate way of measuring the oxygen-carrying capacity of the blood is to determine its hemoglobin content. Oxygen, which combines reversibly with the heme (iron-containing portion) of the hemoglobin molecule, is picked up by the blood cells in the lungs and unloaded in the tissues. Thus, the more hemoglobin molecules the RBCs contain, the more oxygen they will be able to transport. Normal blood contains 12 to 18 g of hemoglobin per 100 ml of blood. Hemoglobin content in men is slightly higher (13 to 18 g) than in women (12 to 16 g).

29

Activity 5

Determining Hemoglobin Concentration

Several techniques have been developed to estimate the hemoglobin content of blood, ranging from the old, rather inaccurate Tallquist method to expensive hemoglobinometers, which are precisely calibrated and yield highly accurate results. Directions for both the Tallquist method and a hemoglobinometer are provided here.

Tallquist Method

1. Obtain a Tallquist hemoglobin scale, test paper, lancets, alcohol swabs, and cotton balls.

2. Use instructor-provided blood or prepare the finger as previously described. (For best results, make sure the alcohol evaporates before puncturing your finger.) Place one good-sized drop of blood on the special absorbent paper provided with the color scale. The blood stain should be larger than the holes on the color scale.

3. As soon as the blood has dried and loses its glossy appearance, match its color, under natural light, with the color standards by moving the specimen under the comparison scale so that the blood stain appears at all the various apertures. (The blood should not be allowed to dry to a brown color, as this will result in an inaccurate reading.) Because the colors on the scale represent 1% variations in hemoglobin content, it may be necessary to estimate the percentage if the color of your blood sample is intermediate between two color standards.

4. On the data sheet (p. 438) record your results as the percentage of hemoglobin concentration and as grams per 100 ml of blood.

Hemoglobinometer Determination

1. Obtain a hemoglobinometer, hemolysis applicator, alcohol swab, and lens paper, and bring them to your bench.

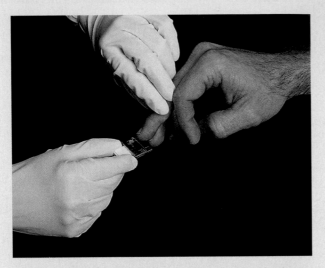

(a) A drop of blood is added to the moat plate of the blood chamber. The blood must flow freely.

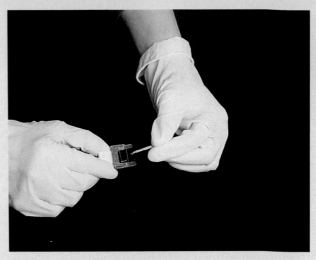

(b) The blood sample is hemolyzed with a wooden hemolysis applicator. Complete hemolysis requires 35 to 45 seconds.

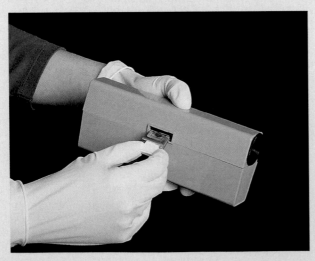

(c) The charged blood chamber is inserted into the slot on the side of the hemoglobinometer.

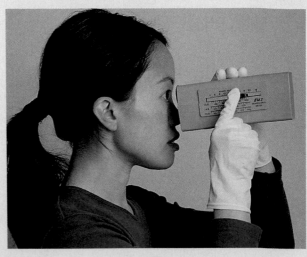

(d) The colors of the green split screen are found by moving the slide with the right index finger. When the two colors match in density, the grams/100 ml and % Hb are read on the scale.

Figure 29.7 Hemoglobin determination using a hemoglobinometer.

Test the hemoglobinometer light source to make sure it is working; if not, request new batteries before proceeding and test it again.

2. Remove the blood chamber from the slot in the side of the hemoglobinometer and disassemble the blood chamber by separating the glass plates from the metal clip. Notice as you do this that the larger glass plate has an H-shaped depression cut into it that acts as a moat to hold the blood, whereas the smaller glass piece is flat and serves as a coverslip.

3. Clean the glass plates with an alcohol swab, and then wipe them dry with lens paper. Hold the plates by their sides to prevent smearing during the wiping process.

4. Reassemble the blood chamber (remember: larger glass piece on the bottom with the moat up), but leave the moat plate about halfway out to provide adequate exposed surface to charge it with blood.

5. Obtain a drop of blood (from the provided sample or from your fingertip as before), and place it on the depressed area of the moat plate that is closest to you (**Figure 29.7a**).

6. Using the wooden hemolysis applicator, stir or agitate the blood to rupture (lyse) the RBCs (Figure 29.7b). This usually takes 35 to 45 seconds. Hemolysis is complete when the blood appears transparent rather than cloudy.

7. Push the blood-containing glass plate all the way into the metal clip and then firmly insert the charged blood chamber back into the slot on the side of the instrument (Figure 29.7c).

8. Hold the hemoglobinometer in your left hand with your left thumb resting on the light switch located on the underside of the instrument. Look into the eyepiece and notice that there is a green area divided into two halves (a split field).

9. With the index finger of your right hand, slowly move the slide on the right side of the hemoglobinometer back and forth until the two halves of the green field match (Figure 29.7d).

10. Note and record on the data sheet (p. 438) the grams of Hb (hemoglobin)/100 ml of blood indicated on the uppermost scale by the index mark on the slide. Also record % Hb, indicated by one of the lower scales.

11. Disassemble the blood chamber once again, and carefully place its parts (glass plates and clip) into a bleach-containing beaker.

Generally speaking, the relationship between the hematocrit and grams of hemoglobin per 100 ml of blood is 3:1—for example, a hematocrit of 36% with 12 g of Hb per 100 ml of blood is a ratio of 3:1. How do your values compare?

Record on the data sheet (p. 438) the value obtained from your data.

Bleeding Time

Normally a sharp prick of the finger or earlobe results in bleeding that lasts from 2 to 7 minutes (Ivy method) or 0 to 5 minutes (Duke method), although other factors such as altitude affect the time. How long the bleeding lasts is referred to as **bleeding time** and tests the ability of platelets to stop bleeding in capillaries and small vessels. Absence of some clotting factors may affect bleeding time, but prolonged bleeding time is most often associated with deficient or abnormal platelets.

Coagulation Time

Blood clotting, or **coagulation,** is a protective mechanism that minimizes blood loss when blood vessels are ruptured.

This process requires the interaction of many substances normally present in the plasma (clotting factors, or procoagulants) as well as some released by platelets and injured tissues. Basically hemostasis proceeds as follows (**Figure 29.8a**, p. 436): The injured tissues and platelets release **tissue factor (TF)** and **PF$_3$ (platelet factor 3)** respectively, which trigger the clotting mechanism, or cascade. Tissue factor and PF$_3$ interact with other blood protein clotting factors and calcium ions to form **prothrombin activator,** which in turn converts **prothrombin** (present in plasma) to **thrombin.** Thrombin then acts enzymatically to polymerize (combine) the soluble **fibrinogen** proteins (present in plasma) into insoluble **fibrin,** which forms a meshwork of strands that traps the RBCs and forms the basis of the clot (Figure 29.8b). Normally, blood removed from the body clots within 2 to 6 minutes.

29

Activity 6

Determining Coagulation Time

1. Obtain a *nonheparinized* capillary tube, a timer (or watch), a lancet, cotton balls, a triangular file, and alcohol swabs.

2. Clean and prick the finger to produce a free flow of blood. Discard the lancet in the disposal container.

3. Place one end of the capillary tube in the blood drop, and hold the opposite end at a lower level to collect the sample.

4. Lay the capillary tube on a paper towel after collecting the sample.

Record the time. _____

5. At 30-second intervals, make a small nick on the tube close to one end with the triangular file, and then carefully break the tube. Slowly separate the ends to see whether a gel-like thread of fibrin spans the gap. When this occurs, record below and on the data sheet (p. 438) the time for coagulation to occur. Are your results within the normal time range?

_____

6. Put used supplies in the autoclave bag and broken capillary tubes into the sharps container.

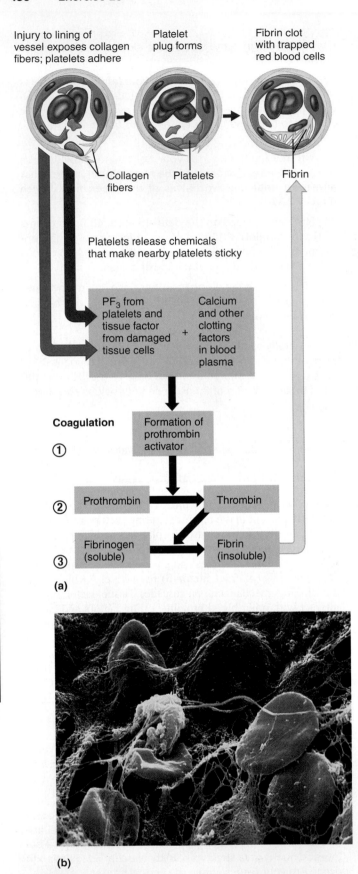

Injury to lining of vessel exposes collagen fibers; platelets adhere

Platelet plug forms

Fibrin clot with trapped red blood cells

Collagen fibers

Platelets

Fibrin

Platelets release chemicals that make nearby platelets sticky

PF₃ from platelets and tissue factor from damaged tissue cells + Calcium and other clotting factors in blood plasma

Coagulation

① Formation of prothrombin activator

② Prothrombin → Thrombin

③ Fibrinogen (soluble) → Fibrin (insoluble)

(a)

(b)

Figure 29.8 Events of hemostasis and blood clotting.
(a) Simple schematic of events. Steps numbered 1–3 represent the major events of coagulation.
(b) Photomicrograph of RBCs trapped in a fibrin mesh (2700×).

Blood Typing

Blood typing is a system of blood classification based on the presence of specific glycoproteins on the outer surface of the RBC plasma membrane. Such proteins are called **antigens,** or **agglutinogens,** and are genetically determined. For ABO blood groups, these antigens are accompanied by plasma proteins, called **antibodies** or **agglutinins.** These antibodies act against RBCs carrying antigens that are not present on the person's own RBCs. If the donor blood type doesn't match, the recipient's antibodies react with the donor's blood antigens, causing the RBCs to clump, agglutinate, and eventually hemolyze. It is because of this phenomenon that a person's blood must be carefully typed before a whole blood or packed cell transfusion.

Several blood typing systems exist, based on the various possible antigens, but the factors routinely typed for are antigens of the ABO and Rh blood groups which are most commonly involved in transfusion reactions. The basis of the ABO typing is shown in **Table 29.2**.

Individuals whose red blood cells carry the Rh antigen are Rh positive (approximately 85% of the U.S. population); those lacking the antigen are Rh negative. Unlike ABO blood groups, the blood of neither Rh-positive (Rh⁺) nor Rh-negative (Rh⁻) individuals carries preformed anti-Rh antibodies. This is understandable in the case of the Rh-positive individual. However, Rh-negative persons who receive transfusions of Rh-positive blood become sensitized by the Rh antigens of the donor RBCs, and their systems begin to produce anti-Rh antibodies. On subsequent exposures to Rh-positive blood, typical transfusion reactions occur, resulting in the clumping and hemolysis of the donor blood cells.

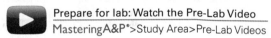

Prepare for lab: Watch the Pre-Lab Video
MasteringA&P®>Study Area>Pre-Lab Videos

Activity 7

Typing for ABO and Rh Blood Groups

Blood may be typed on microscope slides or using blood test cards. Each method is described in this activity. The artificial blood kit does not use any body fluids and produces results similar to but not identical to results for human blood.

Typing Blood Using Glass Slides

1. Obtain two clean microscope slides, a wax marking pencil, anti-A, anti-B, and anti-Rh typing sera, toothpicks, lancets, alcohol swabs, medicine dropper, and the Rh typing box.

2. Divide slide 1 into halves with the wax marking pencil. Label the lower left-hand corner "anti-A" and the lower right-hand corner "anti-B." Mark the bottom of slide 2 "anti-Rh."

3. Place one drop of anti-A serum on the *left* side of slide 1. Place one drop of anti-B serum on the *right* side of slide 1. Place one drop of anti-Rh serum in the center of slide 2.

4. If you are using your own blood, cleanse your finger with an alcohol swab, pierce the finger with a lancet, and wipe away the first drop of blood. Obtain 3 drops of freely flowing blood, placing one drop on each side of slide 1 and a drop on slide 2. Immediately dispose of the lancet in a designated disposal container.

Table 29.2   ABO Blood Typing

ABO blood type	Antigens present on RBC membranes	Antibodies present in plasma	% of U.S. population		
			White	Black	Asian
A	A	Anti-B	40	27	28
B	B	Anti-A	11	20	27
AB	A and B	None	4	4	5
O	Neither	Anti-A and anti-B	45	49	40

If using instructor-provided animal blood or red blood cells treated with EDTA (an anticoagulant), use a medicine dropper to place one drop of blood on each side of slide 1 and a drop of blood on slide 2.

⚠ 5. Quickly mix each blood-antiserum sample with a *fresh* toothpick. Then dispose of the toothpicks and used alcohol swab in the autoclave bag.

6. Place slide 2 on the Rh typing box and rock gently back and forth. (A slightly higher temperature is required for precise Rh typing than for ABO typing.)

7. After 2 minutes, observe all three blood samples for evidence of clumping. The agglutination that occurs in the positive test for the Rh factor is very fine and difficult to interpret; thus if there is any question, observe the slide under the microscope. Record your observations in the Activity 7 chart.

8. Interpret your ABO results (see the examples of each type) in **Figure 29.9**. If clumping was observed on slide 2, you are Rh positive. If not, you are Rh negative.

9. Record your blood type on the data sheet (p. 438).

10. Put the used slides in the bleach-containing bucket at the general supply area; put disposable supplies in the autoclave bag.

Activity 7: Blood Typing

Result	Observed (+)	Not observed (−)
Presence of clumping with anti-A		
Presence of clumping with anti-B		
Presence of clumping with anti-Rh		

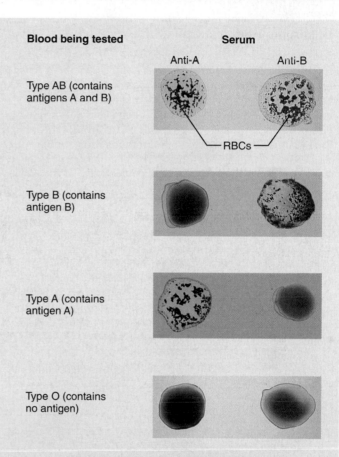

Blood being tested — **Serum** (Anti-A, Anti-B)

Type AB (contains antigens A and B) — RBCs

Type B (contains antigen B)

Type A (contains antigen A)

Type O (contains no antigen)

Figure 29.9 Blood typing of ABO blood types.
When serum containing anti-A or anti-B antibodies (agglutinins) is added to a blood sample, agglutination will occur between the antibody and the corresponding antigen (agglutinogen A or B). As illustrated, agglutination occurs with both sera in blood group AB, with anti-B serum in blood group B, with anti-A serum in blood group A, and with neither serum in blood group O.

Using Blood Typing Cards

1. Obtain a blood typing card marked A, B, and Rh, dropper bottles of anti-A serum, anti-B serum, and anti-Rh serum, toothpicks, lancets, and alcohol swabs.

2. Place a drop of anti-A serum in the spot marked anti-A, place a drop of anti-B serum on the spot marked anti-B, and place a drop of anti-Rh serum on the spot marked anti-Rh (or anti-D).

3. Carefully add a drop of blood to each of the spots marked "Blood" on the card. If you are using your own blood, refer to step 4 in the Activity 7 section Typing Blood Using Glass Slides. Immediately discard the lancet in the designated disposal container.

4. Using a new toothpick for each test, mix the blood sample with the antibody. Dispose of the toothpicks appropriately.

Text continues on next page. →

29

5. Gently rock the card to allow the blood and antibodies to mix.

6. After 2 minutes, observe the card for evidence of clumping. The Rh clumping is very fine and may be difficult to observe. Record your observations in the Activity 7: Blood Typing chart. (Use Figure 29.9 to interpret your results.)

7. Record your blood type on the data sheet below, and discard the card in an autoclave bag.

Activity 8

Observing Demonstration Slides

Look at the slides of *macrocytic hypochromic anemia, microcytic hypochromic anemia, sickle cell anemia, lymphocytic leukemia* (chronic), and *eosinophilia* that have been put on demonstration by your instructor. Record your observations in the appropriate section of the Review Sheet. You can refer to your notes, the text, and other references later to respond to questions about the blood pathologies represented on the slides.

Cholesterol Concentration in Plasma

Atherosclerosis is the disease process in which the body's blood vessels become increasingly occluded, or blocked, by plaques. By narrowing the arteries, the plaques can contribute to hypertensive heart disease. They also serve as starting points for the formation of blood clots (thrombi), which may break away and block smaller vessels farther downstream in the circulatory pathway and cause heart attacks or strokes.

Cholesterol is a major component of the smooth muscle plaques formed during atherosclerosis. No physical examination of an adult is considered complete until cholesterol levels are assessed along with other risk factors. A normal value for total plasma cholesterol in adults ranges from 130 to 200 mg per 100 ml of plasma; you will use blood to make such a determination.

Although the total plasma cholesterol concentration is valuable information, it may be misleading, particularly if a person's high-density lipoprotein (HDL) level is high and low-density lipoprotein (LDL) level is relatively low. Cholesterol, being water insoluble, is transported in the blood complexed to lipoproteins. In general, cholesterol bound into HDLs is destined to be degraded by the liver and then eliminated from the body, whereas that forming part of the LDLs is "traveling" to the body's tissue cells. When LDL levels are excessive, cholesterol is deposited in the blood vessel walls; hence, LDLs are considered to carry the "bad" cholesterol.

Activity 9

Measuring Plasma Cholesterol Concentration

1. Go to the appropriate supply area, and obtain a cholesterol test card and color scale, lancet, and alcohol swab.

2. Clean your fingertip with the alcohol swab, allow it to dry, then prick it with a lancet. Place a drop of blood on the test area of the card. Put the lancet in the designated disposal container.

3. After 3 minutes, remove the blood sample strip from the card and discard in the autoclave bag.

4. Analyze the underlying test spot, using the included color scale. Record the cholesterol level below and on the **Hematologic Test Data Sheet.**

Total cholesterol level _____ mg/dl

⚠ **5.** Before leaving the laboratory, use the spray bottle of bleach solution, and saturate a paper towel to thoroughly wash down your laboratory bench.

Hematologic Test Data Sheet		

Differential WBC count:

WBC	Student blood smear	Unknown sample # ____
% neutrophils	_____	_____
% eosinophils	_____	_____
% basophils	_____	_____
% lymphocytes	_____	_____
% monocytes	_____	_____

Hematocrit:

RBC _____ % of blood volume

WBC _____ % of blood volume   not generally reported

Plasma _____ % of blood

Hemoglobin (Hb) content:

Tallquist method: _____ g/100 ml of blood; _____ % Hb

Hemoglobinometer (type:

_____)

_____ g/100 ml of blood; _____ %Hb

Ratio (hematocrit to grams of Hb per 100 ml of blood):

_____

Coagulation time: _____

Blood typing:

ABO group _____ Rh factor _____

Total cholesterol level: _____ mg/dl of blood

Name_____ Lab Time/Date _____

Composition of Blood

1. What is the blood volume of an average-size adult male? _____ liters; an average adult female? _____ liters

2. What determines whether blood is bright red or a dull brick red? _____

_____

3. Use the key to identify the cell type(s) or blood elements that fit the following descriptive statements. Some terms will be used more than once.

   Key: a. red blood cell   d. basophil   g. lymphocyte
         b. megakaryocyte   e. monocyte   h. formed elements
         c. eosinophil   f. neutrophil   i. plasma

   _____ 1. most numerous leukocyte

   _____ , _____ , and _____ 2. granulocytes (3)

   _____ 3. also called an erythrocyte; anucleate formed element

   _____, _____, _____ 4. phagocytic leukocytes (3)

   _____, _____ 5. agranulocytes

   _____ 6. precursor cell of platelets

   _____ 7. (a) through (g) are all examples of these

   _____ 8. involved in destroying parasitic worms

   _____ 9. releases histamine; promotes inflammation

   _____ 10. produces antibodies

   _____ 11. transports oxygen

   _____ 12. primarily water, noncellular; the fluid matrix of blood

   _____ 13. exits a blood vessel to develop into a macrophage

   _____ , _____ , _____ ,

   _____ , _____ 14. the five types of white blood cells

4. List four classes of nutrients normally found in plasma. _____ ,

_____ , _____ , and _____

Name two gases. _____ and _____

Name three ions. _____ , _____ , and _____

5. Describe the consistency and color of the plasma you observed in the laboratory. _____

_____

6. What is the average life span of a red blood cell? How does its anucleate condition affect this life span?

_____

_____

_____

7. From memory, describe the structural characteristics of each of the following blood cell types as accurately as possible, and note the percentage of each in the total white blood cell population.

eosinophils: _____

_____

neutrophils: _____

_____

lymphocytes: _____

_____

basophils: _____

_____

monocytes: _____

_____

8. Correctly identify the blood pathologies described in column A by matching them with selections from column B:

Column A   **Column B**

_____ 1. abnormal increase in the number of WBCs   a. anemia

_____ 2. abnormal increase in the number of RBCs   b. leukocytosis

_____ 3. condition of too few RBCs or of RBCs with   c. leukopenia
                                          hemoglobin deficiencies

_____ 4. abnormal decrease in the number of WBCs   d. polycythemia

Hematologic Tests

9. Broadly speaking, why are hematologic studies of blood so important in diagnosing disease?

_____

_____

10. In the chart below, record information from the blood tests you read about or conducted. Complete the chart by recording values for healthy male adults and indicating the significance of high or low values for each test.

Test	Student test results	Normal values (healthy male adults)	Significance	
			High values	**Low values**
Total WBC count	No data			
Total RBC count	No data			
Hematocrit				
Hemoglobin determination				
Bleeding time	No data			
Coagulation time				

11. Why is a differential WBC count more valuable than a total WBC count when trying to determine the specific source of

    pathology? _____

_____

_____

12. What name is given to the process of RBC production? (Consult an appropriate reference as necessary.) _____

_____

    What hormone acts as a stimulus for this process? _____

    Why might patients with kidney disease suffer from anemia? _____

_____

    How can such patients be treated? _____

13. Discuss the effect of each of the following factors on RBC count. Consult an appropriate reference as necessary, and explain your reasoning.

    long-term effect of athletic training (for example, running 4 to 5 miles per day over a period of 6 to 9 months).

_____

_____

a permanent move from sea level to a high-altitude area: _____

_____

_____

14. Define *hematocrit*. _____

15. If you had a high hematocrit, would you expect your hemoglobin determination to be high or low? _____

    Why? _____

    _____

16. What is an anticoagulant? _____

    Name two anticoagulants used in conducting the hematologic tests. _____

    and _____

    What is the body's natural anticoagulant? _____

17. If your blood clumped with both anti-A and anti-B sera, your ABO blood type would be _____

    To what ABO blood groups could you give blood? _____

    From which ABO donor types could you receive blood? _____

    Which ABO blood type is most common? _____ Least common? _____

18. What blood type is theoretically considered the universal donor? _____ Why? _____

    _____

19. Assume the blood of two patients has been typed for ABO blood type.

Typing results
Mr. Adams:

Blood drop and   Blood drop and
anti-A serum   anti-B serum

Typing results
Mr. Calhoon:

Blood drop and   Blood drop and
anti-A serum   anti-B serum

On the basis of these results, Mr. Adams has type _____ blood, and Mr. Calhoon has type _____
blood.

20. Explain why an Rh-negative person does not have a transfusion reaction on the first exposure to Rh-positive blood

but *does* have a reaction on the second exposure. _____

_____

What happens when an ABO blood type is mismatched for the first time? _____

_____

21. Record your observations of the five demonstration slides viewed.

    a.   Macrocytic hypochromic anemia: _____

    b.   Microcytic hypochromic anemia: _____

               _____

    c.   Sickle cell anemia: _____

    d.   Lymphocytic leukemia (chronic): _____

               _____

    e.   Eosinophilia: _____

    Which of the slides above (a through e) corresponds with the following conditions?

_____ 1. iron-deficient diet

_____ 2. a type of bone marrow cancer

_____ 3. genetic defect that causes hemoglobin to become sharp/spiky

_____ 4. lack of vitamin B_{12}

_____ 5. a tapeworm infestation in the body

_____ 6. a bleeding ulcer

22. Provide the normal, or at least "desirable," range for plasma cholesterol concentration.

_____ mg/100 ml

23. Describe the relationship between high blood cholesterol levels and cardiovascular diseases such as hypertension, heart attacks, and strokes.

_____

_____

_____

_____

EXERCISE 30

Anatomy of the Heart

Objectives

☐ Describe the location of the heart.

☐ Name and describe the covering and lining tissues of the heart.

☐ Name and locate the major anatomical areas and structures of the heart when provided with an appropriate model, image, or dissected sheep heart, and describe the function of each.

☐ Explain how the atrioventricular and semilunar valves operate.

☐ Distinguish blood vessels carrying oxygen-rich blood from those carrying carbon dioxide–rich blood, and describe the system used to color code them in images.

☐ Explain why the heart is called a double pump, and compare the pulmonary and systemic circuits.

☐ Trace the pathway of blood through the heart.

☐ Trace the functional blood supply of the heart, and name the associated blood vessels.

☐ Describe the histology of cardiac muscle, and state the importance of its intercalated discs and the spiral arrangement of its cells.

Materials

- X-ray film of the human thorax; X-ray viewing box
- Three-dimensional heart model or laboratory chart showing heart anatomy
- Red and blue pencils
- Highlighter
- Three-dimensional models of cardiac and skeletal muscle
- Compound microscope
- Prepared slides of cardiac muscle (l.s.)
- Preserved or fresh sheep hearts, pericardial sacs intact (if possible)
- Dissecting instruments and tray
- Pointed glass rods or blunt probes
- Small plastic metric rulers
- Disposable gloves

Text continues on next page. →

MasteringA&P®

For related exercise study tools, go to the Study Area of **MasteringA&P**. There you will find:

- Practice Anatomy Lab PAL
- A&PFlix *A&PFlix*
- PhysioEx **PEx**
- Practice quizzes, Histology Atlas, eText, Videos, and more!

Pre-Lab Quiz

1. The heart is enclosed in a double-walled sac called the:
   a. apex   **b.** mediastinum   **c.** pericardium   **d.** thorax

2. The heart is divided into _____ chambers.
   a. two   **b.** three   **c.** four   **d.** five

3. What is the name of the two receiving chambers of the heart?

   _____

4. The left ventricle discharges blood into the _____, from which all systemic arteries of the body diverge to supply the body tissues.
   a. aorta   **c.** pulmonary vein
   b. pulmonary artery   **d.** vena cava

5. Circle True or False. Blood flows through the heart in one direction—from the atria to the ventricles.

6. Circle the correct underlined term. The right atrioventricular valve, or <u>tricuspid valve</u> / <u>mitral valve</u>, prevents backflow into the right atrium when the right ventricle is contracting.

7. Circle the correct underlined term. The heart serves as a double pump. The <u>right</u> / <u>left</u> side serves as the pulmonary circuit pump, shunting carbon dioxide–rich blood to the lungs.

8. The blood vessels that supply blood to the heart itself are the:
   a. aortas   **c.** coronary arteries
   b. carotid arteries   **d.** pulmonary trunks

9. Two microscopic features of cardiac cells that help distinguish them from other types of muscle cells are branching of the cells and:
   a. intercalated discs   **c.** sarcolemma
   b. myosin fibers   **d.** striations

10. Circle the correct underlined term. In the heart, the <u>left</u> / <u>right</u> ventricle has thicker walls and a basically circular cavity shape.

- Container for disposal of organic debris
- Laboratory detergent
- Spray bottle with 10% household bleach solution

✂ For instructions on animal dissections, see the dissection exercises (starting on p. 705) in the cat and fetal pig editions of this manual.

The major function of the **cardiovascular system** is transportation. Using blood as the transport vehicle, the system carries oxygen, digested foods, cell wastes, electrolytes, and many other substances vital to the body's homeostasis to and from the body cells. The system's propulsive force is the contracting heart, which can be compared to a muscular pump equipped with one-way valves. As the heart contracts, it forces blood into a closed system of large and small plumbing tubes (blood vessels) within which the blood is confined and circulated.

Gross Anatomy of the Human Heart

The **heart,** a cone-shaped organ approximately the size of a fist, is located within the mediastinum of the thorax. It is flanked laterally by the lungs, posteriorly by the vertebral column, and anteriorly by the sternum (**Figure 30.1**). Its more pointed **apex** extends slightly to the left and rests on the diaphragm, approximately at the level of the fifth intercostal space. Its broader **base,** from which the great vessels emerge, lies beneath the second rib and points toward the right shoulder.

☐ If an X-ray image of a human thorax is available, verify the relationships described above (otherwise, Figure 30.1 should suffice).

The gross anatomy figure (**Figure 30.2**) shows three views of the heart—external anterior and posterior views and a frontal section.

The heart is enclosed within a double-walled sac called the pericardium. The loose-fitting superficial part of the sac is the **fibrous pericardium.** Deep to it is the *serous pericardium,* which lines the fibrous pericardium as the **parietal layer.** At the base of the heart, the parietal layer reflects back to cover the external surface of the heart as the **visceral layer,** or **epicardium.** The epicardium is an integral part of the heart wall. Serous fluid produced by these layers allows the heart to beat in a relatively frictionless environment.

The walls of the heart are composed of three layers:

- **Epicardium:** The outer layer, which is also the visceral pericardium.

- **Myocardium:** The middle layer and thickest layer, which is composed mainly of cardiac muscle. It is reinforced with dense irregular connective tissue, the *cardiac skeleton,* which is thicker around the heart valves and at the base of the great vessels leaving the heart.

- **Endocardium:** The inner lining of the heart, which covers the heart valves and is continuous with the inner lining of the great vessels. It is composed of simple squamous epithelium resting on areolar connective tissue.

Heart Chambers

The heart is divided into four chambers: two superior **atria** (singular: *atrium*) and two inferior **ventricles.** The septum that divides the heart longitudinally is referred to as the **interatrial** or **interventricular septum,** depending on which chambers it partitions. Functionally, the atria are receiving chambers and are relatively ineffective as pumps.

The inferior thick-walled ventricles, which form the bulk of the heart, are the discharging chambers. They force blood out of the heart into the large arteries that emerge from its base.

Heart Valves

Four valves enforce a one-way blood flow through the heart chambers. The **atrioventricular (AV) valves** are located between the atrium and the ventricle on the left and right side of the heart. The **semilunar (SL) valves** are located between a ventricle and a great vessel (**Figure 30.3**, p. 449).

- **Tricuspid valve:** The right AV valve has three flaplike cusps anchored to the **papillary muscles** of the ventricular wall by tiny, white collagenic cords called **chordae tendineae** (literally, "heart strings")

- **Mitral valve** (*bicuspid valve*): The left AV valve has two flaplike cusps anchored to the papillary muscles by chordae tendineae.

The AV valves are open and hang into the ventricles when blood is flowing into the atria and the ventricles are relaxed. When the ventricles contract, the blood in the ventricles is compressed, causing the AV valves to move superiorly and close the opening between the atrium and the ventricle. The chordae tendineae, pulled tight by the contracting papillary muscles, anchor the cusps in the closed position and prevent the backflow of blood from the ventricles into the atria. If unanchored, the cusps would move upward into the atria like an umbrella being turned inside out by a strong wind.

- **Pulmonary (SL) valve:** Has three pocketlike cusps located between the right ventricle and the pulmonary trunk.

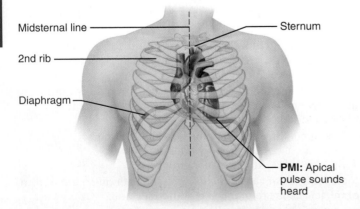

Midsternal line —

2nd rib —

Diaphragm —

Sternum

PMI: Apical pulse sounds heard

Figure 30.1 Location of the heart in the thorax. PMI is the point of maximal intensity where the apical pulse is heard.

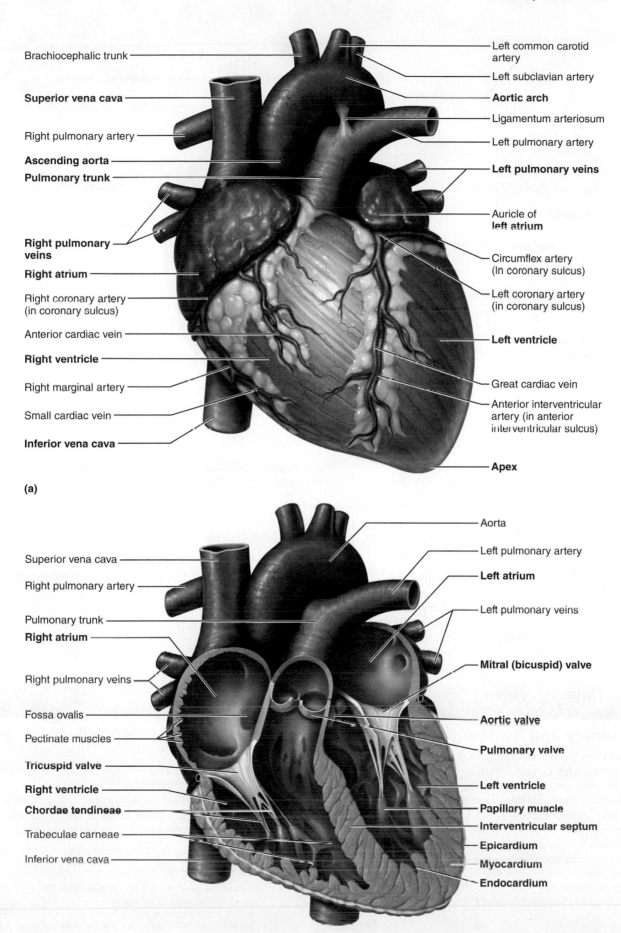

Brachiocephalic trunk

Superior vena cava

Right pulmonary artery

Ascending aorta

Pulmonary trunk

Right pulmonary veins

Right atrium

Right coronary artery (in coronary sulcus)

Anterior cardiac vein

Right ventricle

Right marginal artery

Small cardiac vein

Inferior vena cava

Left common carotid artery

Left subclavian artery

Aortic arch

Ligamentum arteriosum

Left pulmonary artery

Left pulmonary veins

Auricle of **left atrium**

Circumflex artery (In coronary sulcus)

Left coronary artery (in coronary sulcus)

Left ventricle

Great cardiac vein

Anterior interventricular artery (in anterior interventricular sulcus)

Apex

(a)

Superior vena cava

Right pulmonary artery

Pulmonary trunk

Right atrium

Right pulmonary veins

Fossa ovalis

Pectinate muscles

Tricuspid valve

Right ventricle

Chordae tendineae

Trabeculae carneae

Inferior vena cava

Aorta

Left pulmonary artery

Left atrium

Left pulmonary veins

Mitral (bicuspid) valve

Aortic valve

Pulmonary valve

Left ventricle

Papillary muscle

Interventricular septum

Epicardium

Myocardium

Endocardium

30

(b)

Figure 30.2 Gross anatomy of the human heart.
(a) External anterior view. **(b)** Frontal section.

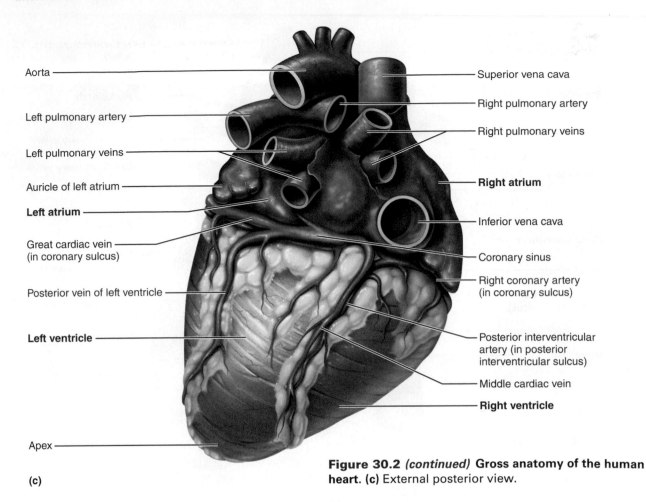

Aorta

Left pulmonary artery

Left pulmonary veins

Auricle of left atrium

Left atrium

Great cardiac vein
(in coronary sulcus)

Posterior vein of left ventricle

Left ventricle

Apex

(c)

Superior vena cava

Right pulmonary artery

Right pulmonary veins

Right atrium

Inferior vena cava

Coronary sinus

Right coronary artery
(in coronary sulcus)

Posterior interventricular
artery (in posterior
interventricular sulcus)

Middle cardiac vein

Right ventricle

Figure 30.2 *(continued)* **Gross anatomy of the human heart. (c)** External posterior view.

• **Aortic (SL) valve:** Has three pocketlike cusps located between the left ventricle and the aorta.

The SL valves are open and flattened against the wall of the vessel when the contraction of the ventricles pushes blood into the great vessels. When the ventricles relax, blood flows backward toward the ventricle and the cusps fill with blood, closing the SL valves. This prevents the backflow of blood from the great vessels into the ventricles.

Activity 1

Using the Heart Model to Study Heart Anatomy

Locate in Figure 30.2 all the structures described above. Then, observe the human heart model and laboratory charts, and identify the same structures without referring to the figure.

Pulmonary, Systemic, and Coronary Circulations

Pulmonary and Systemic Circuits

The heart functions as a double pump. The right side serves as the **pulmonary circuit,** shunting the carbon dioxide–rich blood entering its chambers to the lungs to unload carbon dioxide and pick up oxygen, and then back to the left side of the heart (**Figure 30.4,** p. 450). The function of the pulmonary circuit is strictly to provide for gas exchange. The second circuit, which carries oxygen-rich blood from the left heart through the body tissues and back to the right side of the heart, is called the **systemic circuit.**

The following steps describe the blood flow through the right side of the heart (pulmonary circuit):

1. The right atrium receives oxygen-poor blood from the body via the vena cavae (**superior vena cava** and **inferior vena cava**) and the coronary sinus.

2. From the right atrium, blood flows through the tricuspid valve to the right ventricle.

3. From the right ventricle, blood flows through the pulmonary valve into the **pulmonary trunk.**

4. The pulmonary trunk branches into left and right **pulmonary arteries,** which carry blood to the lungs, where the blood unloads carbon dioxide and picks up oxygen.

5. Oxygen-rich blood returns to the heart via four pulmonary veins.

The remaining steps describe the blood flow through the left side of the heart (systemic circuit):

6. Oxygen-rich blood enters the left atrium via four pulmonary veins.

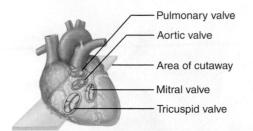

Pulmonary valve
Aortic valve
Area of cutaway
Mitral valve
Tricuspid valve

Figure 30.3 Heart valves. (a) Superior view of the two sets of heart valves (atria removed). **(b)** Photograph of the heart valves, superior view. **(c)** Photograph of the tricuspid valve. This inferior-to-superior view shows the valve as seen from the right ventricle. **(d)** Frontal section of the heart.

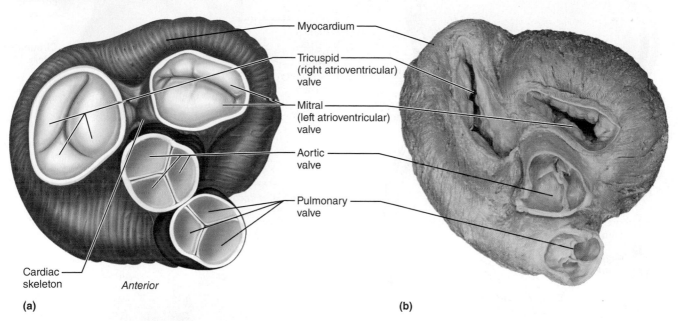

Myocardium

Tricuspid (right atrioventricular) valve

Mitral (left atrioventricular) valve

Aortic valve

Pulmonary valve

Cardiac skeleton

Anterior

(a)

(b)

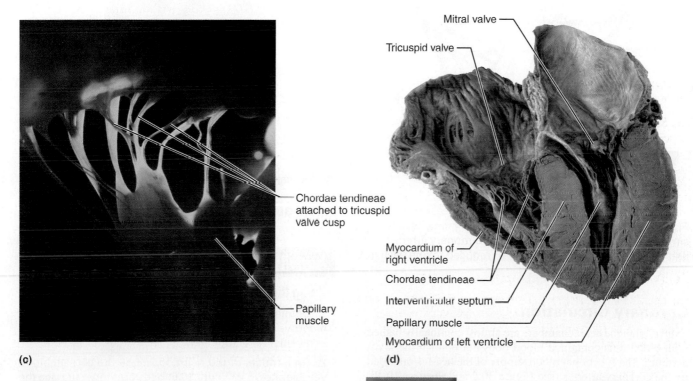

Chordae tendineae attached to tricuspid valve cusp

Papillary muscle

(c)

Mitral valve

Tricuspid valve

Myocardium of right ventricle

Chordae tendineae

Interventricular septum

Papillary muscle

Myocardium of left ventricle

(d)

30

7. From the left atrium, blood flows through the mitral valve to the left ventricle.

8. From the left ventricle, blood flows through the aortic valve to the **aorta.**

9. Oxygen-rich blood is delivered to the body tissues by the systemic arteries.

Activity 2

Tracing the Path of Blood Through the Heart

Use colored pencils to trace the pathway of a red blood cell through the heart by adding arrows to the frontal section diagram (Figure 30.2b). Use red arrows for the oxygen-rich blood and blue arrows for the carbon dioxide–rich blood.

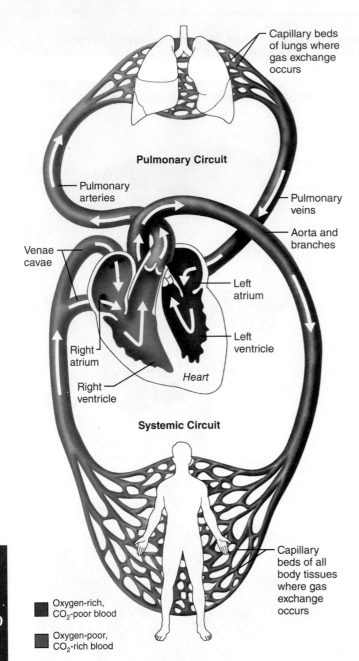

Figure 30.4 labels:
- Capillary beds of lungs where gas exchange occurs
- Pulmonary Circuit
- Pulmonary arteries
- Pulmonary veins
- Aorta and branches
- Venae cavae
- Left atrium
- Left ventricle
- Right atrium
- Right ventricle
- Heart
- Systemic Circuit
- Oxygen-rich, CO₂-poor blood
- Oxygen-poor, CO₂-rich blood
- Capillary beds of all body tissues where gas exchange occurs

Figure 30.4 The systemic and pulmonary circuits. For simplicity, the actual number of two pulmonary arteries and four pulmonary veins has been reduced to one each in this art.

Coronary Circulation

Even though the heart chambers are almost continually bathed with blood, this contained blood does not nourish the myocardium. The functional blood supply of the heart is provided by the coronary arteries (see Figure 30.2 and **Figure 30.5**). Table 30.1 summarizes the arteries that supply blood to the heart and the veins that drain the blood.

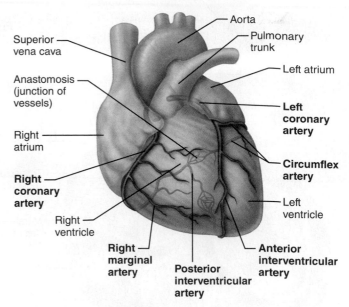

(a) The major coronary arteries

Figure 30.5(a) labels:
- Superior vena cava
- Aorta
- Pulmonary trunk
- Left atrium
- Anastomosis (junction of vessels)
- Left coronary artery
- Right atrium
- Circumflex artery
- Right coronary artery
- Left ventricle
- Right ventricle
- Anterior interventricular artery
- Right marginal artery
- Posterior interventricular artery

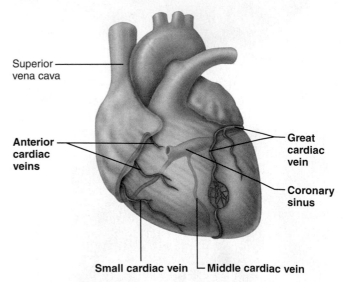

(b) The major cardiac veins

Figure 30.5(b) labels:
- Superior vena cava
- Anterior cardiac veins
- Great cardiac vein
- Coronary sinus
- Small cardiac vein
- Middle cardiac vein

Figure 30.5 Coronary circulation.

Activity 3

Using the Heart Model to Study Cardiac Circulation

1. Obtain a highlighter and highlight all the cardiac blood vessels in Figure 30.2a and c. Note how arteries and veins travel together.

2. On a model of the heart, locate all the cardiac blood vessels shown in Figure 30.5. Use your finger to trace the pathway of blood from the right coronary artery to the lateral aspect of the right side of the heart and back to the right atrium. Name the arteries and veins along the pathway. Trace the pathway of blood from the left coronary artery to the anterior ventricular walls and back to the right atrium. Name the arteries and veins along the pathway. Note that there are multiple different pathways to distribute blood to these parts of the heart.

Microscopic Anatomy of Cardiac Muscle

The cardiac cells, crisscrossed by connective tissue fibers for strength, are arranged in spiral or figure-8-shaped bundles (**Figure 30.6**, overview diagram). When the heart contracts, its internal chambers become smaller, forcing the blood into the large arteries leaving the heart.

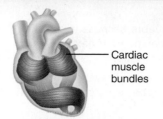

Cardiac muscle bundles

Activity 4

Examining Cardiac Muscle Tissue Anatomy

1. Observe the three-dimensional model of cardiac muscle, examining its branching cells and the areas where the cells interdigitate, the **intercalated discs.**

2. Compare the model of cardiac muscle to the model of skeletal muscle. Note the similarities and differences between the two kinds of muscle tissue.

3. Observe a longitudinal section of cardiac muscle under high power. Identify the nucleus, striations, intercalated discs, and sarcolemma of the individual cells, and compare your observations to Figure 30.6.

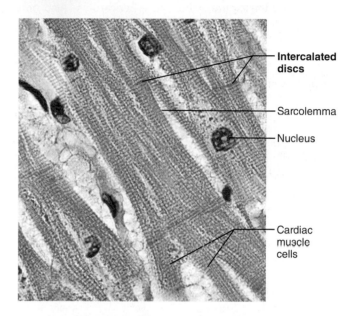

Intercalated discs

Sarcolemma

Nucleus

Cardiac muscle cells

Figure 30.6 Photomicrograph of cardiac muscle (665×). The overview diagram illustrates the circular and spiral arrangement of cardiac muscle bundles in the myocardium of the heart.

Table 30.1	Coronary Circulation (Figure 30.5)	
Arteries	**Description**	**Areas supplied/branches**
Right coronary artery (RCA)	Branches from the ascending aorta just above the aortic valve and encircles the heart in the coronary sulcus.	Its branches include the right marginal artery and the posterior interventricular artery.
Right marginal artery	Branches off the RCA and is located in the lateral portion of the right ventricle.	Supplies the lateral right side of the heart.
Posterior interventricular artery	Branches off the RCA and is located in the posterior interventricular sulcus.	Supplies the posterior walls of the ventricles and the posterior portion of the interventricular septum. Near the apex of the heart it merges (anastomoses) with the anterior interventricular artery.
Left coronary artery (LCA)	Branches from the ascending aorta and passes posterior to the pulmonary trunk.	Its branches include the anterior interventricular artery and the circumflex artery.
Anterior interventricular artery	Branches off the LCA and is located in the anterior interventricular sulcus. This artery is referred to clinically as the left anterior descending artery (LAD).	Supplies the anterior portion of the interventricular septum and the anterior walls of both ventricles.
Circumflex artery	Branches off the LCA; located in the coronary sulcus.	Supplies the left atrium and the posterior portion of the left ventricle.
Veins	**Description**	**Areas drained**
Great cardiac vein	Located in the anterior interventricular sulcus, parallel to the anterior interventricular artery.	Anterior portions of the right and left ventricles.
Middle cardiac vein	Located in the posterior interventricular sulcus, parallel to the posterior interventricular artery.	Posterior portions of the right and left ventricles.
Small cardiac vein	Located on the lateral right ventricle, parallel to the right marginal artery.	Lateral right ventricle.
Coronary sinus	Located in the coronary sulcus on the posterior surface of the heart; drains into the right atrium.	The entire heart; the great, middle and small cardiac veins all drain into the coronary sinus.
Anterior cardiac veins	Located on the anterior surface of the right atrium.	They drain directly into the right atrium.

30

DISSECTION

The Sheep Heart

Dissecting a sheep heart is valuable because it is similar in size and structure to the human heart. (Refer to **Figure 30.7** as you proceed with the dissection.)

1. Obtain a preserved sheep heart, a dissecting tray, dissecting instruments, a glass probe, a plastic ruler, and gloves.

2. Observe the texture of the fibrous pericardium. Also, note its point of attachment to the heart. Where is it attached?

_____

_____

3. If the fibrous pericardial sac is still intact, slit it open and cut it from its attachments. Observe the slippery parietal pericardium that lines the sac and the visceral pericardium (epicardium) that covers the heart wall. Using a sharp scalpel, carefully pull a little of the epicardium away from the myocardium. How do its position, thickness, and apposition to the heart differ from those of the parietal pericardium?

_____

_____

_____

4. Examine the external surface of the heart. Notice the accumulation of adipose tissue, which in many cases marks the separation of the chambers and the location of the coronary arteries. Carefully scrape away some of the fat with a scalpel to expose the coronary blood vessels.

5. Identify the base and apex of the heart, and then identify the two wrinkled **auricles**, earlike flaps of tissue projecting from the atria. The balance of the heart muscle is ventricular tissue. To identify the left ventricle, compress the ventricles on each side of the longitudinal fissures carrying the coronary blood vessels. The side that feels thicker and more solid is the left ventricle. The right ventricle feels much thinner and somewhat flabby when compressed. This difference reflects the greater demand placed on the left ventricle, which must pump blood through the much longer systemic circuit, a pathway with much higher resistance than the pulmonary circuit served by the right ventricle. Hold the heart in its anatomical position (Figure 30.7a), with the anterior surface uppermost. In this position the left ventricle composes the entire apex and the left side of the heart.

6. Identify the pulmonary trunk and the aorta extending from the superior aspect of the heart. The pulmonary trunk is more anterior, and you may see its division into the right and left pulmonary arteries if it has not been cut

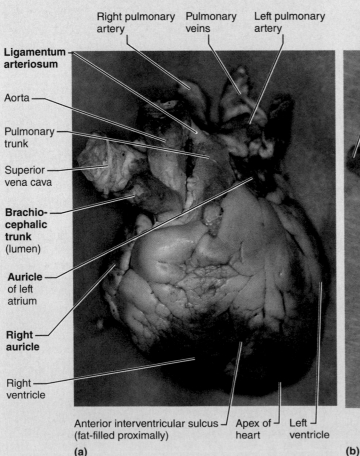

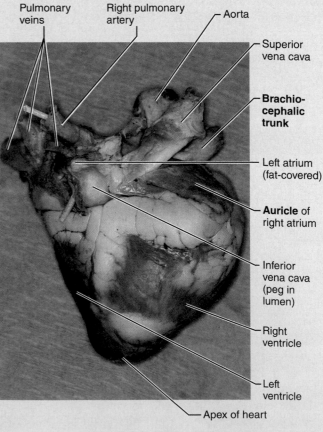

Figure 30.7 Anatomy of the sheep heart. (a) Anterior view. **(b)** Posterior view.

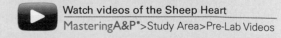

Watch videos of the Sheep Heart

MasteringA&P®>Study Area>Pre-Lab Videos

too closely to the heart. The thicker-walled aorta, which branches almost immediately, is located just beneath the pulmonary trunk. The first observable branch of the sheep aorta, the **brachiocephalic trunk,** is identifiable unless the aorta has been cut immediately as it leaves the heart.

Gently pull on the aorta with your gloved fingers or forceps to stretch it. Repeat with the venae cavae.

Which vessel is easier to stretch? _____

How does the elasticity of each vessel relate to its ability to withstand pressure?

_____

_____

Carefully clear away some of the fat between the pulmonary trunk and the aorta to expose the **ligamentum arteriosum,** a cordlike remnant of the **ductus arteriosus.** (In the fetus, the ductus arteriosus allows blood to pass directly from the pulmonary trunk to the aorta, thus bypassing the nonfunctional fetal lungs.)

7. Cut through the wall of the aorta until you see the aortic valve. Identify the two openings to the coronary arteries just above the valve. Insert a probe into one of these holes to see if you can follow the course of a coronary artery across the heart.

8. Turn the heart to view its posterior surface (compare it to the view in Figure 30.7b). Notice that the right and left ventricles appear equal-sized in this view. Try to identify the four thin-walled pulmonary veins entering the left atrium. Identify the superior and inferior venae cavae entering the right atrium. Because of the way the heart is trimmed, the pulmonary veins and superior vena cava may be very short or missing. If possible, compare the approximate diameter of the superior vena cava with the diameter of the aorta.

Which is larger? _____

Which has thicker walls? _____

Why do you suppose these differences exist?

_____

_____

_____

9. Insert a probe into the superior vena cava, through the right atrium, and out the inferior vena cava. Use scissors to cut along the probe so that you can view the interior of the right atrium. Observe the tricuspid valve.

How many cusps does it have? _____

10. Return to the pulmonary trunk and cut through its anterior wall until you can see the pulmonary valve (**Figure 30.8**, p. 454). How does its action differ from that of the tricuspid valve?

_____

_____

_____

Extend the cut through the pulmonary trunk into the right ventricle. Cut down, around, and up through the tricuspid valve to make the cut continuous with the cut across the right atrium (see Figure 30.8).

11. Reflect the cut edges of the superior vena cava, right atrium, and right ventricle to obtain the view seen in the dissection photo (Figure 30.8). Observe the comblike ridges of muscle throughout most of the right atrium. This is called **pectinate muscle** (*pectin* = comb). Identify, on the ventral atrial wall, the large opening of the inferior vena cava, and follow it to its external opening with a probe. Notice that the atrial walls in the vicinity of the venae cavae are smooth and lack the roughened appearance (pectinate musculature) of the other regions of the atrial walls. Just below the inferior vena caval opening, identify the opening of the **coronary sinus,** which returns venous blood of the coronary circulation to the right atrium. Nearby, locate an oval depression, the **fossa ovalis,** in the interatrial septum. This depression marks the site of an opening in the fetal heart, the **foramen ovale,** which allows blood to pass from the right to the left atrium, thus bypassing the fetal lungs.

12. Identify the papillary muscles in the right ventricle, and follow their attached chordae tendineae to the cusps of the tricuspid valve. Notice the pitted and ridged appearance (**trabeculae carneae**) of the inner ventricular muscle.

13. Identify the **moderator band** (septomarginal band), a bundle of cardiac muscle fibers connecting the interventricular septum to anterior papillary muscles. It contains a branch of the atrioventricular bundle and helps coordinate contraction of the ventricle.

14. Make a longitudinal incision through the left atrium and continue it into the left ventricle. Notice how much thicker the myocardium of the left ventricle is than that of the right ventricle. Measure the thickness of right and left ventricular walls in millimeters and record the numbers.

_____

How do your numbers compare with those of your classmates?

_____

Text continues on next page →

30

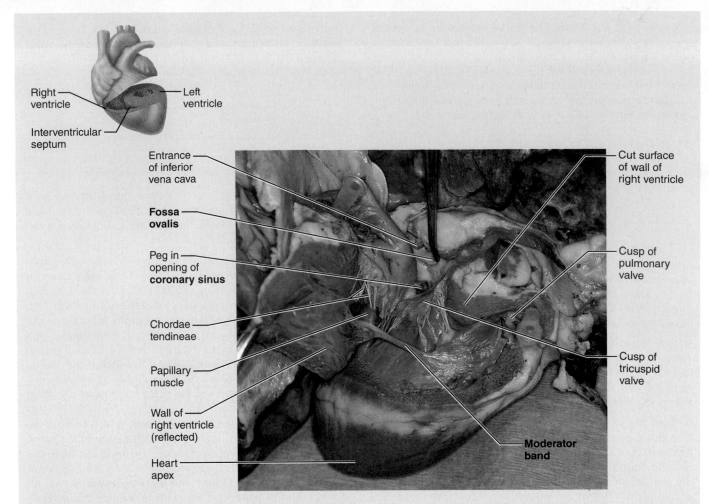

Right ventricle

Left ventricle

Interventricular septum

Entrance of inferior vena cava

Fossa ovalis

Peg in opening of **coronary sinus**

Chordae tendineae

Papillary muscle

Wall of right ventricle (reflected)

Heart apex

Cut surface of wall of right ventricle

Cusp of pulmonary valve

Cusp of tricuspid valve

Moderator band

Figure 30.8 Right side of the sheep heart opened and reflected to reveal internal structures. Overview diagram illustrates the anatomical differences between the right and left ventricles. The left ventricle has thicker walls, and its cavity is basically circular. By contrast, the right ventricle cavity is crescent-shaped and wraps around the left ventricle.

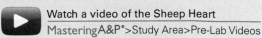

Watch a video of the Sheep Heart

MasteringA&P®>Study Area>Pre-Lab Videos

Compare the *shape* of the left ventricular cavity to the shape of the right ventricular cavity. (See overview diagram in Figure 30.8.)

_____

_____

Are the papillary muscles and chordae tendineae observed in

the right ventricle also present in the left ventricle? _____

Count the number of cusps in the mitral valve. How does this compare with the number seen in the tricuspid valve?

_____

How do the sheep valves compare with their human counterparts?

_____

15. Reflect the cut edges of the atrial wall, and attempt to locate the entry points of the pulmonary veins into the left atrium. Follow the pulmonary veins, if present, to the heart exterior with a probe. Notice how thin-walled these vessels are.

16. Dispose of the organic debris in the designated container, clean the dissecting tray and instruments with detergent and water, and wash the lab bench with bleach solution before leaving the laboratory.

30

REVIEW SHEET
Anatomy of the Heart

Name _____   Lab Time/Date _____

Gross Anatomy of the Human Heart

1. An anterior view of the heart is shown here. Match each structure listed on the left with the correct letter in the figure.

_____ 1. right atrium

_____ 2. right ventricle

_____ 3. left atrium

_____ 4. left ventricle

_____ 5. superior vena cava

_____ 6. inferior vena cava

_____ 7. ascending aorta

_____ 8. aortic arch

_____ 9. brachiocephalic trunk

_____ 10. left common carotid artery

_____ 11. left subclavian artery

_____ 12. pulmonary trunk

_____ 13. right pulmonary artery

_____ 14. left pulmonary artery

_____ 15. ligamentum arteriosum

_____ 16. right pulmonary veins

_____ 17. left pulmonary veins

_____ 18. right coronary artery

_____ 19. anterior cardiac vein

_____ 20. left coronary artery

_____ 21. circumflex artery

_____ 22. anterior interventricular artery

_____ 23. apex of heart

_____ 24. great cardiac vein

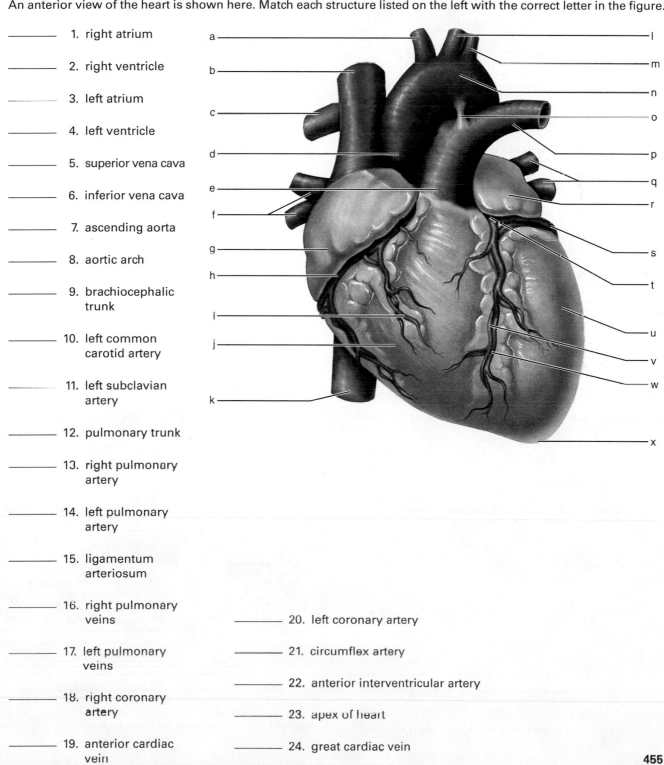

2. What is the function of the fluid that fills the pericardial sac? _____

_____

3. Match the terms in the key to the descriptions provided below. Some terms are used more than once.

_____ 1. location of the heart in the thorax

_____ 2. superior heart chambers

_____ 3. inferior heart chambers

_____ 4. visceral pericardium

_____ 5. receiving chambers of the heart

_____ 6. layer composed of cardiac muscle

_____ 7. provide nutrient blood to the heart muscle

_____ 8. lining of the heart chambers

_____ 9. actual "pumps" of the heart

_____ 10. drains blood into the right atrium

Key:

a. atria

b. coronary arteries

c. coronary sinus

d. endocardium

e. epicardium

f. mediastinum

g. myocardium

h. ventricles

4. What is the function of the valves found in the heart? _____

_____

5. What is the role of the chordae tendineae? _____

_____

Pulmonary, Systemic, and Coronary Circulations

6. A simple schematic of general circulation is shown below. Which circuit is missing from this diagram?

_____ Add to the diagram as best you can to make it depict the two circuits.

Label the two circuits.

7. Differentiate clearly between the roles of the pulmonary and systemic circuits. _____

_____

_____

8. Complete the following scheme of circulation of a red blood cell in the human body.

Right atrium through the tricuspid valve to the _____, through the _____

valve to the pulmonary trunk, to the _____, to the capillary beds of the lungs,

to the _____, to the _____

of the heart, through the _____ valve to the _____, through the _____

_____ valve to the _____, to the systemic arteries, to the _____

of the tissues, to the systemic veins, to the _____, _____

_____, and _____ entering the right atrium of the heart.

9. If the mitral valve does not close properly, which circuit is affected? _____

10. Why might a thrombus (blood clot) in the anterior descending branch of the left coronary artery cause sudden death?

_____

Microscopic Anatomy of Cardiac Muscle

11. How would you distinguish the structure of cardiac muscle from that of skeletal muscle? _____

_____

_____

12. Add the following terms to the photograph of cardiac muscle below.

    a.   intercalated disc   b.   nucleus of cardiac fiber   c.   striations   d.   cardiac muscle fiber

Describe the unique anatomical features of cardiac muscle. What role does the unique structure of cardiac muscle play in its function?

_____

_____

_____

_____

_____

Dissection of the Sheep Heart

13. During the sheep heart dissection, you were asked initially to identify the right and left ventricles without cutting into the heart. During this procedure, what differences did you observe between the two chambers?

_____

_____

When you measured thickness of ventricular walls, was the right or left ventricle thicker? _____

Knowing that structure and function are related, how would you say this structural difference reflects the relative functions

of these two heart chambers? _____

_____

14. Semilunar valves prevent backflow into the _____; mitral and tricuspid valves prevent backflow

into the _____. Using your own observations, explain how the operation of the semilunar

valves differs from that of the atrioventricular valves. _____

_____

_____

_____

15. Compare and contrast the structure of the mitral and tricuspid valves. _____

_____

16. Two remnants of fetal structures are observable in the heart—the ligamentum arteriosum and the fossa ovalis. What were the fetal heart structures called, where was each located, and what common purpose did they serve as functioning fetal structures?

_____

_____

_____

Conduction System of the Heart and Electrocardiography

Objectives

☐ State the function of the intrinsic conduction system of the heart.

☐ List and identify the elements of the intrinsic conduction system, and describe how impulses are initiated and conducted through this system and the myocardium.

☐ Interpret the ECG in terms of depolarization and repolarization events occurring in the myocardium; and identify the P, QRS, and T waves on an ECG recording using an ECG recorder or BIOPAC®.

☐ Define *tachycardia, bradycardia,* and *fibrillation.*

☐ Calculate the heart rate, durations of the QRS complex, P-R interval, and Q-T interval from an ECG obtained during the laboratory period, and recognize normal values for the durations of these events.

☐ Describe and explain the changes in the ECG observed during experimental conditions such as exercise or breath holding.

Materials

• ECG or BIOPAC® equipment:*

ECG recording apparatus, electrode paste, alcohol swabs, rubber straps or disposable electrodes

🌊 **BIOPAC®** BIOPAC BSL System with BSL software version 3.7.5 to 3.7.7 (for Windows 7/Vista/XP or Mac OS X 10.4-10.6), data acquisition unit MP36/35 or MP45, PC or Mac computer, Biopac Student Lab electrode lead set, disposable vinyl electrodes.

Instructors using the MP36/35/30 data acquisition unit with BSL software versions earlier than 3.7.5 (for Windows or Mac) will need slightly different channel settings and collection strategies. Instructions for using

Text continues on next page. →

***Note:** Instructions for using PowerLab® equipment can be found on MasteringA&P.*

MasteringA&P®

For related exercise study tools, go to the Study Area of **MasteringA&P.** There you will find:

• Practice Anatomy Lab PAL
• A&PFlix *A&PFlix*
• PhysioEx **PEx**
• Practice quizzes, Histology Atlas, eText, Videos, and more!

Pre-Lab Quiz

1. Circle True or False. Cardiac muscle cells are electrically connected by gap junctions and behave as a single unit.

2. Because it sets the rate of depolarization for the normal heart, the _____ node is known as the pacemaker of the heart.
   a. atrioventricular   **b.** Purkinje   **c.** sinoatrial

3. Circle True or False. Stimulation by the nerves of the autonomic nervous system is essential for cardiac muscle to contract.

4. Today you will create a graphic recording of the electrical changes that occur during a cardiac cycle. This is known as an:
   a. electrocardiogram
   b. electroencephalogram
   c. electromyogram

5. Circle the correct underlined term. The typical ECG has three / six normally recognizable deflection waves.

6. In a typical ECG, the _____ wave signals the depolarization of the atria immediately before they contract.
   a. P   **c.** R
   b. Q   **d.** T

7. Circle True or False. The repolarization of the atria is usually masked by the large QRS complex.

8. Circle the correct underlined term. A heart rate over 100 beats/minute is known as <u>tachycardia</u> / <u>bradycardia</u>.

9. How many electrodes will you place on your subject for today's activity if you use a standard ECG apparatus?
   a. 3   **c.** 10
   b. 4   **d.** 12

10. Circle True or False. ECG can be used to calculate heart rate.

the older data acquisition unit can be found on MasteringA&P.

- Cot or lab table; pillow (optional)
- Millimeter ruler

Heart contraction results from a series of depolarization waves that travel through the heart preliminary to each beat. Because cardiac muscle cells are electrically connected by gap junctions, the entire myocardium behaves like a single unit, a **functional syncytium.**

The Intrinsic Conduction System

The ability of cardiac muscle to beat is intrinsic—it does not depend on impulses from the nervous system to initiate its contraction and will continue to contract rhythmically even if all nerve connections are severed. The **intrinsic conduction system** of the heart consists of **cardiac pacemaker cells.** The intrinsic conduction system ensures that heart muscle depolarizes in an orderly and sequential manner, from atria to ventricles, and that the heart beats as a coordinated unit.

The components of the intrinsic conduction system include the **sinoatrial (SA) node,** located in the right atrium just inferior to the entrance to the superior vena cava; the **atrioventricular (AV) node** in the lower atrial septum at the junction of the atria and ventricles; the **AV bundle (bundle of His)** and right and left **bundle branches,** located in the interventricular septum; and the **subendocardial conducting network,** also called **Purkinje fibers** (**Figure 31.1**).

Note that the atria and ventricles are separated from one another by a region of electrically inert connective tissue, so the depolarization wave can be transmitted to the ventricles only via the tract between the AV node and AV bundle. Thus, any damage to the AV node-bundle pathway partially or totally insulates the ventricles from the influence of the SA node.

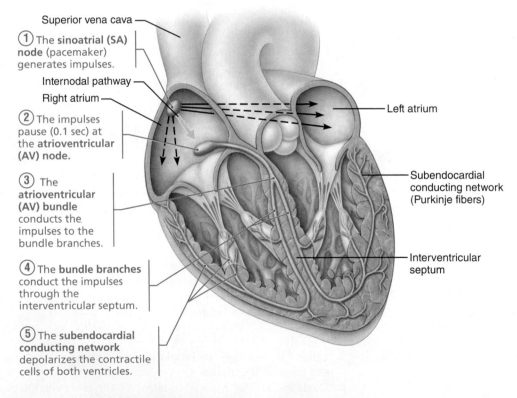

Figure 31.1 The intrinsic conduction system of the heart. Impulses travel through the heart in order ① to ⑤ following the yellow pathway. Dashed-line arrows indicate transmission of the impulse from the SA node through the atria. Solid yellow arrow indicates transmission of the impulse from the SA node to the AV node via the internodal pathway.

Superior vena cava

① The **sinoatrial (SA) node** (pacemaker) generates impulses.

Internodal pathway

Right atrium

② The impulses pause (0.1 sec) at the **atrioventricular (AV) node.**

③ The **atrioventricular (AV) bundle** conducts the impulses to the bundle branches.

④ The **bundle branches** conduct the impulses through the interventricular septum.

⑤ The **subendocardial conducting network** depolarizes the contractile cells of both ventricles.

Left atrium

Subendocardial conducting network (Purkinje fibers)

Interventricular septum

31

Electrocardiography

The conduction of impulses through the heart generates electrical currents that eventually spread throughout the body. These impulses can be detected on the body's surface and recorded with an instrument called an *electrocardiograph*. The graphic recording of the electrical changes occurring during the cardiac cycle is called an **electrocardiogram (ECG or EKG)** (**Figure 31.2**). A typical ECG has three recognizable deflection waves: the P wave, the QRS complex, and the T wave. For analysis, the ECG is divided into segments and intervals. A **segment** is a region between two waves. For example, the S-T segment is the region between the end of the S deflection and the start of the T wave. An **interval** is a region that contains a segment and one or more waves. For example, the Q-T interval includes the S-T segment as well as the QRS complex and the T wave. Boundaries for waves as well as some commonly measured segments and intervals are described in **Table 31.1**. The deflection waves of an ECG correlate to the depolarization and repolarization of the heart's chambers (**Figure 31.3**, p. 462).

Abnormalities of the deflection waves and changes in the time intervals of the ECG are useful in detecting myocardial infarcts (heart attacks) or problems with the conduction system of the heart.

Table 31.1	Boundaries of Each ECG Component
Feature	**Boundaries**
P wave	Start of P deflection to return to baseline
P-R interval	Start of P deflection to start of Q deflection
QRS complex	Start of Q deflection to S return to baseline
S-T segment	End of S deflection to start of T wave
Q-T interval	Start of Q deflection to end of T wave
T wave	Start of T deflection to return to baseline
T-P segment	End of T wave to start of next P wave
R-R interval	Peak of R wave to peak of next R wave

Table 31.2 summarizes some examples of abnormal electrocardiogram tracings and their possible clinical significance.

A heart rate over 100 beats/min is referred to as **tachycardia;** a rate below 60 beats/min is **bradycardia.** Although neither condition is pathological, prolonged tachycardia may progress to **fibrillation,** a condition of rapid uncoordinated heart contractions. Bradycardia in athletes is a positive finding; that is, it indicates an increased efficiency of cardiac functioning. Because *stroke volume* (the amount of blood ejected by a ventricle with each contraction) increases with physical conditioning, the heart can contract more slowly and still meet circulatory demands.

Twelve standard leads are used to record an ECG for diagnostic purposes. Three of these are bipolar leads that measure the voltage difference between the arms, or an arm and a leg, and nine are unipolar leads. Together the 12 leads provide a fairly comprehensive picture of the electrical activity of the heart.

For this investigation, four electrodes are used (**Figure 31.4**, p. 462), and results are obtained from the three *standard limb leads* (also shown in Figure 31.4). Several types of physiographs or ECG recorders are available. Your instructor will provide specific directions on how to set up and use the available

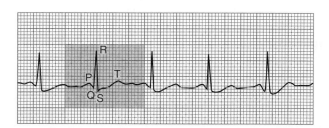

(a)

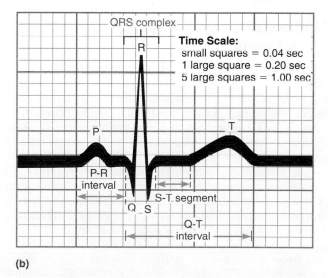

(b)

Figure 31.2 The normal electrocardiogram. (a) Regular sinus rhythm. **(b)** Waves, segments, and intervals of a normal ECG.

Table 31.2	Examples of Abnormal ECGs and Possible Clinical Significance
Finding	**Possible clinical significance**
Enlarged R wave	Enlarged ventricles.
Prolonged P-R interval	First-degree heart block. The signal from the SA node to the AV node is delayed longer than normal.
Prolonged Q-T interval (when compared to the R-R interval)	Increased risk of ventricular arrhythmias. This interval corresponds to the beginning of ventricular depolarization through ventricular repolarization.
S-T segment elevated from baseline	Myocardial infarction (heart attack).

31

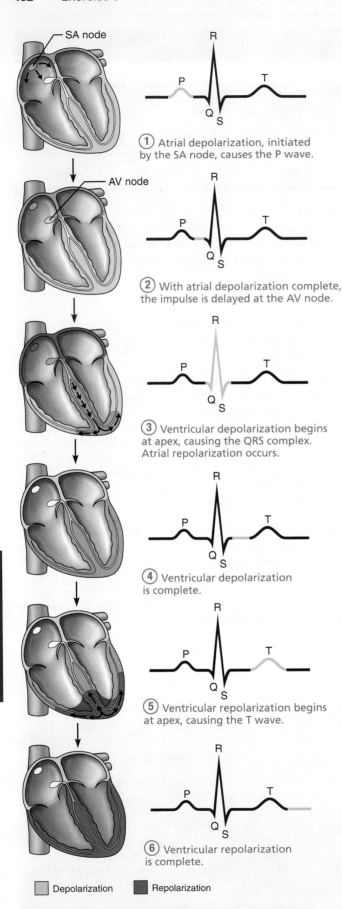

1 Atrial depolarization, initiated by the SA node, causes the P wave.

2 With atrial depolarization complete, the impulse is delayed at the AV node.

3 Ventricular depolarization begins at apex, causing the QRS complex. Atrial repolarization occurs.

4 Ventricular depolarization is complete.

5 Ventricular repolarization begins at apex, causing the T wave.

6 Ventricular repolarization is complete.

☐ Depolarization   ■ Repolarization

Figure 31.3 The sequence of depolarization and repolarization of the heart related to the deflection waves of an ECG tracing.

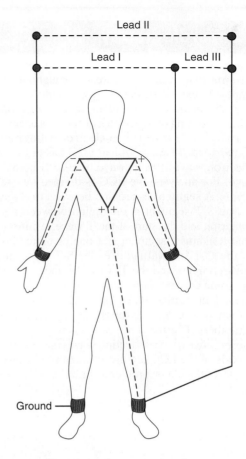

Figure 31.4 ECG recording positions for the standard limb leads.

 Prepare for lab: Watch the Pre-Lab Video
MasteringA&P®>Study Area>Pre-Lab Videos

apparatus if standard ECG apparatus is used (Activity 1A). Instructions for use of BIOPAC® apparatus (Activity 1B) follow (pp. 464–468).

Understanding the Standard Limb Leads

As you might expect, electrical activity recorded by any lead depends on the location and orientation of the recording electrodes. Clinically, it is assumed that the heart lies in the center of a triangle with sides of equal lengths *(Einthoven's triangle)* and that the recording connections are made at the corners of that triangle. But in practice, the electrodes connected to each arm and to the left leg are considered to connect to the triangle corners. The standard limb leads record the voltages generated between any two of the connections. A recording using lead I (RA-LA), which connects the right arm (RA) and the left arm (LA), is most sensitive to electrical activity spreading horizontally across the heart. Lead II (RA-LL) and lead III (LA-LL) record activity along the vertical axis (from the base of the heart to its apex) but from different orientations. The significance of Einthoven's triangle is that the sum of the voltages of leads I and III equals that in lead II (Einthoven's law). Hence, if the voltages of two of the standard leads are recorded, that of the third lead can be determined mathematically.

Activity 1A

Recording ECGs Using a Standard ECG Apparatus

Preparing the Subject

1. If using electrodes that require gel, place the gel on four electrode plates and position each electrode as follows after scrubbing the skin at the attachment site with an alcohol swab. Attach an electrode to the anterior surface of each forearm, about 5 to 8 cm (2 to 3 in.) above the wrist, and secure them with rubber straps. In the same manner, attach an electrode to each leg, approximately 5 to 8 cm above the ankle. Disposable electrodes may be placed directly on the subject in the same areas.

2. Attach the appropriate tips of the patient cable to the electrodes. The cable leads are marked RA (right arm), LA (left arm), LL (left leg), and RL (right leg, the ground).

Making a Baseline Recording

1. Position the subject comfortably in a supine position on a cot, or sitting relaxed on a laboratory chair.

2. Turn on the power switch, and adjust the sensitivity knob to 1. Set the paper speed to 25 mm/sec and the lead selector to the position corresponding to recording from lead I (RA-LA).

3. Set the control knob at the **RUN** position and record the subject's at-rest ECG from lead I for 2 to 3 minutes or until the recording stabilizes. The subject should try to relax and not move unnecessarily, because the skeletal muscle action potentials will also be recorded.

4. Stop the recording and mark it "lead I."

5. Repeat the recording procedure for leads II (RA-LL) and III (LA-LL).

6. Each student should take a representative portion of one of the lead recordings and label the record with the name of the subject and the lead used. Identify and label the P, QRS, and T waves. The calculations you perform for your recording should be based on the following information: Because the paper speed was 25 mm/sec, each millimeter of paper corresponds to a time interval of 0.04 sec. Thus, if an interval requires 4 mm of paper, its duration is 4 mm × 0.04 sec/mm = 0.16 sec.

7. Calculate the heart rate. Obtain a millimeter ruler and measure the R to R interval. Enter this value into the following equation to find the time for one heartbeat.

_____ mm/beat × 0.04 sec/mm = _____ sec/beat

Now find the beats per minute, or heart rate, by using the figure just calculated for seconds per beat in the following equation:

60 sec/min ÷ (sec/beat) = beats/min

Measure the QRS complex, and calculate its duration.

_____

Measure the Q-T interval, and calculate its duration.

_____

Measure the P-R interval, and calculate its duration.

_____

Are the calculated values within normal limits?

_____

8. At the bottom of this page, attach sections of the ECG recordings from leads I through III. Make sure you indicate the paper speed, lead, and subject's name on each tracing. Also record the heart rate on the tracing.

> **WHY THIS MATTERS** | **Atrial Fibrillation (AF)**
>
> During atrial fibrillation, or AF, the atria spasm instead of contracting as a coordinated unit, which leads to pooling of blood in the atria. Atrial fibrillation is a result of damage to the intrinsic conduction system. During afib, the atria generate as many as 500 action potentials per minute, much faster than the usual 100 action potentials per minute generated by the SA node. Multiple signals flood the AV node, but it can't repolarize fast enough to pass on all of these action potentials. The rate of contraction of the ventricles may increase or decrease, contributing to the irregular, but often rapid, heartbeat. The underlying cause of AF is usually other heart conditions, such as hypertension and coronary heart disease. ■

Recording the ECG for Running in Place

1. Make sure the electrodes are securely attached to prevent electrode movement while recording the ECG.

2. Set the paper speed to 25 mm/sec, and prepare to make the recording using lead I.

3. Record the ECG while the subject is running in place for 3 minutes. Then have the subject sit down, but continue to record the ECG for an additional 4 minutes. *Mark the recording* at the end of the 3 minutes of running and at 1 minute after cessation of activity.

4. Stop the recording. Calculate the beats/min during the third minute of running, at 1 minute after exercise, and at 4 minutes after exercise. Record below:

_____ beats/min while running in place

_____ beats/min at 1 minute after exercise

_____ beats/min at 4 minutes after exercise

5. Compare this recording with the previous recording from lead I. Which intervals are shorter in the "running" recording?

_____

Text continues on next page. ➡

31

6. Does the subject's heart rate return to resting level by 4 minutes after exercise?

_____

Recording the ECG During Breath Holding

1. Position the subject comfortably in the sitting position.

2. Using lead I and a paper speed of 25 mm/sec, begin the recording. After approximately 10 seconds, instruct the subject to begin breath holding, and mark the record to indicate the onset of the 1-minute breath-holding interval.

3. Stop the recording after 1 minute, and remind the subject to breathe. Calculate the beats/minute during the 1-minute experimental (breath-holding) period.

Beats/min during breath holding: _____

4. Compare this recording with the lead I recording obtained under resting conditions.

What differences do you see? _____

_____

Attempt to explain the physiological reason for the differences you have seen. (Hint: A good place to start might be to check hypoventilation or the role of the respiratory system in acid-base balance of the blood.)

_____

_____

_____

_____

Activity 1B

Electrocardiography Using BIOPAC®

In this activity, you will record the electrical activity of the heart under three different conditions: (1) while the subject is lying down, (2) after the subject sits up and breathes normally, and (3) after the subject has exercised and is breathing deeply.

In order to obtain a clear ECG, it is important that the subject:

- Remain still during the recording.
- Refrain from laughing or talking during the recording.
- When in the sitting position, keep arms and legs steady and relaxed.
- Remove metal watches and bracelets.

Setting Up the Equipment

1. Connect the BIOPAC® unit to the computer and turn the computer **ON**.

2. Make sure the BIOPAC® unit is **OFF**.

3. Plug in the equipment (as shown in **Figure 31.5**):

- Electrode lead set—CH 1

4. Turn the BIOPAC® unit **ON**.

5. Place the three electrodes on the subject (as shown in **Figure 31.6**), and attach the electrode leads according to the colors indicated. The electrodes should be placed on the medial surface of each leg, 5 to 8 cm (2 to 3 in.) superior to the ankle. The other electrode should be placed on the right anterior forearm 5 to 8 cm above the wrist.

6. The subject should lie down and relax in a comfortable position with eyes closed. A chair or place to sit up should be available nearby.

7. Start the Biopac Student Lab program on the computer by double-clicking the icon on the desktop or by following your instructor's guidance.

8. Select lesson **L05-ECG-1** from the menu, and click **OK**.

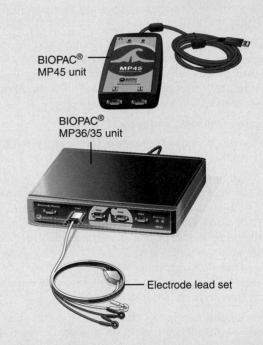

BIOPAC® MP45 unit

BIOPAC® MP36/35 unit

Electrode lead set

Figure 31.5 Setting up the BIOPAC® unit. Plug the electrode lead set into Channel 1. Leads are shown plugged into the MP36/35 unit.

9. Type in a filename that will save this subject's data on the computer hard drive. You may want to use the subject's last name followed by ECG-1 (for example, SmithECG-1), then click **OK**.

10. Because we are not recording all available lesson options, click the File Menu, choose **Lesson Preferences**, choose **Heart Rate Data**, and click **OK**. Choose **Do Not Calculate** and click **OK**. Click the file menu and choose **Lesson Preferences** again; select **Lesson Segments** and click **OK**; click the box for **Deep Breathing** to deselect it; and then click **OK**.

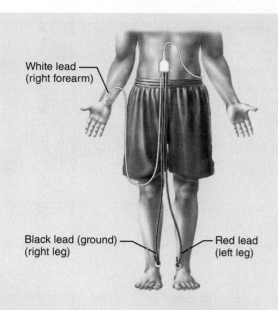

White lead
(right forearm)

Black lead (ground)
(right leg)

Red lead
(left leg)

Figure 31.6 Placement of electrodes and the appropriate attachment of electrode leads by color.

Calibrating the Equipment

- Examine the electrodes and the electrode leads to be certain they are properly attached.

1. The subject must remain supine, still, and relaxed. With the subject in a still position, click **Calibrate**. This will initiate the process whereby the computer will automatically establish parameters to record the data.

2. The calibration procedure will stop automatically after 8 seconds.

3. Observe the recording of the calibration data, which should look similar to the data example (**Figure 31.7**).

- If the data look very different, click **Redo Calibration** and repeat the steps above.

- If the data look similar, proceed to the next section. *Don't* click **Done** until you have completed all 3 segments.

Recording Segment 1: Subject Lying Down

1. To prepare for the recording, remind the subject to remain still and relaxed while lying down.

2. Click **Continue** and when prepared, click **Record** and gather data for 20 seconds. At the end of 20 seconds, click **Suspend**.

3. Observe the data, which should look similar to the data example (**Figure 31.8**).

- If the data look very different, click **Redo** and repeat the steps above. Be certain to check attachment of the electrodes and leads, and remind the subject not to move, talk, or laugh.

- If the data look similar, move on to the next recording segment.

Recording Segment 2: After Subject Sits Up, with Normal Breathing

1. Tell the subject to be ready to sit up in the designated location. With the exception of breathing, the subject

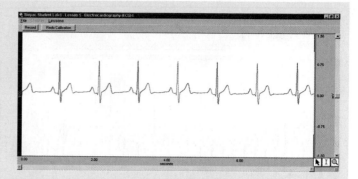

Figure 31.7 Example of calibration data.

should try to remain motionless after assuming the seated position. *If the subject moves too much during recording after sitting up, unwanted skeletal muscle artifacts will affect the recording.*

2. Click **Continue** and when prepared, instruct the subject to sit up. Immediately after the subject assumes a motionless state, click **Record,** and the data will begin recording.

3. At the end of 20 seconds, click **Suspend** to stop recording.

4. Observe the data, which should look similar to the data example (**Figure 31.9**, p. 466).

- If the data look very different, have the subject lie down, then click **Redo**. Be certain to check attachment of the electrodes, then repeat steps 1–4 above. Do not click **Record** until the subject is motionless.

- If the data look similar, move on to the next recording segment.

Recording Segment 3: After Subject Exercises, with Deep Breathing

1. Click **Continue,** but *don't* click **Record** until after the subject has exercised. Remove the electrode pinch connectors from the electrodes on the subject.

2. Have the subject do a brief round of exercise, such as jumping jacks or running in place for 1 minute, in order to elevate the heart rate.

3. As quickly as possible after the exercise, have the subject resume a motionless, seated position. Reattach the pinch

Text continues on next page. →

31

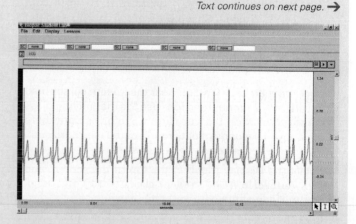

Figure 31.8 Example of ECG data while the subject is lying down.

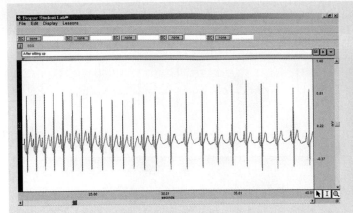

Figure 31.9 Example of ECG data after the subject sits up and breathes normally.

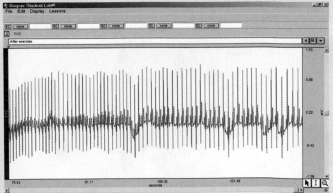

Figure 31.10 Example of ECG data after the subject exercises.

connectors. Once again, if the subject moves too much during recording, unwanted skeletal muscle artifacts will affect the data. After exercise, the subject is likely to be breathing deeply but otherwise should remain as still as possible.

4. Immediately after the subject assumes a motionless, seated state, click **Record,** and the data will begin recording. Record the ECG for 60 seconds in order to observe post-exercise recovery.

5. After 60 seconds, click **Suspend** to stop recording.

6. Observe the data, which should look similar to the data example (**Figure 31.10**).

- If the data look very different, click **Redo** and repeat the steps above. Be certain to check attachment of the electrodes and leads, and remember not to click **Record** until the subject is motionless.

7. When finished, click **Done** and then **Yes.** Remove the electrodes from the subject.

8. A pop-up window will appear. To record from another subject, select **Record from another subject** and return to step 5 under Setting Up the Equipment. If continuing to the Data Analysis section, select **Analyze current data file** and proceed to step 2 of the Data Analysis section.

Data Analysis

1. If just starting the BIOPAC® program to perform data analysis, enter **Review Saved Data** mode and choose the file with the subject's ECG data (for example, SmithECG-1).

2. Use the following tools to adjust the data in order to clearly view and analyze four consecutive cardiac cycles:

- Click the magnifying glass (near the I-beam cursor box) to activate the **zoom** function. Use the magnifying glass cursor to click on the very first waveforms until there are about 4 seconds of data represented (see horizontal time scale at the bottom of the screen).

- Select the **Display** menu at the top of the screen, and click **Autoscale Waveforms** (or click Ctrl + Y). This function will adjust the data for better viewing.

- Click the **Adjust Baseline** button. Two new buttons will appear; simply click these buttons to move the waveforms **Up** or **Down** so they appear clearly in the center of the display window. Once they are centered, click **Exit.**

3. Note that the first two pairs of channel/measurement boxes at the top of the screen are set to Delta T and bpm.

Channel	Measurement	Data
CH 1	Delta T	ECG
CH 1	bpm	ECG

Analysis of Segment 1: Subject Lying Down

1. Use the arrow cursor and click the I-beam cursor box for the "area selection" function.

2. First measure **Delta T** and **bpm** in Segment 1 (approximately seconds 0–20). Using the I-beam cursor, highlight from the peak of one R wave to the peak of the next R wave (as shown in **Figure 31.11**).

3. Observe that the computer automatically calculates the **Delta T** and **bpm** for the selected area. These measurements represent the following:

Delta T (difference in time): Computes the elapsed time between the beginning and end of the highlighted area

bpm (beats per minute): Computes the beats per minute when the area from the R wave of one cycle to the R wave of another cycle is highlighted

4. Record these data in the **Segment 1 Samples chart** under R to R Sample 1 (round to the nearest 0.01 second and 0.1 beat per minute).

5. Using the I-beam cursor, highlight two other pairs of R to R areas in this segment. Record the data in the same chart under Samples 2 and 3.

6. Calculate the means of the data in this chart.

7. Next, use the **zoom, Autoscale Waveforms,** and **Adjust Baseline** tools described in step 2 to focus in on one ECG waveform within Segment 1. (See the example in **Figure 31.12**).

8. Once a single ECG waveform is centered for analysis, click the I-beam cursor box to activate the "area selection" function.

9. Using the highlighting function and **Delta T** computation, measure the duration of every component of the

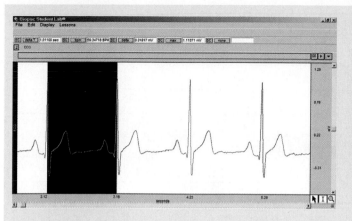

Figure 31.11 Example of highlighting from R wave to R wave.

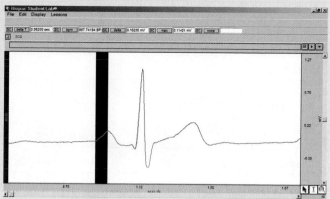

Figure 31.12 Example of a single ECG waveform with the first part of the P wave highlighted.

Segment 1 Samples for Delta T and bpm					
Measure	**Channel**	**R to R Sample 1**	**R to R Sample 2**	**R to R Sample 3**	**Mean**
Delta T	CH 1				
bpm	CH 1				

ECG waveform. (Refer to Figure 31.2b and Table 31.1 for guidance in highlighting each component.)

10. Highlight each component of one cycle. Observe the elapsed time, and record this data under Cycle 1 in the **Segment 1 Elapsed Time chart**.

11. Scroll along the horizontal axis at the bottom of the data to view and analyze two additional cycles in Segment 1. Record the elapsed time for every component of Cycle 2 and Cycle 3 in the Segment 1 Elapsed Time chart.

12. In the same chart, calculate the means for the three cycles of data and record.

Analysis of Segment 2: Subject Sitting Up and Breathing Normally

1. Scroll along the horizontal time bar until you reach the data for Segment 2 (approximately seconds 20–40). A marker with "Seated" should denote the beginning of this data.

2. As in the analysis of Segment 1, use the I-beam tool to highlight and measure the **Delta T** and **bpm** between three different pairs of R waves in this segment, and record the data in the **Segment 2 Samples chart** on p. 468.

Analysis of Segment 3: After Exercise with Deep Breathing

1. Scroll along the horizontal time bar until you reach the data for Segment 3 (approximately seconds 40–100). A marker with "After exercise" should denote the beginning of this data.

2. As before, use the I-beam tool to highlight and measure the **Delta T** and **bpm** between three pairs of R waves in this segment, and record the data in the **Segment 3 Samples chart** on p. 468.

Segment 1 Elapsed Time for ECG Components (seconds)				
Component	**Cycle 1**	**Cycle 2**	**Cycle 3**	**Mean**
P wave				
P-R interval				
QRS complex				
S-T segment				
Q-T interval				
T wave				
T-P segment				
R-R interval				

3. Using the instructions for steps 7–9 in the section Analysis of Segment 1, highlight and observe the elapsed time for each component of one cycle, and record these data under Cycle 1 in the **Segment 3 Elapsed Time chart** on p. 468.

4. When finished, **Exit** from the file menu to quit.

Compare the average **Delta T** times and average **bpm** between the data in Segment 1 (lying down) and the data in Segment 3 (after exercise). Which is greater in each case?

_____

_____

Text continues on next page. →

31

Segment 2 Samples for Delta T and bpm					
Measure	Channel	R to R Sample 1	R to R Sample 2	R to R Sample 3	Mean
Delta T	CH 1				
bpm	CH 1				

Segment 3 Samples for Delta T and bpm					
Measure	Channel	R to R Sample 1	R to R Sample 2	R to R Sample 3	Mean
Delta T	CH 1				
bpm	CH 1				

What is the relationship between the elapsed time (**Delta T**) between R waves and the heart rate?

_____

What event does the period between R waves correspond to?

_____

Is there a change in heart rate when the subject makes the transition from lying down (Segment 1) to a sitting position (Segment 2)?

_____

Examine the average duration of each of the ECG components in Segment 1 and the data in Segment 3. In the Average Duration chart, record the average values for Segment 1 and the data for Segment 3. Draw a circle around those measures that fit within the normal range.

Compare the Q-T intervals in the data while the subject is at rest versus after exercise; this interval corresponds closely to the duration of contraction of the ventricles. Describe and explain any difference.

_____

Compare the duration in the period from the end of each T wave to the next P wave while the subject is at rest versus after exercise. Describe and explain any difference.

_____

_____

A patient presents with a P-R interval three times longer than the normal duration. What might be the cause of this abnormality?

_____

_____

Segment 3 Elapsed Time for ECG Components (seconds)	
Component	Cycle 1
P wave	
P-R interval	
QRS complex	
S-T segment	
Q-T interval	
T wave	
T-P segment	
R-R interval	

Average Duration for ECG Components			
ECG component	Normal duration (seconds)	Segment 1 (lying down)	Segment 3 (post-exercise)
P wave	0.07–0.18		
P-R interval	0.12–0.20		
QRS complex	0.06–0.12		
S-T segment	<0.20		
Q-T interval	0.32–0.38		
T wave	0.10–0.25		
T-P segment	0–0.40		
R-R interval	varies		

31

REVIEW SHEET
Conduction System of the Heart and Electrocardiography

469

Name _____ Lab Time/Date _____

The Intrinsic Conduction System

1. List the elements of the intrinsic conduction system in order, starting from the SA node.

   SA node → _____ → _____ →

   _____ → _____

   At what structure in the transmission sequence is the impulse temporarily delayed? _____

   Why? _____

   _____

2. Even though cardiac muscle has an inherent ability to beat, the intrinsic conduction system plays a critical role in heart physiology.

   What is that role? _____

   _____

Electrocardiography

3. Define *ECG.* _____

   _____

4. Draw an ECG wave form representing one heartbeat. Label the P wave, QRS complex, and T wave; the P-R interval; the S-T segment, and the Q-T interval.

   ┌───┐
   │   │
   │   │
   │   │
   │   │
   │   │
   │   │
   │   │
   └───┘

5. Why does heart rate increase during running? _____

   _____

   _____

6. Describe what happens in the cardiac cycle in the following situations.

   1. immediately before the P wave: _____

   2. during the P wave: _____

   3. immediately after the P wave: _____

   4. during the QRS complex: _____

   5. immediately after the QRS complex (S-T segment): _____

   6. during the T wave: _____

7. Define the following terms.

   1. *tachycardia:* _____

   2. *bradycardia:* _____

   3. *fibrillation:* _____

8. Abnormalities of heart valves can be detected more accurately by auscultation than by electrocardiography. Why is this so?

   _____

   _____

   _____

WHY THIS MATTERS | 9. Given what you know about the correlation between the ECG waves and the electrical events in the heart, what wave of the ECG tracing would you expect to be affected in atrial fibrillation? Explain.

   _____

   10. Which is more serious, atrial fibrillation or ventricular fibrillation? _____

   Why? _____

Anatomy of Blood Vessels

Objectives

☐ Describe the tunics of blood vessel walls, and state the function of each layer.

☐ Correlate differences in artery, vein, and capillary structure with the functions of these vessels.

☐ Recognize a cross-sectional view of an artery and vein when provided with a microscopic view or appropriate image.

☐ List and identify the major arteries arising from the aorta, and indicate the body region supplied by each.

☐ Describe the cerebral arterial circle, and discuss its importance in the body.

☐ List and identify the major veins draining into the superior and inferior venae cavae, and indicate the body regions drained.

☐ Describe these special circulations in the body: pulmonary circulation, hepatic portal system, and fetal circulation, and discuss the important features of each.

Materials

- Compound microscope
- Prepared microscope slides showing cross sections of an artery and vein
- Anatomical charts of human arteries and veins (or a three-dimensional model of the human circulatory system)
- Anatomical charts of the following specialized circulations: pulmonary circulation, hepatic portal circulation, fetal circulation, arterial supply of the brain (or a brain model showing this circulation)

✂ For instructions on animal dissections, see the dissection exercises (starting on p. 705) in the cat and fetal pig editions of this manual.

MasteringA&P®

For related exercise study tools, go to the Study Area of **MasteringA&P.** There you will find:

- Practice Anatomy Lab PAL
- PhysioEx PEx
- A&PFlix **A&PFlix**
- Practice quizzes, Histology Atlas, eText, Videos, and more!

Pre-Lab Quiz

1. Circle the correct underlined term. <u>Arteries</u> / <u>Veins</u> drain tissues and return blood to the heart.

2. Circle True or False. Gas exchange takes place between tissue cells and blood through capillary walls.

3. The _____ is the largest artery of the body.
   - **a.** aorta
   - **b.** carotid artery
   - **c.** femoral artery
   - **d.** subclavian artery

4. Circle the correct underlined term. The largest branch of the abdominal aorta, the <u>renal</u> / <u>superior mesenteric</u> artery, supplies most of the small intestine and the first half of the large intestine.

5. The anterior tibial artery terminates with the _____ artery, which is often palpated in patients with circulatory problems to determine the circulatory efficiency of the lower limb.
   - **a.** dorsalis pedis
   - **b.** external iliac
   - **c.** obturator
   - **d.** tibial

6. Circle the correct underlined term. Veins draining the head and upper extremities empty into the <u>superior</u> / <u>inferior</u> vena cava.

7. Located in the lower limb, the _____ is the longest vein in the body.
   - **a.** external iliac
   - **b.** fibular
   - **c.** great saphenous
   - **d.** internal iliac

8. Circle the correct underlined term. The <u>renal</u> / <u>hepatic</u> veins drain the liver.

9. The function of the _____ is to drain the digestive viscera and carry dissolved nutrients to the liver for processing.
   - **a.** fetal circulation
   - **b.** hepatic portal circulation
   - **c.** pulmonary circulation system

10. Circle the correct underlined term. In the developing fetus, the umbilical <u>artery</u> / <u>vein</u> carries blood rich in nutrients and oxygen to the fetus.

Arteries, carrying blood away from the heart, and veins, which drain the tissues and return blood to the heart, function simply as conducting vessels or conduits. Only the tiny capillaries that connect the arterioles and venules and branch throughout the tissues directly serve the needs of the body's cells. It is through the capillary walls that exchanges are made between tissue cells and blood.

In this exercise you will examine the microscopic structure of blood vessels and identify the major arteries and veins of the systemic circulation and other special circulations.

Microscopic Structure of the Blood Vessels

Except for the microscopic capillaries, the walls of blood vessels are constructed of three coats, or *tunics* (**Figure 32.1**).

• **Tunica intima:** Lines the lumen of a vessel and is composed of a single thin layer of *endothelium,* subendothelial layer, and internal elastic membrane. Its cells fit closely together, forming an extremely smooth blood vessel lining that helps decrease resistance to blood flow.

• **Tunica media:** Middle coat, composed primarily of smooth muscle and elastin. The smooth muscle plays an active role in regulating the diameter of blood vessels, which in turn alters peripheral resistance and blood pressure.

• **Tunica externa:** Outermost tunic, composed of areolar or fibrous connective tissue. Its function is basically supportive and protective. In larger vessels, the tunica externa contains a system of tiny blood vessels, the **vasa vasorum.**

In general, the walls of arteries are thicker than those of veins. The tunica media in particular tends to be much heavier and contains substantially more smooth muscle and elastic tissue. Arteries, which are closer to the pumping action of the heart, must be able to expand as an increased volume of blood is propelled into them during systole and then recoil passively as the blood flows off into the circulation during diastole. The anatomical differences between the different types of vessels reflect their functional differences. **Table 32.1** summarizes the structure and function of various blood vessels.

Valves in veins act to prevent backflow of blood in much the same manner as the semilunar valves of the heart. The

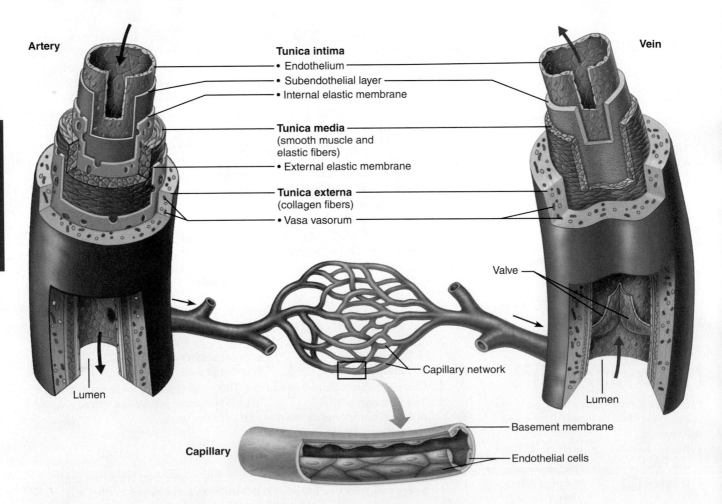

Figure 32.1 Generalized structure of arteries, veins, and capillaries.

32

Table 32.1	Summary of Blood Vessel Anatomy and Physiology				
Type of vessel	Description	Average lumen diameter	Average wall thickness	Function
Elastic (conducting) arteries	Largest, most elastic arteries. Contain more elastic tissue than other arteries.	1.5 cm	1.0 mm	Act as a pressure reservoir, expanding and recoiling for continuous blood flow. Examples: aorta, brachiocephalic artery, and common carotid artery.
Muscular (distributing) arteries	Medium-sized arteries, accounting for most arteries found in the body. They have less elastic tissue and more smooth muscle than other arteries.	0.6 cm	1.0 mm	Better ability to constrict and less stretchable than elastic arteries. They distribute blood to specific areas of the body. Examples: brachial artery and radial artery.
Arterioles	Smallest arteries with a very thin tunica externa and only a few layers of smooth muscle in the tunica media.	37 μm	6 μm	Blood flows from arterioles into a capillary bed. They play a role in regulating the blood flow to specific areas of the body.
Capillaries	Contain only a tunica intima.	9 μm	0.5 μm	Provide for the exchange of materials (gases, nutrients, etc.) between the blood and tissue cells.
Venules	Smallest veins. All tunics are very thin, with at most two layers of smooth muscle and no elastic tissue.	20 μm	1 μm	Drain capillary beds and merge to form veins.
Veins	Contain more fibrous tissue in the tunica externa than corresponding arteries. The tunica media is thinner, with a larger lumen than the corresponding artery.	0.5 cm	0.5 mm	Low-pressure vessels; return blood to the heart. Valves prevent the backflow of the blood.

skeletal muscle "pump" also promotes venous return; as the skeletal muscles surrounding the veins contract and relax, the blood is milked through the veins toward the heart. Anyone who has been standing relatively still for an extended time has experienced swelling in the ankles, caused by blood pooling in their feet during the period of muscle inactivity. Pressure changes that occur in the thorax during breathing also aid the return of blood to the heart.

☐ To demonstrate how efficiently venous valves prevent backflow of blood, perform the following simple experiment.

Allow one hand to hang by your side until the blood vessels on the dorsal aspect become distended. Place two fingertips against one of the distended veins and, pressing firmly, move the superior finger proximally along the vein and then release this finger. The vein will remain flattened and collapsed despite gravity. Then remove the distal fingertip and observe the rapid filling of the vein.

Check the box when you have completed this task.

Activity 1

Examining the Microscopic Structure of Arteries and Veins

1. Obtain a slide showing a cross-sectional view of blood vessels and a microscope.

2. Scan the section to identify a thick-walled artery (use **Figure 32.2** as a guide). Very often, but not always, an arterial lumen will appear scalloped due to the constriction of its walls by the elastic tissue of the media.

3. Identify a vein. Its lumen may be elongated or irregularly shaped and collapsed, and its walls will be considerably thinner. Notice the difference in the relative amount of elastic fibers in the media of the two vessels. Also, note the thinness of the intima layer, which is composed of flat squamous cells.

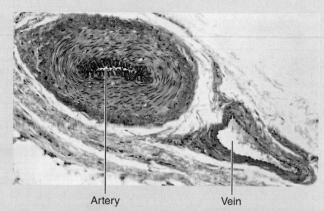

Figure 32.2 Photomicrograph of a muscular artery and the corresponding vein in cross section (76×).

32

Major Systemic Arteries of the Body

The **aorta** is the largest artery of the body. It has three main regions: ascending aorta, aortic arch, and descending aorta. Extending upward as the **ascending aorta** from the left ventricle, it arches posteriorly and to the left (**aortic arch**) and then courses downward as the **descending aorta** through the thoracic cavity. Called the **thoracic aorta** from T_5 to T_{12}, the descending aorta penetrates the diaphragm to enter the abdominal cavity just anterior to the vertebral column. As it enters the abdominal cavity, it becomes the **abdominal aorta.** The branches of the ascending aorta and aortic arch are summarized in **Table 32.2,** p. 476. Branches of the thoracic and abdominal aorta are summarized in Tables 32.3 and 32.4.

As you locate the arteries diagrammed on **Figure 32.3,** be aware of ways in which you can make your memorization task easier. In many cases, the name of the artery reflects the body region it travels through (axillary, subclavian, brachial, popliteal), the organ served (renal, hepatic), or the bone followed (tibial, femoral, radial, ulnar).

Aortic Arch

The **brachiocephalic** (literally, "arm-head") **trunk** is the first branch of the aortic arch (**Figure 32.4**). The other two major arteries branching off the aortic arch are the **left common carotid artery** and the **left subclavian artery.** The brachiocephalic trunk persists briefly before dividing into the **right common carotid artery** and the **right subclavian artery.**

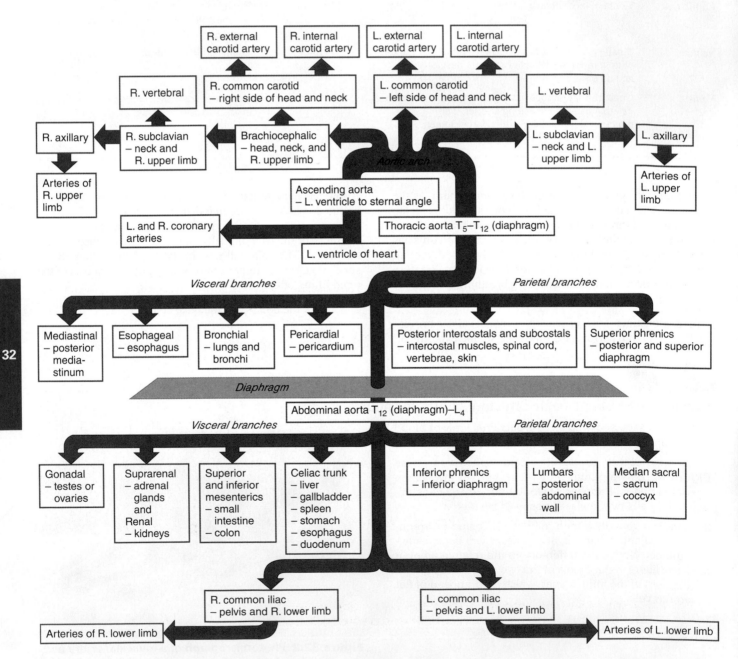

Figure 32.3 Schematic of the systemic arterial circulation. (R. = right, L. = left)

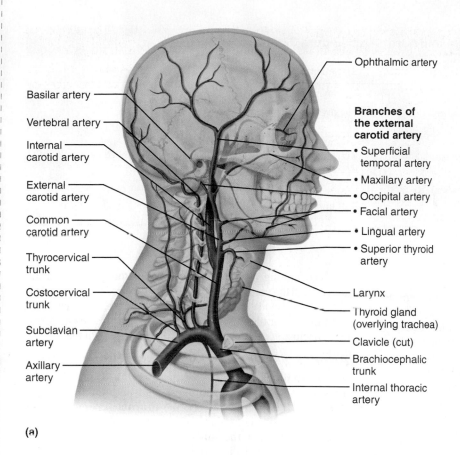

Ophthalmic artery

Basilar artery

Vertebral artery

Internal
carotid artery

External
carotid artery

Common
carotid artery

Thyrocervical
trunk

Costocervical
trunk

Subclavian
artery

Axillary
artery

**Branches of
the external
carotid artery**
- Superficial
  temporal artery
- Maxillary artery
- Occipital artery
- Facial artery
- Lingual artery
- Superior thyroid
  artery

Larynx

Thyroid gland
(overlying trachea)

Clavicle (cut)

Brachiocephalic
trunk

Internal thoracic
artery

(a)

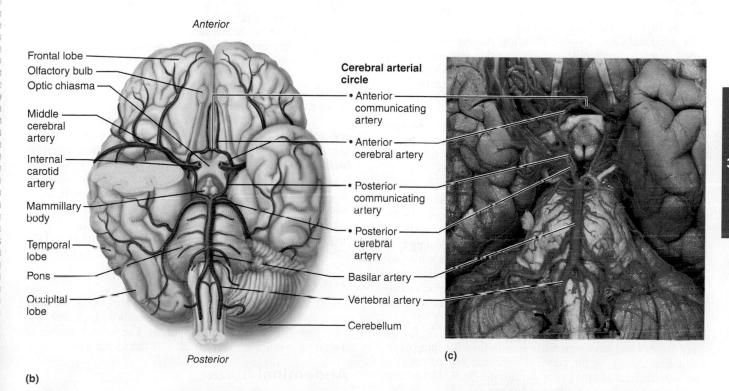

Anterior

Frontal lobe

Olfactory bulb

Optic chiasma

Middle
cerebral
artery

Internal
carotid
artery

Mammillary
body

Temporal
lobe

Pons

Occipital
lobe

**Cerebral arterial
circle**
- Anterior
  communicating
  artery
- Anterior
  cerebral artery
- Posterior
  communicating
  artery
- Posterior
  cerebral
  artery

Basilar artery

Vertebral artery

Cerebellum

Posterior

(b)

(c)

Figure 32.4 Arteries of the head, neck, and brain. (a) Right aspect. **(b)** Drawing of the cerebral arteries. Cerebellum is not shown on the left side of the figure. **(c)** Cerebral arterial circle (circle of Willis) in a human brain.

32

Table 32.2	The Aorta: Ascending Aorta and Aortic Arch (Figure 32.3)
Ascending aorta branches	**Structures served**
Right coronary artery	The myocardium of the heart (see Exercise 30)
Left coronary artery	The myocardium of the heart (see Exercise 30)
Aortic arch branches	**Structures served**
Brachiocephalic trunk (branches into right common carotid and right subclavian arteries)	Right common carotid artery – right side of the head and neck Right subclavian artery – right upper limb
Left common carotid artery	Left side of the head and neck
Left subclavian artery	Left upper limb

Arteries Serving the Head and Neck

The common carotid artery on each side divides to form an internal and an external carotid artery. The **internal carotid artery** serves the brain and gives rise to the **ophthalmic artery** that supplies orbital structures. The **external carotid artery** supplies the tissues external to the skull, largely via its **superficial temporal, maxillary, facial,** and **occipital** arterial branches. (Notice that several arteries are shown in the figure that are not described here.)

The right and left subclavian arteries each give off several branches to the head and neck. The first of these is the **vertebral artery,** which runs up the posterior neck to supply the cerebellum, part of the brain stem, and the posterior cerebral hemispheres. Issuing just lateral to the vertebral artery are the **thyrocervical trunk,** which mainly serves the thyroid gland and some scapular muscles, and the **costocervical trunk,** which supplies deep neck muscles and some of the upper intercostal muscles. In the armpit, the subclavian artery becomes the axillary artery, which serves the upper limb.

Arteries Serving the Brain

The brain is supplied by two pairs of arteries arising from the region of the aortic arch—the internal carotid arteries and the vertebral arteries. (Figure 32.4b is a diagram of the brain's arterial supply.)

Within the cranium, each internal carotid artery divides into **anterior** and **middle cerebral arteries,** which supply the bulk of the cerebrum. The right and left anterior cerebral arteries are connected by a short shunt called the **anterior communicating artery.** This shunt, along with shunts from each of the middle cerebral arteries, called the **posterior communicating arteries,** contribute to the formation of the **cerebral arterial circle** (circle of Willis), an arterial anastomosis at the base of the brain surrounding the pituitary gland and the optic chiasma.

The paired **vertebral arteries** diverge from the subclavian arteries and pass superiorly through the foramina of the transverse process of the cervical vertebrae to enter the skull through the foramen magnum. Within the skull, the vertebral arteries unite to form a single **basilar artery,** which continues superiorly along the ventral aspect of the brain stem, giving off branches to the pons, cerebellum, and inner ear. At the base of the cerebrum, the basilar artery divides to form the **posterior cerebral arteries.** These supply portions of the temporal and occipital lobes of the cerebrum and complete the cerebral arterial circle posteriorly.

The uniting of the blood supply of the internal carotid arteries and the vertebral arteries via the cerebral arterial circle is a protective device that theoretically provides an alternative set of pathways for blood to reach the brain tissue in the case of arterial occlusion or impaired blood flow anywhere in the system.

Arteries Serving the Thorax and Upper Limbs

As the **axillary artery** runs through the axilla, it gives off several branches to the chest wall and shoulder girdle (**Figure 32.5**). These include the **thoracoacromial artery** (to shoulder and pectoral region), the **lateral thoracic artery** (lateral chest wall), the **subscapular artery** (to scapula and dorsal thorax), and the **anterior** and **posterior circumflex humeral arteries** (to the shoulder and the deltoid muscle). At the inferior edge of the teres major muscle, the axillary artery becomes the **brachial artery** as it enters the arm. The brachial artery gives off a major branch, the **deep artery of the arm,** and as it nears the elbow it gives off several small branches. At the elbow, the brachial artery divides into the **radial** and **ulnar arteries,** which follow the same-named bones to supply the forearm and hand.

The **internal thoracic arteries** that arise from the subclavian arteries supply the mammary glands, most of the thorax wall, and anterior intercostal structures via their **anterior intercostal artery** branches. The first two pairs of **posterior intercostal arteries** arise from the costocervical trunk, noted above. The more inferior pairs arise from the thoracic aorta. (Not shown in Figure 32.5 are the small arteries that serve the diaphragm [phrenic arteries], esophagus [esophageal arteries], bronchi [bronchial arteries], and other structures of the mediastinum [mediastinal and pericardial arteries].)

Thoracic Aorta

The thoracic aorta is the superior portion of the descending aorta (Figure 32.5). It begins where the aortic arch ends and ends just as it pierces the diaphragm. The main branches of the thoracic aorta are summarized in **Table 32.3**.

Abdominal Aorta

Although several small branches of the descending aorta serve the thorax, the more major branches of the descending aorta are those serving the abdominal organs and ultimately the lower limbs (**Figure 32.6**, pp. 478–479). Most of the branches of the abdominal aorta serve the abdominal organs. The major branches of the abdominal aorta are summarized in **Table 32.4** on p. 479.

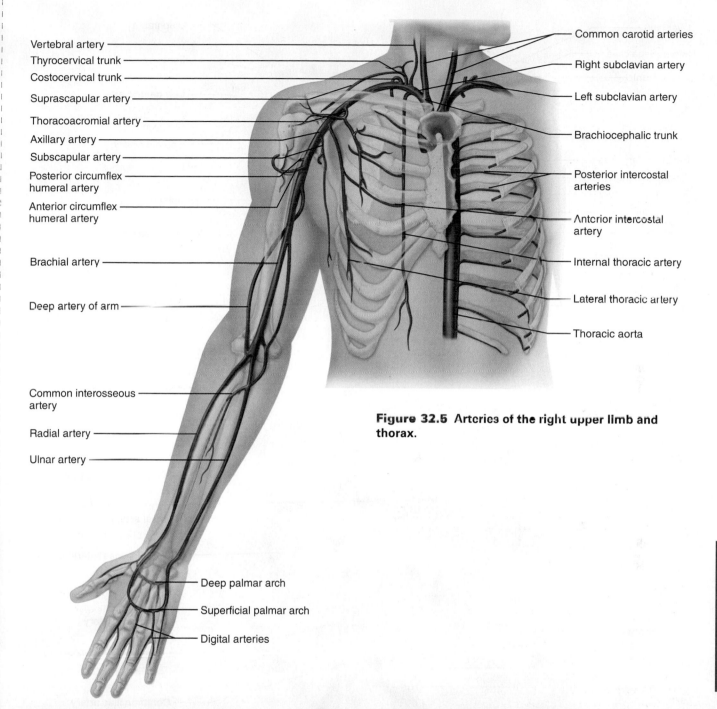

Vertebral artery

Thyrocervical trunk

Costocervical trunk

Suprascapular artery

Thoracoacromial artery

Axillary artery

Subscapular artery

Posterior circumflex
humeral artery

Anterior circumflex
humeral artery

Brachial artery

Deep artery of arm

Common interosseous
artery

Radial artery

Ulnar artery

Common carotid arteries

Right subclavian artery

Left subclavian artery

Brachiocephalic trunk

Posterior intercostal
arteries

Anterior intercostal
artery

Internal thoracic artery

Lateral thoracic artery

Thoracic aorta

Deep palmar arch

Superficial palmar arch

Digital arteries

Figure 32.5 Arteries of the right upper limb and thorax.

Table 32.3	The Aorta: Thoracic Aorta (Figure 32.3) (Note that many of these arteries vary in number from person to person)
Visceral thoracic aorta branches	**Structures served**
Pericardial arteries	Pericardium, the serous membrane of the heart
Bronchial arteries	Bronchi, bronchioles, and lungs
Esophageal arteries	Esophagus
Mediastinal arteries	Posterior mediastinum
Parietal thoracic aorta branches	**Structures served**
Posterior intercostal arteries (inferior pairs)	Intercostal muscles, spinal cord, vertebrae, and skin
Subcostal arteries	Intercostal muscles, spinal cord, vertebrae, and skin
Superior phrenic arteries	Posterior, superior part of the diaphragm

32

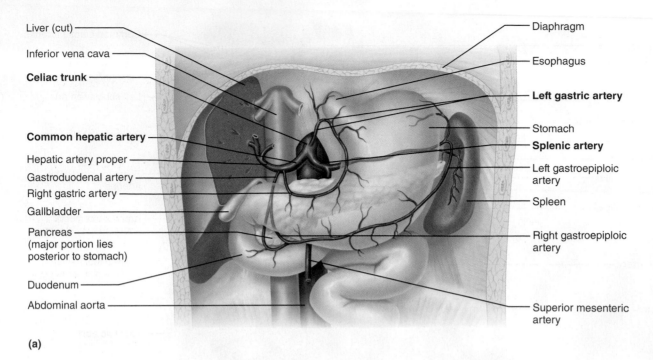

Liver (cut)

Inferior vena cava

Celiac trunk

Common hepatic artery

Hepatic artery proper

Gastroduodenal artery

Right gastric artery

Gallbladder

Pancreas
(major portion lies
posterior to stomach)

Duodenum

Abdominal aorta

Diaphragm

Esophagus

Left gastric artery

Stomach

Splenic artery

Left gastroepiploic
artery

Spleen

Right gastroepiploic
artery

Superior mesenteric
artery

(a)

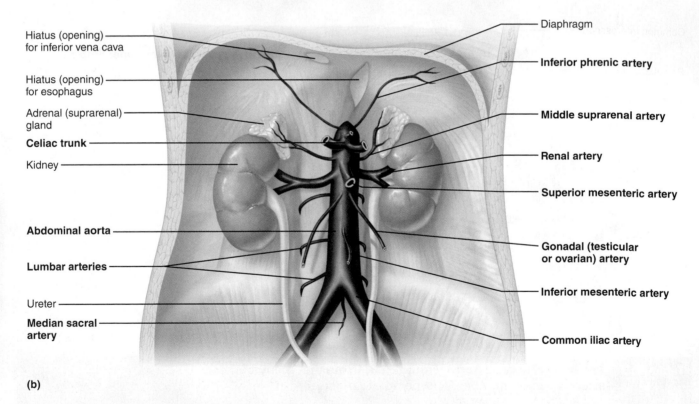

Hiatus (opening)
for inferior vena cava

Hiatus (opening)
for esophagus

Adrenal (suprarenal)
gland

Celiac trunk

Kidney

Abdominal aorta

Lumbar arteries

Ureter

**Median sacral
artery**

Diaphragm

Inferior phrenic artery

Middle suprarenal artery

Renal artery

Superior mesenteric artery

**Gonadal (testicular
or ovarian) artery**

Inferior mesenteric artery

Common iliac artery

(b)

Figure 32.6 Arteries of the abdomen. (a) The celiac trunk and its major branches.
(b) Major branches of the abdominal aorta.

32

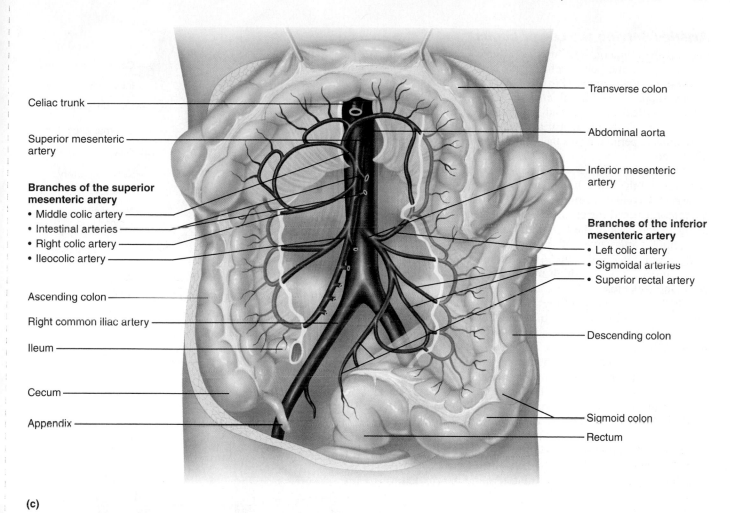

Celiac trunk

Superior mesenteric artery

Branches of the superior mesenteric artery
- Middle colic artery
- Intestinal arteries
- Right colic artery
- Ileocolic artery

Ascending colon

Right common iliac artery

Ileum

Cecum

Appendix

Transverse colon

Abdominal aorta

Inferior mesenteric artery

Branches of the inferior mesenteric artery
- Left colic artery
- Sigmoidal arteries
- Superior rectal artery

Descending colon

Sigmoid colon

Rectum

(c)

Figure 32.6 *(continued)* **(c)** Distribution of the superior and inferior mesenteric arteries, transverse colon pulled superiorly.

Table 32.4	The Aorta: Abdominal Aorta (Figure 32.6)
Branches	**Structures served**
Inferior phrenic arteries	Inferior surface of the diaphragm
Celiac trunk: left gastric artery	Stomach and esophagus
Celiac trunk: splenic artery	Branches to the spleen; short gastric arteries branch to the stomach; and the left gastroepiploic artery branches to the stomach
Celiac trunk: common hepatic artery	Branches into the hepatic artery proper (its branches serve the liver, gallbladder, and stomach) and the gastroduodenal artery (its branches serve the stomach, pancreas, and duodenum)
Superior mesenteric artery	Most of the small intestine and the first part of the large intestine
Middle suprarenal arteries	Adrenal glands that sit on top of the kidneys
Renal arteries	Kidneys
Gonadal arteries	Ovarian arteries (female) – ovaries Testicular arteries (male) – testes
Inferior mesenteric artery	Distal portion of the large intestine
Lumbar arteries	Posterior abdominal wall
Median sacral artery	Sacrum and coccyx
Common iliac arteries	The distal abdominal aorta splits to form the left and right common iliac arteries, which serve the pelvic organs, lower abdominal wall, and the lower limbs

32

Arteries Serving the Lower Limbs

Each of the common iliac arteries extends for about 5 cm (2 inches) into the pelvis before it divides into the internal and external iliac arteries (**Figure 32.7**). The **internal iliac artery** supplies the gluteal muscles via the **superior** and **inferior gluteal arteries,** and the adductor muscles of the medial thigh via the **obturator artery,** as well as the external genitalia and perineum (via the *internal pudendal artery,* not illustrated).

The **external iliac artery** supplies the anterior abdominal wall and the lower limb. As it continues into the thigh, its name changes to **femoral artery.** Proximal branches of the femoral artery, the **circumflex femoral arteries,** supply the head and neck of the femur and the hamstring muscles. The femoral artery gives off a deep branch, the **deep artery of the thigh** (also called the *deep femoral artery*), which is the main supply to the thigh muscles (hamstrings, quadriceps, and adductors). In the knee region, the femoral artery briefly becomes the **popliteal artery;** its subdivisions—the **anterior** and **posterior tibial arteries**—supply the leg, ankle, and foot. The posterior tibial, which supplies flexor muscles, gives off one main branch, the **fibular artery,** that serves the lateral calf (fibular muscles). It then divides into the **lateral** and **medial plantar arteries,** which supply blood to the sole of the foot. The anterior tibial artery supplies the extensor muscles and terminates with the **dorsalis pedis artery.** The dorsalis pedis supplies the dorsum of the foot and continues on as the **arcuate artery,** which issues the **dorsal metatarsal arteries** to the metatarsus of the foot. The dorsalis pedis is often palpated in patients with circulation problems of the leg to determine the circulatory efficiency to the limb as a whole.

☐ Palpate your own dorsalis pedis artery.

Check the box when you have completed this task.

Activity 2

Locating Arteries on an Anatomical Chart or Model

Now that you have identified the arteries in Figures 32.3–32.7, attempt to locate and name them (without a reference) on a large anatomical chart or three-dimensional model of the vascular system.

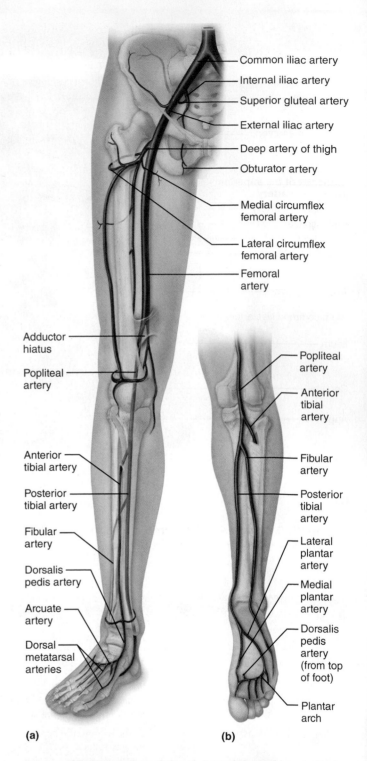

(a)   **(b)**

Figure 32.7 Arteries of the right pelvis and lower limb. (a) Anterior view. **(b)** Posterior view.

Major Systemic Veins of the Body

Arteries are generally located in deep, well-protected body areas. However, many veins follow a more superficial course and are often easily seen and palpated on the body surface. Most deep veins parallel the course of the major arteries, and in many cases the naming of the veins and arteries is identical except for the designation of the vessels as veins. Whereas the major systemic arteries branch off the aorta, the veins tend to converge on the venae cavae, which enter the right atrium of the heart. Veins draining the head and upper extremities empty into the **superior vena cava,** and those draining the lower body empty into the **inferior vena cava. Figure 32.8**, a schematic of the systemic veins and their relationship to the venae cavae, will get you started.

Veins Draining into the Inferior Vena Cava

The inferior vena cava, a much longer vessel than the superior vena cava, returns blood to the heart from all body regions

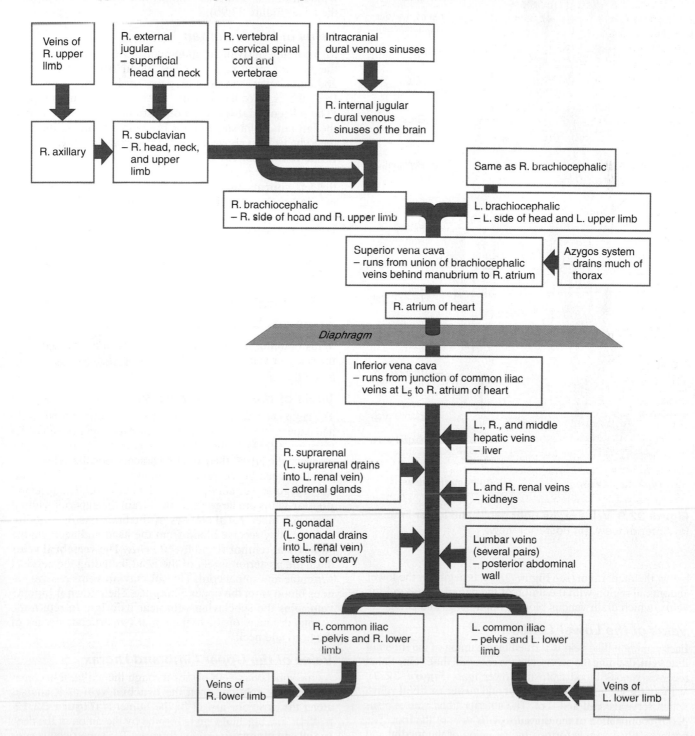

Figure 32.8 Schematic of systemic venous circulation.

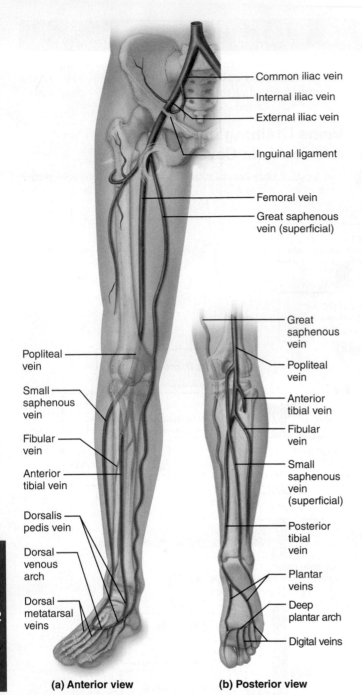

(a) Anterior view   **(b) Posterior view**

Common iliac vein
Internal iliac vein
External iliac vein
Inguinal ligament
Femoral vein
Great saphenous vein (superficial)
Great saphenous vein
Popliteal vein
Anterior tibial vein
Fibular vein
Small saphenous vein (superficial)
Posterior tibial vein
Plantar veins
Deep plantar arch
Digital veins
Popliteal vein
Small saphenous vein
Fibular vein
Anterior tibial vein
Dorsalis pedis vein
Dorsal venous arch
Dorsal metatarsal veins

Figure 32.9 Veins of the right pelvis and lower limb.
(a) Anterior view. **(b)** Posterior view.

below the diaphragm (see Figure 32.8). It begins in the lower abdominal region with the union of the paired **common iliac veins,** which drain venous blood from the legs and pelvis.

Veins of the Lower Limbs

Each common iliac vein is formed by the union of the **internal iliac vein,** draining the pelvis, and the **external iliac vein,** which receives venous blood from the lower limb (**Figure 32.9**). Veins of the leg include the **anterior** and **posterior tibial veins,** which serve the calf and foot. The anterior tibial vein is a superior continuation of the **dorsalis pedis vein** of the foot. The posterior tibial vein is formed by the union of the **medial** and

lateral plantar veins, and ascends deep in the calf muscles. It receives the **fibular vein** in the calf and then joins with the anterior tibial vein at the knee to produce the **popliteal vein,** which crosses the back of the knee. The popliteal vein becomes the **femoral vein** in the thigh; the femoral vein in turn becomes the external iliac vein in the inguinal region.

The **great saphenous vein,** a superficial vein, is the longest vein in the body. Beginning in common with the **small saphenous vein** from the **dorsal venous arch,** it extends up the medial side of the leg, knee, and thigh to empty into the femoral vein. The small saphenous vein runs along the lateral aspect of the foot and through the calf muscle, which it drains, and then empties into the popliteal vein at the knee (Figure 32.9b).

Veins of the Abdomen

Moving superiorly in the abdominal cavity (**Figure 32.10**), the inferior vena cava receives blood from the posterior abdominal wall via several pairs of **lumbar veins,** and from the right ovary or testis via the **right gonadal vein.** (The **left gonadal [ovarian or testicular] vein** drains into the left renal vein superiorly.) The paired **renal veins** drain the kidneys. Just above the right renal vein, the **right suprarenal vein** (receiving blood from the adrenal gland on the same side) drains into the inferior vena cava, but its partner, the **left suprarenal vein,** empties into the left renal vein inferiorly. The **hepatic veins** drain the liver. The unpaired veins draining the digestive tract organs empty into a special vessel, the hepatic portal vein, which carries blood to the liver to be processed before it enters the systemic venous system. (The hepatic portal system is discussed separately on p. 486.)

Veins Draining into the Superior Vena Cava

Veins draining into the superior vena cava are named from the superior vena cava distally, *but remember that the flow of blood is in the opposite direction*.

Veins of the Head and Neck

The **right** and **left brachiocephalic veins** drain the head, neck, and upper extremities and unite to form the superior vena cava (**Figure 32.11**). Notice that although there is only one brachiocephalic artery, there are two brachiocephalic veins.

Branches of the brachiocephalic veins include the internal jugular, vertebral, and subclavian veins. The **internal jugular veins** are large veins that drain the superior sagittal sinus and other **dural sinuses** of the brain. As they run inferiorly, they receive blood from the head and neck via the **superficial temporal** and **facial veins.** The **vertebral veins** drain the posterior aspect of the head including the cervical vertebrae and spinal cord. The **subclavian veins** receive venous blood from the upper extremity. The **external jugular vein** joins the subclavian vein near its origin to return the venous drainage of the extracranial (superficial) tissues of the head and neck.

Veins of the Upper Limb and Thorax

As the subclavian vein passes through the axilla, it becomes the **axillary vein** and then the **brachial vein** as it courses along the posterior aspect of the humerus (**Figure 32.12**, p. 484). The brachial vein is formed by the union of the deep **radial** and **ulnar veins** of the forearm. The superficial venous

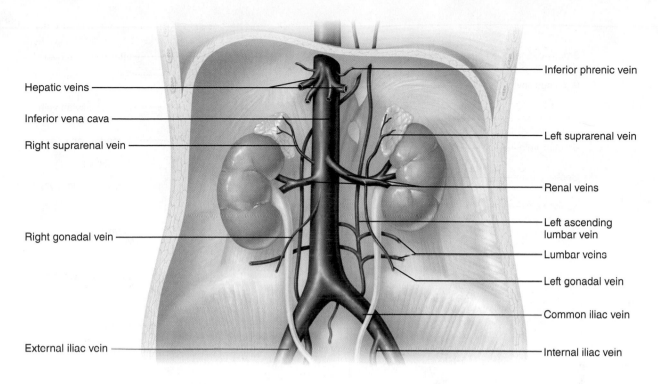

Figure 32.10 Venous drainage of abdominal organs not drained by the hepatic portal vein.

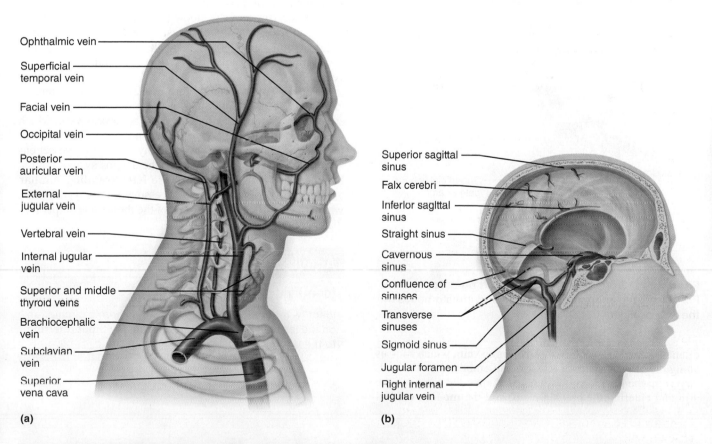

(a)   (b)

Figure 32.11 Venous drainage of the head, neck, and brain. (a) Veins of the head and neck, right superficial aspect. **(b)** Dural sinuses of the brain, right aspect.

32

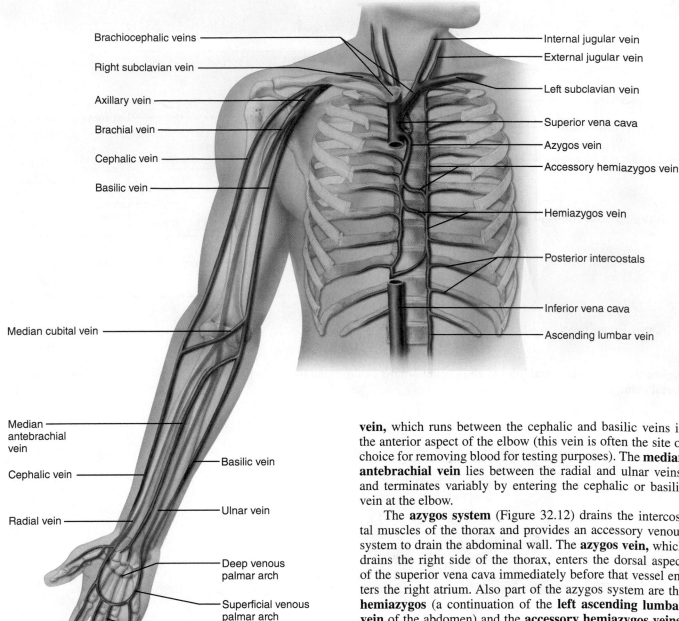

Brachiocephalic veins

Right subclavian vein

Axillary vein

Brachial vein

Cephalic vein

Basilic vein

Internal jugular vein

External jugular vein

Left subclavian vein

Superior vena cava

Azygos vein

Accessory hemiazygos vein

Hemiazygos vein

Posterior intercostals

Inferior vena cava

Ascending lumbar vein

Median cubital vein

Median antebrachial vein

Cephalic vein

Radial vein

Basilic vein

Ulnar vein

Deep venous palmar arch

Superficial venous palmar arch

Digital veins

Figure 32.12 Veins of the thorax and right upper limb. For clarity, the abundant branching and anastomoses of these vessels are not shown.

drainage of the arm includes the **cephalic vein,** which courses along the lateral aspect of the arm and empties into the axillary vein; the **basilic vein,** found on the medial aspect of the arm and entering the brachial vein; and the **median cubital vein,** which runs between the cephalic and basilic veins in the anterior aspect of the elbow (this vein is often the site of choice for removing blood for testing purposes). The **median antebrachial vein** lies between the radial and ulnar veins, and terminates variably by entering the cephalic or basilic vein at the elbow.

The **azygos system** (Figure 32.12) drains the intercostal muscles of the thorax and provides an accessory venous system to drain the abdominal wall. The **azygos vein,** which drains the right side of the thorax, enters the dorsal aspect of the superior vena cava immediately before that vessel enters the right atrium. Also part of the azygos system are the **hemiazygos** (a continuation of the **left ascending lumbar vein** of the abdomen) and the **accessory hemiazygos veins,** which together drain the left side of the thorax and empty into the azygos vein.

Activity 3

Identifying the Systemic Veins

Identify the important veins of the systemic circulation on the large anatomical chart or model without referring to the figures.

Special Circulations

Pulmonary Circulation

The pulmonary circulation (discussed previously in relation to heart anatomy on p. 448) differs in many ways from systemic circulation because it does not serve the metabolic needs of the body tissues with which it is associated (in this case, lung tissue). It functions instead to bring the blood into close contact with the alveoli of the lungs to permit gas exchanges that rid the blood of excess carbon dioxide and replenish its supply of vital oxygen. The arteries of the pulmonary circulation are structurally much like veins, and they create a low-pressure bed in the lungs. (If the arterial pressure in the systemic circulation is 120/80, the pressure in the pulmonary artery is likely to be approximately 24/8.) The functional blood supply of the lungs is provided by the **bronchial arteries** (not shown), which diverge from the thoracic portion of the descending aorta.

Pulmonary circulation begins with the large **pulmonary trunk,** which leaves the right ventricle and divides into the **right** and **left pulmonary arteries** about 5 cm (2 inches) above its origin. The right and left pulmonary arteries plunge into the lungs, where they subdivide into **lobar arteries** (three on the right and two on the left). The lobar arteries accompany the main bronchi into the lobes of the lungs and branch extensively within the lungs to form arterioles, which finally terminate in the capillary networks surrounding the alveolar sacs of the lungs. Diffusion of the respiratory gases occurs across the walls of the alveoli and **pulmonary capillaries.** The pulmonary capillary beds are drained by venules, which converge to form sequentially larger veins and finally the four **pulmonary veins** (two leaving each lung), which return the blood to the left atrium of the heart.

Activity 4

Identifying Vessels of the Pulmonary Circulation

Find the vessels of the pulmonary circulation in **Figure 32.13** and on an anatomical chart (if one is available).

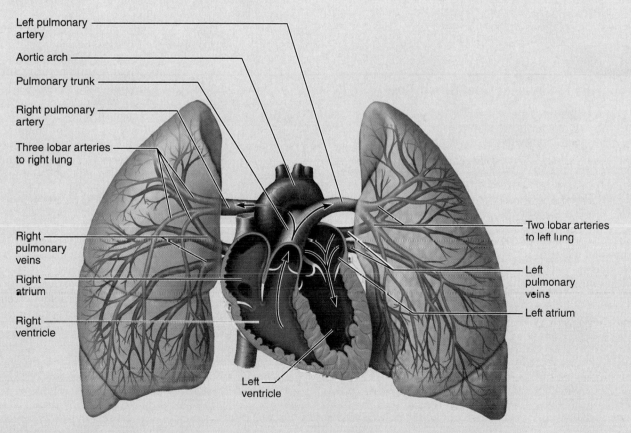

Figure 32.13 The pulmonary circulation. The pulmonary arterial system is shown in blue to indicate that the blood it carries is relatively oxygen-poor. The pulmonary venous drainage is shown in red to indicate that the blood it transports is oxygen-rich.

Group Challenge

Fix the Blood Trace

Several artery or vein sequences are listed below. Working in small groups, decide whether each sequence of blood vessels is correct or whether a blood vessel is missing. If correct, simply write "all correct." If incorrect, list the missing vessel, and draw an insertion mark ("v") on the arrow to indicate where the vessel would be located in the sequence. Refrain from using a figure or other reference to help with your decision. Instead, try to find the answers by working together. Note: The missing vessel will not be at the beginning or end of the sequence.

1. aortic arch → R. subclavian artery → R. axillary artery → R. brachial artery → R. radial artery → R. superficial palmar arch

_____

2. abdominal aorta → R. common iliac artery → R. femoral artery → R. popliteal artery → R. anterior tibial artery → R. dorsalis pedis artery

_____

3. ascending aorta → aortic arch → L. common carotid artery → L. internal carotid artery → L. anterior cerebral artery

_____

4. R. median antebrachial vein → R. basilic vein → R. axillary vein → R. brachiocephalic vein → superior vena cava

_____

Fetal Circulation

In a developing fetus, the lungs and digestive system are not yet functional, and all nutrient, excretory, and gaseous exchanges occur through the placenta (**Figure 32.14a**). Nutrients and oxygen move across placental barriers from the mother's blood into fetal blood, and carbon dioxide and other metabolic wastes move from the fetal blood supply to the mother's blood.

Activity 5

Tracing the Pathway of Fetal Blood Flow

Trace the pathway of fetal blood flow using Figure 32.14a and an anatomical chart (if available). Locate all the named vessels. Identify the named remnants of the foramen ovale and fetal vessels (refer to Figure 32.14b).

Fetal blood travels through the umbilical cord, which contains three blood vessels: one large umbilical vein and two smaller umbilical arteries. The **umbilical vein** carries blood rich in nutrients and oxygen to the fetus; the **umbilical arteries** carry blood laden with carbon dioxide and waste from the fetus to the placenta. The umbilical arteries, which transport blood away from the fetal heart, meet the umbilical vein at the *umbilicus* and wrap around the vein within the cord en route to their placental attachments. Newly oxygenated blood flows in the umbilical vein superiorly toward the fetal heart. Some of this blood perfuses the liver, but the larger proportion is ducted through the relatively nonfunctional liver to the inferior vena cava via a shunt vessel called the **ductus venosus,** which carries the blood to the right atrium of the heart.

Because fetal lungs are nonfunctional and collapsed, two shunting mechanisms ensure that blood almost entirely bypasses the lungs. Much of the blood entering the right atrium is shunted into the left atrium through the **foramen ovale,** a flaplike opening in the interatrial septum. The left ventricle then pumps the blood out the aorta to the systemic circulation. Blood that does enter the right ventricle and is pumped out of the pulmonary trunk encounters a second shunt, the **ductus arteriosus,** a short vessel connecting the pulmonary trunk and the aorta. Because the collapsed lungs present an extremely high-resistance pathway, blood more readily enters the systemic circulation through the ductus arteriosus.

The aorta carries blood to the tissues of the body; this blood ultimately finds its way back to the placenta via the umbilical arteries. The only fetal vessel that carries highly oxygenated blood is the umbilical vein. All other vessels contain varying degrees of oxygenated and deoxygenated blood.

At birth, or shortly after, the foramen ovale closes and becomes the **fossa ovalis,** and the ductus arteriosus collapses and is converted to the fibrous **ligamentum arteriosum** (Figure 32.14b). Lack of blood flow through the umbilical vessels leads to their eventual obliteration, and the circulatory pattern becomes that of the adult. Remnants of the umbilical arteries persist as the **medial umbilical ligaments** on the inner surface of the anterior abdominal wall; the occluded umbilical vein becomes the **ligamentum teres** (or **round ligament** of the liver); and the ductus venosus becomes a fibrous band called the **ligamentum venosum** on the inferior surface of the liver.

Hepatic Portal Circulation

Blood vessels of the hepatic portal circulation drain the digestive viscera, spleen, and pancreas and deliver this blood to the liver for processing via the **hepatic portal vein** (formed by the union of the splenic and superior mesenteric veins). If a meal has recently been eaten, the hepatic portal blood will be nutrient-rich. The liver is the key body organ involved in maintaining proper sugar, fatty acid, and amino acid concentrations in the blood, and this system ensures that these substances pass through the liver before entering the systemic circulation. As blood travels through the liver sinusoids, some of the nutrients are removed to be stored or processed in various ways for release to the general circulation. At the same time, the hepatocytes are detoxifying alcohol and other possibly harmful chemicals present in the blood, and the liver's macrophages are removing bacteria and other debris from the passing blood. The liver in turn is drained by the hepatic veins that enter the inferior vena cava.

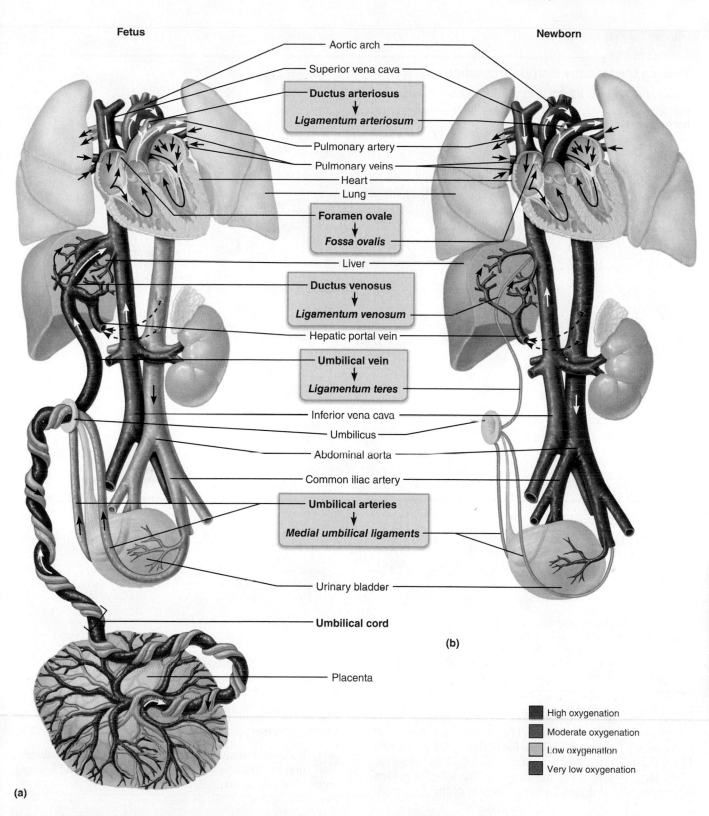

Fetus

Newborn

Aortic arch

Superior vena cava

Ductus arteriosus
↓
Ligamentum arteriosum

Pulmonary artery

Pulmonary veins

Heart

Lung

Foramen ovale
↓
Fossa ovalis

Liver

Ductus venosus
↓
Ligamentum venosum

Hepatic portal vein

Umbilical vein
↓
Ligamentum teres

Inferior vena cava

Umbilicus

Abdominal aorta

Common iliac artery

Umbilical arteries
↓
Medial umbilical ligaments

Urinary bladder

Umbilical cord

(b)

Placenta

High oxygenation

Moderate oxygenation

Low oxygenation

Very low oxygenation

(a)

32

Figure 32.14 Circulation in the fetus and newborn. Arrows indicate direction of
blood flow. Arrows in the blue boxes go from the fetal structure to what it becomes
after birth (the postnatal structure). **(a)** Special adaptations for embryonic and
fetal life. The umbilical vein (red) carries oxygen- and nutrient-rich blood from the
placenta to the fetus. The umbilical arteries (pink) carry waste-laden blood from the
fetus to the placenta. **(b)** Changes in the cardiovascular system at birth. The umbilical
vessels are occluded, as are the liver and lung bypasses (ductus venosus and
arteriosus, and the foramen ovale).

Activity 6

Tracing the Hepatic Portal Circulation

Locate on **Figure 32.15** and on an anatomical chart of the hepatic portal circulation (if available), the vessels named below.

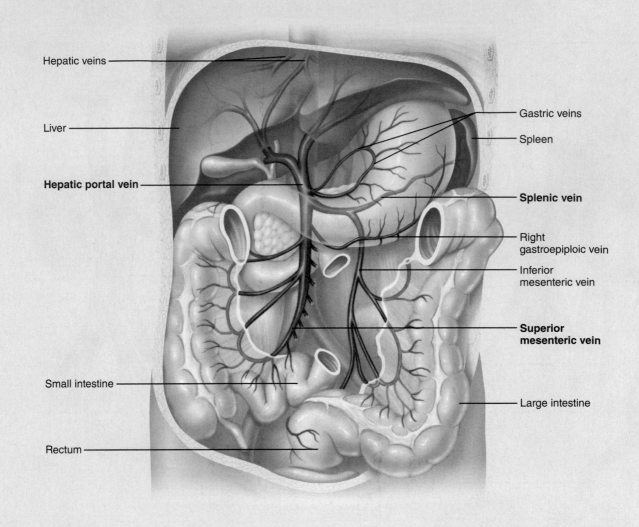

Figure 32.15 Hepatic portal circulation.

The **splenic vein** carries blood from the spleen, parts of the pancreas, and the stomach. The splenic vein unites with the **superior mesenteric vein** to form the hepatic portal vein. The superior mesenteric vein drains the small intestine, part of the large intestine, and the stomach. The **inferior mesenteric vein,** which drains the distal portion of the large intestine and rectum, empties into the splenic vein just before the splenic vein merges with the superior mesenteric vein.

For instructions on animal dissections, see the dissection exercises (starting on p. 705) in the cat and fetal pig editions of this manual.

REVIEW SHEET
Anatomy of Blood Vessels

Name _____ Lab Time/Date _____

Microscopic Structure of the Blood Vessels

1. Cross-sectional views of an artery and of a vein are shown here. Identify each; on the lines to the sides, note the structural details that enabled you to make these identifications:

(vessel type) _____

(a) _____

(b) _____

(vessel type) _____

(a) _____

(b) _____

Now describe each tunic more fully by selecting its characteristics from the key below and placing the appropriate key letters on the answer lines.

Tunica intima _____   Tunica media _____   Tunica externa _____

Key:

a. innermost tunic
b. most superficial tunic
c. thin tunic of capillaries

d. regulates blood vessel diameter
e. contains smooth muscle and elastin
f. has a smooth surface to decrease resistance to blood flow

2. Why are valves present in veins but not in arteries? _____

_____

_____

3. Name two events *occurring within the body* that aid in venous return.

_____ and _____

4. Considering their functional differences, why do you think the walls of arteries are proportionately thicker than those

of the corresponding veins? _____

_____

_____

Major Systemic Arteries and Veins of the Body

5. Use the key on the right to identify the arteries or veins described on the left. Some terms are used more than once.

Key:
a. anterior tibial
b. basilic
c. brachial
d. brachiocephalic
e. celiac trunk
f. cephalic
g. common carotid
h. common iliac
i. coronary
j. deep artery of the thigh
k. dorsalis pedis
l. external carotid
m. femoral
n. fibular
o. great saphenous
p. hepatic
q. inferior mesenteric
r. internal carotid
s. internal iliac
t. phrenic
u. posterior tibial
v. radial
w. renal
x. subclavian
y. superior mesenteric
z. vertebral

_____ 1. the arterial system has one of these; the venous system has two

_____ 2. these arteries supply the myocardium

_____, _____ 3. two paired arteries serving the brain

_____ 4. longest vein in the lower limb

_____ 5. artery on the dorsum of the foot

_____ 6. main artery that serves the thigh muscles

_____ 7. supplies the diaphragm

_____ 8. formed by the union of the radial and ulnar veins

_____, _____ 9. two superficial veins of the arm

_____ 10. artery serving the kidney

_____ 11. veins draining the liver

_____ 12. artery that supplies the distal half of the large intestine

_____ 13. drains the pelvic organs

_____ 14. what the external iliac artery becomes on entry into the thigh

_____ 15. artery that branches into radial and ulnar arteries

_____ 16. supplies most of the small intestine

_____ 17. join to form the inferior vena cava

_____ 18. an arterial trunk that has three major branches, which run to the liver, spleen, and stomach

_____ 19. major artery serving the tissues external to the skull

_____, _____, _____, _____ 20. four veins serving the leg

_____ 21. artery generally used to take the pulse at the wrist

6. What is the function of the cerebral arterial circle?

_____

7. The anterior and middle cerebral arteries arise from the _____ artery.

They serve the _____ of the brain.

8. Trace the pathway of a drop of blood from the aorta to the left occipital lobe of the brain, noting all structures through

which it flows. _____

_____

9. The human arterial and venous systems are diagrammed on this page and the next. Identify all indicated blood vessels.

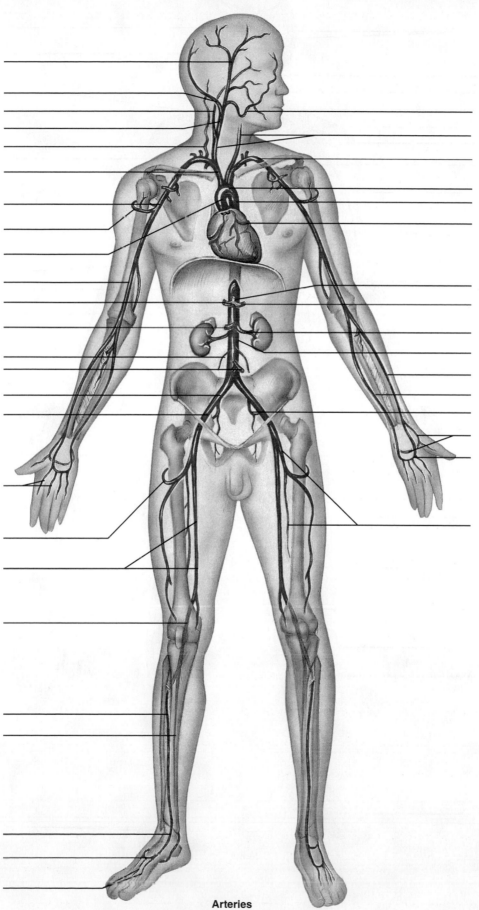

Arteries

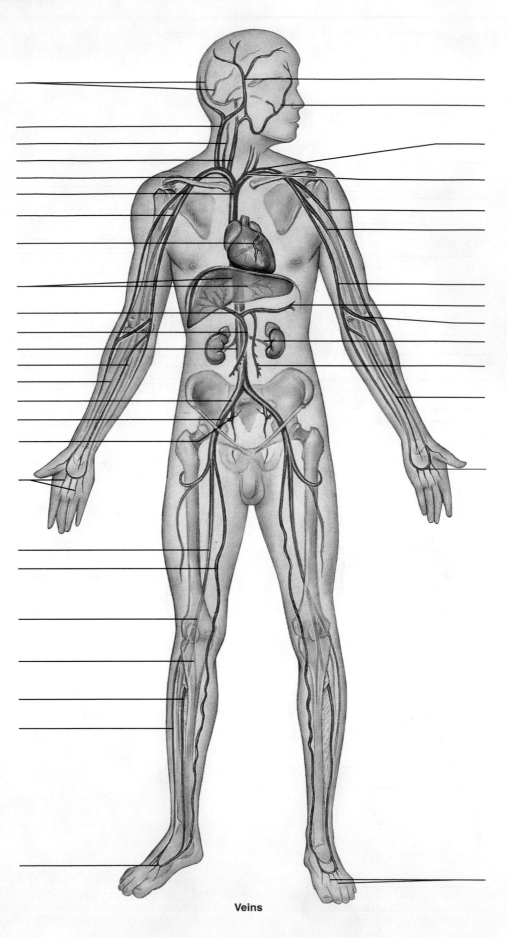

Veins

10. Trace the blood flow for each of the following situations.

    a. from the capillary beds of the left thumb to the capillary beds of the right thumb:_____

_____

_____

_____

_____

    b. from the mitral valve to the tricuspid valve by way of the great toe: _____

_____

_____

_____

_____

Pulmonary Circulation

11. Trace the pathway of a carbon dioxide gas molecule in the blood from the inferior vena cava until it leaves the bloodstream. Name all structures (vessels, heart chambers, and others) passed through en route.

_____

_____

_____

12. Trace the pathway of oxygen gas molecules from an alveolus of the lung to the right ventricle of the heart. Name all

    structures through which it passes. **Circle the areas of gas exchange.** _____

_____

_____

_____

13. Most arteries of the adult body carry oxygen-rich blood, and the veins carry oxygen-poor blood.

    How does this differ in the pulmonary arteries and veins? _____

_____

14. How do the arteries of the pulmonary circulation differ structurally from the systemic arteries? What condition is indicated

    by this anatomical difference? _____

_____

Fetal Circulation

15. For each of the following structures, first indicate its function in the fetus; and then note its fate (what happens to it or what it is converted to after birth). **Circle the blood vessel that carries the most oxygen-rich blood.**

Structure	Function in fetus	Fate and postnatal structure
Umbilical artery		
Umbilical vein		
Ductus venosus		
Ductus arteriosus		
Foramen ovale		

16. What organ serves as a respiratory/digestive/excretory organ for the fetus? _____

Hepatic Portal Circulation

17. What is the source of blood in the hepatic portal system? _____

_____

18. Why is this blood carried to the liver before it enters the systemic circulation? _____

_____

19. The hepatic portal vein is formed by the union of the _____ and the _____.

The _____ vein carries blood from the _____, _____ and _____.

The _____ vein drains the _____, _____, and _____.

The _____ vein empties into the splenic vein and drains the _____ and _____.

20. Trace the flow of a drop of blood from the small intestine to the right atrium of the heart, noting all structures encountered

or passed through on the way. _____

_____

EXERCISE

33

Human Cardiovascular Physiology: Blood Pressure and Pulse Determinations

Objectives

☐ Define *systole*, *diastole*, and *cardiac cycle*.

☐ Indicate the normal length of the cardiac cycle, the relative pressure changes occurring within the atria and ventricles, the timing of valve closure, and the volume changes occurring in the ventricles during the cycle.

☐ Correlate the events of the ECG with the events of the cardiac cycle.

☐ Use the stethoscope to auscultate heart sounds, relate heart sounds to cardiac cycle events, and describe the clinical significance of heart murmurs.

☐ Demonstrate thoracic locations where the first and second heart sounds are most accurately auscultated.

☐ Define *pulse*, *pulse pressure*, *pulse deficit*, *blood pressure*, *sounds of Korotkoff*, and *mean arterial pressure (MAP)*.

☐ Determine a subject's apical and radial pulse.

☐ Determine a subject's blood pressure with a sphygmomanometer, and relate systolic and diastolic pressures to events of the cardiac cycle.

☐ Compare the value of venous pressure to systemic blood pressure, and describe how venous pressure is measured.

☐ Investigate the effects of exercise on blood pressure, pulse, and cardiovascular fitness.

☐ Indicate factors affecting blood flow and skin color.

Materials

- Recording of "Interpreting Heart Sounds" (if available on free loan from the local chapters of the American Heart Association) or any of the suitable Internet sites on heart sounds
- Stethoscope
- Alcohol swabs
- Watch (or clock) with second hand

Text continues on next page. →

MasteringA&P®

For related exercise study tools, go to the Study Area of **MasteringA&P**. There you will find:

- Practice Anatomy Lab PAL
- A&PFlix *A&PFlix*
- PhysioEx PEx
- Practice quizzes, Histology Atlas, eText, Videos, and more!

Pre-Lab Quiz

1. Circle the correct underlined term. According to general usage, <u>systole</u> / <u>diastole</u> refers to ventricular relaxation.

2. A graph illustrating the pressure and volume changes during one heartbeat is called the:
   - a. blood pressure
   - b. cardiac cycle
   - c. conduction system of the heart
   - d. electrical events of the heartbeat

3. Circle True or False. When ventricular systole begins, intraventricular pressure increases rapidly, closing the atrioventricular (AV) valves.

4. The average heart beats approximately _____ times per minute.
   - a. 50
   - b. 75
   - c. 100
   - d. 125

5. Circle the correct underlined term. Abnormal heart sounds called <u>murmurs</u> / <u>stroke</u> can indicate valve problems.

6. The term _____ refers to the alternating surges of pressure in an artery that occur with each contraction and relaxation of the left ventricle.
   - a. diastole
   - b. pulse
   - c. murmur
   - d. systole

7. Circle the correct underlined term. The pulse is most often taken at the lateral aspect of the wrist, above the thumb, by compressing the <u>popliteal</u> / <u>radial</u> artery.

8. What device will you use today to measure your subject's blood pressure?

   _____

Text continues on next page. →

BIOPAC® BIOPAC® BSL System with BSL software version 3.7.6 to 3.7.7 (for Windows 7/Vista/XP or Mac OS X 10.4–10.6), data acquisition unit MP36/35 or MP45, PC or Mac computer, BIOPAC pulse plethysmograph.

Instructors using the MP36/35/30 data acquisition unit with BSL software versions earlier than 3.7.6 (for Windows or Mac) will need slightly different channel settings and collection strategies. Instructions for using the older data acquisition unit can be found on MasteringA&P.

- Sphygmomanometer
- Felt marker
- Meter stick
- Cot (if available)
- Step stools (0.4 m [16 in.] and 0.5 m [20 in.] in height)
- Small basin suitable for the immersion of one hand
- Ice
- Laboratory thermometer

PEx PhysioEx™ 9.1 Computer Simulation Ex. 5 on p. PEx-75

Note: *Instructions for using PowerLab® equipment can be found on MasteringA&P.*

9. In reporting a blood pressure of 120/90, which number represents the *diastolic* pressure? _____

10. The _____ are characteristic sounds that indicate the resumption of blood flow to the artery being occluded when taking blood pressure.
   a. ectopic heartbeats   c. heart rhythms
   b. heart murmurs   d. sounds of Korotkoff

During this lab, we will conduct investigations of a few phenomena such as pulse, heart sounds, and blood pressure, all of which reflect the heart in action and the function of blood vessels. A discussion of the cardiac cycle will provide a basis for understanding and interpreting the various physiological measurements taken.

Cardiac Cycle

In a healthy heart, the two atria contract simultaneously. As they begin to relax, the ventricles contract simultaneously. According to general usage, the terms **systole** and **diastole** refer to events of ventricular contraction and relaxation, respectively. The **cardiac cycle** is equivalent to one complete heartbeat—during which both atria and ventricles contract and then relax. It is marked by a succession of changes in blood volume and pressure within the heart.

The events of the cardiac cycle for the left side of the heart can be graphed in several ways (**Figure 33.1**). Although pressure changes in the right side are less dramatic than those in the left, the same relationships apply.

The events of the cardiac cycle are summarized in **Table 33.1**. We begin with the heart in total relaxation, mid-to-late diastole; however, you could start anywhere within the cycle.

The average heart beats approximately 75 beats per minute, and so the length of the cardiac cycle is about 0.8 second. Of this time period, atrial contraction occupies the first 0.1 second, which is followed by atrial relaxation and ventricular contraction for the next 0.3 second. The remaining 0.4 second is a period of total heart relaxation, the **quiescent period.**

Notice that two different types of events control the movement of blood through the heart: the alternate contraction and relaxation of the chambers, and the opening and closing of valves, which is entirely dependent on the pressure changes within the heart chambers.

33

Table 33.1	**Events of the Cardiac Cycle**					
Phase	**Description of events**	**Status of ventricles**	**Status of atria**	**Status of AV valves**	**Status of SL valves**
Mid-to-late diastole ① Ventricular filling	When atrial pressure is greater than ventricular pressure, the AV valves are forced open, and blood flows passively into the atria and on through to the ventricles.	Relaxed (diastole)	Relaxed (atrial diastole)	Opened	Closed
Mid-to-late diastole ① Ventricular filling with atrial contraction	The atria contract to complete the filling of the ventricles. Ventricular diastole ends, and so the end diastolic volume (EDV) of the ventricles is achieved.	Relaxed (diastole)	Contracted (atrial systole)	Opened	Closed
Ventricular systole ②a Isovolumetric contraction	The contraction of the ventricles begins, and ventricular pressure increases, closing the AV valves.	Contracted (systole)	Relaxed (atrial diastole)	Closed	Closed
Ventricular systole ②b Ventricular ejection	Ventricular pressure continues to rise; when the pressure in the ventricles exceeds the pressure in the great vessels exiting the heart, the SL valves open, and blood is ejected.	Contracted (systole)	Relaxed (atrial diastole)	Closed	Opened
Early diastole ③ Isovolumetric relaxation	The ventricles relax, decreasing the pressure in the ventricles; the decrease in pressure causes the SL valves to close. The **dicrotic notch** is the result of a pressure fluctuation that occurs when the aortic valve snaps shut.	Relaxed (diastole)	Relaxed (atrial diastole)	Closed	Closed

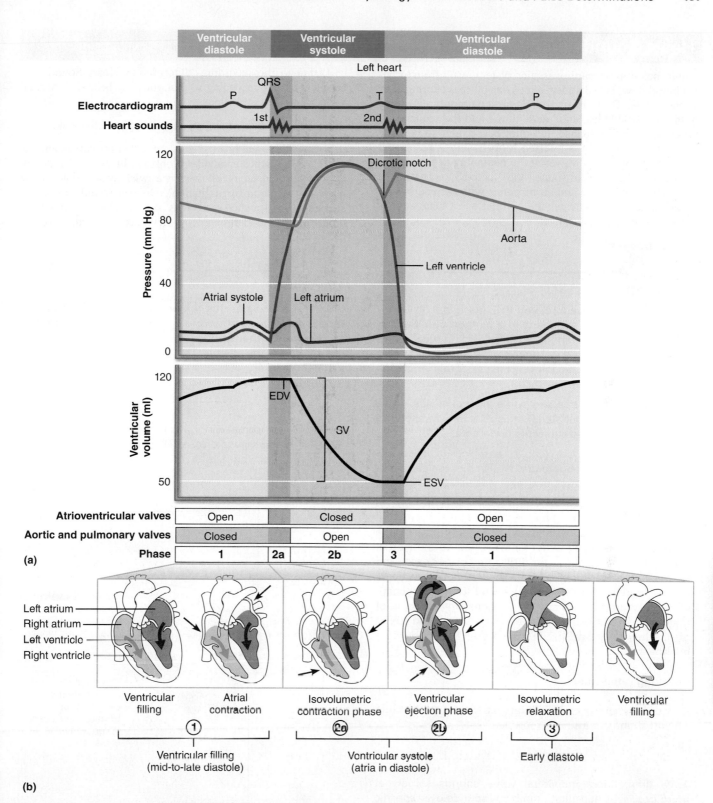

Figure 33.1 Summary of events occurring in the heart during the cardiac cycle.
(a) An ECG tracing is superimposed on the graph (top) so that electrical events can
be related to pressure and volume changes (center) in the left side of the heart.
Pressures are lower in the right side of the heart. Time occurrence of heart sounds is
also indicated. (EDV = end diastolic volume, SV = stroke volume, ESV = end systolic
volume) (b) Events of phases 1 through 3 of the cardiac cycle are diagrammed.

33

Heart Sounds

Sounds heard in the cardiovascular system result from turbulent blood flow. Two distinct sounds can be heard during each cardiac cycle. These heart sounds are commonly described by the monosyllables "lub" and "dup"; and the sequence is designated lub-dup, pause, lub-dup, pause, and so on. The first heart sound (lub) is referred to as S_1 and is associated with closure of the AV valves at the beginning of ventricular systole. The second heart sound (dup), called S_2, occurs as the semilunar valves close and corresponds with the end of systole. (Figure 33.1a indicates the correlation of heart sounds with events of the cardiac cycle.)

☐ Listen to the recording "Interpreting Heart Sounds" or another suitable recording so that you may hear both normal and abnormal heart sounds.

Check the box when you have completed the task.

Abnormal heart sounds are called **murmurs** and often indicate valvular problems. In valves that do not close tightly, closure is followed by a swishing sound due to the backflow of blood (regurgitation). Distinct sounds, often described as high-pitched screeching, are associated with the tortuous flow of blood through constricted, or stenosed, valves. ✚

Activity 1

Auscultating Heart Sounds

In the following procedure, you will auscultate (listen to) your partner's heart sounds with a stethoscope. Several more sophisticated heart-sound amplification systems are on the market, and your instructor may prefer to use one if it is available. If so, directions for the use of this apparatus will be provided by the instructor.

1. Obtain a stethoscope and some alcohol swabs. Heart sounds are best auscultated if the subject's outer clothing is removed, so a male subject is preferable.

2. With an alcohol swab, clean the earpieces of the stethoscope. Allow the alcohol to dry. Notice that the earpieces are angled. For comfort and best auscultation, the earpieces should be angled in a *forward* direction when placed into the ears.

3. Don the stethoscope. Place the diaphragm of the stethoscope on your partner's thorax, just to the sternal side of the left nipple at the fifth intercostal space, and listen carefully for heart sounds. The first sound will be a longer, louder (more booming) sound than the second, which is short and sharp. After listening for a couple of minutes, try to time the pause between the second sound of one heartbeat and the first sound of the subsequent heartbeat.

How long is this interval? _____ sec

How does it compare to the interval between the first and second sounds of a single heartbeat?

_____

4. To differentiate individual valve sounds somewhat more precisely, auscultate the heart sounds over specific thoracic regions. (Refer to **Figure 33.2** for positioning of the stethoscope.)

Auscultation of AV Valves

As a rule, the mitral valve closes slightly before the tricuspid valve. You can hear the mitral valve more clearly if you place the stethoscope over the apex of the heart, which is at the fifth intercostal space, approximately in line with the middle region of the left clavicle. Listen to

the heart sounds at this region; then move the stethoscope medially to the right margin of the sternum to auscultate the tricuspid valve. Can you detect the slight lag between the closure of the mitral and tricuspid valves?

_____

There are normal variations in the site for "best" auscultation of the tricuspid valve. These range from the right sternal margin over the fifth intercostal space (depicted in

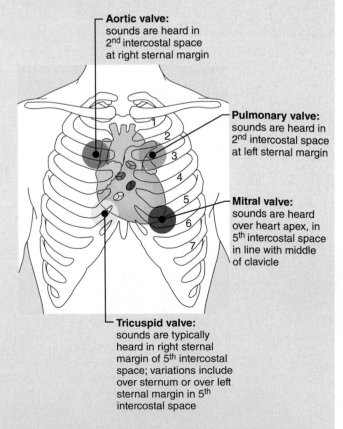

Aortic valve: sounds are heard in 2nd intercostal space at right sternal margin

Pulmonary valve: sounds are heard in 2nd intercostal space at left sternal margin

Mitral valve: sounds are heard over heart apex, in 5th intercostal space in line with middle of clavicle

Tricuspid valve: sounds are typically heard in right sternal margin of 5th intercostal space; variations include over sternum or over left sternal margin in 5th intercostal space

Figure 33.2 Areas of the thorax where heart sounds can best be detected.

Figure 33.2) to over the sternal body in the same plane, to the left sternal margin over the fifth intercostal space. If you have difficulty hearing closure of the tricuspid valve, try one of these other locations.

Auscultation of Semilunar Valves

Again there is a slight dissynchrony of valve closure; the aortic valve normally snaps shut just ahead of the pulmonary valve. If the subject inhales deeply but gently, filling of the right ventricle (due to decreased intrapulmonary pressure) and closure of the pulmonary valve will be delayed slightly. The two sounds can therefore be heard more distinctly.

Position the stethoscope over the second intercostal space, just to the *right* of the sternum. The aortic valve is best heard at this position. As you listen, have your partner take a deep breath. Then move the stethoscope to the *left* side of the sternum in the same line, and auscultate the pulmonary valve. Listen carefully; try to hear the "split" between the closure of these two valves in the second heart sound.

Although at first it may seem a bit odd that the pulmonary valve issuing from the *right* heart is heard most clearly to the *left* of the sternum and the aortic valve of the left heart is best heard at the right sternal border, this is easily explained by reviewing heart anatomy.

The Pulse

The term **pulse** refers to the alternating surges of pressure (expansion and then recoil) in an artery that occur with each contraction and relaxation of the left ventricle. This difference between systolic and diastolic pressure is called the **pulse pressure.** (See p. 502.) Normally the pulse rate (pressure surges per minute) equals the heart rate (beats per minute), and the pulse averages 70 to 76 beats per minute in the resting state.

Parameters other than pulse rate are also useful clinically. You may also assess the regularity (or rhythmicity) of the pulse, and its amplitude and/or tension—does the blood vessel expand and recoil (sometimes visibly) with the pressure waves? Can you feel it strongly, or is it difficult to detect? Is it regular like the ticking of a clock, or does it seem to skip beats?

Activity 2

Palpating Superficial Pulse Points

The pulse may be felt easily on any superficial artery when the artery is compressed over a bone or firm tissue. Palpate the following pulse or pressure points on your partner by placing the fingertips of the first two or three fingers of one hand over the artery. It helps to compress the artery firmly as you begin your palpation and then immediately ease up on the pressure slightly. In each case, notice the regularity of the pulse, and assess the degree of tension or amplitude. (**Figure 33.3**, illustrates the superficial pulse points to be palpated.) Check off the boxes as you locate each pulse point.

☐ **Superficial temporal artery:** Anterior to the ear, in the temple region.

☐ **Facial artery:** Clench the teeth, and palpate the pulse just anterior to the masseter muscle on the mandible (in line with the corner of the mouth).

☐ **Common carotid artery:** At the side of the neck.

☐ **Brachial artery:** In the cubital fossa, at the point where it bifurcates into the radial and ulnar arteries.

☐ **Radial artery:** At the lateral aspect of the wrist, above the thumb.

☐ **Femoral artery:** In the groin.

☐ **Popliteal artery:** At the back of the knee.

☐ **Posterior tibial artery:** Just above the medial malleolus.

☐ **Dorsalis pedis artery:** On the dorsum of the foot.

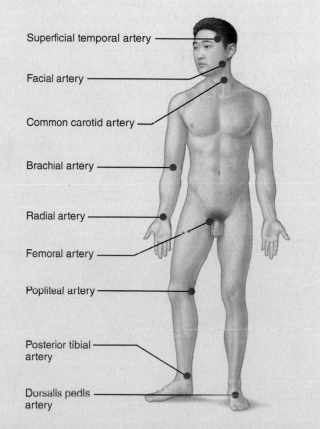

Superficial temporal artery

Facial artery

Common carotid artery

Brachial artery

Radial artery

Femoral artery

Popliteal artery

Posterior tibial artery

Dorsalis pedis artery

Figure 33.3 Body sites where the pulse is most easily palpated.

Text continues on next page. →

Which pulse point had the greatest amplitude?

_____

Which had the least? _____

Can you offer any explanation for this? _____

_____

_____

_____

Because of its easy accessibility, the pulse is most often taken on the radial artery. With your partner sitting quietly, practice counting the radial pulse for 1 minute. Make three counts, and average the results.

count 1 _____ count 2 _____

count 3 _____ average _____

Activity 3

Measuring Pulse Using BIOPAC®

Because of the elasticity of the arteries, blood pressure decreases and smooths out as blood moves farther away from the heart. A pulse, however, can still be felt in the fingers. A device called a plethysmograph or a piezoelectric pulse transducer can measure this pulse.

Setting Up the Equipment

1. Connect the BIOPAC® unit to the computer, and turn the computer **ON**.

2. Make sure the BIOPAC® unit is **OFF**.

3. Plug in the equipment (as shown in **Figure 33.4**).

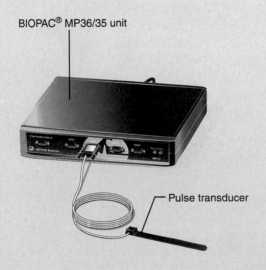

BIOPAC® MP36/35 unit

Pulse transducer

Figure 33.4 Setting up the BIOPAC®equipment. Plug the pulse tranducer into Channel 2.

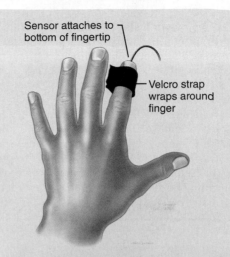

Sensor attaches to bottom of fingertip

Velcro strap wraps around finger

Figure 33.5 Placement of the pulse transducer around the tip of the index finger.

• Pulse transducer (plethysmograph)—CH 2

4. Turn the BIOPAC® unit **ON**.

5. Wrap the pulse transducer around the tip of the index finger (as shown in **Figure 33.5**). Wrap the Velcro around the finger gently (if wrapped too tightly, it will reduce circulation to the finger and obscure the recording). *Do not wiggle the finger or move the plethysmograph cord during recording.*

6. Have the subject sit down with the forearms supported and relaxed.

7. Start the Biopac Student Lab program on the computer by double-clicking the icon on the desktop or by following your instructor's guidance.

8. Select lesson **L07-ECG&P-1** from the menu, and click **OK**.

9. Type in a filename that will save this subject's data on the computer hard drive. You may want to use the subject's last name followed by Pulse (for example, Smith-Pulse-1), then click **OK**.

Calibrating the Equipment

1. When the subject is relaxed, click **Calibrate**. Click **Ignore** when prompted for SS2L on Ch.1. The calibration will stop automatically after 8 seconds.

2. Observe the recording of the calibration data, which should look like the waveforms in the example (**Figure 33.6**). ECG data is not recorded in this activity.

• If the data look very different, click **Redo Calibration** and repeat the steps above. If there is no signal, you may need to loosen the Velcro on the finger and check all attachments.

• If the data look similar, proceed to the next section.

Recording the Data

1. When the subject is ready, click **Record** to begin recording the pulse. After 30 seconds, click **Suspend**.

2. Observe the data, which should look similar to the pulse data example (**Figure 33.7**).

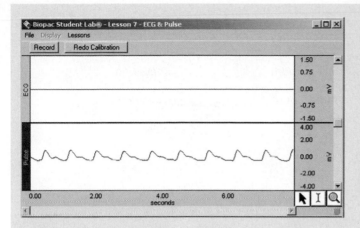

Figure 33.6 Example of waveforms during the calibration procedure.

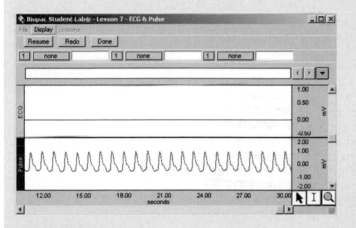

Figure 33.7 Example of waveforms during the recording of data.

- If the data look very different, click **Redo** and repeat the steps above.
- If the data look similar, go to the next step.

3. When you are finished, click **Done** and then click **Yes**. A pop-up window will appear; to record from another subject select **Record from another subject** and return to step 5 under Setting Up the Equipment. If continuing to the Data Analysis section, select **Analyze current data file** and proceed to step 2 in the Data Analysis section.

Data Analysis

1. If just starting the BIOPAC® program to perform data analysis, enter **Review Saved Data** mode and choose the file with the subject's pulse data (for example, SmithPulse-1).

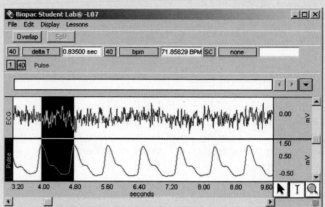

Figure 33.8 Using the I-beam cursor to highlight data for analysis.

2. Observe that pulse data is in the lower scale.

3. To analyze the data, set up the first channel/measurement box at the top of the screen.

Channel	Measurement	Data
CH 40	Delta T	pulse

Delta T: Measures the time elapsed in the selected area

4. Use the arrow cursor and click the I-beam cursor box on the lower right side of the screen to activate the "area selection" function. Using the activated I-beam cursor, highlight from the peak of one pulse to the peak of the next pulse (as shown in **Figure 33.8**).

Observe the elapsed time between heartbeats and record here (to the nearest 0.01 second): _____

5. Calculate the beats per minute by inserting the elapsed time into this formula:

(1 beat/ _____ sec) × (60 sec/min)

= _____ beats/min

Optional Activity with BIOPAC®
Pulse Measurement

To expand the experiment, you can measure the effects of heat and cold on pulse rate. To do this, you can submerge the subject's hand (the hand without the plethysmograph!) in hot and/or ice water for 2 minutes, and then record the pulse. Alternatively, you can investigate change in pulse rate after brief exercise such as jogging in place.

Apical-Radial Pulse

The correlation between the apical and radial pulse rates can be determined by simultaneously counting them. The **apical pulse** (actually the counting of heartbeats) may be slightly faster than the radial because of a slight lag in time as the blood rushes from the heart into the large arteries where it can be palpated. A difference between the values observed is referred to as a **pulse deficit.** A large difference may indicate cardiac impairment (a weakened heart that is unable to pump blood into the arterial tree to a normal extent), low cardiac output, or abnormal heart rhythms. Apical pulse counts are routinely ordered for patients with cardiac decompensation.

33

Activity 4

Taking an Apical Pulse

With the subject sitting quietly, one student, using a stethoscope, should determine the apical pulse rate while another simultaneously counts the radial pulse rate. The stethoscope should be positioned over the fifth left intercostal space. The person taking the radial pulse should determine the starting point for the count and give the stop-count signal exactly 1 minute later. Record your values at right.

apical count _____ beats/min

radial count _____ pulses/min

pulse deficit _____ pulses/min

Blood Pressure Determinations

Blood pressure (BP) is defined as the pressure the blood exerts against any unit area of the blood vessel walls, and it is generally measured in the arteries. Because the heart alternately contracts and relaxes, the resulting rhythmic flow of blood into the arteries causes the blood pressure to rise and fall during each beat. Thus you must take two blood pressure readings: the **systolic pressure,** which is the pressure in the arteries at the peak of ventricular contraction, and the **diastolic pressure,** which reflects the pressure during ventricular relaxation. Blood pressures are reported in millimeters of mercury (mm Hg), with the systolic pressure appearing first; 120/80 translates to 120 over 80, or a systolic pressure of 120 mm Hg and a diastolic pressure of 80 mm Hg. Normal blood pressure varies considerably from one person to another.

In this procedure, you will measure arterial pressure by indirect means and under various conditions. You will investigate and demonstrate factors affecting blood pressure, and the rapidity of blood pressure changes.

 Prepare for lab: Watch the Pre-Lab Video
MasteringA&P®>Study Area>Pre-Lab Videos

Activity 5

Using a Sphygmomanometer to Measure Arterial Blood Pressure Indirectly

The **sphygmomanometer,** commonly called a *blood pressure cuff,* is an instrument used to obtain blood pressure readings by the auscultatory method (**Figure 33.9**). It consists of an inflatable cuff with an attached pressure gauge. The cuff is placed around the arm and inflated to a pressure higher than systolic pressure to occlude circulation to the forearm. As cuff pressure is gradually released, the examiner listens with a stethoscope for characteristic sounds called the **sounds of Korotkoff,** which indicate the resumption of blood flow into the forearm.

1. Work in pairs to obtain radial artery blood pressure readings. Obtain a felt marker, stethoscope, alcohol swabs, and a sphygmomanometer. Clean the earpieces of the stethoscope with the alcohol swabs, and check the cuff for the presence of trapped air by compressing it against the laboratory table. (A partially inflated cuff will produce erroneous measurements.)

2. The subject should sit in a comfortable position with one arm resting on the laboratory table (approximately at heart level if possible). Wrap the cuff around the subject's arm, just above the elbow, with the inflatable area on the medial arm surface. The cuff may be marked with an arrow; if so, the arrow should be positioned over the brachial artery (Figure 33.9). Secure the cuff by tucking the distal end under the wrapped portion or by bringing the Velcro areas together.

3. Palpate the brachial pulse, and lightly mark its position with a felt pen. Don the stethoscope, and place its diaphragm over the pulse point.

⚠ *Do not keep the cuff inflated for more than 1 minute.* If you have any trouble obtaining a reading

within this time, deflate the cuff, wait 1 or 2 minutes, and try again. (A prolonged interference with blood pressure homeostasis can lead to fainting.)

4. Inflate the cuff to approximately 160 mm Hg pressure, and slowly release the pressure valve. Watch the pressure gauge as you listen carefully for the first soft thudding sounds of the blood spurting through the partially occluded artery. Mentally note this pressure (systolic pressure), and continue to release the cuff pressure. You will notice first an increase, then a muffling, of the sound. For the diastolic pressure, note the pressure at which the sound becomes muffled or disappears. Make two blood pressure determinations, and record your results below.

First trial:	**Second trial:**
systolic pressure _____	systolic pressure _____
diastolic pressure _____	diastolic pressure _____

5. Compute the **pulse pressure** for each trial. The pulse pressure is the difference between the systolic and diastolic pressures, and it indicates the amount of blood forced from the heart during systole, or the actual "working" pressure. A narrowed pulse pressure (less than 30 mm Hg) may be a signal of severe aortic stenosis, constrictive pericarditis, or tachycardia. A widened pulse

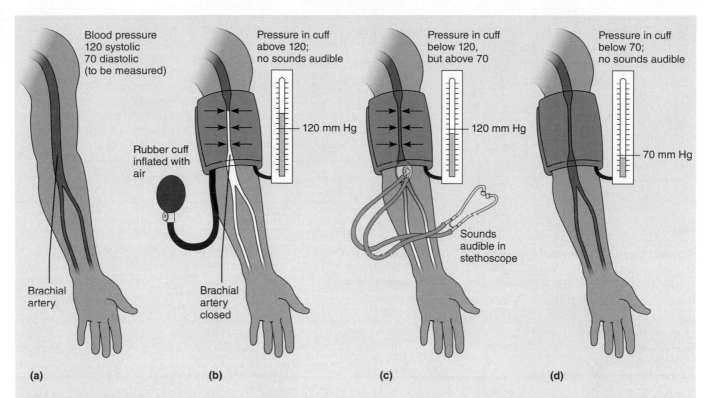

(a)

Blood pressure
120 systolic
70 diastolic
(to be measured)

Rubber cuff
inflated with
air

Brachial
artery

Brachial
artery
closed

(b)

Pressure in cuff
above 120;
no sounds audible

— 120 mm Hg

(c)

Pressure in cuff
below 120,
but above 70

— 120 mm Hg

Sounds
audible in
stethoscope

(d)

Pressure in cuff
below 70;
no sounds audible

— 70 mm Hg

Figure 33.9 Procedure for measuring blood pressure. (a) The course of the brachial artery of the arm. Assume a blood pressure of 120/70. **(b)** The blood pressure cuff is wrapped snugly around the arm just above the elbow and inflated until blood flow into the forearm is stopped and no brachial pulse can be felt or heard.

(c) Pressure in the cuff is gradually reduced while the examiner uses a stethoscope to listen (auscultate) for sounds (of Korotkoff) in the brachial artery. The pressure, read as the first soft tapping sounds are heard (the first point at which a small amount of blood is spurting through the constricted artery), is recorded as the

systolic pressure. **(d)** As the pressure is reduced still further, the sounds become louder and more distinct, but when the artery is no longer restricted and blood flows freely, the sounds can no longer be heard. The pressure at which the sounds disappear is routinely recorded as the diastolic pressure.

pressure (over 40 mm Hg) is common in hypertensive individuals.

Pulse pressure:

first trial _____   second trial _____

6. Compute the **mean arterial pressure (MAP)** for each trial using the following equation:

$$\text{MAP} = \text{diastolic pressure} + \frac{\text{pulse pressure}}{3}$$

first trial _____   second trial _____

33

Activity 6

Estimating Venous Pressure

It is not possible to measure venous pressure with the sphygmomanometer. The methods available for measuring it produce estimates at best, because venous pressures are so much lower than arterial pressures. The difference in pressure becomes obvious when these vessels are cut. If a vein is cut, the blood flows evenly from the cut. A lacerated artery produces rapid spurts of blood. In this activity, you will estimate venous pressures.

1. Obtain a meter stick, and ask your lab partner to stand with his or her right side toward the blackboard, arms hanging freely at the sides. On the board, mark the approximate level of the right atrium. (This will be just slightly higher than the point at which you auscultated the apical pulse.)

2. Observe the superficial veins on the dorsum of the right hand as the subject alternately raises and lowers it. Notice the collapsing and filling of the veins as internal pressures change. Have the subject repeat this action until you can determine the point at which the veins have just collapsed. Mark this hand level on the board. Then measure, in millimeters, the distance in the vertical plane from this point to the level of the right atrium (previously marked). Record this value. Distance of right arm from right atrium at point of venous collapse:

_____ mm

Text continues on next page. →

3. Compute the venous pressure (P$_v$), in millimeters of mercury, with the following formula:

$$P_v = \frac{1.056 \text{ (specific gravity of blood)} \times \text{mm (measured)}}{13.6 \text{ (specific gravity of Hg)}}$$

Venous pressure computed: _____ mm Hg

Normal venous pressure varies from approximately 30 to 90 mm Hg. That of the hand ranges between 30 and 40 mm Hg. How does your computed value compare?

_____

4. Because venous walls are so thin, pressure within them is readily affected by external factors such as muscle activity, deep pressure, and pressure changes occurring in the thorax during breathing. The Valsalva maneuver, which increases intrathoracic pressure, is used to demonstrate the effect of thoracic pressure changes on venous pressure.

To perform this maneuver take a deep breath, and then mimic the motions of exhaling forcibly, but without actually exhaling. In reaction to this, the glottis will close;

and intrathoracic pressure will increase. (Most of us have performed this maneuver unknowingly in acts of defecation in which we are "bearing down.") Have the same subject again stand next to the blackboard mark for the level of his or her right atrium. While the subject performs the Valsalva maneuver and raises and lowers one hand, determine the point of venous collapse and mark it on the board. Measure the distance of that mark from the right atrium level, and record it below. Then compute the venous pressure, and record it.

_____ mm   Venous pressure: _____ mm Hg

How does this value compare with the venous pressure measurement computed for the relaxed state?

_____

Explain: _____

_____

_____

Activity 7

Observing the Effect of Various Factors on Blood Pressure and Heart Rate

Arterial blood pressure is directly proportional to cardiac output (CO, amount of blood pumped out of the left ventricle per unit time) and peripheral resistance (PR) to blood flow, that is,

$$BP = CO \times PR$$

Peripheral resistance is increased by blood vessel constriction (most importantly the arterioles), by an increase in blood viscosity, and by a loss of elasticity of the arteries (seen in arteriosclerosis). Any factor that increases either the cardiac output or the peripheral resistance causes an almost immediate reflex rise in blood pressure. A close examination of these relationships reveals that many factors—age, weight, time of day, exercise, body position, emotional state, and various drugs, for example—alter blood pressure. The influence of a few of these factors is investigated here.

The following tests are done most efficiently if students work in groups of four: one student acts as the subject; two are examiners (one takes the radial pulse and the other auscultates the brachial blood pressure); and a fourth student collects and records data. The sphygmomanometer cuff should be left on the subject's arm throughout the experiments (in a deflated state, of course) so that, at the proper times, the blood pressure can be taken quickly. In each case, take the measurements at least twice. For each of the following tests, students should formulate hypotheses, collect data, and write lab reports. (See Getting Started, on MasteringA&P.) Conclusions should be shared with the class.

Posture

To monitor circulatory adjustments to changes in position, take blood pressure and pulse measurements under the conditions noted in the **Activity 7 Posture chart**. Record your results on the chart.

Activity 7: Posture				
	Trial 1		Trial 2	
	BP	Pulse	BP	Pulse
Sitting quietly				
Reclining (after 2 to 3 min)				
Immediately on standing from the reclining position ("at attention" stance)				
After standing for 3 min				

33

Activity 7: Exercise										
Harvard step test for 5 min at 30/min	Baseline		Interval following test							
			Immediately		1 min		2 min		3 min	
	BP	P	BP	P	BP	P	BP	P	BP	P
Well-conditioned individual	___	___	___	___	___	___	___	___	___	___
Poorly conditioned individual	___	___	___	___	___	___	___	___	___	___

Exercise

Blood pressure and pulse changes during and after exercise provide a good yardstick for measuring one's overall cardiovascular fitness. Although there are more sophisticated and more accurate tests that evaluate fitness according to a specific point system, the *Harvard step test* described here is a quick way to compare the relative fitness level of a group of people.

You will be working in groups of four, duties assigned as indicated above, except that student 4, in addition to recording the data, will act as the timer and call the cadence.

! Any student with a known heart problem should refuse to participate as the subject.

All four students may participate as the subject in turn, if desired, but the bench stepping is to be performed *at least twice* in each group—once with a well-conditioned person acting as the subject, and once with a poorly conditioned subject.

Bench stepping is the following series of movements repeated sequentially:

1. Place one foot on the step.

2. Step up with the other foot so that both feet are on the platform. Straighten the legs and the back.

3. Step down with one foot.

4. Bring the other foot down.

The pace for the stepping will be set by the "timer" (student 4), who will repeat "Up 2 3 4, up 2 3 4" at such a pace that each "up-2-3-4" sequence takes 2 sec (30 cycles/min).

1. Student 4 should obtain the step (0.5 m [20-in.] tall for male subject or 0.4 m [16-in.] for a female subject) while baseline measurements are being obtained on the subject.

2. Once the baseline pulse and blood pressure measurements have been recorded on the **Activity 7 Exercise chart** (above), the subject is to stand quietly at attention for 2 minutes to allow his or her blood pressure to stabilize before beginning to step.

3. The subject is to perform the bench stepping for as long as possible, up to a maximum of 5 minutes, according to the cadence called by the timer. The subject must keep their posture erect; the other group members should watch for and work against crouching. If he or she is unable to keep the pace for a span of 15 seconds, the test is to be terminated.

4. When the subject is stopped by the timer for crouching, stops voluntarily because he or she is unable to continue, or has completed 5 minutes of bench stepping, he or she is to sit down. Record the duration of exercise (in seconds), and measure the blood pressure and pulse immediately and thereafter at 1-minute intervals for 3 minutes post-exercise.

Duration of exercise: _____ sec

5. Calculate the subject's *index of physical fitness* using the following formula:

$$\text{Index} = \frac{\text{duration of exercise in second} \times \text{mm } 100}{2 \times \text{sum of the three pulse counts in recovery}}$$

Interpret scores according to the following scale:

below 55	poor physical condition
55 to 62	low average
63 to 71	average
72 to 79	high average
80 to 89	good
90 and over	excellent

6. Record the test values on the Activity 7 Exercise chart above, and repeat the testing and recording procedure with the second subject.

When did you notice a greater elevation of blood pressure and pulse?

_____

Explain: _____

_____

Was there a sizable difference between the after-exercise values for well-conditioned and poorly conditioned individuals?

_____ Explain: _____

_____

Did the diastolic pressure also increase? _____

Text continues on next page. →

33

Activity 7: Stimulus							
Baseline		**1 min**		**2 min**		**3 min**	
BP	P	BP	P	BP	P	BP	P

Explain: _____

_____

A Noxious Sensory Stimulus (Cold)

Blood pressure can be affected by emotions and pain. This lability of blood pressure will be investigated through use of the **cold pressor test,** in which one hand will be immersed in unpleasantly (even painfully) cold water.

1. Measure the blood pressure and pulse of the subject as he or she sits quietly. Record these as the baseline values on the **Activity 7 Stimulus chart** (above).

2. Obtain a basin and thermometer, fill the basin with ice cubes, and add water. When the temperature of the ice bath has reached 5°C, immerse the subject's other hand (the noncuffed limb) in the ice water. With the hand still immersed, take blood pressure and pulse readings at 1-minute intervals for a period of 3 minutes, and record the values on the chart.

How did the blood pressure change during cold exposure?

_____

_____

Was there any change in pulse? _____

3. Subtract the respective baseline readings of systolic and diastolic blood pressure from the highest single reading of systolic and diastolic pressure obtained during cold immersion. (For example, if the highest experimental reading is 140/88 and the baseline reading is 120/70, then the differences in blood pressure would be systolic pressure, 20 mm Hg, and diastolic pressure, 18 mm Hg.) These differences are called the index of response. According to their index of response, subjects can be classified as follows:

Hyporeactors (stable blood pressure): Exhibit a rise of diastolic and/or systolic pressure ranging from 0 to 22 mm Hg or a drop in pressures

Hyperreactors (labile blood pressure): Exhibit a rise of 23 mm Hg or more in the diastolic and/or systolic blood pressure

Is the subject tested a hypo- or hyperreactor?

_____

_____

Skin Color as an Indicator of Local Circulatory Dynamics

Skin color reveals with surprising accuracy the state of the local circulation and allows inferences concerning the larger blood vessels and the circulation as a whole. The Activity 8 experiments on local circulation illustrate a number of factors that affect blood flow to the tissues.

Clinical expertise often depends upon good observation skills, accurate recording of data, and logical interpretation of the findings. One of the earliest compensatory responses of the body to impaired blood flow to the brain is constriction of cutaneous blood vessels. This reduces blood flow to the skin and diverts it into the circulatory mainstream to serve other, more vital tissues. As a result, the skin of the face and particularly of the extremities becomes pale, cold, and eventually moist with perspiration. Therefore, pale, cold, clammy skin should immediately prompt a suspicion that the circulation is dangerously inefficient.

Activity 8

Examining the Effect of Local Chemical and Physical Factors on Skin Color

The local blood supply to the skin (indeed, to any tissue) is influenced by (1) local metabolites, (2) oxygen supply, (3) local temperature, (4) autonomic nervous system impulses, (5) local vascular reflexes, (6) certain hormones, and (7) substances released by injured tissues. A number of these factors are examined in the simple experiments that follow. Each experiment should be conducted by students in groups of three or four. One student will act as the subject; the others will conduct the tests and make and record observations.

Vasodilation and Flushing of the Skin Due to Local Metabolites

1. Obtain a sphygmomanometer (blood pressure cuff) and stethoscope. You will also need a watch or clock with a second hand.

2. The subject should bare both arms by rolling up the sleeves as high as possible and then lay the forearms side by side on the bench top.

33

3. Observe the general color of the subject's forearm skin, and the normal contour and size of the veins. Notice whether skin color is bilaterally similar. Record your observations:

_____

_____

_____

4. Apply the blood pressure cuff to one arm, and inflate it to 250 mm Hg. Keep it inflated for 1 minute. During this period, repeat the observations made above, and record the results:

_____

_____

5. Release the pressure in the cuff (leaving the deflated cuff in position), and again record the forearm skin color and the condition of the forearm veins. Make this observation immediately after deflation and then again 30 seconds later.

Immediately after deflation: _____

_____

30 sec after deflation: _____

_____

The above observations constitute your baseline information. Now conduct the following tests.

6. Instruct the subject to raise the cuffed arm above his or her head and to clench the fist as tightly as possible. While the hand and forearm muscles are tightly contracted, rapidly inflate the cuff to 240 mm Hg or more. This maneuver partially empties the hand and forearm of blood and stops most blood flow to the hand and forearm. Once the cuff has been inflated, the subject is to relax the fist and return the forearm to the bench top so that it can be compared to the other forearm.

7. Leave the cuff inflated for exactly 1 minute. During this interval, compare the skin color in the "ischemic" (blood-deprived) hand to that of the "normal" (noncuffed-limb) hand. Quickly release the pressure immediately after 1 minute.

What are the subjective effects (sensations felt by the subject, such as pain, cold, warmth, tingling, weakness) of stopping blood flow to the arm and hand for 1 minute? These sensations are "symptoms" of a change in function.

_____

_____

What are the objective effects (color of skin and condition of veins)?

_____

_____

How long does it take for the subject's ischemic hand to regain its normal color?

_____

Effects of Venous Congestion

1. Again, but with a different subject, observe and record the appearance of the skin and veins on the forearms resting on the bench top. This time, pay particular attention to the color of the fingers, particularly the distal phalanges, and the nail beds. Record this information:

_____

_____

2. Wrap the blood pressure cuff around one of the subject's arms, and inflate it to 40 mm Hg. Maintain this pressure for 5 minutes. Make a record of the subjective and objective findings just before the 5 minutes are up, and then again immediately after release of the pressure at the end of 5 minutes.

Subjective (arm cuffed): _____

_____

Objective (arm cuffed): _____

_____

Subjective (pressure released): _____

_____

Objective (pressure released): _____

_____

3. With still another subject, conduct the following simple experiment: Raise one arm above the head, and let the other hang by the side for 1 minute. After 1 minute, quickly lay both arms on the bench top, and compare their color.

Color of raised arm: _____

Color of dependent arm: _____

From this and the two preceding observations, analyze the factors that determine tint of color (pink or blue) and

Text continues on next page. →

33

intensity of skin color (deep pink or blue as opposed to light pink or blue). Record your conclusions.

_____

_____

_____

_____

_____

Collateral Blood Flow

In some diseases, blood flow to an organ through one or more arteries may be completely and irreversibly obstructed. Fortunately, in most cases a given body area is supplied both by one main artery and by anastomosing channels connecting the main artery with one or more neighboring blood vessels. Consequently, an organ may remain viable even though its main arterial supply is occluded, as long as the **collateral vessels** are still functional.

The effectiveness of collateral blood flow in preventing ischemia can be easily demonstrated.

1. Check the subject's hands to be sure they are _warm_ to the touch. If not, choose another subject, or warm the subject's hands in 35°C water for 10 minutes before beginning.

2. Palpate the subject's radial and ulnar arteries approximately 2.5 cm (1 in.) above the wrist flexure, and mark their locations with a felt marker.

3. Instruct the subject to supinate one forearm and to hold it in a partially flexed (about a 30° angle) position, with the elbow resting on the bench top.

4. Face the subject and grasp his or her forearm with both of your hands, the thumb and fingers of one hand compressing the marked radial artery and the thumb and fingers of the other hand compressing the ulnar artery. Maintain the pressure for 5 minutes, noticing the progression of the subject's hand to total ischemia.

5. At the end of 5 minutes, release the pressure abruptly. Record the subject's sensations, as well as the intensity and duration of the flush in the previously occluded hand. (Use the other hand as a baseline for comparison.)

_____

_____

_____

6. Allow the subject to relax for 5 minutes; then repeat the maneuver, but this time _compress only the radial artery._ Record your observations.

_____

_____

How do the results of the first test differ from those of the second test with respect to color changes during compression and to the intensity and duration of reactive hyperemia (redness of the skin)?

_____

_____

_____

_____

7. Once again allow the subject to relax for 5 minutes. Then repeat the maneuver, _compressing only the ulnar artery._ Record your observations:

_____

_____

What can you conclude about the relative sizes of, and hand areas served by, the radial and ulnar arteries?

_____

_____

Effect of Mechanical Stimulation of Blood Vessels of the Skin

With moderate pressure, draw the blunt end of your pen across the skin of a subject's forearm. Wait 3 minutes to observe the effects, and then repeat with firmer pressure.

What changes in skin color do you observe with light-to-moderate pressure?

_____

With heavy pressure? _____

The redness, or _flare,_ observed after mechanical stimulation of the skin results from a local inflammatory response promoted by chemical mediators released by injured tissues. These mediators stimulate increased blood flow into the area and leaking of fluid (from the capillaries) into the local tissues.

REVIEW SHEET

Human Cardiovascular Physiology: Blood Pressure and Pulse Determinations

Name _____   Lab Time/Date _____

Cardiac Cycle

1. Using the grouped sets of terms to the right of the diagram, correctly identify each trace, valve closings and openings, and each time period of the cardiac cycle.

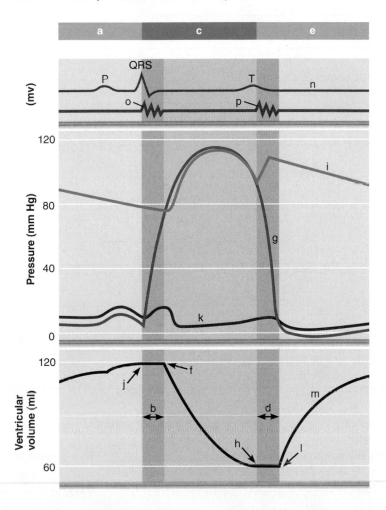

_____ 1. aortic pressure

_____ 2. atrial pressure (left)

_____ 3. ECG

_____ 4. first heart sound

_____ 5. second heart sound

_____ 6. ventricular pressure (left)

_____ 7. ventricular volume

_____ 8. aortic (semilunar) valve closes

_____ 9. aortic (semilunar) valve opens

_____ 10. AV and semilunar valves are closed (2 letters)

_____ 11. AV valve closes

_____ 12. AV valve opens

_____ 13. ventricular diastole (2 letters)

14. ventricular systole

2. Define the following terms.

systole: _____

diastole: _____

cardiac cycle: _____

3. Answer the following questions concerning events of the cardiac cycle.

When are the AV valves closed? _____

What event within the heart causes the AV valves to open? _____

When are the semilunar valves closed? _____

What event causes the semilunar valves to open? _____

Are both sets of valves closed during any part of the cycle? _____

If so, when? _____

Are both sets of valves open during any part of the cycle? _____

At what point in the cardiac cycle is the pressure in the heart highest? _____

Lowest? _____

_____

What event results in the pressure deflection called the dicrotic notch? _____

_____

4. Using the key below, indicate the time interval occupied by the following events of the cardiac cycle.

   Key: a. 0.8 sec   b. 0.4 sec   c. 0.3 sec   d. 0.1 sec

   _____ 1. the length of the normal cardiac cycle   _____ 3. the quiescent period

   _____ 2. the time interval of atrial contraction   _____ 4. the time interval of ventricular contraction

5. If an individual's heart rate is 80 beats/min, what is the length of the cardiac cycle? _____ What portion of the

   cardiac cycle decreases with a more rapid heart rate? _____

   _____

6. What two factors promote the movement of blood through the heart? _____

   _____ and _____

   _____

Heart Sounds

7. Complete the following statements.

    The two monosyllables describing the heart sounds are
_1_. The first heart sound is a result of closure of the _2_ valves,
whereas the second is a result of closure of the _3_ valves. The
heart chambers that have just been filled when you hear the first
heart sound are the _4_, and the chambers that have just emptied are the _5_. Immediately after the second heart sound, both
the _6_ and _7_ are filling with blood.

1. _____

2. _____

3. _____

4. _____

5. _____

6. _____

7. _____

8. As you listened to the heart sounds during the laboratory session, what differences in pitch, length, and amplitude

(loudness) of the two sounds did you observe? _____

_____

_____

9. In order to auscultate most accurately, indicate where you would place your stethoscope for the following sounds:

closure of the tricuspid valve: _____

closure of the aortic valve: _____

apical heartbeat: _____

Which valve is heard most clearly when the apical heartbeat is auscultated?_____

10. No one expects you to be a full-fledged physician on such short notice, but on the basis of what you have learned
about heart sounds, give an example of how abnormal sounds might be used to diagnose a heart problem.

_____

_____

_____

The Pulse

11. Define *pulse*. _____

12. Describe the procedure used to take the pulse. _____

_____

_____

13. Identify the artery palpated at each of the pressure points listed.

at the wrist: _____

in front of the ear: _____

on the dorsum of the foot: _____

at the side of the neck: _____

14. When you were palpating the various pulse or pressure points, which appeared to have the greatest amplitude or tension?

_____ Why do you think this was so? _____

_____

15. Assume someone has been injured in an auto accident and is hemorrhaging badly. What pressure point would you compress to help stop bleeding from each of the following areas?

the thigh: _____   the calf: _____

the forearm: _____   the thumb: _____

16. How could you tell by simple observation whether bleeding is arterial or venous? _____

_____

17. You may sometimes observe a slight difference between the value obtained from an apical pulse (beats/min) and that from an arterial pulse taken elsewhere on the body. What is this difference called?

_____

Blood Pressure Determinations

18. Define *blood pressure.* _____

19. Identify the phase of the cardiac cycle to which each of the following apply.

systolic pressure: _____   diastolic pressure: _____

20. What is the name of the instrument used to compress the artery and record pressures in the auscultatory method of

determining blood pressure? _____

21. What are the sounds of Korotkoff? _____

_____

What causes the systolic sound? _____

_____

What causes the disappearance of the sound? _____

22. Interpret the pressure reading for each of the numbers listed: 145/85. _____

_____

_____

23. Define *pulse pressure.* _____

Why is this measurement important? _____

_____

24. Explain why *pulse pressure* is different from *pulse rate*. _____

_____

_____

25. How do venous pressures compare to arterial pressures? _____

Why?_____

26. What maneuver to increase the thoracic pressure illustrates the effect of external factors on venous pressure? _____

_____ How is it performed? _____

_____

27. What might an abnormal increase in venous pressure indicate? (Think!) _____

_____

Observing the Effect of Various Factors
on Blood Pressure and Heart Rate

28. What effect do the following have on blood pressure? (Indicate increase by ↑ and decrease by ↓.)

_____ 1. increased diameter of the arterioles   _____ 4. hemorrhage

_____ 2. increased blood viscosity   _____ 5. arteriosclerosis

_____ 3. increased cardiac output   _____ 6. increased pulse rate

29. In which position (sitting, reclining, or standing) is the blood pressure normally the highest?

_____The lowest? _____

What immediate changes in blood pressure did you observe when the subject stood up after being in the sitting or

reclining position? _____

_____

What changes in the blood vessels might account for the change? _____

_____

_____

After the subject stood for 3 minutes, what changes in blood pressure did you observe? _____

_____

How do you account for this change? _____

30. What was the effect of exercise on blood pressure? _____

On pulse rate? _____ Do you think these effects reflect changes in cardiac output *or* in

peripheral resistance? _____

_____

_____

Why are there normally no significant increases in diastolic pressure after exercise? _____

_____

31. What effects of cold temperature did you observe on blood pressure in the laboratory? _____

_____

What do you think the effect of heat would be? _____

Why? _____

32. Differentiate between a hypo- and a hyperreactor relative to the cold pressor test. _____

_____

Skin Color as an Indicator of Local Circulatory Dynamics

33. Describe normal skin color and the appearance of the veins in the subject's forearm before any testing was conducted.

_____

34. What changes occurred when the subject emptied the forearm of blood (by raising the arm and making a fist) and the

flow was occluded with the cuff? _____

What changes occurred during venous congestion? _____

_____

35. What is the importance of collateral blood supplies? _____

_____

36. Explain the mechanism by which mechanical stimulation of the skin produced a flare. _____

_____

34 Frog Cardiovascular Physiology

Objectives

☐ Describe the properties of automaticity and rhythmicity as they apply to cardiac muscle.

☐ Discuss the anatomical differences between frog and human hearts.

☐ Compare the intrinsic rate of contraction of the "pacemaker" of the frog heart (sinus venosus) to that of the atria and ventricle.

☐ Define *extrasystole,* and explain when an extrasystole can be induced.

☐ Explain why it is important that cardiac muscle cannot be tetanized.

☐ Describe the effects of the following on heart rate: cold, heat, vagal stimulation, pilocarpine, atropine sulfate, epinephrine, digitalis, and potassium, sodium, and calcium ions.

☐ Define *ectopic pacemaker, vagal escape,* and *partial* and *total heart block.*

☐ Name the blood vessels associated with a capillary bed and describe microcirculation.

☐ Identify an arteriole, venule, and capillaries in a frog's web, and cite the differences between relative size of these vessels and the rate of blood flow through them.

☐ Discuss the effect of heat, cold, local irritation, and histamine on blood flow in capillaries, and explain how these responses help maintain homeostasis.

Materials

- Dissecting instruments and tray
- Disposable gloves
- Petri dishes
- Medicine dropper
- Millimeter ruler
- Disposal container for organic debris
- Frog Ringer's solutions (at room temperature, 5°C, and 32°C)
- Frogs*
- Thread

Instructor will double-pith frogs as required for student experimentation.

Text continues on next page. →

MasteringA&P®

For related exercise study tools, go to the Study Area of **MasteringA&P**. There you will find:

- Practice Anatomy Lab **PAL**
- A&PFlix **A&PFlix**
- PhysioEx **PEx**
- Practice quizzes, Histology Atlas, eText, Videos, and more!

Pre-Lab Quiz

1. Circle True or False. Heart muscle can depolarize spontaneously in the absence of any external stimulation.

2. Spontaneous depolarization-repolarization events occur in a regular and continuous manner in cardiac muscle, a property known as:
   a. automaticity
   b. rhythmicity
   c. synchronicity

3. How many chambers does the frog heart have?
   a. two   b. three   c. four

4. Circle True or False. Heart rate can be modified by extrinsic impulses from the autonomic nerves.

5. What is an extrasystole? _____

_____

6. Which chemical agent will you use to modify the frog heart rate?
   a. caffeine   c. magnesium solution
   b. digitalis   d. Ringer's solution

7. The _____ nerve carries parasympathetic impulses to the heart.
   a. cardiac   c. phrenic
   b. olfactory   d. vagus

8. Circle the correct underlined term. The phenomenon of <u>vagal escape</u> / <u>heart block</u> occurs when the heart stops momentarily, and then begins to beat again.

9. Circle True or False. The flow of blood through capillary beds is slow and intermittent.

Text continues on next page. →

- Large rubber bands
- Fine common pins
- Frog board
- Cotton balls
- Physiograph or BIOPAC® equipment:

Physiograph (polygraph), physiograph paper and ink, force transducer, transducer cable, transducer stand, stimulator output extension cable, electrodes

BIOPAC® BIOPAC® BSL System with BSL software version 3.7.7 (for Windows 7/Vista/XP) or BSL 3.7.3 (Mac OS X 10.4–10.6), data acquisition unit MP36/35, PC or Mac computer, BIOPAC® HDW100A tension adjuster or equivalent, BIOPAC® SS12LA force transducer with S-hook, small hook with thread, and transducer (or ring) stand.

Instructors using the MP36/35/30 data acquisition unit with BSL software versions earlier than 3.7.7 (for Windows or Mac) and will need slightly different channel settings and collection strategies. Instructions for using the older data acquisition unit can be found on MasteringA&P.

- Dropper bottles of freshly prepared solutions (using frog Ringer's solution as the solvent) of the following:

  2.5% pilocarpine

  5% atropine sulfate

  1% epinephrine

  2% digitalis

  2% calcium chloride ($CaCl_2$)

  0.7% sodium chloride (NaCl)

  5% potassium chloride (KCl)

  0.01% histamine

  0.01 N HCl

- Dissecting pins
- Paper towels
- Compound microscope

PEx PhysioEx™ 9.1 Computer Simulation Ex. 6 on p. PEx-93

Note: *Instructions for using PowerLab® equipment can be found on MasteringA&P.*

34

---

10. _____ causes extensive vasodilation when applied to the frog web.
   a. Calcium
   b. Epinephrine
   c. Histamine
   d. Ringer's solution

Investigations of human cardiovascular physiology are very interesting, but many areas obviously do not lend themselves to experimentation. However, investigation of frog cardiovascular physiology can be done and provides valuable data because the physiological mechanisms in these animals are similar, if not identical, to those in humans.

In this exercise, you will conduct cardiac investigations. In addition, you will observe the microcirculation in a frog's web and subject it to various chemical and thermal agents to demonstrate their influence on local blood flow.

Special Electrical Properties of Cardiac Muscle: Automaticity and Rhythmicity

Cardiac muscle differs from skeletal muscle both functionally and in its fine structure. Skeletal muscle must be electrically stimulated to contract. In contrast, heart muscle can and does depolarize spontaneously in the absence of external stimulation. This property, called **automaticity,** is due to plasma membranes that have reduced permeability to potassium ions but still allow sodium ions to slowly leak into the cells. This leakage causes the muscle cells to gradually depolarize until the action potential threshold is reached. Shortly thereafter, contraction occurs. Also, the spontaneous depolarization-repolarization events occur in a regular, continuous manner in cardiac muscle, a property called **rhythmicity.**

In the following experiment, you will observe these properties of cardiac muscle in vitro (that is, removed from the body). Work together in groups of three or four.

Activity 1

Investigating the Automaticity and Rhythmicity of Heart Muscle

1. Obtain a dissecting tray and instruments, disposable gloves, two petri dishes, frog Ringer's solution, a metric ruler, and a medicine dropper, and bring them to your laboratory bench.

⚠ **2.** Don the gloves, and then request and obtain a doubly pithed frog from your instructor. Quickly open the thoracic cavity and observe the heart rate in situ (at the site or within the body).

Record the heart rate: _____ beats/min

3. Dissect out the heart and the gastrocnemius muscle of the calf, and place the removed organs in separate petri dishes containing frog Ringer's solution. (**Note:** The procedure for removing the gastrocnemius muscle is provided on pp. 241–242 in Exercise 14. The extreme care used in that procedure for the removal of the gastrocnemius muscle need not be exercised here.)

4. Observe the activity of the two organs for a few seconds.

Which is contracting? _____

At what rate? _____ beats/min

Is the contraction rhythmic? _____

5. Sever the sinus venosus from the heart (**Figure 34.1**). The **sinus venosus** of the frog's heart corresponds to the SA node of the human heart.

Does the sinus venosus continue to beat? _____

If not, lightly touch it with a probe to stimulate it. Record its rate of contraction.

Rate: _____ beats/min

6. Sever the right atrium from the heart; then remove the left atrium. Does each atrium continue to beat?

_____ Rate: _____ beats/min

Does the ventricle continue to beat? _____

Rate: _____ beats/min

7. Notice that frogs have a single ventricle (Figure 34.1). Fragment the ventricle to determine how small the ventricular fragments must be before the automaticity of ventricular muscle is abolished. Measure these fragments and record their approximate size.

_____ mm × _____ mm × _____ mm

Which portion of the heart exhibited the most marked automaticity?

_____

Which showed the least? _____

8. Properly dispose of the frog and heart fragments in the appropriate container before continuing.

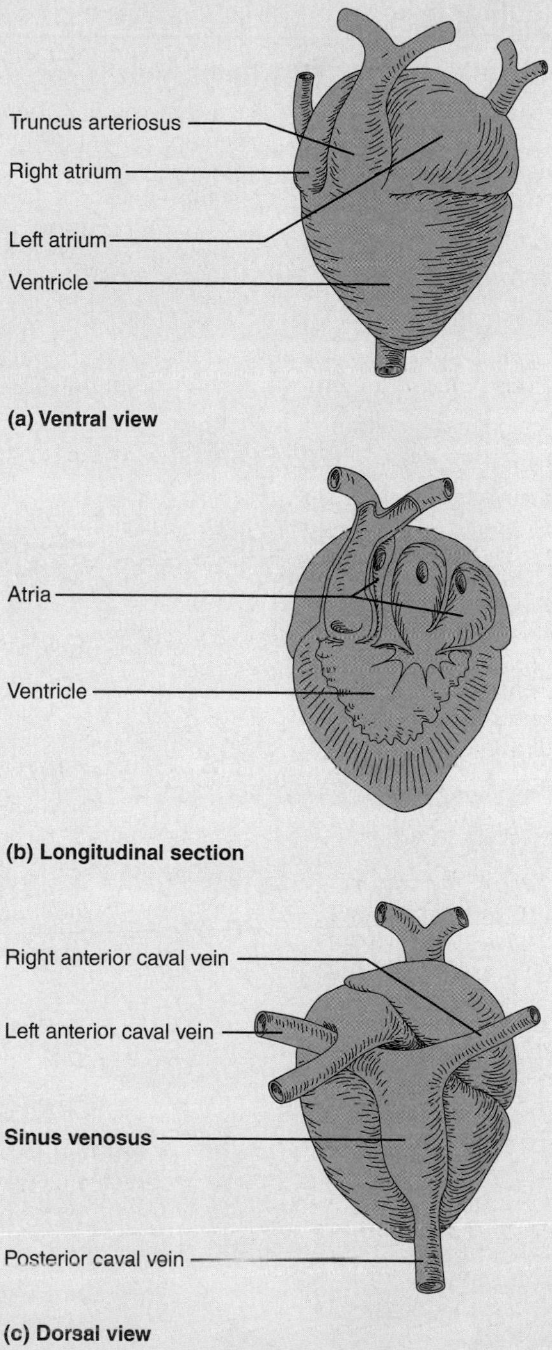

(a) Ventral view

Truncus arteriosus
Right atrium
Left atrium
Ventricle

(b) Longitudinal section

Atria
Ventricle

(c) Dorsal view

Right anterior caval vein
Left anterior caval vein
Sinus venosus
Posterior caval vein

Figure 34.1 Anatomy of the frog heart.

34

Baseline Frog Heart Activity

The heart's effectiveness as a pump is dependent both on intrinsic (within the heart) and extrinsic (external to the heart) controls. In this activity, you will investigate some of these factors.

The nodal system, in which the "pacemaker" imposes its depolarization rate on the rest of the heart, is one intrinsic factor that influences the heart's pumping action. If its impulses fail to reach the ventricles (as in heart block), the ventricles continue to beat but at their own inherent rate, which is much slower than that usually imposed on them. Although heart contraction does not depend on nerve impulses, its rate can be modified by extrinsic impulses reaching it through the autonomic nerves. Additionally, cardiac activity is modified by various chemicals, hormones, ions, and metabolites. The effects of several of these chemical factors are examined in the next experimental series.

The frog heart has two atria and a single, incompletely divided ventricle (see Figure 34.1). The pacemaker is located in the sinus venosus, an enlarged region between the venae cavae and the right atrium.

Activity 2

Recording Baseline Frog Heart Activity

To record baseline frog heart activity, work in groups of four—two students handling the equipment setup and two preparing the frog for experimentation. Two sets of instructions are provided for apparatus setup—one for the physiograph (**Figure 34.2**), the other for BIOPAC® (**Figure 34.3**).

Apparatus Setups

Physiograph Apparatus Setup

1. Obtain a force transducer, transducer cable, and transducer stand, and bring them to the recording site.

2. Attach the force transducer to the transducer stand (as shown in Figure 34.2).

3. Then attach the transducer cable to the transducer coupler (input) on the channel amplifier of the physiograph and to the force transducer.

4. Attach the stimulator output extension cable to output on the stimulator panel (red to red, black to black).

BIOPAC® Apparatus Setup

1. Connect the BIOPAC® apparatus to the computer and turn the computer **ON**.

2. Make sure the BIOPAC® unit is **OFF**.

3. Set up the equipment (as shown in Figure 34.3).

4. Turn the BIOPAC® unit **ON**.

5. Launch the BIOPAC® BSL *PRO* software by clicking the icon on the desktop or by following your instructor's guidance.

6. Open the Frog Heart template. MP36/35 users: Go to the **File** menu at the top of the screen and choose **Open > Files of Type = Graph Template (GTL) > FrogHeart.gtl**. MP45 users: Click the **BSL *PRO*** tab and navigate to look in the *PRO* lessons. Select **a04.gtl** and click **OK**.

7. Put the tension adjuster (BIOPAC® HDW100A, or equivalent) on the transducer stand, and attach the BIOPAC® SS12LA force transducer with the hook holes pointing down. Level the force transducer both horizontally and vertically.

8. Set the tension adjuster to approximately one-quarter of its full range. (**Note:** *Do not firmly tighten any of the thumbscrews at this stage.*) Select a force range of 0 to 50 grams for this experiment.

9. Select and attach the small S-hook to the force transducer.

Preparation of the Frog

1. Obtain room-temperature frog Ringer's solution, a medicine dropper, dissecting instruments and tray, disposable

Figure 34.2 Physiograph setup for recording the activity of the frog heart.

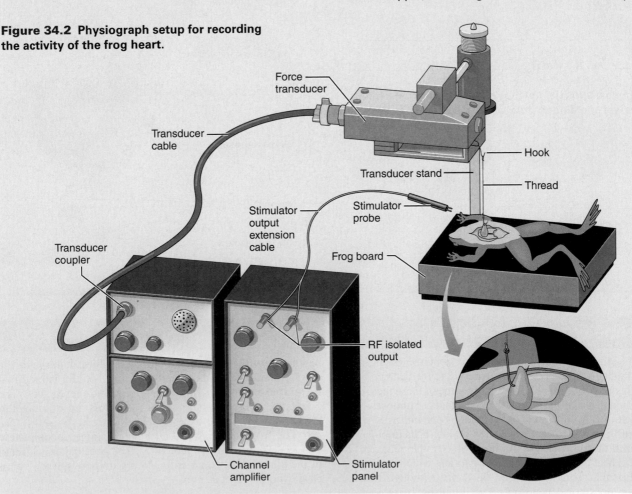

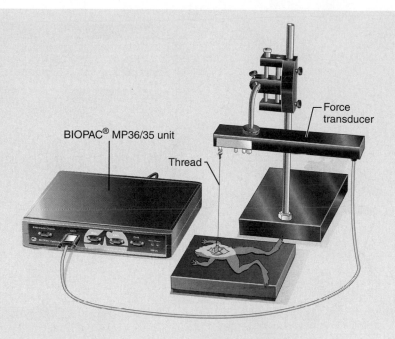

Force
transducer

BIOPAC® MP36/35 unit

Thread

Figure 34.3 BIOPAC® setup for recording the activity of the frog heart. Plug the force transducer into channel 1. Transducer is shown plugged into the MP36/35 unit.

gloves, fine common pins (physiograph) or small hook (BIOPAC®), cotton ball, frog board, large rubber bands, and some thread, and bring them to your bench.

 2. Don the gloves, and obtain a doubly pithed frog from your instructor.

3. Make a longitudinal incision through the abdominal and thoracic walls with scissors, and then cut through the sternum to expose the heart.

4. Grasp the pericardial sac with forceps, and cut it open so that the beating heart can be observed.

5. Locate the vagus nerve, which runs down the lateral aspect of the neck and parallels the trachea and carotid artery. (In good light, it appears to be striated.) Slip an 18-inch length of thread under the vagus nerve so that it can later be lifted away from the surrounding tissues by the thread. Then place a Ringer's solution–soaked cotton ball over the nerve to keep it moistened until you are ready to stimulate it later in the procedure.

6. Using a medicine dropper, flush the heart with Ringer's solution. _From this point on, the heart must be kept continually moistened with room-temperature Ringer's solution unless other solutions are being used for the experimentation._

7. Attach the frog to the frog board using large rubber bands.

Physiograph Frog Heart Preparation

1. Bend a common pin to a 90° angle, and tie to its head a thread 0.46–0.5 m (18–20 inches) long. Take care not to penetrate the ventricular chamber as you force the pin through the apex of the heart until the apex is well secured in the angle of the pin.

2. Tie the thread from the heart to the hook on the force transducer. Do not pull the thread too tightly. It should

be taut enough to lift the heart apex upward, away from the thorax, but should *not* stretch the heart. Adjust the force transducer as necessary. (See Figure 34.2.)

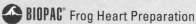 **BIOPAC®** Frog Heart Preparation

1. Attach a small hook tied with thread to the frog heart, following the instructions in step 1 of the physiograph instructions above to insert it through the apex of the heart. Confirm that the prepared frog is firmly attached to the frog board, positioned below the ring stand with the line running vertically from the frog heart to the transducer.

2. Slide the tension adjuster/force transducer assembly down the ring stand until you can hang the loop loosely from the S-hook, then slide it back up until the line is taut but the heart muscle is not stretched (be careful not to tear the heart).

3. Position the tension adjuster and/or force transducer so that the top is level, with approximately 10 cm (4 inches) of line from the heart to the S-hook. Adjust the assembly so that the thread line runs vertically; for a true reflection of the muscle's contractile force, the muscle must not be pulled at an angle.

4. Use the tension adjuster knob to make the line taut and tighten all thumbscrews to secure positioning of the assembly. Let the setup sit for a minute, then recheck the tension to make sure nothing has slipped or stretched.

5. Data may be distorted if the transducer line is not pulling directly vertical from the frog heart to the S-hook. Once again, align the frog as described above and make sure that the heart is not twisting the thread. If it is twisting, you will need to *carefully* remove the hook and repeat the setup.

6. BIOPAC® calculates *rate* data, which always trails the actual rate by one cycle. Data collection is a sensitive

34

Text continues on next page. →

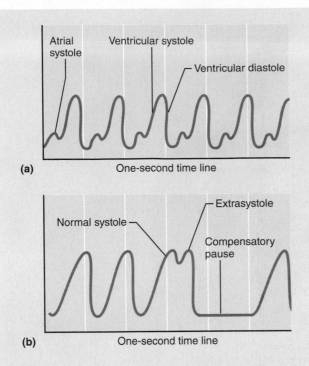

(a) One-second time line

(b) One-second time line

Figure 34.4 Physiograph recording of contractile activity of a frog heart. (a) Normal heartbeat. **(b)** Induction of an extrasystole.

process and may display artifacts from table movement, heart movement (for instance, from breathing on the heart), and chemicals touching the heart. To get the best data, keep the experimental area stable, clean, and clear of obstructions. Hints for obtaining best data:

- Make sure the SS12LA force transducer is level on the horizontal and vertical planes.

- Set up the tension adjuster and force transducer in positions that minimize their movement when tension is applied. Keep the point of S-hook attachment as close as possible to the ring stand support.

- Position the tension adjuster so that you will not bump the cables or frog board when using the adjustment knob.

- Position and/or tape the force transducer cables where they will not be pulled or bumped easily.

- Make sure the frog board is on a stable surface.

- Make sure the frog is firmly attached to the frog board so it will not rise up when tension is applied.

Making the Baseline Recording

Using the Physiograph

1. Turn the amplifier on, and balance the apparatus according to instructions provided by your instructor. Set the paper speed at 0.5 cm/sec. Press the record and paper advance buttons.

2. Set the signal magnet or time marker at 1/sec.

3. Record 12 to 15 normal heartbeats. Be sure you can distinguish atrial and ventricular contractions (**Figure 34.4**). Then adjust the paper or scroll speed so that the peaks of ventricular contractions are approximately 2 cm apart. (Peaks

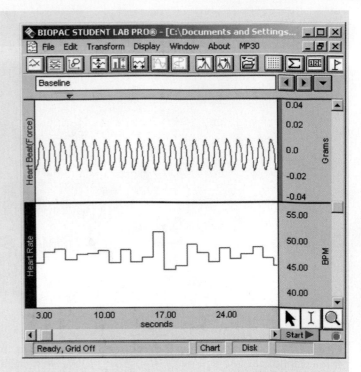

Figure 34.5 Example of baseline frog heart rate data.

indicate systole; troughs indicate diastole.) Pay attention to the relative force of heart contractions while recording.

Using BIOPAC®

1. Click **Start** to begin recording.

2. Observe at least 5 heart rate cycles, then click **Stop** to stop recording. Your data should look like that in the example (**Figure 34.5**).

3. Choose **Save** from the File menu, and type in a filename to save the recorded data. You may want to save by your team's name followed by FrogHeart-1 (for example, Smith-FrogHeart-1).

Analyzing the Baseline Data

Count the number of ventricular contractions per minute from your physiograph or BIOPAC® data, and record:

_____ beats/min

Compute the A–V interval (period from the beginning of atrial contraction to the beginning of ventricular contraction).

_____ sec

How do the two tracings compare in time?

_____

_____

Mark the atrial and ventricular systoles on the record. _Remember to keep the heart moistened with Ringer's solution._

Activity 3

Investigating the Refractory Period of Cardiac Muscle Using the Physiograph

Repeated rapid stimuli can cause skeletal muscle to remain in a contracted state (as demonstrated in Exercise 14). In other words, the muscle can be tetanized. This is possible because of the relatively short refractory period of skeletal muscle. In this experiment, you will use the physiograph to investigate the refractory period of cardiac muscle and its response to stimulation.* During the procedure, one student should keep the stimulating electrodes in constant contact with the frog heart ventricle while another student operates the stimulator panel.

1. Using the physiograph, set the stimulator to deliver 20-V shocks of 2-msec duration, and begin recording.

2. Deliver single shocks at the beginning of ventricular contraction, the peak of ventricular contraction, and then later and later in the cardiac cycle.

3. Observe the recording for **extrasystoles,** which are extra beats that show up riding on the ventricular contraction peak. Also note the **compensatory pause,** which allows the heart to get back on schedule after an extrasystole. (See Figure 34.4b.)

During which portion of the cardiac cycle was it possible to induce an extrasystole?

_____

_____

4. Attempt to tetanize the heart by stimulating it at the rate of 20 to 30 impulses per second. What is the result?

_____

_____

Considering the function of the heart, why is it important that heart muscle cannot be tetanized?

_____

_____

*BIOPAC® users may investigate the refractory period of cardiac muscle by using PhysioEx Exercise 6.

Activity 4

Assessing Physical and Chemical Modifiers of Heart Rate

Now that you have observed normal frog heart activity, you will have an opportunity to investigate the effects of various factors that modify heart activity. In each case, record a few normal heartbeats before introducing the modifying factor. After removing the agent, allow the heart to return to its normal rate before continuing with the testing. On each record, indicate the point of introduction and removal of the modifying agent.

For each physical agent or solution that is applied:
If using the physiograph, increase the scroll or paper speed so that heartbeats appear as spikes 4 to 5 mm apart.

If using BIOPAC®, after applying each solution click the **Start** button. When the effect is observed, record the effect for five cycles, then click **Stop.** Choose **Save** from the File menu to save your data.

(Note: Repeat these steps for each physical agent or chemical solution that is being applied.)

Temperature

1. Obtain 5°C and 32°C frog Ringer's solutions and medicine droppers.

2. Bathe the heart with 5°C Ringer's solution, and continue to record until the recording indicates a change in cardiac activity and five cardiac cycles have been recorded.

3. Stop recording, pipette off the cold Ringer's solution and flood the heart with room-temperature Ringer's solution.

4. Start recording again to determine the resumption of the normal heart rate. When this has been achieved, flood the heart with 32°C Ringer's solution, and again record five cardiac cycles after a change is noted.

5. Stop the recording, pipette off the warm Ringer's solution, and bathe the heart with room-temperature Ringer's solution once again.

What change occurred with the cold (5°C) Ringer's solution?

_____

What change occurred with the warm (32°C) Ringer's solution?

_____

6. Count the heart rate at the two temperatures, and record the data below.

_____ beats/min at 5°C; _____ beats/min at 32°C

Chemical Agents

Pilocarpine

Flood the heart with a 2.5% solution of pilocarpine. Record until a change in the pattern of the ECG is noticed. Pipette off the excess pilocarpine solution, and proceed immediately to the next test. What happened when the heart was bathed in the pilocarpine solution?

_____

Pilocarpine simulates the effect of parasympathetic (vagal) nerve stimulation by enhancing acetylcholine release; such drugs are called parasympathomimetic drugs.

Text continues on next page. →

34

Is pilocarpine an agonist or an antagonist of acetylcholine?

_____

_____

Atropine Sulfate

Apply a few drops of atropine sulfate to the frog's heart, and observe the recording. If no changes are observed within 2 minutes, apply a few more drops. When you observe a response, pipette off the excess atropine sulfate and flood the heart with room-temperature Ringer's solution. What happens when the atropine sulfate is added?

_____

_____

Atropine is a drug that blocks the effect of the neurotransmitter acetylcholine, which is liberated by the parasympathetic nerve endings. Do your results accurately reflect this effect of atropine?

_____

_____

Is atropine an agonist or an antagonist of acetylcholine?

_____

Epinephrine

Flood the frog heart with epinephrine solution, and continue to record until a change in heart activity is noted.

What are the results? _____

_____

Which division of the autonomic nervous system does its effect mimic?

_____

_____

Digitalis

Pipette off the excess epinephrine solution, and rinse the heart with room-temperature Ringer's solution. Continue recording, and when the heart rate returns to baseline values, bathe it in digitalis solution. What is the effect of digitalis on the heart?

_____

_____

Digitalis is a drug commonly prescribed for heart patients with congestive heart failure. It slows heart rate, providing more time for venous return and decreasing the work of the weakened heart. These effects are thought to be due to inhibition of the sodium-potassium pump and enhancement of Ca^{2+} entry into the myocardial fibers.

Various Ions

To test the effect of various ions on the heart, apply the designated solution until you observe a change in heart rate or in strength of contraction. Pipette off the solution, flush with room-temperature Ringer's solution, and allow the heart to resume its normal rate before continuing. _Do not allow the heart to stop._ If the rate should decrease dramatically, flood the heart with room-temperature Ringer's solution.

Effect of Ca^{2+} (use 2% $CaCl_2$) _____

_____

Effect of Na^+ (use 0.7% NaCl) _____

_____

Effect of K^+ (use 5% KCl) _____

_____

Potassium ion concentration is normally higher within cells than in the extracellular fluid. _Hyperkalemia_ decreases the resting potential of plasma membranes, thus decreasing the force of heart contraction. In some cases, the conduction rate of the heart is so depressed that **ectopic pacemakers** (pacemakers appearing erratically and at abnormal sites in the heart muscle) appear in the ventricles, and fibrillation may occur. Was there any evidence of premature beats in the recording of

potassium ion effects? _____

Was arrhythmia produced with any of the ions tested?

_____ If so, which? _____

Vagus Nerve Stimulation

The vagus nerve carries parasympathetic impulses to the heart, which modify heart activity. If you are using the physiograph, you can test this by stimulating the vagus nerve.*

1. Remove the cotton placed over the vagus nerve. Using the previously tied thread, lift the nerve away from the tissues, and place the nerve on the stimulating electrodes.

2. Using a duration of 0.5 msec at a voltage of 1 mV, stimulate the nerve at a rate of 50/sec. Continue stimulation until the heart stops momentarily and then begins to beat again **(vagal escape)**. If no effect is observed, increase stimulus intensity and try again. If no effect is observed after a substantial increase in stimulus voltage, reexamine your "vagus nerve" to make sure that it is not simply strands of connective tissue.

3. Discontinue stimulation after you observe vagal escape, and flush the heart with room-temperature Ringer's solution until the normal heart rate resumes. What is the effect of vagal stimulation on heart rate?

_____

_____

*BIOPAC® users may observe the effects of vagal stimulation by using PhysioEx Exercise 6.

Intrinsic Conduction System Disturbance (Heart Block)

1. Moisten a 25-cm (10-inch) length of thread, and make a Stannius ligature (loop the thread around the heart at the junction of the atria and ventricle).

2. If using a physiograph, decrease the scroll or paper speed to achieve intervals of approximately 2 cm between the ventricular contractions, and record a few normal heartbeats.

3. Tighten the ligature in a stepwise manner while observing the atrial and ventricular contraction curves. As heart block occurs, the atria and ventricle will no longer show a 1:1 contraction ratio. Record a few beats each time you observe a different degree of heart block—a 2:1 ratio of atrial to ventricular contractions, 3:1, 4:1, and so on. As long as you can continue to count a whole-number ratio

between the two chamber types, the heart is in **partial heart block.** When you can no longer count a whole number ratio, the heart is in **total,** or **complete, heart block.**

4. When total heart block occurs, release the ligature to see if the normal A–V rhythm is reestablished. What is the result?

_____

_____

5. Attach properly labeled recordings (or copies of the recordings) made during this procedure to the last page of this exercise for future reference.

6. Dispose of the frog remains and gloves in appropriate containers, and dismantle the experimental apparatus before continuing.

The Microcirculation and Local Blood Flow

The thin web of a frog's foot provides an opportunity to observe the flow of blood to, from, and within the capillary beds. The flow of blood through a capillary bed is called the **microcirculation.** Arterioles carry blood to the capillary bed; venules carry blood away. Most capillary beds consist of a vascular shunt, called the **metarteriole–thoroughfare channel,** and true capillaries, the actual exchange vessels (**Figure 34.6**).

The total cross-sectional area of the capillaries in the body is much greater than that of the veins and arteries combined. Thus, the velocity of flow through the capillary beds is quite slow. Capillary flow is also intermittent, because if all capillary beds were filled with blood at the same time, there would be no blood at all in the large vessels. The flow of blood into the capillary beds is regulated by the activity of muscular *terminal arterioles,* which feed the beds, and by *precapillary sphincters* at entrances to the true capillaries. The amount of blood flowing into the true capillaries of the bed is regulated most importantly by local chemical controls. Thus a capillary bed may be flooded with blood or almost entirely bypassed depending on what is happening within the body or in a particular body region at any one time. You will investigate some of the local controls in the next group of experiments.

Activity 5

Investigating the Effect of Various Factors on the Microcirculation

1. Obtain a frog board (with a hole at one end), dissecting pins, disposable gloves, frog Ringer's solution (room temperature, 5°C, and 32°C), 0.01 N HCl, 0.01% histamine solution, 1% epinephrine solution, a large rubber band, and some paper towels.

2. Put on the gloves and obtain a frog (alive and hopping, *not* pithed). Moisten several paper towels with room-temperature Ringer's solution, and wrap the frog's

Text continues on next page. →

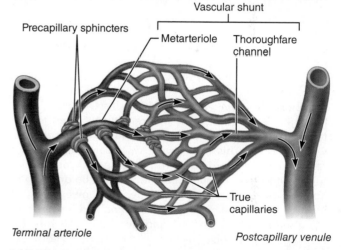

Precapillary sphincters — Metarteriole · Thoroughfare channel · Vascular shunt · True capillaries

Terminal arteriole   *Postcapillary venule*

(a) Sphincters open

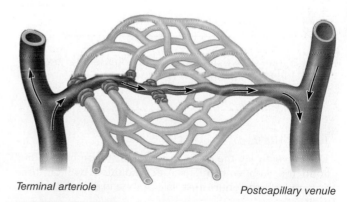

Terminal arteriole   *Postcapillary venule*

(b) Sphincters closed

Figure 34.6 Anatomy of a capillary bed. The composite metarteriole—thoroughfare channels act as shunts to bypass the true capillaries when precapillary sphincters controlling blood entry into the true capillaries are constricted.

34

body securely with them. One hind leg should be left unsecured and extending beyond the paper cocoon.

3. Attach the frog to the frog board (or other supporting structure) with a large rubber band and then carefully spread (but do not stretch) the web of the exposed hindfoot over the hole in the support. Have your partners hold the edges of the web firmly for viewing. Alternatively, secure the toes to the board with dissecting pins.

4. Obtain a compound microscope, and observe the web under low power to find a capillary bed. Focus on the vessels in high power. Keep the web moistened with Ringer's solution as you work. If the circulation seems to stop during your observations, massage the hind leg of the frog gently to restore blood flow.

5. Observe the red blood cells of the frog. Notice that unlike human RBCs, they are nucleated. Watch their movement through the smallest vessels — the capillaries. Do they move in single file, or do they flow through two or three cells abreast?

_____

Are they flexible? _____ Explain. _____

Can you see any white blood cells in the capillaries?

_____ If so, which types? _____

6. Notice the relative speed (velocity) of blood flow through the blood vessels. Differentiate between the arterioles, which feed the capillary bed, and the venules, which drain it. This may be tricky, because images are reversed in the microscope. Thus, the vessel that appears to feed into the capillary bed will actually be draining it. You can distinguish between the vessels, however, if you consider that the flow is more pulsating and turbulent in the arterioles and smoother and steadier in the venules. How does the velocity of flow in the arterioles compare with that in the venules?

_____

In the capillaries? _____

What is the relative difference in the diameter of the arterioles and capillaries?

_____

Temperature

1. To investigate the effect of temperature on blood flow, flood the web with 5°C Ringer's solution two or three times to chill the entire area. Is a change in vessel diameter noticeable?

_____ Which vessels are affected? _____

How? _____

2. Blot the web gently with a paper towel, and then bathe the web with warm (32°C) Ringer's solution. Record your observations.

_____

Inflammation

1. Pipette 0.01 N HCl onto the frog's web. Hydrochloric acid will act as an irritant and cause a localized inflammatory response. Is there an increase or decrease in the blood flow into the capillary bed following the application of HCl?

_____

What purpose do these local changes serve during a localized inflammatory response?

_____

_____

_____

2. Flush the web with room-temperature Ringer's solution and blot.

Histamine

1. Histamine, which is released in large amounts during allergic responses, causes extensive vasodilation. Investigate this effect by adding a few drops of histamine solution to the frog web. What happens?

_____

_____

How does this response compare to that produced by HCl?

_____

2. Blot the web and flood with 32°C Ringer's solution as before. Now add a few drops of 1% epinephrine solution, and observe the web. What are epinephrine's effects on the blood vessels?

_____

3. Return the dropper bottles to the supply area and the frog to the terrarium. Properly clean your work area before leaving the lab.

Name _____ Lab Time/Date _____

Special Electrical Properties of Cardiac Muscle: Automaticity and Rhythmicity

1. Define the following terms.

   automaticity: _____

   rhythmicity: _____

2. Discuss the anatomical differences between frog and human hearts. _____

   _____

   _____

3. Which region of the dissected frog heart had the highest intrinsic rate of contraction? _____

   The greatest automaticity? _____

   The greatest regularity or rhythmicity? _____ How do these properties correlate with

   the duties of a pacemaker? _____

   _____

   Is this region the pacemaker of the frog heart? _____

   Which region had the lowest intrinsic rate of contraction? _____

Investigating the Refractory Period of Cardiac Muscle

4. Define *extrasystole.* _____

5. Respond to the following questions if you used a physiograph. _____

   What was the effect of stimulation of the heart during ventricular contraction? _____

   During ventricular relaxation (first portion)? _____

   During the pause interval? _____

   What does this indicate about the refractory period of cardiac muscle? _____

Assessing Physical and Chemical Modifiers of Heart Rate

6. Describe the effect of thermal factors on the frog heart.

cold: _____ heat: _____

7. Once again refer to your recordings. Did the administration of the following produce any changes in force of contraction (shown by peaks of increasing or decreasing height)? If so, explain the mechanism.

epinephrine: _____

_____

pilocarpine: _____

_____

calcium ions: _____

_____

8. Excessive amounts of each of the following ions would most likely interfere with normal heart activity. Note the type of changes caused in each case.

K^+: _____

Ca^{2+}: _____

Na^+: _____

9. Respond to the following questions if you used a physiograph. What was the effect of vagal stimulation on heart rate?

_____

Which of the following factors cause the same (or very similar) heart rate–reducing effects: epinephrine, acetylcholine, atropine sulfate, pilocarpine, sympathetic nervous system activity, digitalis, potassium ions?

_____

Which of the factors listed above would reverse or antagonize vagal effects? _____

_____

10. What is vagal escape? _____

_____

Why is vagal escape valuable in maintaining homeostasis? _____

_____

11. How does the Stannius ligature used in the laboratory produce heart block? _____

_____

_____

12. Define *partial heart block,* and describe how it was recognized in the laboratory. _____

_____

_____

13. Define *total heart block,* and describe how it was recognized in the laboratory. _____

_____

_____

_____

14. What do your heart block experiment results indicate about the spread of impulses from the atria to the ventricles?

_____

_____

Observing the Microcirculation Under Various Conditions

15. In what way are the red blood cells of the frog different from those of the human? _____

_____

On the basis of this one factor, would you expect their life spans to be longer or shorter? _____

16. The following statements refer to your observation of one or more of the vessel types observed in the microcirculation in the frog's web. Characterize each statement by choosing the best response from the key.

Key: a. arterioles   b. venules   c. capillaries

_____ 1. smallest vessels observed

_____ 2. vessels in which blood flow is rapid, pulsating

_____ 3. vessels in which blood flow is least rapid

_____ 4. red blood cells pass through these vessels in single file

_____ 5. blood flow is smooth and steady

_____ 6. most numerous vessels

_____ 7. vessels that deliver blood to the capillary bed

_____ 8. vessels that serve the needs of the tissues via exchanges

_____ 9. vessels that drain the capillary beds

17. Which of the vessel diameters changed most? _____

What division of the nervous system controls the vessels? _____

18. Discuss the effects of the following on blood vessel diameter (state specifically the blood vessels involved) and rate of blood flow. Then explain the importance of the reaction observed to the general well-being of the body.

local application of cold: _____

_____

local application of heat: _____

_____

inflammation (or application of HCl): _____

_____

histamine: _____

_____

The Lymphatic System and Immune Response

Objectives

☐ State the function of the lymphatic system, name its components, and compare its function to that of the blood vascular system.

☐ Describe the formation and composition of lymph, and discuss how it is transported through the lymphatic vessels.

☐ Relate immunological memory, specificity, and differentiation of self from nonself to immune function.

☐ Differentiate between the roles of B cells and T cells in the immune response.

☐ Describe the structure and function of lymph nodes, and indicate the location of T cells, B cells, and macrophages in a typical lymph node.

☐ Describe the major microanatomical features of the spleen and tonsils.

☐ Draw or describe the structure of the antibody monomer, and name the five immunoglobulin subclasses.

☐ Differentiate between antigen and antibody.

☐ Explain how the Ouchterlony test detects antigens by using the antigen-antibody reaction.

Materials

- Large anatomical chart of the human lymphatic system
- Prepared slides of lymph node, spleen, and tonsil
- Compound microscope
- Wax marking pencil
- Petri dish containing simple saline agar
- Medicine dropper
- Dropper bottles of red and green food color
- Dropper bottles of goat antibody to horse serum albumin, goat antibody to bovine serum albumin, goat antibody to swine serum albumin, horse serum

Text continues on next page. →

MasteringA&P®

For related exercise study tools, go to the Study Area of **MasteringA&P**. There you will find:

- Practice Anatomy Lab **PAL**
- PhysioEx **PEx**
- A&PFlix **A&PFlix**
- Practice quizzes, Histology Atlas, eText, Videos, and more!

Pre-Lab Quiz

1. Circle True or False. The lymphatic system protects the body by removing foreign material such as bacteria from the lymphatic stream.

2. Lymph is:
   a. excess blood that has escaped from veins
   b. excess tissue fluid that has leaked out of capillaries
   c. excess tissue fluid that has escaped from arteries

3. Circle True or False. Collecting lymphatic vessels have three tunics and are equipped with valves like veins.

4. _____, which serve as filters for the lymphatic system, occur at various points along the lymphatic vessels.
   a. Glands   b. Lymph nodes   c. Valves

5. Circle True or False. The immune response is a systemic response that occurs when the body recognizes a substance as foreign and acts to destroy or neutralize it.

6. Three characteristics of the immune response are the ability to distinguish self from nonself, memory, and
   a. autoimmunity   b. specificity   c. susceptibility

7. Circle the correct underlined term. <u>B cells</u> / <u>T cells</u> differentiate in the thymus.

8. Circle the correct underlined term. T cells mediate <u>humoral</u> / <u>cellular</u> immunity because they destroy cells infected with viruses and certain bacteria and parasites.

9. Circle True or False. Antibodies are produced by plasma cells in response to antigens and are found in all body secretions.

10. Antibodies that have only one structural unit (monomer) consist of _____ protein chains, connected by disulfide bonds.
    a. two   c. four
    b. three   d. six

albumin diluted to 20% with physiological saline, unknown albumin sample diluted to 20% (prepared from horse, swine, and/or bovine albumin)

• Colored pencils

For instructions on animal dissections, see the dissection exercises (starting on p. 705) in the cat and fetal pig editions of this manual.

PEx PhysioEx™ 9.1 Computer Simulation Ex. 12 on p. PEx-177

The lymphatic system has two chief functions: (1) it transports tissue fluid (lymph) to the blood vessels, and (2) it protects the body by removing foreign material such as bacteria from the lymphatic stream and by serving as a site for lymphocyte "policing" of body fluids and for lymphocyte multiplication.

The Lymphatic System

The **lymphatic system** consists of a network of lymphatic vessels (lymphatics), lymphoid tissue, lymph nodes, and a number of other lymphoid organs, such as the tonsils, thymus, and spleen. We will focus on the lymphatic vessels and lymph nodes in this section. The white blood cells that are the central actors in body immunity are described later in this exercise.

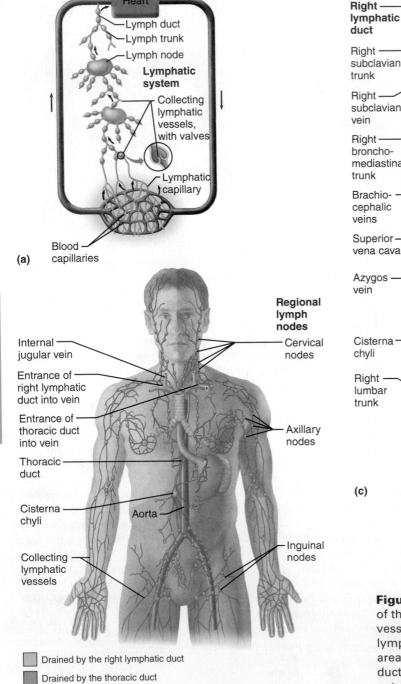

Figure 35.1 Lymphatic system. (a) Simplified scheme of the relationship of lymphatic vessels to blood vessels of the cardiovascular system. **(b)** Distribution of lymphatic vessels and lymph nodes. The green-shaded area represents body area drained by the right lymphatic duct. **(c)** Major veins in the superior thorax showing entry points of the thoracic and right lymphatic ducts. The major lymphatic trunks are also identified.

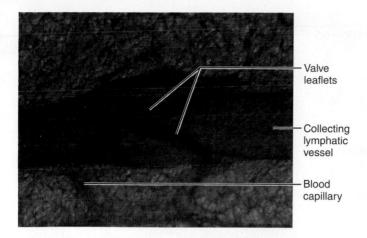

Figure 35.2 Collecting lymphatic vessel (700×).

Distribution and Function of Lymphatic Vessels and Lymph Nodes

As blood circulates through the body, the hydrostatic and osmotic pressures operating at the capillary beds result in fluid outflow at the arterial end of the bed and in its return at the venous end. However, not all of the lost fluid is returned to the bloodstream by this mechanism. It is the microscopic, blind-ended **lymphatic capillaries (Figure 35.1a)**, which branch through nearly all the tissues of the body, that pick up this leaked fluid and carry it through successively larger vessels—**collecting lymphatic vessels** to **lymphatic trunks**—until the lymph finally returns to the blood vascular system through one of the two large ducts in the thoracic region (Figure 35.1b). The **right lymphatic duct** drains lymph from the right upper extremity, head, and thorax delivered by the jugular, subclavian, and bronchomediastinal trunks. The large **thoracic duct** receives lymph from the rest of the body (see Figure 35.1c). The enlarged terminus of the thoracic duct is the **cisterna chyli,** which receives lymph from the digestive organs. In humans, both ducts empty the lymph into the venous circulation at the junction of the internal jugular vein and the subclavian vein, on their respective sides of the body. Notice that the lymphatic system, lacking both a contractile "heart" and arteries, is a one-way system; it carries lymph only toward the heart.

Like veins of the blood vascular system, the collecting lymphatic vessels have three tunics and are equipped with valves **(Figure 35.2)**. However, lymphatics tend to be thinner-walled, to have *more* valves, and to anastomose (form branching networks) more than veins. Since the lymphatic system is a pumpless system, lymph transport depends largely on the milking action of the skeletal muscles and on pressure changes within the thorax that occur during breathing.

As lymph is transported, it filters through bean-shaped **lymph nodes,** which cluster along the lymphatic vessels of

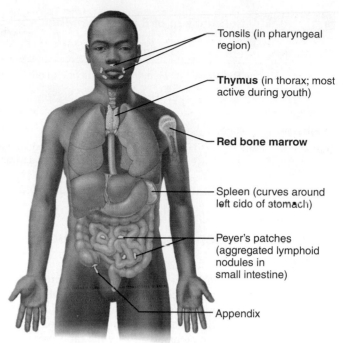

Figure 35.3 Lymphoid organs. Locations of the tonsils, thymus, spleen, appendix, and Peyer's patches. Primary lymphoid organs are in bold.

the body. There are thousands of lymph nodes, but because they are usually embedded in connective tissue, they are not ordinarily seen. Particularly large collections of lymph nodes are found in the inguinal, axillary, and cervical regions of the body.

Other lymphoid organs—the tonsils, thymus, and spleen **(Figure 35.3)**—resemble the lymph nodes histologically, and they house similar cell populations (lymphocytes and macrophages).

35

Activity 1

Identifying the Organs of the Lymphatic System

Study the large anatomical chart to observe the general plan of the lymphatic system. Notice the distribution of lymph nodes, various lymphatics, the lymphatic trunks, and the location of the right lymphatic duct, the thoracic duct, and the cisterna chyli.

For instructions on animal dissections, see the dissection exercises (starting on p. 705) in the cat and fetal pig editions of this manual.

The Immune Response

The **adaptive immune system** is a functional system that recognizes something as foreign and acts to destroy or neutralize it. This response is known as the **immune response.** It is a systemic response and is not restricted to the initial infection site. When operating effectively, the immune response protects us from bacterial and viral infections, bacterial toxins, and cancer.

Table 35.1	Comparison of B and T Lymphocytes	
Properties	**B lymphocytes**	**T lymphocytes**
Type of Immune Response	Humoral immunity (antibody-mediated immunity)	Cellular immunity (cell-mediated immunity)
Site of Maturation	Red bone marrow	Thymus
Effector Cells	Plasma cells (antibody-secreting cells)	Cytotoxic T cells, helper T cells, regulatory T cells
Memory Cell Formation	Yes	Yes
Functions	Plasma cells produce antibodies that inactivate antigen and tag antigen for destruction.	Cytotoxic T cells attack infected cells and tumor cells. Helper T cells activate B cells and other T cells.

Major Characteristics of the Immune Response

The most important characteristics of the immune response are its (1) **memory,** (2) **specificity,** and (3) **self-tolerance.** Not only does the immune system have a "memory" for previously encountered foreign antigens (the chicken pox virus, for example), but this memory is also remarkably accurate and highly specific.

An almost limitless variety of things are *antigens*—that is, anything capable of provoking an immune response and reacting with the products of the response. Nearly all foreign proteins, many polysaccharides, bacteria and their toxins, viruses, mismatched RBCs, cancer cells, and many small molecules (haptens) can be antigenic. The cells that recognize antigens and initiate the immune response are lymphocytes, the second most numerous members of the leukocyte, or white blood cell (WBC), population. Each immunocompetent lymphocyte has receptors on its surface allowing it to bind with only one or a few very similar antigens, thus providing specificity.

As a rule, our own proteins are tolerated, a fact that reflects the ability of the immune system to distinguish our own tissues (self) from foreign antigens (nonself). Nevertheless, an inability to recognize self can and does occasionally happen, and the immune system then attacks the body's own tissues. This phenomenon is called *autoimmunity.* Autoimmune diseases include multiple sclerosis (MS), myasthenia gravis, Graves' disease, rheumatoid arthritis (RA), and diabetes mellitus.

Organs, Cells, and Cell Interactions of the Immune Response

The immune system uses as part of its arsenal the **lymphoid organs** and **lymphoid tissues,** which include the thymus, lymph nodes, spleen, tonsils, appendix, and red bone marrow. Of these, the thymus and red bone marrow are considered to be the *primary lymphoid organs.* The others are *secondary lymphoid organs and tissues.*

The primary cells that provide for the immune response are the **B** and **T lymphocytes,** also known as **B** and **T cells.** The B and T cells originate in the red bone marrow. Each cell type must go through a maturation process where they become **immunocompetent** and **self-tolerant.** Immunocompetence involves the addition of receptors on the cell surface that recognize and bind to a specific antigen. Self-tolerance involves the cell's ability to distinguish self from nonself. B cells mature in the red bone marrow. T cells travel to the thymus for their maturation process.

After maturation, the B and T cells leave the bone marrow and thymus, respectively; enter the bloodstream; and travel to peripheral (secondary) lymphoid organs, where clonal selection occurs. **Clonal selection** is triggered when an antigen binds to the specific cell-surface receptors of a T or B cell. This event causes the lymphocyte to proliferate rapidly, forming a clone of like cells, all bearing the same antigen-specific receptors. Then, in the presence of certain regulatory signals, the members of the clone specialize, or differentiate—some form memory cells, and others become effector or regulatory cells. Upon subsequent meetings with the same antigen, the immune response proceeds considerably faster because the troops are already mobilized and awaiting further orders, so to speak. Additional characteristics of B and T cells are compared in **Table 35.1.**

Absence or failure of thymic differentiation of T lymphocytes results in a marked depression of both antibody and cell-mediated immune functions. Additionally, the observation that the thymus naturally shrinks with age has been correlated with the relatively immune-deficient status of elderly individuals. ✛

Activity 2

Studying the Microscopic Anatomy of a Lymph Node, the Spleen, and a Tonsil

1. Obtain a compound microscope and prepared slides of a lymph node, spleen, and a tonsil. As you examine the lymph node slide, notice the following anatomical features (depicted in **Figure 35.4**). The node is enclosed within a fibrous **capsule,** from which connective tissue septa (**trabeculae**) extend inward to divide the node into several compartments. Very fine strands of reticular connective tissue issue from the trabeculae, forming the stroma of the gland within which cells are found.

In the outer region of the node, the **cortex,** some of the cells are arranged in globular masses, referred to as germinal centers. The **germinal centers** contain rapidly dividing B cells. The rest of the cortical cells are primarily T cells that circulate continuously, moving from the blood into the node and then exiting from the node in the lymphatic stream.

In the internal portion of the lymph node, the **medulla,** the cells are arranged in cordlike fashion. Most of the medullary cells are macrophages. Macrophages are important not only for their phagocytic function but also

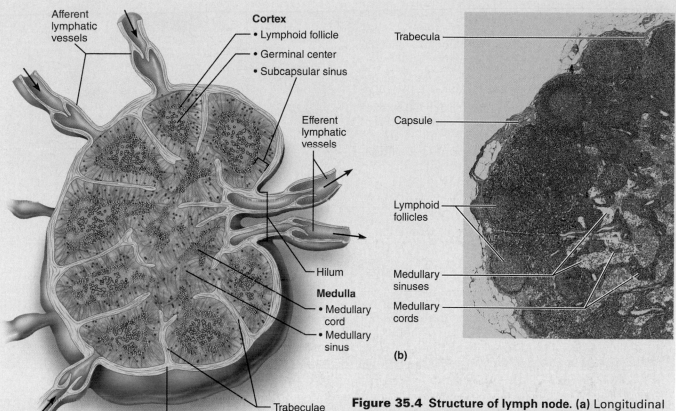

(a)

(b)

Figure 35.4 Structure of lymph node. (a) Longitudinal view of the internal structure of a lymph node and associated lymphatics. The arrows indicate the direction of the lymph flow. **(b)** Photomicrograph of part of a lymph node (20×).

because they play an essential role in "presenting" the antigens to the T cells.

Lymph enters the node through a number of *afferent vessels,* circulates through *lymph sinuses* within the node, and leaves the node through *efferent vessels* at the **hilum.** Since each node has fewer efferent than afferent vessels, the lymph flow stagnates somewhat within the node. This allows time for the generation of an immune response and for the macrophages to remove

debris from the lymph before it reenters the blood vascular system.

2. As you observe the slide of the spleen, look for the areas of lymphocytes suspended in reticular fibers, the **white pulp,** clustered around central arteries (**Figure 35.5**). The remaining tissue in the spleen is the **red pulp,** which is composed of splenic sinusoids and areas of reticular tissue and macrophages called the **splenic cords.**

Text continues on next page. →

35

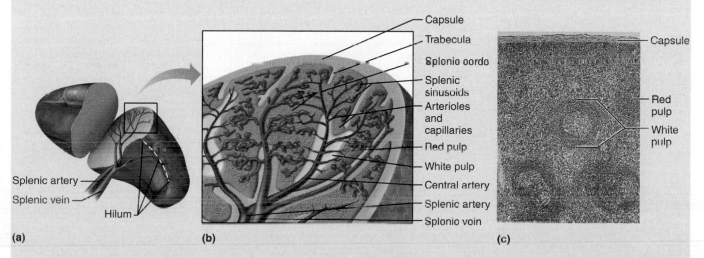

(a)   (b)   (c)

Figure 35.5 The spleen. (a) Gross structure. **(b)** Diagram of the histological structure. **(c)** Photomicrograph of spleen tissue showing white and red pulp regions (30×).

The white pulp, composed primarily of lymphocytes, is responsible for the immune functions of the spleen. Macrophages remove worn-out red blood cells, debris, bacteria, viruses, and toxins from blood flowing through the sinuses of the red pulp.

3. As you examine the tonsil slide, notice the **lymphoid follicles** containing **germinal centers** surrounded by scattered lymphocytes. The characteristic **tonsillar crypts** (invaginations of the mucosal epithelium) of the tonsils trap bacteria and other foreign material (**Figure 35.6**). Eventually the bacteria work their way into the lymphoid tissue and are destroyed.

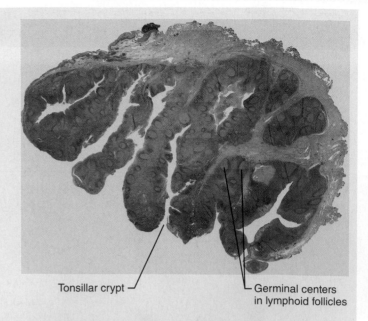

Tonsillar crypt

Germinal centers in lymphoid follicles

Figure 35.6 Histology of a palatine tonsil. The luminal surface is covered with epithelium that invaginates deeply to form crypts (10×).

Group Challenge

Compare and Contrast Lymphoid Organs and Tissues

Work in groups of three to discuss the characteristics of each lymphoid structure listed in the **Group Challenge chart** below. On a separate piece of paper, one student will record the characteristics for each structure for the group. The group will then consider each pair of structures listed in the chart, discuss the similarities and differences for each pair, and complete the chart based on consensus answers.

Use your textbook or another appropriate reference for comparing Peyer's patches and the thymus with other structures. Remember to consider both structural and functional similarities and differences.

Group Challenge: Comparing Lymphoid Structures		
Lymphoid pair	**Similarities**	**Differences**
Lymph node Spleen		
Lymph node Tonsil		
Peyer's patches Tonsils		
Tonsil Spleen		
Thymus Spleen		

Antibodies and Tests for Their Presence

Antibodies, or **immunoglobulins (Igs),** are produced by sensitized B cells and their plasma cell offspring in response to an antigen. They are a heterogeneous group of proteins that make up the general class of plasma proteins called **gamma globulins.** Antibodies are found not only in plasma but also (to greater or lesser extents) in all body secretions. Five major classes of immunoglobulins have been identified: IgM, IgG, IgD, IgA, and IgE. The immunoglobulin classes share a common basic structure but differ functionally and in their localization in the body.

All Igs are composed of one or more structural units called **antibody monomers.** A monomer consists of four protein chains bound together by disulfide bridges (**Figure 35.7**). Two of the chains are quite large and have a high molecular weight; these are the **heavy chains.** The other two chains are only half as long and have a low molecular weight. These are called **light chains.** The two heavy chains have a *constant (C) region,* in which the amino acid sequence is the same in a class of immunoglobulins, and a *variable (V) region,* in which the amino acid sequence varies considerably between antibodies. The same is true of the two light chains; each has a constant and a variable region.

The intact Ig molecule has a three-dimensional shape that is generally Y shaped. Together, the variable regions of the light and heavy chains in each "arm" construct one **antigen-binding** site uniquely shaped to "fit" a specific *antigenic determinant* (portion) of an antigen. Thus, each Ig monomer bears two identical sites that bind to identical antigenic determinants. Binding of the immunoglobulins to their complementary antigen(s) effectively immobilizes the antigens until they can be phagocytized or lysed by complement fixation.

The antigen-antibody reaction is used diagnostically in a variety of ways. One of the most familiar is blood typing. (See Exercise 29 for instructions on ABO and Rh blood typing.) The antigen-antibody test that we will use in this laboratory session is the Ouchterlony technique, which is used mainly for rapid screening of suspected antigens.

Ouchterlony Double-Gel Diffusion, an Immunological Technique

The Ouchterlony double-gel diffusion technique was developed in 1948 to detect the presence of particular antigens in sera or extracts. Antigens and antibodies are placed in wells in a gel and allowed to diffuse toward each other. If an antigen reacts with an antibody, a thin white line called a *precipitin line* forms. In the following activity, the double-gel diffusion technique will be used to identify antigens. Work in groups of no more than three.

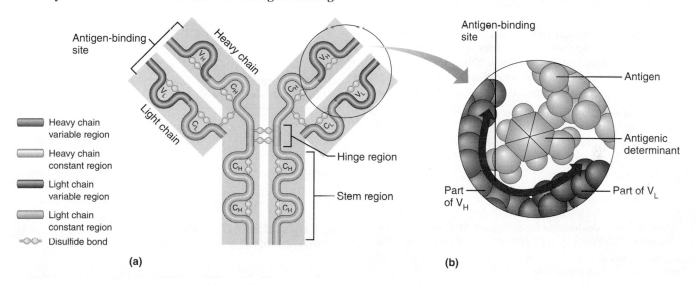

(a)   (b)

Figure 35.7 Antibody structure. (a) Schematic antibody structure consists of two heavy chains and two light chains connected by disulfide bonds. **(b)** Enlargement of an antigen-binding site of an immunoglobulin.

Activity 3

Using the Ouchterlony Technique to Identify Antigens

1. Obtain one each of the materials for conducting the Ouchterlony test: petri dish containing saline agar; medicine dropper; wax marking pencil; and dropper bottles of red and green food dye, horse serum albumin, an unknown serum albumin sample, and antibodies to horse, bovine, and swine albumin. Put your initials and the number of the unknown albumin sample used on the bottom of the petri dish near the edge.

2. Use the wax marking pencil and the template (**Figure 35.8**, p. 536) to divide the dish into three sections, and mark them I, II, and III.

3. Prepare sample wells (again using the template in Figure 35.8). Squeeze the medicine dropper bulb, and gently touch the tip to the surface of the agar. While releasing the bulb, push the tip down through the agar to the bottom of the dish. Lift the dropper vertically; this should leave a straight-walled well in the agar.

4. Repeat step 3 so that section I has two wells and sections II and III have four wells each.

Text continues on next page. →

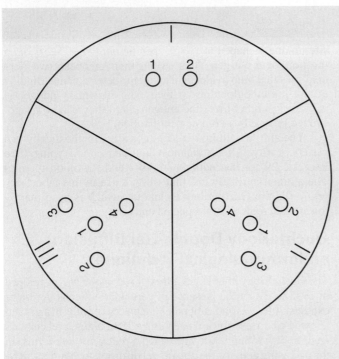

Figure 35.8 Template for well preparation for the Ouchterlony double-gel diffusion experiment.

5. To observe diffusion through the gel, nearly fill one well in section I with red dye and the other well with green dye. Be careful not to overfill the wells. Observe periodically for 30 to 45 minutes as the dyes diffuse through the agar. Draw your results as instructed in the Results section.

6. To demonstrate positive and negative results, fill the wells in section II as instructed (Table 35.2). A precipitin line should form only between wells 1 and 2.

7. To test the unknown sample, fill the wells in section III as instructed (Table 35.3).

8. Replace the cover on the petri dish, and incubate at room temperature for at least 16 hours. Make arrangements to observe the agar for precipitin lines after 16 hours. *The lines may begin to fade after 48 hours.* Draw the results as instructed in the following section, parts 1–3, indicating the location of all precipitin lines that form.

Results

1. For the demonstration of diffusion using dye, draw on the template (Figure 35.8) the appearance of section I of your dish after 30 to 45 minutes. Use colored pencils.

2. Draw section II on the template (Figure 35.8) as it appears after incubation, 16 to 48 hours. Be sure the wells are numbered.

Table 35.2	Section II
Well	**Solution**
1	Horse serum albumin
2	Goat anti–horse albumin
3	Goat anti–bovine albumin
4	Goat anti–swine albumin

Table 35.3	Section III
Well	**Solution**
1	Unknown # _____
2	Goat anti–horse albumin
3	Goat anti–bovine albumin
4	Goat anti–swine albumin

3. Draw section III on the template as it appears after incubation, 16 to 48 hours. Be sure the wells are numbered.

Unknown # _____

4. What evidence for diffusion did you observe in section I?

_____

_____

5. Is there any evidence of a precipitate in section I?

_____

6. Which of the sera functioned as an antigen in section II?

_____

7. Which antibody reacted with the antigen in section II?

_____

How do you know? (Be specific about your observations.)

_____

_____

8. If swine albumin had been placed in well 1, what would you expect to happen? Explain.

_____

_____

9. If chicken albumin had been placed in well 1, what would you expect to happen? Explain.

_____

_____

10. What antigens were present in the unknown solution?

_____

How do you know? (Be specific about your observations.)

_____

_____

35

REVIEW SHEET
The Lymphatic System and Immune Response

Name _____ Lab Time/Date _____

The Lymphatic System

1. Match the terms below with the correct letters on the diagram.

_____ 1. axillary lymph nodes

_____ 2. cervical lymph nodes

_____ 3. cisterna chyli

_____ 4. inguinal lymph nodes

_____ 5. lymphatic vessels

_____ 6. Peyer's patches (in small intestine)

_____ 7. red bone marrow

_____ 8. right lymphatic duct

_____ 9. spleen

_____ 10. thoracic duct

_____ 11. thymus

_____ 12. tonsils

2. Explain why the lymphatic system is a one-way system, whereas the blood vascular system is a two-way system.

_____

_____

3. How do lymphatic vessels resemble veins? _____

   How do lymphatic capillaries differ from blood capillaries? _____

_____

4. What is the function of the lymphatic vessels? _____

_____

5. What is lymph? _____

6. What factors are involved in the flow of lymphatic fluid? _____

_____

7. What name is given to the terminal duct draining most of the body? _____

8. What is the cisterna chyli? _____

_____

How does the composition of lymph in the cisterna chyli differ from lymph composition in the general lymphatic stream? Use your textbook or other reference if necessary.

_____

9. Which portion of the body is drained by the right lymphatic duct? _____

10. Note three areas where lymph nodes are densely clustered: _____,

_____, and _____

11. What are the two major functions of the lymph nodes? _____

and _____

12. The radical mastectomy is an operation in which a cancerous breast, surrounding tissues, and the underlying muscles of the anterior thoracic wall, plus the axillary lymph nodes, are removed. After such an operation, the arm usually swells, or becomes edematous, and is very uncomfortable—sometimes for months. Why?

_____

The Immune Response

13. What is the function of B cells in the immune response? _____

_____

14. What is the function of T cells in the immune response? _____

_____

_____

15. Define the following terms related to the operation of the immune system.

immunological memory: _____

_____

specificity: _____

_____

self-tolerance _____

_____

autoimmune disease: _____

_____

Studying the Microscopic Anatomy of a Lymph Node, the Spleen, and a Tonsil

16. In the box below, make a rough drawing of the structure of a lymph node. Identify the cortex area, germinal centers, and medulla. For each identified area, note the cell type (T cell, B cell, or macrophage) most likely to be found there.

_____

17. What structural characteristic ensures a *slow* flow of lymph through a lymph node? _____

_____

Why is this desirable? _____

_____

18. What similarities in structure and function are found in the lymph nodes, spleen, and tonsils? _____

_____

_____

Antibodies and Tests for Their Presence

19. Distinguish between antigen and antibody. _____

_____

20. Describe the structure of the immunoglobulin monomer, and label the diagram with the choices given in the key. ____

_____

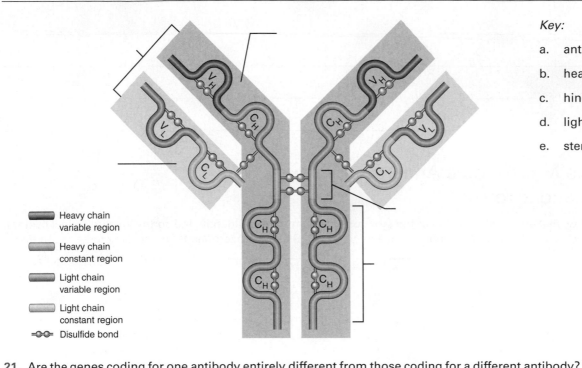

Key:

a. antigen-binding site

b. heavy chain

c. hinge region

d. light chain

e. stem region

Heavy chain
variable region

Heavy chain
constant region

Light chain
variable region

Light chain
constant region

Disulfide bond

21. Are the genes coding for one antibody entirely different from those coding for a different antibody? _____

Explain your answer. _____

_____

_____

22. In the Ouchterlony test, what happened when the antibody to horse serum albumin mixed with horse serum albumin?

_____

_____

_____

23. If the unknown antigen contained bovine and swine serum albumin, what would you expect to happen in the Ouchterlony

test, and why? _____

_____

_____

36

Anatomy of the Respiratory System

Objectives

☐ State the major functions of the respiratory system.

☐ Define the following terms: *pulmonary ventilation, external respiration, and internal respiration.*

☐ Identify the major respiratory system structures on models or appropriate images, and describe the function of each.

☐ Describe the difference between the conducting and respiratory zones.

☐ Name the serous membrane that encloses each lung, and describe its structure.

☐ Demonstrate lung inflation in a fresh sheep pluck or preserved tissue specimen.

☐ Recognize the histologic structure of the trachea and lung tissue microscopically.

Materials

- Resin cast of the respiratory tree
- Human torso model
- Thoracic cavity structures model and/or chart of the respiratory system
- Larynx model (if available)
- Preserved inflatable lung preparation or sheep pluck fresh from the slaughterhouse
- Source of compressed air
- Dissecting tray
- Disposable gloves
- Disposable autoclave bag
- Prepared slides of the following (if available): trachea (cross section), lung tissue, both normal and pathological specimens
- Compound and stereomicroscopes

For instructions on animal dissections, see the dissection exercises (starting on p. 705) in the cat and fetal pig editions of this manual.

MasteringA&P®

For related exercise study tools, go to the Study Area of **MasteringA&P**. There you will find:

- Practice Anatomy Lab **PAL**
- PhysioEx **PEx**
- A&PFlix **A&PFlix**
- Practice quizzes, Histology Atlas, e Text, Videos, and more!

Pre-Lab Quiz

1. The major role of the respiratory system is to:
   a. dispose of waste products in a solid form
   b. permit the flow of nutrients through the body
   c. supply the body with carbon dioxide and dispose of oxygen
   d. supply the body with oxygen and dispose of carbon dioxide

2. Circle True or False. Four processes—pulmonary ventilation, external respiration, transport of respiratory gases, and internal respiration—must all occur in order for the respiratory system to function fully.

3. The upper respiratory structures include the nose, the larynx, and the:
   a. epiglottis   c. pharynx
   b. lungs   d. trachea

4. Circle the correct underlined term. The thyroid cartilage / arytenoid cartilage is the largest and most prominent of the laryngeal cartilages.

5. Circle True or False. The epiglottis forms a lid over the larynx when we swallow food: it closes off the respiratory passageway to incoming food or drink.

6. Air flows from the larynx to the trachea, and then enters the:
   a. left and right lungs   c. pharynx
   b. left and right main bronchi   d. segmental bronchi

7. Circle the correct underlined term. The lining of the trachea is: pseudostratified ciliated columnar epithelium / transitional epithelium, which propels dust particles, bacteria, and other debris away from the lungs.

8. Circle True or False. All but the smallest branches of the bronchial tree have cartilaginous reinforcements in their walls.

9. _____, tiny balloonlike structures, are composed of a single thin layer of squamous epithelium. They are the main structural and functional units of the lung and the actual sites of gas exchange.

10. Circle the correct underlined term. Fissures divide the lungs into lobes, three on the right and two / three on the left.

Body cells require an abundant and continuous supply of oxygen. The major role of the **respiratory system** is to supply the body with oxygen and dispose of carbon dioxide. To fulfill this role, at least four distinct processes, collectively referred to as **respiration,** must occur:

Pulmonary ventilation: The tidelike movement of air into and out of the lungs that allows the gases to be continuously changed and refreshed. Also more simply called *breathing.*

External respiration: The gas exchange between the blood and the air-filled chambers of the lungs.

Transport of respiratory gases: The transport of respiratory gases between the lungs and tissue cells of the body using blood as the transport vehicle.

Internal respiration: Exchange of gases between systemic blood and tissue cells.

Upper Respiratory System Structures

The upper respiratory system structures—the external nose, nasal cavity, pharynx, paranasal sinuses and larynx—are summarized in **Table 36.1** and illustrated in **Figure 36.1** and **Figure 36.2** on p. 544. As you read through the descriptions in the table, identify each structure in the figures. Note that different sources divide the upper and lower respiratory systems slightly differently.

Table 36.1A	Structures of the Upper Respiratory System (Figure 36.1)		
Structure	**Description**	**Function**	
External Nose	Externally visible, its inferior surface has nostrils (nares). Supported by bone and cartilage and covered with skin.	The nostrils provide an entrance for air into the respiratory system.	
Nasal Cavity (includes the structures listed below)	Lined with respiratory mucosa composed of pseudostratified ciliated columnar epithelium. The floor of the cavity is formed by the hard and soft palates.	Functions to filter, warm, and moisten incoming air; resonance chambers for voice production.	
Nasal vestibule	Anterior portion of the nasal cavity; contains sebaceous and sweat glands and numerous hair follicles.	Filters coarse particles from the air.	
Nasal septum	Formed by the vomer, perpendicular plate of the ethmoid bone, and septal cartilage.	Divides the nasal cavity into left and right sides.	
Superior, middle, and inferior nasal conchae	Turbinates that project medially from the lateral walls of the cavity. Each concha has a corresponding meatus beneath it.	Increase the surface area of the mucosa, which enhances air turbulence and aids in trapping large particles in the mucus.	
Posterior nasal apertures	Posterior openings of the nasal cavity.	Provide an exit for the air into the nasopharynx.	
Pharynx (3 subdivisions: nasopharynx, oropharynx, and laryngopharynx—listed below)			
Nasopharynx	Superior portion of the pharynx located posterior to the nasal cavity; lined with pseudostratified ciliated columnar epithelium. The pharyngeal tonsil and openings of the pharyngotympanic tubes (surrounded by tubal tonsils) are located in this region.	Provides for the passage of air from the nasal cavity. Tonsils in the region provide protection against pathogens.	
Oropharynx	Located posterior to the oral cavity and extends from the soft palate to the epiglottis; lined with stratified squamous epithelium. Its lateral walls contain the palatine tonsils. The lingual tonsils are in the anterior oropharynx at the base of the tongue.	Provides for the passage of air and swallowed food. Tonsils provide protection against pathogens.	
Laryngopharynx	Extends from the epiglottis to the larynx; lined with stratified squamous epithelium. It diverges into respiratory and digestive branches.	Provides for the passage of air and swallowed food.	
Pharyngotympanic Tube	Tube that opens into the lateral walls of the nasopharynx and connects the nasopharynx to the middle ear.	Allows the middle ear pressure to equalize with the atmospheric pressure.	
Paranasal Sinuses	Surround the nasal cavity and are named for the bones in which they are located. Lined with pseudostratified ciliated columnar epithelium.	Act as resonance chambers for speech; warm and moisten incoming air.	

Lower Respiratory System Structures

Air entering the **trachea,** or windpipe, from the larynx travels down its length (about 11.0 cm or 4 inches) to the level of the *sternal angle.* There the passageway divides into the right and left **main (primary) bronchi (Figure 36.3,** p. 545). The right main bronchus is wider, shorter, and more vertical than the left, and foreign objects that enter the respiratory passageways are more likely to become lodged in it.

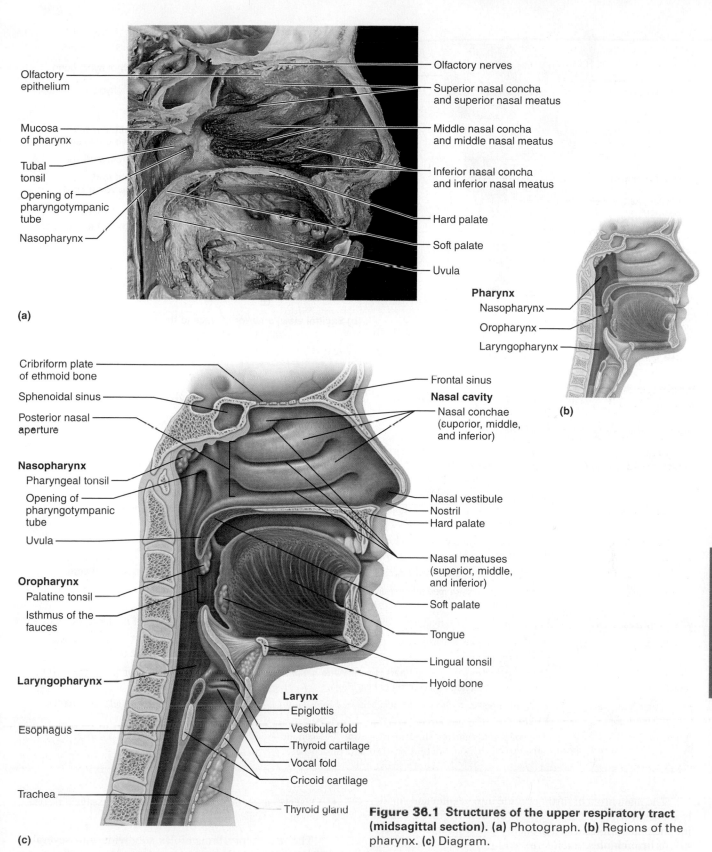

(a)

Olfactory epithelium

Mucosa of pharynx

Tubal tonsil

Opening of pharyngotympanic tube

Nasopharynx

Olfactory nerves

Superior nasal concha and superior nasal meatus

Middle nasal concha and middle nasal meatus

Inferior nasal concha and inferior nasal meatus

Hard palate

Soft palate

Uvula

Pharynx
Nasopharynx
Oropharynx
Laryngopharynx

(b)

Cribriform plate of ethmoid bone

Sphenoidal sinus

Posterior nasal aperture

Nasopharynx
Pharyngeal tonsil
Opening of pharyngotympanic tube
Uvula

Oropharynx
Palatine tonsil
Isthmus of the fauces

Laryngopharynx

Esophagus

Trachea

(c)

Frontal sinus

Nasal cavity

Nasal conchae (superior, middle, and inferior)

Nasal vestibule
Nostril
Hard palate

Nasal meatuses (superior, middle, and inferior)

Soft palate

Tongue

Lingual tonsil

Hyoid bone

Larynx
Epiglottis
Vestibular fold
Thyroid cartilage
Vocal fold
Cricoid cartilage
Thyroid gland

Figure 36.1 Structures of the upper respiratory tract (midsagittal section). (a) Photograph. **(b)** Regions of the pharynx. **(c)** Diagram.

36

The trachea is lined with a ciliated, mucus-secreting, pseudostratified columnar epithelium. The cilia propel mucus laden with dust particles and other debris away from the lungs and toward the throat, where it can be expectorated or swallowed. The walls of the trachea are reinforced with C-shaped cartilaginous rings (see Figure 36.6, p. 547). These C-shaped cartilages serve a double function: The incomplete parts allow the esophagus to expand anteriorly when a large food bolus is swallowed. The solid portions reinforce the trachea walls to maintain its open passageway when pressure changes occur during breathing.

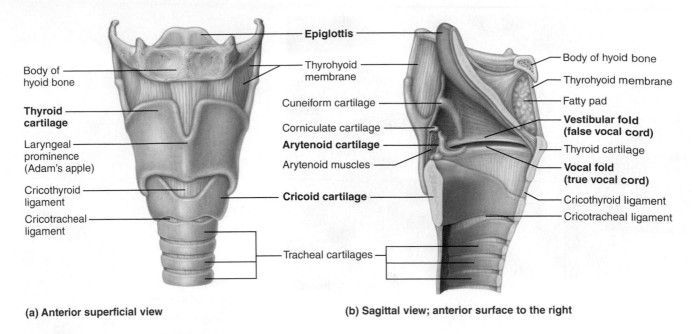

(a) Anterior superficial view

(b) Sagittal view; anterior surface to the right

Figure 36.2 Structure of the larynx.

Table 36.1B	Structures of the Upper Respiratory System (Figures 36.1 and 36.2)	
Structure	**Description**	**Function**
Larynx (includes the structures listed below)	Tube connecting the laryngopharynx and the trachea. Nine cartilages are present. Epithelium superior to the vocal folds is stratified squamous. Epithelium inferior to the vocal folds is pseudostratified ciliated columnar.	Air passageway; prevents food from entering the lower respiratory tract. Responsible for voice production.
Thyroid cartilage	Large cartilage made up of hyaline cartilage. Its laryngeal prominence is commonly referred to as the Adam's apple.	Forms the framework of the larynx.
Cricoid cartilage	Single ring of hyaline cartilage located inferior to the thyroid cartilage and superior to the trachea.	Attaches the larynx to the trachea via the cricotracheal ligament.
Arytenoid cartilage	Paired pyramid-shaped hyaline cartilages.	Anchor the vocal folds (true vocal cords).
Corniculate cartilage	Paired small horn-shaped hyaline cartilages located atop the arytenoid cartilages.	Form part of the posterior wall of the larynx.
Cuneiform cartilage	Paired wedge-shaped hyaline cartilages.	Form the lateral aspect of the laryngeal wall.
Epiglottis	Single flap of elastic cartilage anchored to the inner rim of the thyroid cartilage.	"Guardian of the airways" forms a lid over the larynx during swallowing.
Vocal folds (true vocal cords)	Composed of mostly elastic fibers covered with mucous membrane; attached to the arytenoid cartilage.	Vibrate with expired air for sound production.
Vestibular folds (false vocal cords)	Located superior to the vocal folds. Mucosal folds similar in composition to the vocal folds.	Protect the vocal folds and help to close the glottis when we swallow.
Glottis	The vocal folds and the slitlike passageway between the vocal folds.	Plays a role in the Valsalva maneuver.

The main bronchi further divide into smaller and smaller branches (the secondary, tertiary, on down), finally becoming the **bronchioles.** Each bronchiole divides into many **terminal bronchioles.** Each terminal bronchiole branches into two or more **respiratory bronchioles** (Figure 36.3). All but the smallest branches have cartilaginous reinforcements in their walls. As the respiratory tubes get smaller and smaller, the relative amount of smooth muscle in their walls increases as the amount of cartilage declines and finally disappears. The continuous branching of the respiratory passageways in the lungs is often referred to as the **bronchial tree.**

• Observe a resin cast of respiratory passages if one is available.

The respiratory bronchioles subdivide into several **alveolar ducts,** which terminate in alveolar sacs that resemble clusters of grapes. **Alveoli,** tiny balloonlike expansions along the alveolar sacs, are composed of a single thin layer of squamous epithelium overlying a basal lamina. The external surfaces of the alveoli are covered with a network of pulmonary capillaries (**Figure 36.4**). Together, the alveolar and capillary walls and their fused basement membranes

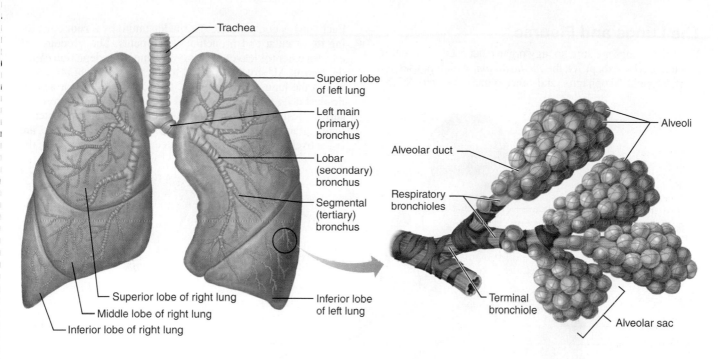

Figure 36.3 Structures of the lower respiratory tract.

form the **respiratory membrane,** also called the *blood air barrier.*

Because gas exchanges occur across the respiratory membrane, the alveolar sacs, alveolar ducts, and respiratory bronchioles are referred to collectively as **respiratory zone** **structures.** All other respiratory passageways (from the nasal cavity to the terminal bronchioles) simply serve as access or exit routes to and from these gas exchange chambers and are called **conducting zone structures** or *anatomical dead space.*

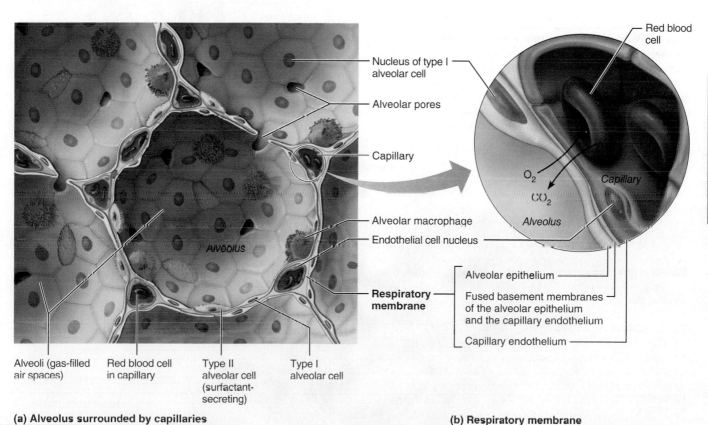

(a) Alveolus surrounded by capillaries

(b) Respiratory membrane

Figure 36.4 Diagram of the relationship between the alveoli and pulmonary capillaries involved in gas exchange.

The Lungs and Pleurae

The paired lungs are soft, spongy organs that occupy the entire thoracic cavity except for the *mediastinum,* which houses the heart, bronchi, esophagus, and other organs (**Figure 36.5**).

Each lung is connected to the mediastinum by a **root** containing its vascular and bronchial attachments. The structures of the root enter (or leave) the lung via a medial indentation called the **hilum.** All structures distal to the main bronchi are found within the lung. A lung's **apex,** the narrower superior aspect, lies just deep to the clavicle, and its **base,** the inferior concave surface, rests on the diaphragm. Anterior, lateral, and posterior lung surfaces are in close contact with the ribs and, hence, are collectively called the **costal surface.** The medial surface of the

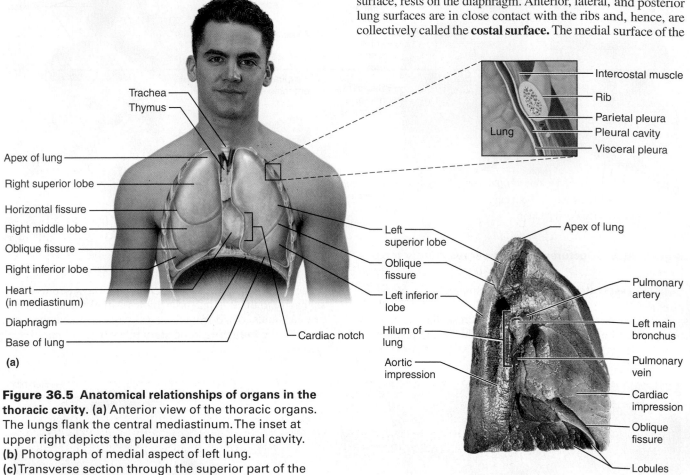

(a)

Figure 36.5 Anatomical relationships of organs in the thoracic cavity. (a) Anterior view of the thoracic organs. The lungs flank the central mediastinum. The inset at upper right depicts the pleurae and the pleural cavity. **(b)** Photograph of medial aspect of left lung. **(c)** Transverse section through the superior part of the thorax, showing the lungs and the main organs in the mediastinum.

(b)

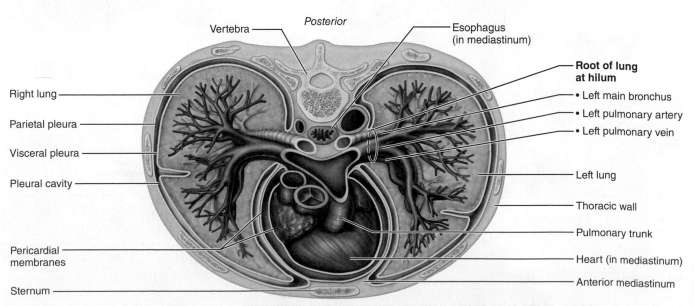

(c)

36

left lung exhibits a concavity called the **cardiac notch,** which accommodates the heart where it extends left from the body midline. Fissures divide the lungs into a number of **lobes**—two in the left lung and three in the right. Other than the respiratory passageways and air spaces that make up the bulk of their volume, the lungs are mostly elastic connective tissue, which allows them to recoil passively during expiration.

Each lung is enclosed in a double-layered sac of serous membrane called the **pleura.** The outer layer, the **parietal pleura,** is attached to the thoracic walls and the **diaphragm;** the inner layer, covering the lung tissue, is the **visceral pleura.** The two pleural layers are separated by the **pleural cavity,** which is filled with a thin film of *pleural fluid.* Produced by the pleurae, this fluid allows the lungs to glide without friction over the thoracic wall during breathing.

Activity 1

Identifying Respiratory System Organs

Before proceeding, be sure to locate on the torso model, thoracic cavity structures model, larynx model, or an anatomical chart all the respiratory structures described— both upper and lower respiratory system organs.

For instructions on animal dissections, see the dissection exercises (starting on p. 705) in the cat and fetal pig editions of this manual.

Activity 2

Demonstrating Lung Inflation in a Sheep Pluck

A *sheep pluck* includes the larynx, trachea with attached lungs, the heart and pericardium, and portions of the major blood vessels found in the mediastinum.

⚠ Don disposable gloves, obtain a dissecting tray and a fresh sheep pluck (or a preserved pluck of another animal), and identify the lower respiratory system organs. Once you have completed your observations, insert a hose from an air compressor (vacuum pump) into the trachea,

and allow air to flow alternately in and out of the lungs. Notice how the lungs inflate. This observation is educational in a preserved pluck, but it is a spectacular sight in a fresh one. When a fresh pluck used, the lung pluck changes color (becomes redder) as hemoglobin in trapped RBCs becomes loaded with oxygen.

⚠ Dispose of the gloves in the autoclave bag immediately after use.

Activity 3

Examining Prepared Slides of Trachea and Lung Tissue

1. Obtain a compound microscope and a slide of a cross section of the tracheal wall. Identify the smooth muscle layer, the hyaline cartilage supporting rings, and the pseudostratified

ciliated epithelium (use **Figure 36.6** as a guide). Also try to identify a few goblet cells in the epithelium. (See Figure 6.3c and d on p. 71.)

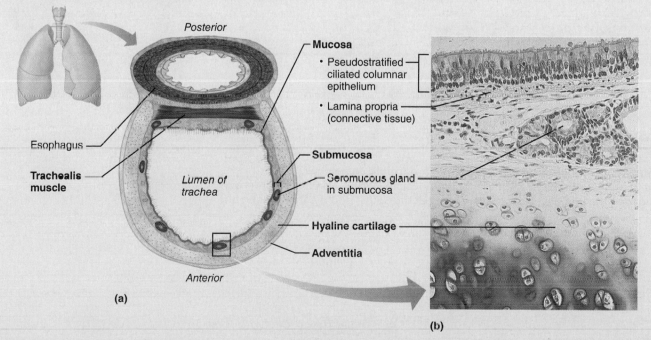

(a)

(b)

Figure 36.6 Tissue composition of the tracheal wall. (a) Cross-sectional view of the trachea. **(b)** Photomicrograph of a portion of the tracheal wall (125×).

Text continues on next page. →

2. Obtain a slide of lung tissue for examination. The alveolus is the main structural and functional unit of the lung and is the actual site of gas exchange. Identify a bronchiole (**Figure 36.7a**) and the simple squamous epithelium of the alveolar walls (Figure 36.7b).

3. Examine slides of pathological lung tissues, and compare them to the normal lung specimens. Record your observations in the review sheet at the end of this exercise.

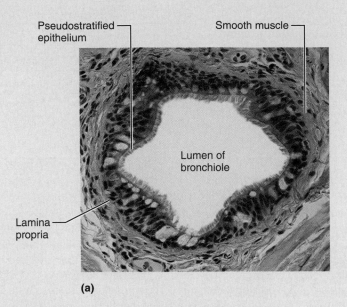

Pseudostratified epithelium

Smooth muscle

Lumen of bronchiole

Lamina propria

(a)

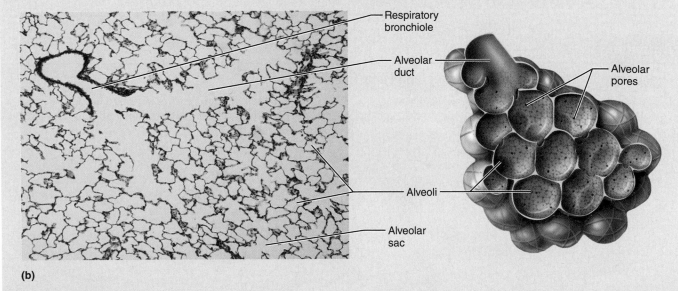

Respiratory bronchiole

Alveolar duct

Alveolar pores

Alveoli

Alveolar sac

(b)

Figure 36.7 Microscopic structure of a bronchiole and alveoli.
(a) Photomicrograph of a section of a bronchiole (180×). **(b)** Photomicrograph showing the final divisions of the bronchial tree (50×) and diagram of alveoli.

36

REVIEW SHEET
Anatomy of the Respiratory System

Name _____ Lab Time/Date _____

Upper and Lower Respiratory System Structures

1. Complete the labeling of the diagram of the upper respiratory structures (sagittal section).

Opening of
pharyngotympanic
tube

Nasopharynx

Hard palate

Hyoid bone

Thyroid cartilage

Cricoid cartilage

Thyroid gland

2. Two pairs of vocal folds are found in the larynx. Which pair are the true vocal cords (superior or inferior)?

_____

3. Name the specific cartilages in the larynx that correspond to the following descriptions.

   forms the Adam's apple: _____   shaped like a ring: _____

   a "lid" for the larynx: _____   vocal cord attachment: _____

4. Why is it important that the human trachea is reinforced with cartilaginous rings?

_____

   Why is it important that the rings are incomplete posteriorly?

_____

5. What is the function of the pleural fluid? _____

_____

_____

6. Name two functions of the nasal conchae: _____

and _____

7. The following questions refer to the main bronchi.

Which is longer? _____ Larger in diameter? _____ More horizontal? _____

Which more commonly traps a foreign object that has entered the respiratory passageways? _____

8. Appropriately label all structures provided with leader lines on the diagrams below.

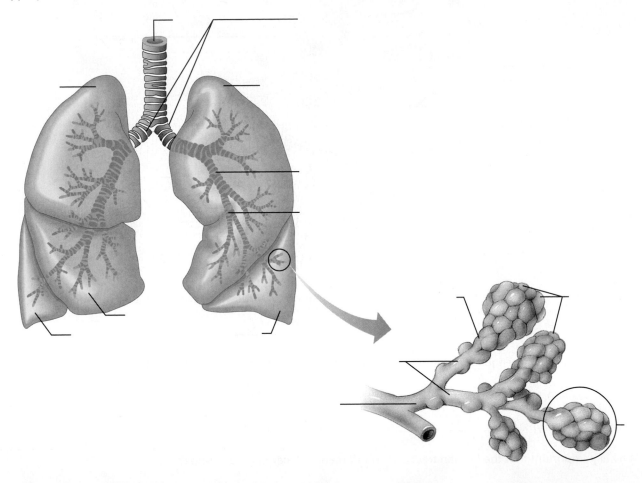

9. Trace a molecule of oxygen from the nostrils to the pulmonary capillaries of the lungs: Nostrils →

_____

_____

_____

10. Match the terms in column B to the descriptions in column A.

Column A

_____ 1. connects the larynx to the main bronchi

_____ 2. includes terminal and respiratory as subtypes

_____ 3. food passageway posterior to the trachea

_____ 4. covers the glottis during swallowing of food

_____ 5. contains the vocal cords

_____ 6. nerve that activates the diaphragm during inspiration

_____ 7. pleural layer lining the walls of the thorax

_____ 8. site from which oxygen enters the pulmonary blood

_____ 9. connects the middle ear to the nasopharynx

_____ 10. contains opening between the vocal folds

_____ 11. increases air turbulence in the nasal cavity

_____ 12. separates the oral cavity from the nasal cavity

Column B

a. alveolus

b. bronchiole

c. conchae

d. epiglottis

e. esophagus

f. glottis

g. larynx

h. palate

i. pharyngotympanic tube

j. parietal pleura

k. phrenic nerve

l. trachea

m. vagus nerve

n. visceral pleura

11. What portions of the respiratory system are referred to as anatomical dead space? _____

_____

_____

Why? _____

12. Define the following terms.

external respiration: _____

_____

internal respiration: _____

_____

Demonstrating Lung Inflation in a Sheep Pluck

13. Does the lung inflate part by part or as a whole, like a balloon? _____

14. What happened when the pressure was released? _____

15. What type of tissue ensures this phenomenon? _____

Examining Prepared Slides of Trachea and Lung Tissue

16. What structural characteristics of the alveoli make them an ideal site for the diffusion of gases?

_____

Why does oxygen move from the alveoli into the pulmonary capillary blood? _____

_____

_____

17. If you observed pathological lung sections, record your observations. Also record how the tissue differed from normal lung tissue. Complete the table below using your answers.

Slide type	Observations	Comparison to normal lung tissue

37

Respiratory System Physiology

Objectives

☐ Define the following and provide volume figures if applicable:

inspiration	inspiratory reserve volume (IRV)
expiration	minute respiratory volume (MRV)
tidal volume (TV)	forced vital capacity (FVC)
vital capacity (VC)	forced expiratory volume (FEV$_T$)
expiratory reserve volume (ERV)	

☐ Explain the role of muscles and volume changes in the mechanical process of breathing.

☐ Describe bronchial and vesicular breathing sounds.

☐ Demonstrate proper usage of a spirometer or an airflow transducer and associated BIOPAC® equipment.

☐ Discuss the relative importance of various mechanical and chemical factors in producing respiratory variations.

☐ Explain the importance of the carbonic acid–bicarbonate buffer system in maintaining blood pH.

Materials

- Model lung (bell jar demonstrator)
- Tape measure with centimeter divisions (cloth or plastic)
- Stethoscope
- Alcohol swabs
- Spirometer or BIOPAC® equipment:

Spirometer, disposable cardboard mouthpieces, nose clips, table (on board) for recording class data, disposable autoclave bag, battery jar containing 70% ethanol solution

Text continues on next page. →

MasteringA&P®

For related exercise study tools, go to the Study Area of **MasteringA&P**. There you will find:

- Practice Anatomy Lab **PAL**
- A&PFlix **A&PFlix**
- PhysioEx **PEx**
- Practice quizzes, Histology Atlas, e text, Videos, and more!

Pre-Lab Quiz

1. Circle the correct underlined term. <u>Inspiration</u> / <u>Expiration</u> is the phase of pulmonary ventilation when air passes out of the lungs.

2. Which of the following processes does *not* occur during inspiration?
   a. diaphragm moves to a flattened position
   b. gas pressure inside the lungs is lowered
   c. inspiratory muscles relax
   d. size of thoracic cavity increases

3. Circle True or False. Vesicular breathing sounds are produced by air rushing through the trachea and bronchi.

4. During normal quiet breathing, about _____ ml of air moves into and out of the lungs with each breath.
   a. 250
   b. 500
   h. 1000
   d. 2000

5. Circle the correct underlined term. <u>Tidal volume</u> / <u>Vital capacity</u> is the maximum amount of air that can be exhaled after a maximal inspiration.

6. Circle True or False. The neural centers that control respiratory rhythm and maintain a rate of 12–18 respirations per minute are located in the medulla and thalamus.

7. Circle the correct underlined term. Changes in pH and oxygen concentrations in the blood are monitored by chemoreceptor regions in the <u>medulla</u> / <u>aortic and carotid bodies</u>.

8. The carbonic acid–bicarbonate buffer system stabilizes arterial blood pH at:
   a. 2.0 ± 1.00
   b. 7.4 ± 0.02
   c. 6.2 ± 0.07
   d. 9.5 ±1.15

Text continues on next page. →

BIOPAC® BIOPAC® BSL System with BSL software version 3.7.5 to 3.7.7 (for Windows 7/Vista/XP or Mac OS X 10.4–10.6), data acquisition unit MP36/35 or MP45, PC or Mac computer, BIOPAC® airflow transducer, BIOPAC® calibration syringe, disposable mouthpiece, nose clip, and bacteriological filter.

Instructors using the MP36/35/30 data acquisition unit with BSL software versions earlier than 3.7.5 (for Windows or Mac) will need slightly different channel settings and collection strategies. Instructions for using the older data acquisition unit can be found on MasteringA&P.

- Paper bag
- 0.05 *M* NaOH
- Phenol red in a dropper bottle
- 100-ml beakers
- Distilled water
- Straws
- Concentrated HCl and NaOH in dropper bottles
- 250- and 50-ml beakers
- Plastic wash bottles containing distilled water
- pH meter (standardized with buffer of pH 7)
- Buffer solution (pH 7)
- Graduated cylinder (100 ml)
- Glass stirring rod
- Animal plasma
- 0.01 *M* HCl in dropper bottles

PEx PhysioEx™ 9.1 Computer Simulation Ex. 7 on p. PEx-105

Note: *Instructions for using PowerLab® equipment can be found on MasteringA&P.*

37

9. Circle the correct underlined term. <u>Acids</u> / <u>Bases</u> released into the blood by the body cells tend to lower the pH of the blood and cause it to become acidic.

10. Circle True or False. Rate and depth of breathing, hyperventilation, and hypoventilation should have little or no effect on the acid-base balance of blood.

The body's trillions of cells require O_2 and give off CO_2 as a waste the body must get rid of. The **respiratory system** provides the link with the external environment for both taking in O_2 and eliminating CO_2, but it doesn't work alone. The cardiovascular system via its contained blood provides the watery medium for transporting O_2 and CO_2 in the body. Let's look into how the respiratory system carries out its role.

Mechanics of Respiration

Pulmonary ventilation, or **breathing,** consists of two phases: **inspiration,** during which air is taken into the lungs, and **expiration,** during which air passes out of the lungs. As the inspiratory muscles (external intercostals and diaphragm) contract during inspiration, the size of the thoracic cavity increases. The diaphragm moves from its relaxed dome shape to a flattened position, increasing the vertical dimension. The external intercostals lift the rib cage, increasing the anterior-posterior and lateral dimensions (**Figure 37.1**). Because the lungs adhere to the thoracic walls like flypaper owing to the presence of serous fluid in the pleural cavity, the intrapulmonary volume (volume within the lungs) also increases, lowering the air (gas) pressure inside the lungs. The gases then expand to fill the available space, creating a partial vacuum that causes air to flow into the lungs—constituting the act of inspiration. During expiration, the inspiratory muscles relax, and the natural tendency of the elastic lung tissue to recoil decreases the intrathoracic and intrapulmonary volumes. As the gas molecules within the lungs are forced closer together, the intrapulmonary pressure rises to a point higher than atmospheric pressure. This causes gases to flow out of the lungs to equalize the pressure inside and outside the lungs—the act of expiration.

Activity 1

Operating the Model Lung

Observe the model lung, which demonstrates the principles involved in gas flows into and out of the lungs. It is a simple apparatus with a bottle "thorax," a rubber membrane "diaphragm," and balloon "lungs."

1. Go to the demonstration area and work the model lung by moving the rubber diaphragm up and down. Notice the *relative* changes in balloon (lung) size as the volume of the thoracic cavity is alternately increased and decreased.

2. Check the appropriate columns in the chart concerning these observations in the review sheet at the end of this exercise.

3. A pneumothorax is a condition in which air has entered the pleural cavity, as with a puncture wound. Simulate a pneumothorax: Inflate the balloon lungs by pulling down on the diaphragm. Ask your lab partner to let air into the bottle "thorax" by loosening the rubber stopper.

What happens to the balloon lungs?

_____

4. After observing the operation of the model lung, conduct the following tests on your lab partner. Use the tape measure to determine his or her

Inspiration   Expiration

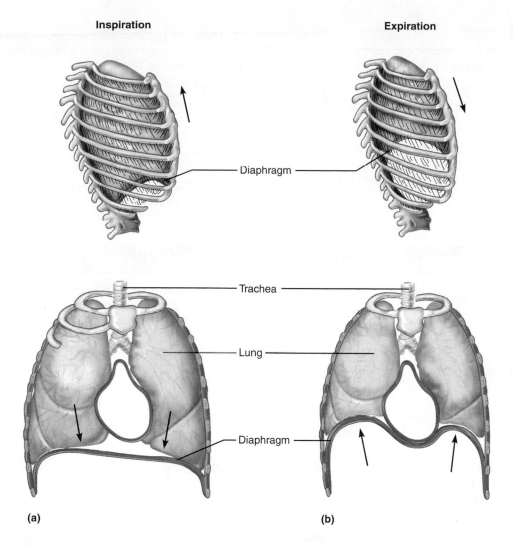

Figure 37.1 Rib cage and diaphragm positions during breathing. (a) At the end of a normal inspiration; chest expanded, diaphragm depressed. **(b)** At the end of a normal expiration; chest depressed, diaphragm elevated.

chest circumference by placing the tape around the chest as high up under the armpits as possible. Record the measurements in centimeters in the appropriate space below for each of the conditions.

Quiet breathing:

Inspiration _____ cm   Expiration _____ cm

Forced breathing:

Inspiration _____ cm   Expiration _____ cm

Do the results coincide with what you expected on the basis of what you have learned thus far? _____

How does the structural relationship between the balloon-lungs and bottle-thorax differ from that seen in the human lungs and thorax?

_____

_____

_____

Respiratory Sounds

As air flows in and out of the bronchial tree, it produces two characteristic sounds that can be auscultated with a stethoscope. The **bronchial sounds** are produced by air rushing through the large respiratory passageways (the trachea and the bronchi). The second sound type, **vesicular breathing sounds,** apparently results from air filling the alveolar sacs and resembles the sound of a rustling of leaves.

Activity 2

Auscultating Respiratory Sounds

1. Obtain a stethoscope, and clean the earpieces with an alcohol swab. Allow the alcohol to dry before donning the stethoscope.

2. Place the diaphragm of the stethoscope on the throat of the test subject just below the larynx. Listen for bronchial sounds on inspiration and expiration. Move the stethoscope down toward the bronchi until you can no longer hear sounds.

3. Place the stethoscope over the following chest areas and listen for vesicular sounds during respiration (heard primarily during inspiration).

- At various intercostal spaces
- At the *triangle of auscultation* (a small depressed area of the back where the muscles fail to cover the rib cage; located just medial to the inferior part of the scapula)
- Inferior to the clavicle

Diseased respiratory tissue, mucus, or pus can produce abnormal chest sounds such as rales (a rasping sound) and wheezing (a whistling sound). ✚

Respiratory Volumes and Capacities—Spirometry

A person's size, sex, age, and physical condition produce variations in respiratory volumes. Normal quiet breathing moves about 500 ml of air in and out of the lungs with each breath. As you have seen in the first activity, a person can usually forcibly inhale or exhale much more air than is exchanged in normal quiet breathing. The terms used for the

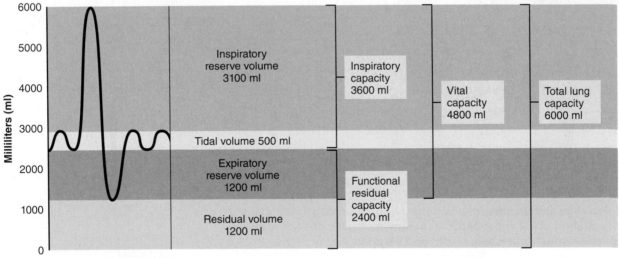

(a) Spirographic record for a male

	Measurement	Adult male average value	Adult female average value	Description
Respiratory volumes	Tidal volume (TV)	500 ml	500 ml	Amount of air inhaled or exhaled with each breath under resting conditions
	Inspiratory reserve volume (IRV)	3100 ml	1900 ml	Amount of air that can be forcefully inhaled after a normal tidal volume inspiration
	Expiratory reserve volume (ERV)	1200 ml	700 ml	Amount of air that can be forcefully exhaled after a normal tidal volume expiration
	Residual volume (RV)	1200 ml	1100 ml	Amount of air remaining in the lungs after a forced expiration
Respiratory capacities	Total lung capacity (TLC)	6000 ml	4200 ml	Maximum amount of air contained in lungs after a maximum inspiratory effort: TLC = TV + IRV + ERV + RV
	Vital capacity (VC)	4800 ml	3100 ml	Maximum amount of air that can be expired after a maximum inspiratory effort: VC = TV + IRV + ERV
	Inspiratory capacity (IC)	3600 ml	2400 ml	Maximum amount of air that can be inspired after a normal tidal volume expiration: IC = TV + IRV
	Functional residual capacity (FRC)	2400 ml	1800 ml	Volume of air remaining in the lungs after a normal tidal volume expiration: FRC = ERV + RV

(b) Summary of respiratory volumes and capacities for males and females

Figure 37.2 Respiratory volumes and capacities.

37

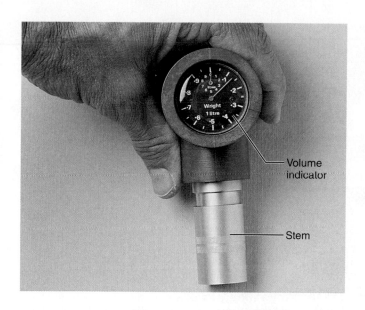

Volume indicator

Stem

Figure 37.3 The Wright handheld dry spirometer. Reset to zero prior to each test.

measurable respiratory volumes and capacities are defined and illustrated with an idealized tracing in **Figure 37.2**.

Respiratory volumes can be measured, as in Activities 3 and 4, with an apparatus called a **spirometer.** There are two major types of spirometers, which give comparable results—the handheld dry, or wheel, spirometers (such as the Wright spirometer illustrated in **Figure 37.3**) and "wet" spirometers, such as the Phipps and Bird spirometer and the Collins spirometer (which is available in both recording and nonrecording varieties). The somewhat more sophisticated wet spirometer consists of a plastic or metal *bell* within a rectangular or cylindrical tank that air can be added to or removed from (**Figure 37.4**, p. 558).

In nonrecording spirometers, an indicator moves as air is *exhaled,* and only expired air volumes can be measured directly. By contrast, recording spirometers allow both inspired and expired gas volumes to be measured.

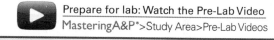

Prepare for lab: Watch the Pre-Lab Video
MasteringA&P*>Study Area>Pre-Lab Videos

Activity 3

Measuring Respiratory Volumes Using Spirometers

The steps for using a nonrecording spirometer and a wet recording spirometer are given separately below.

Using a Nonrecording Spirometer

1. Before using the spirometer, count and record the subject's normal respiratory rate. The subject should face away from you as you make the count.

Respirations per minute: _____

Now identify the parts of the spirometer you will be using by comparing it to the illustration of a similar spirometer (in Figure 37.3 or 37.4a). Examine the spirometer volume indicator *before beginning* to make sure you know how to read the scale. Work in pairs, with one person acting as the subject while the other records the data of the volume determinations. *Reset the indicator to zero before beginning each trial.*

Obtain a disposable cardboard mouthpiece. Prior to inserting the cardboard mouthpiece, clean the valve assembly with an alcohol swab. Then insert the mouthpiece in the open end of the valve assembly (attached to the flexible tube) of the wet spirometer or over the fixed stem of the handheld dry spirometer. Before beginning, the subject should practice exhaling through the mouthpiece without exhaling through the nose, or prepare to use the nose clips. If you are using the handheld spirometer, make sure its dial faces upward so that the volumes can be easily read during the tests.

2. The subject should stand erect during testing. Conduct the test three times for each required measurement. Record the data where indicated in this section, and then find the average volume figure for that respiratory measurement. After you have completed the trials and computed

the averages, enter the average values on the table prepared on the board for tabulation of class data,* and copy all averaged data onto the review sheet at the end of the exercise.

3. **Measuring tidal volume (TV).** The TV, or volume of air inhaled and exhaled with each quiet, normal respiration, is approximately 500 ml. To conduct the test, inhale a normal breath, and then exhale a normal breath of air into the spirometer mouthpiece. (Do not force the expiration!) Record the volume and repeat the test twice.

trial 1: _____ ml   trial 2: _____ ml

trial 3: _____ ml   average TV: _____ ml

4. Compute the subject's **minute respiratory volume (MRV)** using the following formula:

$$\text{MRV} = \text{TV} \times \text{respirations/min} = \underline{\hspace{1cm}} \text{ml/min}$$

5. **Measuring expiratory reserve volume (ERV).** The ERV is the volume of air that can be forcibly exhaled after a normal expiration. Normally it ranges between 700 and 1200 ml.

Inhale and exhale normally two or three times, then insert the spirometer mouthpiece and exhale forcibly as

*****Note to the Instructor:** The format of class data tabulation can be similar to that shown here. However, it would be interesting to divide the class into smokers and nonsmokers and then compare the mean average VC and ERV for each group. Such a comparison might help to determine whether smokers are handicapped in any way. It also might be a good opportunity for an informal discussion of the early warning signs of chronic bronchitis and emphysema, which are primarily smokers' diseases.

37

Text continues on p. 559. →

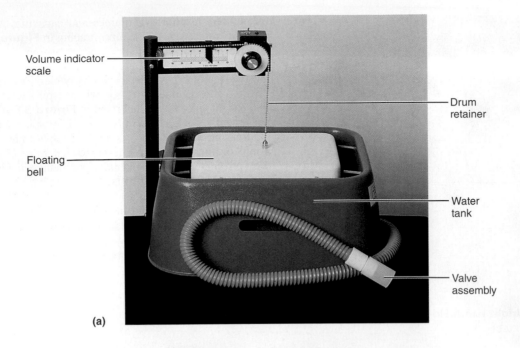

Volume indicator scale

Drum retainer

Floating bell

Water tank

Valve assembly

(a)

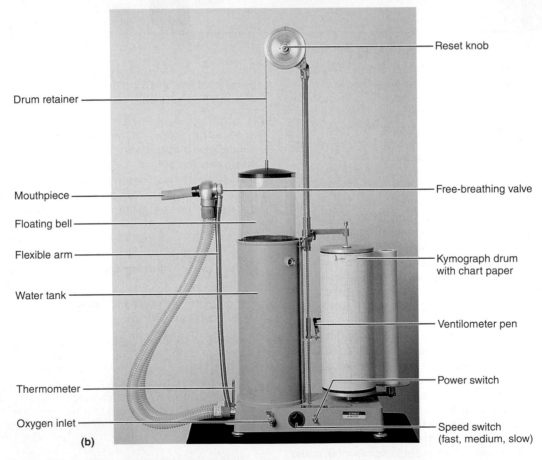

Drum retainer

Reset knob

Mouthpiece

Free-breathing valve

Floating bell

Flexible arm

Kymograph drum with chart paper

Water tank

Ventilometer pen

Thermometer

Power switch

Oxygen inlet

Speed switch (fast, medium, slow)

(b)

Figure 37.4 Wet spirometers. (a) The Phipps and Bird wet spirometer.
(b) The Collins-9L wet recording spirometer.

37

much of the additional air as you can. Record your results, and repeat the test twice again.

trial 1: _____ ml   trial 2: _____ ml

trial 3: _____ ml   average ERV: _____ ml

ERV is dramatically reduced in conditions in which the elasticity of the lungs is decreased by a chronic obstructive pulmonary disease (COPD) such as **emphysema**. Since energy must be used to *deflate* the lungs in such conditions, expiration is physically exhausting to individuals suffering from COPD. ✚

6. Measuring vital capacity (VC). The VC, or total exchangeable air of the lungs (the sum of TV + IRV + ERV), normally ranges from 3100 ml to 4800 ml.

Breathe in and out normally two or three times, and then bend forward and exhale all the air possible. Then, as you raise yourself to the upright position, inhale as fully as possible. It is important to *strain* to inhale the maximum amount of air that you can. Quickly insert the mouthpiece, and exhale as forcibly as you can. Record your results and repeat the test twice again.

trial 1: _____ ml   trial 2: _____ ml

trial 3: _____ ml   average VC: _____ ml

7. The inspiratory reserve volume (IRV), or volume of air that can be forcibly inhaled following a normal inspiration, can now be computed using the average values obtained for TV, ERV, and VC and plugging them into the equation:

$$IRV = VC - (TV + ERV)$$

Record your calculated IRV: _____ ml

The normal IRV range is substantial, ranging from 1900 to 3100 ml. How does your calculated value compare?

_____

Steps 8–10, which provide common directions for use of both nonrecording and recording spirometers, continue (on p. 561) after the wet recording spirometer directions.

WHY THIS MATTERS | Incentive Spirometry

If you or a family member has had surgery, chances are you have been introduced to the incentive spirometer placed on the hospital bedside table. The device used may vary from model to model but usually requires the patient to perform a sustained maximal inspiration (SMI) while providing a visual cue to track his or her efforts. This slow, deep inspiration is similar to a yawn and ventilates all of the alveoli (not the case in quiet breathing). With flow-oriented devices, the inspiratory effort is visualized by an uplifted ball. With volume-oriented devices, the inspiratory effort is visualized by a raised piston. In either case, incentive spirometry has been found to reduce postoperative pulmonary complications. ■

Using a Wet Recording Spirometer

1. In preparation for recording, familiarize yourself with the spirometer by comparing it to the illustrated equipment (Figure 37.4b).

2. Examine the chart paper, noting that its horizontal lines represent milliliter units. To apply the chart paper to the recording drum, first lift the drum retainer and then remove the kymograph drum. Wrap a sheet of chart paper around the drum, *making sure that the right edge overlaps the left.* Fasten it with tape, and then replace the kymograph drum and lower the drum retainer into its original position in the hole in the top of the drum.

3. Raise and lower the floating bell several times, noting as you do that the *ventilometer pen* moves up and down on the drum. This pen, which writes in black ink, will be used for recording and should be adjusted so that it records in the approximate middle of the chart paper. This adjustment is made by repositioning the floating bell using the *reset knob* on the metal pulley at the top of the spirometer apparatus. The other pen, the respirometer pen, which records in red ink, will not be used for these tests and should be moved away from the drum's recording surface.

4. Recording your normal respiratory rate. Clean the nose clips with an alcohol swab. While you wait for the alcohol to air dry, count and record your normal respiratory rate.

Respirations per minute: _____

5. Recording tidal volume. After the alcohol has air dried, apply the nose clips to your nose. This will enforce mouth breathing.

Open the *free-breathing valve.* Insert a disposable cardboard mouthpiece into the end (valve assembly) of the breathing tube, and then insert the mouthpiece into your mouth. Practice breathing for several breaths to get used to the apparatus. At this time, you are still breathing room air.

Set the spirometer switch to **SLOW** (32 mm/min). Close the free-breathing valve, and breathe in a normal manner for 2 minutes to record your tidal volume—the amount of air inspired or expired with each normal respiratory cycle. This recording should show a regular pattern of inspiration-expiration spikes and should gradually move upward on the chart paper. (A downward slope indicates that there is an air leak somewhere in the system—most likely at the mouthpiece.) Notice that on an apparatus using a counter-weighted pen (such as the Collins-9L Ventilometer shown in Figure 37.4b), inspirations are recorded by upstrokes and expirations are recorded by downstrokes.*

6. Recording vital capacity. To record your vital capacity, take the deepest possible inspiration you can and then exhale to the greatest extent possible—really *push* the air out. (The recording obtained should resemble that shown in Figure 37.5). Repeat the vital capacity measurement twice again. Then turn off the spirometer and remove the chart paper from the kymograph drum.

*If a Collins survey spirometer is used, the situation is exactly opposite: Upstrokes are expirations, and downstrokes are inspirations.

text continues on next page. →

37

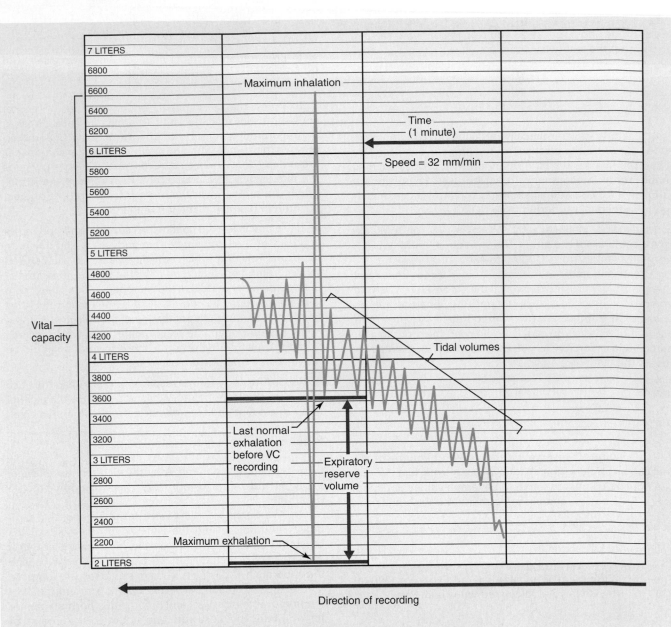

Figure 37.5 A typical spirometry recording of tidal volume, inspiratory capacity, expiratory reserve volume, and vital capacity. At a drum speed of 32 mm/min, each vertical column of the chart represents a time interval of 1 minute. (Note that downstrokes represent exhalations, and upstrokes represent inhalations.)

7. Determine and record your measured, averaged, and corrected respiratory volumes. Because the pressure and temperature inside the spirometer are influenced by room temperature and differ from those in the body, all measured values are to be multiplied by a **BTPS** (body temperature, atmospheric pressure, and water saturation) **factor.** At room temperature, the BTPS factor is typically 1.1 or very close to that value. Hence, you will multiply your average measured values by 1.1 to obtain your corrected respiratory volume values. Copy the averaged and corrected values onto the review sheet at the end of this exercise.

- Tidal volume (TV). Select a typical resting tidal breath recording. Subtract the millimeter value of the trough (exhalation) from the millimeter value of the peak (inspiration). Record this value below as *measured*

TV 1. Select two other TV tracings to determine the TV values for the TV 2 and TV 3 measurements. Then, determine your average TV and multiply it by 1.1 to obtain the BTPS-corrected average TV value.

measured TV 1: _____ ml   average TV: _____ ml

measured TV 2: _____ ml   corrected average TV:

measured TV 3: _____ ml   _____ ml

Also compute your **minute respiratory volume (MRV)** using the following formula:

MRV = TV × respirations/min = _____ ml/min

- Inspiratory capacity (IC). In the first vital capacity recording, find the expiratory trough immediately preceding the maximal inspiratory peak achieved during vital capacity determination. Subtract the milliliter value of that expiration from the value corresponding to the peak of the maximal inspiration that immediately follows. For example, according to our typical recording (**Figure 37.5**), these values would be

$$6600 - 3650 = 2950 \text{ ml}$$

Record your computed value and the results of the two subsequent tests on the appropriate lines below. Then calculate the measured and corrected inspiratory capacity averages, and record.

measured IC 1: _____ ml   average IC: _____ ml

measured IC 2: _____ ml   corrected
                                 average IC: _____ ml
measured IC 3: _____ ml

- Inspiratory reserve volume (IRV). Subtract the corrected average tidal volume from the corrected average for the inspiratory capacity and record below.

IRV = corrected average IC – corrected average TV

corrected average IRV: _____ ml

- Expiratory reserve volume (ERV). Subtract the number of milliliters corresponding to the trough of the maximal expiration obtained during the vital capacity recording from milliliters corresponding to the last *normal* expiration before the VC maneuver is performed. For example, according to our typical recording (Figure 37.5), these values would be

$$3650 \text{ ml} - 2050 \text{ ml} = 1600 \text{ ml}$$

Record your measured and averaged values (three trials) below.

measured ERV 1: _____ ml   average ERV: _____ ml

measured ERV 2: _____ ml   corrected average ERV:

measured ERV 3: _____ ml   _____ ml

- Vital capacity (VC). Add your corrected values for ERV and IC to obtain the corrected average VC. Record below and on the review sheet at the end of this exercise.

corrected average VC: _____ ml

Now continue with step 8 (below) whether you are following the procedure for the nonrecording or recording spirometer.

8. Figure out how closely your measured average vital capacity volume compares with the *predicted values* for someone your age, sex, and height. Obtain the predicted value either from the following equation or the appropriate table (see your instructor for the printed table). Notice that you will have to convert your height in inches to centimeters (cm) to find the corresponding value. This is easily done by multiplying your height in inches by 2.54.

Computed height: _____ cm

Male VC = (0.052) H − (0.022) A − 3.60

Female VC = (0.041) H − (0.018) A − 2.69

Note: (VC) = vital capacity in liters, (H) = height in centimeters, and (A) = age in years.

Predicted VC (obtained from the equation or appropriate table):

_____ ml

Use the following equation to compute your VC as a percentage of the predicted VC value:

$$\% \text{ of predicted VC} = \left(\frac{\text{average VC}}{\text{predicted VC}} \right) \times 100$$

% predicted VC value: _____%

9. Computing residual volume. A respiratory volume that cannot be experimentally demonstrated here is the residual volume (RV). RV is the amount of air remaining in the lungs after a maximal expiratory effort. The presence of residual air (usually about 1200 ml) that cannot be voluntarily flushed from the lungs is important because it allows gas exchange to go on continuously — even between breaths.

Although the residual volume cannot be measured directly, it can be approximated by using one of the following factors:

For ages 16–34   Factor = 0.250

For ages 35–49   Factor = 0.305

For ages 50–69   Factor = 0.445

Compute your predicted RV using the following equation.

$$RV = VC \times \text{factor}$$

 10. Recording is finished for this subject. Before continuing with the next member of your group:

- Dispose of used cardboard mouthpieces in the autoclave bag.

- Swish the valve assembly (if removable) in the 70% ethanol solution, then rinse with tap water.

- Put a fresh mouthpiece into the valve assembly (or on the stem of the handheld spirometer). Using the procedures outlined above, measure and record the respiratory volumes for all members of your group.

37

37

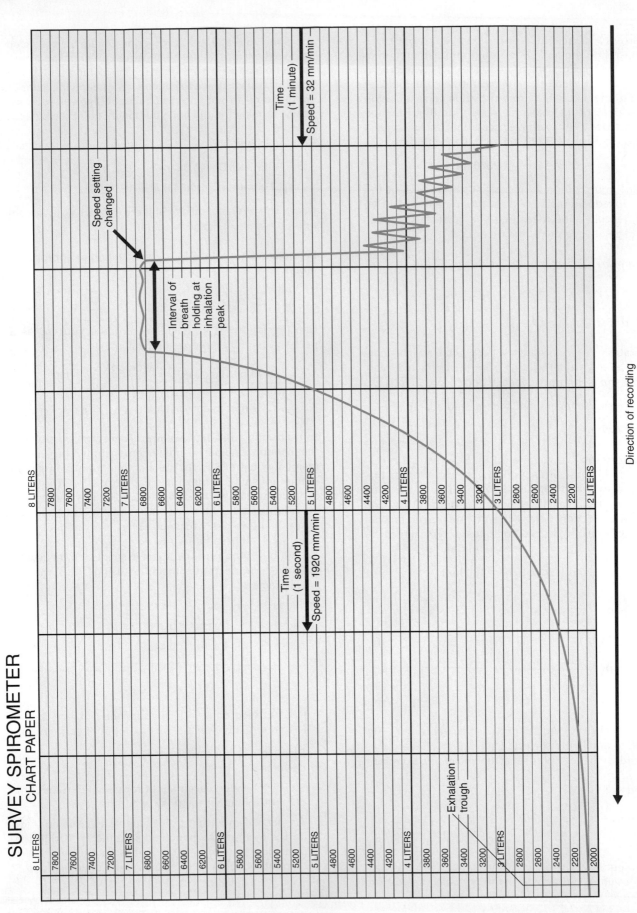

Figure 37.6 A recording of the forced vital capacity (FVC) and forced expiratory volume (FEV) or timed vital capacity test.

Forced Expiratory Volume (FEV$_T$) Measurement

Though not really diagnostic, pulmonary function tests can help the clinician distinguish between obstructive and restrictive pulmonary diseases. (In obstructive disorders, like chronic bronchitis and asthma, airway resistance is increased, whereas in restrictive diseases, such as polio and tuberculosis, total lung capacity declines.) Two highly useful pulmonary function tests used for this purpose are the FVC and the FEV$_T$ (**Figure 37.6**).

The **FVC** (forced vital capacity) measures the amount of gas expelled when the subject takes the deepest possible breath and then exhales forcefully and rapidly. This volume is reduced in those with restrictive pulmonary disease. The **FEV$_T$** (forced expiratory volume) involves the same basic testing procedure, but it specifically looks at the percentage of the vital capacity that is exhaled during specific time intervals of the FVC test. FEV$_1$, for instance, is the amount exhaled during the first second. Healthy individuals can expire 75% to 85% of their FVC in the first second. The FEV$_1$ is low in those with obstructive disease.

Activity 4

Measuring the FVC and FEV$_1$

Directions provided here for the FEV$_T$ determination apply only to the recording spirometer.

1. Prepare to make your recording as described for the recording spirometer, steps 1–5 (on p. 559).

2. At a signal agreed upon by you and your lab partner, take the deepest inspiration possible and hold it for 1 to 2 seconds. As the inspiratory peak levels off, your partner is to change the drum speed to **FAST** (1920 mm/min) so that the distance between the vertical lines on the chart represents 1 second.

3. Once the drum speed is changed, exhale as much air as you can as rapidly and forcibly as possible.

4. When the tracing plateaus (bottoms out), stop recording and determine your FVC. Subtract the milliliter reading in the expiration trough (the bottom plateau) from the preceding inhalation peak (the top plateau). Record this value.

FVC: _____ ml

5. Prepare to calculate the FEV$_1$. Draw a vertical line intersecting with the spirogram tracing at the precise point that

exhalation began. Identify this line as *line 1*. From line 1, measure 32 mm horizontally to the left, and draw a second vertical line. Label this as *line 2*. The distance between the two lines represents 1 second, and the volume exhaled in the first second is read where line 2 intersects the spirogram tracing. Subtract that milliliter value from the milliliter value of the inhalation peak (at the intersection of line 1), to determine the volume of gas expired in the first second. According to the values given in the example (Figure 37.6), that figure would be 3400 ml (6800 ml − 3400 ml). Record your measured value below.

Milliliters of gas expired in second 1: _____ ml

6. To compute the FEV$_1$ use the following equation:

$$FEV_1 = \frac{\text{volume expired in second 1}}{\text{FVC volume}} \times 100\%$$

Record your calculated value below and on the review sheet at the end of this exercise.

FEV$_1$: _____ % of FVC

Activity 5

Measuring Respiratory Volumes Using BIOPAC®

In this activity, you will measure respiratory volumes using the BIOPAC® airflow transducer. An example of these volumes is demonstrated in the computer-generated spirogram (**Figure 37.7**, p. 564). Since it is not possible to measure **residual volume (RV)** using the airflow transducer, assume that it is 1.0 liter for each subject, which is a reasonable estimation. Or enter a volume between 1 and 5 liters via Preferences. It is also important to estimate the **predicted vital capacity** of the subject for comparison to the measured value. A rough estimate of the vital capacity in liters (VC) of a subject can be calculated using the following formulas based on height in centimeters (*H*) and age in years (*A*).

Male VC = (0.052)*H* − (0.022)*A* − 3.60

Female VC = (0.041)*H* − (0.018)*A* − 2.69

Because many factors besides height and age influence vital capacity, it should be assumed that measured values

up to 20% above or below the calculated predicted value are normal.

Setting Up the Equipment

1. Connect the BIOPAC® unit to the computer and turn the computer **ON**.

2. Make sure the BIOPAC® unit is **OFF**.

3. Plug in the equipment (as shown in **Figure 37.8**, p. 564).

- Airflow transducer—CH 1

4. Turn the BIOPAC® unit **ON**.

5. Place a *clean* bacteriological filter onto the end of the BIOPAC® calibration syringe (as shown in **Figure 37.9**, p. 564). Since the subject will be blowing through a filter, it is necessary to use a filter for calibration.

Text continues on next page. →

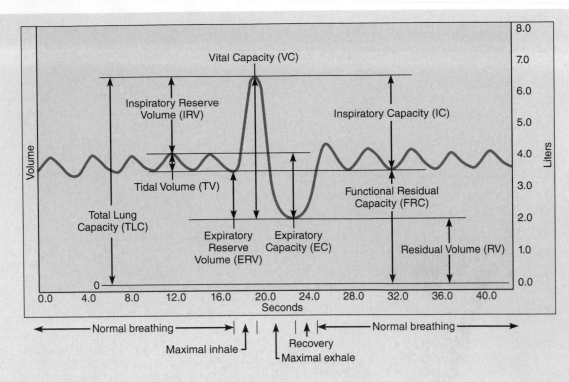

Figure 37.7 Example of a computer-generated spirogram.

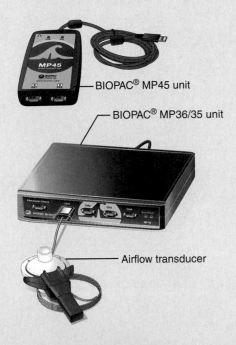

Figure 37.8 Setting up the BIOPAC® equipment.
Plug the airflow transducer into Channel 1. Transducer is shown plugged into the MP36/35 unit.

6. Insert the calibration syringe and filter assembly into the airflow transducer on the side labeled **Inlet**.

7. Start the Biopac Student Lab program on the computer by double-clicking the icon on the desktop or by following your instructor's guidance.

8. Select lesson **L12-Pulmonary Functions-1** from the menu, and click **OK**.

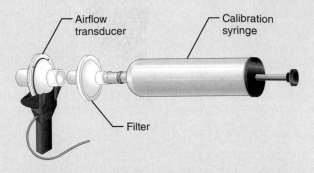

Figure 37.9 Placement of the calibration syringe and filter assembly onto the airflow transducer for calibration.

9. Type in a filename that will save this subject's data on the computer hard drive. You may want to use the subject's last name followed by Pulmonary Functions (PF)-1 (for example, SmithPF-1), then click **OK**.

Calibrating the Equipment

Two precautions must be followed:

• The airflow transducer is sensitive to gravity, so it must be held directly parallel to the ground during calibration and recording.

• Do not hold onto the airflow transducer when it is attached to the calibration syringe and filter assembly—the syringe tip is likely to break. (See **Figure 37.10** for the proper handling of the calibration assembly). The size of the calibration syringe can be altered via Preferences.

1. Make sure the plunger is pulled all the way out. While the assembly is held in a steady position parallel to the ground, click **Calibrate** and then **OK** after you have read

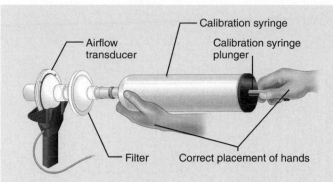

Figure 37.10 Proper handling of the calibration assembly.

the alert box. This part of the calibration will terminate automatically with an alert box ensuring that you have read the on-screen instructions and those indicated in step 2, below.

2. The final part of the calibration involves simulating five breathing cycles using the calibration syringe. A single cycle consists of:

- Pushing the plunger in (taking 1 second for this stroke)
- Waiting for 2 seconds
- Pulling the plunger out (taking 1 second for this stroke)
- Waiting 2 seconds

Remember to hold the airflow transducer directly parallel to the ground during calibration and recording.

3. When ready to perform this second stage of the calibration, click **Yes.** After you have completed five cycles, click **End Calibration.**

4. Observe the data, which should look similar to that in the example **(Figure 37.11)**.

- If the data look very different, click **Redo Calibration** and repeat the steps above.
- If the data look similar, gently remove the calibration syringe, leaving the air filter attached to the transducer. Proceed to the next section.

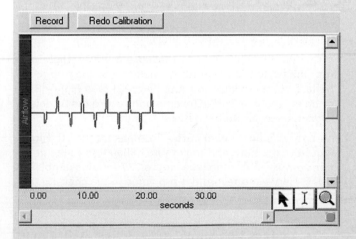

Figure 37.11 Example of calibration data.

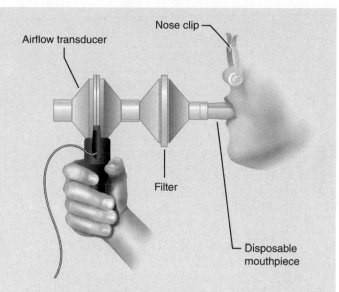

Figure 37.12 Proper equipment setup for recording data.

Recording the Data

Follow these procedures precisely, because the airflow transducer is very sensitive. Hints to obtain the best data:

- Always insert air filter on, and breathe through, the transducer side labeled **Inlet.**
- Keep the airflow transducer upright at all times.
- The subject should not look at the computer screen during the recording of data.
- The subject must keep a nose clip on throughout the experiment.

1. Insert a clean mouthpiece into the air filter that is already attached to the airflow transducer. *Be sure that the filter is attached to the **Inlet** side of the airflow transducer.*

2. Write the name of the subject on the mouthpiece and air filter. For safety purposes, each subject must use his or her own air filter and mouthpiece.

3. The subject should now place the nose clip on the nose (or hold the nose very tightly with finger pinch), wrap the lips tightly around the mouthpiece, and begin breathing normally through the airflow transducer (as shown in **Figure 37.12**)

4. When prepared, the subject will complete the following unbroken series with nose plugged and lips tightly sealed around the mouthpiece:

- Take five normal breaths (1 breath = inhale + exhale).
- Inhale as much air as possible.
- Exhale as much air as possible.
- Take five normal breaths.

5. When the subject is prepared to proceed, click **Record** on the first normal inhalation and proceed. When the subject finishes the last exhalation at the end of the series, click **Stop.**

Text continues on next page. →

37

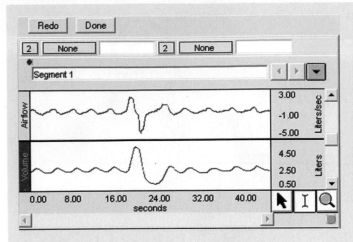

Figure 37.13 Example of pulmonary data.

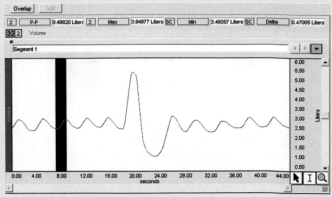

Figure 37.14 Highlighting data for the inhalation of the third breath.

6. Observe the data, which should look similar to that in the example (**Figure 37.13**).

- If the data look very different, click **Redo** and repeat the steps above. Be certain that the lips are sealed around the mouthpiece, the nose is completely plugged, and the transducer is upright.

- If the data look similar, proceed to step 7.

7. When finished, click **Done**. A pop-up window will appear.

- Click **Yes** if you are done and want to stop recording.

- To record from another subject, select **Record from another Subject** and return to step 1 under Recording the Data. You will not need to redo the calibration procedure for the second subject.

- If continuing to the Data Analysis section, select **Analyze current data file** and proceed to step 2 of the Data Analysis section.

Data Analysis

1. If just starting the BIOPAC® program to perform data analysis, enter **Review Saved Data** mode and choose the file with the subject's PF data (for example, SmithPF-1).

2. Observe how the channel numbers are designated: CH 1— Airflow; CH 2—Volume.

3. To set up the display for optimal viewing, hide CH 1— Airflow. To do this, hold down the Ctrl key (PC) or Option key (Mac) while using the cursor to click the Channel box 1 (the small box with a 1 at the upper left of the screen).

4. To analyze the data, set up the first pair of channel/measurement boxes at the top of the screen by selecting the following channel and measurement type from the drop-down menu:

Channel	Measurement	Data
CH 2	p-p	volume

5. Take two measures for an averaged TV calculation: Use the arrow cursor and click the I-beam cursor box on the lower right side of the screen to activate the "area selection" function. Using the activated I-beam cursor, highlight the inhalation of cycle 3 (as shown in **Figure 37.14**).

Pulmonary Measurements	
Volumes	**Measurements (liters)**
Tidal volume (TV)	
Inspiratory reserve volume (IRV)	
Expiratory reserve volume (ERV)	
Vital capacity (VC)	
Residual volume (RV)	1.00 (assumed)

6. The computer automatically calculates the **p-p** value for the selected area. This measure is the difference between the highest and lowest values in the selected area. Note the value. Use the I-beam cursor to select the exhalation of cycle 3 and note the **p-p** value.

7. Calculate the average of the two **p-p** values. This represents the **tidal volume** (in liters). Record the value in the **Pulmonary Measurements chart** above.

8. Use the I-beam cursor to measure the IRV: Highlight from the peak of maximum inhalation to the peak of the last normal inhalation just before it (see Figure 37.7 for an example of IRV). Observe and record the Δ **(delta)** value in the chart (to the nearest 0.01 liter).

9. Use the I-beam cursor to measure the ERV: Highlight from the trough of maximum exhalation to the trough of the last normal exhalation just before it (see Figure 37.7 for an example of ERV). Observe and record the Δ **(delta)** value in the chart (to the nearest 0.01 liter).

10. Last, use the I-beam cursor to measure the VC: Highlight from the trough of maximum exhalation to the peak of maximum inhalation (see Figure 37.7 for an example of VC). Observe and record the **p-p** value in the chart (to the nearest 0.01 liter).

11. When finished, choose **File menu** and **Quit** to close the program.

Using the measured data, calculate the capacities listed in the **Calculated Pulmonary Capacities chart**.

Calculated Pulmonary Capacities		
Capacity	Formula	Calculation (liters)
Inspiratory capacity (IC)	= TV + IRV	
Functional residual capacity (FRC)	= ERV + RV	
Total lung capacity (TLC)	= TV + RV + IRV + ERV	

Use the formula in the introduction of this activity (p. 563) to calculate the predicted vital capacity of the subject based on height and age.

Predicted VC: _____ liters

How does the measured vital capacity compare to the predicted vital capacity?

_____

_____

_____

_____

Describe why height and weight might correspond with a subject's VC.

_____

_____

_____

What other factors might influence the VC of a subject?

_____

_____

_____

_____

Factors Influencing Rate and Depth of Respiration

The neural centers that control respiratory rhythm and maintain a rate of 12 to 18 respirations/min are located in the medulla and pons. On occasion, input from the stretch receptors in the lungs (via the vagus nerve to the medulla) modifies the respiratory rate, as in cases of extreme overinflation of the lungs (Hering-Breuer reflex).

Death occurs when medullary centers are completely suppressed, as from an overdose of sleeping pills or gross overindulgence in alcohol, and respiration ceases completely. ✚

Although the neural centers initiate the basic rhythm of breathing, there is no question that physical phenomena such as talking, yawning, coughing, and exercise can modify the rate and depth of respiration. So, too, can chemical factors such as changes in oxygen or carbon dioxide concentrations in the blood or fluctuations in blood pH. This is especially important in initiating breathing in a newborn. The buildup of carbon dioxide in the blood triggers the baby's first breath. The experimental sequence in Activity 6 is designed to test the relative importance of various physical and chemical factors in the process of respiration.

Activity 6

Visualizing Respiratory Variations

In this activity, you will count the respiratory rate of the subject visually by observing the movement of the chest or abdomen.

1. Record quiet breathing for 1 minute with the subject in a sitting position.

Breaths per minute: _____

2. Record the subject's breathing as he or she performs activities from the following list. Record your results on the review sheet at the end of this exercise.

talking

yawning

laughing

standing

doing a math problem
   (concentrating)

swallowing water

coughing

lying down

running in place

3. Without recording, have the subject breathe normally for 2 minutes, then inhale deeply and hold his or her breath for as long as he or she can.

Breath-holding interval: _____ sec

As the subject exhales, record the recovery period (time to return to normal breathing—usually slightly over 1 minute):

Time of recovery period: _____ sec

Did the subject have the urge to inspire *or* expire during

breath holding? _____

Without recording, repeat the above experiment, but this time exhale completely and forcefully *after* taking the deep breath.

Breath-holding interval _____ sec

Text continues on next page. →

Time of recovery period _____ sec

Did the subject have the urge to inspire *or* expire? _____

Explain the results. (Hint: The vagus nerve is the sensory nerve of the lungs and plays a role here.)

_____

_____

_____

4. During the next task, a sensation of dizziness may develop. As the carbon dioxide is washed out of the blood by hyperventilation, the blood pH increases, leading to a decrease in blood pressure and reduced cerebral circulation.

⚠ If you have a history of dizzy spells or a heart condition, do not perform this task.

The subject may experience a lack of desire to breathe after forced breathing is stopped. If the period of breathing cessation—apnea—is extended, cyanosis of the lips may occur.

Have the subject hyperventilate (breathe deeply and forcefully at the rate of 1 breath/4 sec) for about 30 sec.

Is the respiratory rate after hyperventilation faster *or* slower than during normal quiet breathing?

_____

5. Repeat the hyperventilation step. After hyperventilation, the subject is to hold his or her breath as long as possible.

Breath-holding interval: _____

Can the breath be held for a longer or shorter time after hyperventilating?

_____

6. Without recording, have the subject breathe into a paper bag for 3 minutes, then record his or her breathing movements.

⚠ During the bag-breathing exercise, the subject's partner should watch the subject carefully.

Is the breathing rate faster *or* slower than that recorded during normal quiet breathing?

_____

After hyperventilating? _____.

7. Run in place for 2 minutes, and then have your partner determine how long you can hold your breath.

Breath-holding interval: _____ sec

8. To prove that respiration has a marked effect on circulation, conduct the following test. Have your lab partner record the rate and relative force of your radial pulse before you begin.

Rate: _____ beats/min   Relative force: _____

Inspire forcibly. Immediately close your mouth and nose to retain the inhaled air, and then make a forceful and prolonged expiration. Your lab partner should observe and record the condition of the blood vessels of your neck and face, and again immediately palpate the radial pulse.

Observations: _____

Radial pulse: _____ beats/min   Relative force: _____

Explain the changes observed. _____

_____

_____

⚠ Dispose of the paper bag in the autoclave bag. Observation of the test results should enable you to determine which chemical factor, carbon dioxide or oxygen, has the greatest effect on modifying the respiratory rate and depth.

Role of the Respiratory System in Acid-Base Balance of Blood

Blood pH must be relatively constant for the cells of the body to function optimally. The carbonic acid–bicarbonate buffer system of the blood is extremely important because it helps stabilize arterial blood pH at 7.4 ± 0.02.

When carbon dioxide diffuses into the blood from the tissue cells, much of it enters the red blood cells, where it combines with water to form carbonic acid (**Figure 37.15**):

$$H_2O + CO_2 \xrightarrow[\text{enzyme present in RBC}]{\text{carbonic anhydrase}} H_2CO_3 \text{ (carbonic acid)}$$

Some carbonic acid is also formed in the plasma, but that reaction is very slow because of the lack of the carbonic anhydrase enzyme. Shortly after it forms, carbonic acid dissociates to release bicarbonate (HCO_3^-) and hydrogen ions (H^+). The hydrogen ions that remain in the cells are neutralized when they combine with hemoglobin molecules. If they were not neutralized, the intracellular pH would become very acidic as H^+ ions accumulated. The bicarbonate ions diffuse out of the red blood cells into the plasma, where they become part of the carbonic acid–bicarbonate buffer system. As HCO_3^- follows its concentration gradient into the plasma, an electrical imbalance develops in the RBCs that draws Cl^- into them from the plasma. This exchange phenomenon is called the *chloride shift*.

Acids (more precisely, H^+) released into the blood by the body cells tend to lower the pH of the blood and to cause it to

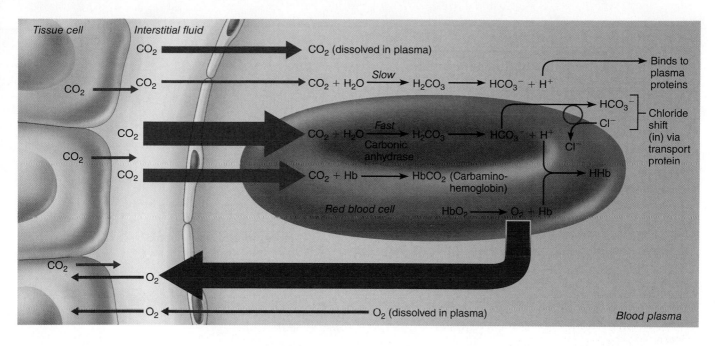

Figure 37.15 Oxygen release and carbon dioxide pickup at the tissues.

become acidic. On the other hand, basic substances that enter the blood tend to cause the blood to become more alkaline and the pH to rise. Both of these tendencies are resisted in large part by the carbonic acid–bicarbonate buffer system. If the H^+ concentration in the blood begins to increase, the H^+ ions combine with bicarbonate ions to form carbonic acid (a weak acid that does not tend to dissociate at physiological or acid pH) and are thus removed.

$$H^+ + HCO_3^- \rightarrow H_2CO_3$$

Likewise, as blood H^+ concentration drops below what is desirable and blood pH rises, H_2CO_3 dissociates to release bicarbonate ions and H^+ ions to the blood.

$$H_2CO_3 \rightarrow H^+ + HCO_3^-$$

The released H^+ lowers the pH again. The bicarbonate ions, being *weak* bases, are poorly functional under alkaline conditions and have little effect on blood pH unless and until blood pH drops toward acid levels.

In the case of excessively slow or shallow breathing (hypoventilation) or fast deep breathing (hyperventilation), the amount of carbonic acid in the blood can be greatly modified—increasing dramatically during hypoventilation and decreasing substantially during hyperventilation. In either situation, if the buffering ability of the blood is inadequate, respiratory acidosis or alkalosis can result. Therefore, maintaining the normal rate and depth of breathing is important for proper control of blood pH.

Activity 7

Demonstrating the Reaction Between Carbon Dioxide (in Exhaled Air) and Water

1. Fill a beaker with 100 ml of distilled water.

2. Add 5 ml of 0.05 M NaOH and five drops of phenol red. Phenol red is a pH indicator that turns yellow in acidic solutions.

3. Blow through a straw into the solution.

What do you observe?

_____

_____

What chemical reaction is taking place in the beaker?

_____

_____

4. Discard the straw in the autoclave bag.

37

Activity 8

Observing the Operation of Standard Buffers

1. A **buffer** is a molecule or molecular system that stabilizes the pH of a solution. To observe the action of a buffer system, obtain five 250-ml beakers and a wash bottle containing distilled water. Set up the following experimental samples:

Beaker 1:
(150 ml distilled water)   pH _____

Beaker 2:
(150 ml distilled water and
1 drop concentrated HCl)   pH _____

Beaker 3:
(150 ml distilled water and
1 drop concentrated NaOH)   pH _____

Beaker 4:
(150 ml standard buffer solution
[pH 7] and 1 drop concentrated HCl)   pH _____

Beaker 5:
(150 ml standard buffer solution
[pH 7] and 1 drop concentrated NaOH)   pH _____

2. Using a pH meter standardized with a buffer solution of pH 7, determine the pH of the contents of each beaker and record above. After *each and every* pH recording, turn the pH meter switch to **STANDBY,** and rinse the electrodes thoroughly with a stream of distilled water from the wash bottle.

3. Add 3 more drops of concentrated HCl to beaker 4, stir,

and record the pH:_____

4. Add 3 more drops of concentrated NaOH to beaker 5, stir,

and record the pH:_____

How successful was the buffer solution in resisting pH changes when a strong acid (HCl) or a strong base (NaOH) was added?

_____

Activity 9

Exploring the Operation of the Carbonic Acid–Bicarbonate Buffer System

To observe the ability of the carbonic acid–bicarbonate buffer system of blood to resist pH changes, perform the following simple experiment.

1. Obtain two small beakers (50 ml), animal plasma, graduated cylinder, glass stirring rod, and a dropper bottle of 0.01 *M* HCl. Using the pH meter standardized with the buffer solution of pH 7.0, measure the pH of the animal plasma. Use only enough plasma to allow immersion of the electrodes and measure the volume used carefully.

pH of the animal plasma: _____

2. Add 2 drops of the 0.01 *M* HCl solution to the plasma; stir and measure the pH again.

pH of plasma plus 2 drops of HCl: _____

3. Turn the pH meter switch to **STANDBY,** rinse the electrodes, and then immerse them in a quantity of distilled

water (pH 7) exactly equal to the amount of animal plasma used. Measure the pH of the distilled water.

pH of distilled water: _____

4. Add 2 drops of 0.01 *M* HCl, swirl, and measure the pH again.

pH of distilled water plus the two drops of HCl: _____

Is the plasma a good buffer? _____

What component of the plasma carbonic acid–bicarbonate buffer system was acting to counteract a change in pH when HCl was added?

_____

EXERCISE

37

REVIEW SHEET

Respiratory System Physiology

Name _____ Lab Time/Date _____

Mechanics of Respiration

1. For each of the following cases, check the column appropriate to your observations on the operation of the model lung.

Change	Diaphragm pushed up		Diaphragm pulled down	
	Increased	Decreased	Increased	Decreased
In internal volume of the bell jar (thoracic cage)				
In internal pressure				
In the size of the balloons (lungs)				

2. Base your answers to the following on your observations in question 1.

   Under what internal conditions does air tend to flow into the lungs? _____

   _____

   Under what internal conditions does air tend to flow out of the lungs? Explain why this is so. _____

   _____

3. Activation of the diaphragm and the external intercostal muscles begins the inspiratory process. What effect does

   contraction of these muscles have on thoracic volume, and how is this accomplished? _____

   _____

   _____

4. What was the approximate increase in diameter of chest circumference during a quiet inspiration? _____ cm

   During forced inspiration? _____ cm

   What temporary physiological advantage is created by the substantial increase in chest circumference during forced

   inspiration? _____

   _____

5. The presence of a partial vacuum between the pleural membranes is integral to normal breathing movements. What
   would happen if an opening were made into the chest cavity, as with a puncture wound?

   _____

   What must be done to treat this condition medically? _____

Respiratory Sounds

6. Which of the respiratory sounds is heard during both inspiration and expiration? _____

   Which is heard primarily during inspiration? _____

7. Where did you best hear the vesicular respiratory sounds? _____

Respiratory Volumes and Capacities—Spirometry or BIOPAC®

8. Write the respiratory volume term and the normal value that is described by the following statements.

   Volume of air present in the lungs after a forceful expiration: _____

   Volume of air that can be expired forcibly after a normal expiration: _____

   Volume of air that is breathed in and out during a normal respiration: _____

   Volume of air that can be inspired forcibly after a normal inspiration: _____

   Volume of air corresponding to TV + IRV + ERV: _____

9. For the spirometer activities, record experimental respiratory volumes as determined in the laboratory. (Corrected values and FEV_1 are for the recording spirometer only.)

   Average TV: _____ ml

   Corrected value for TV: _____ ml

   Average IRV: _____ ml

   Corrected value for IRV: _____ ml

   MRV: _____ ml/min

   Average ERV: _____ ml

   Corrected value for ERV: _____ ml

   Average VC: _____ ml

   Corrected value for VC: _____ ml

   % predicted VC: _____ %

   FEV_1: _____ % FVC

   For the BIOPAC® activity, record the following experimental respiratory volumes as determined in the laboratory.

   TV: _____ L   IRV: _____ L

   ERV: _____ L   VC: _____ L

WHY THIS MATTERS 10. With incentive spirometry, the patient is instructed to take a normal quiet breath and then inhale from the mouthpiece of the incentive spirometer as slowly and completely as possible. What respiratory volume/

   capacity measurement do you think this maneuver approximates? _____

   _____

11. Explain how you would obtain this respiratory volume/capacity value (from question 10) when using the

   nonrecording dry spirometer. _____

   _____

   _____

12. Which respiratory ailments can respiratory volume tests be used to detect?

_____

_____

Factors Influencing Rate and Depth of Respiration

13. Where are the neural control centers of respiratory rhythm? _____ and _____

For questions 14–21, use your Activity 6 data.

14. In your data, what was the rate of quiet breathing?

Initial testing _____ breaths/min

Test performed	Observations (breaths per minute)
Talking	
Yawning	
Laughing	
Standing	
Concentrating	
Swallowing water	
Coughing	
Lying down	
Running in place	

15. Record student data below.

Breath-holding interval after a deep Inhalation: _____ sec   length of recovery period: _____ sec

Breath-holding interval after a forceful expiration: _____ sec   length of recovery period: _____ sec

After breathing quietly and taking a deep breath (which you held), was your urge to inspire *or* expire? _____

After exhaling and then holding one's breath, was the desire for inspiration *or* expiration? _____

Explain these results. (Hint: What reflex is involved here?) _____

_____

_____

16. Observations after hyperventilation: _____

_____

17. Breath-holding interval after hyperventilation: _____ sec

    Why does hyperventilation produce apnea or a reduced respiratory rate? _____

    _____

18. Observations for rebreathing air: _____

    _____

    Why does rebreathing air produce an increased respiratory rate? _____

    _____

19. What was the effect of running in place (exercise) on the duration of breath holding? _____

    Explain this effect. _____

    _____

20. Record student data from the test illustrating the effect of respiration on circulation.

    Radial pulse before beginning test: _____ /min   Radial pulse after testing: _____ /min

    Relative pulse force before beginning test: _____   Relative force of radial pulse after testing: _____

    Condition of neck and facial veins after testing: _____

    Explain these data. _____

    _____

21. Do the following factors generally increase (indicate ↑) or decrease (indicate ↓) the respiratory rate and depth?

    increase in blood CO_2: _____   increase in blood pH: _____

    decrease in blood O_2: _____   decrease in blood pH: _____

    Did it appear that CO_2 or O_2 had a more marked effect on modifying the respiratory rate? _____

22. Where are sensory receptors sensitive to changes in blood pressure located? _____

    _____

23. Where are sensory receptors sensitive to changes in O_2 levels in the blood located? _____

    _____

24. What is the primary factor that initiates breathing in a newborn infant? _____

    _____

25. Which, if any, of the measurable respiratory volumes would likely be increased in a person who is cardiovascularly fit, such as a runner or a swimmer?

_____

_____

Which, if any, of the measurable respiratory volumes would likely be decreased in a person who has smoked a lot for over 20 years?

_____

_____

26. Blood CO_2 levels and blood pH are related. When blood CO_2 levels increase, does the pH increase or decrease?

_____ Explain why. _____

_____

Role of the Respiratory System in Acid-Base Balance of Blood

27. Define *buffer*. _____ _____

28. How successful was the laboratory buffer (pH 7) in resisting changes in pH when the acid was added? _____

_____

When the base was added? _____

How successful was the buffer in resisting changes in pH when the additional drops of the acid and base were added

to the original samples? _____

29. What buffer system operates in blood plasma? _____

Which component of the buffer system resists a *drop* in pH? _____ Which resists a *rise* in pH? _____

30. Explain how the carbonic acid–bicarbonate buffer system of the blood operates. _____

_____

_____

_____

31. What happened when the carbon dioxide in exhaled air mixed with water? _____

_____

What role does exhalation of carbon dioxide play in maintaining relatively constant blood pH? _____

_____

_____

Objectives

- ☐ State the overall function of the digestive system.

- ☐ Describe the general histologic structure of the alimentary canal wall, and identify the following structures on an appropriate image of the wall: mucosa, submucosa, muscularis externa, and serosa or adventitia.

- ☐ Identify on a model or image the organs of the alimentary canal, and name their subdivisions, if any.

- ☐ Describe the general function of each of the digestive system organs or structures.

- ☐ List and explain the specializations of the structure of the stomach and small intestine that contribute to their functional roles.

- ☐ Name and identify the accessory digestive organs, listing a function for each.

- ☐ Describe the anatomy of the generalized tooth, and name the human deciduous and permanent teeth.

- ☐ List the major enzymes or enzyme groups produced by the salivary glands, stomach, small intestine, and pancreas.

- ☐ Recognize microscopically or in an image the histologic structure of the following organs:

  small intestine   tooth   liver
  salivary glands   stomach

Materials

- Dissectible torso model
- Anatomical chart of the human digestive system
- Prepared slides of the liver and mixed salivary glands; of longitudinal sections of the gastroesophageal junction and a tooth; and of cross sections of the stomach, duodenum, ileum, and large intestine
- Compound microscope
- Three-dimensional model of a villus (if available)

Text continues on next page. →

MasteringA&P®

For related exercise study tools, go to the Study Area of **MasteringA&P**. There you will find:

- Practice Anatomy Lab **PAL**
- A&PFlix **A&PFlix**
- PhysioEx **PEx**
- Practice quizzes, Histology Atlas, eText, Videos, and more!

Pre-Lab Quiz

1. The digestive system:
   a. eliminates undigested food
   b. provides the body with nutrients
   c. provides the body with water
   d. all of the above

2. Circle the correct underlined term. <u>Digestion</u> / <u>Absorption</u> occurs when small molecules pass through epithelial cells into the blood for distribution to the body cells.

3. The _____ abuts the lumen of the alimentary canal and consists of epithelium, lamina propria, and muscularis mucosae.
   a. mucosa   b. serosa   c. submucosa

4. Circle the correct underlined term. Approximately 25 cm long, the <u>esophagus</u> / <u>alimentary canal</u> conducts food from the pharynx to the stomach.

5. Wavelike contractions of the digestive tract that propel food along are called:
   a. digestion   c. ingestion
   b. elimination   d. peristalsis

6. The _____ is located on the left side of the abdominal cavity and is hidden by the liver and diaphragm.
   a. gallbladder   c. small intestine
   b. large intestine   d. stomach

Text continues on next page. →

- Jaw model or human skull
- Three-dimensional model of liver lobules (if available)

✂ For instructions on animal dissections, see the dissection exercises (starting on p. 705) in the cat and fetal pig editions of this manual.

7. Circle True or False. Nearly all nutrient absorption occurs in the small intestine.

8. Circle the correct underlined term. The <u>ascending colon</u> / <u>descending colon</u> traverses down the left side of the abdominal cavity and becomes the sigmoid colon.

9. A tooth consists of two major regions, the crown and the:
   a. dentin   c. gingiva
   b. enamel   d. root

10. Located inferior to the diaphragm, the _____ is the largest gland in the body.
    a. gallbladder   c. pancreas
    b. liver   d. thymus

The **digestive system** provides the body with the nutrients, water, and electrolytes essential for health. The organs of this system ingest, digest, and absorb food and eliminate the undigested remains as feces.

The digestive system consists of a hollow tube extending from the mouth to the anus, into which various accessory organs or glands empty their secretions (**Figure 38.1**). For ingested food to become available to the body cells, it must first be broken down into its smaller diffusible molecules—a process called **digestion.** The digested end products can then pass through the epithelial cells lining the tract into the blood for distribution to the body cells—a process termed **absorption.**

The organs of the digestive system are traditionally separated into two major groups: the **alimentary canal,** or **gastrointestinal (GI) tract,** and the **accessory digestive organs.** The alimentary canal consists of the mouth, pharynx, esophagus, stomach, and small and large intestines. The accessory structures include the teeth, which physically break down foods, and the salivary glands, gallbladder, liver, and pancreas, which secrete their products into the alimentary canal.

General Histological Plan of the Alimentary Canal

From the esophagus to the anal canal, the basic structure of the alimentary canal is similar. As we study individual parts of the alimentary canal, we will note how this basic plan is modified to provide the unique digestive functions of each subsequent organ.

Essentially the alimentary canal wall has four basic layers or tunics. From the lumen outward, these are the *mucosa,* the *submucosa,* the *muscularis externa,* and either a *serosa* or *adventitia* (**Figure 38.2**, p. 580). Each of these layers has a predominant tissue type and a specific function in the digestive process.

Table 38.1 on p. 581 summarizes the characteristics of the layers of the wall of the alimentary canal.

Organs of the Alimentary Canal

Activity 1

Identifying Alimentary Canal Organs

The sequential pathway and fate of food as it passes through the alimentary canal are described in the next sections. Identify each structure in Figure 38.1 and on the torso model or anatomical chart of the digestive system as you work.

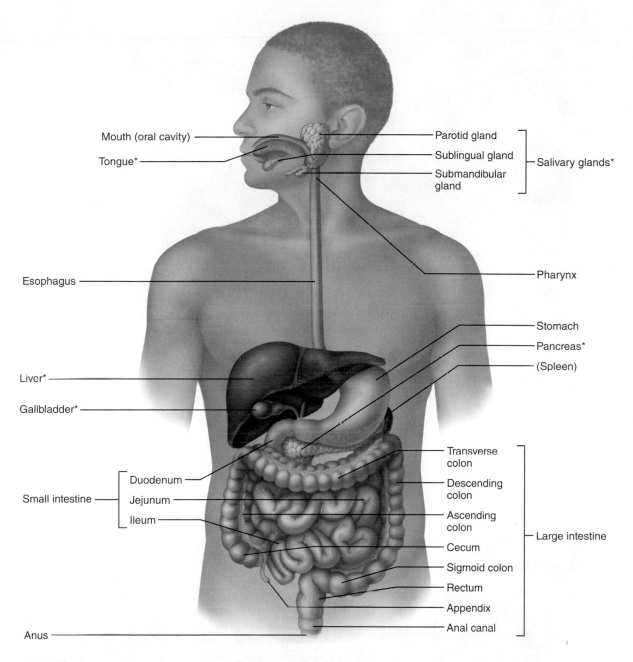

Figure 38.1 The human digestive system: alimentary tube and accessory organs. Organs marked with asterisks are accessory organs. Those without asterisks are alimentary canal organs (except the spleen, an organ of the lymphatic system).

Oral Cavity or Mouth

Food enters the digestive tract through the **oral cavity,** or **mouth (Figure 38.3,** p. 580). Within this mucous membrane–lined cavity are the gums, teeth, tongue, and openings of the ducts of the salivary glands. The **lips (labia)** protect the opening of the chamber anteriorly, the **cheeks** form its lateral walls, and the **palate,** its roof. The anterior portion of the palate is referred to as the **hard palate** because the palatine processes of the maxillae and horizontal plates of the palatine bones underlie it. The posterior **soft palate** is a fibromuscular structure that is unsupported by bone. The

uvula, a fingerlike projection of the soft palate, extends inferiorly from its posterior margin. The floor of the oral cavity is occupied by the muscular **tongue,** which is largely supported by the *mylohyoid muscle* (**Figure 38.4,** p. 581) and attaches to the hyoid bone, mandible, styloid processes, and pharynx. A membrane called the **lingual frenulum** secures the inferior midline of the tongue to the floor of the mouth. The space between the teeth and cheeks (or lips) is the **oral vestibule;** the area that lies within the teeth and gums is the **oral cavity proper.** (The teeth and gums are discussed in more detail on pp. 588–590.)

On each side of the mouth at its posterior end are masses of lymphoid tissue, the **palatine tonsils** (see Figure 38.3). Each

38

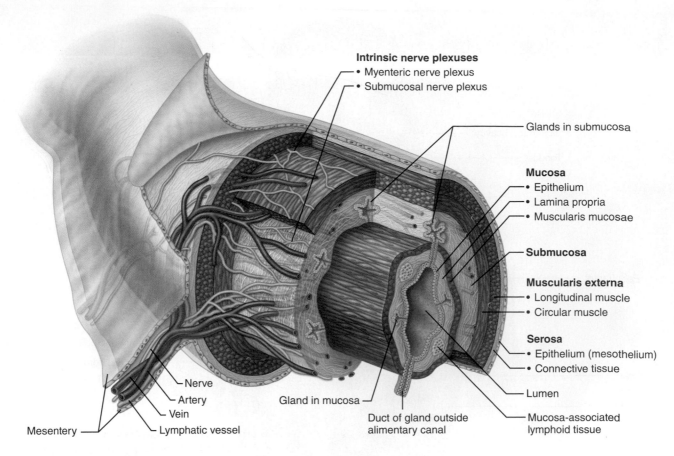

Figure 38.2 Basic structural pattern of the alimentary canal wall.

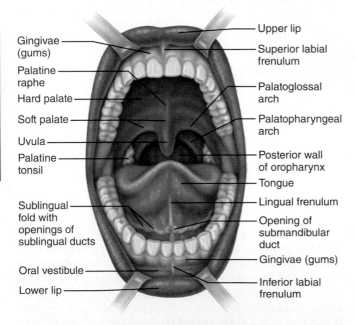

Figure 38.3 Anterior view of the oral cavity.

lies in a concave area bounded anteriorly and posteriorly by membranes, the **palatoglossal arch** and the **palatopharyngeal arch,** respectively. Another mass of lymphoid tissue, the **lingual tonsil** (see Figure 38.4), covers the base of the tongue,

posterior to the oral cavity proper. The tonsils, in common with other lymphoid tissues, are part of the body's defense system.

Very often in young children, the palatine tonsils become inflamed and enlarge, partially blocking the entrance to the pharynx posteriorly and making swallowing difficult and painful. This condition is called **tonsillitis. ✚**

Three pairs of salivary glands duct their secretion, saliva, into the oral cavity. One component of saliva, salivary amylase, begins the digestion of starchy foods within the oral cavity. (The salivary glands are discussed in more detail on p. 590.)

As food enters the mouth, it is mixed with saliva and masticated (chewed). The cheeks and lips help hold the food between the teeth during mastication, and the highly mobile tongue manipulates the food during chewing and initiates swallowing. Thus the mechanical and chemical breakdown of food begins before the food has left the oral cavity.

Pharynx

When the tongue initiates swallowing, the food passes posteriorly into the pharynx, a common passageway for food, fluid, and air (see Figure 38.4). The pharynx is subdivided anatomically into three parts—the **nasopharynx** (behind the nasal cavity), the **oropharynx** (behind the oral cavity extending from the soft palate to the epiglottis), and the **laryngopharynx** (extending from the epiglottis to the base of the larynx).

The walls of the pharynx consist largely of two layers of skeletal muscle: an inner layer of longitudinal muscle and an outer layer of circular constrictor muscles. Together these initiate wavelike contractions that propel the food inferiorly

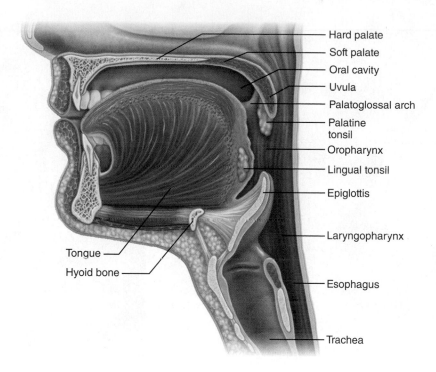

Figure 38.4 Sagittal view of the head showing oral cavity and pharynx.

into the esophagus. The mucosa of the oropharynx and laryngopharynx, like that of the oral cavity, contains a protective stratified squamous epithelium.

Esophagus

The **esophagus,** or gullet, extends from the pharynx through the diaphragm to the gastroesophageal sphincter in the superior aspect of the stomach. Approximately 25 cm long in humans, it is essentially a food passageway that conducts food to the stomach in a wavelike peristaltic motion. The esophagus has no digestive or absorptive function. The walls at its superior end contain skeletal muscle, which is replaced by smooth muscle in the area nearing the stomach. The **gastroesophageal sphincter,** a slight thickening of the smooth muscle layer at the esophagus-stomach junction, controls food passage into the stomach (Figure 38.6c).

Table 38.1	**Alimentary Canal Wall Layers (Figure 38.1)**		
Layer	**Subdivision of the layer**	**Tissue type**	**Major functions (generalized for the layer)**
Mucosa	Epithelium	Stratified squamous epithelium in the mouth, esophagus, and anus; simple columnar epithelium in the remainder of the canal	Secretion of mucus, digestive enzymes, and hormones; absorption of end products into the blood; protection against infectious disease.
	Lamina propria	Areolar connective tissue with blood vessels; many lymphoid follicles, especially as tonsils and mucosa-associated lymphoid tissue (MALT)	
	Muscularis mucosae	A thin layer of smooth muscle	
Submucosa	N/A	Areolar and dense irregular connective tissue containing blood vessels, lymphatic vessels, and nerve fibers (submucosal nerve plexus)	Blood vessels absorb and transport nutrients. Elastic fibers help maintain the shape of each organ.
Muscularis externa	Circular layer	Inner layer of smooth muscle	Segmentation and peristalsis of digested food along the tract are regulated by the myenteric nerve plexus.
	Longitudinal layer	Outer layer of smooth muscle	
Serosa* (visceral peritoneum)	Connective tissue	Areolar connective tissue	Reduces friction as the digestive system organs slide across one another.
	Epithelium (mesothelium)	Simple squamous epithelium	

*Since the esophagus is outside the peritoneal cavity, the serosa is replaced by an adventitia made of aerolar connective tissue that binds the esophagus to surrounding tissues.

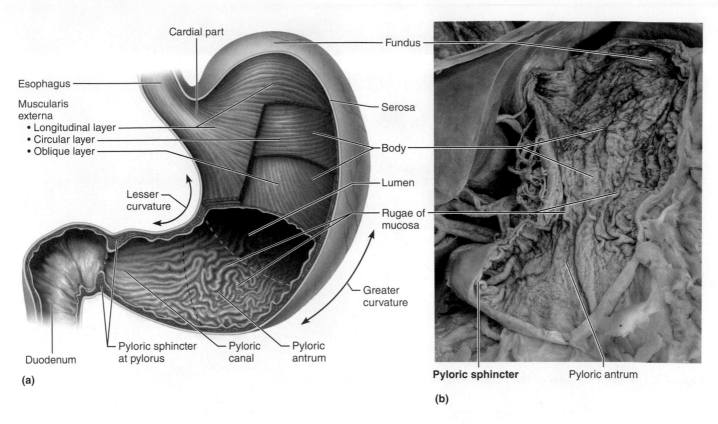

(a)

(b)

Pyloric sphincter   Pyloric antrum

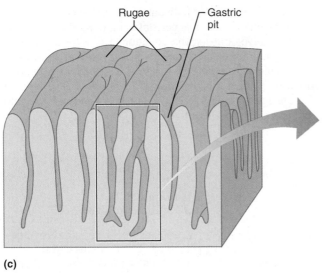

(c)

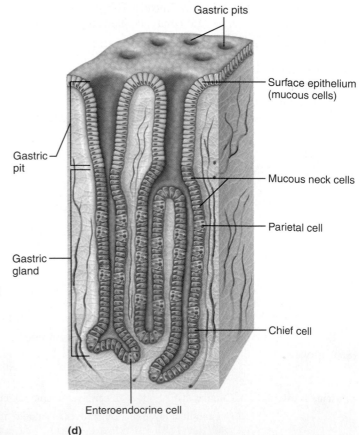

(d)

Figure 38.5 Anatomy of the stomach. (a) Gross internal and external anatomy. **(b)** Photograph of internal aspect of stomach. **(c, d)** Section of the stomach wall showing rugae and gastric pits.

Stomach

The **stomach** (**Figure 38.5**) is on the left side of the abdominal cavity and is hidden by the liver and diaphragm. The stomach is made up of several regions, summarized in **Table 38.2** on p. 584. *Mesentery* is the general term that refers to the double layer of the peritoneum that extends from the digestive organs to the body wall. There are two mesenteries, the **greater omentum** and **lesser omentum,** that connect to the stomach. The lesser omentum extends from the liver to the **lesser curvature** of the stomach. The greater omentum extends from the **greater curvature** of the stomach, reflects downward, and covers most of the abdominal organs in an apronlike fashion. (Figure 38.7 on p. 585 illustrates the omenta as well as the other peritoneal attachments of the abdominal organs.)

The stomach is a temporary storage region for food as well as a site for mechanical and chemical breakdown of food. It contains a third (innermost) *obliquely* oriented layer of smooth muscle in its muscularis externa that allows it to churn, mix, and pummel the food, physically reducing it to smaller fragments. **Gastric glands** of the mucosa secrete hydrochloric acid (HCl) and hydrolytic enzymes. The *mucosal glands* also secrete a viscous mucus that helps prevent the stomach itself from being digested by the proteolytic enzymes. Most digestive activity occurs in the pyloric part of the stomach. After the food is processed in the stomach, it resembles a creamy mass called **chyme,** which enters the small intestine through the pyloric sphincter.

Activity 2

Studying the Histologic Structure of the Stomach and the Gastroesophageal Junction

1. Stomach: View the stomach slide first. Refer to **Figure 38.6a** as you scan the tissue under low power to locate the muscularis externa; then move to high power to more closely examine this layer. Try to pick out the three smooth muscle layers. How does the extra oblique layer of smooth muscle found in the stomach correlate with the stomach's churning movements?

_____

_____

_____

Identify the gastric glands and the gastric pits (see Figures 38.5 and 38.6b). If the section is taken from the stomach fundus and is differentially stained, you can identify, in the gastric glands, the blue-staining **chief cells,** which produce pepsinogen, and the red-staining **parietal cells,** which secrete HCl. The enteroendocrine cells that release hormones are indistinguishable. Draw a small section of the stomach wall, and label it appropriately.

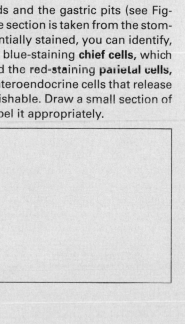

Text continues on next page. →

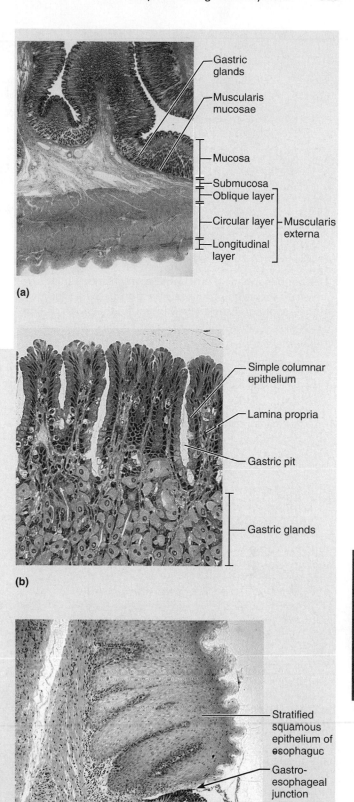

(a)

— Gastric glands

— Muscularis mucosae

— Mucosa

— Submucosa
— Oblique layer
— Circular layer ⎱ Muscularis externa
— Longitudinal layer

(b)

— Simple columnar epithelium

— Lamina propria

— Gastric pit

— Gastric glands

(c)

— Stratified squamous epithelium of esophagus

— Gastroesophageal junction

— Simple columnar epithelium of stomach

Figure 38.6 Histology of selected regions of the stomach and gastroesophageal junction. (a) Stomach wall (12×). **(b)** Gastric pits and glands (130×). **(c)** Gastroesophageal junction, longitudinal section (60×).

38

Table 38.2	Parts of the Stomach (Figure 38.5)
Structure	**Description**
Cardial part (cardia)	The area surrounding the cardial orifice through which food enters the stomach
Fundus	The dome-shaped area that is located superior and lateral to the cardial part
Body	Midportion of the stomach and largest region
Pyloric part:	Funnel-shaped pouch that forms the distal stomach
Pyloric antrum	Wide superior portion of the pyloric part
Pyloric canal	Narrow tubelike portion of the pyloric part
Pylorus	Distal end of the pyloric part that is continuous with the small intestine
Pyloric sphincter	Valve that controls the emptying of the stomach into the small intestine

2. Gastroesophageal junction: Scan the slide under low power to locate the mucosal junction between the end of the esophagus and the beginning of the stomach, the gastroesophageal junction. Draw a small section of the junction and label it appropriately.

Compare your observations to Figure 38.6c. What is the functional importance of the epithelial differences seen in the two organs?

_____

_____

_____

Small Intestine

The **small intestine** is a convoluted tube, 6 to 7 meters (about 20 feet) long in a cadaver but only about 2 m (6 feet) long during life because of its muscle tone. It extends from the pyloric sphincter to the ileocecal valve. The small intestine is suspended by a double layer of peritoneum, the fan-shaped **mesentery,** from the posterior abdominal wall (**Figure 38.7**), and it lies, framed laterally and superiorly by the large intestine, in the abdominal cavity. The small intestine has three subdivisions (see Figure 38.1):

1. The **duodenum** extends from the pyloric sphincter for about 25 cm (10 inches) and curves around the head of the pancreas; most of the duodenum lies in a retroperitoneal position.

2. The **jejunum,** continuous with the duodenum, extends for 2.5 m (about 8 feet). Most of the jejunum occupies the umbilical region of the abdominal cavity.

3. The **ileum,** the terminal portion of the small intestine, is about 3.6 m (12 feet) long and joins the large intestine at the **ileocecal valve.** It is located inferiorly and somewhat to the right in the abdominal cavity, but its major portion lies in the hypogastric region.

In the small intestine, enzymes from two sources complete the digestion process: **brush border enzymes,** which are hydrolytic enzymes bound to the microvilli of the columnar epithelial cells; and, more important, enzymes produced by the pancreas and ducted into the duodenum largely via the **main pancreatic duct.** Bile (formed in the liver) also enters the duodenum via the **bile duct** in the same area. At the duodenum, the ducts join to form the bulblike **hepatopancreatic ampulla** and empty their products into the duodenal lumen through the **major duodenal papilla,** an orifice controlled by a muscular valve called the **hepatopancreatic sphincter** (see Figure 38.15 on p. 591).

Nearly all nutrient absorption occurs in the small intestine, where three structural modifications increase the absorptive surface of the mucosa: the microvilli, villi, and circular folds (**Figure 38.8**, p. 586).

- **Microvilli:** Microscopic projections of the surface plasma membrane of the columnar epithelial lining cells of the mucosa.

- **Villi:** Fingerlike projections of the mucosa tunic that give it a velvety appearance and texture.

- **Circular folds:** Deep, permanent folds of the mucosa and submucosa layers that force chyme to spiral through the intestine, mixing it and slowing its progress. These structural modifications decrease in frequency and size toward the end of the small intestine. Any residue remaining undigested and unabsorbed at the terminus of the small intestine enters the large intestine through the ileocecal valve. The amount of lymphoid tissue in the submucosa of the small intestine (especially the aggregated lymphoid nodules called **Peyer's patches, Figure 38.9b**, p. 587) increases along the length of the small intestine and is very apparent in the ileum.

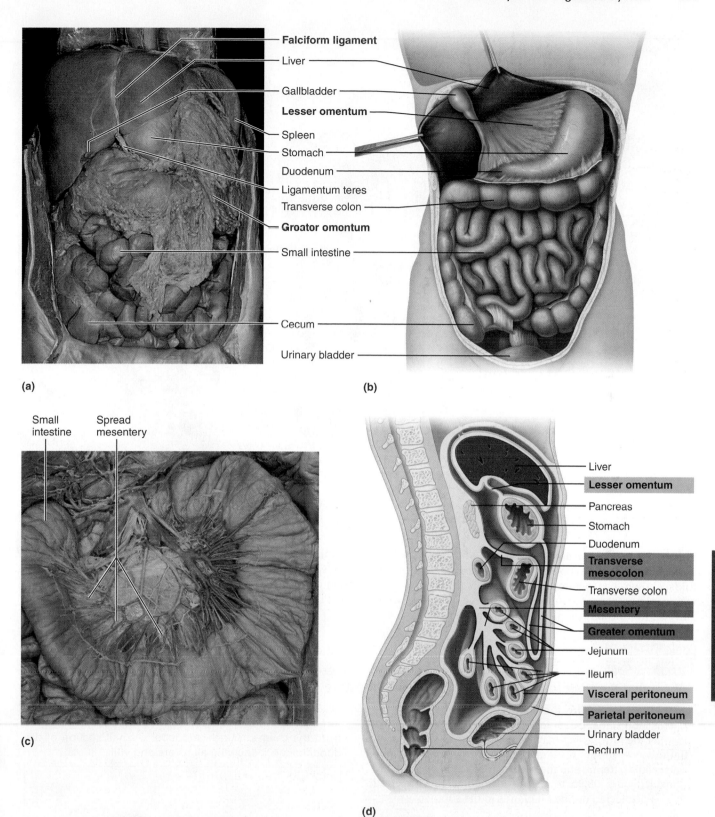

Figure 38.7 Peritoneal attachments of the abdominal organs. Superficial anterior views of abdominal cavity: **(a)** photograph with the greater omentum in place and **(b)** diagram showing greater omentum removed and liver and gallbladder reflected superiorly. **(c)** Mesentery of the small intestine. **(d)** Sagittal view of a male torso. Mesentery labels appear in colored boxes.

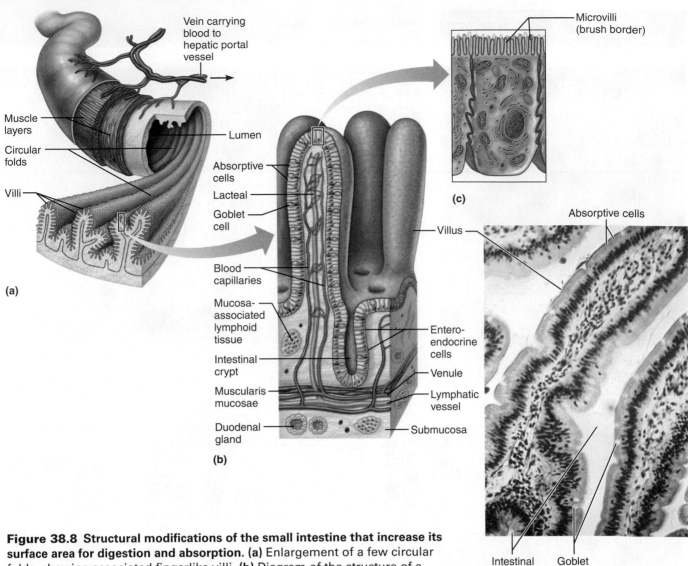

Figure 38.8 Structural modifications of the small intestine that increase its surface area for digestion and absorption. (a) Enlargement of a few circular folds, showing associated fingerlike villi. **(b)** Diagram of the structure of a villus. **(c)** Two absorptive cells that exhibit microvilli on their free (luminal) surface. **(d)** Photomicrograph of the mucosa showing villi (105×).

38

Activity 3

Observing the Histologic Structure of the Small Intestine

1. Duodenum: Secure the slide of the duodenum to the microscope stage. Observe the tissue under low power to identify the four basic tunics of the intestinal wall—that is, the **mucosa** and its three sublayers, the **submucosa,** the **muscularis externa,** and the **serosa,** or *visceral peritoneum.* Consult Figure 38.9a to help you identify the scattered mucus-producing **duodenal glands** in the submucosa.

What type of epithelium do you see here? _____

_____

Examine the large leaflike *villi,* which increase the surface area for absorption. Notice the scattered mucus-producing goblet cells in the epithelium of the villi. Note also the **intestinal crypts** (see also Figure 38.8), invaginated areas of the mucosa between the villi containing the cells that

produce intestinal juice, a watery mucus-containing mixture that serves as a carrier fluid for absorption of nutrients from the chyme. Sketch and label a small section of the duodenal wall, showing all layers and villi.

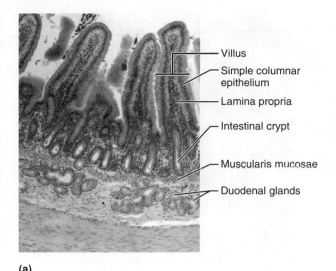

Villus

Simple columnar epithelium

Lamina propria

Intestinal crypt

Muscularis mucosae

Duodenal glands

(a)

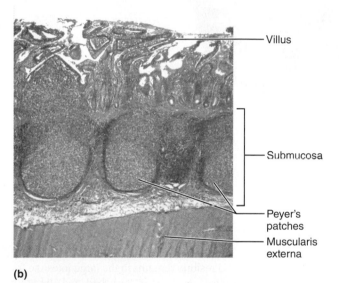

Villus

Submucosa

Peyer's patches

Muscularis externa

(b)

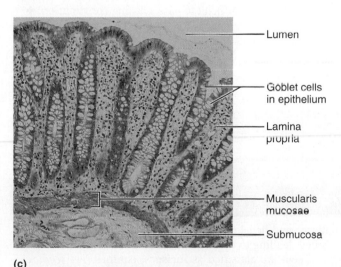

Lumen

Goblet cells in epithelium

Lamina propria

Muscularis mucosae

Submucosa

(c)

Figure 38.9 Histology of selected regions of the small and large intestines. Cross-sectional views. **(a)** Duodenum of the small intestine (95×). **(b)** Ileum of the small intestine (20×). **(c)** Large intestine (80×).

2. Ileum: The structure of the ileum resembles that of the duodenum, except that the villi are less elaborate (because most of the absorption has occurred by the time that chyme reaches the ileum). Secure a slide of the ileum to the microscope stage for viewing. Observe the villi, and identify the four layers of the wall and the large, generally spherical Peyer's patches (Figure 38.9b). What tissue type are Peyer's patches?

3. If a villus model is available, identify the following cells or regions before continuing: absorptive epithelium, goblet cells, lamina propria, the muscularis mucosae, capillary bed, and lacteal. If possible, also identify the intestinal crypts.

Large Intestine

The **large intestine** (**Figure 38.10**, p. 588) is about 1.5 m (5 feet) long and extends from the ileocecal valve to the anus. It encircles the small intestine on three sides and consists of the following subdivisions: **cecum, appendix, colon, rectum,** and **anal canal.**

The blind wormlike appendix, which hangs from the cecum, is a trouble spot in the large intestine. Since it is generally twisted, it provides an ideal location for bacteria to accumulate and multiply. Inflammation of the appendix, or appendicitis, is the result. ✚

The colon is divided into several distinct regions. The **ascending colon** travels up the right side of the abdominal cavity and makes a right-angle turn at the **right colic (hepatic) flexure** to cross the abdominal cavity as the **transverse colon.** It then turns at the **left colic (splenic) flexure** and continues down the left side of the abdominal cavity as the **descending colon,** where it takes an S-shaped course as the **sigmoid colon.** The sigmoid colon, rectum, and the anal canal lie in the pelvis anterior to the sacrum and thus are not considered abdominal cavity structures. Except for the transverse and sigmoid colons, the colon is retroperitoneal.

The anal canal terminates in the **anus,** the opening to the exterior of the body. The anal canal has two sphincters, a voluntary *external anal sphincter* composed of skeletal muscle, and an involuntary *internal anal sphincter* composed of smooth muscle. The sphincters are normally closed except during defecation, when undigested food and bacteria are eliminated from the body as feces.

In the large intestine, the longitudinal muscle layer of the muscularis externa is reduced to three longitudinal muscle bands called the **teniae coli.** Since these bands are shorter than the rest of the wall of the large intestine, they cause the wall to pucker into small pocketlike sacs called **haustra.** Fat-filled pouches of visceral peritoneum, called *epiploic appendages,* hang from the colon's surface.

The major function of the large intestine is to consolidate and propel the unusable fecal matter toward the anus and eliminate it from the body. While it does this task, it (1) provides a site where intestinal bacteria manufacture vitamins B and K; and (2) reclaims most of the remaining water from undigested food, thus conserving body water.

38

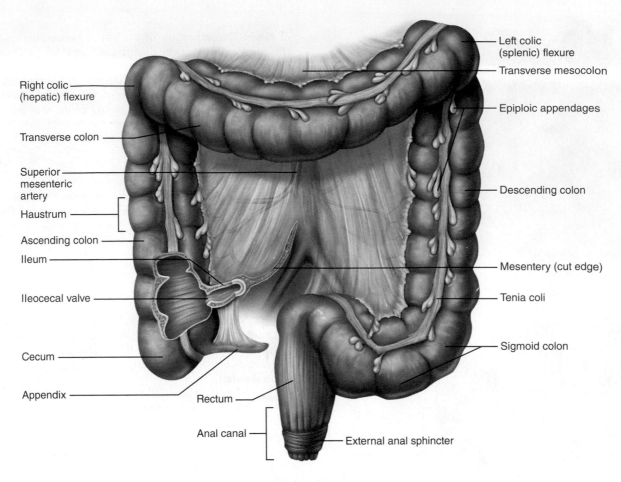

Figure 38.10 The large intestine. (Section of the cecum removed to show the ileocecal valve.)

Watery stools, or **diarrhea,** result from any condition that rushes undigested food residue through the large intestine before it has had sufficient time to absorb the water.

Conversely, when food residue remains in the large intestine for extended periods, excessive water is absorbed and the stool becomes hard and difficult to pass, causing **constipation. +**

Activity 4

Examining the Histologic Structure of the Large Intestine

Large intestine: Secure a slide of the large intestine to the microscope stage for viewing. Observe the numerous goblet cells in the epithelium (Figure 38.9c). Why do you think the large intestine produces so much mucus?

Accessory Digestive Organs

Teeth

By the age of 21, two sets of teeth have developed (**Figure 38.11**). The initial set, called the **deciduous** (or **milk**) **teeth,** normally appears between the ages of 6 months and 2½ years. The first of these to erupt are the lower central incisors. The child begins to shed the deciduous teeth around the age of 6, and a second set of teeth, the **permanent teeth,** gradually replaces them. As the deeper permanent teeth progressively

enlarge and develop, the roots of the deciduous teeth are resorbed, leading to their final shedding.

Teeth are classified as **incisors, canines** (*eye teeth, cuspids*), **premolars** (*bicuspids*), and **molars.** The incisors are chisel shaped and exert a shearing action used in biting. Canines are cone-shaped teeth used for tearing food. The premolars have two *cusps* (grinding surfaces); the molars have broad crowns with rounded cusps specialized for the fine grinding of food.

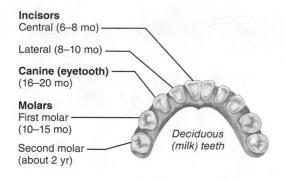

Incisors
Central (6–8 mo)
Lateral (8–10 mo)
Canine (eyetooth)
(16–20 mo)
Molars
First molar
(10–15 mo)
Second molar
(about 2 yr)

*Deciduous
(milk) teeth*

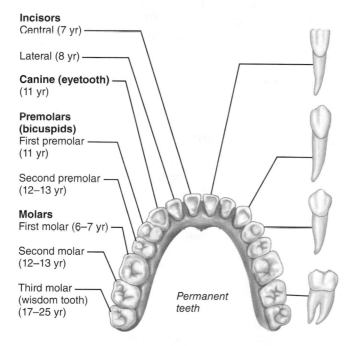

Incisors
Central (7 yr)
Lateral (8 yr)
Canine (eyetooth)
(11 yr)
**Premolars
(bicuspids)**
First premolar
(11 yr)
Second premolar
(12–13 yr)
Molars
First molar (6–7 yr)
Second molar
(12–13 yr)
Third molar
(wisdom tooth)
(17–25 yr)

*Permanent
teeth*

Figure 38.11 Human deciduous teeth and permanent teeth. (Approximate time of teeth eruption shown in parentheses.)

Dentition is described by means of a **dental formula,** which designates the numbers, types, and position of the teeth in one side of the jaw. Because tooth arrangement is bilaterally symmetrical, it is necessary to designate one only side of the jaw. The complete dental formula for the deciduous teeth from the medial aspect of each jaw and proceeding posteriorly is as follows:

$$\frac{\text{Upper teeth: 2 incisors, 1 canine, 0 premolars, 2 molars}}{\text{Lower teeth: 2 incisors, 1 canine, 0 premolars, 2 molars}} \times 2$$

This formula is generally abbreviated to read as follows:

$$\frac{2,1,0,2}{2,1,0,2} \times 2 \text{ (20 deciduous teeth)}$$

The permanent teeth are then described by the following dental formula:

$$\frac{2,1,2,3}{2,1,2,3} \times 2 \text{ (32 permanent teeth)}$$

Although 32 is designated as the normal number of permanent teeth, not everyone develops a full set. In many people, the third molars, commonly called *wisdom teeth,* never erupt.

Activity 5

Identifying Types of Teeth

Identify the four types of teeth (incisors, canines, premolars, and molars) on the jaw model or human skull.

A tooth consists of two major regions, the **crown** and the **root.** These two regions meet at the **neck** near the gum line. A longitudinal section made through a tooth shows the following basic anatomical plan (**Figure 38.12**). The crown is the superior portion of the tooth visible above the **gingiva,** or **gum,** which surrounds the tooth. The surface of the crown is covered by **enamel.** Enamel consists of 95% to 97% inorganic calcium salts and thus is heavily mineralized. The crevice between the end of the crown and the upper margin of the gingiva is referred to as the *gingival sulcus.*

That portion of the tooth embedded in the bone is the root. The outermost surface of the root is covered by **cement,** which is similar to bone in composition and less brittle than enamel. The cement attaches the tooth to the **periodontal ligament,** which holds the tooth in the tooth socket and exerts a cushioning effect. **Dentin,** which composes the bulk of the tooth, is the bonelike material interior to the enamel and cement.

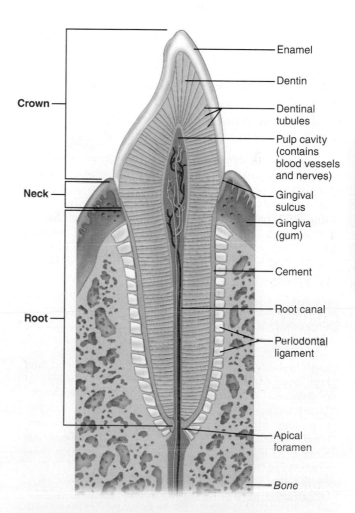

Crown
Neck
Root

Enamel
Dentin
Dentinal
tubules
Pulp cavity
(contains
blood vessels
and nerves)
Gingival
sulcus
Gingiva
(gum)
Cement
Root canal
Periodontal
ligament
Apical
foramen
Bone

Figure 38.12 Longitudinal section of human canine tooth within its bony socket (alveolus).

38

The **pulp cavity** occupies the central portion of the tooth. **Pulp,** connective tissue liberally supplied with blood vessels, nerves, and lymphatics, occupies this cavity and provides for tooth sensation and supplies nutrients to the tooth tissues. **Odontoblasts,** specialized cells in the outer margins of the pulp cavity, produce the dentin. Odontoblasts have slender processes that extend into the *dentinal tubules* of the dentin. The pulp cavity extends into distal portions of the root and becomes the **root canal.** An opening at the root apex, the **apical foramen,** provides a route of entry into the tooth for blood vessels, nerves, and other structures from the tissues beneath.

Activity 6

Studying Microscopic Tooth Anatomy

Observe a slide of a longitudinal section of a tooth, and compare your observations with the structures detailed in Figure 38.12. Identify as many of these structures as possible.

Salivary Glands

Three pairs of major **salivary glands** (see Figure 38.1) empty their secretions into the oral cavity.

Parotid glands: Large glands located anterior to the ear and ducting into the mouth over the second upper molar through the parotid duct.

Submandibular glands: Located along the medial aspect of the mandibular body in the floor of the mouth, and ducting under the tongue to the base of the lingual frenulum.

Sublingual glands: Small glands located most anteriorly in the floor of the mouth and emptying under the tongue via several small ducts.

Food in the mouth and mechanical pressure stimulate the salivary glands to secrete saliva. Saliva consists primarily of a glycoprotein called *mucin,* which moistens the food and helps to bind it together into a mass called a **bolus,** and a clear serous fluid containing the enzyme *salivary amylase.* Salivary amylase begins the digestion of starch. Parotid gland secretion is mainly serous; the submandibular is a mixed gland that produces both mucin and serous components; and the sublingual gland is a mixed gland that produces mostly mucin.

Activity 7

Examining Salivary Gland Tissue

Examine salivary gland tissue under low power and then high power to become familiar with the appearance of a glandular tissue. Notice the clustered arrangement of the cells around their ducts. The cells are basically triangular, with their pointed ends facing the duct opening. Differentiate between mucus-producing cells, which have a clear cytoplasm, and serous cells, which have granules in their cytoplasm. The serous cells often form *demilunes* (caps) around the more central mucous cells. (**Figure 38.13** may be helpful in this task.)

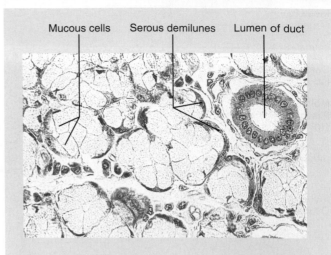

Figure 38.13 Histology of a mixed salivary gland. Sublingual gland (170×).

Liver and Gallbladder

The **liver** (see Figure 38.1), the largest gland in the body, is located inferior to the diaphragm, more to the right than the left side of the body. The human liver has four lobes and is suspended from the diaphragm and anterior abdominal wall by the **falciform ligament** (**Figure 38.14**).

The liver performs many metabolic roles. However, its digestive function is to produce bile, which leaves the liver through the **common hepatic duct** and then enters the duodenum through the **bile duct** (**Figure 38.15**). Bile has no enzymatic action but emulsifies fats, breaking up fat globules into small droplets. Without bile, very little fat digestion or absorption occurs.

When digestive activity is not occurring in the digestive tract, bile backs up into the **cystic duct** and enters the **gallbladder,** a small, green sac on the inferior surface of the liver. Bile is stored there until needed for the digestive process.

If the common hepatic or bile duct is blocked (for example, by wedged gallstones), bile is prevented from entering the small intestine, accumulates, and eventually backs up into the liver. This exerts pressure on the liver cells, and bile begins to enter the bloodstream. As the bile circulates through the body, the tissues become yellow, or jaundiced.

Blockage of the ducts is just one cause of jaundice. More often it results from actual liver problems such as **hepatitis,** (which is any inflammation of the liver,) or **cirrhosis,** a condition in which the liver is severely damaged and becomes hard and fibrous. ✚

As demonstrated by its highly organized anatomy, the liver (**Figure 38.16**, p. 592) is very important in the initial processing of the nutrient-rich blood draining the digestive organs. Its structural and functional units are called **lobules.** Each lobule is a basically hexagonal structure consisting of cordlike arrays of **hepatocytes** or *liver cells,* which radiate outward from a central vein running upward in the longitudinal axis of the lobule. At each of the six corners of the lobule is a **portal triad,** so named because three basic structures are always present there: a *portal arteriole* (a branch of the *hepatic artery,* the functional blood supply of the liver), a *portal venule* (a branch of the *hepatic portal vein* carrying nutrient-rich blood from the digestive viscera), and a *bile duct.* Between the liver cells are blood-filled spaces, or **sinusoids,** through which

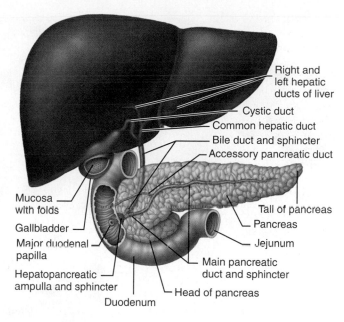

Right and left hepatic ducts of liver
Cystic duct
Common hepatic duct
Bile duct and sphincter
Accessory pancreatic duct
Tail of pancreas
Pancreas
Jejunum
Main pancreatic duct and sphincter
Head of pancreas
Duodenum
Hepatopancreatic ampulla and sphincter
Major duodenal papilla
Gallbladder
Mucosa with folds

Figure 38.15 Ducts of accessory digestive organs.

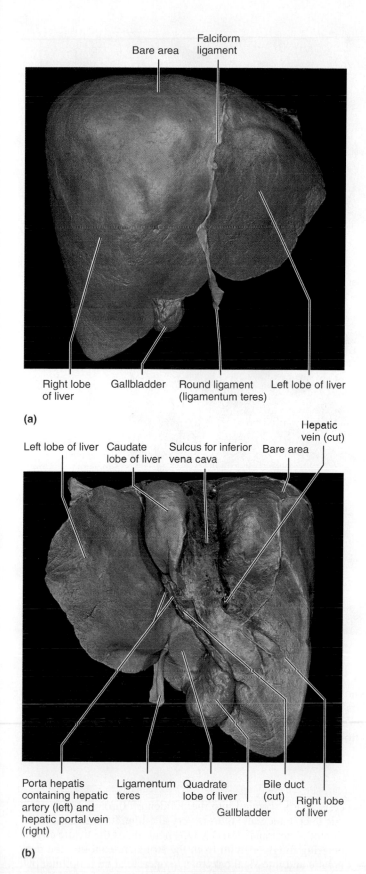

Falciform ligament
Bare area
Right lobe of liver
Gallbladder
Round ligament (ligamentum teres)
Left lobe of liver

(a)

Left lobe of liver
Caudate lobe of liver
Sulcus for inferior vena cava
Bare area
Hepatic vein (cut)
Porta hepatis containing hepatic artery (left) and hepatic portal vein (right)
Ligamentum teres
Quadrate lobe of liver
Gallbladder
Bile duct (cut)
Right lobe of liver

(b)

Figure 38.14 Gross anatomy of the human liver.
(a) Anterior view. **(b)** Posteroinferior aspect. The four liver lobes are separated by a group of fissures in this view.

blood from the hepatic portal vein and hepatic artery percolates. **Stellate macrophages,** special phagocytic cells, also called **hepatic macrophages,** line the sinusoids and remove debris such as bacteria from the blood as it flows past, while the hepatocytes pick up oxygen and nutrients. The sinusoids empty into the central vein, and the blood ultimately drains from the liver via the *hepatic veins.*

Bile is continuously being made by the hepatocytes. It flows through tiny canals, the **bile canaliculi,** which run between adjacent cells toward the bile duct branches in the triad regions, where the bile eventually leaves the liver.

Activity 8

Examining the Histology of the Liver

Examine a slide of liver tissue and identify as many as possible of the structural features (see Figure 38.16). Also examine a three-dimensional model of liver lobules if this is available. Reproduce a small pie-shaped section of a liver lobule in the space below. Label the hepatocytes, the stellate macrophages, sinusoids, a portal triad, and a central vein.

38

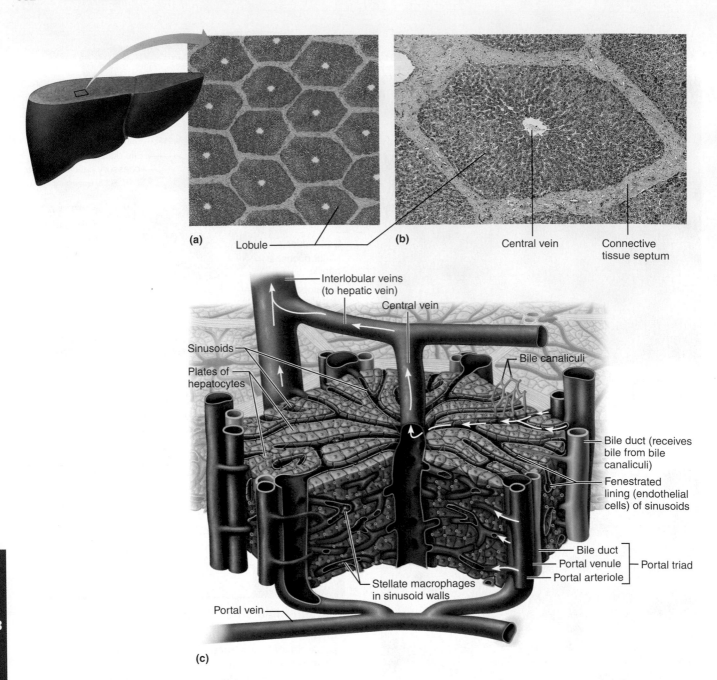

Figure 38.16 Microscopic anatomy of the liver. (a) Schematic view of the cut surface of the liver showing the hexagonal nature of its lobules. **(b)** Photomicrograph of one liver lobule (55×). **(c)** Enlarged three-dimensional diagram of one liver lobule. Arrows show direction of blood flow. Bile flows in the opposite direction toward the bile ducts.

Pancreas

The **pancreas** is a soft, triangular gland that extends horizontally across the posterior abdominal wall from the spleen to the duodenum (see Figure 38.1). Like the duodenum, it is a retroperitoneal organ (see Figure 38.7). The pancreas has both an endocrine function, producing the hormones insulin and glucagon, and an exocrine function. Its exocrine secretion includes many hydrolytic enzymes produced by the acinar

cells and is secreted into the duodenum through the pancreatic duct. Pancreatic juice is very alkaline. Its high concentration of bicarbonate ion (HCO_3^-) neutralizes the acidic chyme entering the duodenum from the stomach, enabling the pancreatic and intestinal enzymes to operate at their optimal pH, which is slightly alkaline. (See Figure 27.3c.)

For instructions on animal dissections, see the dissection exercises (starting on p. 705) in the cat and fetal pig editions of this manual.

REVIEW SHEET
Anatomy of the Digestive System

Name _____ Lab Time/Date _____

General Histological Plan of the Alimentary Canal

1. The general anatomical features of the alimentary canal are listed below. Fill in the table to complete the information.

Wall layer	Subdivisions of the layer (if applicable)	Major functions
Mucosa		
Submucosa		
Muscularis externa		
Serosa or adventitia		

Organs of the Alimentary Canal

2. The tubelike digestive system canal that extends from the mouth to the anus is known as the _____ canal or the _____ tract.

3. How is the muscularis externa of the stomach modified? _____

_____

How does this modification relate to the function of the stomach? _____

_____

4. What transition in epithelial type exists at the gastroesophageal junction? _____

_____

How do the epithelia of these two organs relate to their specific functions?

_____

5. Differentiate the colon from the large intestine. _____

_____

6. Match the items in column B with the descriptive statements in column A.

Column A

_____ 1. structure that suspends the small intestine from the posterior body wall

_____ 2. fingerlike extensions of the intestinal mucosa that increase the surface area for absorption

_____ 3. large collections of lymphoid tissue found in the submucosa of the small intestine

_____ 4. deep folds of the mucosa and submucosa that extend completely or partially around the circumference of the small intestine

_____ 5. mobile organ that manipulates food in the mouth and initiates swallowing

_____ 6. conduit for both air and food

_____ 7. the "gullet"; no digestive/absorptive function

_____ 8. folds of the gastric mucosa

_____ 9. pocketlike sacs of the large intestine

_____ 10. projections of the plasma membrane of a mucosal epithelial cell

_____ 11. valve at the junction of the small and large intestines

_____ 12. primary region of food and water absorption

_____ 13. membrane securing the tongue to the floor of the mouth

_____ 14. absorbs water and forms feces

_____ 15. area between the teeth and lips/cheeks

_____ 16. wormlike sac that outpockets from the cecum

_____ 17. initiates protein digestion

_____ 18. structure attached to the lesser curvature of the stomach

_____ 19. covers most of the abdominal organs like an apron

_____ 20. valve controlling food movement from the stomach into the duodenum

_____ 21. posterosuperior boundary of the oral cavity

_____ 22. region containing two sphincters through which feces are expelled from the body

_____ 23. bone-supported anterosuperior boundary of the oral cavity

Column B

a. anus

b. appendix

c. circular folds

d. esophagus

e. frenulum

f. greater omentum

g. hard palate

h. haustra

i. ileocecal valve

j. large intestine

k. lesser omentum

l. mesentery

m. microvilli

n. oral vestibule

o. Peyer's patches

p. pharynx

q. pyloric valve

r. rugae

s. small intestine

t. soft palate

u. stomach

v. tongue

w. villi

7. Correctly identify all organs depicted in the diagram below.

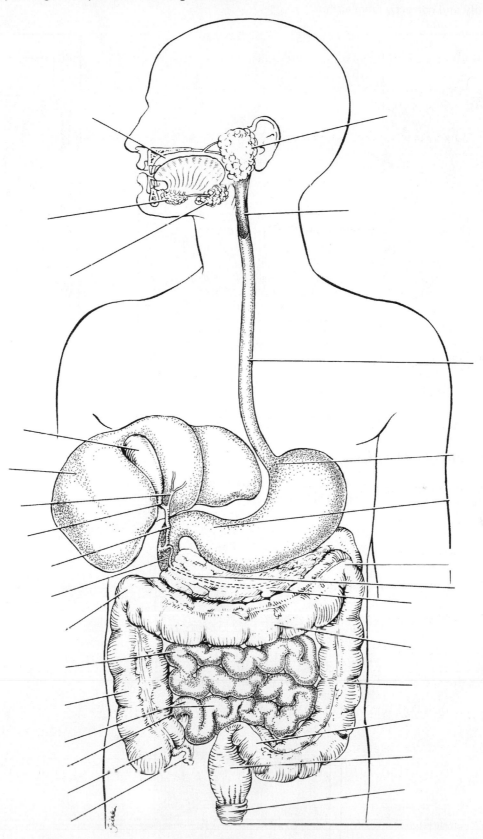

8. You have studied the histologic structure of a number of organs in this laboratory. Three of these are diagrammed below. Identify and correctly label each.

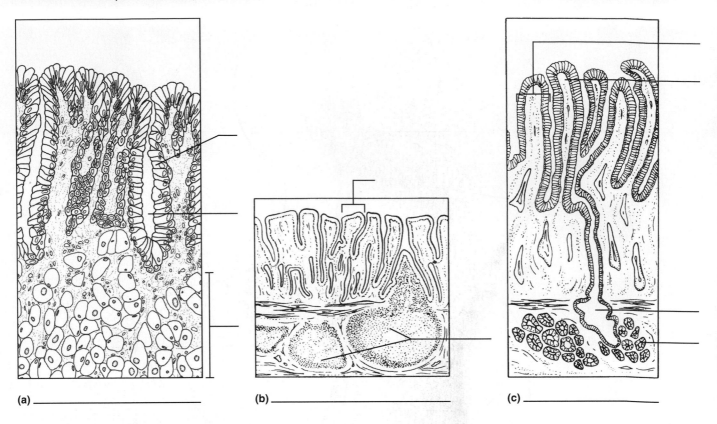

(a) _____   (b) _____   (c) _____

Accessory Digestive Organs

9. Correctly label all structures provided with leader lines in the diagram of a molar below. (Note: Some of the terms in the key for question 10 may be helpful in this task.)

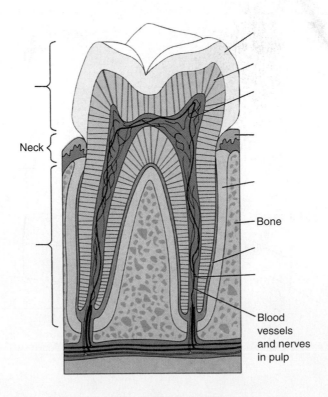

10. Use the key to identify each tooth area described below.   *Key:*   a. cement

    _____ 1. visible portion of the tooth in situ

    _____ 2. material covering the tooth root   b. crown

    _____ 3. hardest substance in the body   c. dentin

    _____ 4. attaches the tooth to the tooth socket   d. enamel

    _____ 5. portion of the tooth embedded in bone   e. gingival sulcus

    _____ 6. forms the major portion of tooth structure; similar to bone   f. odontoblast

    _____ 7. produces the dentin   g. periodontal ligament

    _____ 8. site of blood vessels, nerves, and lymphatics   h. pulp

    _____ 9. narrow gap between the crown and the gum   i. root

11. In the human, the number of deciduous teeth is _____; the number of permanent teeth is _____.

12. The dental formula for permanent teeth is $\dfrac{2,1,2,3}{2,1,2,3} \times 2$

    Explain what this means. _____

    _____

    What is the dental formula for the deciduous teeth? _____ × _____ (_____ deciduous teeth)

13. Which teeth are the "wisdom teeth"? _____

14. Various types of glands form a part of the alimentary canal wall or duct their secretions into it. Match the glands listed in column B with the function/locations described in column A.

    Column A   **Column B**

    _____ 1. produce(s) mucus; found in the submucosa of the small intestine   a. duodenal glands

    _____ 2. produce(s) a product containing amylase that begins starch breakdown in the mouth   b. gastric glands

    c. intestinal crypts

    _____ 3. produce(s) many enzymes and an alkaline fluid that is secreted into the duodenum   d. liver

    _____ 4. produce(s) bile that it secretes into the duodenum via the bile duct   e. pancreas

    _____ 5. produce(s) HCl and pepsinogen   f. salivary glands

    _____ 6. found in the mucosa of the small intestine; produce(s) intestinal juice

15. Which of the salivary glands produces a secretion that is mainly serous? _____

16. What is the role of the gallbladder? _____

17. Name three structures always found in the portal triad regions of the liver. _____,

    _____, and _____

18. Where would you expect to find the stellate macrophages of the liver? _____

    What is their function? _____

19. Why is the liver so dark red in the living animal? _____

    _____

20. The pancreas has two major populations of secretory cells—those in the islets and the acinar cells. Which population

    serves the digestive process? _____

Digestive System Processes: Chemical and Physical

Objectives

☐ List the digestive system enzymes involved in the digestion of proteins, fats, and carbohydrates; state their site of origin; and summarize the conditions promoting their optimal functioning.

☐ Name the end products of protein, fat, and carbohydrate digestion.

☐ Define *enzyme, catalyst, control, substrate,* and *hydrolase.*

☐ Describe the different types of enzyme assays, and the appropriate chemical tests to determine if digestion of a particular foodstuff has occurred.

☐ Discuss the role of temperature and pH in the regulation of enzyme activity.

☐ State the function of bile in the digestive process.

☐ Explain why swallowing is both a voluntary and a reflex activity, and discuss the role of the tongue, larynx, and gastroesophageal sphincter in swallowing.

☐ Compare and contrast segmentation and peristalsis as mechanisms of mixing and propulsion in digestive tract organs.

Materials

Part I: Enzyme Action

General Supply Area

- Hot plates
- 250-ml beakers
- Boiling chips
- Test tubes and test tube rack
- Wax markers
- Water bath set at 37°C (if not available, incubate at room temperature and double the time)
- Ice water bath
- Chart on board for recording class results

Activity 1: Starch Digestion

- Dropper bottle of distilled water

Text continues on next page. →

MasteringA&P®

For related exercise study tools, go to the Study Area of **MasteringA&P**. There you will find:

- Practice Anatomy Lab **PAL**
- PhysioEx **PEx**
- A&PFlix **A&PFlix**
- Practice quizzes, Histology Atlas, eText, Videos, and more!

Pre-Lab Quiz

1. Circle the correct underlined term. Enzymes are <u>catalysts</u> / <u>substrates</u> that increase the rate of chemical reactions without becoming a part of the product.

2. Circle True or False. Breakdown products of fats are absorbed by the lymphatic system and are then transported into the systemic circulation by lymph.

3. A(n) _____ is a specimen or standard against which all experimental samples are compared.
   a. assay   b. control   c. substrate   d. trial

4. One enzyme that you will be studying today, produced by the salivary glands and secreted into the mouth, hydrolyzes starch to maltose. It is
   _____.

5. Circle True or False. When you use iodine to test for starch, a color change to blue-black indicates a positive starch test.

6. If Benedict's test in the starch assay produces a _____ precipitate, then your test will be recorded as positive for maltose.
   a. blue to black   b. green to orange   c. white

7. The enzyme _____, produced by the pancreas, is responsible for breaking down proteins.
   a. amylase   b. kinase   c. lipase   d. trypsin

8. Circle the correct underlined term. The enzyme <u>pancreatic lipase</u> / <u>pepsin</u> hydrolyzes neutral fats to their component monoglycerides and fatty acids.

9. Circle True or False. Both smooth and skeletal muscles are involved in the propulsion of foodstuffs along the alimentray canal.

10. _____ movements are local contractions that mix foodstuffs with digestive juices and increase the rate of absorption.
   a. Deglutition   b. Elimination   c. Peristaltic   d. Segmental

- Dropper bottles of the following:
  1% alpha-amylase solution*
  1% boiled starch solution, freshly prepared†
  1% maltose solution
  Lugol's iodine solution (IKI)
  Benedict's solution
- Spot plate

Activity 2: Protein Digestion

- Dropper bottles of 1% trypsin and 0.01% BAPNA solution

Activity 3: Bile Action and Fat Digestion

- Dropper bottles of 1% pancreatin solution, litmus cream (fresh cream to which powdered litmus is added to achieve a deep blue color), 0.1 *N* HCl, and vegetable oil
- Bile salts (sodium taurocholate)
- Parafilm® (small squares to cover the test tubes)

Part II: Physical Processes

Activity 5: Observing Digestive Movements

- Water pitcher
- Paper cups
- Stethoscope
- Alcohol swab
- Disposable autoclave bag
- Watch, clock, or timer

Activity 6: Video Viewing

- Television and VCR or DVD player for independent viewing of video by student
- *Interactive Physiology*®, Digestive System
- **PEx** PhysioEx™ 9.1 Computer Simulation Ex. 8 on p. PEx-119

* The alpha-amylase must be a low-maltose preparation for good results.

†Prepare by adding 1 g starch to 100 ml distilled water; boil and cool; add a pinch of salt (NaCl). Prepare fresh daily.

39

The food we eat must be processed so that its nutrients can reach the cells of our body. First the food is mechanically broken down into small particles, and then the particles are chemically (enzymatically) digested into the molecules that can be absorbed. Food digestion is a prerequisite to food absorption. (You have already studied mechanisms of passive and active absorption in Exercise 5 and/or PhysioEx Exercise 1. Before proceeding, review that material.)

Digestion of Foodstuffs: Enzymatic Action

Enzymes are large protein molecules produced by body cells. They are biological **catalysts,** meaning that they increase the rate of a chemical reaction without themselves becoming part of the product. The digestive enzymes are hydrolytic enzymes, or **hydrolases.** Their **substrates,** or the molecules on which they act, are organic food molecules which they break down by adding water to the molecular bonds, thus cleaving the bonds between the chemical building blocks, or monomers.

Each enzyme hydrolyzes only one or a small group of substrate molecules, and specific environmental conditions are necessary for it to function optimally. Since digestive enzymes actually function outside the body cells in the digestive tract, their hydrolytic activity can also be studied in a test tube.

Figure 39.1 is a flowchart depicting the progressive digestion of carbohydrates, proteins, fats, and nucleic acids. It summarizes the specific enzymes involved, their site of formation, and their site of action. Acquaint yourself with the flowchart before beginning this experiment, and refer to it as necessary during the laboratory session.

General Instructions for Activities 1–3

Work in groups of four, with each group taking responsibility for setting up and conducting one of the following experiments. In each activity, you are directed to boil the contents of one or more test tubes. To do this, obtain a 250-ml beaker, boiling chips, and a hot plate from the general supply area. Place a few boiling chips into the beaker, add about 125 ml of water, and bring to a boil. Place the test tube for each specimen in the water for the number of minutes specified in the directions. You will also be using a 37°C bath and an ice water bath for parts of these experiments.

Upon completion of the experiments, each group should communicate its results to the rest of the class by recording them in a chart on the board. Each assay contains one or more **controls,** the specimens against which experimental samples are compared. All members of the class should observe the controls as well as the experimental results and be able to explain the tests used and the results observed and anticipated for each experiment.

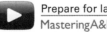

Prepare for lab: Watch the Pre-Lab Video
MasteringA&P®>Study Area > Pre-Lab Videos

Activity 1

Assessing Starch Digestion by Salivary Amylase

1. From the general supply area, obtain a test tube rack, 10 test tubes, and a wax marking pencil. From the Activity 1 supply area, obtain a dropper bottle of distilled water and dropper bottles of maltose, amylase, and starch solutions.

2. In this experiment you will investigate the hydrolysis of starch to maltose by **salivary amylase.** You will need to be able to identify the presence of starch and maltose, the breakdown product of starch, to determine to what extent the enzymatic activity has occurred. Thus controls must be prepared to provide a known standard against which comparisons can be made. Starch decreases and sugar increases as hydrolysis occurs, according to the following formula:

$$\text{Starch} + \text{water} \xrightarrow{\text{amylase}} \text{maltose}$$

Text continues on p. 602. ➡

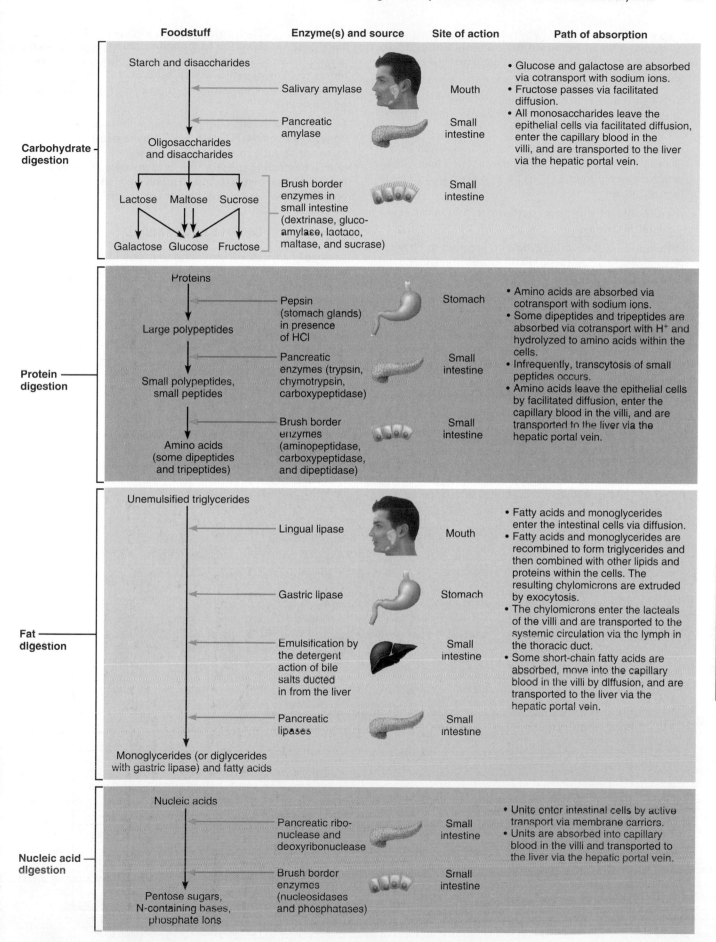

Figure 39.1 Flowchart of digestion and absorption of foodstuffs.

Two students should prepare the controls (tubes 1A to 3A) while the other two prepare the experimental samples (tubes 4A to 6A).

- Mark each tube with a wax pencil and load the tubes as indicated in the **Activity 1 chart** below, using 3 drops (gtt) of each indicated substance.

- Place tubes in the incubation conditions listed in the **Activity 1 chart** below for approximately 1 hour. Shake the rack gently from time to time to keep the contents evenly mixed.

- At the end of the hour, perform the amylase assay described below.

- While these tubes are incubating, proceed to Physical Processes: Mechanisms of Food Propulsion and Mixing (p. 606). Be sure to monitor the time so as to complete this activity as needed.

Amylase Assay

1. After one hour, obtain a spot plate and dropper bottles of Lugol's iodine solution (for the IKI, or iodine, test) and Benedict's solution from the Activity 1 supply area. Set up your boiling water bath using a hot plate, boiling chips, and a 250-ml beaker.

2. While the water is heating, mark six depressions of the spot plate 1A–6A (*A* for amylase) for sample identification.

3. Using a pipet, transfer a drop of the sample from each of the tubes 1A–6A into the appropriately numbered spot. Into each sample drop, place a drop of Lugol's iodine (IKI) solution. A blue-black color indicates the presence of starch and is referred to as a **positive starch test**. If starch is not present, the mixture will not turn blue, which is referred to as a **negative starch test**. Record your results (+ for positive, – for negative) in the Activity 1 chart and on the board.

4. Into the remaining mixture in each tube, place 3 drops of Benedict's solution. Put each tube into the beaker of boiling water for about 5 minutes. If a green-to-orange precipitate forms, maltose is present; this is a **positive sugar test**. A **negative sugar test** is indicated by no color change. Record your results in the Activity 1 chart and on the board.

WHY THIS MATTERS | Lactose Intolerance

Lactose is a disaccharide found in milk and, to a lesser extent, in other dairy products. Lactose is hydrolyzed into glucose and galactose by the enzyme lactase. Infants usually produce lactase, but many individuals become lactose intolerant (lactase nonpersistent) as they mature. When lactose-intolerant individuals ingest dairy products, the bacteria in their colon react with lactose, producing gas, which causes bloating, flatulence, abdominal cramps, and diarrhea. Individuals who continue to produce lactase into adulthood are lactase-persistent and are able to metabolize lactose normally. Ingestion of probiotic bacteria has been found to reduce the symptoms of lactose intolerance, possibly because probiotic bacteria that produce their own lactase replace the gas-producing bacterial flora in the colon. ∎

Activity 1: Salivary Amylase Digestion of Starch

Tube no.	1A	2A	3A	4A		5A	6A
Additives (3 gtt ea)	amylase, water	starch, water	maltose, water	amylase	Boil amylase 4 min, then add starch → boiled amylase, starch	amylase, starch	amylase, starch
Incubation condition	37°C	37°C	37°C	37°C		37°C	0°C
IKI test (color change)							
Result: (+) or (−)							
Benedict's test (color change)							
Result: (+) or (−)							

Additive key:

■ = Amylase   ■ = Starch   ☐ = Maltose   ☐ = Water

Protein Digestion by Trypsin

Trypsin, an enzyme produced by the pancreas, hydrolyzes proteins to small peptides. BAPNA (*N*-alpha-benzoyl-L-arginine-*p*-nitroanilide) is a synthetic trypsin substrate consisting of a dye covalently bound to an amino acid. Trypsin hydrolysis of BAPNA cleaves the dye molecule from the amino acid, causing the solution to change from colorless to bright yellow. The color change from clear to yellow is direct evidence of hydrolysis by trypsin.

Activity 2

Assessing Protein Digestion by Trypsin

1. From the general supply area, obtain five test tubes and a test tube rack, and from the Activity 2 supply area get a dropper bottle of trypsin and one of BAPNA. Bring these items to your bench.

2. Two students should prepare the controls (tubes 1T and 2T) while the other two prepare the experimental samples (tubes 3T to 5T).

- Mark each tube with a wax pencil, and load the tubes as indicated in the Activity 2 chart, using 3 drops (gtt) of each indicated substance.

- Place all tubes in a rack in the appropriate water bath for approximately 1 hour. Shake the rack occasionally to keep the contents well mixed.

- At the end of the hour, examine the tubes for the results of the trypsin assay (detailed below).

Trypsin Assay

Since BAPNA is a synthetic color-producing substrate, the presence of yellow color indicates a **positive hydrolysis test**; the dye molecule has been cleaved from the amino acid. If the sample mixture remains clear, a **negative hydrolysis test** has occurred.

Record the results in the Activity 2 chart and on the board.

Activity 2: Trypsin Digestion of Protein						
Tube no.	1T	2T	3T		4T	5T
Additives (3 gtt ea)	trypsin, water	BAPNA, water	trypsin	Boil trypsin 4 min, then add BAPNA. → boiled trypsin, BAPNA	trypsin, BAPNA	trypsin, BAPNA
Incubation condition	37°C	37°C	37°C		37°C	0°C
Color change						
Result: (+) or (−)						

Additive key:

■ = Trypsin   ■ = BAPNA   □ = Water

Pancreatic Lipase Digestion of Fats and the Action of Bile

The treatment that fats and oils go through during digestion in the small intestine is a bit more complicated than that of carbohydrates or proteins—pretreatment with bile to physically emulsify the fats is required. Hence, two sets of reactions occur.

First:

$$\text{Fats/oils} \xrightarrow[\text{(emulsification)}]{\text{bile}} \text{minute fat/oil droplets}$$

Then:

$$\text{Fat/oil droplets} \xrightarrow[\text{(digestion)}]{\text{lipase}} \text{monoglycerides and fatty acids}$$

The term **pancreatin** describes the enzymatic product of the pancreas, which includes enzymes that digest proteins, carbohydrates, nucleic acids, and fats. It is used here to investigate the properties of **pancreatic lipase,** which hydrolyzes fats and oils to their component monoglycerides and two fatty acids.

Since fatty acids are organic acids, they acidify solutions, decreasing the pH. An easy way to recognize that digestion is ongoing or completed is to test pH. You will be using a pH indicator called *litmus blue* to follow these changes; it changes from blue to pink as the test tube contents become acidic.

Activity 3

Demonstrating the Emulsification Action of Bile and Assessing Fat Digestion by Lipase

1. From the general supply area, obtain nine test tubes and a test tube rack, plus one dropper bottle of each of the solutions in the Activity 3 supply area.

2. Although *bile,* a secretory product of the liver, is not an enzyme, it is important to fat digestion because of its emulsifying action. It physically breaks down large fat particles into smaller ones. Emulsified fats provide a larger surface area for enzymatic activity. To demonstrate the action of bile on fats, prepare two test tubes and mark them 1E and 2E (*E* for emulsified fats).

- To tube 1E, add 20 drops of water and 4 drops of vegetable oil.

- To tube 2E, add 20 drops of water, 4 drops of vegetable oil, and a pinch of bile salts.

- Cover each tube with a small square of Parafilm, shake vigorously, and allow the tubes to stand at room temperature.

After 10 to 15 minutes, observe both tubes. If emulsification has not occurred, the oil will be floating on the surface of the water. If emulsification has occurred, the fat droplets will be suspended throughout the water, forming an emulsion.

In which tube has emulsification occurred? _____

3. Two students should prepare the controls (1L and 2L, *L* for lipase) while the other two students in the group set up the experimental samples (3L to 5L, 4B, and 5B, where *B* is for bile), as illustrated in the **Activity 3 chart**.

- Mark each tube with a wax pencil and load the tubes using 5 drops (gtt) of each indicated solution.

- Place a pinch of bile salts in tubes 4B and 5B.

- Cover each tube with a small square of Parafilm, and shake to mix the contents of the tube.

- Remove the Parafilm, and place all tubes in a rack in the appropriate water bath for approximately 1 hour. Shake the test tube rack from time to time to keep the contents well mixed.

- At the end of the hour, perform the lipase assay below.

Lipase Assay

Fresh cream provides the fat substrate for this assay; add litmus powder to it to make litmus cream. The basis of this assay is a pH change that is detected by the litmus powder indicator. Alkaline or neutral solutions containing litmus are blue but will turn reddish in the presence of acid. If digestion occurs, the fatty acids produced will turn the litmus cream from blue to pink. Because the effect of

Activity 3: Pancreatic Lipase Digestion of Fats

Tube no.	1L	2L	3L	4L	5L	4B	5B
Additives (5 gtt ea)	pancreatin, water	litmus cream, water	pancreatin → Boil pancreatin 4 min, then add litmus cream. boiled pancreatin, litmus cream	pancreatin, litmus cream	pancreatin, litmus cream	pancreatin, litmus cream, bile salts	pancreatin, litmus cream, bile salts
Incubation condition	37°C	37°C	37°C	37°C	0°C	37°C	0°C
Color change							
Result: (+) or (−)							

Additive key:

☐ = Pancreatin   ☐ = Litmus cream   ☐ = Water   △ = Pinch bile salts

39

hydrolysis by lipase is directly seen, additional assay reagents are not necessary.

1. To prepare a color control, add 0.1 *N* HCl drop by drop to tubes 1L and 2L (covering the tubes with a square of

Parafilm after each addition and shaking to mix) until the cream turns pink.

2. Record the color of the tubes in the Activity 3 chart and on the board.

Activity 4

Reporting Results and Conclusions

1. Share your results with the class as directed in the General Instructions (p. 600).

2. Suggest additional experiments, and carry out experiments if time permits.

3. Prepare a lab report for the experiments on digestion. (See Getting Started, on MasteringA&P.)

 Group Challenge

Odd Enzyme Out

The following boxes each contain four digestive enzymes. One of the listed enzymes does not share a characteristic that the other three do. Working in groups of three, discuss the characteristics of the enzymes in each group. On a separate piece of paper, one student will record the characteristics for each enzyme for the group. Discuss the possible candidates for the "odd enzyme". Once the group has come to a consensus, circle the enzyme that doesn't belong with the others, and explain why it is singled out. Sometimes there may be multiple reasons why the enzyme doesn't belong with the others. Include as many as you can think of, but make sure it does not have the key characteristic shared by the other three.

1. Which is the "odd enzyme"?	Why is it the odd one out?
Trypsin	
Carboxypeptidase	
Pepsin	
Chymotrypsin	

2. Which is the "odd enzyme"?	Why is it the odd one out?
Lactase	
Pepsin	
Aminopeptidase	
Trypsin	

3. Which is the "odd enzyme"?	Why is it the odd one out?
Maltase	
Pancreatic lipase	
Nucleosidase	
Dipeptidase	

4. Which is the "odd enzyme"?	Why is it the odd one out?
Sucrase	
Dextrinase	
Glucoamylase	
Chymotrypsin	

39

Physical Processes: Mechanisms of Food Propulsion and Mixing

Although enzyme activity is a very important part of the overall digestion process, foods must also be processed physically (by chewing and churning) and moved by mechanical means along the tract if digestion and absorption are to be completed. Muscles are involved in producing the movements of foodstuffs along the gastrointestinal tract. Although we tend to think only of smooth muscles when visceral activities are involved, both skeletal and smooth muscles are involved in the physical processes. This fact is demonstrated by the simple activities that follow.

Deglutition (Swallowing)

Swallowing, or **deglutition,** is largely the result of skeletal muscle activity and occurs in two phases: *buccal* (mouth) and *pharyngeal-esophageal*. The buccal phase (**Figure 39.2** step ①) is voluntarily controlled and initiated by the tongue. Once begun, the process continues involuntarily in the pharynx and esophagus through peristalsis, resulting in the delivery of the swallowed contents to the stomach (Figure 39.2 steps ②–③).

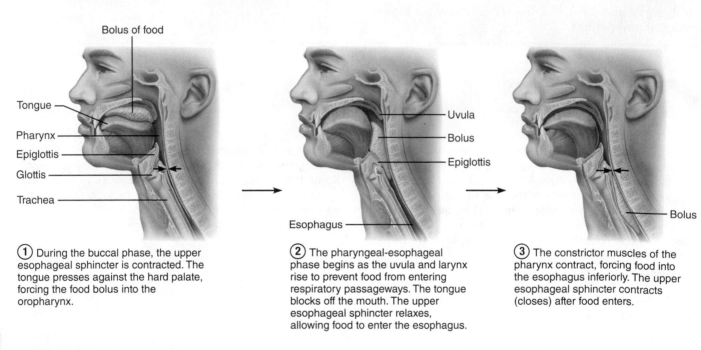

① During the buccal phase, the upper esophageal sphincter is contracted. The tongue presses against the hard palate, forcing the food bolus into the oropharynx.

② The pharyngeal-esophageal phase begins as the uvula and larynx rise to prevent food from entering respiratory passageways. The tongue blocks off the mouth. The upper esophageal sphincter relaxes, allowing food to enter the esophagus.

③ The constrictor muscles of the pharynx contract, forcing food into the esophagus inferiorly. The upper esophageal sphincter contracts (closes) after food enters.

Figure 39.2 Swallowing. The process of swallowing consists of voluntary (buccal) (step ①) and involuntary (pharyngeal-esophageal) phases (steps ②–③).

Activity 5

Observing Movements and Sounds of the Digestive System

1. Obtain a pitcher of water, a stethoscope, a paper cup, an alcohol swab, and an autoclave bag in preparation for making the following observations.

2. While swallowing a mouthful of water, consciously note the movement of your tongue during the process. Record your observations.

_____

_____

_____

3. Repeat the swallowing process while your laboratory partner watches the externally visible movements of your larynx. (This movement is more obvious in a male, since males have a larger Adam's apple.) Record your observations.

_____

_____

What do these movements accomplish? _____

_____

_____

4. Before donning the stethoscope, your lab partner should clean the earpieces with an alcohol swab. Then, he or she should place the diaphragm of the stethoscope over your abdominal wall, approximately 2.5 cm (1 inch) below the xiphoid process and slightly to the left, to listen for sounds as you again take two or three swallows of water. There should be two audible sounds—one when the water splashes against the gastroesophageal sphincter, and the second when the peristaltic wave of the esophagus arrives at the sphincter and the sphincter opens, allowing water to gurgle into the stomach. Determine, as accurately as possible, the time interval between these two sounds and record it below.

Interval between arrival of water at the sphincter and the

opening of the sphincter: _____ sec

This interval gives a fair indication of the time it takes for the peristaltic wave to travel down the 25 cm (10 inches) of the esophagus. (Actually the time interval is slightly less than it seems, because pressure causes the sphincter to relax before the peristaltic wave reaches it.)

 Dispose of the used paper cup in the autoclave bag.

Segmentation and Peristalsis

Although several types of movements occur in the digestive tract organs, peristalsis and segmentation are most important as mixing and propulsive mechanisms (**Figure 39.3**).

Peristaltic movements are the major means of propelling food through most of the alimentary canal. Essentially, they are waves of contraction followed by waves of relaxation that squeeze foodstuffs through the alimentary canal, and they are superimposed on segmental movements.

Segmental movements are local constrictions of the organ wall that occur rhythmically. They serve mainly to mix the foodstuffs with digestive juices and to increase the rate of absorption by continually moving different portions of the chyme over adjacent regions of the intestinal wall. However, segmentation is also an important means of food propulsion in the small intestine, and slow segmenting movements called haustral contractions are frequently seen in the large intestine.

Activity 6

Viewing Segmental and Peristaltic Movements

If a video showing some of the propulsive movements is available, go to a viewing station to view it before leaving the laboratory. Alternatively, use the *Interactive Physiology*® module on the Digestive System to observe gut motility.

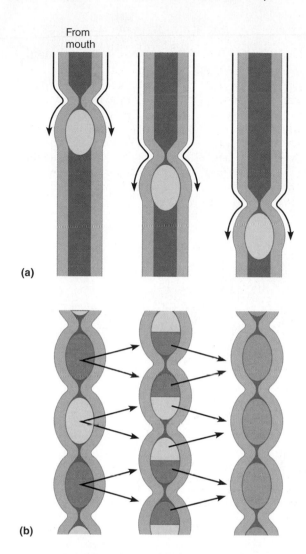

(a)

(b)

Figure 39.3 Peristaltic and segmental movements of the digestive tract. (a) Peristalsis: neighboring segments of the intestine alternately contract and relax, moving food along the tract. **(b)** Segmentation: single segments of intestine alternately contract and relax. Because inactive segments exist between active segments, food mixing occurs to a greater degree than food movement. Peristalsis is superimposed on segmentation movements.

39

EXERCISE

39

REVIEW SHEET
Digestive System Processes:
Chemical and Physical

Name _____   Lab Time/Date _____

Digestion of Foodstuffs: Enzymatic Action

1. Match the following definitions with the proper choices from the key.

   Key: a. catalyst   b. control   c. enzyme   d. substrate

   _____ 1. substance on which a catalyst works

   _____ 2. biologic catalyst; protein in nature

   _____ 3. increases the rate of a chemical reaction without becoming part of the product

   _____ 4. provides a standard of comparison for test results

2. List the three characteristics of enzymes. _____

_____

_____

3. The enzymes of the digestive system are classified as hydrolases. What does this mean?

_____

_____

4. Fill in the following chart about the various digestive system enzymes encountered in this exercise.

Enzyme	Organ producing it	Site of action	Substrate(s)	Optimal pH
Salivary amylase				
Trypsin				
Lipase (pancreatic)				

5. Name the end products of digestion for the following types of foods.

   proteins: _____   carbohydrates: _____

   fats: _____ and _____

WHY THIS MATTERS | 6. How does the substrate for amylase differ from the substrate for lactase? _____

_____

How are the substrates similar? _____

7. Where does lactose hydrolysis occur for lactase-persistent individuals? _____

Where does lactose hydrolysis occur for lactose-intolerant individuals who have consumed probiotic bacte-

rial microflora? _____

8. You used several indicators or tests in the laboratory to determine the presence or absence of certain substances. Choose the correct test or indicator from the key to correspond to the condition described below.

Key: a. Lugol's iodine (IKI)   b. Benedict's solution   c. litmus   d. BAPNA

_____ 1. used to test for protein hydrolysis, which was indicated by a yellow color

_____ 2. used to test for the presence of starch, which was indicated by a blue-black color

_____ 3. used to test for the presence of fatty acids, which was evidenced by a color change from blue to pink

_____ 4. used to test for the presence of reducing sugars (maltose, glucose) as indicated by a blue to green or orange color change

9. What conclusions can you draw when an experimental sample gives both a positive starch test and a positive maltose

test after incubation? _____

Why was 37°C the optimal incubation temperature? _____

Why did very little, if any, starch digestion occur in test tube 4A? _____

_____

When starch was incubated with amylase at 0°C, did you see any starch digestion? _____

Why or why not? _____

_____

Assume you have said to a group of your peers that amylase is capable of starch hydrolysis to maltose. If you had

not done control tube 1A, what objection to your statement could be raised? _____

_____

What if you had not done tube 2A? _____

10. In the exercise concerning trypsin function, why was an enzyme assay such as Benedict's or Lugol's iodine (IKI), which

test for the presence of a reaction product, not necessary? _____

Why was tube 1T necessary? _____

Why was tube 2T necessary? _____

Trypsin is a protease similar to pepsin, the protein-digesting enzyme in the stomach. Would trypsin work well in the

stomach? _____ Why? _____

11. In the procedure concerning pancreatic lipase digestion of fats and the action of bile salts, how did the appearance

of tubes 1E and 2E differ? _____

_____

Explain the reason for the difference. _____

_____

Why did the litmus indicator change from blue to pink during fat hydrolysis? _____

_____

Why is bile not considered an enzyme? _____

How did the tubes containing bile compare with those not containing bile? _____

What role does bile play in fat digestion? _____

_____

12. The three-dimensional structure of a functional protein is altered by intense heat or nonphysiological pH even though peptide bonds may not break. Such inactivation is called denaturation, and denatured enzymes are nonfunctional. Explain why.

_____

_____

What specific experimental conditions resulted in denatured enzymes? _____

_____

13. Pancreatic and intestinal enzymes operate optimally at a pH that is slightly alkaline, yet the chyme entering the duodenum from the stomach is very acid. How is the proper pH for the functioning of the pancreatic-intestinal enzymes ensured?

_____

_____

14. Assume you have been chewing a piece of bread for 5 or 6 minutes. How would you expect its taste to change during

this interval? _____

Why? _____

15. Note the mechanism of absorption (passive or active transport) of the following food breakdown products, and indicate by a check mark (✓) whether the absorption would result in their movement into the blood capillaries or the lymphatic capillaries (lacteals).

Substance	Mechanism of absorption	Blood	Lymph
Monosaccharides			
Fatty acids and monoglycerides			
Amino acids			
Water			
Na^+, Cl^-, Ca^{2+}			

16. People on a strict diet to lose weight begin to metabolize stored fats at an accelerated rate. How does this condition affect blood pH? _____

17. Using a flowchart, trace the pathway of a ham sandwich (ham = protein and fat; bread = starch) from the mouth to the site of absorption of its breakdown products, noting where digestion occurs and what specific enzymes are involved.

18. Some of the digestive organs have groups of secretory cells that liberate hormones into the blood. These exert an effect on the digestive process by acting on other cells or structures and causing them to release digestive enzymes, expel bile, or increase the motility of the digestive tract. For each hormone below, note the organ producing the hormone and its effects on the digestive process. Include the target organs affected.

Hormone	Produced by	Target organ(s) and effects
Secretin		
Gastrin		
Cholecystokinin		

Physical Processes: Mechanisms of Food Propulsion and Mixing

19. Complete the following statements.

Swallowing, or __1__, occurs in two phases—the __2__ and __3__. One of these phases, the __4__ phase, is voluntary. During the voluntary phase, the __5__ is used to push the food into the back of the throat. During swallowing, the __6__ rises to ensure that its passageway is covered by the epiglottis so that the ingested substances don't enter the respiratory passageways. It is possible to swallow water while standing on your head because the water is carried along the esophagus involuntarily by the process of __7__. The pressure exerted by the foodstuffs on the __8__ sphincter causes it to open, allowing the foodstuffs to enter the stomach.

The two major types of propulsive movements that occur in the small intestine are __9__ and __10__. One of these movements, __11__, acts to continually mix the foods and to increase the absorption rate by moving different parts of the chyme mass over the intestinal mucosa, but it has less of a role in moving foods along the digestive tract.

1. _____

2. _____

3. _____

4. _____

5. _____

6. _____

7. _____

8. _____

9. _____

10. _____

11. _____

Anatomy of the Urinary System

Objectives

- ☐ List the functions of the urinary system.
- ☐ Identify, on a model or image, the urinary system organs, and state the general function of each.
- ☐ Compare the course and length of the urethra in males and females.
- ☐ Identify these regions of the dissected kidney (longitudinal section): hilum, cortex, medulla, medullary pyramids, major and minor calyces, pelvis, renal columns, and fibrous and perirenal fat capsules.
- ☐ Trace the blood supply of the kidney from the renal artery to the renal vein.
- ☐ Define *nephron,* and describe its anatomy.
- ☐ Define *glomerular filtration, tubular reabsorption,* and *tubular secretion,* and indicate the nephron areas involved in these processes.
- ☐ Define *micturition,* and explain the differences in the control of the internal and external urethral sphincters.
- ☐ Recognize the histologic structure of the kidney and ureter microscopically or in an image.

Materials

- Human dissectible torso model, three-dimensional model of the urinary system, and/or anatomical chart of the human urinary system
- Dissecting instruments and tray
- Pig or sheep kidney, doubly or triply injected
- Disposable gloves
- Three-dimensional models of the cut kidney and of a nephron (if available)
- Compound microscope
- Prepared slides of a longitudinal section of kidney and cross sections of the bladder
- Post-it® Notes
- ✂ For instructions on animal dissections, see the dissection exercises (starting on p. 705) in the cat and fetal pig editions of this manual.

MasteringA&P®

For related exercise study tools, go to the Study Area of **MasteringA&P**. There you will find:

- Practice Anatomy Lab **PAL**
- A&PFlix *A&PFlix*
- PhysioEx **PEx**
- Practice quizzes, Histology Atlas, eText, Videos, and more!

Pre-Lab Quiz

1. Circle the correct underlined term. In its excretory role, the urinary system is primarily concerned with the removal of <u>carbon-containing</u> / <u>nitrogenous</u> wastes from the body.

2. The _____ perform(s) the excretory and homeostatic functions of the urinary system.
   - **a.** kidneys
   - **b.** ureters
   - **c.** urinary bladder
   - **d.** all of the above

3. Circle the correct underlined term. The <u>cortex</u> / <u>medulla</u> of the kidney is segregated into triangular regions with a striped appearance.

4. Circle the correct underlined term. As the renal artery approaches a kidney, it is divided into branches known as the <u>segmental arteries</u> / <u>afferent arterioles</u>.

5. What do we call the anatomical units responsible for the formation of urine? _____

6. This knot of coiled capillaries, found in the kidneys, forms the filtrate. It is the:
   - **a.** arteriole
   - **b.** glomerulus
   - **c.** podocyte
   - **d.** tubule

7. The section of the renal tubule closest to the glomerular capsule is the:
   - **a.** collecting duct
   - **b.** distal convoluted tubule
   - **c.** nephron loop
   - **d.** proximal convoluted tubule

8. Circle the correct underlined term. The <u>afferent</u> / <u>efferent</u> arteriole drains the glomerular capillary bed.

9. Circle True or False. During tubular reabsorption, components of the filtrate move from the bloodstream into the tubule.

10. Circle the correct underlined term. The <u>internal</u> / <u>external</u> urethral sphincter consists of skeletal muscle and is voluntarily controlled.

Metabolism of nutrients by the body produces wastes including carbon dioxide, nitrogenous wastes, and ammonia that must be eliminated from the body if normal function is to continue. Excretory processes involve multiple organ systems, with the **urinary system** primarily responsible for the removal of nitrogenous wastes from the body. In addition to this excretory function, the kidney maintains the electrolyte, acid-base, and fluid balances of the blood and is thus a major, if not *the* major, homeostatic organ of the body.

To perform its functions, the kidney acts first as a blood filter, and then as a filtrate processor. It allows toxins, metabolic wastes, and excess ions to leave the body in the urine, while retaining needed substances and returning them to the blood. Malfunction of the urinary system, particularly of the kidneys, leads to a failure in homeostasis which, unless corrected, is fatal.

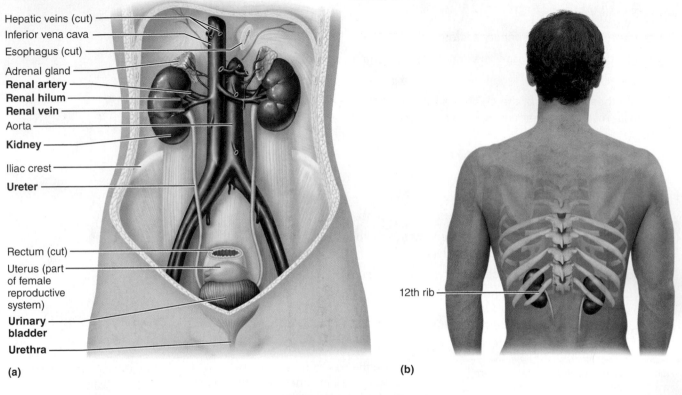

(a)

Hepatic veins (cut)
Inferior vena cava
Esophagus (cut)
Adrenal gland
Renal artery
Renal hilum
Renal vein
Aorta
Kidney
Iliac crest
Ureter
Rectum (cut)
Uterus (part of female reproductive system)
Urinary bladder
Urethra

(b)

12th rib

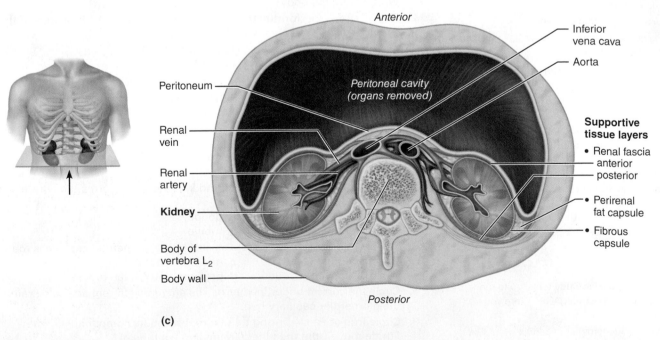

(c)

Anterior

Peritoneum
Renal vein
Renal artery
Kidney
Body of vertebra L₂
Body wall

Peritoneal cavity (organs removed)

Inferior vena cava
Aorta

Supportive tissue layers
• Renal fascia
  anterior
  posterior
• Perirenal fat capsule
• Fibrous capsule

Posterior

Figure 40.1 Organs of the urinary system. (a) Anterior view of the female urinary organs. Most unrelated abdominal organs have been removed. **(b)** Posterior in situ view showing the position of the kidneys relative to the twelfth ribs. **(c)** Cross section of the abdomen viewed from inferior direction. Note the retroperitoneal position and supportive tissue layers of the kidneys.

40

Gross Anatomy of the Human Urinary System

The urinary system (**Figure 40.1**) consists of the paired kidneys and ureters and the single urinary bladder and urethra. The **kidneys** perform the functions described above and manufacture urine in the process. The remaining organs of the system provide temporary storage reservoirs or transportation channels for urine.

Activity 1

Identifying Urinary System Organs

Examine the human torso model, a large anatomical chart, or a three-dimensional model of the urinary system to locate and study the anatomy and relationships of the urinary organs.

1. Locate the paired kidneys on the dorsal body wall in the superior lumbar region. Notice that they are not positioned at exactly the same level. Because it is crowded by the liver, the right kidney is slightly lower than the left kidney. Three layers of support tissue surround each kidney. Beginning with the innermost layer they are: (1) a transparent *fibrous capsule*, (2) a *perirenal fat capsule*, and (3) the fibrous *renal fascia* that holds the kidneys in place in a retroperitoneal position.

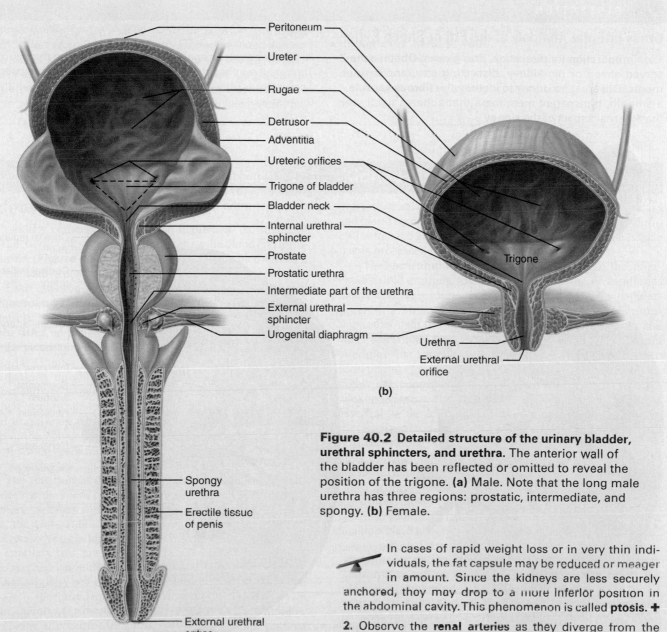

Labels (left figure, a):
- Peritoneum
- Ureter
- Rugae
- Detrusor
- Adventitia
- Ureteric orifices
- Trigone of bladder
- Bladder neck
- Internal urethral sphincter
- Prostate
- Prostatic urethra
- Intermediate part of the urethra
- External urethral sphincter
- Urogenital diaphragm
- Spongy urethra
- Erectile tissue of penis
- External urethral orifice

(a)

Labels (right figure, b):
- Trigone
- Urethra
- External urethral orifice

(b)

Figure 40.2 Detailed structure of the urinary bladder, urethral sphincters, and urethra. The anterior wall of the bladder has been reflected or omitted to reveal the position of the trigone. **(a)** Male. Note that the long male urethra has three regions: prostatic, intermediate, and spongy. **(b)** Female.

In cases of rapid weight loss or in very thin individuals, the fat capsule may be reduced or meager in amount. Since the kidneys are less securely anchored, they may drop to a more inferior position in the abdominal cavity. This phenomenon is called **ptosis. +**

2. Observe the **renal arteries** as they diverge from the descending aorta and plunge into the indented medial region, called the **hilum,** of each kidney. Note also the **renal**

Table 40.1	Structures of the Nephron (Figure 40.4)		
Structure	**Description**	**Epithelium**	**Function**
Structures Within the Renal Corpuscle			
Glomerulus	A cluster of capillaries supplied by the afferent arteriole and drained by the efferent arteriole	Fenestrated endothelium (simple squamous)	Forms part of the filtration membrane
Visceral layer of the glomerular capsule	Podocytes that branch into foot processes	Simple squamous epithelium	Forms part of the filtration membrane. Spaces between the foot processes form filtration slits.
Parietal layer of the glomerular capsule	Outer impermeable wall of the glomerular capsule	Simple squamous epithelium	Forms the outside of the cuplike glomerular capsule. Plays no role in filtration.
Structures Within the Renal Tubule			
Proximal convoluted tubule (PCT)	Highly coiled first section of the renal tubule	Simple cuboidal with many microvilli and many mitochondria	Primary site of tubular reabsorption of water and solutes. Some secretion also occurs.
Descending limb of the nephron loop	First portion of the nephron loop	Simple cuboidal with some microvilli	Tubular reabsorption and secretion of water and solutes.
Descending thin limb of the nephron loop	A continuation of the descending limb	Simple squamous epithelium	Very permeable to water. Water is reabsorbed, but no solutes are reabsorbed.
Thick ascending limb of the nephron loop	In most nephrons the ascending limb is thick	Cuboidal or low columnar, with very few aquaporins	Not permeable to water. Solutes are reabsorbed actively and passively.
Distal convoluted tubule (DCT)	Coiled distal portion of the tubule	Simple cuboidal with few microvilli but many mitochondria	Some reabsorption of water and solutes and secretion, which are regulated to meet the body's needs
Collecting duct	Receives filtrate from the DCT of multiple nephrons	Simple cuboidal epithelium with two specialized cell types: principal cells and intercalated cells	Some reabsorption and secretion to conserve body fluids, maintain blood pH, and regulate solute concentrations.

reabsorption is passive, such as that of water which passes by osmosis, but the reabsorption of most substances depends on active transport processes and is highly selective. Substances that are almost entirely reabsorbed from the filtrate include water, glucose, and amino acids. Various ions are selectively reabsorbed or allowed to go out in the urine according to what is required to maintain appropriate blood pH and electrolyte composition. Waste products including urea, creatinine, uric acid, and drug metabolites are reabsorbed to a much lesser degree or not at all. Most (75% to 80%) of tubular reabsorption occurs in the proximal convoluted tubule.

Tubular secretion is essentially the reverse process of tubular reabsorption. Substances such as hydrogen and potassium ions and creatinine move from the blood of the peritubular capillaries through the tubular cells into the filtrate to be disposed of in the urine.

Activity 2

Studying Nephron Structure

1. Begin your study of nephron structure by identifying the glomerular capsule, proximal and distal convoluted tubule regions, and the nephron loop on a model of the nephron. Then, obtain a compound microscope and a prepared slide of kidney tissue to continue with the microscope study of the kidney.

2. Hold the longitudinal section of the kidney up to the light to identify cortical and medullary areas. Then secure the slide on the microscope stage, and scan the slide under low power.

3. Move the slide so that you can see the cortical area. Identify a glomerulus, which appears as a ball of tightly packed material containing many small nuclei (Figure 40.6). It is usually surrounded by a vacant-appearing region corresponding to the space between the visceral and parietal layers of the glomerular capsule that surrounds it.

4. Notice that the renal tubules are cut at various angles. Try to differentiate between the fuzzy cuboidal epithelium of the proximal convoluted tubule, which has dense microvilli, and that of the distal convoluted tubule with sparse microvilli. Also identify the thin-walled nephron loop.

40

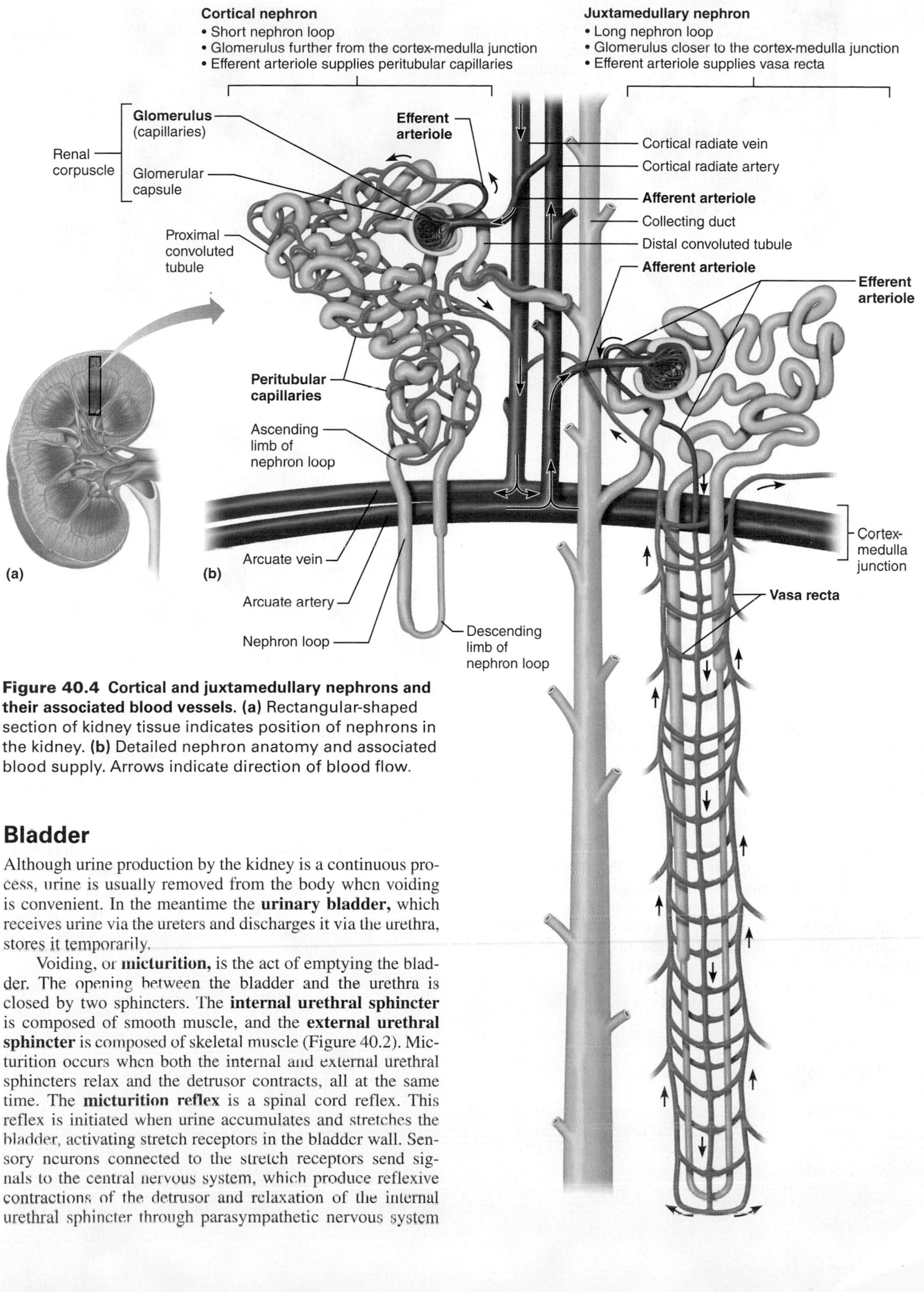
Cortical nephron
• Short nephron loop
• Glomerulus further from the cortex-medulla junction
• Efferent arteriole supplies peritubular capillaries
Juxtamedullary nephron
• Long nephron loop
• Glomerulus closer to the cortex-medulla junction
• Efferent arteriole supplies vasa recta
Renal corpuscle
Glomerulus (capillaries)
Glomerular capsule
Efferent arteriole
Proximal convoluted tubule
Cortical radiate vein
Cortical radiate artery
Afferent arteriole
Collecting duct
Distal convoluted tubule
Afferent arteriole
Efferent arteriole
Peritubular capillaries
Ascending limb of nephron loop
Arcuate vein
Arcuate artery
Nephron loop
Descending limb of nephron loop
Cortex-medulla junction
Vasa recta
(a)
(b)

Figure 40.4 Cortical and juxtamedullary nephrons and their associated blood vessels. (a) Rectangular-shaped section of kidney tissue indicates position of nephrons in the kidney. **(b)** Detailed nephron anatomy and associated blood supply. Arrows indicate direction of blood flow.

Bladder

Although urine production by the kidney is a continuous process, urine is usually removed from the body when voiding is convenient. In the meantime the **urinary bladder,** which receives urine via the ureters and discharges it via the urethra, stores it temporarily.

Voiding, or **micturition,** is the act of emptying the bladder. The opening between the bladder and the urethra is closed by two sphincters. The **internal urethral sphincter** is composed of smooth muscle, and the **external urethral sphincter** is composed of skeletal muscle (Figure 40.2). Micturition occurs when both the internal and external urethral sphincters relax and the detrusor contracts, all at the same time. The **micturition reflex** is a spinal cord reflex. This reflex is initiated when urine accumulates and stretches the bladder, activating stretch receptors in the bladder wall. Sensory neurons connected to the stretch receptors send signals to the central nervous system, which produce reflexive contractions of the detrusor and relaxation of the internal urethral sphincter through parasympathetic nervous system

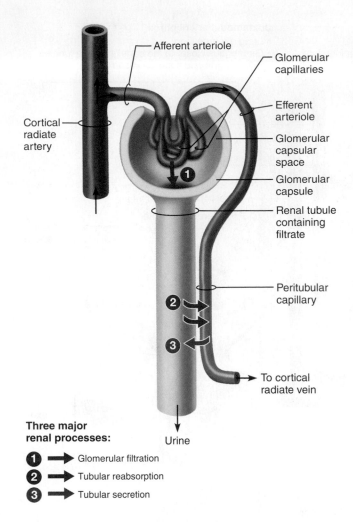

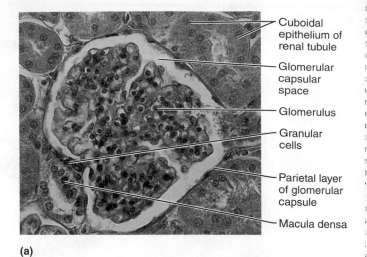

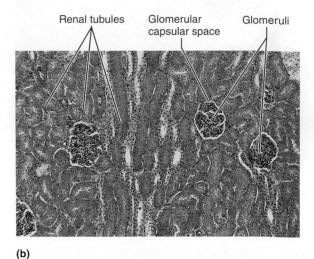

(a)

(b)

Three major renal processes:

1 ➡ Glomerular filtration

2 ➡ Tubular reabsorption

3 ➡ Tubular secretion

Figure 40.5 A schematic, uncoiled nephron. A kidney actually has millions of nephrons acting in parallel. The three major renal processes by which the kidneys adjust the composition of plasma are depicted. Black arrows show the path of blood flow through the renal microcirculation.

Figure 40.6 Microscopic structure of kidney tissue.
(a) Detailed structure of the glomerulus (225×).
(b) Low-power view of the renal cortex (67×).

pathways. Somatic motor neurons leading to the external urethral sphincter are inhibited, causing the skeletal muscle to relax. With both sphincters open and the bladder contracting, urine is voided. Higher brain centers allow or inhibit the micturition reflex, depending on the convenience and desire to urinate. Inhibition of the micturition reflex relies in part on control of the external urethral sphincter.

Lack of voluntary control over the external urethral sphincter is referred to as **incontinence.** Incontinence is normal in children under 2 years old; in older children and adults, it can result from spinal cord injuries or urinary tract pathology.

Activity 3

Studying Bladder Structure

1. Return the kidney slide to the supply area, and obtain a slide of bladder tissue. Scan the bladder tissue. Identify its three layers: mucosa, muscular layer, and fibrous adventitia.

2. Study the highly specialized transitional epithelium of the mucosa. The plump, transitional epithelial cells have the ability to slide over one another, thus decreasing the thickness of the mucosa layer as the bladder fills and stretches to accommodate the increased urine volume. Depending on the degree of stretching of the bladder, the mucosa may be three to eight cell layers thick. (Compare the transitional epithelium of the mucosa to that shown in Figure 6.3h on p. 73).

3. Examine the heavy muscular wall (detrusor), which consists of three irregularly arranged muscular layers. The innermost and outermost muscle layers are arranged longitudinally; the middle layer is arranged circularly. Attempt to differentiate the three muscle layers.

4. Draw a small section of the bladder wall, and label all regions or tissue areas.

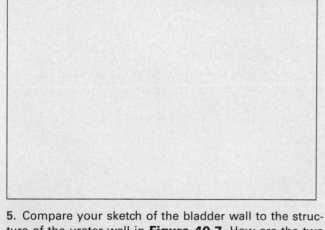

5. Compare your sketch of the bladder wall to the structure of the ureter wall in **Figure 40.7**. How are the two organs similar histologically?

_____

_____

_____

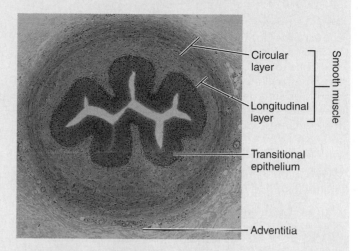

Figure 40.7 Structure of the ureter wall. Cross section of ureter (35×).

What is/are the most obvious differences?

_____

_____

WHY THIS MATTERS | Elimination Communication

Although babies do not have the ability to inhibit the micturition reflex, they can be conditioned to void at certain times. Elimination Communication (EC) is a new technique that takes advantage of this conditioned response. To achieve EC, a watchful parent must pick up on clues that indicate when a baby needs to void or defecate. Parents condition the baby to eliminate (urine or feces) when prompted with a verbal cue. Although the baby has not fully achieved control of the micturition reflex, this ability to briefly delay voiding enables parents to reduce the use of diapers and cut down on the waste that goes into landfills. ■

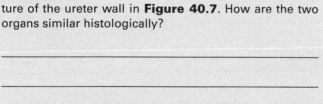

For instructions on animal dissections, see the dissection exercises (starting on p. 705) in the cat and fetal pig editions of this manual.

40

Group Challenge

Urinary System Sequencing

Arrange the following sets of urinary structures in the correct order for the flow of urine, filtrate, or blood. Work in small groups, but refrain from using a figure or other reference to determine the sequence. Within your group, assign a facilitator and a recorder. The facilitator will list each term in a given set on a separate Post-it® Note. All members of the group will discuss the correct order for the structures and arrange them accordingly. After all of the Post-it® Notes have been arranged, the recorder will write down the terms in the appropriate order.

1. renal pelvis, minor calyx, renal papilla, urinary bladder, ureter, major calyx, and urethra

_____

_____

2. distal convoluted tubule, ascending limb of the nephron loop, glomerulus, collecting duct, descending limb of the nephron loop, proximal convoluted tubule, and glomerular capsule

_____

_____

3. segmental artery, afferent arteriole, cortical radiate artery, glomerulus, renal artery, interlobar artery, and arcuate artery _____

_____

4. arcuate vein, inferior vena cava, peritubular capillaries, renal vein, interlobar vein, cortical radiate vein, and efferent arteriole _____

_____

REVIEW SHEET
Anatomy of the Urinary System

Name _____ Lab Time/Date _____

Gross Anatomy of the Human Urinary System

1. Complete the following statements.

   The kidney is referred to as an excretory organ because it excretes __1__ wastes. It is also a major homeostatic organ because it maintains the electrolyte, __2__, and __3__ balance of the blood.

   Urine is continuously formed by the structural and functional units of the kidneys, the __4__, and is routed down the __5__ by the mechanism of __6__ to a storage organ called the __7__. Eventually, the urine is conducted to the body __8__ by the urethra. In the male, the urethra is __9__ centimeters long and transports both urine and __10__. The female urethra is __11__ centimeters long and transports only urine.

   Voiding or emptying the bladder is called __12__. Voiding has both voluntary and involuntary components. The voluntary sphincter is the __13__ sphincter. An inability to control this sphincter is referred to as __14__.

1. _____

2. _____

3. _____

4. _____

5. _____

6. _____

7. _____

8. _____

9. _____

10. _____

11. _____

12. _____

13. _____

14. _____

2. What is the function of the fat cushion that surrounds the kidneys in life? _____

_____

3. Define *ptosis.* _____

_____

4. Complete the labeling of the diagram to correctly identify the urinary system organs.

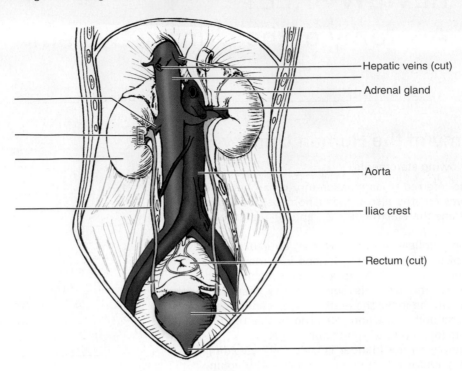

Hepatic veins (cut)

Adrenal gland

Aorta

Iliac crest

Rectum (cut)

Gross Internal Anatomy of the Pig or Sheep Kidney

5. Match the appropriate structure in column B to its description in column A. The items in column B may be used more than once.

Column A

_____ 1. smooth membrane, tightly adherent to the kidney surface

_____ 2. portion of the kidney containing mostly collecting ducts

_____ 3. portion of the kidney containing the bulk of the nephron structures

_____ 4. superficial region of kidney tissue

_____ 5. basinlike area of the kidney, continuous with the ureter

_____ 6. a cup-shaped extension of the pelvis that encircles the apex of a pyramid

_____ 7. area of cortical tissue running between the medullary pyramids

Column B

a.   cortex

b.   fibrous capsule

c.   medulla

d.   minor calyx

e.   renal column

f.   renal pelvis

Functional Microscopic Anatomy of the Kidney and Bladder

6. Label the blood vessels and parts of the nephron by selecting the letter for the correct structure from the key below. The items in the key may be used more than once.

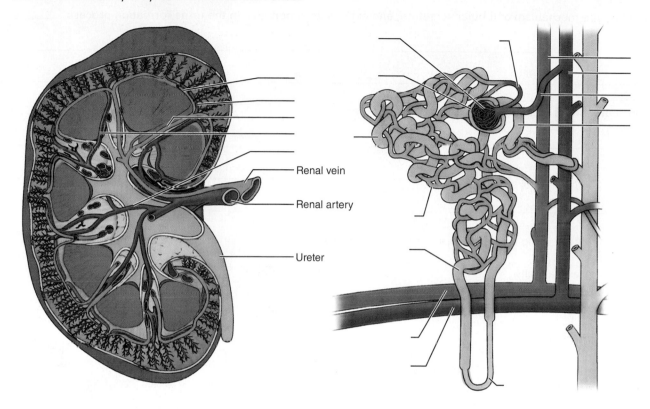

Renal vein

Renal artery

Ureter

Key:

a. afferent arteriole

b. arcuate artery

c. arcuate vein

d. collecting duct

e. cortical radiate artery

f. cortical radiate vein

g. distal convoluted tubule

h. efferent arteriole

i. glomerular capsule

j. glomerulus

k. interlobar artery

l. interlobar vein

m. nephron loop—ascending limb

n. nephron loop—descending limb

o. peritubular capillary

p. proximal convoluted tubule

q. segmental artery

7. For each of the following descriptions of a structure, find the matching name in the question 6 key.

_____ 1. capillary specialized for filtration

_____ 2. capillary specialized for reabsorption

_____ 3. cuplike part of the renal corpuscle

_____ 4. location of macula densa

_____ 5. primary site of tubular reabsorption

_____ 6. receives urine from many nephrons

8. Explain *why* the glomerulus is such a high-pressure capillary bed. _____

_____

How does its high-pressure condition aid its function of filtrate formation? _____

_____

9. What structural modification of certain tubule cells enhances their ability to reabsorb substances from the filtrate?

_____

10. Explain the mechanism of tubular secretion, and explain its importance in the urine formation process. _____

_____

_____

11. Compare and contrast the composition of blood plasma and glomerular filtrate. _____

_____

12. Define *juxtaglomerular complex.* _____

_____

13. Label the figure using the key letters of the correct terms.

    Key: a. granular cells

             b. cuboidal epithelium

             c. macula densa

             d. glomerular capsule (parietal layer)

             e. ascending limb of the nephron loop

14. What is important functionally about the specialized epithelium (transitional epithelium) in the bladder?

_____

_____

WHY THIS MATTERS | 15. Name and describe the sphincter that provides conscious control over the micturition reflex. _____

_____

16. The practice of Elimination Communication seems to indicate that babies can gain the ability to inhibit the micturition reflex. Hypothesize how this control might be achieved and which nervous system (central, autonomic, or somatic) is most likely involved. _____

_____

_____

Urinalysis

Objectives

☐ List the physical characteristics of urine, and indicate the normal pH and specific gravity ranges.

☐ List substances that are normal urinary constituents.

☐ Conduct various urinalysis tests and procedures, and use them to determine the substances present in a urine specimen.

☐ Define the following urinary conditions:

calculi	albuminuria	hemoglobinuria
casts	glycosuria	ketonuria
	hematuria	pyuria

☐ Discuss the possible causes and implications of the conditions listed above.

Materials

- Disposable gloves
- Student urine samples collected at the beginning of the laboratory or "normal" artificial urine provided by the instructor*
- Numbered "pathological" urine specimens provided by the instructor*
- Wide-range pH paper
- Dipsticks: individual (Clinistix®, Ketostix®, Albustix®, Hemastix®) or combination (Chemstrip® or Multistix®)
- Urinometer
- Test tubes, test tube rack, and test tube holders
- 10-cc graduated cylinders

*Directions for making artificial urine are provided in the Instructor's Guide for this manual.

Text continues on next page. →

MasteringA&P®

For related exercise study tools, go to the Study Area of **MasteringA&P.** There you will find:

- Practice Anatomy Lab **PAL**
- A&PFlix **A&PFlix**
- PhysioEx **PEx**
- Practice quizzes, Histology Atlas, eText, Videos, and more!

Pre-Lab Quiz

1. Normal urine is usually pale yellow to amber in color because of the presence of:
   a. hemochrome
   b. melanin
   c. urochrome

2. Circle the correct underlined term. The average pH value of urine is 6.0 / 11.0.

3. Circle True or False. Glucose can usually be found in all normal urine.

4. _____, like other blood proteins, is/are too large to pass through the glomerular filtration membrane and is/are normally not found in urine.
   a. Albumin   **c.** Nitrates
   b. Chloride   **d.** Sulfate

5. Circle the correct underlined term. Hematuria / Ketonuria, the appearance of red blood cells in the urine, almost always indicates pathology of the urinary system.

6. The appearance of bile pigments in the urine, a condition known as _____, can be an indication of liver disease.
   a. albuminuria   **c.** ketonuria
   b. bilirubinuria   **d.** pyuria

7. Circle the correct underlined term. Proteinuria / Pyuria, the presence of white blood cells or pus in the urine, is consistent with inflammation of the urinary tract.

8. Circle the correct underlined term. Casts / Calculi are hardened cell fragments formed in the distal convoluted tubules and collecting ducts and flushed out of the urinary tract.

9. Circle True or False. When testing the pH of the urine, you will use the same piece of wide-range pH paper for each test.

10. When determining the presence of inorganic constituents such as sulfates, phosphates, and chlorides, you will be looking for the "formation of a precipitate." What is a precipitate? _____

- Test reagents for sulfates: 10% barium chloride solution, dilute hydrochloric acid (HCl)
- Hot plate
- 500-ml beaker
- Test reagent for phosphates: dilute nitric acid (HNO_3), dilute ammonium molybdate
- Glass stirring rod
- Test reagent for chloride: 3.0% silver nitrate solution ($AgNO_3$), freshly prepared
- Clean microscope slide and coverslip
- Compound microscope
- Test reagent for urea: concentrated nitric acid in dropper bottles
- Test reagent for glucose: Clinitest® tablets; Clinitest color chart
- Medicine droppers
- Timer (watch or clock with a second hand)
- Ictotest® reagent tablets and test mat
- Flasks and laboratory buckets containing 10% bleach solution
- Disposable autoclave bags
- *Demonstration:* Instructor-prepared specimen of urine sediment set up for microscopic analysis
- **PEx** PhysioEx™ 9.1 Computer Simulation Ex. 9 on p. PEx-131.

B lood composition depends on three major factors: diet, cellular metabolism, and urinary output. In 24 hours, the kidneys' 2 million nephrons filter 150 to 180 liters of blood plasma through their glomeruli into the tubules, where it is selectively processed by tubular reabsorption and secretion. In the same period, urinary output, which contains by-products of metabolism and excess ions, is 1.0 to 1.8 liters. In healthy people, the kidneys can maintain blood constancy despite wide variations in diet and metabolic activity.

Characteristics of Urine

Color and Transparency

Freshly voided urine is generally clear and pale yellow to amber in color. This normal yellow color is due to *urochrome*, a pigment metabolite that arises from the body's destruction of hemoglobin and travels to the kidney as bilirubin or bile pigments. As a rule, color variations from pale yellow to deeper amber indicate the relative concentration of solutes to water in the urine. The greater the solute concentration, the deeper the color. Abnormal urine color may be due to certain foods, such as beets, various drugs, bile, or blood. Cloudy urine may indicate a urinary tract infection.

Odor

The odor of freshly voided urine is slightly aromatic, but bacterial action gives it an ammonia-like odor when left standing. Some drugs, vegetables (such as asparagus), and various disease processes (such as diabetes mellitus) alter the characteristic odor of urine. For example, the urine of a person with uncontrolled diabetes mellitus (and elevated levels of ketones) smells fruity or acetone-like.

pH

The pH of urine ranges from 4.5 to 8.0, but its average value, 6.0, is slightly acidic. Diet may markedly influence the pH of the urine. For example, a diet high in protein (meat, eggs, cheese) and whole wheat products increases the acidity of urine. Conversely, a vegetarian diet usually increases the alkalinity of the urine. A bacterial infection of the urinary tract may also cause the urine to become more alkaline.

Specific Gravity

Specific gravity is the relative weight of a specific volume of liquid compared with an equal volume of distilled water. The specific gravity of distilled water is 1.000, because 1 ml weighs 1 g. Since urine contains dissolved solutes, it weighs more than water, and its customary specific gravity ranges from 1.001 to 1.030. Urine with a specific gravity of 1.001 contains few solutes and is considered very dilute. Dilute urine commonly results when a person drinks excessive amounts of water, uses diuretics, or suffers from diabetes insipidus or chronic renal failure. Conditions that produce urine with a high specific gravity include limited fluid intake, fever, diabetes mellitus, gonorrhea, and kidney inflammation, called *pyelonephritis.* If urine becomes excessively concentrated, some of the substances normally held in solution begin to precipitate or crystallize, forming **kidney stones,** or **renal calculi.**

Water is the largest component of urine, accounting for 95% of its volume. The second largest component of urine is urea. Nitrogenous wastes in the urine include urea, uric acid, and creatinine. *Urea* comes from the breakdown of proteins. *Uric acid* is a breakdown product from nucleic acids. *Creatinine* is a metabolite produced from the metabolism of creatine phosphate in muscle tissue.

Normal solute constituents of urine, in order of decreasing concentration, include urea, sodium, potassium, phosphate, and sulfate ions; creatinine; and uric acid. Much smaller but highly variable amounts of calcium, magnesium, and bicarbonate ions are also found in the urine. Abnormally high concentrations of any of these urinary constituents may indicate a pathological condition.

Abnormal Urinary Constituents

Abnormal urinary constituents are substances not normally present in the urine when the body is operating properly. +

When certain pathological conditions are present, urine composition often changes dramatically. **Table 41.1** identifies substances that are not normally found in the urine and describes their characteristics.

Casts

Any complete discussion of the varieties and implications of casts is beyond the scope of this exercise. However, because they always represent a pathological condition of the kidney or urinary tract, they should at least be mentioned. **Casts** are hardened cell fragments, usually cylindrical, which are formed in the distal convoluted tubules and collecting ducts and then flushed out of the urinary tract. Hyaline casts are formed from a mucoprotein secreted by tubule cells (Figure 41.1b, p. 632). These casts form when the filtrate flow rate is slow, the pH is low, or the salt concentration is high, all conditions that cause protein to denature. Red blood cell casts are typical in glomerulonephritis, as red blood cells leak through the filtration membrane and stick together in the tubules. White blood cell casts form when the kidney is inflamed, which is typically a result of pyelonephritis (a type of urinary tract infection) but sometimes occurs with glomerulonephritis. Degenerated renal tubule cells form granular casts (Figure 41.1b).

Activity 1

Analyzing Urine Samples

In this part of the exercise, you will use prepared dipsticks and perform chemical tests to determine the characteristics of normal urine as well as to identify abnormal urinary components. You will investigate two or more urine samples. The first, designated as the *standard urine specimen* in the **Activity 1 chart** (p. 631), will be either yours or a "standard" sample provided by your instructor. The second will be an unknown urine specimen provided by your instructor. Make the following determinations on both samples, and record your results by circling the appropriate item or description or by adding data to complete the chart. If you have more than one unknown sample, accurately identify each sample by number.

Text continues on next page. →

Table 41.1	**Abnormal Urinary Constituents**			
Abnormal urinary constituent	**Clinical term**	**Description**		**Possible conditions**
Glucose	Glycosuria (glucosuria)	High blood sugar levels due to inadequate insulin levels; or can result when active transport mechanisms for glucose are exceeded temporarily		Pathological: uncontrolled diabetes mellitus Nonpathological: Excessive carbohydrate intake
Protein	Proteinuria (albuminuria)	Increased permeability of the glomerular filtration membrane (proteins are usually too large to pass through); albumin is the most abundant blood protein		Pathological: hypertension, glomerulonephritis, ingestion of poisons, bacterial toxins, kidney trauma Nonpathological: excessive physical exertion, pregnancy
Ketone bodies	Ketonuria	Excessive production of intermediates of fat metabolism, which may result in acidosis		Uncontrolled diabetes mellitus, starvation, low-carbohydrate diets
Erythrocytes (RBCs)	Hematuria	Irritation of the urinary tract organs that results in bleeding; or a result of leakage of RBCs through a damaged filtration membrane		Bleeding in the tract: kidney stones, urinary tract tumors, trauma to urinary tract organs Damaged filtration membrane: glomerulonephritis
Hemoglobin	Hemoglobinuria	Fragmentation of erythrocytes, resulting in the release of hemoglobin into the plasma and subsequently into the filtrate		Hemolytic anemia, transfusion reactions, severe burns, poisonous snake bites, renal disease
Nitrites	Nitrituria	Results when gram-negative bacteria such as *E. coli* reduce nitrates to form nitrites		Urinary tract infections (UTIs)
Bile pigments	Bilirubinuria	Increased levels of bilirubin in the urine as a result of liver damage or blockage of the bile duct		Hepatitis, cirrhosis of the liver, gallstones
Leukocytes (WBCs)	Pyuria	Presence of WBCs or pus in the urine caused by inflammation of the urinary tract		Urinary tract infections (including pyelonephritis), gonorrhea

41

⚠️ *Obtain and wear disposable gloves throughout this laboratory session.* Although the instructor-provided urine samples are actually artificial urine (concocted in the laboratory to resemble real urine), you should still observe the techniques of safe handling of body fluids as part of your learning process. When you have completed the laboratory procedures: (1) dispose of the gloves, used pH paper strips, and dipsticks in the autoclave bag; (2) put used glassware in the bleach-containing laboratory bucket; (3) wash the lab bench down with 10% bleach solution.

Determination of the Physical Characteristics of Urine

1. Determine the color, transparency, and odor of your "standard" sample and one of the numbered pathological samples, and circle the appropriate descriptions in the Activity 1 chart.

2. Obtain a roll of wide-range pH paper to determine the pH of each sample. Use a fresh piece of paper for each test, and dip the strip into the urine to be tested two or three times before comparing the color obtained with the chart on the dispenser. Record your results in the chart. (If you will be using one of the combination dipsticks—Chemstrip or Multistix—you can use these dipsticks to determine pH.)

3. To determine specific gravity, obtain a urinometer cylinder and float. Mix the urine well, and fill the urinometer cylinder about two-thirds full with urine.

4. Examine the urinometer float to determine how to read its markings. In most cases, the scale has numbered lines separated by a series of unnumbered lines. The numbered lines give the reading for the first two decimal places. You must determine the third decimal place by reading the lower edge of the meniscus—the curved surface representing the urine-air junction—on the stem of the float.

5. Carefully lower the urinometer float into the urine. Make sure it is floating freely before attempting to take the reading. Record the specific gravity of both samples in the chart. _Do not dispose of this urine if the samples that you have are less than 200 ml in volume_ because you will need to make several more determinations.

Determination of Inorganic Constituents in Urine

Sulfates

Using a 10-cc graduated cylinder, add 5 ml of urine to a test tube, and then add a few drops of dilute hydrochloric acid and 2 ml of 10% barium chloride solution. The appearance of a white precipitate (barium sulfate) indicates the presence of sulfates in the sample. Clean the graduated cylinder and the test tubes well after use. Record your results.

Phosphates

Obtain a hot plate and a 500-ml beaker. To prepare the hot water bath, half fill the beaker with tap water and heat it on the hot plate. Add 5 ml of urine to a test tube, and then add three or four drops of dilute nitric acid and 3 ml of ammonium molybdate. Mix well with a glass stirring rod, and then heat gently in a hot water bath. Formation of a yellow precipitate indicates the presence of phosphates in the sample. Record your results.

Chlorides

Place 5 ml of urine in a test tube, and add several drops of silver nitrate. The appearance of a white precipitate (silver chloride) is a positive test for chlorides. Record your results.

Nitrites

Use a combination dipstick to test for nitrites. Record your results.

Determination of Organic Constituents in Urine

Individual dipsticks or combination dipsticks (Chemstrip or Multistix) may be used for many of the tests in this section. If you are using combination dipsticks, be prepared to take the readings on several factors (pH, protein [albumin], glucose, ketones, blood/hemoglobin, leukocytes, urobilinogen, bilirubin, and nitrites) at the same time. Generally speaking, results for all of these tests may be read *during* the second minute after immersion, but readings taken after 2 minutes have passed should be considered invalid. Pay careful attention to the directions for method and time of immersion and disposal of excess urine from the strip, regardless of the dipstick used. Identify the dipsticks that you use in the chart. If you are testing your own urine and get an unanticipated result, it is helpful to know that most of the combination dipsticks produce false positive or negative results for certain solutes when the subject is taking vitamin C, aspirin, or certain drugs.

Urea

Put two drops of urine on a clean microscope slide and *carefully* add one drop of concentrated nitric acid to the urine. Slowly warm the mixture on a hot plate until it begins to dry at the edges, but do not allow it to boil or to evaporate to dryness. When the slide has cooled, examine the edges of the preparation under low power to identify the rhombic or hexagonal crystals of urea nitrate, which form when urea and nitric acid react chemically. Keep the light low for best contrast. Record your results.

Glucose

Use a combination dipstick or obtain a vial of Clinistix, and conduct the dipstick test according to the instructions on the vial. Record your results in the Activity 1 chart.

Because the Clinitest reagent is routinely used in clinical agencies for glucose determinations in pediatric patients, it is worthwhile to conduct this test as well. Obtain the Clinitest tablets and the associated color chart. You will need a timer (watch or clock with a second hand) for this test. Using a medicine dropper, put 5 drops of urine into a test tube; then rinse the dropper and add 10 drops of water to the tube. Add a Clinitest tablet. Wait 15 seconds and then compare the color obtained to the color chart. Record your results.

Protein

Use a combination dipstick or obtain the Albustix dipsticks, and conduct the determinations as indicated on the vial. Record your results.

Ketones

Use a combination dipstick or obtain the Ketostix dipsticks. Conduct the determinations as indicated on the vial. Record your results.

Blood/Hemoglobin

Test your urine samples for the presence of hemoglobin by using a Hemastix dipstick or a combination dipstick according to the directions on the vial. Usually a short drying period is required before making the reading, so read the directions carefully. Record your results.

Bilirubin

Using a combination dipstick, determine if there is any bilirubin in your urine samples. Record your results.

Also conduct the Ictotest for the presence of bilirubin. Using a medicine dropper, place one drop of urine in the center of one of the special test mats provided with the Ictotest reagent tablets. Place one of the reagent tablets over the drop of urine, and then add two drops of water directly to the tablet. If the mixture turns purple when you add water, bilirubin is present. Record your results.

Leukocytes

Use a combination dipstick to test for leukocytes. Record your results.

Urobilinogen

Use a combination dipstick to test for urobilinogen. Record your results.

Clean up your area following the procedures described at the beginning of this activity.

Activity 1: Urinalysis Results					
Observation or test	**Normal values**	**Standard urine specimen**		**Unknown specimen (# _____)**	
Physical Characteristics					
Color	Pale yellow	Yellow: pale medium dark		Yellow: pale medium dark	
		other _____		other _____	
Transparency	Clear	Clear Slightly cloudy Cloudy		Clear Slightly cloudy Cloudy	
Odor	Aromatic	Describe:		Describe:	
		_____		_____	
pH	4.5–8.0	_____		_____	
Specific gravity	1.001–1.030	_____		_____	
Inorganic Components					
Sulfates	Present	Present	Absent	Present	Absent
Phosphates	Present	Present	Absent	Present	Absent
Chlorides	Present	Present	Absent	Present	Absent
Nitrites	Absent	Present	Absent	Present	Absent
Dipstick: _____					
Organic Components					
Urea	Present	Present	Absent	Present	Absent
Glucose					
Dipstick: _____	Negative	Record results:		Record results:	
		_____		_____	
Clinitest	Negative	_____		_____	
Protein					
Dipstick: _____	Negative	_____		_____	
Ketone bodies					
Dipstick: _____	Negative	_____		_____	
RBCs/hemoglobin					
Dipstick: _____	Negative	_____		_____	
Bilirubin					
Dipstick: _____	Negative	_____		_____	
Ictotest	Negative (no color change)	Negative	Positive (purple)	Negative	Positive (purple)
Leukocytes	Absent	Present	Absent	Present	Absent
Dipstick: _____					
Urobilinogen	Present	Present	Absent	Present	Absent
Dipstick: _____					

41

Activity 2

Analyzing Urine Sediment Microscopically (Optional)

If your instructor so indicates, conduct a microscopic analysis of urine sediment in "real" urine. The urine sample to be analyzed microscopically has been centrifuged to spin the more dense urine components to the bottom of a tube, and some of the sediment has been mounted on a slide and stained with Sedi-stain to make the components more visible.

Go to the demonstration microscope to conduct this study. Using the lowest light source possible, examine the slide under low power to determine whether you can see any common sediments (**Figure 41.1**).

Unorganized sediments: Chemical substances that form crystals or precipitate from solution; for example, calcium oxalates, carbonates, and phosphates; uric acid; ammonium ureates; and cholesterol. Also, if one has been taking antibiotics or certain drugs such as sulfa drugs, these may be detectable in the urine in crystalline form. Normal urine contains very small amounts of crystals, but conditions such as urinary retention or urinary tract infection may cause the appearance of much larger amounts. The high-power lens may be needed to view the various crystals, which tend to be much more minute than the organized cellular sediments.

Organized sediments: Include epithelial cells (rarely of any pathological significance), white blood cells, red blood cells, and casts. The presence of white blood cells, red blood cells, and casts other than trace amounts always indicates kidney pathology. Note that red blood cells, white blood cells, and epithelial cells can also form casts.

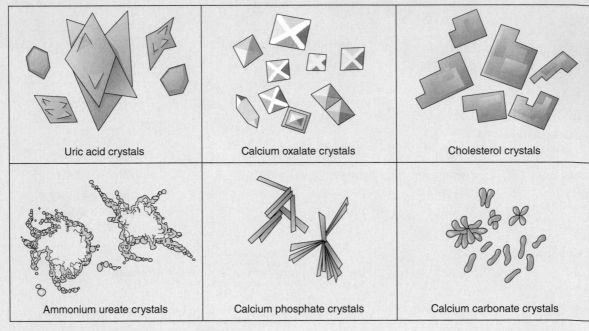

Uric acid crystals   Calcium oxalate crystals   Cholesterol crystals

Ammonium ureate crystals   Calcium phosphate crystals   Calcium carbonate crystals

(a) Unorganized sediments

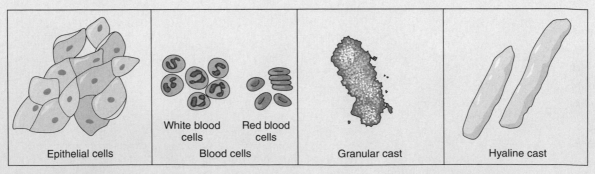

Epithelial cells   White blood cells   Red blood cells   Granular cast   Hyaline cast
                     Blood cells

(b) Organized sediments

Figure 41.1 Examples of sediments.

Name _____ LabTime/Date _____

Characteristics of Urine

1. What is the normal volume of urine excreted in a 24-hour period? _____

2. Assuming normal conditions, note whether each of the following substances would be (a) in greater relative concentration in the urine than in the glomerular filtrate, (b) in lesser concentration in the urine than in the glomerular filtrate, or (c) absent from both the urine and the glomerular filtrate. Use an appropriate reference as needed.

_____ 1. water   _____ 6. amino acids   _____ 11. uric acid

_____ 2. phosphate ions   _____ 7. glucose   _____ 12. creatinine

_____ 3. sulfate ions   _____ 8. protein   _____ 13. white blood cells

_____ 4. potassium ions   _____ 9. red blood cells   _____ 14. nitrites

_____ 5. sodium ions   _____ 10. urea

3. Explain why urinalysis is a routine part of any good physical examination. _____

_____

4. What substance is responsible for the normal yellow color of urine? _____

5. Which has a greater specific gravity: 1 ml of urine or 1 ml of distilled water? _____ Explain your answer. _____

_____

6. Explain the relationship between the color, specific gravity, and volume of urine. _____

_____

Abnormal Urinary Constituents

7. A microscopic examination of urine may reveal the presence of certain abnormal urinary constituents.

   Name three constituents that might be present if a urinary tract infection exists.   _____,

   _____, and _____

8. How does a urinary tract infection influence urine pH? _____

   How does starvation influence urine pH? _____

9. All urine specimens become alkaline and cloudy on standing at room temperature. Explain why. _____

_____

10. Several specific terms have been used to indicate the presence of abnormal urine constituents. Identify each of the abnormalities described below by inserting a term from the key at the right that names the condition.

_____ 1. presence of erythrocytes in the urine

_____ 2. presence of hemoglobin in the urine

_____ 3. presence of glucose in the urine

_____ 4. presence of protein in the urine

_____ 5. presence of ketone bodies in the urine

_____ 6. presence of white blood cells in the urine

Key:

a.   glycosuria

b.   hematuria

c.   hemoglobinuria

d.   ketonuria

e.   proteinuria

f.   pyuria

11. What are renal calculi, and what conditions favor their formation? _____

_____

12. Glucose and protein are both normally absent in the urine, but the reason for their exclusion differs. Explain the reason for

the absence of glucose. _____

Explain the reason for the absence of protein. _____

13. The presence of abnormal constituents or conditions in urine may be associated with diseases, disorders, or other causes listed in the key. Select and list all conditions associated with each numbered item. Some numbered items will have multiple letters.

_____ 1. low specific gravity

_____ 2. high specific gravity

_____ 3. glucose

_____ 4. protein

_____ 5. blood cells

_____ 6. hemoglobin

_____ 7. bile pigments

_____ 8. ketone bodies

_____ 9. casts

_____ 10. pus

Key:

a.   cirrhosis of the liver

b.   diabetes insipidus (uncontrolled)

c.   diabetes mellitus (uncontrolled)

d.   eating a 2-lb box of sweets for lunch

e.   glomerulonephritis

f.   gonorrhea

g.   hemolytic anemias

h.   hepatitis

i.   kidney stones

j.   pregnancy, exertion

k.   pyelonephritis

l.   starvation

14. Name the three major nitrogenous wastes found in the urine. _____,

_____, and _____

15. Explain the difference between organized and unorganized sediments. _____

_____

_____

42 Anatomy of the Reproductive System

Objectives

☐ Discuss the general function of the reproductive system.

☐ Identify the structures of the male and female reproductive systems on an appropriate model or image, and list the general function of each.

☐ Define *semen,* state its composition, and name the organs involved in its production.

☐ Trace the pathway followed by a sperm from its site of formation to the external environment.

☐ Define *erection* and *ejaculation.*

☐ Define *gonad,* and name the gametes and endocrine products of the testes and ovaries, indicating the cell types or structures responsible for the production of each.

☐ Describe the microscopic structure of the penis, seminal glands, epididymis, uterine wall and uterine tube, and relate structure to function.

☐ Explain the role of the fimbriae and ciliated epithelium of the uterine tubes in the movement of the egg from the ovary to the uterus.

☐ Identify the fundus, body, and cervical regions of the uterus.

☐ Define *endometrium, myometrium,* and *ovulation.*

☐ Describe the anatomy and discuss the reproduction-related function of female mammary glands.

Materials

* Three-dimensional models or large laboratory charts of the male and female reproductive tracts
* Prepared slides of cross sections of the penis, seminal glands, epididymis, uterus showing endometrium (secretory phase), and uterine tube
* Compound microscope

For instructions on animal dissections, see the dissection exercises (starting on p. 705) in the cat and fetal pig editions of this manual.

MasteringA&P®

For related exercise study tools, go to the Study Area of **MasteringA&P.** There you will find:

* Practice Anatomy Lab **PAL**
* PhysioEx **PEx**
* A&PFlix **A&PFlix**
* Practice quizzes, Histology Atlas, eText, Videos, and more!

Pre-Lab Quiz

1. The essential organs of reproduction are the _____, which produce the sex cells.
   a. accessory male glands   c. seminal glands
   b. gonads   d. uterus

2. Circle the correct underlined term. The paired oval testes lie in the scrotum / prostate outside the abdominopelvic cavity, where they are kept slightly cooler than body temperature.

3. After sperm are produced, they enter the first part of the duct system, the:
   a. ductus deferens   c. epididymis
   b. ejaculatory duct   d. urethra

4. The prostate, seminal glands, and bulbo-urethral glands produce _____, the liquid medium in which sperm leaves the body.
   a. seminal fluid   b. testosterone   c. urine   d. water

5. Circle the correct underlined term. The interstitial endocrine cells / seminiferous tubules produce testosterone, the hormonal product of the testis.

6. The endocrine products of the ovaries are estrogen and:
   a. luteinizing hormone   c. prolactin
   b. progesterone   d. testosterone

7. Circle the correct underlined term. The labia majora / clitoris are/is homologous to the penis.

8. The _____ is a pear-shaped organ that houses the embryo or fetus during its development.
   a. bladder   b. cervix   c. uterus   d. vagina

Text continues on next page. →

9. Circle the correct underlined term. The <u>endometrium</u> / <u>myometrium</u>, the thick mucosal lining of the uterus, has a superficial layer that sloughs off periodically.
10. Circle the correct underlined term. A developing egg is ejected from the ovary at the appropriate stage of maturity in an event known as <u>menstruation</u> / <u>ovulation</u>.

Most organ systems of the body function from the time they are formed to sustain the existing individual. However, the **reproductive system** begins its biological function, the production of offspring, at puberty.

The essential organs of reproduction are the **gonads,** the testes and the ovaries, which produce the sex cells, or **gametes,** and the sex hormones. The reproductive role of the male is to manufacture sperm and to deliver them to the female reproductive tract. The female, in turn, produces eggs. If the time is suitable, the combination of sperm and egg produces a fertilized egg, which is the first cell of a new individual. Once fertilization has occurred, the female uterus provides a nurturing, protective environment in which the embryo, later called the fetus, develops until birth.

Gross Anatomy of the Human Male Reproductive System

The primary reproductive organs of the male are the **testes,** which produce sperm and the male sex hormones. All other reproductive structures are ducts or sources of secretions, which aid in the safe delivery of the sperm to the body exterior or female reproductive tract.

Activity 1

Identifying Male Reproductive Organs

As the following organs and structures are described, locate them on **Figure 42.1** and then identify them on a three-dimensional model of the male reproductive system or on a large laboratory chart.

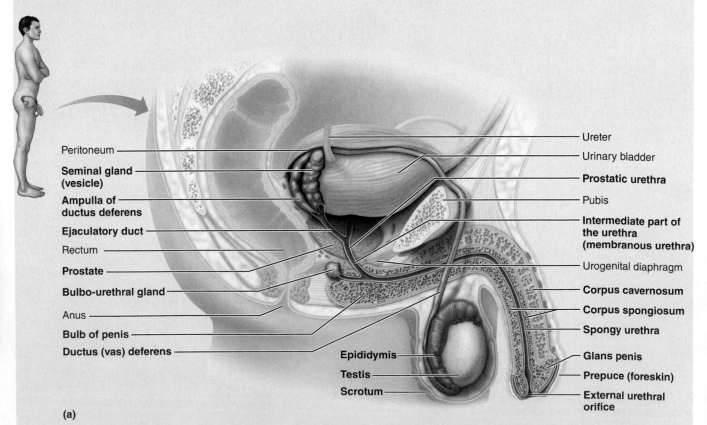

Peritoneum
Seminal gland (vesicle)
Ampulla of ductus deferens
Ejaculatory duct
Rectum
Prostate
Bulbo-urethral gland
Anus
Bulb of penis
Ductus (vas) deferens
Epididymis
Testis
Scrotum
Ureter
Urinary bladder
Prostatic urethra
Pubis
Intermediate part of the urethra (membranous urethra)
Urogenital diaphragm
Corpus cavernosum
Corpus spongiosum
Spongy urethra
Glans penis
Prepuce (foreskin)
External urethral orifice

(a)

Figure 42.1 Reproductive organs of the human male. (a) Sagittal view.

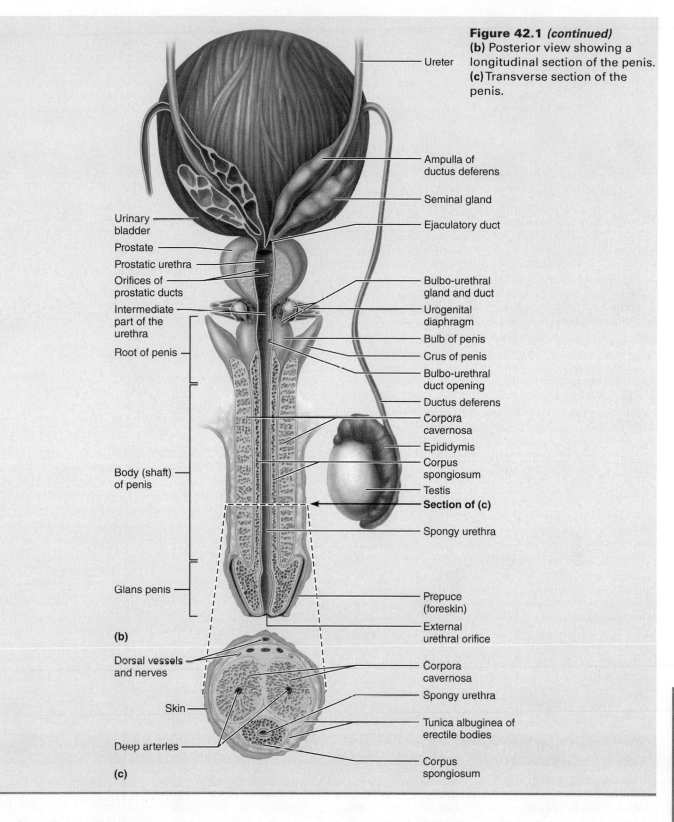

Figure 42.1 *(continued)*
(b) Posterior view showing a longitudinal section of the penis.
(c) Transverse section of the penis.

Ureter

Ampulla of ductus deferens

Seminal gland

Ejaculatory duct

Urinary bladder

Prostate

Prostatic urethra

Orifices of prostatic ducts

Bulbo-urethral gland and duct

Intermediate part of the urethra

Urogenital diaphragm

Root of penis

Bulb of penis

Crus of penis

Bulbo-urethral duct opening

Ductus deferens

Corpora cavernosa

Epididymis

Corpus spongiosum

Body (shaft) of penis

Testis

Section of (c)

Spongy urethra

Glans penis

Prepuce (foreskin)

External urethral orifice

(b)

Dorsal vessels and nerves

Corpora cavernosa

Spongy urethra

Skin

Tunica albuginea of erectile bodies

Deep arteries

Corpus spongiosum

(c)

42

The paired oval testes lie in the **scrotum** outside the abdominopelvic cavity. The temperature there (approximately 94°F, or 34°C) is slightly lower than body temperature, a requirement for producing viable sperm.

The accessory structures forming the *duct system* are the epididymis, the ductus deferens, the ejaculatory duct, and the urethra. The **epididymis** is an elongated structure running up the posterolateral aspect of the testis and capping its superior

Table 42.1	Accessory Glands of the Male Reproductive System (Figure 42.1)	
Accessory gland	Location	Secretion
Seminal glands	Paired glands located posterior to the urinary bladder. The duct of each gland merges with a ductus deferens to form the ejaculatory duct.	A thick, light yellow, alkaline secretion containing fructose and citric acid, which nourish the sperm, and prostaglandins for enhanced sperm motility. Its secretion has the largest contribution to the volume of semen.
Prostate	Single gland that encircles the prostatic urethra inferior to the bladder.	A milky, slightly acidic fluid that contains citric acid, several enzymes, and prostate-specific antigen (PSA). Its secretion plays a role in activating the sperm.
Bulbo-urethral glands	Paired tiny glands that drain into the intermediate part of the urethra.	A clear alkaline mucus that lubricates the tip of the penis for copulation and neutralizes traces of acidic urine in the urethra prior to ejaculation.

aspect. The epididymis forms the first portion of the duct system and provides a site for immature sperm entering it from the testis to complete their maturation process. The **ductus deferens,** or **vas deferens** (sperm duct), arches superiorly from the epididymis, passes through the inguinal canal into the pelvic cavity, and courses over the superior aspect of the urinary bladder. In life, the ductus deferens is enclosed along with blood vessels and nerves in a connective tissue sheath called the **spermatic cord** (Figure 42.2). The terminus of the ductus deferens enlarges to form the region called the **ampulla,** which empties into the **ejaculatory duct.** During **ejaculation,** contraction of the ejaculatory duct propels the sperm through the prostate to the **prostatic urethra,** which in turn empties into the **intermediate part of the urethra** and then into the **spongy urethra,** which runs through the length of the penis to the body exterior.

The *accessory glands* include the prostate, the seminal glands, and the bulbo-urethral glands. These glands produce **seminal fluid,** the liquid medium in which sperm leave the body.

The location and secretion of the accessory glands are summarized in **Table 42.1**.

Semen consists of sperm and seminal fluid. Seminal fluid is overall alkaline, which buffers the sperm against the acidity of the female vagina.

The **penis,** part of the external genitalia of the male along with the scrotal sac, is the copulatory organ of the male. Designed to deliver sperm into the female reproductive tract, it consists of a body, or shaft, which terminates in an enlarged tip, the **glans penis** (Figure 42.1a and b). The skin covering the penis is loosely applied, and it reflects downward to form

WHY THIS MATTERS | **Benign Prostatic Hyperplasia (BPH)**

Benign has become synonymous with "noncancerous" and *hyperplasia* means an increase in the number of cells. In benign prostatic hyperplasia (BPH), an increased number of cells results in an enlarged prostate. As men age, normal hormonal changes can cause an increased proliferation of cells, which can enlarge the prostate and lead to urinary dysfunction. An enlarged prostate can also be a result of prostate cancer. Previously, an elevated blood level of prostate-specific antigen (PSA), a secretion of the prostate, was correlated to prostate cancer. However, research has shown that elevated PSA levels can have other causes, including BPH. A more accurate predictor of prostate cancer appears to be regular monitoring of the PSA level over time to see whether it continues to rise. ■

a circular fold of skin, the **prepuce,** or **foreskin,** around the proximal end of the glans. The foreskin may be removed in the surgical procedure called *circumcision.* Internally, the penis consists primarily of three elongated cylinders of erectile tissue, which engorge with blood during sexual excitement. This causes the penis to become rigid and enlarged so that it may more adequately serve as a penetrating device. This event is called **erection.** The paired dorsal cylinders are the **corpora cavernosa.** The single ventral **corpus spongiosum** surrounds the spongy urethra (Figure 42.1c).

Microscopic Anatomy of Selected Male Reproductive Organs

Each **testis** is covered by a dense connective tissue capsule called the **tunica albuginea** (literally, "white tunic"). Extensions of this sheath enter the testis, dividing it into a number of lobes, each of which houses one to four highly coiled **seminiferous tubules,** the sperm-forming factories (**Figure 42.2**). The seminiferous tubules of each lobe converge to empty the sperm into another set of tubules, the **rete testis,** at the posterior of the testis. Sperm traveling through the rete testis then enter the epididymis, located on the exterior aspect of the testis, as previously described. Lying between the seminiferous tubules and softly padded with connective tissue are the **interstitial endocrine cells,** which produce testosterone, the main hormonal product of the testis. Microscopic study of the testis is not included here (it is in Exercise 43) unless your instructor adds it.

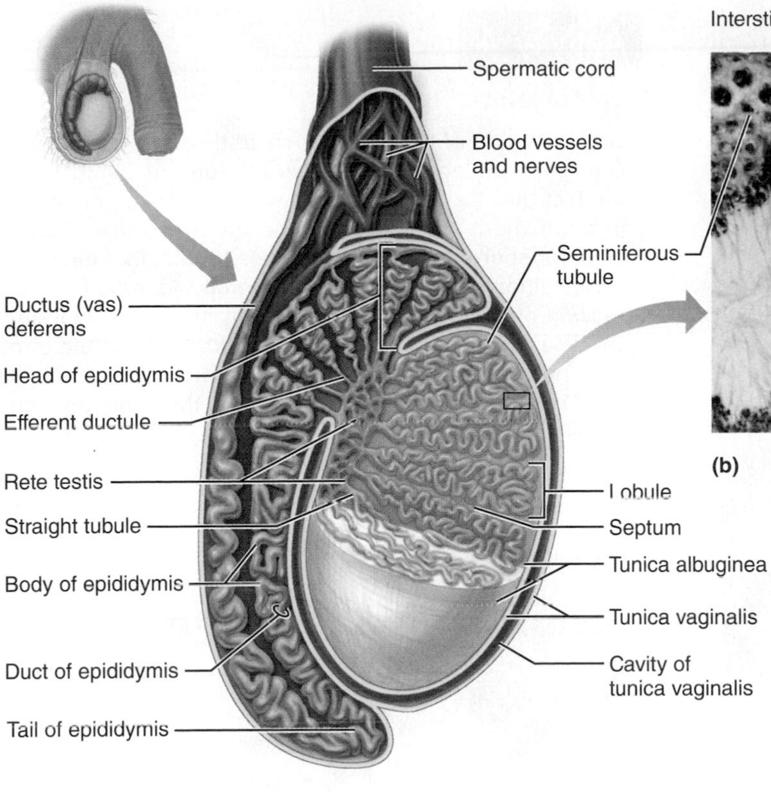

Spermatic cord

Blood vessels and nerves

Seminiferous tubule

Ductus (vas) deferens

Head of epididymis

Efferent ductule

Rete testis

Straight tubule

Body of epididymis

Duct of epididymis

Tail of epididymis

Lobule

Septum

Tunica albuginea

Tunica vaginalis

Cavity of tunica vaginalis

(a)

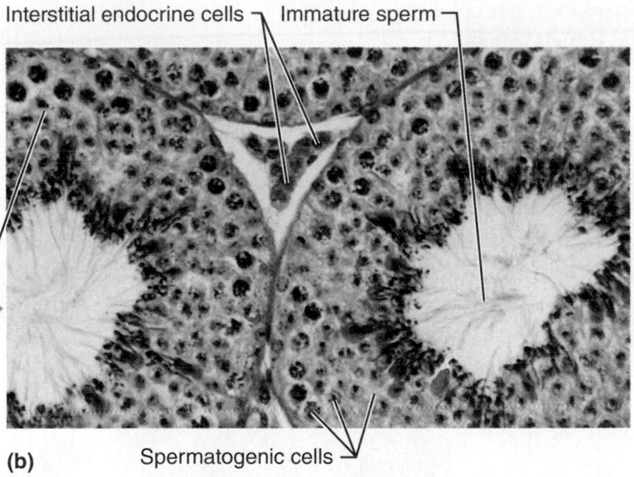

Interstitial endocrine cells   Immature sperm

(b)

Spermatogenic cells

Spermatic cord

Epididymis

Testis

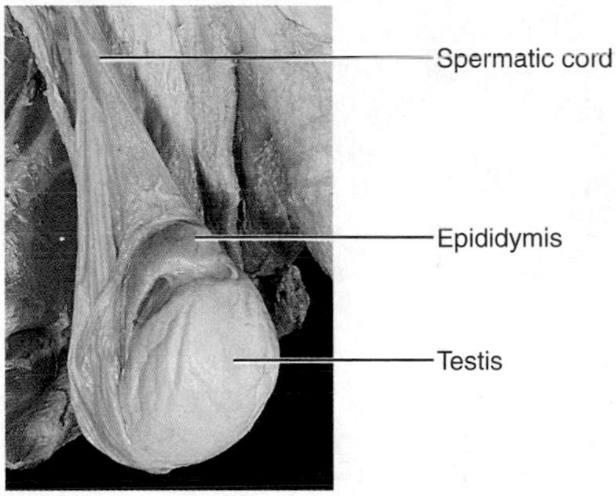

(c)

Figure 42.2 Structure of the testis. (a) Partial sagittal section of the testis and associated epididymis. **(b)** Cross-sectional view of portions of the seminiferous tubules, showing the spermatogenic (sperm-forming) cells, which make up the epithelium of the tubule walls, and the interstitial endocrine cells in the loose connective tissue between the tubules (200×). **(c)** External view of a testis from a cadaver; same orientation as in (a).

Activity 2

Penis

Obtain a slide of a cross section of the penis. Scan the tissue under low power to identify the urethra and the cavernous bodies. Compare your observations to Figure 42.1c and **Figure 42.3**. Observe the lumen of the urethra carefully. What type of epithelium do you see?

_____

_____

Explain the function of this type of epithelium.

_____

_____

_____

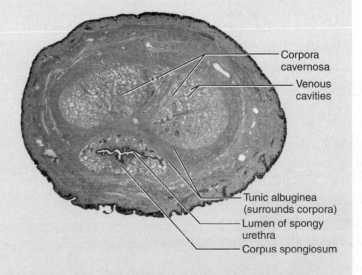

Corpora cavernosa

Venous cavities

Tunic albuginea (surrounds corpora)

Lumen of spongy urethra

Corpus spongiosum

Figure 42.3 Transverse section of the penis (3×).

42

Activity 3

Seminal Gland

Obtain a slide showing a cross-sectional view of the seminal gland. Examine the slide at low magnification to get an overall view of the highly folded mucosa of this gland. Switch to higher magnification, and notice that the folds of the gland protrude into the lumen where they divide further, giving the lumen a honeycomb look (**Figure 42.4**). Notice that the loose connective tissue lamina propria is underlain by smooth muscle fibers—first a circular layer, and then a longitudinal layer.

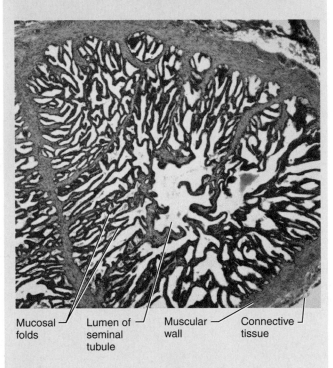

Mucosal folds · Lumen of seminal tubule · Muscular wall · Connective tissue

Figure 42.4 Cross-sectional view of a seminal gland with its elaborate network of mucosal folds. Glandular secretion is seen in the lumen (25×).

Activity 4

Epididymis

Obtain a slide of a cross section of the epididymis. Notice the abundant tubule cross sections resulting from the fact that the coiling epididymis tubule has been cut through many times in the specimen (**Figure 42.5**). Look for sperm in the lumen of the tubule. Examine the composition of the tubule wall carefully. Identify the *stereocilia* of the pseudostratified columnar epithelial lining. These nonmotile microvilli absorb excess fluid and pass nutrients to the sperm in the lumen. Now identify the smooth muscle layer. What do you think the function of the smooth muscle is?

_____

_____

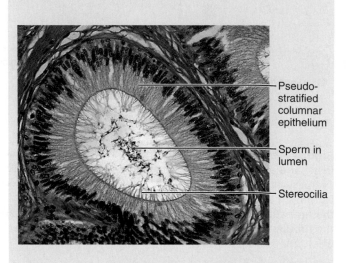

Pseudo-stratified columnar epithelium

Sperm in lumen

Stereocilia

Figure 42.5 Cross section of epididymis (120×).

Gross Anatomy of the Human Female Reproductive System

The **ovaries** are the primary reproductive organs of the female. Like the testes of the male, the ovaries produce gametes (in this case eggs, or ova) and also sex hormones (estrogens and progesterone). The other accessory structures of the female reproductive system transport, house, nurture, or otherwise serve the needs of the reproductive cells and/or the developing fetus.

Activity 5

Identifying Female Reproductive Organs

As you read the descriptions of these structures, locate them in Figure 42.6 and Figure 42.7 and then on the female reproductive system model or large laboratory chart.

External Genitalia

The **external genitalia (vulva)** consist of the mons pubis, the labia majora and minora, the clitoris, the external urethral and vaginal orifices, the hymen, and the greater vestibular glands. **Table 42.2** summarizes the structures of the female external genitalia (**Figure 42.6**).

The diamond-shaped region between the anterior end of the labial folds, the ischial tuberosities laterally, and the anus posteriorly is called the **perineum.**

Internal Organs

The internal female organs include the vagina, uterus, uterine tubes, ovaries, and the ligaments and supporting structures that suspend these organs in the pelvic cavity (**Figure 42.7** p. 642). The **vagina** extends for approximately 10 cm (4 inches) from the vestibule to the uterus superiorly. It serves as a copulatory organ and birth canal and permits passage of the menstrual flow. The

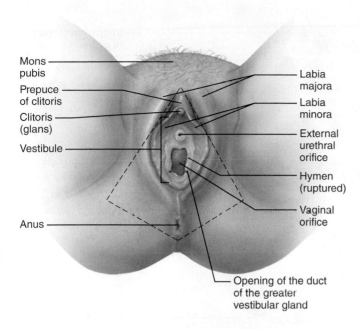

Mons pubis
Prepuce of clitoris
Clitoris (glans)
Vestibule
Anus
Labia majora
Labia minora
External urethral orifice
Hymen (ruptured)
Vaginal orifice
Opening of the duct of the greater vestibular gland

Figure 42.6 External genitalia (vulva) of the human female. The region enclosed by dashed lines is the perineum.

pear-shaped **uterus,** situated between the bladder and the rectum, is a muscular organ with its narrow end, the **cervix,** directed inferiorly. The major portion of the uterus is referred to as the **body;** its superior rounded region above the entrance of the uterine tubes is called the **fundus.** A fertilized egg is implanted in the uterus, which houses the embryo or fetus during its development.

In some cases, the fertilized egg may implant in a uterine tube or even on the abdominal viscera, creating an **ectopic pregnancy.** Such implantations are usually unsuccessful and may even endanger the mother's life because the uterine tubes cannot accommodate the increasing size of the fetus. ✚

The **endometrium,** the thick mucosal lining of the uterus, has a superficial **functional layer,** or **stratum functionalis,** that sloughs off periodically (about every 28 days) in response to cyclic changes in the levels of ovarian hormones in the woman's blood. This sloughing-off process, which is accompanied by bleeding, is referred to as **menstruation,** or **menses.** The deeper **basal layer,** or **stratum basalis** (Figure 43.6b), forms a new stratum functionalis after menstruation ends.

The **uterine,** or **fallopian, tubes** are about 10 cm (4 inches) long and extend from the ovaries in the peritoneal cavity to the superolateral region of the uterus. The distal ends of the tubes are funnel-shaped and have fingerlike projections called **fimbriae.** Unlike in the male duct system, there is no actual contact between the female gonad and the initial part of the female duct system—the uterine tube.

Because of this open passageway between the female reproductive organs and the peritoneal cavity, reproductive system infections, such as gonorrhea and other **sexually transmitted infections (STIs),** also called *sexually transmitted diseases (STDs),* can cause widespread inflammations of the pelvic viscera, a condition called **pelvic inflammatory disease (PID).** ✚

The internal female organs are all retroperitoneal, except the ovaries. They are supported and suspended somewhat freely by ligamentous folds of peritoneum. The supporting structures for the uterus, uterine tubes, and ovaries are summarized in **Table 42.3** on p. 643.

Within the ovaries, the female gametes, or eggs, begin their development in saclike structures called *follicles.* The growing follicles also produce *estrogens.* When a developing egg has reached the appropriate stage of maturity, it is ejected from the ovary in an event called **ovulation.** The ruptured follicle is then converted to a second type of endocrine structure called a *corpus luteum,* which secretes progesterone and some estrogens.

The flattened almond-shaped ovaries lie adjacent to the uterine tubes but are not connected to them; consequently, an ovulated "egg," actually a secondary oocyte (see Exercise 43), enters the pelvic cavity. The waving fimbriae of the uterine tubes create fluid currents that, if successful, draw the egg into the lumen of the uterine tube. There the egg begins its passage to the uterus, propelled by the cilia of the tubule walls. The usual and most desirable site of fertilization is the uterine tube, because the journey to the uterus takes about 3 to 4 days and an egg is viable only for up to 24 hours after it is expelled from the ovary. Thus, sperm must swim upward through the vagina and uterus and into the uterine tubes to reach the egg. This must be an arduous journey, because they must swim against the downward current created by ciliary action—rather like swimming upstream!

Table 42.2	External Genitalia (Vulva) of the Human Female (Figure 42.6)
Structure	**Description**
Mons pubis	Rounded fatty eminence that cushions the pubic symphysis; covered with coarse pubic hair after puberty.
Labia majora (singular: *labium majus*)	Two elongated hair-covered skin folds that extend from the mons pubis. They contain sebaceous glands, apocrine glands, and adipose. They are homologous to the scrotum.
Labia minora (singular: *labium minus*)	Two smaller folds located medial to the labia majora. They don't have hair or adipose but they do have many sebaceous glands.
Vestibule	Region located between the two labia minora. From anterior to posterior, it contains the clitoris, the external urethral orifice, and the vaginal orifice.
Clitoris	Small mass of erectile tissue located where the labia minora meet anteriorly. It is homologous to the penis.
Prepuce of the clitoris	Skin folds formed by the union of the labia minora; they serve to hood the clitoris.
External urethral orifice	Serves as the outlet for the urinary system. It has no reproductive function in the female.
Hymen	A thin fold of vascular mucous membrane that may partially cover the vaginal opening.
Greater vestibular glands	Pea-sized mucus-secreting glands located on either side of the hymen. They lubricate the distal end of the vagina during coitus. They are homologous to the bulbo-urethral glands of males.

42

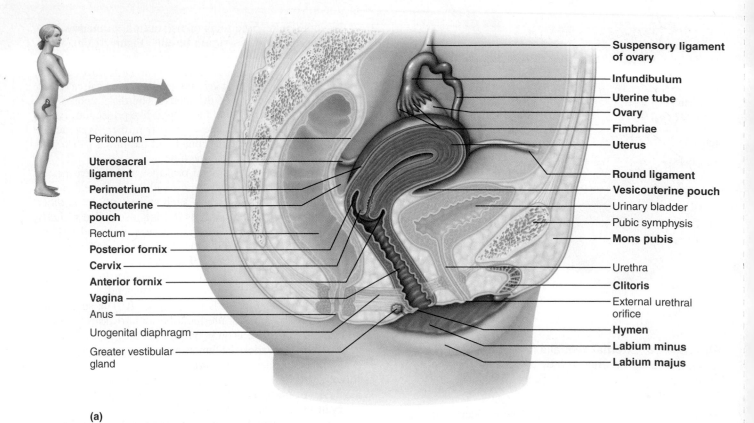

Peritoneum

Uterosacral ligament

Perimetrium

Rectouterine pouch

Rectum

Posterior fornix

Cervix

Anterior fornix

Vagina

Anus

Urogenital diaphragm

Greater vestibular gland

Suspensory ligament of ovary

Infundibulum

Uterine tube

Ovary

Fimbriae

Uterus

Round ligament

Vesicouterine pouch

Urinary bladder

Pubic symphysis

Mons pubis

Urethra

Clitoris

External urethral orifice

Hymen

Labium minus

Labium majus

(a)

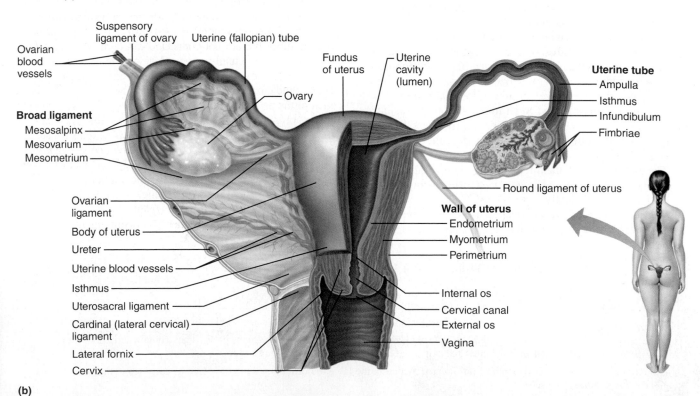

Ovarian blood vessels

Suspensory ligament of ovary

Uterine (fallopian) tube

Ovary

Broad ligament

Mesosalpinx

Mesovarium

Mesometrium

Ovarian ligament

Body of uterus

Ureter

Uterine blood vessels

Isthmus

Uterosacral ligament

Cardinal (lateral cervical) ligament

Lateral fornix

Cervix

Fundus of uterus

Uterine cavity (lumen)

Uterine tube

Ampulla

Isthmus

Infundibulum

Fimbriae

Round ligament of uterus

Wall of uterus

Endometrium

Myometrium

Perimetrium

Internal os

Cervical canal

External os

Vagina

(b)

Figure 42.7 Internal reproductive organs of the human female. (a) Midsagittal section of the human female reproductive system. **(b)** Posterior view. The posterior walls of the vagina, uterus, and uterine tubes, and the broad ligament have been removed on the right side to reveal the shape of the lumen of these organs.

Table 42.3	Supporting Structures for the Uterus, Uterine Tubes, and Ovaries (Figure 42.7)
Structure	**Description**
Broad ligament	Fold of the peritoneum that drapes over the superior uterus to enclose the uterus and uterine tubes and anchors them to the lateral body walls
Mesometrium	Portion of the broad ligament that supports the uterus laterally (mesentery of the uterus)
Round ligaments	Anchor the uterus to the anterior pelvic wall by descending through the mesometrium and the inguinal canal; attach to the skin of one of the labia majora
Uterosacral ligaments	Secure the inferior uterus to the sacrum posteriorly
Cardinal (lateral cervical) ligaments	Connect the cervix and vagina to the pelvic wall laterally
Mesosalpinx	Portion of the broad ligament that anchors the uterine tube (mesentery of the uterine tube)
Mesovarium	Posterior fold of the broad ligament that supports the ovaries (mesentery of the ovaries)
Suspensory ligaments	A lateral continuation of the broad ligament that attaches the ovaries to the lateral pelvic wall
Ovarian ligaments	Anchors the ovaries to the uterus medially and is enclosed within the broad ligament

Microscopic Anatomy of Selected Female Reproductive Organs

Activity 6

Wall of the Uterus

Obtain a slide of a cross-sectional view of the uterine wall. Identify the three layers of the uterine wall—the endometrium, myometrium, and serosa. **Figure 42.8**, a photomicrograph that includes the secretory endometrium, will help with this study.

As you study the slide, notice that the bundles of smooth muscle are oriented in several different directions. What is the function of the **myometrium** (smooth muscle layer) during the birth process?

_____

_____

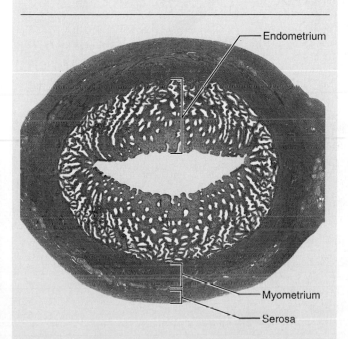

Figure 42.8 Cross-sectional view of the uterine wall. The mucosa is in the secretory phase. (3×).

Activity 7

Uterine Tube

Obtain a slide of a cross-sectional view of a uterine tube for examination. Notice that the mucosal folds nearly fill the tubule lumen (**Figure 42.9**). Then switch to high power to examine the ciliated secretory epithelium.

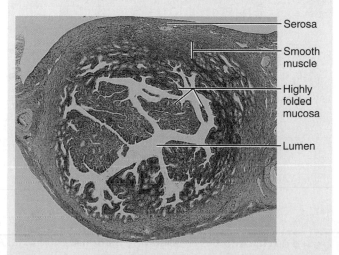

Figure 42.9 Cross-sectional view of the uterine tube (12×).

The Mammary Glands

The **mammary glands** exist within the breasts in both sexes, but they normally have a reproduction-related function only in females. Since the function of the mammary glands is to produce milk to nourish the newborn infant, their importance is more closely associated with events that occur when reproduction has already been accomplished. Periodic stimulation by the female sex hormones, especially estrogens, increases the size of the female mammary glands at puberty. During this period, the duct system becomes more elaborate, and fat is deposited—fat deposition is the more important contributor to increased breast size.

The rounded, skin-covered mammary glands lie anterior to the pectoral muscles of the thorax, attached to them by connective tissue. Slightly below the center of each breast is a pigmented area, the **areola,** which surrounds a centrally protruding **nipple** (**Figure 42.10**).

Internally each mammary gland consists of 15 to 25 **lobes** that radiate around the nipple and are separated by fibrous connective tisse and adipose, or fatty, tissue. Within each lobe are smaller chambers called **lobules,** containing the glandular **alveoli** that produce milk during lactation. The alveoli of each lobule pass the milk into a number of **lactiferous ducts,** which join to form an expanded storage chamber, the **lactiferous sinus,** as they approach the nipple. The sinuses open to the outside at the nipple.

For instructions on animal dissections, see the dissection exercises (starting on p. 705) in the cat and fetal pig editions of this manual.

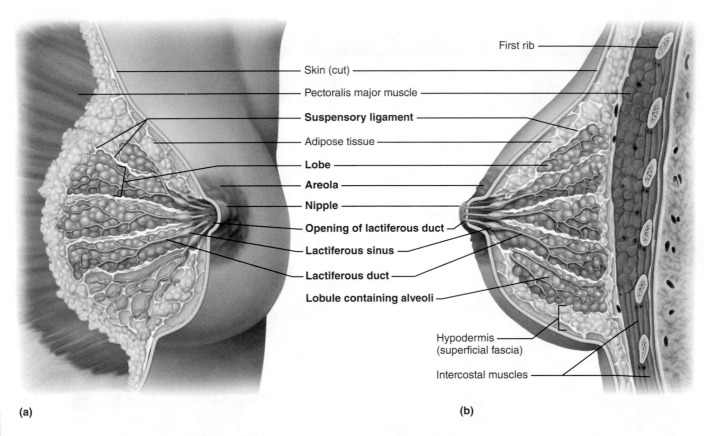

First rib

Skin (cut)

Pectoralis major muscle

Suspensory ligament

Adipose tissue

Lobe

Areola

Nipple

Opening of lactiferous duct

Lactiferous sinus

Lactiferous duct

Lobule containing alveoli

Hypodermis (superficial fascia)

Intercostal muscles

(a)

(b)

Figure 42.10 Anatomy of lactating mammary gland. (a) Anterior view of partially dissected breast. **(b)** Sagittal section of the breast.

REVIEW SHEET
Anatomy of the Reproductive System

Name_____   Lab Time/Date _____

Gross Anatomy of the Human Male Reproductive System

1. List the two principal functions of the testis. _____

   and _____

2. Identify all indicated structures or portions of structures on the diagrammatic view of the male reproductive system below.

3. Why are the testes located in the scrotum rather than inside the ventral body cavity? _____

_____

WHY THIS MATTERS   4. A screening tool for benign prostatic hyperplasia and prostate cancer includes palpation of the prostate.

   Explain how this is accomplished. _____

_____

_____

5. Would you expect a male with benign prostatic hyperplasia to have difficulty with ejaculation? Why or why not?

_____

_____

6. Match the terms in column B to the descriptive statements in column A.

Column A

   _____   1.   copulatory organ/penetrating device

   _____   2.   muscular passageway conveying sperm to the ejaculatory duct; in the spermatic cord

   _____   3.   distal urethra that transports both sperm and urine

   _____   4.   sperm maturation site

   _____   5.   location of the testis in adult males

   _____   6.   loose fold of skin encircling the glans penis

   _____   7.   portion of the urethra that is located in the urogenital diaphragm

   _____   8.   empties a secretion into the prostatic urethra

   _____   9.   empties a secretion into the intermediate part of the urethra

Column B

a. bulbo-urethral glands

b. ductus (vas) deferens

c. epididymis

d. intermediate part of the urethra

e. penis

f. prepuce

g. prostate

h. prostatic urethra

i. scrotum

j. seminal gland

k. spongy urethra

7. Describe the composition of semen, and name all structures contributing to its formation. _____

_____

_____

8. Of what importance is the fact that seminal fluid is alkaline? _____

_____

9. What structures compose the spermatic cord? _____

_____

   Where is it located? _____

10. Using the following terms, trace the pathway of sperm from the testes to the urethra: rete testis, epididymis, seminiferous tubule, ductus deferens.

_____ → _____ → _____ → _____

Gross Anatomy of the Human Female Reproductive System

11. Name the structures composing the external genitalia, or vulva, of the female. _____

_____

_____

12. On the diagram below of a frontal section of a portion of the female reproductive system, identify all indicated structures.

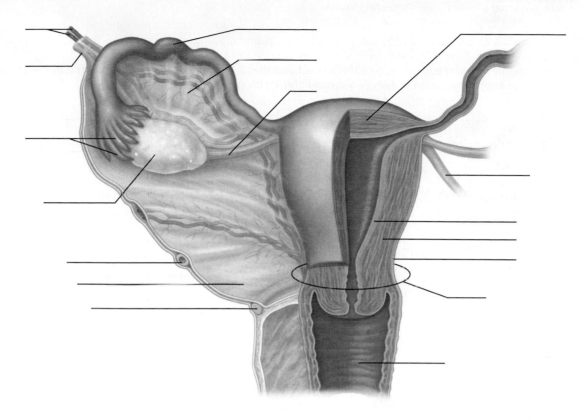

13. Identify the female reproductive system structures described below.

_____   1.   site of fetal development

_____   2.   copulatory canal

_____   3.   egg typically fertilized here

_____   4.   becomes erect during sexual excitement

_____   5.   duct extending from ovaries to the uterus

_____   6.   partially closes the vaginal opening; a membrane

_____   7.   produces oocytes, estrogens, and progesterone

_____   8.   fingerlike ends of the uterine tube

14. Do any sperm enter the pelvic cavity of the female? Why or why not? _____

_____

15. What is an ectopic pregnancy, and how can it happen? _____

_____

_____

16. Put the following vestibular-perineal structures in their proper order from the anterior to the posterior aspect: vaginal orifice, anus, external urethral opening, and clitoris.

   Anterior limit: _____ → _____ → _____ → _____

17. Assume that a couple has just consummated the sex act and that the sperm have been deposited in the vagina. Trace the pathway of the sperm through the female reproductive tract.

   _____

18. Define *ovulation*. _____

Microscopic Anatomy of Selected Male and Female Reproductive Organs

19. The testis is divided into a number of lobes by connective tissue. Each of these lobes contains one to four _____

   _____, which converge to empty sperm into another set of tubules called the

   _____.

20. What is the function of the cavernous bodies seen in the penis? _____

   _____

21. Name the three layers of the uterine wall from the inside out.

   _____, _____, _____

   Which of these is sloughed during menses? _____

   Which contracts during childbirth? _____

22. Describe the epithelium found in the uterine tube. _____

   _____

23. Describe the arrangement of the layers of smooth muscle in the seminal gland. _____

   _____

24. What is the function of the stereocilia exhibited by the epithelial cells of the mucosa of the epididymis? _____

   _____

25. On the diagram showing the sagittal section of the human testis, correctly identify all structures provided with leader lines.

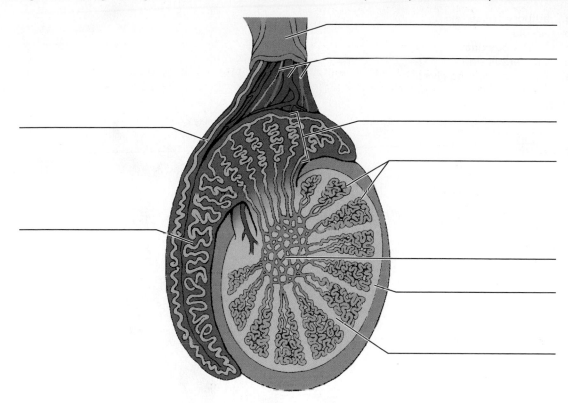

The Mammary Glands

26. Match the key term with the correct description.

_____ glands that produce milk during lactation

_____ subdivision of mammary lobes that contains alveoli

_____ enlarged storage chamber for milk

_____ duct connecting alveoli to the storage chambers

_____ pigmented area surrounding the nipple

_____ releases milk to the outside

Key:

a. alveoli

b. areola

c. lactiferous duct

d. lactiferous sinus

e. lobule

f. nipple

27. Using the key terms, correctly identify breast structures.

Key: a.   adipose tissue
         b.   areola
         c.   lactiferous duct
         d.   lactiferous sinus
         e.   lobule containing alveoli
         f.   nipple

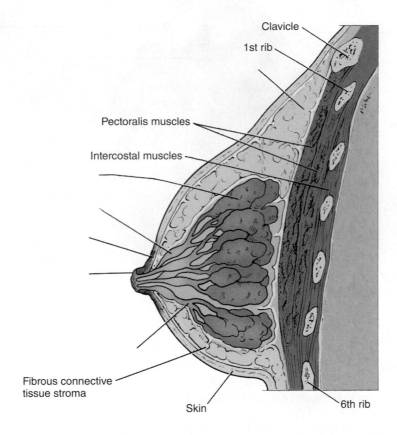

28. Describe the procedure for self-examination of the breasts. (Men are not exempt from breast cancer, you know!)

_____

_____

_____

_____

Physiology of Reproduction: Gametogenesis and the Female Cycles

Objectives

☐ Define *meiosis, gametogenesis, oogenesis, spermatogenesis, synapsis, haploid, zygote,* and *diploid.*

☐ Cite similarities and differences between mitosis and meiosis.

☐ Describe the stages of spermatogenesis, and relate each to the cross-sectional structure of the seminiferous tubule.

☐ Define *spermiogenesis,* and relate the anatomy of sperm to their function.

☐ Describe the effects of FSH and LH on testicular function.

☐ Discuss the microscopic structure of the ovary; identify primary, secondary, and vesicular follicles, and the corpus luteum; list the hormones produced by the follicles and the corpus luteum.

☐ Relate the stages of oogenesis to follicle development in the ovary.

☐ Compare and contrast spermatogenesis and oogenesis.

☐ Discuss the effect of FSH and LH on the ovary, and describe the feedback relationship between anterior pituitary gonadotropins and ovarian hormones.

☐ List the phases of the menstrual cycle, and discuss the hormonal control of each.

Materials

- Three-dimensional models illustrating meiosis, spermatogenesis, and oogenesis
- Sets of "pop it" beads in two colors with magnetic centromeres, available in Chromosome Simulation Lab Activity from Ward's Natural Science
- Compound microscope
- Prepared slides of testis and human sperm
- *Demonstration:* microscopes set up to demonstrate the following stages of oogenesis in *Ascaris megalocephala:*

  Slide 1: Primary oocyte with fertilization membrane, sperm nucleus, and aligned tetrads apparent

  Slide 2: Formation of the first polar body

Text continues on next page. →

MasteringA&P®

For related exercise study tools, go to the Study Area of **MasteringA&P**. There you will find:

- Practice Anatomy Lab **PAL**
- PhysioEx **PEx**
- A&PFlix **A&PFlix**
- Practice quizzes, Histology Atlas, eText, Videos, and more!

Pre-Lab Quiz

1. Human gametes contain _____ chromosomes.
   a. 13   b. 23   c. 36   d. 46

2. The end product of meiosis is:
   a. two diploid daughter cells
   b. two haploid daughter cells
   c. four diploid daughter cells
   d. four haploid daughter cells

3. Circle the correct underlined terms. A grouping of four chromatids, known as a dyad / tetrad, occurs only during mitosis / meiosis.

4. _____ extend inward from the periphery of the seminiferous tubule and provide nourishment to the spermatids as they begin their transformation into sperm.
   a. Interstitial endocrine cells
   b. Granulosa cells
   c. Sustentocytes
   d. Follicle cells

5. Circle the correct underlined term. The acrosome / midpiece of the sperm contains enzymes involved in the penetration of the egg.

6. Circle the correct underlined term. Within each ovary, the immature ovum develops in a saclike structure called a corpus / follicle.

7. As the primordial follicle grows and its epithelium changes from squamous to cuboidal cells, it becomes a(n) _____ and begins to produce estrogens.
   a. oocyte
   b. oogonia
   c. primary follicle
   d. primary ovum

8. Circle True or False. A sudden release of luteinizing hormone by the anterior pituitary triggers ovulation.

9. Circle the correct underlined term. The corpus luteum / corpus albicans is a solid glandular structure with a scalloped lumen that develops from a ruptured follicle.

Text continues on next page. →

Slide 3: Secondary oocyte with dyads
aligned

Slide 4: Formation of the ovum and second
polar body

Slide 5: Fusion of the male and female
pronuclei to form the fertilized egg

- Prepared slides of ovary and uterine
endometrium (showing menstrual,
proliferative, and secretory phases)

10. The _____ phase of the female cycle occurs from days 1–5
and is signaled by the sloughing off of the thick functional layer of the
endometrium.
   a. endometrial
   b. menstrual
   c. proliferative
   d. secretory

H uman beings develop from the union of egg and sperm. Each of these gametes is a unique cell produced either in the ovary or testis. Unlike all other body cells, gametes have only half the normal chromosome number, and they are produced by a special type of nuclear division called meiosis.

Meiosis

The normal number of chromosomes in most human body cells is 46, the **diploid,** or **2n,** chromosomal number. This number is made up of two sets of similar chromosomes, one set of 23 from each parent. Thus each body cell contains 23 pairs of similar chromosomes called **homologous chromosomes** or homologues (**Figure 43.1**). Each member of a homologous pair contains genes that code for the same traits.

Gametes contain only one member of each homologous pair of chromosomes. Therefore each human gamete contains a total of 23 chromosomes, the **haploid,** or **n,** chromosomal number. When egg and sperm fuse, they form a **zygote** that restores the diploid number of chromosomes. The zygote divides by the process of **mitosis** to produce the multicellular human body. Mitosis is the process by which most body cells divide. It produces two diploid daughter cells, each containing 46 chromosomes that are identical to those of the mother cell (see Exercise 4).

Gametogenesis is the process of gamete formation. It involves nuclear division by **meiosis,** which reduces the number of chromosomes by half. Before meiosis begins, the chromosomes in the *mother cells,* or stem cells, are replicated just as they are before mitosis. The identical copies remain together as *sister chromatids.* They are held together by a centromere, forming a structure called a **dyad** (Figure 43.1).

Two nuclear divisions, called meiosis I and meiosis II, occur during meiosis. Each has the same phases as mitosis—prophase, metaphase, anaphase, and telophase. Meiosis I is the *reduction division.* During prophase of meiosis I, the homologous chromosomes pair up in a process called **synapsis** that forms little groups of four chromatids, called **tetrads** (Figure 43.1). During synapsis, the free ends of adjacent maternal and paternal chromatids wrap around each other at one or more points, forming **crossovers,** or **chiasmata.** Crossovers allow maternal and paternal chromosomes to exchange genetic material. The tetrads align randomly at the metaphase plate so that either the maternal or paternal chromosome may be on a given side of the plate. Then the two homologous chromosomes, each still composed of two sister chromatids, are pulled to opposite ends of the cell. At the end of meiosis I, each haploid daughter cell contains one member of each original homologous pair.

Meiosis II begins immediately without replication of the chromosomes. The dyads align on the metaphase plate, and the two sister chromatids are pulled apart, each now becoming a full chromosome. The net result of meiosis is four haploid daughter cells, each containing 23 chromosomes. The events of crossover and the random alignment of tetrads during meiosis I introduce great genetic variability. As a result, it is unlikely that any gamete is exactly like another.

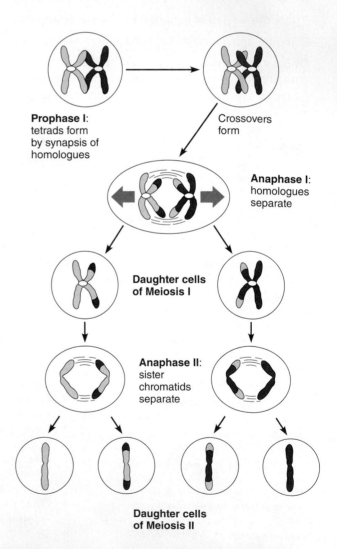

Prophase I:
tetrads form
by synapsis of
homologues

Crossovers
form

Anaphase I:
homologues
separate

**Daughter cells
of Meiosis I**

Anaphase II:
sister
chromatids
separate

**Daughter cells
of Meiosis II**

Figure 43.1 Events of meiosis involving one pair of homologous chromosomes. Male homologue is purple; female homologue is pink.

Activity 1

Identifying Meiotic Phases and Structures

1. Obtain a model depicting the events of meiosis, and follow the sequence of events during meiosis I and II. Identify prophase, metaphase, anaphase, and telophase in each nuclear division. Also identify tetrads and chiasmata during meiosis I and dyads during meiosis II. Note ways in which the daughter cells resulting from meiosis I differ from the mother cell and how the gametes differ from both cell populations. Use the key on the model, your textbook, or an appropriate reference as necessary to aid you in these observations.

2. Using strings of colored "pop it" beads with magnetic centromeres, demonstrate the phases of meiosis, including crossing over, for a cell with a diploid (2n) number of 4. Use one bead color for the male chromosomes and another color for the female chromosomes.

3. Ask your instructor to verify the accuracy of your "creation" before returning the beads to the supply area.

Spermatogenesis

Human sperm production, or **spermatogenesis,** begins at puberty and continues without interruption throughout life. The average male ejaculate contains about a quarter billion sperm. Because only one sperm fertilizes an ovum, the perpetuation of the species will not be endangered by lack of sperm.

Spermatogenesis, the process of gametogenesis in males, occurs in the seminiferous tubules of the testes (**Figure 43.2**, p. 654, and Exercise 42). The primitive stem cells, or **spermatogonia,** found at the tubule periphery, divide extensively to build up the stem cell line. Before puberty, all divisions are mitotic divisions that produce more spermatogonia. At puberty, however, under the influence of follicle-stimulating hormone

(FSH) secreted by the anterior pituitary gland, each mitotic division of a spermatogonium produces one spermatogonium and one **primary spermatocyte,** which is destined to undergo meiosis. As meiosis occurs, the dividing cells approach the lumen of the tubule. Thus the progression of meiotic events can be followed from the tubule periphery to the lumen. It is important to recognize that **spermatids,** haploid cells that are the actual product of meiosis, are not functional gametes. They are nonmotile cells and have too much excess baggage to function well in a reproductive capacity. A subsequent process, called **spermiogenesis,** strips away the extraneous cytoplasm from the spermatid, converting it to a motile, streamlined **sperm.**

Activity 2

Examining Events of Spermatogenesis

1. Obtain a slide of the testis and a microscope. Examine the slide under low power to identify the cross-sectional views of the cut seminiferous tubules. Then rotate the high-power lens into position, and observe the wall of one of the cut tubules (Figure 43.2).

2. Scrutinize the cells at the periphery of the tubule. The cells in this area are the spermatogonia. About half of these will form primary spermatocytes, which begin meiosis. These are recognizable by their pale-staining nuclei with centrally located nucleoli. The remaining daughter cells resulting from mitotic divisions of spermatogonia stay at the tubule periphery to maintain the germ cell line.

3. Observe the cells in the middle of the tubule wall. Here you should see a large number of spermatocytes that are obviously undergoing a nuclear division process. Look for coarse clumps of threadlike chromatin or chromosomes that have the appearance of coiled springs. Attempt to differentiate between the larger primary spermatocytes and the somewhat smaller secondary spermatocytes. Once formed, the secondary spermatocytes quickly undergo division and so are more difficult to find.

Can you see tetrads? _____

Is there evidence of crossover? _____

In which location would you expect to see cells containing tetrads, closer to the spermatogonia or closer to the lumen?

_____

Would these cells be primary or secondary spermatocytes?

_____

4. Examine the cells at the tubule lumen. Identify the small round-nucleated spermatids, many of which may appear lopsided and look as though they are starting to lose their cytoplasm. See if you can find a spermatid embedded in an elongated cell type—a **sustentocyte,** or *Sertoli cell*—which extends inward from the periphery of the tubule. The sustentocytes nourish the spermatids as they begin their transformation into sperm. Also in the adluminal area (area toward the lumen), locate immature sperm, which can be identified by their tails. The sperm develop directly from the spermatids by the loss of extraneous cytoplasm and the development of a propulsive tail.

5. Identify the **interstitial endocrine cells,** also called *Leydig cells,* lying external to and between the seminiferous tubules. LH (luteinizing hormone) prompts these cells to produce

Text continues on next page. →

43

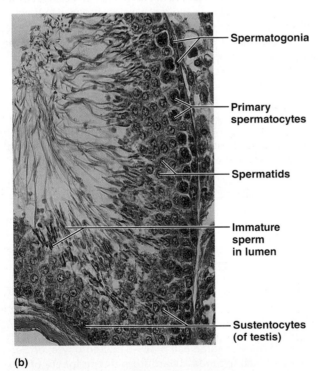

Figure 43.2 Spermatogenesis. (a) Flowchart of meiotic events and spermiogenesis. **(b)** Micrograph of an active seminiferous tubule (275×).

43

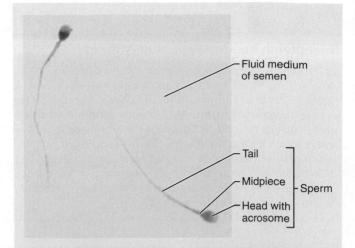

Figure 43.3 Sperm in semen. (1000×).

testosterone, which acts synergistically with FSH (follicle-stimulating hormone) to stimulate sperm production. Both LH and FSH are named for their effects on the female gonad.

In the next stage of sperm development, spermiogenesis, all the superficial cytoplasm is sloughed off, and the remaining cell organelles are compacted into the three regions of the mature sperm. At the risk of oversimplifying, these anatomical regions are the *head*, the *midpiece*, and the *tail*, which correspond roughly to the activating and genetic region, the metabolic region (rich in mitochondria for ATP production), and the locomotor region (a typical flagellum powered by ATP), respectively. The mature sperm is a streamlined cell equipped with an organ of locomotion and a high rate of metabolism that enable it to move long distances quickly to get to the egg. It is a prime example of the correlation of form and function.

The pointed sperm head contains the DNA, or genetic material, of the chromosomes. Essentially it is the nucleus of the spermatid. Anterior to the nucleus is the **acrosome,** which contains enzymes necessary for penetration of the egg.

6. Obtain a prepared slide of human sperm, and view it with the oil immersion lens. Identify the head, acrosome, and tail regions of the sperm (**Figure 43.3**). Deformed sperm, for example, sperm with multiple heads or tails, are sometimes present in such preparations. Did you observe any?

_____ If so, describe them. _____

_____

7. Examine the model of spermatogenesis to identify the spermatogonia, the primary and secondary spermatocytes, the spermatids, and the functional sperm.

Demonstration of Oogenesis in *Ascaris* (Optional)

Oogenesis (the process of producing an egg) in mammals is difficult to demonstrate. However, the process may be studied rather easily in the transparent eggs of *Ascaris megalocephala*, an invertebrate roundworm parasite found in the intestine of mammals. Since its diploid chromosome number is 4, the chromosomes are easily counted.

Activity 3

Examining Meiotic Events Microscopically

Go to the demonstration area where the slides are set up, and make the following observations:

1. Scan the first demonstration slide to identify a *primary oocyte*, the cell type that begins the meiotic process. It will have what appears to be a relatively thick cell membrane; this is the *fertilization membrane* that the oocyte produces after sperm penetration. Find and study a primary oocyte that is undergoing meiosis I. Look for a barrel-shaped spindle with two tetrads (two groups of four beadlike chromosomes) in it. Most often the spindle is located at the periphery of the cell. The sperm nucleus may or may not be seen, depending on how the cell was cut.

2. Observe slide 2. Locate a cell in which half of each tetrad (a dyad) is being extruded from the cell surface into a smaller cell called the *first polar body*.

3. On slide 3, attempt to locate a *secondary oocyte* (a daughter cell produced during meiosis I) undergoing meiosis II.

In this view, you should see two dyads (each with two beadlike chromosomes) on the spindle.

4. On slide 4, locate a cell in which the *second polar body* is being formed. In this case, both it and the ovum will now contain two chromosomes, the haploid number for *Ascaris*.

5. On slide 5, identify a *fertilized egg*, or a cell in which the sperm and ovum nuclei (actually *pronuclei*) are fusing to form a single nucleus containing four chromosomes.

Human Oogenesis and the Ovarian Cycle

Once the adult ovarian cycle is established, gonadotropic hormones produced by the anterior pituitary influence the development of ova in the ovaries and their cyclic production of female sex hormones. Within an ovary, each immature ovum develops within a saclike structure called a *follicle,* where it is encased by one or more layers of smaller cells. The surrounding cells are called **follicle cells** if one layer is present and **granulosa cells** when more than one layer is present.

The process of **oogenesis,** or female gamete formation, which occurs in the ovary, is similar to spermatogenesis occurring in the testis, but there are some important differences. Oogenesis begins with primitive stem cells called **oogonia,** located in the cortex of the ovaries of the developing female fetus (**Figure 43.4**, p. 656). During fetal development, the oogonia undergo mitosis thousands of times until their number reaches 2 million or more. They then become encapsulated by a single layer of squamouslike follicle cells and form the **primordial follicles** of the ovary. By the time the female child is born, most of her oogonia have increased in size and have become **primary oocytes,** which are in the prophase stage of meiosis I. Thus at birth, the female is presumed to have her lifetime supply of primary oocytes.

From birth until puberty, the primary oocytes are quiescent. Then, under the influence of FSH, one or sometimes more of the follicles begin to undergo maturation approximately every 28 days. The stages of maturation that follicles undergo are described in **Table 43.1** on p. 657 and illustrated in Figure 43.4 and **Figure 43.5** on p. 657.

In the female, meiosis produces only one functional gamete, in contrast to the four produced in the male. Another major difference is in the relative size and structure of the functional gametes. Sperm are tiny and equipped with tails for locomotion. They have few organelles and virtually no nutrient-containing cytoplasm; hence the nutrients contained in semen are essential to their survival. In contrast, the egg is a relatively large nonmotile cell, well stocked with cytoplasmic reserves that nourish the developing embryo until implantation can be accomplished. Essentially all the zygote's organelles are provided by the egg.

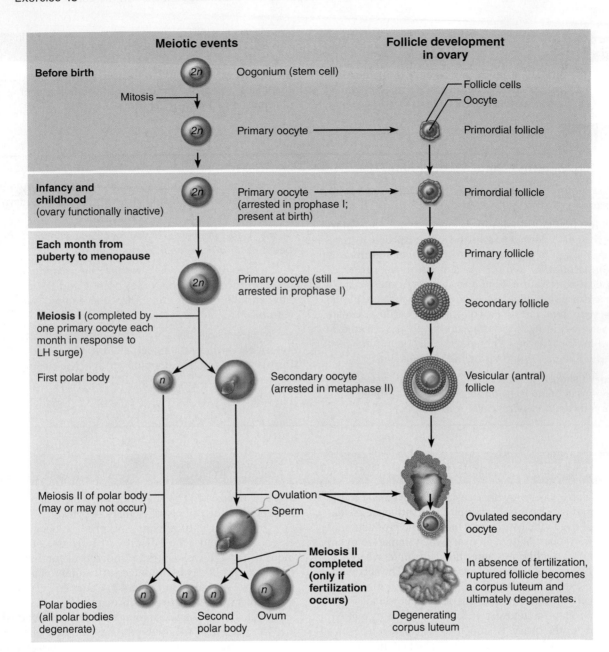

Figure 43.4 Oogenesis. Left, flowchart of meiotic events. Right, correlation with follicular development and ovulation in the ovary.

Activity 4

Examining Oogenesis in the Ovary

Because many different stages of ovarian development exist within the ovary at any one time, a single microscopic preparation will contain follicles at many different stages of development (Figure 43.5). Obtain a cross section of ovary tissue, and identify the following structures: germinal epithelium, primary follicle, late secondary follicle, vesicular follicle, and corpus luteum. Use Table 43.1 and Figure 43.5 to guide your observations.

Activity 5

Comparing and Contrasting Oogenesis and Spermatogenesis

Examine the model of oogenesis, and compare it with the spermatogenesis model. Note differences in the number, size, and structure of the functional gametes.

Table 43.1 — Maturation Stages of Ovarian Follicles (Figure 43.4)

Follicle stage	Description
The follicles below (primordial through late secondary follicle) are primary oocytes arrested in prophase I.	
Primordial follicle	A primary oocyte surrounded by a single layer of squamous follicle cells. They are produced in the fetus and are present at birth.
Primary follicle	This stage begins at puberty. The squamous follicle cells become cuboidal and surround the primary oocyte.
Secondary follicle	The follicle cells divide and form a stratified cuboidal epithelium around the oocyte. The follicle cells are now called *granulosa cells*.
Late secondary follicle	The oocyte secretes a glycoprotein to form its extracellular membrane, the zona pellucida. This thick membrane is now surrounded by a layer of connective tissue and epithelial cells called the *theca folliculi*. Clear liquid has accumulated in between the granulosa cells.
The follicles below (vesicular and the ruptured follicle) are secondary oocytes arrested in metaphase II.	
Vesicular (antral) follicle	The liquid continues to accumulate between the granulosa cells and forms a large fluid-filled cavity called an *antrum*. There are now several layers of granulosa cells called the *corona radiata*. The primary oocyte completes the first meiotic division to produce a *secondary oocyte* and a *polar body*.
Ruptured follicle	The follicle ruptures, and the secondary oocyte is released with its corona radiata to begin its journey down the uterine tube. Note that meiosis II will be completed to produce an ovum only if fertilization occurs. If no fertilization occurs, the secondary oocyte degenerates.
The structures below (corpus luteum and corpus albicans) are remnants of the ruptured follicle.	
Corpus luteum	The "yellow body" formed by granulosa cells where the follicle ruptured. It secretes the steroid hormones estrogen and progesterone.
Corpus albicans	The "white body" that represents the further deterioration of the corpus luteum.

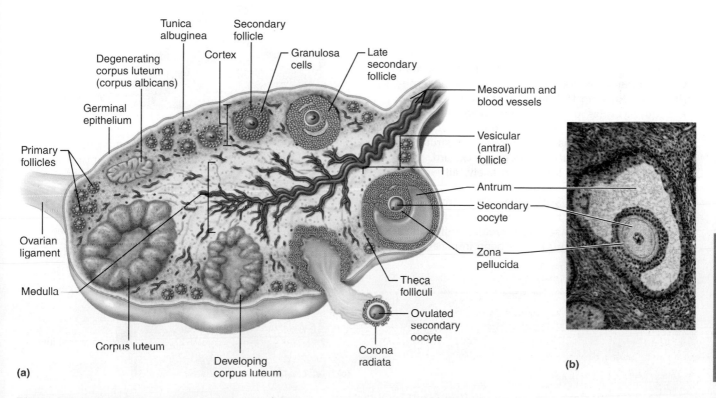

Figure 43.5 Anatomy of the human ovary. (a) The ovary has been sectioned to reveal the follicles in its interior. Note that not all the structures would appear in the ovary at the same time. **(b)** Photomicrograph of a vesicular (antral) follicle (75×).

43

The Menstrual Cycle

The **uterine cycle,** or **menstrual cycle,** is hormonally controlled by estrogens and progesterone secreted by the ovary. It is normally divided into three stages: menstrual, proliferative, and secretory. Notice how the endometrial changes (**Figure 43.6**) correlate with hormonal and ovarian changes (**Figure 43.7**).

If fertilization has occurred, the embryo will produce a hormone much like LH, which will maintain the function of the corpus luteum. Otherwise, as the corpus luteum begins to deteriorate, lack of ovarian hormones in the blood causes blood vessels supplying the endometrium to kink and become spastic, setting the stage for menstruation to begin by the 28th day.

Although 28 days is a common length for the menstrual cycle (Figure 43.7d), its length is highly variable, sometimes as short as 21 days or as long as 38.

Activity 6

Observing Histological Changes in the Endometrium During the Menstrual Cycle

Obtain slides showing the menstrual, secretory, and proliferative phases of the endometrium of the uterus. Observe each carefully, comparing their relative thicknesses and vascularity. As you work, refer to the corresponding photomicrographs (Figure 43.6).

WHY THIS MATTERS | **Endometriosis**

Endometriosis is a disorder characterized by endometrial tissue located in places outside the uterine cavity, most commonly attached to the pelvic peritoneum, on the ovaries, and in the uterine tubes. For women with endometriosis, fragments of displaced endometrial tissue respond to hormonal changes, resulting in frequent cycles of severe pain even after menstruation has ended. The most commonly accepted cause of endometriosis is retrograde menstruation, literally menstruation that proceeds backward, or in the reverse direction. More than likely, all women experience some backflow of endometrial cells during menstruation, so it is likely that other contributing factors ultimately lead to endometriosis. Differences in the immune response and structural differences in the peritoneum have recently been implicated as contributing factors. ■

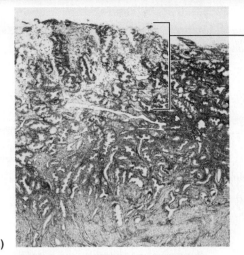

(a)

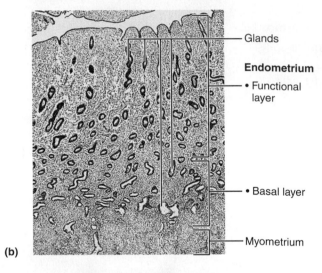

(b)

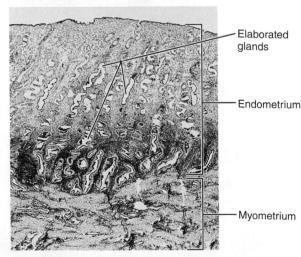

(c)

Figure 43.6 Endometrial changes during the menstrual cycle. (a) Onset of menstruation (8×). **(b)** Early proliferative phase (10×). **(c)** Early secretory phase (8×).

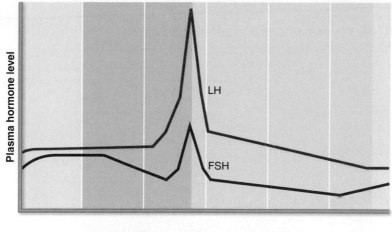

(a) Fluctuation of gonadotropin levels: Fluctuating levels of pituitary gonadotropins (follicle-stimulating hormone and luteinizing hormone) in the blood regulate the events of the ovarian cycle.

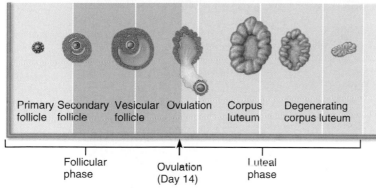

Primary follicle   Secondary follicle   Vesicular follicle   Ovulation   Corpus luteum   Degenerating corpus luteum

Follicular phase   Ovulation (Day 14)   Luteal phase

(b) Ovarian cycle: Structural changes in the ovarian follicles during the ovarian cycle are correlated with (d) changes in the endometrium of the uterus during the uterine cycle.

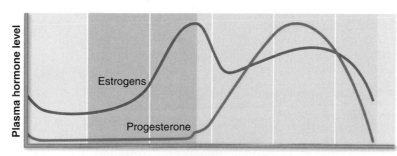

Estrogens

Progesterone

(c) Fluctuation of ovarian hormone levels: Fluctuating levels of ovarian hormones (estrogens and progesterone) cause the endometrial changes of the uterine cycle. The high estrogen levels are also responsible for the LH/FSH surge in (a).

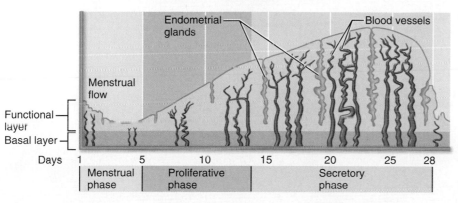

Endometrial glands   Blood vessels

Menstrual flow

Functional layer
Basal layer

Days   1   5   10   15   20   25   28

Menstrual phase | Proliferative phase | Secretory phase

(d) The three phases of the uterine cycle:
- Menstrual: The functional layer of the endometrium is shed. Approximately days 1–5.
- Proliferative: The functional layer of the endometrium is rebuilt under influence of estrogens. Approximately days 6–14. Ovulation occurs at the end of this phase.
- Secretory: Begins immediately after ovulation under the influence of progesterone. Enrichment of the blood supply and glandular secretion of nutrients prepare the endometrium to receive an embryo.

43

Figure 43.7 Correlation of anterior pituitary and ovarian hormones with structural changes in the ovary and uterus. The time bar applies to all parts of the figure.

EXERCISE 43

REVIEW SHEET
Physiology of Reproduction: Gametogenesis and the Female Cycles

Name _____ Lab Time/Date _____

Meiosis

1. The following statements refer to events occurring during mitosis and/or meiosis. For each statement, decide whether the event occurs in (a) mitosis only, (b) meiosis only, or (c) both mitosis and meiosis.

_____ 1. dyads are visible

_____ 2. tetrads are visible

_____ 3. product is two diploid daughter cells genetically identical to the mother cell

_____ 4. product is four haploid daughter cells quantitatively and qualitatively different from the mother cell

_____ 5. involves the phases prophase, metaphase, anaphase, and telophase

_____ 6. occurs throughout the body

_____ 7. occurs only in the ovaries and testes

_____ 8. provides cells for growth and repair

_____ 9. homologues synapse; crossovers are seen

_____ 10. chromosomes are replicated before the division process begins

_____ 11. provides cells for perpetuation of the species

_____ 12. consists of two consecutive nuclear divisions, without chromosomal replication occurring before the second division

2. Describe the process of synapsis. _____

_____

3. How does crossover introduce variability in the daughter cells? _____

_____

4. Define *homologous chromosomes*. _____

_____

Spermatogenesis

5. The cell types seen in the seminiferous tubules are listed in the key. Match the correct cell type or types with the descriptions given below.

Key: a. primary spermatocyte   c. spermatogonium   e. spermatid
         b. secondary spermatocyte   d. sustentocyte   f. sperm

_____ 1. primitive stem cell   _____ 4. products of meiosis II

_____ 2. haploid (3 responses)   _____ 5. product of spermiogenesis

_____ 3. provides nutrients to   _____ 6. product of meiosis I
                         developing sperm

6. Why are spermatids not considered functional gametes? _____

_____

7. Differentiate *spermatogenesis* from *spermiogenesis*. _____

_____

_____

8. Draw a sperm below, and identify the acrosome, head, midpiece, and tail. Then beside each label, note the composition and function of each of these sperm structures.

_____

9. The life span of a sperm is very short. What anatomical characteristics might lead you to suspect this even if you don't

   know its life span? _____

Oogenesis, the Ovarian Cycle, and the Menstrual Cycle

10. The sequence of events leading to germ cell formation in the female begins during fetal development. By the time the child

    is born, all viable oogonia have been converted to _____. _____.

    In view of this fact, how does the total germ cell potential of the female compare to that of the male?

    _____

11. The female gametes develop in structures called *follicles*. What is a follicle? _____

_____

How are primary and vesicular follicles anatomically different? _____

_____

_____

_____

What is a corpus luteum? _____

_____

12. What are the hormones produced by the corpus luteum? _____   _____

13. Use the key to identify the cell type you would expect to find in the following structures. The items in the key may be used once, more than once, or not at all.

Key: a.   oogonium   b.   primary oocyte   c.   secondary oocyte   d.   ovum

_____ 1.   forming part of the primary follicle in the   _____ 3.   in the vesicular follicle of the ovary
ovary

_____ 4.   in the uterine tube shortly after fertilization

_____ 2.   in the uterine tube before fertilization

14. The cellular product of spermatogenesis is four _____; the final product of oogenesis is one

_____ and three _____. What is the function of this unequal cytoplasmic

division seen during oogenesis in the female? _____

_____

_____

What is the fate of the polar bodies produced during oogenesis? _____

Why? _____

WHY THIS
MATTERS | 15. Which layer of the endometrium, the functional or basal layer, would you expect to be displaced with endometriosis?

Why? _____

_____

16. Explain the path that endometrial cells would travel to reach the peritoneum when retrograde menstruation occurs.

_____

_____

17. For each statement below dealing with plasma hormone levels during the female ovarian and menstrual cycles, decide whether the condition in column A is usually (a) greater than, (b) less than, or (c) essentially equal to the condition in column B.

Column A   **Column B**

_____ 1. amount of LH in the blood . . . . amount of LH in the blood at ovulation
                during menstruation

_____ 2. amount of FSH in the blood . . . . amount of FSH in the blood on day 20 of the cycle
                on day 6 of the cycle

_____ 3. amount of estrogen in the blood . . . . amount of estrogen in the blood at ovulation
                during menstruation

_____ 4. amount of progesterone in the blood. . . . . amount of progesterone in the blood on day 23
                on day 14

_____ 5. amount of estrogen in the blood . . . . amount of progesterone in the blood on day 10
                on day 10

18. What uterine tissue undergoes dramatic changes during the menstrual cycle? _____

19. When during the female menstrual cycle would fertilization be unlikely? Explain why. _____

_____

20. Assume that a woman could be an "on demand" ovulator like the rabbit, in which copulation stimulates the hypothalamic-pituitary-gonadal axis and causes LH release, and that an oocyte was ovulated and fertilized on day 26 of her 28-day cycle. Why would a successful pregnancy be unlikely at this time?

_____

_____

_____

21. The menstrual cycle depends on events within the female ovary. The stages of the menstrual cycle are listed below. For each, note its approximate time span and the related events in the uterus; and then to the right, record the ovarian events occurring simultaneously. Pay particular attention to hormonal events.

Menstrual cycle stage	Uterine events	Ovarian events
Menstruation		
Proliferative		
Secretory		

44 Survey of Embryonic Development

Objectives

☐ Define *fertilization* and *zygote*.

☐ Define and discuss the function of *cleavage* and *gastrulation*.

☐ Differentiate between the blastula and gastrula forms of the sea urchin and human using appropriate models or diagrams.

☐ Define *blastocyst*.

☐ Identify the following structures: inner cell mass, trophoblast, chorionic villi, amnion, yolk sac and allantois, and state the function of each.

☐ Describe the process and timing of implantation in the human.

☐ Define *decidua basalis* and *decidua capsularis*.

☐ Name the three primary germ layers, and list the organs or organ systems that arise from each.

☐ Describe the gross anatomy and general function of the human placenta.

Materials

- Prepared slides of sea urchin development (zygote through larval stages)
- Compound microscope
- Three-dimensional human development models or plaques (if available)
- *Demonstration:* Phases of human development in *Life Before Birth: Normal Fetal Development,* Second Edition (by Marjorie A. England, 1996, Mosby-Wolfe)
- Disposable gloves
- *Demonstration:* Pregnant cat, rat, or pig uterus with uterine wall dissected
- Three-dimensional model of pregnant human torso
- *Demonstration:* Fresh or formalin-preserved human placenta
- Prepared slide of placenta tissue

MasteringA&P®

For related exercise study tools, go to the Study Area of **MasteringA&P.** There you will find:

- Practice Anatomy Lab **PAL**
- PhysioEx **PEx**
- A&PFlix **A&PFlix**
- Practice quizzes, Histology Atlas, eText, Videos, and more!

Pre-Lab Quiz

1. Circle the correct underlined term. The fertilized egg, or <u>zygote</u> / <u>embryo</u>, appears as a single cell surrounded by a fertilization membrane and a jellylike membrane.

2. The uniting of the egg and sperm nuclei is known as:
   a. embryogenesis
   b. fertilization
   c. implantation
   d. mitosis

3. Circle True or False. Cleavage is a series of mitotic divisions without any intervening growth periods and results in a multicellular embryonic body.

4. Circle the correct underlined term. As a result of gastrulation, a <u>three-</u> / <u>four</u>-layered embryo forms, with each layer corresponding to a primary germ layer.

5. The _____ implants in the uterine wall.
   a. zygote
   b. morula
   c. blastocyst
   d. gastrula

6. The _____ gives rise to the epidermis of the skin and the nervous system.
   a. ectoderm
   b. endoderm
   c. mesoderm

7. By the ninth week of development, the embryo is referred to as a:
   a. blastocyst
   b. blastomere
   c. fetus
   d. gastrula

8. Circle True or False. The placenta is composed solely of embryonic membranes.

9. Circle the correct underlined term. The <u>allantois</u> / <u>amnion</u> encases the young embryonic body in a fluid-filled chamber that acts to protect the developing embryo against trauma.

10. What is the function of the placenta? _____

Early development in all animals involves three basic types of activities, which are integrated to ensure the formation of a viable offspring: (1) an increase in cell number and subsequent cell growth; (2) cellular specialization; and (3) morphogenesis, the formation of functioning organ systems. This exercise first provides a rather broad overview of the changes in structure that take place during embryonic development in sea urchins. The pattern of changes in this marine animal provides a basis of comparison with developmental events in the human.

Developmental Stages of Sea Urchins and Humans

Activity 1

Microscopic Study of Sea Urchin Development

1. Obtain a compound microscope and a set of slides depicting embryonic development of the sea urchin.

2. Observe the fertilized egg, or **zygote,** which appears as a single cell immediately surrounded by a jellylike membrane and a fertilization membrane. After an egg is penetrated by a sperm, the egg and the sperm nuclei fuse to form a single nucleus. This process is called **fertilization.** Within 2 to 5 minutes after sperm penetration, a fertilization membrane forms beneath the jelly coat to prevent the entry of additional sperm. Draw the zygote and label the fertilization and jelly membranes.

Zygote

3. Observe the cleavage stages. Once fertilization has occurred, the zygote begins to divide, forming a mass of successively smaller and smaller cells, called **blastomeres.** This series of mitotic divisions without intervening growth periods is referred to as **cleavage,** and it results in a multicellular embryonic body. The cleavage stage of embryonic development provides a large number of building blocks (cells) with which to fashion the forming body.

As the division process continues, a solid ball of cells forms. At the 32-cell stage, it is called the **morula,** and the embryo resembles a berry in form. Then the cell mass hollows out to become the embryonic form called the **blastula,** which is a ball of cells surrounding a central cavity. The blastula is the final product of cleavage.

Identify and sketch the blastula stage of cleavage—a ball of cells with an apparently lighter center due to the presence of the central cavity.

Blastula

4. Identify the **early gastrula** form, which follows the blastula in the developmental sequence. The gastrula looks as though one end of the blastula had been indented or pushed into the central cavity, forming a two-layered embryo. In time, a third layer of cells appears between the initial two cell layers. Thus, as a result of **gastrulation,** a three-layered embryo forms, each layer corresponding to a **primary germ layer** from which all body tissues develop. The innermost layer, the **endoderm,** and the middle layer, the **mesoderm,** form the internal organs; the outermost layer, the **ectoderm,** forms the surface tissues of the body.

Draw a gastrula below. Label the ectoderm and endoderm. If you can see the third layer of cells, the mesoderm, budding off between the other two layers, label that also.

Gastrula

5. Gastrulation in the sea urchin is followed by the appearance of the free-swimming larval form, in which the three germ layers have differentiated into the various tissues and organs of the animal's body. If time allows, observe the larval form on the prepared slides.

Activity 2

Examining the Stages of Human Development

Use **Figure 44.1** and **Figure 44.2** on p. 668 and models or plaques of human development to identify the various stages of development and to respond to the questions posed below.

1. Observe the fertilized egg, or zygote, which appears as a single cell immediately surrounded by a jellylike *zona pellucida* and then a crown of granulosa cells (the *corona radiata*).

2. Next, observe the cleavage stages.

How is the human cleavage process similar to that in the sea urchin?

_____

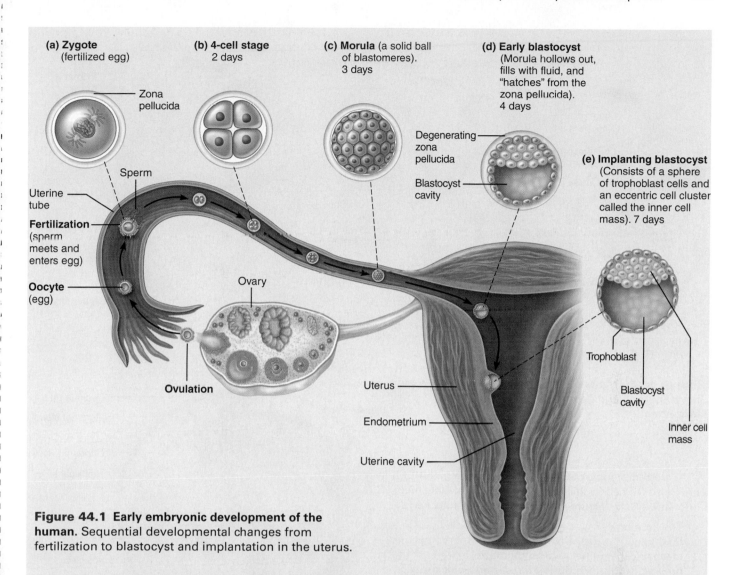

Figure 44.1 Early embryonic development of the human. Sequential developmental changes from fertilization to blastocyst and implantation in the uterus.

Observe the blastula, the final product of cleavage, which is called the **blastocyst** in the human. Unlike the sea urchin, only a portion of the blastocyst cells in the human contributes to the formation of the embryonic body—those seen at one side of the blastocyst (Figure 44.1e) forming the **inner cell mass (ICM)**. The rest of the blastocyst that enclosing the central cavity and overriding the ICM—is referred to as the **trophoblast**. The trophoblast becomes an extraembryonic membrane called the **chorion**, which forms the fetal portion of the **placenta**. The ICM becomes the **embryonic disc,** which forms the embryo proper.

3. Observe the *implanting* blastocyst shown on the model or in the figures. By approximately day 7 after ovulation, a developing human embryo (blastocyst) is floating free in the uterine cavity. About that time, it adheres to the uterine wall over the ICM area, and implantation begins.

The trophoblast cells secrete enzymes that break down the endometrium to reach the blood vessels beneath, assisting in implantation. Implantation is completed by day 14 after fertilization, and the ICM is imbedded in the endometrium. The portion of the endometrium that lies beneath the ICM and eventually lies beneath the embryo is the **decidua basalis,** and the portion of the endometrium that is on the luminal face of the uterus is the **decidua capsularis.**

By the time implantation has been completed, embryonic development has progressed to the **gastrula stage,** and the three primary germ layers are present and are beginning to differentiate (Figure 44.2). Within the next 6 weeks, virtually all of the body organ systems will have been laid down at least in rudimentary form by the germ layers. The three primary germ layers and their major derivatives are summarized in **Table 44.1** on p. 668.

By the ninth week of development, the embryo is referred to as **a fetus,** and from this point on, the major activities are growth and tissue and organ specialization.

4. Again observe the blastocyst to follow the formation of the embryonic membranes and the placenta (Figure 44.2). Notice the villus extensions of the trophoblast. By the time implantation is complete, the trophoblast has differentiated into the chorion, and its large elaborate villi are lying in the blood-filled sinusoids in the uterine tissue. This composite of the decidua basalis and **chorionic villi** is called the **placenta** (Figure 44.3), and all exchanges to and from the embryo occur through the chorionic membranes.

Text continues on next page. →

44

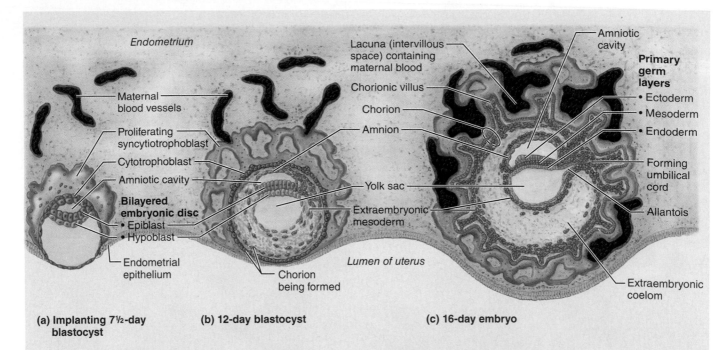

(a) Implanting 7½-day blastocyst

(b) 12-day blastocyst

(c) 16-day embryo

Figure 44.2 Events of placentation, early embryonic development, and extraembryonic membrane formation.

Three embryonic membranes (originating in the ICM) have also formed by this time—the amnion, the allantois, and the yolk sac (Figure 44.2). Attempt to identify each.

- The **amnion** encases the young embryonic body in a fluid-filled chamber that protects the embryo against mechanical trauma and temperature extremes and prevents adhesions during rapid embryonic growth.

- The **yolk sac** in humans has lost its original function, which was to pass nutrients to the embryo after digesting the yolk mass. The placenta has taken over that task; also, the human egg has very little yolk. However, the yolk sac is not totally useless. The embryo's first blood cells originate here, and the

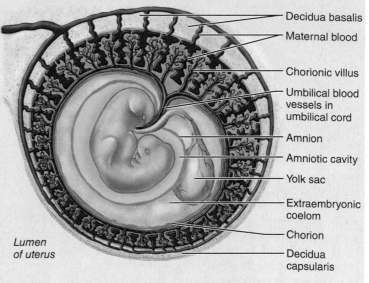

(d) 4½-week embryo

Table 44.1	Major Derivatives of the Three Primary Germ Layers (Figure 44.2c)
Primary germ layer	**Major derivatives**
Ectoderm (outermost layer)	Epidermis and accessory skin structures; nervous system and organs of the special senses; epithelia of the oral cavity, the nasal cavity, and the anal canal; adrenal medulla, pituitary, and pineal glands
Mesoderm (middle layer)	Dermis; skeleton; skeletal, cardiac, and most smooth muscle; cartilage; blood vessels; kidneys and ureters; most lymphoid organs and tissues; internal reproductive organs; serous membranes; adrenal cortex
Endoderm (inner most layer)	Epithelial lining of the respiratory tract, GI tract, urinary tract, and reproductive tract; liver, gallbladder, and pancreas; thymus, parathyroid glands, and thyroid glands

primordial germ cells migrate from it into the embryo's body to seed the gonadal tissue.

- The **allantois,** which protrudes from the posterior end of the yolk sac, is also largely redundant in humans because of the placenta. In birds and reptiles, it is a repository for embryonic wastes. In humans, it is the structural basis on which the mesoderm migrates to form the **umbilical cord,** which attaches the embryo to the placenta.

5. Go to the demonstration area to view the photographic series, *Life Before Birth: Normal Fetal Development*. After viewing them, respond to the following questions.

What organs or organ systems appear *very* early in embryonic development?

_____

_____

Does development occur in a rostral to caudal (head to toe) direction, or vice versa?

_____

Does development occur in a distal to proximal direction, or vice versa?

_____

What is vernix caseosa? _____

_____

_____

What is lanugo? _____

_____

_____

In Utero Development

Activity 3

Identifying Fetal Structures

1. Put on disposable gloves. Go to the appropriate demonstration area and observe the fetuses in the Y-shaped animal (cat, rat, or pig) uterus. Identify the following fetal or fetal-related structures:

- **Placenta** (a composite structure formed from the uterine mucosa and the fetal chorion).

Describe its appearance. _____

_____

_____

- **Umbilical cord.** Describe its relationship to the placenta and fetus.

_____

_____

- **Amniotic sac.** Identify the transparent amnion surrounding a fetus. Open one amniotic sac, and note the amount, color, and consistency of the fluid.

_____

_____

_____

Remove a fetus, and observe the degree of development of the head, body, and extremities. Is the skin thick or thin?

_____

2. Observe the model of a pregnant human torso. Identify the placenta. How does it differ in shape from the animal placenta observed?

_____

_____

_____

Identify the umbilical cord. In what region of the uterus does implantation usually occur, as indicated by the position of the placenta?

_____

_____

What might be the consequence if it occurred lower?

_____

_____

44

Gross and Microscopic Anatomy of the Placenta

The placenta is a remarkable temporary organ. Composed of maternal and fetal tissues, it is responsible for providing nutrients and oxygen to the embryo and fetus while removing carbon dioxide and metabolic wastes.

Activity 4

Studying Placental Structure

1. Notice that the human placenta on display has two very different-appearing surfaces—one smooth and the other spongy, roughened, and torn-looking.

Which is the fetal side? _____

What is the basis for your conclusion? _____

_____

_____

Identify the umbilical cord. Within the cord, identify the umbilical vein and two umbilical arteries. What is the function of the umbilical vein?

_____

The umbilical arteries? _____

_____

2. Obtain a microscope slide of placental tissue. Observe the tissue carefully and identify the *intervillous spaces* (lacunae), which are blood-filled in life (**Figure 44.3**). Identify the villi, and notice their rich vascular supply.

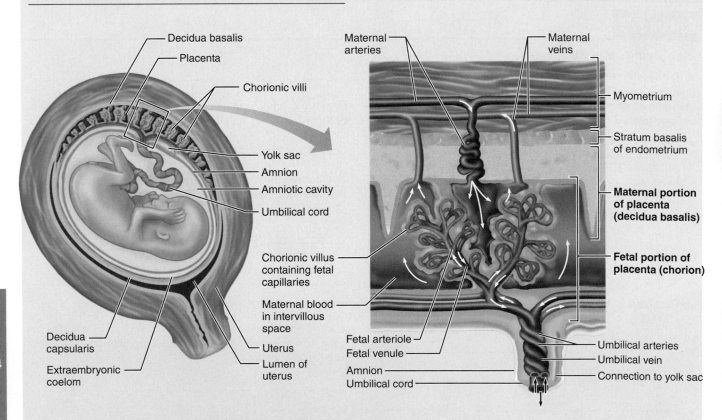

Figure 44.3 Diagrammatic representation of the structure of the placenta for a 13-week fetus.

REVIEW SHEET
Survey of Embryonic Development

Name _____ LabTime/Date _____

Developmental Stages of Sea Urchins and Humans

1. Define *zygote*. _____

2. Describe how you were able to tell by observation when a sea urchin egg was fertilized. _____

_____

3. Use the key choices to identify the embryonic stage or process described below.

   Key: a. blastocyst (blastula in sea urchins)   c. fertilization   e. morula
         b. cleavage   d. gastrulation   f. zygote

   _____ 1. process of male and female pronuclei fusion

   _____ 2. solid ball of embryonic cells

   _____ 3. process of rapid mitotic cell division without intervening growth periods

   _____ 4. cell resulting from combination of egg and sperm

   _____ 5. process involving cell rearrangements to form the three primary germ layers

   _____ 6. embryonic stage in which the embryo consists of a hollow ball of cells

4. What is the importance of cleavage in embryonic development? _____

_____

How is cleavage different from mitotic cell division, which occurs later in life? _____

_____

_____

5. The cells of the human blastocyst have various fates. Which blastocyst derivatives have the following fates?

   _____ 1. forms the embryo proper

   _____ 2. becomes the extraembryonic membrane called the chorion

   _____ 3. produces the amnion, yolk sac, and allantois

   _____ 4. produces the primordial germ cells

   _____ 5. an embryonic membrane that provides the structural basis for the umbilical cord

6. Using the letters on the diagram, correctly identify each of the following maternal or embryonic structures.

_____ amnion   _____ chorionic villi   _____ decidua capsularis   _____ forming umbilical cord

_____ chorion   _____ decidua basalis

_____ ectoderm   _____ mesoderm

_____ endoderm   _____ uterine cavity

7. Explain the process and importance of gastrulation. _____

_____

_____

8. What is the function of the amnion and the amniotic fluid? _____

_____

_____

9. Describe the process of implantation, noting the role of the trophoblast cells. _____

_____

_____

10. How many days after fertilization is implantation generally completed? _____ What event in the female menstrual

cycle ordinarily occurs just about this time if implantation does *not* occur? _____

_____

11. Referring to the illustrations and text of *Life Before Birth: Normal Fetal Development,* answer the following:

Which two organ systems are extensively developed in the *very young* embryo?

_____ and _____

Describe the direction of development by circling the correct descriptions below:

proximal-distal   distal-proximal   caudal-rostral   rostral-caudal

Does body control during infancy develop in the same directions? Think! Can an infant pick up a common pin (pincer grasp) or wave his arms earlier? Is arm-hand or leg-foot control achieved earlier?

_____

12. Note whether each of the following organs or organ systems develops from the (a) ectoderm, (b) endoderm, or (c) mesoderm.

_____ 1. skeletal muscle   _____ 4. respiratory mucosa   _____ 7. nervous system

_____ 2. skeleton   _____ 5. circulatory system   _____ 8. serous membrane

_____ 3. lining of the GI tract   _____ 6. epidermis of skin   _____ 9. liver, pancreas

In Utero Development

13. Make the following comparisons between a human and the dissected structures of another pregnant mammal.

Comparison object	Human	Dissected animal
Shape of the placenta		
Shape of the uterus		

14. Where in the human uterus do implantation and placentation ordinarily occur? _____

_____

15. Describe the function(s) of the placenta. _____

_____

16. Which two embryonic membranes has the placenta more or less "put out of business"? _____

_____

17. When does the human embryo come to be called a fetus? _____

18. What is the usual and most desirable fetal position in utero? _____

Why is this the most desirable position? _____

_____

_____

Gross and Microscopic Anatomy of the Placenta

19. Describe fully the gross structure of the human placenta as observed in the laboratory. _____

_____

_____

20. What is the tissue origin of the placenta: fetal, maternal, or both? _____

21. What placental barriers must be crossed to exchange materials? _____

_____

Objectives

☐ Define *allele, heterozygous, homozygous, dominance, recessiveness, genotype, phenotype,* and *incomplete dominance.*

☐ Work simple genetics problems using a Punnett square.

☐ State the basic laws of probability.

☐ Observe selected human phenotypes, and determine their genotype basis.

☐ Separate variants of hemoglobin using agarose gel electrophoresis.

Materials

- Pennies (for coin tossing)
- PTC (phenylthiocarbamide) taste strips
- Sodium benzoate taste strips
- Chart drawn on board for tabulation of class results of human phenotype/ genotype determinations

Activity 5: Blood typing

- Anti-A and Anti-B sera
- Clean microscope slides
- Toothpicks
- Wax pencils
- Sterile lancets
- Alcohol swabs
- Beaker containing 10% bleach solution
- Disposable autoclave bag

Activity 6: Hemoglobin phenotyping

- Electrophoresis equipment and power supply
- 1.2% agarose gels
- 1X TBE (Tris-Borate/EDTA) buffer pH 8.4
- Micropipette or variable autvmicropipette with tips
- Marking pen

Text continues on next page. →

MasteringA&P®

For related exercise study tools, go to the Study Area of **MasteringA&P.** There you will find:

- Practice Anatomy Lab **PAL**
- PhysioEx **PEx**
- A&PFlix **A&PFlix**
- Practice quizzes, Histology Atlas, eText, Videos, and more!

Pre-Lab Quiz

1. Circle the correct underlined term. Genes that code for the same genetic trait are <u>alleles</u> / <u>sister chromatids</u>.

2. Circle True or False. A heterozygous individual will have two of the same alleles in a chromosome pair.

3. The allele with less potency, which is present but not expressed, is the _____ allele.
   a. dominant
   b. genotypic
   c. homozygous
   d. recessive

4. Circle the correct underlined term. The physical appearance of an individual's genetic makeup, the characteristics that we can see, is that person's <u>genotype</u> / <u>phenotype</u>.

5. A _____ is used to demonstrate the genotypes of potential offspring that might result from mating.
   a. karyotype
   b. phenotype cross
   c. Punnett square

6. A condition known as _____ can result when heterozygous individuals exhibit a phenotype intermediate between homozygous individuals and both alleles are expressed in the offspring.
   a. incomplete dominance
   b. sex-linked inheritance
   c. total dominance
   d. total heterozygosity

7. Circle the correct underlined term. Possession of the <u>X</u> / <u>Y</u> chromosome determines maleness.

8. Circle True or False. The Y chromosome is only about one-third the size of the X chromosome, and it lacks many of the genes that are found on the X chromosome.

9. Circle True or False. Males are more likely to inherit hemophilia because it is a result of receiving the sex-linked gene from the father.

10. Circle True or False. The inheritance of ABO blood type involves three possible alleles.

- Safety goggles (student-provided)
- Metric ruler
- Plastic baggies
- Disposable gloves
- Coomassie blue protein stain solution
- Coomassie blue de-stain solution
- Distilled water
- Staining tray
- 100-ml graduated cylinder
- Hemoglobin samples dissolved in TBE solubilizing buffer with bromophenol blue: HbA, labeled A; HbS, labeled S; HbA + HbS, labeled AS; and unknown samples of each

The field of genetics is bristling with excitement. Complex gene-splicing techniques have allowed researchers to precisely isolate genes coding for specific proteins and then to use those genes to harvest large amounts of specific proteins and even to cure some dreaded human diseases. At present, growth hormone, insulin, erythropoietin, and interferon produced by these genetic engineering techniques are available for clinical use, and the list is growing daily.

Introduction to the Language of Genetics

In humans all cells, except eggs and sperm, contain 46 chromosomes, that is, the diploid number. The diploid chromosomal number actually represents two complete (or nearly complete) sets of genetic instructions—one from the mother and the other from the father—or 23 pairs of *homologous chromosomes*.

Genes coding for the same traits on each pair of homologous chromosomes are called **alleles.** The alleles may be identical or different. For example, the pair of alleles coding for hairline shape on your forehead may specify either straight across or widow's peak. When both alleles in a homologous chromosome pair have the same expression, the individual is **homozygous** for that trait. When the alleles differ in their expression, the individual is **heterozygous** for the given trait; and often only one of the alleles, called the **dominant gene,** exerts its effects. The allele with less potency, the **recessive gene,** is present but suppressed. Whereas dominant genes, or alleles, exert their effects in both homozygous and heterozygous conditions, as a rule recessive alleles *must* be present in double dose (homozygosity) to exert their influence.

An individual's actual genetic makeup, that is, whether the person is homozygous or heterozygous for the various alleles, is called the **genotype.** The expression of the genotype, for example, the presence or absence of a widow's peak (Figure 45.2) is referred to as a **phenotype.**

The complete story of heredity is much more complex than just outlined, and in actuality the expression of many traits (for example, eye color) is determined by the interaction of many allele pairs. However, our emphasis here will be to investigate only the less complex aspects of genetics.

Dominant-Recessive Inheritance

One of the best ways to master the terminology and learn the principles of heredity is to work out the solutions to some genetic crosses in much the same way Gregor Mendel did in his classic experiments on pea plants.

To work out the various simple monohybrid (one pair of alleles) crosses in this exercise, you will be given the genotype of the parents. You will then determine the possible genotypes of their offspring by using a grid called the *Punnett square,* and you will record the percentages of both genotype and phenotype. To illustrate the procedure, an example of one of Mendel's pea plant crosses is outlined next.

Alleles: *T* (determines *tallness;* dominant)

   t (determines *dwarfism;* recessive)

Genotypes of parents: *TT* (♂) × *tt* (♀)

Phenotypes of parents: Tall × dwarf

To use the Punnett, or checkerboard, square, write the alleles (actually gametes) of one parent across the top and the gametes of the other parent down the left side. Then combine the gametes across and down to achieve all possible combinations (possible genotypes of their offspring), as follows:

Results: Genotypes 100% *Tt* (all heterozygous)

   Phenotypes 100% tall (because *T*, which determines tallness, is dominant, and all contain the *T* allele)

Activity 1

Working Out Crosses Involving Dominant and Recessive Genes

For each of the following crosses, draw your own Punnett square and use the technique outlined above to determine the genotypes and phenotypes of the offspring.

1. Genotypes of parents: *Tt* (♂) × *tt* (♀)

45

% of each genotype: _____

% of each phenotype: _____% tall, _____% dwarf

2. Genotypes of parents: *Tt* (♂) × *Tt* (♀)

% of each genotype: _____

% of each phenotype: _____% tall, _____% dwarf

3. Genotypes of parents: *TT* (♂) × *Tt* (♀)

% of each genotype: _____

% of each phenotype: _____% tall, _____% dwarf

Incomplete Dominance

The concepts of dominance and recessiveness are somewhat arbitrary and artificial in some instances because so-called dominant genes may be expressed differently in homozygous and heterozygous individuals. This produces a condition called **incomplete dominance,** or *intermediate inheritance.*

In such cases, both alleles express themselves in the offspring. The crosses are worked out in the same manner as indicated previously, but heterozygous offspring exhibit a phenotype intermediate between that of the homozygous individuals. Some examples follow.

Activity 2

Working Out Crosses Involving Incomplete Dominance

1. The inheritance of flower color in snapdragons illustrates the principle of incomplete dominance. The genotype *RR* is expressed as a red flower, *Rr* yields pink flowers, and *rr* produces white flowers. Work out the following crosses to determine the expected phenotypes and both genotypic and phenotypic percentages.

a. Genotypes of parents: *RR* × *rr*

% of each genotype: _____

% of each phenotype: _____

b. Genotypes of parents: *Rr* × *rr*

% of each genotype: _____

% of each phenotype: _____

_____

c. Genotypes of parents: *Rr* × *Rr*

% of each genotype: _____

% of each phenotype: _____

_____

2. In humans, the inheritance of sickle cell anemia/trait is determined by a single pair of alleles that exhibit incomplete

dominance. Individuals homozygous for the sickling gene *(s)* have *sickle cell anemia.* In double dose *(ss),* the sickling gene causes production of a very abnormal hemoglobin, which crystallizes and becomes sharp and spiky under conditions of oxygen deficit. Heterozygous individuals *(Ss)* have the *sickle cell trait;* they make both normal and sickling hemoglobin. Usually these individuals are healthy, but prolonged decreases in blood oxygen levels can lead to a sickle cell crisis. Individuals with the genotype *SS* form normal hemoglobin. Work out the following crosses:

a. Parental genotypes: *SS* × *ss*

% of each genotype: _____

% of each phenotype: _____

b. Parental genotypes: *Ss* × *Ss*

% of each genotype: _____

% of each phenotype: _____

_____

c. Parental genotypes: *ss* × *Ss*

% of each genotype: _____

% of each phenotype: _____

45

Sex-Linked Inheritance

A cell's chromosomes can be stained, photographed, and digitally rearranged to produce an image called a *karyotype*, which shows the complete human diploid chromosomal complement displayed in homologous pairs (**Figure 45.1**).

Of the 23 pairs of homologous chromosomes, 22 pairs are referred to as **autosomes.** The autosomes guide the expression of most body traits. The 23rd pair, the **sex chromosomes,** determine the sex of an individual, that is, whether an individual will be male or female. Normal females possess two sex chromosomes that look alike, the X chromosomes. Males possess two dissimilar sex chromosomes, referred to as X and Y. Possession of the Y chromosome determines maleness. The Y sex chromosome is only about a third the size of the X sex chromosome, and it lacks many of the genes that are found on the X.

Inherited traits determined by genes on the sex chromosomes are said to be *sex-linked,* and genes present *only* on the X sex chromosome are said to be *X-linked.* Some examples of X-linked genes include those that determine normal color vision (or, conversely, color blindness), and normal clotting ability (as opposed to hemophilia). The alleles that determine color blindness and hemophilia are recessive alleles. In females, *both* X chromosomes must carry the recessive alleles for a woman to express either of these conditions, and thus they tend to be infrequently seen. However, should a male receive an X-linked recessive allele for these conditions, he will exhibit the recessive phenotype because his Y chromosome does not contain alleles for that gene.

The critical point to understand about X-linked inheritance is the *absence* of male to male (that is, father to son) transmission of sex-linked genes. The X of the father *will* pass to each of his daughters but to none of his sons. Males always inherit sex-linked conditions from their mothers via the X chromosome.

Activity 3

Working Out Crosses Involving Sex-Linked Inheritance

1. A heterozygous woman carrying the recessive gene for color blindness marries a man who is color-blind. Assume the dominant gene is X^c (allele for normal color vision) and the recessive gene is X^c (determines color blindness). The mother's genotype is $X^C X^c$ and the father's $X^c Y$. Do a Punnett square to determine the answers to the following questions.

According to the laws of probability, what percentage of all their children will be color-blind?

_____ %

What is the percentage of color-blind individuals by sex?

_____% males; _____% females

What percentage of all children will be carriers? _____%

What is the sex of the carriers? _____

2. A heterozygous woman carrying the recessive gene for hemophilia marries a man who is not a hemophiliac. Assume the dominant gene is X^H and the recessive gene is X^h. The woman's genotype is $X^H X^h$, and her husband's genotype is $X^H Y$. What is the potential percentage and sex of their offspring who will be hemophiliacs?

_____% males; _____% females

What percentage can be expected to lack the allele for hemophilia?

_____%

What is the anticipated sex and percentage of individuals who will be carriers for hemophilia?

_____%; _____sex

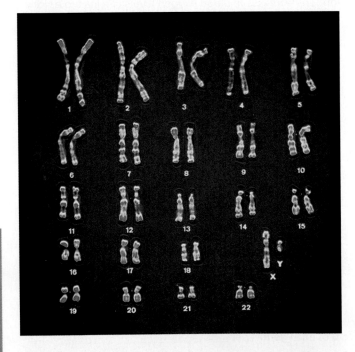

Figure 45.1 Karyotype (chromosomal complement) of human male. Each pair of homologous chromosomes is numbered except the sex chromosomes, which are identified by their letters, X and Y.

Probability

Parceling out of chromosomes to gametes during meiosis and the combination of egg and sperm are random events. Hence, the possibility that certain genomes will arise and be expressed is based on the laws of probability. The randomness of gene recombination from each parent determines individual uniqueness and explains why siblings, however similar, never have totally corresponding traits (unless, of course, they are identical twins). The Punnett square method that you have been using to work out the genetics problems actually provides information on the *probability* of the appearance of certain genotypes considering all possible events. Probability *(P)* is defined as:

$$P = \frac{\text{number of specific events or cases}}{\text{total number of events or cases}}$$

If an event is certain to happen, its probability is 1. If it happens one out of every two times, its probability is ½; if one out of four times, its probability is ¼, and so on.

When figuring the probability of separate events occurring together (or consecutively), the probability of each event must be multiplied together to get the final probability figure. For example, the probability of a penny coming up "heads" in each toss is ½ (because it has two sides—heads and tails). But the probability of a tossed penny coming up heads four times in a row is ½ × ½ × ½ × ½ = ¹⁄₁₆.

Activity 4

Exploring Probability

1. Obtain two pennies and perform the following simple experiment to explore the laws of probability.

a. Toss one penny into the air ten times, and record the number of heads (H) and tails (T) observed.

_____ heads _____ tails

Probability: _____/10 tails; _____/10 heads

b. Now simultaneously toss two pennies into the air for 24 tosses, and record the results of each toss below. In each case, report the probability in the lowest fractional terms.

\# of HH: _____   Probability: _____

\# of HT: _____   Probability: _____

\# of TT: _____   Probability: _____

Does the first toss have any influence on the second?

_____

c. Do a Punnett square using HT for the "alleles" of one "parental" coin and HT for the "alleles" of the other.

Probability of HH: _____

Probability of HT: _____

Probability of TT: _____

How closely do your coin-tossing results correlate with the percentages obtained from the Punnett square results?

_____

_____

2. Determine the probability of having a boy or girl offspring for each conception.

Parental genotypes: XY × XX

Probability of males: _____%

Probability of females: _____%

3. Dad wants an all-male baseball team! What are the chances of his having nine sons in a row?

_____

Genetic Determination of Selected Human Characteristics

Most human traits are determined by multiple alleles or the interaction of several gene pairs. However, a few visible human traits or phenotypes can be traced to a single gene pair. It is some of these that will be investigated here in our "genetics sampler."

Activity 5

Using Phenotype to Determine Genotype

For each of the characteristics described here, determine (as best you can) both your own phenotype and

Text continues on next page. ➜

genotype, and record this information on the **Activity 5 chart**. Since it is difficult to know if you are homozygous or heterozygous for a nondetrimental trait when you exhibit its dominant expression, you are to record your genotype as A—(or B—, and so on, depending on the letter used to indicate the alleles) in such cases. However, if you exhibit the recessive trait, you are homozygous for the recessive allele and should record it accordingly as *aa* (*bb, cc,* and so on). When you have completed your observations, also record your data in the chart on the board for tabulation of class results.

Activity 5: Record of Human Genotypes/Phenotypes

Characteristic	Phenotype	Genotype
PTC taste *(P,p)*		
Sodium benzoate taste *(S,s)*		
Sex (X,Y)		
Dimples *(D,d)*		
Widow's peak *(W,w)*		
Proximal finger hair *(H,h)*		
Freckles *(F,f)*		
Blaze *(B,b)*		
ABO blood type *(I^A, I^B, i)*		

PTC taste: Obtain a PTC taste strip. PTC, or phenylthiocarbamide, is a harmless chemical that some people can taste and others find tasteless. Chew the strip. If it tastes slightly bitter, you are a "taster" and possess the dominant gene *(P)* for this trait. If you cannot taste anything, you are a nontaster and are homozygous recessive *(pp)* for the trait. Approximately 70% of the people in the United States are tasters.

Sodium benzoate taste: Obtain a sodium benzoate taste strip and chew it. A different pair of alleles (from that determining PTC taste) determines the ability to taste sodium benzoate. If you can taste it, you have at least one of the dominant alleles *(S)*. If not, you are homozygous recessive *(ss)* for the trait. Also record whether sodium benzoate tastes salty, bitter, or sweet to you (if a taster). Even though PTC and sodium benzoate taste are inherited independently, they interact to determine a person's taste sensations. Individuals who find PTC bitter and sodium benzoate salty tend to like sauerkraut, buttermilk, spinach, and other slightly bitter or salty foods.

Sex: The genotype XX usually determines the female phenotype, whereas XY usually determines the male phenotype.

Dimpled cheeks: The presence of dimples in one or both cheeks is due to a dominant gene *(D)*. Absence of dimples indicates the homozygous recessive condition *(dd)*.

Widow's peak: A distinct downward V-shaped hairline at the middle of the forehead is referred to as a "widow's peak". It is determined by a dominant allele *(W)*, whereas the

Widow's peak   Straight hairline

Proximal finger hair   No finger hair

Freckles   No freckles

Figure 45.2 Selected examples of human phenotypes.

straight or continuous forehead hairline is determined by the homozygous recessive condition *(ww)* (**Figure 45.2**).

Proximal finger hair: Critically examine the dorsum of the proximal phalanx of fingers 3 and 4. If no hair is obvious, you are recessive *(hh)* for this condition. If hair is visible, you have the dominant gene *(H)* for this trait (which, however, is determined by multigene inheritance) (Figure 45.2).

Freckles: Freckles are the result of a dominant gene. Use *F* as the dominant allele and *f* as the recessive allele (Figure 45.2).

Blaze: A lock of hair different in color from the rest of scalp hair is called a blaze; it is determined by a dominant gene. Use *B* for the dominant gene and *b* for the recessive gene.

Blood type: Some genes exhibit more than two allele forms, leading to a phenomenon called **multiple-allele inheritance**. Inheritance of the ABO blood type is based on the existence of three alleles designated as I^A, I^B, and *i*. Both I^A and I^B are dominant over *i*, but neither is dominant over the other. The I^A and I^B alleles are *codominant*. Thus the possession of I^A and I^B will yield type AB blood, whereas the possession of the I^A and *i* alleles will yield type A blood, and so on (as explained in Exercise 29). The four ABO blood groups, or phenotypes, are A, B, AB, and O. Their correlation to genotype is indicated in **Table 45.1**.

If you have previously typed your blood, record your phenotype and genotype in the Activity 5 chart. If not, type your blood following your instructor's instructions (see Exercise 29, pp. 436–438), and then enter your results in the table.

⚠ Dispose of any blood-soiled supplies by placing the glassware in the bleach-containing beaker and all other items in the autoclave bag.

Once class data have been tabulated, scrutinize the results. Is there a single trait that is expressed in an identical manner by all members of the class?

Table 45.1	Blood Groups	
ABO blood group (phenotype)		**Genotype**
A		$I^A I^A$ or $I^A i$
B		$I^B I^B$ or $I^B i$
AB		$I^A I^B$
O		ii

Hemoglobin Phenotype Identification Using Agarose Gel Electrophoresis

Agarose gel electrophoresis separates molecules based on size and charge. In the appropriate buffer with an alkaline pH, hemoglobin molecules will move toward the anode of the apparatus at different speeds, based on the number of negative charges on the molecules.

Sickle cell anemia and sickle cell trait are discussed in Activity 2 of this exercise. The beta chains of hemoglobin S (HbS) contain a base substitution where a valine replaces glutamic acid. As a result of the substitution, HbS has fewer negative charges than the predominant form of adult hemoglobin (HbA) and can be separated from HbA using agarose gel electrophoresis.

Activity 6

Using Agarose Gel Electrophoresis to Identify Normal Hemoglobin, Sickle Cell Anemia, and Sickle Cell Trait

1. You will need an electrophoresis unit and power supply, a 1.2% agarose gel with eight wells, 1XTBE buffer, micropipettes or a variable automatic micropipette (2–20 µl) with tips, samples of hemoglobin (marked A, AS, S, and unknown #_____) dissolved in TBE solubilizing buffer containing bromophenol blue, a marking pen, safety goggles, metric ruler, plastic bag, and disposable gloves.

2. Record the number of your unknown sample. _____

3. Place the agarose gel into the electrophoresis unit.

4. Using a micropipette, carefully add 15 µl of hemoglobin sample A to wells number 1 and 5, AS to wells 2 and 6, S to wells 3 and 7, and the unknown samples to wells 4 and 8.

5. Slowly add electrophoresis buffer until the gels are covered with about 0.25 cm of buffer.

⚠ 6. The electrophoresis unit runs with high voltage. Do not attempt to open it while the power supply is attached. Close and lock the electrophoresis unit, and connect the unit to the power source, red to red and black to black.

7. Run the unit about 50 minutes at 120 volts until the bromophenol blue is about 0.25 cm from the anode.

8. Turn the power supply **OFF**. Disconnect the cables.

9. Open the electrophoresis unit, carefully remove the gel, and slide the gel into a plastic bag. You should be able to see the hemoglobin bands on the gel. Mark each of the bands on the gel with the marking pen.

Alternatively, the gels may be stained with Coomassie blue. Obtain a flask of Coomassie blue stain, a flask of destaining solution, a flask of distilled water, a staining tray, and a 100-ml graduated cylinder, and do the following:

a. Carefully remove the gel from the plastic plate, and place the gel into a staining dish.

b. Add about 30 ml of stain (enough stain to cover the gel), and be sure the agarose is not stuck to the dish.

c. Allow the gel to remain in the stain for at least an hour (more time might be necessary), and then remove the stain and pour into an appropriate waste container. Rinse the gel and dish with distilled water.

d. Add about 100 ml of de-staining solution. Change the solution after a day. If the background stain has been reduced enough to see the bands, place the staining dish over a light source, and observe the bands. If the stain is still too dark, repeat the de-staining process until the bands can be observed.

e. To store the gels, refrigerate in a bag with a small amount of de-staining solution or dry on a glass plate.

10. Draw the banding patterns for samples 1 through 8 in the figure for question 16 in the Review Sheet (p. 686). Based on the banding patterns of the known samples, what are the genotypes of your unknown samples? Record on the Review Sheet chart.

11. Rinse the electrophoresis unit in distilled or de-ionized water, and clean the glass plates with soap and water.

45

Group Challenge

Odd Phenotype Out

The following boxes each contain four phenotypes. One of the listed traits in each case does not belong with the other three for some reason. Working in groups of three, discuss the characteristics of the phenotypes in each set. On a separate piece of paper, one student will record the characteristics for each phenotype in the set. For each set of phenotypes, discuss the possible candidates for the "odd phenotype" and which characteristic it lacks based on your notes. Once you have come to a consensus within your group, circle the trait that doesn't belong with the others, and explain how it is different from the others.

1. Which is the "odd phenotype"?	Why is it the odd one out?
Freckles Widow's peak Sickle cell trait Sodium benzoate "taster"	
2. Which is the "odd phenotype"?	**Why is it the odd one out?**
Blaze Type AB blood PTC "taster" Widow's peak	
3. Which is the "odd phenotype"?	**Why is it the odd one out?**
No proximal finger hair Straight hairline Color blindness Sodium benzoate "nontaster"	
4. Which is the "odd phenotype"?	**Why is it the odd one out?**
Dimpled cheek Straight hairline Freckles Blaze	

EXERCISE

45

REVIEW SHEET
Principles of Heredity

Name _____ Lab Time/Date _____

Introduction to the Language of Genetics

1. Match the key choices with the definitions given below.

   Key: a. alleles   d. genotype   g. phenotype
         b. autosomes   e. heterozygous   h. recessive
         c. dominant   f. homozygous   i. sex chromosomes

   _____ 1. actual genetic makeup

   _____ 2. chromosomes determining maleness/femaleness

   _____ 3. situation in which an individual has identical alleles for a particular trait

   _____ 4. genes not expressed unless they are present in homozygous condition

   _____ 5. expression of a genetic trait

   _____ 6. situation in which an individual has different alleles making up his or her genotype for a
                         particular trait

   _____ 7. genes for the same trait that may have different expressions

   _____ 8. chromosomes regulating most body characteristics

   _____ 9. the more potent gene allele; masks the expression of the less potent allele

Dominant-Recessive Inheritance

2. In humans, farsightedness is inherited by possession of a dominant allele *(A)*. If a man who is homozygous for normal
   vision *(aa)* marries a woman who is heterozygous for farsightedness *(Aa)*, what percentage of their children would be
   expected to be farsighted?

   _____%

3. A metabolic disorder called phenylketonuria (PKU) is due to an abnormal recessive gene *(p)*. Only homozygous recessive individuals exhibit this disorder. What percentage of the offspring will be anticipated to have PKU if the parents
   are *Pp* and *pp?*

   _____%

4. A man obtained 32 spotted and 10 solid-color rabbits from a mating of two spotted rabbits.

   Which trait is dominant? _____ Recessive? _____

   If the dominant allele is *S,* what is the probable genotype of the rabbit parents? _____ × _____

5. Assume that the allele controlling brown eyes *(B)* is dominant over that controlling blue eyes *(b)* in human beings. (In actuality, eye color in humans is an example of polygenic inheritance, which is much more complex than this.) A blue-eyed man marries a brown-eyed woman, and they have six children, all brown-eyed. What is the most likely genotype of the father?

_____ Of the mother? _____ If the seventh child had *blue* eyes, given this new information what could you conclude about the parents' genotypes?

_____

_____

Incomplete Dominance

6. Tail length on a bobcat is controlled by incomplete dominance. The alleles are *T* for normal tail length and *t* for tail-less.

What name could/would you give to the tails of heterozygous *(Tt)* bobcats? _____

How would their tail length compare with that of *TT* or *tt* bobcats? _____

7. If curly-haired individuals are genotypically *CC*, straight-haired individuals are *cc*, and wavy-haired individuals are heterozygotes *(Cc)*, what percentage of the various phenotypes would be anticipated from a cross between a *CC* woman and a *cc* man?

_____% curly   _____% wavy   _____% straight

Sex-Linked Inheritance

8. What does it mean when someone says a particular characteristic is sex-linked? _____

_____

9. You are a male, and you have been told that hemophilia "runs in your genes." Whose ancestors, your mother's or your

father's, should you investigate? _____ Why? _____

_____

_____

10. An $X^C X^c$ female marries an $X^C Y$ man. Do a Punnett square for this match.

What is the probability of producing a color-blind son? _____

A color-blind daughter? _____

A daughter who is a carrier for the color-blind allele? _____

11. Why are consanguineous marriages (marriages between blood relatives) prohibited in most cultures?

_____

_____

Probability

12. What is the probability of having three daughters in a row? _____

13. A man and a woman, each of seemingly normal intellect, marry. Although neither is aware of the fact, each is a heterozygote

for the allele for mental retardation. Is the allele for mental retardation dominant or recessive? _____

What is the probability of their having one mentally retarded child? _____

What is the probability that all their children (they plan a family of four) will be mentally retarded? _____

Genetic Determination of Selected Human Characteristics

14. Look back at your data to complete this section. For each of the situations described here, determine if an offspring with the characteristics noted is possible with the parental genotypes listed. Check (✓) the appropriate column.

Parental genotypes	Phenotype of child	Possibility	
		Yes	**No**
$Ff \times ff$	Freckles		
$dd \times dd$	Dimples		
$HH \times Hh$	Proximal finger hair		
$I^A i \times I^B i$	Type O blood		
$I^A I^B \times ii$	Type O blood		

15. You have dimples, and you would like to know whether you are homozygous or heterozygous for this trait. You have six brothers and sisters. By observing your siblings, how could you tell, with some degree of certainty, that you are a heterozygote?

_____

_____

_____

Using Agarose Gel Electrophoresis to Identify Hemoglobin Phenotypes

16. Draw the banding patterns you obtained on the figure below.

Sample	Well	Banding pattern
1. A	1. ☐	
2. AS	2. ☐	
3. S	3. ☐	
4. Unknown	4. ☐	
5. A	5. ☐	
6. AS	6. ☐	
7. S	7. ☐	
8. Unknown	8. ☐	

17. What is the genotype of sickle cell anemia? _____ Sickle cell trait? _____

18. Why does sickle cell hemoglobin behave differently from normal hemoglobin during agarose gel electrophoresis?

_____

_____

46 Surface Anatomy Roundup

Objectives

☐ Define *surface anatomy,* and explain why it is an important field of study; define *palpation.*

☐ Describe and palpate the major surface features of the cranium, face, and neck.

☐ Describe the easily palpated bony and muscular landmarks of the back, and locate the vertebral spines on the living body.

☐ List the bony surface landmarks of the thoracic cage, explain how they relate to the major soft organs of the thorax, and explain how to find the second to eleventh ribs.

☐ Name and palpate the important surface features on the anterior abdominal wall, and explain how to palpate a full bladder.

☐ Define and explain the following: *linea alba, umbilical hernia,* examination for an inguinal hernia, *linea semilunaris,* and *McBurney's point.*

☐ Locate and palpate the main surface features of the upper limb.

☐ Explain the significance of the cubital fossa, pulse points in the distal forearm, and the anatomical snuff box.

☐ Describe and palpate the surface landmarks of the lower limb.

☐ Explain exactly where to administer an injection in the gluteal region and in the other major sites of intramuscular injection.

Materials

- Articulated skeletons
- Three-dimensional models or charts of the skeletal muscles of the body
- Hand mirror
- Stethoscope
- Alcohol swabs
- Washable markers

MasteringA&P®

For related exercise study tools, go to the Study Area of **MasteringA&P.** There you will find:

- Practice Anatomy Lab **PAL**
- A&PFlix **A&PFlix**
- PhysioEx **PEx**
- Practice quizzes, Histology Atlas, eText, Videos, and more!

Pre-Lab Quiz

1. Why is it useful to study surface anatomy?
   a. You can easily locate deep muscle insertions.
   b. You can relate external surface landmarks to the location of internal organs.
   c. You can study cadavers more easily.
   d. You really can't learn that much by studying surface anatomy; it's a gimmick.

2. Circle the correct underlined term. <u>Palpation</u> / <u>Dissection</u> allows you to feel internal structures through the skin.

3. The epicranial aponeurosis binds to the subcutaneous tissue of the cranium to form the:
   a. mastoid process
   b. occipital protuberance
   c. true scalp
   d. xiphoid process

4. The _____ is the most prominent neck muscle and also the neck's most important landmark.
   a. buccinator
   b. epicranius
   c. masseter
   d. sternocleidomastoid

5. The three boundaries of the _____ are the trapezius medially, the latissimus dorsi inferiorly, and the scapula laterally.
   a. torso triangle
   b. triangle of auscultation
   c. triangle of back muscles
   d. triangle of McBurney

6. Circle True or False. The lungs do not fill the inferior region of the pleural cavity.

Text continues on next page →

7. Circle True or False. With the exception of a full bladder, most internal pelvic organs are not easily palpated through the skin of the body surface.

8. On the dorsum of your hand is a grouping of superficial veins known as the _____, which provides a site for drawing blood and inserting intravenous catheters.
   a. anatomical snuff box
   b. dorsal venous network
   c. radial and ulnar veins
   d. palmar arches

9. Circle True or False. To avoid harming major nerves and blood vessels, clinicians who administer intramuscular injections in the gluteal region of adults use the gluteus medius muscle.

10. The large femoral artery and vein descend vertically through the _____, formed by the border of the inguinal ligament, the medial border of the adductor longus muscle, and the medial border of the sartorius muscle.
    a. femoral triangle
    b. lateral condyle
    c. medial condyle
    d. quadriceps

Surface anatomy is a valuable branch of anatomical and medical science. True to its name, **surface anatomy** does indeed study the *external surface* of the body, but more importantly, it also studies *internal* organs as they relate to external surface landmarks and as they are seen and felt through the skin. Feeling internal structures through the skin with the fingers is called **palpation** (literally, "touching").

Surface anatomy is living anatomy, better studied in live people than in cadavers. It can provide a great deal of information about the living skeleton (almost all bones can be palpated) and about the muscles and blood vessels that lie near the body surface. Furthermore, a skilled examiner can learn a good deal about the heart, lungs, and other deep organs by performing a surface assessment. Thus, surface anatomy serves as the basis of the standard physical examination. For those planning a career in the health sciences or physical education, a study of surface anatomy will show you where to take pulses, where to insert tubes and needles, where to locate broken bones and inflamed muscles, and where to listen for the sounds of the lungs, heart, and intestines.

We will take a regional approach to surface anatomy, exploring the head first and proceeding to the trunk and the limbs. You will be observing and palpating your own body as you work through the exercise, because your body is the best learning tool of all. To aid your exploration of living anatomy, skeletons and muscle models or charts are provided around the lab so that you can review the bones and muscles you will encounter. For skin sites you are asked to mark that you cannot reach on your own body, it probably would be best to choose a male student as a subject.

Activity 1

Palpating Landmarks of the Head

The head (**Figure 46.1** and **Figure 46.2**) is divided into the cranium and the face.

Cranium

1. Run your fingers over the superior surface of your head. Notice that the underlying cranial bones lie very near the surface. Proceed to your forehead and palpate the **superciliary arches** (brow ridges) directly superior to your orbits (Figure 46.1).

2. Move your hand to the posterior surface of your skull, where you can feel the knoblike **external occipital protuberance**. Run your finger directly laterally from this projection to feel the ridgelike *superior nuchal line* on the occipital bone. This line, which marks the superior extent of the muscles of the posterior neck, serves as the boundary between the head and the neck.

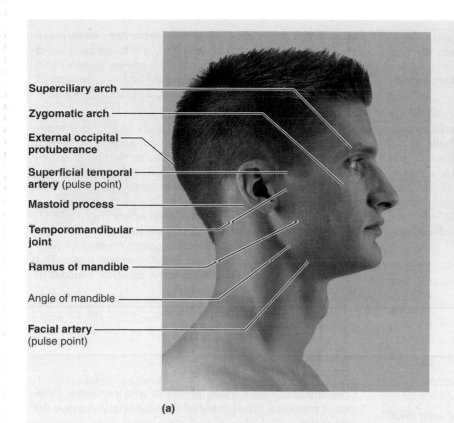

Superciliary arch

Zygomatic arch

External occipital protuberance

Superficial temporal artery (pulse point)

Mastoid process

Temporomandibular joint

Ramus of mandible

Angle of mandible

Facial artery (pulse point)

(a)

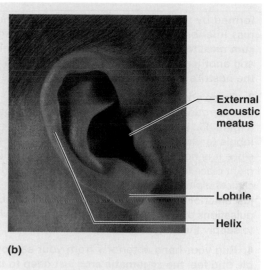

External acoustic meatus

Lobule

Helix

(b)

Figure 46.1 Surface anatomy of the head. (a) Lateral aspect. **(b)** Close-up of an auricle.

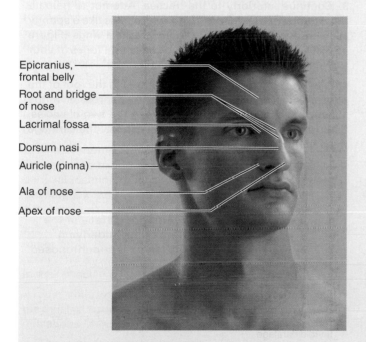

Epicranius, frontal belly

Root and bridge of nose

Lacrimal fossa

Dorsum nasi

Auricle (pinna)

Ala of nose

Apex of nose

Figure 46.2 Surface structures of the face.

Now feel the prominent **mastoid process** on each side of the cranium just posterior to your ear.

3. The **frontal belly** of the epicranius (Figure 46.2) inserts superiorly onto the broad aponeurosis called the *epicranial aponeurosis* (Table 13.1, p. 204) that covers the superior surface of the cranium. This aponeurosis binds tightly to the overlying subcutaneous tissue and skin to form the true **scalp**. Push on your scalp, and confirm that it slides freely over the underlying cranial bones. Because the scalp is only loosely bound to the skull, people can easily be "scalped" (in industrial accidents, for example). The scalp is richly vascularized by a large number of arteries running through its subcutaneous tissue. Most arteries of the body constrict and close after they are cut or torn, but those in the scalp are unable to do so because they are held open by the dense connective tissue surrounding them.

What do these facts suggest about the amount of bleeding that accompanies scalp wounds?

_____

_____

Face

The surface of the face is divided into many different regions, including the *orbital, nasal, oral* (mouth), and *auricular* (ear) areas.

1. Trace a finger around the entire margin of the bony orbit. The **lacrimal fossa**, which contains the tear-gathering lacrimal sac, may be felt on the medial side of the eye socket.

2. Touch the most superior part of your nose, its **root**, which lies between the eyebrows (Figure 46.2). Just inferior to this, between your eyes, is the **bridge** of the nose

Text continues on next page. →

46

Activity 3

Palpating Landmarks of the Trunk

The trunk of the body consists of the thorax, abdomen, pelvis, and perineum. The *back* includes parts of all of these regions, but for convenience it is treated separately.

The Back

Bones

1. The vertical groove in the center of the back is called the **posterior median furrow (Figure 46.5)**. The *spinous processes* of the vertebrae are visible in the furrow when the spinal column is flexed.

- Palpate a few of these processes on your partner's back (C_7 and T_1 are the most prominent and the easiest to find).

- Also palpate the posterior parts of some ribs, as well as the prominent **spine of the scapula** and the scapula's long **medial border.**

    The scapula lies superficial to ribs 2 to 7; its **inferior angle** is at the level of the spinous process of vertebra T_7. The medial end of the scapular spine lies opposite the T_3 spinous process.

2. Now feel the **iliac crests** (superior margins of the iliac bones) in your own lower back. You can find these crests

effortlessly by resting your hands on your hips. Locate the most superior point of each crest, a point that lies roughly halfway between the posterior median furrow and the lateral side of the body (Figure 46.5). A horizontal line through these two superior points, the **supracristal line,** intersects L_4, providing a simple way to locate that vertebra. The ability to locate L_4 is essential for performing a *lumbar puncture,* a procedure in which the clinician inserts a needle into the vertebral canal of the spinal column directly superior or inferior to L_4 and withdraws cerebrospinal fluid.

3. The *sacrum* is easy to palpate just superior to the cleft in the buttocks. You can feel the *coccyx* in the extreme inferior part of that cleft, just posterior to the anus.

Muscles

The largest superficial muscles of the back are the **trapezius** superiorly and **latissimus dorsi** inferiorly (Figure 46.5). Furthermore, the deeper **erector spinae** muscles are very evident in the lower back, flanking the vertebral column like thick vertical cords.

1. Shrug your shoulders to feel the trapezius contracting just deep to the skin.

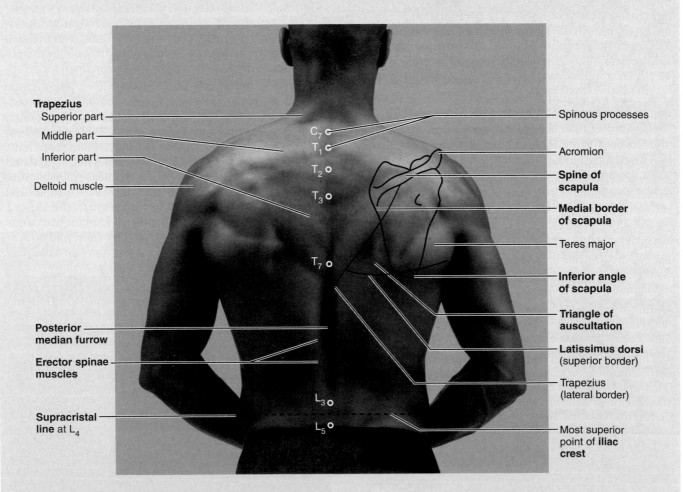

Figure 46.5 Surface anatomy of the back.

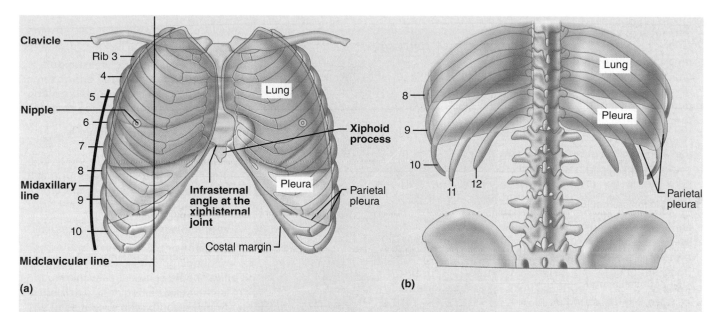

Figure 46.6 The bony rib cage as it relates to the underlying lungs and pleural cavities. Both the pleural cavities (blue) and the lungs (pink) are outlined. **(a)** Anterior view. **(b)** Posterior view.

2. Feel your partner's erector spinae muscles contract and bulge as he straightens his spine from a slightly bent-over position.

The superficial muscles of the back fail to cover a small area of the rib cage called the **triangle of auscultation** (Figure 46.5). This triangle lies just medial to the inferior part of the scapula. Its three boundaries are formed by the trapezius medially, the latissimus dorsi inferiorly, and the scapula laterally. The physician places a stethoscope over the skin of this triangle to listen for lung sounds (*auscultation* = listening). To hear the lungs clearly, the doctor first asks the patient to fold the arms together in front of the chest and then flex the trunk.

What do you think is the precise reason for having the patient take this action?

_____

_____

3. Have your partner assume the position just described. After cleaning the earpieces with an alcohol swab, use the stethoscope to auscultate the lung sounds. Compare the clarity of the lung sounds heard over the triangle of auscultation to that over other areas of the back.

The Thorax

Bones

1. Start exploring the anterior surface of your partner's bony *thoracic cage* (**Figure 46.6** and **Figure 46.7**, p. 694) by defining the extent of the *sternum*. Use a finger to trace the sternum's triangular *manubrium* inferior to the jugular notch, its flat *body,* and the tongue-shaped **xiphoid process.** Now palpate the ridgelike **sternal angle,** where the manubrium meets the body of the sternum. Locating the sternal angle is important because it directs you

to the second ribs (which attach to it). Once you find the second rib, you can count down to identify every other rib in the thorax (except the first and sometimes the twelfth rib, which lie too deep to be palpated). The sternal angle is a highly reliable landmark—it is easy to locate, even in overweight people.

2. By locating the individual ribs, you can mentally "draw" a series of horizontal lines of "latitude" that you can use to map and locate the underlying visceral organs of the thoracic cavity. Such mapping also requires lines of "longitude," so let us construct some vertical lines on the wall of your partner's trunk. As he lifts an arm straight up in the air, extend a line inferiorly from the center of the axilla onto his lateral thoracic wall. This is the **midaxillary line** (Figure 46.6a). Now estimate the midpoint of his **clavicle,** and run a vertical line inferiorly from that point toward the groin. This is the **midclavicular line,** and it will pass about 1 cm medial to the nipple.

3. Next, feel along the V-shaped inferior edge of the rib cage, the **costal margin**. At the **infrasternal angle,** the superior angle of the costal margin, lies the **xiphisternal joint.** The heart lies on the diaphragm deep to the xiphisternal joint.

4. The thoracic cage provides many valuable landmarks for locating the vital organs of the thoracic and abdominal cavities. On the anterior thoracic wall, ribs 2–6 define the superior-to-inferior extent of the female breast, and the fourth intercostal space indicates the location of the **nipple** in men, children, and small-breasted women. The right costal margin runs across the anterior surface of the liver and gallbladder. Surgeons must be aware of the inferior margin of the *pleural cavities* because if they accidentally cut into one of these cavities, a lung collapses. The inferior pleural margin lies adjacent to vertebra T_{12} near the posterior

Text continues on next page. ➡

Activity 5

Palpating Landmarks of the Upper Limb

Axilla

The **base of the axilla** is the groove in which the underarm hair grows (Figure 46.7). Deep to this base lie the axillary *lymph nodes* (which swell and can be palpated in breast cancer), the large *axillary vessels* serving the upper limb, and much of the brachial plexus. The base of the axilla forms a "valley" between two thick, rounded ridges, the **axillary folds.** Just anterior to the base, clutch your **anterior axillary fold,** formed by the pectoralis major muscle. Then grasp your **posterior axillary fold.** This fold is formed by the latissimus dorsi and teres major muscles of the back as they course toward their insertions on the humerus.

Shoulder

1. Again locate the prominent spine of the scapula posteriorly (Figure 46.5). Follow the spine to its lateral end, the flattened **acromion** on the shoulder's summit. Then, palpate the **clavicle** anteriorly, tracing this bone from the sternum to the shoulder (**Figure 46.9**). Notice the clavicle's curved shape.

2. Now locate the junction between the clavicle and the acromion on the superolateral surface of your shoulder, at the **acromioclavicular joint.** To find this joint, thrust your arm anteriorly repeatedly until you can palpate the precise point of pivoting action.

3. Next, place your fingers on the **greater tubercle** of the humerus. This is the most lateral bony landmark on the superior surface of the shoulder. It is covered by the thick **deltoid muscle,** which forms the rounded superior part of the shoulder. Intramuscular injections are often given into the deltoid, about 5 cm (2 inches) inferior to the greater tubercle (Figure 46.17a, p. 700).

Arm

Remember, according to anatomists, the arm runs only from the shoulder to the elbow, and not beyond.

1. In the arm, palpate the humerus along its entire length, especially along its medial and lateral sides.

2. Feel the **biceps brachii** muscle contract on your anterior arm when you flex your forearm against resistance. The medial boundary of the biceps is represented by the **medial bicipital furrow** (Figure 46.9). This groove contains the large *brachial artery,* and by pressing on it with your fingertips you can feel your *brachial pulse.* Recall that the brachial artery is the artery routinely used in measuring blood pressure with a sphygmomanometer.

3. All three heads of the **triceps brachii** muscle (lateral, long, and medial) are visible through the skin of a muscular person (**Figure 46.10**).

Elbow Region

1. In the distal part of your arm, near the elbow, palpate the two projections of the humerus, the **lateral** and **medial epicondyles** (Figures 46.9 and 46.10). Midway between the epicondyles, on the posterior side, feel the **olecranon,** which forms the point of the elbow.

2. Confirm that the two epicondyles and the olecranon all lie in the same horizontal line when the elbow is extended.

Deltoid

Biceps brachii

Medial bicipital furrow

Medial epicondyle of the humerus

Olecranon

Clavicle

Acromioclavicular joint

Greater tubercle of the humerus

Cephalic vein

Figure 46.9 Shoulder and arm.

46

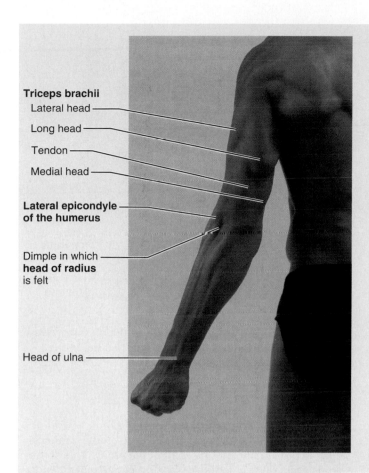

Triceps brachii
Lateral head
Long head
Tendon
Medial head
Lateral epicondyle of the humerus
Dimple in which **head of radius** is felt
Head of ulna

Figure 46.10 Surface anatomy of the upper limb, posterior view.

If these three bony processes do not line up, the elbow is dislocated.

3. Now feel along the posterior surface of the medial epicondyle. You are palpating your ulnar nerve.

4. On the anterior surface of the elbow is a triangular depression called the **cubital fossa** (**Figure 46.11**). The triangle's superior *base* is formed by a horizontal line between the humeral epicondyles; its two inferior sides are defined by the **brachioradialis** and **pronator teres** muscles (Figure 46.11b). Try to define these boundaries on your own limb. To find the brachioradialis muscle, flex your forearm against resistance, and watch this muscle bulge through the skin of your lateral forearm. To feel your pronator teres contract, palpate the cubital fossa as you pronate your forearm against resistance. (Have your partner provide the resistance.)

Superficially, the cubital fossa contains the **median cubital vein** (Figure 46.11a). Clinicians often draw blood from this superficial vein and insert intravenous (IV) catheters into it to administer medications, transfused blood, and nutrient fluids. The large **brachial artery** lies just deep to the median cubital vein (Figure 46.11b), so a needle must be inserted into the vein from a shallow angle (almost parallel to the skin) to avoid puncturing the artery. Tendons and nerves are also found deep in the fossa (Figure 46.11b).

5. The median cubital vein interconnects the larger **cephalic** and **basilic veins** of the upper limb. These veins are visible through the skin of lean people (Figure 46.11a). Examine your arm to see if your cephalic and basilic veins are visible.

Forearm and Hand

The two parallel bones of the forearm are the medial *ulna* and the lateral *radius*.

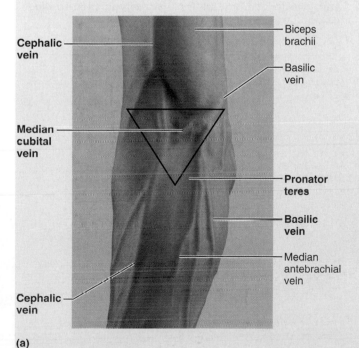

Cephalic vein
Biceps brachii
Basilic vein
Median cubital vein
Pronator teres
Basilic vein
Median antebrachial vein
Cephalic vein

(a)

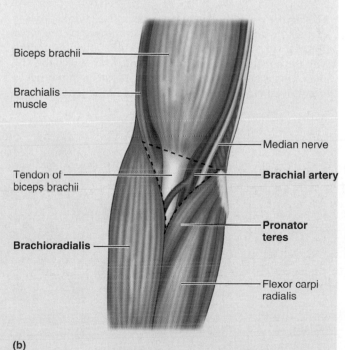

Biceps brachii
Brachialis muscle
Tendon of biceps brachii
Brachioradialis
Median nerve
Brachial artery
Pronator teres
Flexor carpi radialis

(b)

46

Figure 46.11 The cubital fossa on the anterior surface of the right elbow (outlined by the triangle). (a) Photograph. **(b)** Diagram of deeper structures in the fossa.

Text continues on next page. →

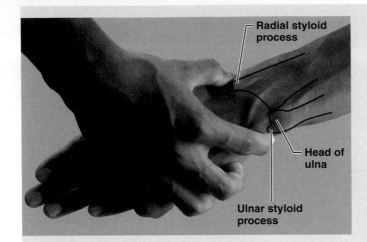

Figure 46.12 A way to locate the ulnar and radial styloid processes. The right hand is palpating the left hand in this picture. Note that the head of the ulna is not the same as the ulnar styloid process. The radial styloid process lies about 1 cm distal to the ulnar styloid process.

1. Feel the ulna along its entire length as a sharp ridge on the posterior forearm (confirm that this ridge runs inferiorly from the olecranon). As for the radius, you can feel its distal half, but most of its proximal half is covered by muscle. You can, however, feel the rotating **head** of the radius. To do this, extend your forearm, and note that a dimple forms on the posterior lateral surface of the elbow region (Figure 46.10). Press three fingers into this dimple, and rotate your free hand as if you were turning a doorknob. You will feel the head of the radius rotate as you perform this action.

2. Both the radius and ulna have a knoblike **styloid process** at their distal ends. Palpate these processes at the wrist (**Figure 46.12**). Do not confuse the ulnar styloid process with the conspicuous **head of the ulna,** from which the styloid process stems. Confirm that the radial styloid process lies about 1 cm (0.4 inch) distal to that of the ulna.

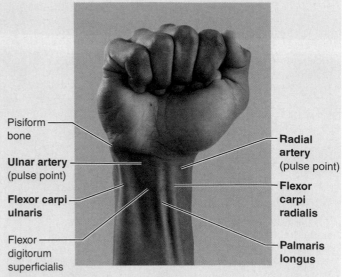

Figure 46.13 The anterior surface of the distal forearm and fist. The tendons of the flexor muscles guide the clinician to several sites for pulse taking.

Colles' fracture of the wrist is an impacted fracture in which the distal end of the radius is pushed proximally into the shaft of the radius. This sometimes occurs when someone falls on outstretched hands, and it most often happens to elderly women with osteoporosis. Colles' fracture bends the wrist into curves that resemble those on a fork. ✚

Can you deduce how physicians use palpation to diagnose a Colles' fracture?

_____

_____

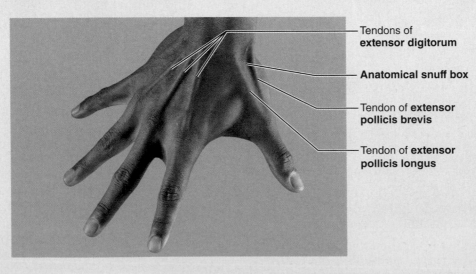

Figure 46.14 The dorsum of the hand. Note especially the anatomical snuff box and dorsal venous network.

3. Next, feel the major groups of muscles within your forearm. Flex your hand and fingers against resistance, and feel the anterior *flexor muscles* contract. Then extend your hand at the wrist, and feel the tightening of the posterior *extensor muscles*.

4. Near the wrist, the anterior surface of the forearm reveals many significant features (**Figure 46.13**). Flex your fist against resistance; the tendons of the main wrist flexors will bulge the skin of the distal forearm. The tendons of the **flexor carpi radialis** and **palmaris longus** muscles are most obvious. The palmaris longus, however, is absent from at least one arm in 30% of all people, so your forearm may exhibit just one prominent tendon instead of two. The **radial artery** lies just lateral to (on the thumb side of) the flexor carpi radialis tendon, where the pulse is easily detected (Figure 46.13). Feel your radial pulse here. The *median nerve,* which innervates the thumb, lies deep to the palmaris longus tendon. Finally, the **ulnar artery** lies on the medial side of the forearm, just lateral to the tendon of the **flexor carpi ulnaris**. Locate and feel your ulnar arterial pulse (Figure 46.13).

5. Extend your thumb and point it posteriorly to form a triangular depression in the base of the thumb on the back of your hand. This is the **anatomical snuff box** (**Figure 46.14**). Its two elevated borders are defined by the tendons of the thumb extensor muscles, **extensor pollicis brevis** and **extensor pollicis longus**. The radial artery runs within the snuff box, so this is another site for taking a radial pulse. The main bone on the floor of the snuff box is the scaphoid bone of the wrist, but the radial styloid process is also present here. If displaced by a bone fracture, the radial styloid process will be felt outside of the snuff box rather than within it. The "snuff box" took its name from the fact that people once put snuff (tobacco for sniffing) in this hollow before lifting it up to the nose.

6. On the dorsum of your hand, observe the superficial veins just deep to the skin. This is the **dorsal venous network,** which drains superiorly into the cephalic vein. This venous network provides a site for drawing blood and inserting intravenous catheters and is preferred over the median cubital vein for these purposes. Next, extend your hand and fingers, and observe the tendons of the **extensor digitorum** muscle.

7. The anterior surface of the hand also contains some features of interest (**Figure 46.15**). These features include the *epidermal ridges* (fingerprints) and many **flexion creases** in the skin. Grasp your **thenar eminence** (the bulge on the palm that contains the thumb muscles) and your **hypothenar eminence** (the bulge on the medial palm that contains muscles that move the little finger).

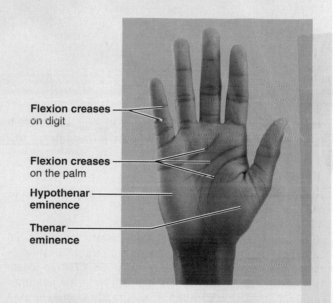

Flexion creases on digit

Flexion creases on the palm

Hypothenar eminence

Thenar eminence

Figure 46.15 The palmar surface of the hand.

Activity 6

Palpating Landmarks of the Lower Limb

Gluteal Region

Dominating the gluteal region are the two *prominences* (cheeks) of the buttocks (**Figure 46.16**, p. 700). These are formed by subcutaneous fat and by the thick **gluteus maximus** muscles. The midline groove between the two prominences is called the **natal cleft** (*natal* = rump) or **gluteal cleft**. The inferior margin of each prominence is the horizontal **gluteal fold,** which roughly corresponds to the inferior margin of the gluteus maximus.

1. Try to palpate your **ischial tuberosity** just above the medial side of each gluteal fold (it will be easier to feel if you sit down or flex your thigh first). The ischial tuberosities are the robust inferior parts of the ischial bones, and they support the body's weight during sitting.

2. Next, palpate the **greater trochanter** of the femur on the lateral side of your hip. This trochanter lies just anterior to a hollow and about 10 cm (one hand's breadth, or 4 inches) inferior to the iliac crest. To confirm that you have found the greater trochanter, alternately flex and extend

your thigh. Because this trochanter is the most superior point on the lateral femur, it moves with the femur as you perform this movement.

3. To palpate the sharp **posterior superior iliac spine,** locate your iliac crests again, and trace each to its most posterior point. You may have difficulty feeling this spine, but it is indicated by a distinct dimple in the skin that is easy to find. This dimple lies two to three finger breadths lateral to the midline of the back. The dimple also indicates the position of the *sacroiliac joint,* where the hip bone attaches to the sacrum of the spinal column. You can check *your* "dimples" out in the privacy of your home.

The gluteal region is a major site for administering intramuscular injections. When such injections are given, extreme care must be taken to avoid piercing the major nerve that lies just deep to the gluteus maximus muscle.

This thick *sciatic nerve* innervates much of the lower limb. Furthermore, the needle must avoid the gluteal

Text continues on next page. →

46

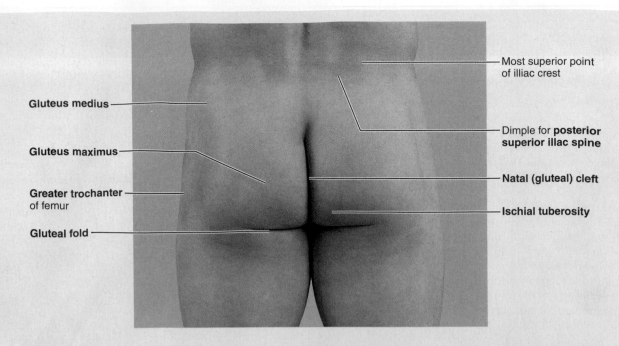

Figure 46.16 The gluteal region. The region extends from the iliac crests superiorly to the gluteal folds inferiorly. Therefore, it includes more than just the prominences of the buttock.

nerves and gluteal blood vessels, which also lie deep to the gluteus maximus.

To avoid harming these structures, the injections are most often applied to the gluteus *medius* (not maximus) muscle superior to the cheeks of the buttocks, in a safe area called the **ventral gluteal site** (**Figure 46.17b**). To locate this site, mentally draw a line laterally from the posterior superior iliac spine (dimple) to the greater trochanter; the injection would be given 5 cm (2 inches) superior to the midpoint of that line. Another safe way to locate the ventral gluteal site is to approach the lateral side of the patient's

left hip with your extended right hand (or the right hip with your left hand). Then, place your thumb on the anterior superior iliac spine and your index finger as far posteriorly on the iliac crest as it can reach. The heel of your hand comes to lie on the greater trochanter, and the needle is inserted in the angle of the V formed between your thumb and index finger about 4 cm (1.5 inches) inferior to the iliac crest.

Gluteal injections are not given to small children because their "safe area" is too small to locate with certainty and because the gluteal muscles are thin at this age. Instead, infants

Text continues on page 702. →

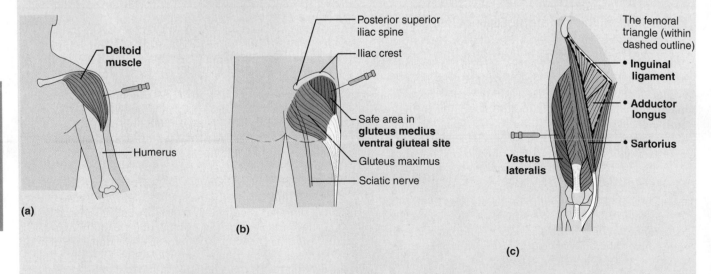

Figure 46.17 Three major sites of intramuscular injections. (a) Deltoid muscle of the arm. **(b)** Ventral gluteal site (gluteus medius). **(c)** Vastus lateralis in the lateral thigh. The femoral triangle is also shown.

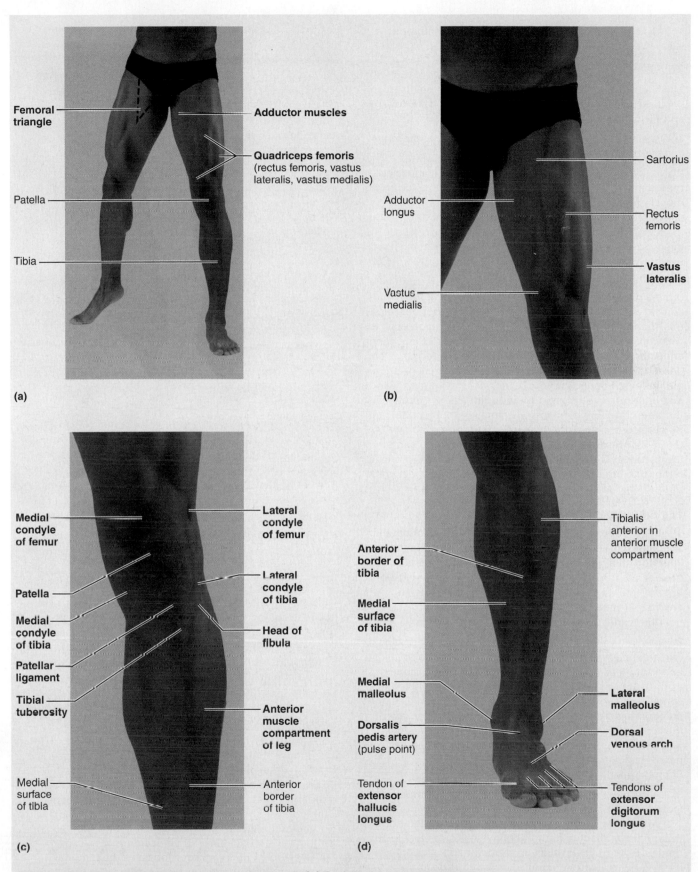

Figure 46.18 Anterior surface of the lower limb. (a) Both limbs, with the right limb revealing its medial aspect. The femoral triangle is outlined on the right limb. **(b)** Enlarged view of the left thigh. **(c)** The left knee region. **(d)** The dorsum of the left foot.

46

and toddlers receive intramuscular shots in the prominent **vastus lateralis** muscle of the thigh (Figure 46.17c).

Thigh

Much of the femur is clothed by thick muscles, so the thigh has few palpable bony landmarks (**Figure 46.18**, p. 701 and **Figure 46.19**).

1. Distally, feel the **medial** and **lateral condyles of the femur** and the **patella** anterior to the condyles (Figure 46.18c and a).

2. Next, palpate your three groups of thigh muscles—the **quadriceps femoris muscles** anteriorly, the **adductor muscles** medially, and the **hamstrings** posteriorly (Figures 46.18a and b and 46.19). The **vastus lateralis,** the lateral muscle of the quadriceps group, is a site for intramuscular injections. Such injections are administered about halfway down the length of this muscle (Figure 46.17c).

3. The anterosuperior surface of the thigh exhibits a three-sided depression called the **femoral triangle** (Figure 46.18a). As shown in Figure 46.17c, the superior border of this triangle is formed by the **inguinal ligament,** and its two inferior borders are defined by the **sartorius** and **adductor longus** muscles. The large *femoral artery* and *vein* descend vertically through the center of the femoral triangle. To feel the pulse of your femoral artery, press inward just inferior to your midinguinal point (halfway between the anterior superior iliac spine and the pubic tubercle). Be sure to push hard, because the artery lies somewhat deep. By pressing very hard on this point, one can stop the bleeding from a hemorrhage in the lower limb. The femoral triangle also contains most of the *inguinal lymph nodes,* which are easily palpated if swollen.

Leg and Foot

1. Locate your patella again, then follow the thick **patellar ligament** inferiorly from the patella to its insertion on the superior tibia (Figure 46.18c). Here you can feel a rough projection, the **tibial tuberosity.** Continue running your fingers inferiorly along the tibia's sharp **anterior border** and its flat **medial surface**—bony landmarks that lie very near the surface throughout their length.

2. Now, return to the superior part of your leg, and palpate the expanded **lateral** and **medial condyles of the tibia** just inferior to the knee. You can distinguish the tibial condyles from the femoral condyles because you can feel the tibial condyles move with the tibia during knee flexion. Feel the bulbous **head of the fibula** in the superolateral region of the leg (Figure 46.18c). Try to feel the *common fibular nerve* where it wraps around the fibula's *neck* just inferior to its head. This nerve, which serves the anterior leg and foot, is often bumped against the bone here and damaged.

3. In the most distal part of the leg, feel the **lateral malleolus** of the fibula as the lateral prominence of the ankle (Figure 46.18d). Notice that this lies slightly inferior to the **medial malleolus** of the tibia, which forms the ankle's medial prominence. Place your finger just posterior to the medial malleolus to feel the pulse of your *posterior tibial artery.*

4. On the posterior aspect of the knee is a diamond-shaped hollow called the **popliteal fossa** (Figure 46.19). Palpate the large muscles that define the four borders of this fossa: The

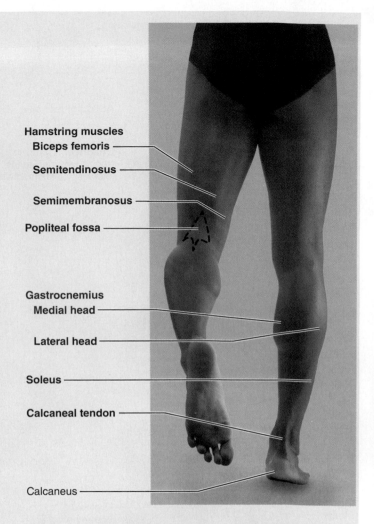

Figure 46.19 Posterior surface of the lower limb. Notice the diamond-shaped popliteal fossa posterior to the knee.

Labels: Hamstring muscles — Biceps femoris; Semitendinosus; Semimembranosus; Popliteal fossa; Gastrocnemius — Medial head; Lateral head; Soleus; Calcaneal tendon; Calcaneus

biceps femoris forming the superolateral border, the **semitendinosus** and **semimembranosus** defining the superomedial border, and the two heads of the **gastrocnemius** forming the inferior border. The main vessels to the leg, the *popliteal artery* and *vein,* lie deep within this fossa. To feel a popliteal pulse, flex your leg at the knee and push your fingers firmly into the popliteal fossa. If a physician is unable to feel a patient's popliteal pulse, the femoral artery may be narrowed by atherosclerosis.

5. Observe the dorsum (superior surface) of your foot. You may see the superficial **dorsal venous arch** overlying the proximal part of the metatarsal bones (Figure 46.18d). This arch gives rise to both saphenous veins (the main superficial veins of the lower limb). Visible in lean people, the *great saphenous vein* ascends along the medial side of the entire limb (Figure 32.9, p. 482). The *small saphenous vein* ascends through the center of the calf.

As you extend your toes, observe the tendons of the **extensor digitorum longus** and **extensor hallucis longus** muscles on the dorsum of the foot. Finally, place a finger on the extreme proximal part of the space between the first and second metatarsal bones. Here you should be able to feel the pulse of the **dorsalis pedis artery.**

REVIEW SHEET
Surface Anatomy Roundup

Name _____ Lab Time/Date _____

_____ 1. A blow to the cheek is most likely to break what superficial bone or bone part? (a) superciliary arches, (b) mastoid process, (c) zygomatic arch, (d) ramus of the mandible

_____ 2. Rebound tenderness (a) occurs in appendicitis, (b) is whiplash of the neck, (c) is a sore foot from playing basketball, (d) occurs when the larynx falls back into place after swallowing.

_____ 3. The anatomical snuff box (a) is in the nose, (b) contains the radial styloid process, (c) is defined by tendons of the flexor carpi radialis and palmaris longus, (d) cannot really hold snuff.

_____ 4. Some landmarks on the body surface can be seen or felt, but others are abstractions that you must construct by drawing imaginary lines. Which of the following pairs of structures is abstract and invisible? (a) umbilicus and costal margin, (b) anterior superior iliac spine and natal cleft, (c) linea alba and linea semilunaris, (d) McBurney's point and midaxillary line, (e) lacrimal fossa and sternocleidomastoid

_____ 5. Many pelvic organs can be palpated by placing a finger in the rectum or the vagina, but only one pelvic organ is readily palpated through the skin. This is the (a) nonpregnant uterus, (b) prostate, (c) full bladder, (d) ovaries, (e) rectum.

_____ 6. Contributing to the posterior axillary fold is/are (a) pectoralis major, (b) latissimus dorsi, (c) trapezius, (d) infraspinatus, (e) pectoralis minor, (f) a and e.

_____ 7. Which of the following is *not* a pulse point? (a) anatomical snuff box, (b) inferior margin of mandible anterior to masseter muscle, (c) center of distal forearm at palmaris longus tendon, (d) medial bicipital furrow on arm, (e) dorsum of foot between the first two metatarsals

_____ 8. Which pair of ribs inserts on the sternum at the sternal angle? (a) first, (b) second, (c) third, (d) fourth, (e) fifth

_____ 9. The inferior angle of the scapula is at the same level as the spinous process of which vertebra? (a) C_5, (b) C_7, (c) T_3, (d) T_7, (e) L_4

_____ 10. An important bony landmark that can be recognized by a distinct dimple in the skin is the (a) posterior superior iliac spine, (b) ulnar styloid process, (c) shaft of the radius, (d) acromion.

_____ 11. A nurse missed a patient's median cubital vein while trying to withdraw blood and then inserted the needle far too deeply into the cubital fossa. This error could cause any of the following problems, *except* this one: (a) paralysis of the ulnar nerve, (b) paralysis of the median nerve, (c) bruising the insertion tendon of the biceps brachii muscle, (d) blood spurting from the brachial artery.

_____ 12. Which of these organs is almost impossible to study with surface anatomy techniques? (a) heart, (b) lungs, (c) brain, (d) nose

_____ 13. A preferred site for inserting an intravenous medication line into a blood vessel is the (a) medial bicipital furrow on arm, (b) external carotid artery, (c) dorsal venous network of hand, (d) popliteal fossa.

_____ 14. One listens for bowel sounds with a stethoscope placed (a) on the four quadrants of the abdominal wall; (b) in the triangle of auscultation; (c) in the right and left midaxillary line, just superior to the iliac crests; (d) inside the patient's bowels (intestines), on the tip of an endoscope.

_____ 15. A stab wound in the posterior triangle of the neck could damage any of the following structures *except* the (a) accessory nerve, (b) phrenic nerve, (c) external jugular vein, (d) external carotid artery.

PhysioEx™ 9.1

PhysioEx™ 9.1 by
Peter Zao, North Idaho College
Timothy Stabler, Indiana University Northwest
Lori Smith, American River College
Andrew Lokuta, University of Wisconsin–Madison
Edwin Griff, University of Cincinnati

Cell Transport Mechanisms and Permeability

PRE-LAB QUIZ

1. Circle the correct underlined term: In <u>active transport</u> / <u>passive transport</u> processes, the cell must provide energy in the form of ATP to power the process.

2. The movement of particles from an area of greater concentration to an area of lesser concentration is:
   a. diffusion
   b. osmosis
   c. active transport
   d. kinetic energy

3. All of the following are true of active transport *except:*
   a. ATP is used to power active transport.
   b. Solutes are moving with their concentration gradient.
   c. It uses a membrane-bound carrier protein.
   d. It can only occur in certain animals.

4. Circle the correct underlined term: In Exercise 1, the dialysis tubing will mimic the <u>nucleus</u> / <u>plasma membrane</u> of a cell.

5. Circle the correct underlined term: The larger the <u>molecular weight</u> / <u>concentration</u> of a compound, the larger the pore size required for passive transport of that compound.

Exercise Overview

The molecular composition of the plasma membrane allows it to be selective about what passes through it. It allows nutrients and appropriate amounts of ions to enter the cell and keeps out undesirable substances. For that reason, we say the plasma membrane is **selectively permeable.** Valuable cell proteins and other substances are kept within the cell, and metabolic wastes pass to the exterior.

Transport through the plasma membrane occurs in two basic ways: either passively or actively. In **passive processes,** the transport process is driven by concentration or pressure differences *(gradients)* between the interior and exterior of the cell. In **active processes**, the cell provides energy (ATP) to power the transport.

Two key passive processes of membrane transport are **diffusion** and **filtration.** Diffusion is an important transport process for every cell in the body. **Simple diffusion** occurs without the assistance of membrane proteins, and **facilitated diffusion** requires a membrane-bound carrier protein that assists in the transport.

In both simple and facilitated diffusion, the substance being transported moves *with* (or *along* or *down)* the *concentration gradient* of the solute (from a region of its higher concentration to a region of its lower concentration). The process does not require energy from the cell. Instead, energy in the form of **kinetic energy** comes from the constant motion of the molecules. The movement of solutes continues until the solutes are evenly dispersed throughout the solution. At this point, the solution has reached **equilibrium.**

A special type of diffusion across a membrane is **osmosis.** In osmosis, water moves with its concentration gradient, from a higher concentration of water to a lower concentration of water. It moves in response to a higher concentration of solutes on the other side of a membrane.

In the body, the other key passive process, **filtration,** usually occurs only across capillary walls. Filtration depends upon a *pressure gradient* as its driving force. It is not a selective process. It is dependent upon the size of the pores in the filter.

The two key active processes (recall that active processes require energy) are **active transport** and **vesicular transport.** Like facilitated diffusion, active transport uses a membrane-bound carrier protein. Active transport differs from facilitated diffusion because the solutes move *against* their concentration gradient and because ATP is used to power the transport. Vesicular transport includes phagocytosis, endocytosis, pinocytosis, and exocytosis. These processes are not covered in this exercise. The activities in this exercise will explore the cell transport mechanisms individually.

ACTIVITY 1

Simulating Dialysis (Simple Diffusion)

OBJECTIVES

1. To understand that diffusion is a passive process dependent upon a solute concentration gradient.

2. To understand the relationship between molecular weight and molecular size.

3. To understand how solute concentration affects the rate of diffusion.

4. To understand how molecular weight affects the rate of diffusion.

Introduction

Recall that all molecules possess *kinetic energy* and are in constant motion. As molecules move about randomly at high speeds, they collide and bounce off one another, changing direction with each collision. For a given temperature, all matter has about the same average kinetic energy. Smaller molecules tend to move faster than larger molecules because kinetic energy is directly related to both mass and velocity ($KE = \frac{1}{2} mv^2$).

When a **concentration gradient** (difference in concentration) exists, the net effect of this random molecular movement is that the molecules eventually become evenly distributed throughout the environment—in other words, diffusion occurs. **Diffusion** is the movement of molecules from a region of their higher concentration to a region of their lower concentration. The driving force behind diffusion is the kinetic energy of the molecules themselves.

The diffusion of particles into and out of cells is modified by the plasma membrane, which is a physical barrier. In general, molecules diffuse passively through the plasma membrane if they are small enough to pass through its pores (and are aided by an electrical and/or concentration gradient) or if they can dissolve in the lipid portion of the membrane (as in the case of CO_2 and O_2). A membrane is called *selectively permeable, differentially permeable,* or *semipermeable* if it allows some solute particles (molecules) to pass but not others.

The diffusion of *solute particles* dissolved in water through a selectively permeable membrane is called **simple diffusion.** The diffusion of *water* through a differentially permeable membrane is called **osmosis.** Both simple diffusion and osmosis involve movement of a substance from an area of its higher concentration to an area of its lower concentration, that is, *with* (or *along* or *down*) its concentration gradient.

This activity provides information on the passage of water and solutes through selectively permeable membranes. You can apply what you learn to the study of transport mechanisms in living, membrane-bounded cells. The dialysis membranes used each have a different *molecular weight cutoff (MWCO),* indicated by the number below it. You can think of MWCO in terms of pore size: the larger the MWCO number, the larger the pores in the membrane. The molecular weight of a solute is the number of grams per mole, where a mole is the constant Avogadro's number 6.02×10^{23} molecules/mole. The larger the molecular weight, the larger the mass of the molecule. The term molecular mass is sometimes used instead of molecular weight.

> **EQUIPMENT USED** The following equipment will be depicted on-screen: left and right beakers—used for diffusion of solutes; dialysis membranes with various molecular weight cutoffs (MWCOs).

Experiment Instructions

Go to the home page in the PhysioEx software and click **Exercise 1: Cell Transport Mechanisms and Permeability.** Click **Activity 1: Simulating Dialysis (Simple Diffusion),** and take the online **Pre-lab Quiz** for Activity 1.

After you take the online Pre-lab Quiz, click the **Experiment** tab and begin the experiment. The experiment instructions are reprinted here for your reference. The opening screen for the experiment is shown below.

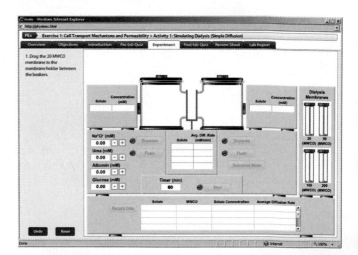

1. Drag the 20 MWCO membrane to the membrane holder between the beakers.

2. Increase the Na^+Cl^- concentration to be dispensed to the left beaker to 9.00 m*M* by clicking the + button beside the Na^+Cl^- display. Click **Dispense** to fill the left beaker with 9.00 m*M* Na^+Cl^- solution.

3. Note that the concentration of Na^+Cl^- in the left beaker is displayed in the concentration window to the left of the beaker. Click **Deionized Water** and then click **Dispense** to fill the right beaker with deionized water.

4. After you start the run, the barrier between the beakers will descend, allowing the solutions in each beaker to have access to the dialysis membrane separating them. You will be

able to determine the amount of solute that passes through the membrane by observing the concentration display to the side of each beaker. A level above zero in Na^+Cl^- concentration in the right beaker indicates that Na^+ and Cl^- ions are diffusing from the left beaker into the right beaker through the selectively permeable dialysis membrane. Note that the timer is set to 60 minutes. The simulation compresses the 60-minute time period into 10 seconds of real time. Click **Start** to start the run and watch the concentration display to the side of each beaker for any activity.

5. Click **Record Data** to display your results in the grid (and record your results in Chart 1).

CHART 1	Dialysis Results (average diffusion rate in mM/min)			
	Membrane MWCO			
Solute	**20**	**50**	**100**	**200**
Na^+Cl^-				
Urea				
Albumin				
Glucose				

> **? PREDICT Question 1**
> The molecular weight of urea is 60.07. Do you think urea will diffuse through the 20 MWCO membrane?

6. Click **Flush** beneath each of the beakers to prepare for the next run.

7. Increase the urea concentration to be dispensed to the left beaker to 9.00 mM by clicking the + button beside the urea display. Click **Dispense** to fill the left beaker with 9.00 mM urea solution.

8. Click **Deionized Water** and then click **Dispense** to fill the right beaker with deionized water.

9. Click **Start** to start the run and watch the concentration display to the side of each beaker for any activity.

10. Click **Record Data** to display your results in the grid (and record your results in Chart 1).

11. Click the 20 MWCO membrane in the membrane holder to automatically return it to the membrane cabinet and then click **Flush** beneath each beaker to prepare for the next run.

12. Drag the 50 MWCO membrane to the membrane holder between the beakers. Increase the Na^+Cl^- concentration to be dispensed to the left beaker to 9.00 mM. Click **Dispense** to fill the left beaker with 9.00 mM Na^+Cl^- solution.

13. Click **Deionized Water** and then click **Dispense** to fill the right beaker with deionized water.

14. Click **Start** to start the run and watch the concentration display to the side of each beaker for any activity.

15. Click **Record Data** to display your results in the grid (and record your results in Chart 1).

16. Click **Flush** beneath each of the beakers to prepare for the next run.

17. Increase the Na^+Cl^- concentration to be dispensed to the left beaker to 18.00 mM. Click **Dispense** to fill the left beaker with 18.00 mM Na^+Cl^- solution.

18. Click **Deionized Water** and then click **Dispense** to fill the right beaker with deionized water.

19. Click **Start** to start the run and watch the concentration display to the side of each beaker for any activity.

20. Click **Record Data** to display your results in the grid (and record your results in Chart 1).

21. Click the 50 MWCO membrane in the membrane holder to automatically return it to the membrane cabinet and then click **Flush** beneath each beaker to prepare for the next run.

22. Drag the 100 MWCO membrane to the membrane holder between the beakers. Increase the Na^+Cl^- concentration to be dispensed to the left beaker to 9.00 mM. Click **Dispense** to fill the left beaker with 9.00 mM Na^+Cl^- solution.

23. Click **Deionized Water** and then click **Dispense** to fill the right beaker with deionized water.

24. Click **Start** to start the run and watch the concentration display to the side of each beaker for any activity.

25. Click **Record Data** to display your results in the grid (and record your results in Chart 1).

26. Click **Flush** beneath each of the beakers to prepare for the next run.

27. Increase the urea concentration to be dispensed to the left beaker to 9.00 mM. Click **Dispense** to fill the left beaker with 9.00 mM urea solution.

28. Click **Deionized Water** and then click **Dispense** to fill the right beaker with deionized water.

29. Click **Start** to start the run and watch the concentration display to the side of each beaker for any activity.

30. Click **Record Data** to display your results in the grid (and record your results in Chart 1).

31. Click the 100 MWCO membrane in the membrane holder to automatically return it to the membrane cabinet and then click **Flush** beneath each beaker to prepare for the next run.

> **? PREDICT Question 2**
> Recall that glucose is a monosaccharide, albumin is a protein with 607 amino acids, and the average molecular weight of a single amino acid is 135 g/mole. Will glucose or albumin be able to diffuse through the 200 MWCO membrane?

32. Drag the 200 MWCO membrane to the membrane holder between the beakers. Increase the glucose concentration to be dispensed to the left beaker to 9.00 m*M*. Click **Dispense** to fill the left beaker with 9.00 m*M* glucose solution.

33. Click **Deionized Water** and then click **Dispense** to fill the right beaker with deionized water.

34. Click **Start** to start the run and watch the concentration display to the side of each beaker for any activity.

35. Click **Record Data** to display your results in the grid (and record your results in Chart 1).

36. Click **Flush** beneath each of the beakers to prepare for the next run.

37. Increase the albumin concentration to be dispensed to the left beaker to 9.00 m*M*. Click **Dispense** to fill the left beaker with 9.00 m*M* albumin solution.

38. Click **Deionized Water** and then click **Dispense** to fill the right beaker with deionized water.

39. Click **Start** to start the run and watch the concentration display to the side of each beaker for any activity.

40. Click **Record Data** to display your results in the grid (and record your results in Chart 1).

After you complete the experiment, take the online **Post-lab Quiz** for Activity 1.

Activity Questions

1. Did any solutes move through the 20 MWCO membrane? Why or why not?

_____

_____

2. Did Na$^+$Cl$^-$ move through the 50 MWCO membrane?

_____

_____

3. Describe how the size of a molecule (molecular weight) affects its rate of diffusion.

_____

_____

4. What happened to the rate of diffusion when you increased the Na$^+$Cl$^-$ solute concentration?

_____

_____

Simulated Facilitated Diffusion

OBJECTIVES

1. To understand that some solutes require a carrier protein to pass through a membrane because of size or solubility limitations.
2. To observe how the concentration of solutes affects the rate of facilitated diffusion.
3. To observe how the number of transport proteins affects the rate of facilitated diffusion.
4. To understand how transport proteins can become saturated.

Introduction

Some molecules are lipid insoluble or too large to pass through pores in the cell's plasma membrane. Instead, they pass through the membrane by a passive transport process called **facilitated diffusion.** For example, sugars, amino acids, and ions are transported by facilitated diffusion. In this form of transport, solutes combine with carrier-protein molecules in the membrane and are then transported *with* (or *along* or *down*) their concentration gradient. The carrier-protein molecules in the membrane might have to change shape slightly to accommodate the solute, but the cell does not have to expend the energy of ATP.

Because facilitated diffusion relies on carrier proteins, solute transport varies with the number of available carrier-protein molecules in the membrane. The carrier proteins can become saturated if too much solute is present and the maximum transport rate is reached. The carrier proteins are embedded in the plasma membrane and act like a shield, protecting the hydrophilic solute from the lipid portions of the membrane.

Facilitated diffusion typically occurs in one direction for a given solute. The greater the concentration difference between one side of the membrane and the other, the greater the rate of facilitated diffusion.

> **EQUIPMENT USED** The following equipment will be depicted on-screen: left and right beakers—used for diffusion of solutes; dialysis membranes with various molecular weight cutoffs (MWCOs); membrane builder—used to build membranes with different numbers of glucose protein carriers.

Experiment Instructions

Go to the home page in the PhysioEx software and click **Exercise 2: Cell Transport Mechanisms and Permeability.** Click **Activity 2: Simulated Facilitated Diffusion,** and take the online **Pre-lab Quiz** for Activity 2.

After you take the online Pre-lab Quiz, click the **Experiment** tab and begin the experiment. The experiment instructions are reprinted here for your reference. The opening screen for the experiment is shown on the following page.

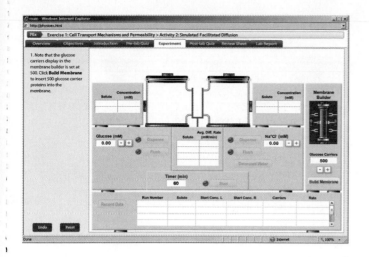

1. Note that the glucose carriers display in the membrane builder is set at 500. Click **Build Membrane** to insert 500 glucose carrier proteins into the membrane.

2. Drag the membrane to the membrane holder between the beakers.

3. Increase the glucose concentration to be dispensed to the left beaker to 2.00 mM by clicking the + button beside the glucose display. Click **Dispense** to fill the left beaker with 2.00 mM glucose solution.

4. Note that the concentration of glucose in the left beaker is displayed in the concentration window to the left of the beaker. Click **Deionized Water** and then click **Dispense** to fill the right beaker with deionized water.

5. After you start the run, the barrier between the beakers will descend, allowing the solutions in each beaker to have access to the dialysis membrane separating them. You will be able to determine the amount of solute that passes through the membrane by observing the concentration display to the side of each beaker. A level above zero in glucose concentration in the right beaker indicates that glucose is diffusing from the left beaker into the right beaker through the selectively permeable dialysis membrane. Note that the timer is set to 60 minutes. The simulation compresses the 60-minute time period into 10 seconds of real time. Click **Start** to start the run and watch the concentration display to the side of each beaker for any activity.

6. Click **Record Data** to display your results in the grid (and record your results in Chart 2).

CHART 2	**Facilitated Diffusion Results** (glucose transport rate, mM/min)		
	Number of glucose carrier proteins		
Glucose concentration	500	700	100
2 mM			
8 mM			
10 mM			
2 mM w/2.00 mM Na⁺Cl⁻			

7. Click **Flush** beneath each of the beakers to prepare for the next run.

8. Increase the glucose concentration to be dispensed to the left beaker to 8.00 mM by clicking the + button beside the glucose display. Click **Dispense** to fill the left beaker with 8.00 mM glucose solution.

9. Click **Deionized Water** and then click **Dispense** to fill the right beaker with deionized water.

10. Click **Start** to start the run and watch the concentration display to the side of each beaker for any activity.

11. Click **Record Data** to display your results in the grid (and record your results in Chart 2).

12. Click the membrane in the membrane holder to automatically return it to the membrane builder and then click **Flush** beneath each beaker to prepare for the next run.

> **? PREDICT Question 1**
> What effect do you think increasing the number of protein carriers will have on the glucose transport rate?
> _____

13. Increase the number of glucose carriers to 700 by clicking the + button beneath the glucose carriers display. Click **Build Membrane** to insert 700 glucose carrier proteins into the membrane.

14. Drag the membrane to the membrane holder between the beakers. Increase the glucose concentration to be dispensed to the left beaker to 2.00 mM. Click **Dispense** to fill the left beaker with 2.00 mM glucose solution.

15. Click **Deionized Water** and then click **Dispense** to fill the right beaker with deionized water.

16. Click **Start** to start the run and watch the concentration display to the side of each beaker for any activity.

17. Click **Record Data** to display your results in the grid (and record your results in Chart 2).

18. Click **Flush** beneath each of the beakers to prepare for the next run.

19. Increase the glucose concentration to be dispensed to the left beaker to 8.00 mM. Click **Dispense** to fill the left beaker with 8.00 mM glucose solution.

20. Click **Deionized Water** and then click **Dispense** to fill the right beaker with deionized water.

21. Click **Start** to start the run and watch the concentration display to the side of each beaker for any activity.

22. Click **Record Data** to display your results in the grid (and record your results in Chart 2).

23. Click the membrane in the membrane holder to automatically return it to the membrane builder and then click **Flush** beneath each beaker to prepare for the next run.

24. Decrease the number of glucose carriers to 100 by clicking the − button beneath the glucose carriers display. Click **Build Membrane** to insert 100 glucose carrier proteins into the membrane.

25. Drag the membrane to the membrane holder between the beakers. Increase the glucose concentration to be dispensed to the left beaker to 10.00 m*M*. Click **Dispense** to fill the left beaker with 10.00 m*M* glucose solution.

26. Click **Deionized Water** and then click **Dispense** to fill the right beaker with deionized water.

27. Click **Start** to start the run and watch the concentration display to the side of each beaker for any activity.

28. Click **Record Data** to display your results in the grid (and record your results in Chart 2).

29. Click the membrane in the membrane holder to automatically return it to the membrane builder and then click **Flush** beneath each beaker to prepare for the next run.

30. Increase the number of glucose carriers to 700. Click **Build Membrane** to insert 700 glucose carrier proteins into the membrane.

> **? PREDICT Question 2**
> What effect do you think adding Na⁺Cl⁻ will have on the glucose transport rate?

31. Increase the glucose concentration to be dispensed to the left beaker to 2.00 m*M*. Click **Dispense** to fill the left beaker with 2.00 m*M* glucose solution.

32. Increase the Na⁺Cl⁻ concentration to be dispensed to the right beaker to 2.00 m*M*. Click **Dispense** to fill the right beaker with 2.00 m*M* Na⁺Cl⁻ solution.

33. Click **Start** to start the run and watch the concentration display to the side of each beaker for any activity.

34. Click **Record Data** to display the results in the grid (and record your results in Chart 2).

After you complete the experiment, take the online **Post-lab Quiz** for Activity 2.

Activity Questions

1. Are the solutes moving with or against their concentration gradient in facilitated diffusion?

2. What happened to the rate of facilitated diffusion when the number of carrier proteins was increased?

3. Explain why equilibrium was not reached with 10 m*M* glucose and 100 membrane carriers.

4. In the simulation you added Na⁺Cl⁻ to test its effect on glucose diffusion. Explain why there was no effect.

ACTIVITY 3

Simulating Osmotic Pressure

OBJECTIVES

1. To explain how osmosis is a special type of diffusion.
2. To understand that osmosis is a passive process that depends upon the concentration gradient of water.
3. To explain how tonicity of a solution relates to changes in cell volume.
4. To understand conditions that affect osmotic pressure.

Introduction

A special form of diffusion, called **osmosis,** is the diffusion of water through a selectively permeable membrane. (A membrane is called *selectively permeable, differentially permeable,* or *semipermeable* if it allows some molecules to pass but not others.) Because water can pass through the pores of most membranes, it can move from one side of a membrane to the other relatively freely. Osmosis takes place whenever there is a difference in water concentration between the two sides of a membrane.

If we place distilled water on both sides of a membrane, *net* movement of water does not occur. Remember, however, that water molecules would still move between the two sides of the membrane. In such a situation, we would say that there is no *net* osmosis.

The concentration of water in a solution depends on the number of solute particles present. For this reason, increasing the solute concentration coincides with decreasing the water concentration. Because water moves down its concentration gradient (from an area of its higher concentration to an area of its lower concentration), it always moves *toward* the solution with the highest concentration of solutes. Similarly, solutes also move down their concentration gradients.

If we position a *fully* permeable membrane (permeable to solutes and water) between two solutions of differing concentrations, then all substances—solutes and water—diffuse freely, and an equilibrium will be reached between the two sides of the membrane. However, if we use a selectively permeable membrane that is impermeable to the solutes, then we have established a condition where water moves but solutes do not. Consequently, water moves toward the more concentrated solution, resulting in a *volume increase* on that side of the membrane.

By applying this concept to a closed system where volumes cannot change, we can predict that the *pressure* in the more concentrated solution will rise. The force that would need to be applied to oppose the osmosis in a closed system is the **osmotic pressure.** Osmotic pressure is measured in *millimeters of mercury (mm Hg).* In general, the more impermeable the solutes, the higher the osmotic pressure.

Osmotic changes can affect the volume of a cell when it is placed in various solutions. The concept of **tonicity** refers to the way a solution affects the volume of a cell. The tonicity of a solution tells us whether or not a cell will shrink or swell. If the concentration of impermeable solutes is the *same* inside and outside of the cell, the solution is **isotonic.** If there is a *higher* concentration of impermeable solutes *outside* the cell than in the cell's interior, the solution is **hypertonic.** Because the net movement of water would be out of the cell, the cell would *shrink* in a hypertonic solution. Conversely, if the concentration of impermeable solutes is *lower* outside of the cell than in the cell's interior, then the solution is **hypotonic.** The net movement of water would be into the cell, and the cell would *swell* and possibly burst.

EQUIPMENT USED The following equipment will be depicted on-screen: left and right beakers—used for diffusion of solutes; dialysis membranes with various molecular weight cutoffs (MWCOs).

Experiment Instructions

Go to the home page in the PhysioEx software and click **Exercise 1: Cell Transport Mechanisms and Permeability.** Click **Activity 3: Simulating Osmotic Pressure,** and take the online **Pre-lab Quiz** for Activity 3.

After you take the online Pre-lab Quiz, click the **Experiment** tab and begin the experiment. The experiment instructions are reprinted here for your reference. The opening screen for the experiment is shown below.

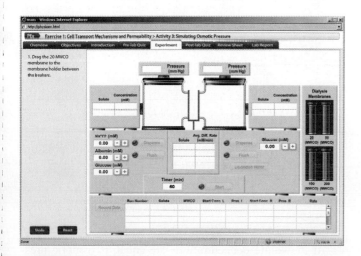

1. Drag the 20 MWCO membrane to the membrane holder between the beakers.

2. Increase the Na$^+$Cl$^-$ concentration to be dispensed to the left beaker to 5.00 mM by clicking the + button beside the Na$^+$Cl$^-$ display. Click **Dispense** to fill the left beaker with 5.00 mM Na$^+$Cl$^-$ solution.

3. Note that the concentration of Na$^+$Cl$^-$ in the left beaker is displayed in the concentration window to the left of the beaker. Click **Deionized Water** and then click **Dispense** to fill the right beaker with deionized water.

4. After you start the run, the barrier between the beakers will descend, allowing the solutions in each beaker to have access to the dialysis membrane separating them. You can observe the changes in pressure in the two beakers by watching the pressure display above each beaker. You will also be able to determine the amount of solute that passes through the membrane by observing the concentration display to the side of each beaker. A level above zero in Na$^+$Cl$^-$ concentration in the right beaker indicates that Na$^+$ and Cl$^-$ ions are diffusing from the left beaker into the right beaker through the selectively permeable dialysis membrane. Note that the timer is set to 60 minutes. The simulation compresses the 60-minute time period into 10 seconds of real time. Click **Start** to start the run and watch the pressure display above each beaker for any activity.

5. Click **Record Data** to display your results in the grid (and record your results in Chart 3).

CHART 3	Osmosis Results		
Solute	Membrane (MWCO)	Pressure on left (mm Hg)	Diffusion rate (mM/min)
Na$^+$Cl$^-$			
Na$^+$Cl$^-$			
Na$^+$Cl$^-$			
Glucose			
Glucose			
Glucose			
Albumin w/glucose			

6. Click **Flush** beneath each of the beakers to prepare for the next run.

7. Increase the Na$^+$Cl$^-$ concentration to be dispensed to the left beaker to 10.00 mM by clicking the + button beside the Na$^+$Cl$^-$ display. Click **Dispense** to fill the left beaker with 10.00 mM Na$^+$Cl$^-$ solution.

8. Click **Deionized Water** and then click **Dispense** to fill the right beaker with deionized water.

? PREDICT Question 1
What effect do you think increasing the Na$^+$Cl$^-$ concentration will have?

9. Click **Start** to start the run and watch the pressure display above each beaker for any activity.

10. Click **Record Data** to display your results in the grid (and record your results in Chart 3).

11. Click the 20 MWCO membrane in the membrane holder to automatically return it to the membrane cabinet and then click **Flush** beneath each beaker to prepare for the next run.

12. Drag the 50 MWCO membrane to the membrane holder between the beakers. Increase the Na^+Cl^- concentration to be dispensed to the left beaker to 10.00 m*M*. Click **Dispense** to fill the left beaker with 10.00 m*M* Na^+Cl^- solution.

13. Click **Deionized Water** and then click **Dispense** to fill the right beaker with deionized water.

14. Click **Start** to start the run and watch the pressure display above each beaker for any activity.

15. Click **Record Data** to display your results in the grid (and record your results in Chart 3).

16. Click the 50 MWCO membrane in the membrane holder to automatically return it to the membrane cabinet and then click **Flush** beneath each beaker to prepare for the next run.

17. Drag the 100 MWCO membrane to the membrane holder between the beakers. Increase the glucose concentration to be dispensed to the left beaker to 8.00 m*M* by clicking the + button beside the glucose display beneath the left beaker. Click **Dispense** to fill the left beaker with 8.00 m*M* glucose solution.

18. Click **Deionized Water** and then click **Dispense** to fill the right beaker with deionized water.

19. Click **Start** to start the run and watch the pressure display above each beaker for any activity.

20. Click **Record Data** to display your results in the grid (and record your results in Chart 3).

21. Click **Flush** beneath each of the beakers to prepare for the next run.

22. Increase the glucose concentration to be dispensed to the left beaker to 8.00 m*M*. Click **Dispense** to fill the left beaker with 8.00 m*M* glucose solution.

23. Increase the glucose concentration to be dispensed to the right beaker to 8.00 m*M* by clicking the + button beside the glucose display beneath the right beaker. Click **Dispense** to fill the right beaker with 8.00 m*M* glucose solution.

24. Click **Start** to start the run and watch the pressure display above each beaker for any activity.

25. Click **Record Data** to display your results in the grid (and record your results in Chart 3).

26. Click the 100 MWCO membrane in the membrane holder to automatically return it to the membrane cabinet and then click **Flush** beneath each beaker to prepare for the next run.

27. Drag the 200 MWCO membrane to the membrane holder between the beakers. Increase the glucose concentration to be dispensed to the left beaker to 8.00 m*M*. Click **Dispense** to fill the left beaker with 8.00 m*M* glucose solution.

28. Click **Deionized Water** and then click **Dispense** to fill the right beaker with deionized water.

29. Click **Start** to start the run and watch the pressure display above each beaker for any activity.

30. Click **Record Data** to display your results in the grid (and record your results in Chart 3).

31. Click **Flush** beneath each of the beakers to prepare for the next run.

32. Increase the albumin concentration to be dispensed to the left beaker to 9.00 m*M*. Click **Dispense** to fill the left beaker with 9.00 m*M* albumin solution.

33. Increase the glucose concentration to be dispensed to the right beaker to 10.00 m*M*. Click **Dispense** to fill the right beaker with 10.00 m*M* glucose solution.

> **? PREDICT Question 2**
> What do you think will be the pressure result of the current experimental conditions?
> _____

34. Click **Start** to start the run and watch the pressure display above each beaker for any activity.

35. Click **Record Data** to display your results in the grid (and record your results in Chart 3).

After you complete the experiment, take the online **Post-lab Quiz** for Activity 3.

Activity Questions

1. Which membrane resulted in the greatest pressure with Na^+Cl^- as the solute? Why?

_____

_____

2. Explain what happens to the osmotic pressure with increasing solute concentration.

_____

_____

3. If the solutes are allowed to diffuse, is osmotic pressure generated?

_____

_____

4. If the solute concentrations are equal, is osmotic pressure generated? Why or why not?

_____

_____

ACTIVITY 4

Simulating Filtration

OBJECTIVES

1. To understand that filtration is a passive process dependent upon a pressure gradient.

2. To understand that filtration is not a selective process.

3. To explain that the size of the membrane pores will determine what passes through.

4. To explain the effect that increasing the hydrostatic pressure has on the filtration rate and how this correlates to events in the body.

5. To understand the relationship between molecular weight and molecular size.

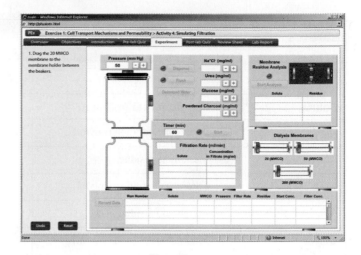

Introduction

Filtration is the process by which water and solutes pass through a membrane (such as a dialysis membrane) from an area of higher hydrostatic (fluid) pressure into an area of lower hydrostatic pressure. Like diffusion, filtration is a passive process. For example, fluids and solutes filter out of the capillaries in the kidneys into the kidney tubules because blood pressure in the capillaries is greater than the fluid pressure in the tubules. So, if blood pressure increases, the rate of filtration increases.

Filtration is not a selective process. The amount of *filtrate*—the fluids and solutes that pass through the membrane—depends almost entirely on the *pressure gradient* (the difference in pressure between the solutions on the two sides of the membrane) and on the *size* of the *membrane pores*. Solutes that are too large to pass through are retained by the capillaries. These solutes usually include blood cells and proteins. Ions and smaller molecules, such as glucose and urea, can pass through.

In this activity the pore size is measured as a *molecular weight cutoff (MWCO)*, which is indicated by the number below the filtration membrane. You can think of MWCO in terms of pore size: the larger the MWCO number, the larger the pores in the filtration membrane. The molecular weight of a solute is the number of grams per mole, where a mole is the constant Avogadro's number 6.02×10^{23} molecules/mole. You will also analyze the filtration membrane for the presence or absence of solutes that might be left sticking to the membrane.

> **EQUIPMENT USED** The following equipment will be depicted on-screen: top and bottom beakers—used for filtration of solutes; dialysis membranes with various molecular weight cutoffs (MWCOs); membrane residue analysis station—used to analyze the filtration membrane.

Experiment Instructions

Go to the home page in the PhysioEx software and click **Exercise 1: Cell Transport Mechanisms and Permeability.** Click **Activity 4: Simulating Filtration** and take the online **Pre-lab Quiz** for Activity 4.

After you take the online Pre-lab Quiz, click the **Experiment** tab and begin the experiment. The experiment instructions are reprinted here for your reference. The opening screen for the experiment is shown above.

1. Drag the 20 MWCO membrane to the membrane holder between the beakers.

2. Increase the concentration of Na^+Cl^-, urea, glucose, and powdered charcoal to be dispensed to 5.00 mg/ml by clicking the + button beside the display for each solute. Click **Dispense** to fill the top beaker.

3. After you start the run, the membrane holder below the top beaker retracts, and the solution will filter through the membrane into the beaker below. You will be able to determine whether solute particles are moving through the filtration membrane by observing the concentration displays beside the bottom beaker. A rise in detected solute concentration indicates that the solute particles are moving through the filtration membrane. Note that the pressure is set at 50 mm Hg and the timer is set to 60 minutes. The simulation compresses the 60-minute time period into 10 seconds of real time. Click **Start** to start the run and watch the concentration displays beside the bottom beaker for any activity.

4. Drag the 20 MWCO membrane to the holder in the membrane residue analysis unit. Click **Start Analysis** to begin analysis (and cleaning) of the membrane.

5. Click **Record Data** to display your results in the grid (and record your results in Chart 4, p. PEx-12).

6. Click the 20 MWCO membrane in the membrane holder to automatically return it to the membrane cabinet and then click **Flush** to prepare for the next run.

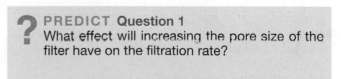

> **? PREDICT Question 1**
> What effect will increasing the pore size of the filter have on the filtration rate?

7. Drag the 50 MWCO membrane to the membrane holder between the beakers. With the concentration of Na^+Cl^-, urea, glucose, and powdered charcoal still set to 5.00 mg/ml, click **Dispense** to fill the top beaker.

8. Click **Start** to start the run and watch the concentration displays beside the bottom beaker for any activity.

CHART 4	Filtration Results				
		Membrane (MWCO)			
		20	50	200	200
Solute	**Filtration rate (ml/min)**				
Na^+Cl^-	Filter concentration (mg/ml)				
	Membrane residue				
Urea	Filter concentration (mg/ml)				
	Membrane residue				
Glucose	Filter concentration (mg/ml)				
	Membrane residue				
Powdered charcoal	Filter concentration (mg/ml)				
	Membrane residue				

9. Drag the 50 MWCO membrane to the holder in the membrane residue analysis unit. Click **Start Analysis** to begin analysis (and cleaning) of the membrane.

10. Click **Record Data** to display your results in the grid (and record your results in Chart 4).

11. Click the 50 MWCO membrane in the membrane holder to automatically return it to the membrane cabinet and then click **Flush** to prepare for the next run.

12. Drag the 200 MWCO membrane to the membrane holder between the beakers. With the concentration of Na^+Cl^-, urea, glucose, and powdered charcoal still set to 5.00 mg/ml, click **Dispense** to fill the top beaker.

13. Click **Start** to start the run and watch the concentration displays beside the bottom beaker for any activity.

14. Drag the 200 MWCO membrane to the holder in the membrane residue analysis unit. Click **Start Analysis** to begin analysis (and cleaning) of the membrane.

15. Click **Record Data** to display your results in the grid (and record your results in Chart 4).

16. Click the 200 MWCO membrane in the membrane holder to automatically return it to the membrane cabinet and then click **Flush** to prepare for the next run.

PREDICT Question 2
What will happen if you increase the pressure above the beaker (the driving pressure)?

17. Increase the pressure to 100 mm Hg by clicking on the + button beside the pressure display above the top beaker.

18. Drag the 200 MWCO membrane to the membrane holder between the beakers. With the concentration of Na^+Cl^-, urea, glucose, and powdered charcoal still set to 5.00 mg/ml, click **Dispense** to fill the top beaker.

19. Click **Start** to start the run and watch the concentration displays beside the bottom beaker for any activity.

20. Drag the 200 MWCO membrane to the holder in the membrane residue analysis unit. Click **Start Analysis** to begin analysis (and cleaning) of the membrane.

21. Click **Record Data** to display your results in the grid (and record your results in Chart 4).

After you complete the Experiment, take the online **Post-lab Quiz** for Activity 4.

Activity Questions

1. Explain your results with the 20 MWCO filter. Why weren't any of the solutes present in the filtrate?

2. Describe two variables that affected the rate of filtration in your experiments.

3. Explain how you can increase the filtration rate through living membranes.

_____

_____

_____

4. Judging from the filtration results, indicate which solute has the largest molecular weight.

_____ ▄▄▄

Simulating Active Transport

OBJECTIVES

1. To understand that active transport requires cellular energy in the form of ATP.

2. To explain how the balance of sodium and potassium is maintained by the Na⁺-K⁺ pump, which moves both ions against their concentration gradients.

3. To understand coupled transport and be able to explain how the movement of sodium and potassium is independent of other solutes, such as glucose.

Introduction

Whenever a cell uses cellular energy (ATP) to move substances across its membrane, the process is an *active transport process*. Substances moved across cell membranes by an active transport process are generally unable to pass by diffusion. There are several reasons why a substance might not be able to pass through a membrane by diffusion: it might be too large to pass through the membrane pores, it might not be lipid soluble, or it might have to move *against*, rather than with, a concentration gradient.

In one type of active transport, substances move across the membrane by combining with a carrier-protein molecule. This kind of process resembles an enzyme-substrate interaction. ATP hydrolysis provides the driving force, and, in many cases, the substances move *against* concentration gradients or electrochemical gradients or both. The carrier proteins are commonly called **solute pumps.** Substances that are moved into cells by solute pumps include amino acids and some sugars. Both of these kinds of solutes are necessary for the life of the cell, but they are lipid insoluble and too large to pass through membrane pores.

In contrast, sodium ions (Na⁺) are ejected from the cells by active transport. There is more Na⁺ outside the cell than inside the cell, so Na⁺ tends to remain in the cell unless actively transported out. In the body, the most common type of solute pump is the Na⁺-K⁺ (sodium-potassium) pump, which moves Na⁺ and K⁺ in opposite directions across cellular membranes. Three Na⁺ ions are ejected from the cell for every two K⁺ ions entering the cell. Note that there is more K⁺ inside the cell than outside the cell, so K⁺ tends to remain outside the cell unless actively transported in.

Membrane carrier proteins that move more than one substance, such as the Na⁺-K⁺ pump, participate in *coupled transport*. If the solutes move in the same direction, the carrier is a *symporter*. If the solutes move in opposite directions, the carrier is an *antiporter*. A carrier that transports only a single solute is a *uniporter*.

> **EQUIPMENT USED** The following equipment will be depicted on-screen: Simulated cell inside a large beaker.

Experiment Instructions

Go to the home page in the PhysioEx software and click **Exercise 1: Cell Transport Mechanisms and Permeability.** Click **Activity 5: Simulating Active Transport** and take the online **Pre-lab Quiz** for Activity 5.

After you take the online Pre-lab Quiz, click the **Experiment** tab and begin the experiment. The experiment instructions are reprinted here for your reference. The opening screen for the experiment is shown below.

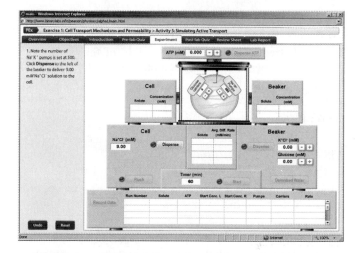

1. Note the number of Na⁺-K⁺ pumps is set at 500. Click **Dispense** to the left of the beaker to deliver 9.00 mM Na⁺Cl⁻ solution to the cell.

2. Increase the K⁺Cl⁻ concentration to be delivered to the beaker to 6.00 mM by clicking the + button beside the K⁺Cl⁻ display. Click **Dispense** to the right of the beaker to deliver 6.00 mM K⁺Cl⁻ solution to the beaker.

3. Increase the ATP concentration to 1.00 mM by clicking the + button beside the ATP display above the beaker. Click **Dispense ATP** to deliver 1.00 mM ATP solution to both sides of the membrane.

4. After you start the run, the solutes will move across the cell membrane, simulating active transport. You will be able to determine the amount of solute that is transported across the membrane by observing the concentration displays on both sides of the beaker (the display on the left shows the concentrations inside the cell and the display on the right shows the concentrations inside the beaker). Note that the timer is set to 60 minutes. The simulation compresses the 60-minute time period into 10 seconds of real time. Click **Start** to start the run and watch the concentration displays on both sides of the beaker for any activity.

5. Click **Record Data** to display your results in the grid.

6. Click **Flush** to reset the beaker and simulated cell.

7. Click **Dispense** to the left of the beaker to deliver 9.00 mM Na$^+$Cl$^-$ solution to the cell.

8. Increase the K$^+$Cl$^-$ concentration to be delivered to the beaker to 6.00 mM by clicking the + button beside the K$^+$Cl$^-$ display. Click **Dispense** to the right of the beaker to deliver 6.00 mM K$^+$Cl$^-$ solution to the beaker.

9. Increase the ATP concentration to 3.00 mM by clicking the + button beside the ATP display above the beaker. Click **Dispense ATP** to deliver 3.00 mM ATP solution to both sides of the membrane.

10. Click **Start** to start the run and watch the concentration displays on both sides of the beaker for any activity.

11. Click **Record Data** to display your results in the grid.

12. Click **Flush** to reset the beaker and simulated cell.

13. Click **Dispense** to the left of the beaker to deliver 9.00 mM Na$^+$Cl$^-$ solution to the cell.

14. Click **Deionized Water** to the right of the beaker and then click **Dispense** to deliver deionized water to the beaker.

15. Increase the ATP concentration to 3.00 mM. Click **Dispense ATP** to deliver 3.00 mM ATP solution to both sides of the membrane.

> **? PREDICT Question 1**
> What do you think will result from these experimental conditions?
>
> _____

16. Click **Start** to start the run and watch the concentration displays on both sides of the beaker for any activity.

17. Click **Record Data** to display your results in the grid.

18. Click **Flush** to reset the beaker and simulated cell.

19. Increase the number of Na$^+$-K$^+$ pumps to 800 by clicking the + button beneath the Na$^+$-K$^+$ pump display. Click **Dispense** to the left of the beaker to deliver 9.00 mM Na$^+$Cl$^-$ solution to the cell.

20. Increase the K$^+$Cl$^-$ concentration to be delivered to the beaker to 6.00 mM. Click **Dispense** to the right of the beaker to deliver 6.00 mM K$^+$Cl$^-$ solution to the beaker.

21. Increase the ATP concentration to 3.00 mM. Click **Dispense ATP** to deliver 3.00 mM ATP solution to both sides of the membrane.

22. Click **Start** to start the run and watch the concentration displays on both sides of the beaker for any activity.

23. Click **Record Data** to display your results in the grid.

24. Click **Flush** to reset the beaker and simulated cell.

25. With the number of Na$^+$-K$^+$ pumps still set to 800, increase the number of glucose carriers to 400 by clicking the + button beneath the glucose carriers display. Click **Dispense** to the left of the beaker to deliver 9.00 mM Na$^+$Cl$^-$ solution to the cell.

> **? PREDICT Question 2**
> Do you think the addition of glucose carriers will affect the transport of sodium or potassium?
>
> _____

26. Increase the K$^+$Cl$^-$ concentration to be delivered to the beaker to 6.00 mM. Increase the glucose concentration to be delivered to the beaker to 10.00 mM. Click **Dispense** to the right of the beaker to deliver 6.00 mM K$^+$Cl$^-$ and 10.00 mM glucose solution to the beaker.

27. Increase the ATP concentration to 3.00 mM. Click **Dispense ATP** to deliver 3.00 mM ATP solution to both sides of the membrane.

28. Click **Start** to start the run and watch the concentration displays on both sides of the beaker for any activity.

29. Click **Record Data** to display your results in the grid.

After you complete the experiment, take the online **Post-lab Quiz** for Activity 5.

Activity Questions

1. In the initial trial the number of Na$^+$-K$^+$ pumps is set to 500, the Na$^+$Cl$^-$ concentration is set to 9.00 mM, the K$^+$Cl$^-$ concentration is set to 6.00 mM, and the ATP concentration is set to 1.00 mM. Explain what happened and why. What would happen if no ATP had been dispensed?

_____

_____

2. Why was there no transport when you dispensed only Na$^+$Cl$^-$, even though ATP was present?

_____

_____

3. What happens to the rate of transport of Na$^+$ and K$^+$ when you increase the number of Na$^+$-K$^+$ pumps?

_____

_____

4. Explain why the Na$^+$ and K$^+$ transports were unaffected by the addition of glucose.

_____

_____

Cell Transport Mechanisms and Permeability

NAME _____

LAB TIME/DATE _____

ACTIVITY 1 Simulating Dialysis (Simple Diffusion)

1. Describe two variables that affect the rate of diffusion. _____

_____

2. Why do you think the urea was not able to diffuse through the 20 MWCO membrane? How well did the results compare

with your prediction? _____

3. Describe the results of the attempts to diffuse glucose and albumin through the 200 MWCO membrane. How well did the

results compare with your prediction? _____

_____

4. Put the following in order from smallest to largest molecular weight: glucose, sodium chloride, albumin, and urea. _____

_____

ACTIVITY 2 Simulated Facilitated Diffusion

1. Explain one way in which facilitated diffusion is the same as simple diffusion and one way in which it differs. _____

_____

2. The larger value obtained when more glucose carriers were present corresponds to an increase in the rate of glucose transport.

Explain why the rate increased. How well did the results compare with your prediction? _____

_____

3. Explain your prediction for the effect Na^+Cl^- might have on glucose transport. In other words, explain why you picked the

choice that you did. How well did the results compare with your prediction? _____

ACTIVITY 3 Simulating Osmotic Pressure

1. Explain the effect that increasing the Na^+Cl^- concentration had on osmotic pressure and why it has this effect. How well did

the results compare with your prediction? _____

_____

2. Describe one way in which osmosis is similar to simple diffusion and one way in which it is different. _____

_____

3. Solutes are sometimes measured in milliosmoles. Explain the statement, "Water chases milliosmoles." _____

_____

4. The conditions were 9 mM albumin in the left beaker and 10 mM glucose in the right beaker with the 200 MWCO membrane

in place. Explain the results. How well did the results compare with your prediction? _____

_____

ACTIVITY 4   Simulating Filtration

1. Explain in your own words why increasing the pore size increased the filtration rate. Use an analogy to support your state-

ment. How well did the results compare with your prediction? _____

_____

2. Which solute did not appear in the filtrate using any of the membranes? Explain why. _____

_____

3. Why did increasing the pressure increase the filtration rate but not the concentration of solutes? How well did the results

compare with your prediction? _____

_____

ACTIVITY 5   Simulating Active Transport

1. Describe the significance of using 9 mM sodium chloride inside the cell and 6 mM potassium chloride outside the cell,

instead of other concentration ratios. _____

_____

2. Explain why there was no sodium transport even though ATP was present. How well did the results compare with your

prediction? _____

3. Explain why the addition of glucose carriers had no effect on sodium or potassium transport. How well did the results

compare with your prediction? _____

4. Do you think glucose is being actively transported or transported by facilitated diffusion in this experiment? Explain your

answer. _____

_____

Skeletal Muscle Physiology

PRE-LAB QUIZ

1. Circle the correct underlined term: <u>Tendons</u> / <u>Ligaments</u> attach skeletal muscle to the periosteum of bones.

2. A motor unit consists of:
   a. a specialized region of the motor end plate
   b. a motor neuron and all of the muscle cells it innervates
   c. a motor fiber and a neuromuscular junction
   d. a muscle fiber's plasma membrane

3. During the _____ phase of a muscle twitch, chemical changes—such as the release of calcium—are occurring intracellularly as the muscle prepares for contraction.
   a. latent   c. coupling
   b. excitation   d. relaxation

4. Circle the correct underlined term: An action potential in a motor neuron triggers the release of <u>acetylcholine</u> / <u>magnesium</u>.

5. Circle True or False: When a weak muscle contraction occurs, each motor unit still develops its maximum tension even though fewer motor units are activated.

6. Circle True or False: During wave summation, muscle twitches follow each other closely, resulting in a step-like increase in force, so that each successive twitch peaks slightly higher than the one before, resulting in a staircase effect.

7. The smallest stimulus required to induce an action potential in a muscle fiber's plasma membrane is known as the:
   a. maximal voltage   c. strong voltage
   b. stimulus voltage   d. threshold voltage

8. Muscles do not shorten even though they are actively contracting during a(n) _____ contraction.
   a. isotonic   c. tetanic
   b. isometric   d. treppe

Exercise Overview

Humans make voluntary decisions to walk, talk, stand up, and sit down. Skeletal muscles, which are usually attached to the skeleton, make these actions possible (view Figure 2.1, p. PEx-18). Skeletal muscles characteristically span two joints and attach to the skeleton via **tendons,** which attach to the periosteum of a bone. Skeletal muscles are composed of hundreds to thousands of individual cells called **muscle fibers,** which produce **muscle tension** (also referred to as **muscle force**). Skeletal muscles are remarkable machines. They provide us with the manual dexterity to create magnificent works of art and can generate the brute force needed to lift a 45-kilogram sack of concrete.

When a skeletal muscle is isolated from an experimental animal and mounted on a **force transducer,** you can generate **muscle contractions** with controlled **electrical stimulation.** Importantly, the contractions of this isolated muscle are known to mimic those of working muscles in the body. That is, in vitro experiments reproduce in vivo functions. Therefore, the activities you perform in this exercise will give you valuable insight into skeletal muscle physiology.

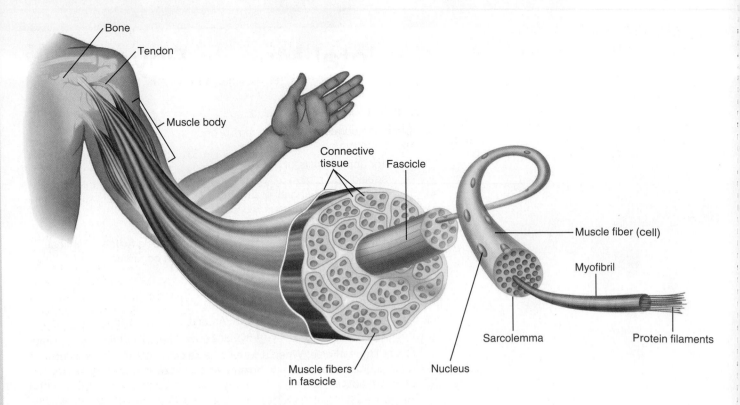

FIGURE 2.1 Structure of a skeletal muscle.

The Muscle Twitch and the Latent Period

OBJECTIVES

1. To understand the terms *excitation-contraction coupling, electrical stimulus, muscle twitch, latent period, contraction phase,* and *relaxation phase.*

2. To initiate muscle twitches with electrical stimuli of varying intensity.

3. To identify and measure the duration of the latent period.

Introduction

A **motor unit** consists of a **motor neuron** and all of the **muscle fibers** it innervates. The motor neuron and a muscle fiber intersect at the **neuromuscular junction.** Specifically, the neuromuscular junction is the location where the axon terminal of the neuron meets a specialized region of the muscle fiber's plasma membrane. This specialized region is called the **motor end plate.**

An action potential in a motor neuron triggers the release of acetylcholine from its terminal. Acetylcholine then diffuses onto the muscle fiber's plasma membrane (or **sarcolemma**) and binds to receptors in the motor end plate, initiating a change in ion permeability that results in a *graded depolarization* of the muscle plasma membrane (the end-plate potential). The events that occur at the neuromuscular junction lead to the **end-plate potential.** The end-plate potential triggers a series of events that results in the contraction of a muscle cell. This entire process is called **excitation-contraction coupling.**

You will be simulating excitation-contraction coupling in this and subsequent activities, but you will be using electrical pulses, rather than acetylcholine, to trigger action potentials. The pulses will be administered by an electrical stimulator that can be set for the precise voltage, frequency, and duration of shock desired. When applied to a muscle that has been surgically removed from an animal, a single electrical stimulus will result in a **muscle twitch**—the mechanical response to a single action potential. A muscle twitch has three phases: the *latent period,* the *contraction phase,* and the *relaxation phase.*

1. The **latent period** is the period of time that elapses between the generation of an action potential in a muscle cell and the start of muscle contraction. Although no force is generated during the latent period, chemical changes (including the release of calcium from the sarcoplasmic reticulum) occur intracellularly in preparation for contraction.

2. The **contraction phase** starts at the end of the latent period and ends when muscle tension peaks.

3. The **relaxation phase** is the period of time from peak tension until the end of the muscle contraction

EQUIPMENT USED   The following equipment will be depicted on-screen: intact, viable skeletal muscle dissected off the leg of a frog; electrical stimulator—delivers the desired amount and duration of stimulating voltage to the muscle via electrodes resting on the muscle; mounting stand—includes a force transducer to measure the amount of force, or tension, developed by the muscle; oscilloscope—displays the stimulated muscle twitch and the amount of active, passive, and total force developed by the muscle.

Experiment Instructions

Go to the home page in the PhysioEx software and click **Exercise 2: Skeletal Muscle Physiology.** Click **Activity 1: The Muscle Twitch and the Latent Period,** and take the online **Pre-lab Quiz** for Activity 1.

After you take the online Pre-lab Quiz, click the **Experiment** tab and begin the experiment. The experiment instructions are reprinted here for your reference. The opening screen for the experiment is shown below.

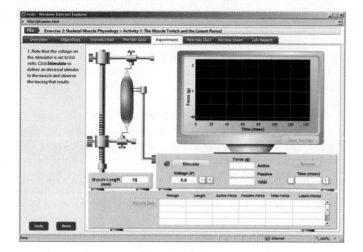

1. Note that the voltage on the stimulator is set to 0.0 volts. Click **Stimulate** to deliver an electrical stimulus to the muscle and observe the tracing that results.

2. The tracing on the oscilloscope indicates active muscle force. Note whether any muscle force developed with the voltage set to zero. Click **Record Data** to display your results in the grid (and record your results in Chart 1).

CHART 1	Latent Period Results	
Voltage	**Active force (g)**	**Latent period (msec)**

3. Increase the voltage to 3.0 volts by clicking the + button beside the voltage display.

4. Click **Stimulate** and observe the tracing that results.

5. Note the muscle force that developed. Click **Record Data** to display your results in the grid (and record your results in Chart 1).

6. Click **Clear Tracings** to remove the tracings from the oscilloscope.

7. Increase the voltage to 4.0 volts by clicking the + button beside the voltage display.

8. Click **Stimulate** and observe the tracing that results. Note that the trace starts at the left side of the screen and stays flat for a short period of time. Remember that the X-axis displays elapsed time in milliseconds. Also note how the force during the twitch also changes.

9. Click **Measure** on the stimulator. A thin, vertical yellow line appears at the far left side of the oscilloscope screen. To measure the length of the latent period, you measure the time between the application of the stimulus and the beginning of the first observable response (here, an increase in force). Click the + button beside the time display. You will see the vertical yellow line start to move across the screen. Watch what happens in the time (msec) display as the line moves across the screen. Keep clicking the + button until the yellow line reaches the point in the tracing where the graph stops being a flat line and begins to rise (this is the point at which muscle tension starts to develop). If the yellow line moves past the desired point, click the − button to move it backward.

When the yellow line is positioned correctly, click **Record Data** to display the latent period in the grid (and record your results in Chart 1).

10. Click **Clear Tracings** to remove the tracings from the oscilloscope.

> **? PREDICT Question 1**
> Will changes to the stimulus voltage alter the duration of the latent period? Explain.

11. You will now gradually increase the voltage to observe how changes to the stimulus voltage alter the duration of the latent period.

- Increase the voltage by 2.0 volts.
- Click **Stimulate** and observe the tracing that results.
- Click **Measure** on the stimulator and then click the + button until the yellow line reaches the point in the tracing where the graph stops being a flat line and begins to rise.
- Click **Record Data** (and record your results in Chart 1).

Repeat this step until you reach 10.0 volts.

After you complete the experiment, take the online **Post-lab Quiz** for Activity 1.

Activity Questions

1. Draw a graph that depicts a single skeletal muscle twitch, placing time on the X-axis and force on the Y-axis. Label the phases of this muscle twitch and describe what is happening in the muscle during each phase.

2. During the latent period of a skeletal muscle twitch, there is an apparent lack of muscle activity. Describe the electrical and chemical changes that occur in the muscle during this period.

_____

_____

_____

The Effect of Stimulus Voltage on Skeletal Muscle Contraction

OBJECTIVES

1. To understand the terms *motor neuron, muscle twitch, motor unit, recruitment, stimulus voltage, threshold stimulus,* and *maximal stimulus.*

2. To understand how motor unit recruitment can increase the tension a whole muscle develops.

3. To identify a threshold stimulus voltage.

4. To observe the effect of increases in stimulus voltage on a whole muscle.

5. To understand how increasing stimulus voltage to an isolated muscle in an experiment mimics motor unit recruitment in the body.

Introduction

A skeletal muscle produces **tension** (also known as **muscle force**) when nervous or electrical stimulation is applied. The force generated by a whole muscle reflects the number of active **motor units** at a given moment. A strong muscle contraction implies that many motor units are activated, with each unit developing its maximal tension, or force. A weak muscle contraction implies that fewer motor units are activated, but each motor unit still develops its maximal tension. By increasing the number of active motor units, we can produce a steady increase in muscle force, a process called **motor unit recruitment.**

Regardless of the number of **motor units** activated, a single stimulated contraction of whole skeletal muscle is called a **muscle twitch.** A tracing of a muscle twitch is divided into three phases: the latent period, the contraction phase, and the relaxation phase. The latent period is a short period between the time of muscle stimulation and the beginning of a muscle response. Although no force is generated during this interval, chemical changes occur intracellularly in preparation for contraction (including the release of calcium from the sarcoplasmic reticulum). During the contraction phase, the myofilaments utilize the cross-bridge cycle and the muscle develops tension. Relaxation takes place when the contraction has ended and the muscle returns to its normal resting state and length.

In this activity you will stimulate an isometric, or fixed-length, contraction of an isolated skeletal muscle. This activity allows you to investigate how the strength of an electrical stimulus affects whole-muscle function. Note that these simulations involve indirect stimulation by an electrode placed on the surface of the muscle. Indirect stimulation differs from the situation in vivo, where each fiber in the muscle receives direct stimulation via a nerve ending. Nevertheless, increasing the intensity of the electrical stimulation mimics how the nervous system increases the number of activated motor units.

The **threshold voltage** is the smallest stimulus required to induce an action potential in a muscle fiber's plasma membrane, or sarcolemma. As the **stimulus voltage** to a muscle is increased beyond the threshold voltage, the amount of force produced by the whole muscle also increases. This result occurs because, as more voltage is delivered to the whole muscle, more muscle fibers are activated and, thus, the total force produced by the muscle increases. Maximal tension in the whole muscle occurs when all the muscle fibers have been activated by a sufficiently strong stimulus (referred to as the **maximal voltage**). Stimulation with voltages greater than the maximal voltage will not increase the force of contraction. This experiment is analogous to, and accurately mimics, muscle activity in vivo, where the recruitment of additional motor units increases the total muscle force produced. This phenomenon is called *motor unit recruitment.*

> **EQUIPMENT USED** The following equipment will be depicted on-screen: intact, viable skeletal muscle dissected off the leg of a frog; electrical stimulator—delivers the desired amount and duration of stimulating voltage to the muscle via electrodes resting on the muscle; mounting stand—includes a force transducer to measure the amount of force, or tension, developed by the muscle; oscilloscope—displays the stimulated muscle twitch and the amount of active, passive, and total force developed by the muscle.

Experiment Instructions

Go to the home page in the PhysioEx software and click **Exercise 2: Skeletal Muscle Physiology.** Click **Activity 2: The Effect of Stimulus Voltage on Skeletal Muscle Contraction,** and take the online **Pre-lab Quiz** for Activity 2.

After you take the online Pre-lab Quiz, click the **Experiment** tab and begin the experiment. The experiment instructions are reprinted here for your reference. The opening screen for the experiment is shown on the following page.

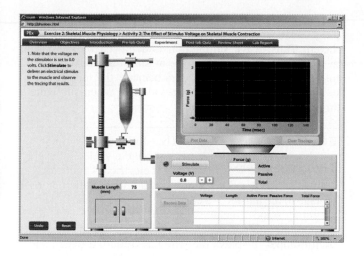

CHART 2	Effect of Stimulus Voltage on Skeletal Muscle Contraction	
Voltage	**Active force (g)**	

1. Note that the voltage on the stimulator is set to 0.0 volts. Click **Stimulate** to deliver an electrical stimulus to the muscle and observe the tracing that results.

2. Note the active force display and then click **Record Data** to display your results in the grid (and record your results in Chart 2).

3. Increase the voltage to 0.2 volts by clicking the + button beside the voltage display. Click **Stimulate** to deliver an electrical stimulus to the muscle and observe the tracing that results.

4. Note the active force display and then click **Record Data** to display your results in the grid (and record your results in Chart 2).

5. You will now gradually increase the voltage and stimulate the muscle to determine the minimum voltage required to generate active force.

- Increase the voltage by 0.1 volts and then click **Stimulate.**

- If no active force is generated, increase the voltage by 0.1 volts and stimulate the muscle again. When active force is generated, click **Record Data** to display your results in the grid (and record your results in Chart 2).

6. Enter the threshold voltage for this experiment in the field below and then click **Submit** to record your answer in the lab report. _____ volts

7. Click **Clear Tracings** to clear the tracings on the oscilloscope.

PREDICT Question 1
As the stimulus voltage is increased from 1.0 volt up to 10 volts, what will happen to the amount of active force generated with each stimulus?

_____

8. Increase the voltage on the stimulator to 1.0 volt and then click **Stimulate.**

9. Note the active force display and then click **Record Data** to display your results in the grid (and record your results in Chart 2).

10. You will now gradually increase the voltage and stimulate the muscle to determine the maximal voltage.

- Increase the voltage by 0.5 volts.

- Click **Stimulate** and observe the tracing that results.

- Note the active force display and then click **Record Data** to display your results in the grid (and record your results in Chart 2).

Repeat this step until you reach 10.0 volts.

11. Click **Plot Data** to view a summary of your data on a plotted grid. Click **Submit** to record your plot in the lab report.

12. Enter the maximal voltage for this experiment in the field below and then click **Submit** to record your answer in the lab report. _____ volts

After you complete the experiment, take the online **Post-lab Quiz** for Activity 2.

Activity Questions

1. For a single skeletal muscle twitch, explain the effect of increasing stimulus voltage.

_____

_____

2. How is this effect achieved in vivo?

_____

_____

The Effect of Stimulus Frequency on Skeletal Muscle Contraction

OBJECTIVES

1. To understand the terms *stimulus frequency, wave summation,* and *treppe.*
2. To observe the effect of an increasing stimulus frequency on the force developed by an isolated skeletal muscle.
3. To understand how increasing stimulus frequency to an isolated skeletal muscle induces the summation of twitch force.

Introduction

As demonstrated in Activity 2, increasing the stimulus voltage to an isolated skeletal muscle (up to a maximal value) results in an increase of force produced by the whole muscle. This experimental result is analogous to motor unit recruitment in the body. Importantly, this result relies on being able to increase the single stimulus intensity in the experiment. You will now explore another way to increase the force produced by an isolated skeletal muscle.

When a muscle first contracts, the force it is able to produce is less than the force it is able to produce with subsequent stimulations within a relatively short time span. **Treppe** is the progressive increase in force generated when a muscle is stimulated in succession, such that muscle twitches follow one another closely, with each successive twitch peaking slightly higher than the one before. This step-like increase in force is why treppe is also known as the staircase effect. For the first few twitches, each successive twitch produces slightly more force than the previous twitch as long as the muscle is allowed to fully relax between stimuli and the stimuli are delivered relatively close together.

When a skeletal muscle is stimulated repeatedly, such that the stimuli arrive one after another within a short period of time, muscle twitches can overlap with each other and result in a stronger muscle contraction than a stand-alone twitch. This phenomenon is known as wave summation.

Wave summation occurs when muscle fibers that are developing tension are stimulated again before the fibers have relaxed. Thus, wave summation is achieved by increasing the **stimulus frequency,** or rate of stimulus delivery to the muscle. Wave summation occurs because the muscle fibers are already in a partially contracted state when subsequent stimuli are delivered.

> **EQUIPMENT USED** The following equipment will be depicted on-screen: intact, viable skeletal muscle dissected off the leg of a frog; an electrical stimulator—delivers the desired amount and duration of stimulating voltage to the muscle via electrodes resting on the muscle; mounting stand—includes a force transducer to measure the amount of force, or tension, developed by the muscle; oscilloscope—displays the stimulated muscle twitch and the amount of active, passive, and total force developed by the muscle.

Experiment Instructions

Go to the home page in the PhysioEx software and click **Exercise 2: Skeletal Muscle Physiology.** Click **Activity 3: The Effect of Stimulus Frequency on Skeletal Muscle Contraction,** and take the online **Pre-lab Quiz** for Activity 3.

After you take the online Pre-lab Quiz, click the **Experiment** tab and begin the experiment. The experiment instructions are reprinted here for your reference. The opening screen for the experiment is shown below.

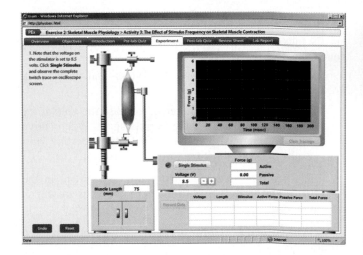

1. Note that the voltage on the stimulator is set to 8.5 volts. Click **Single Stimulus** and observe the tracing that results on the oscilloscope.

2. Note the active force display and then click **Record Data** to display your results in the grid (and record your results in Chart 3).

3. Click **Single Stimulus** and allow the trace to rise and completely fall. *Immediately after* the trace has returned to baseline, click **Single Stimulus** again.

CHART 3	Effect of Stimulus Frequency on Skeletal Muscle Contraction	
Voltage	Stimulus	Active force (g)

4. Note the active force for the second muscle twitch and click **Record Data** to display your results in the grid (and record your results in Chart 3).

5. You should have observed an increase in active force generated by the muscle with the immediate second stimulus. This increase demonstrates the phenomenon of treppe. Click **Clear Tracings** to clear the tracings on the oscilloscope.

6. You will now investigate the process of wave summation. Click **Single Stimulus** and watch the trace rise and begin to fall. *Before* the trace falls completely back to the baseline, click **Single Stimulus** again. (You can simply click **Single Stimulus** twice in quick succession in order to achieve this.)

7. Note the active force for the second muscle twitch and click **Record Data** to display your results in the grid (and record your results in Chart 3).

? PREDICT Question 1
As the stimulus frequency increases, what will happen to the muscle force generated with each successive stimulus? Will there be a limit to this response?

_____

_____

_____

8. Now stimulate the muscle at a higher frequency by clicking **Single Stimulus** four times in rapid succession.

9. Note the active force display and then click **Record Data** to display your results in the grid (and record your results in Chart 3).

10. Click **Clear Tracings** to clear the tracings on the oscilloscope.

? PREDICT Question 2
In order to produce sustained muscle contractions with an active force value of 5.2 grams, do you think you need to increase the stimulus voltage?

_____

11. Increase the voltage to 10.0 volts by clicking the + button beside the voltage display. After setting the voltage, click **Single Stimulus** four times in rapid succession.

12. Note the active force display and then click **Record Data** to display your results in the grid (and record your results in Chart 3).

13. Click **Clear Tracings** to clear the tracings on the oscilloscope.

14. Return the voltage to 8.5 volts by clicking the − button beside the voltage display. After setting the voltage, click **Single Stimulus** as many times as you can in rapid succession. Note the active force display. If you did not achieve an active force of 5.2 grams, click **Clear Tracings** and then click **Single Stimulus** even more rapidly. Repeat this step until you achieve an active force of 5.2 grams.
When you achieve an active force of 5.2 grams, click **Record Data** to display your results in the grid (and record your results in Chart 3).

After you complete the experiment, take the online **Post-lab Quiz** for Activity 3.

Activity Questions

1. Why is treppe also known as the staircase effect?

_____

_____

2. What changes are thought to occur in the skeletal muscle to allow treppe to be observed?

_____

_____

3. How does the frequency of stimulation affect the amount of force generated by a skeletal muscle?

_____

_____

4. Explain how wave summation is achieved in vivo.

_____

_____

Tetanus in Isolated Skeletal Muscle

OBJECTIVES

1. To understand the terms *stimulus frequency, unfused tetanus, fused tetanus,* and *maximal tetanic tension.*

2. To observe the effect of an increasing stimulus frequency on an isolated skeletal muscle.

3. To understand how increasing the stimulus frequency to an isolated skeletal muscle leads to unfused or fused tetanus.

Introduction

As demonstrated in Activity 3, increasing the **stimulus frequency** to an isolated skeletal muscle results in an increase in force produced by the whole muscle. Specifically, you observed that, if electrical stimuli are applied to a skeletal muscle in quick succession, the overlapping twitches generated more force with each successive stimulus. However, if stimuli continue to be applied frequently to a muscle over a prolonged period of time, the maximum possible muscle force from each stimulus will eventually reach a plateau—a state known as **unfused tetanus.** If stimuli are then applied with even greater frequency, the twitches will begin to fuse so that the peaks and valleys of each twitch become indistinguishable from one another—this state is known as **complete (fused) tetanus.** When the stimulus frequency reaches a value beyond which no further increases in force are generated by the muscle, the muscle has reached its **maximal tetanic tension.**

EQUIPMENT USED The following equipment will be depicted on-screen: intact, viable skeletal muscle dissected off the leg of a frog; electrical stimulator—delivers the desired amount and duration of stimulating voltage to the muscle via electrodes resting on the muscle; mounting stand—includes a force transducer to measure the amount of force, or tension, developed by the muscle; oscilloscope—displays the stimulated muscle twitch and the amount of active, passive, and total force developed by the muscle.

Experiment Instructions

Go to the home page in the PhysioEx software and click **Exercise 2: Skeletal Muscle Physiology.** Click **Activity 4: Tetanus in Isolated Skeletal Muscle** and take the online **Pre-lab Quiz** for Activity 4.

After you take the online Pre-lab Quiz, click the **Experiment** tab and begin the experiment. The experiment instructions are reprinted here for your reference. The opening screen for the experiment is shown above.

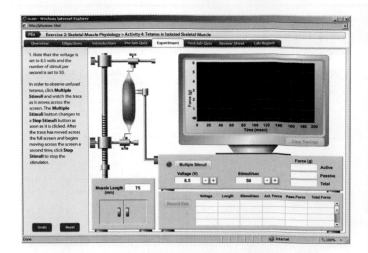

1. Note that the voltage is set to 8.5 volts and the number of stimuli per second is set to 50. To observe *unfused* tetanus, click **Multiple Stimuli** and watch the trace as it moves across the screen. The **Multiple Stimuli** button changes to a **Stop Stimuli** button after it is clicked. After the trace has moved across the full screen and begins moving across the screen a second time, click **Stop Stimuli** to stop the stimulator.

2. Click **Record Data** to display your results in the grid (and record your results in Chart 4).

CHART 4	Tetanus in Isolated Skeletal Muscle
Stimuli/second	**Active force (g)**

? PREDICT Question 1
As the stimulus frequency increases further, what will happen to the muscle tension and twitch appearance with each successive stimulus? Will there be a limit to this response?

_____

_____

3. In order to observe *fused* tetanus, increase the stimuli/sec setting to 130 by clicking the + button beside the stimuli/sec display. Click **Multiple Stimuli** and observe the resulting trace. After the trace has moved across the full screen and begins moving across the screen a second time, click **Stop Stimuli.**

4. Note the fused tetanus and click **Record Data** to display your results in the grid (and record your results in Chart 4).

5. Click **Clear Tracings** to clear the oscilloscope screen.

6. Increase the stimuli/sec setting to 140 by clicking the + button beside the stimuli/sec display. Click **Multiple Stimuli** and observe the resulting trace. After the trace has moved across the full screen and begins moving across the screen a second time, click **Stop Stimuli.**

7. Note the fused tetanus and click **Record Data** to display your results in the grid (and record your results in Chart 4).

8. Click **Clear Tracings** to clear the oscilloscope screen.

9. You will now observe the effect of incremental increases in the number of stimuli per second above 140 stimuli per second.

- Increase the stimuli/sec setting by 2.
- Click **Multiple Stimuli** and observe the resulting trace. After the trace has moved across the full screen and begins moving across the screen a second time, click **Stop Stimuli.**
- Click **Record Data** to display your results in the grid (and record your results in Chart 4).
- Click **Clear Tracings** to clear the oscilloscope screen.

Repeat this step until you reach 150 stimuli per second.

After you complete the experiment, take the online **Post-lab Quiz** for Activity 4.

Activity Questions

1. Explain what you think is being summated in the skeletal muscle to allow a high stimulus frequency to induce a smooth, continuous skeletal muscle contraction.

_____

_____

2. Why do many toddlers receive a tetanus shot (and then subsequent booster shots, as needed, later in life)? How does the condition known as "lockjaw" relate to tetanus shots?

_____

_____

Fatigue in Isolated Skeletal Muscle

OBJECTIVES

1. To understand the terms *stimulus frequency, complete (fused) tetanus, fatigue,* and *rest period.*

2. To observe the development of skeletal muscle fatigue.

3. To understand how the length of intervening rest periods determines the onset of fatigue.

Introduction

As demonstrated in Activities 3 and 4, increasing the stimulus frequency to an isolated skeletal muscle induces an increase of force produced by the whole muscle. Specifically, if voltage stimuli are applied to a muscle frequently in quick succession, the skeletal muscle generates more force with each successive stimulus.

However, if stimuli continue to be applied frequently to a muscle over a prolonged period of time, the maximum force of each twitch eventually reaches a plateau—a state known as *unfused tetanus.* If stimuli are then applied with even greater frequency, the twitches begin to fuse so that the peaks and valleys of each twitch become indistinguishable from one another—this state is known as **complete (fused) tetanus.** When the **stimulus frequency** reaches a value beyond which no further increase in force is generated by the muscle, the muscle has reached its **maximal tetanic tension.**

In this activity you will observe the phenomena of skeletal muscle *fatigue.* Fatigue refers to a decline in a skeletal muscle's ability to maintain a constant level of force, or tension, after prolonged, repetitive stimulation. You will also demonstrate how intervening **rest periods** alter the onset of fatigue in skeletal muscle. The causes of fatigue are still being investigated and multiple molecular events are thought to be involved, though the accumulations of lactic acid, ADP, and P_i in muscles are thought to be the major factors causing fatigue in the case of high-intensity exercise.

Common definitions for **fatigue** are:

- The failure of a muscle fiber to produce tension because of previous contractile activity.

- A decline in the muscle's ability to maintain a constant force of contraction after prolonged, repetitive stimulation.

EQUIPMENT USED The following equipment will be depicted on-screen: intact, viable skeletal muscle dissected off the leg of a frog; electrical stimulator—delivers the desired amount and duration of stimulating voltage to the muscle via electrodes resting on the muscle; mounting stand—includes a force transducer to measure the amount of force, or tension, developed by the muscle; oscilloscope—displays the stimulated muscle twitch and the amount of active, passive, and total force developed by the muscle.

Experiment Instructions

Go to the home page in the PhysioEx software and click **Exercise 2: Skeletal Muscle Physiology.** Click **Activity 5: Fatigue in Isolated Skeletal Muscle,** and take the online **Pre-lab Quiz** for Activity 5.

After you take the online Pre-lab Quiz, click the **Experiment** tab and begin the experiment. The experiment instructions are reprinted here for your reference. The opening screen for the experiment is shown on the following page.

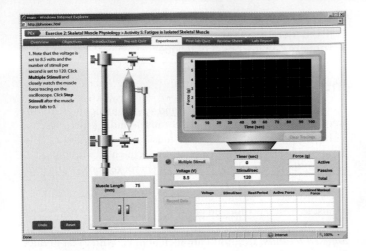

1. Note that the voltage is set to 8.5 volts and the number of stimuli per second is set to 120. Click **Multiple Stimuli** and closely watch the muscle force tracing on the oscilloscope. Click **Stop Stimuli** after the muscle force falls to 0.

2. Click **Record Data** to display your results in the grid (and record your results in Chart 5).

CHART 5	Fatigue Results	
Rest period (sec)	Active force (g)	Sustained maximal force (sec)

3. Click **Clear Tracings** to clear the oscilloscope screen.

> **? PREDICT Question 1**
> If the stimulator is briefly turned off for defined periods of time, what will happen to the length of time that the muscle is able to sustain maximal developed tension when the stimulator is turned on again?
> _____
> _____

4. To demonstrate the onset of fatigue after a variable rest period, you will be clicking the **Multiple Stimuli** button on and off three times. Read through the steps below before proceeding. Watch the timer closely to help you determine when to turn the stimulator back on.

- Click **Multiple Stimuli.**
- After the muscle force falls to 0, click **Stop Stimuli** to turn off the stimulator.
- Wait 10 seconds, then click **Multiple Stimuli** to turn the stimulator back on.
- Click **Stop Stimuli** after the muscle force falls to 0.

- Wait 20 seconds, then click **Multiple Stimuli** to turn the stimulator back on.
- Click **Stop Stimuli** after the muscle force falls to 0.

5. Click **Record Data** to display your results in the grid (and record your results in Chart 5).

After you complete the experiment, take the online **Post-lab Quiz** for Activity 5.

Activity Questions

1. What proposed mechanisms most likely explain why fatigue develops?

_____

_____

2. What would you recommend to an interested friend as the best ways to delay the onset of fatigue?

_____

_____

The Skeletal Muscle Length-Tension Relationship

OBJECTIVES

1. To understand the terms *isometric contraction, active force, passive force, total force,* and *length-tension relationship.*

2. To understand the effect that resting muscle length has on tension development when the muscle is maximally stimulated in an isometric experiment.

3. To explain the molecular basis of the skeletal muscle length-tension relationship.

Introduction

Skeletal muscle contractions are either isometric or isotonic. When a muscle attempts to move a load that is equal to the force generated by the muscle, the muscle contracts isometrically. During an **isometric** contraction, the muscle stays at a fixed length (*isometric* means "same length"). An example of isometric muscle contraction is when you stand in a doorway and push on the doorframe. The load that you are attempting to move (the doorframe) can easily equal the force generated by your muscles, so your muscles do not shorten even though they are actively contracting.

Isometric contractions are accomplished experimentally by keeping both ends of the muscle in a fixed position while electrically stimulating the muscle. Resting length (the length of the muscle before stimulation) is an important factor in determining the amount of force that a muscle can develop when stimulated. **Passive force** is generated by stretching the muscle and results from the elastic recoil of the tissue itself. This passive force is largely caused by the protein titin, which acts as a molecular bungee cord. **Active force** is generated when myosin thick filaments bind to actin thin filaments, thus engaging the cross bridge cycle and ATP hydrolysis. Think of the skeletal muscle as having two force properties: it exerts passive force when it is stretched (like a rubber band exerts passive force) and active force when it is stimulated. **Total force** is the sum of passive and active forces.

This activity allows you to set and hold constant the length of the isolated skeletal muscle and subsequently stimulate it with individual maximal voltage stimuli. A graph relating the three forces generated and the fixed length of the muscle will be automatically plotted after you stimulate the muscle. In muscle physiology this graph is known as the **isometric length-tension relationship.** The results of this simulation can be applied to human muscles to understand how optimum resting length will result in maximum force production.

To understand why muscle tissue behaves as it does, you must understand tension at the cellular level. If you have difficulty understanding the results of this activity, review the sliding filament model of muscle contraction. Think of the length-tension relationship in terms of those sarcomeres that are too short, those that are too long, and those that have the ideal amount of thick and thin filament overlap.

EQUIPMENT USED The following equipment will be depicted on-screen: intact, viable skeletal muscle dissected off the leg of a frog; electrical stimulator—delivers the desired amount and duration of stimulating voltage to the muscle via electrodes resting on the muscle; mounting stand—includes (1) a force transducer to measure the amount of force, or tension, developed by the muscle and (2) a gearing system that allows the hook through the muscle's lower tendon to be moved up or down, thus altering the fixed length of the muscle; oscilloscope—displays the stimulated muscle twitch and the amount of active, passive, and total force developed by the muscle.

Experiment Instructions

Go to the home page in the PhysioEx software and click **Exercise 2, Skeletal Muscle Physiology.** Click **Activity 6, The Skeletal Muscle Length-Tension Relationship,** and take the online **Pre-lab Quiz** for Activity 6.

After you take the online Pre-lab Quiz, click the **Experiment** tab and begin the experiment. The experiment instructions are reprinted here for your reference. The opening screen for the experiment is shown below.

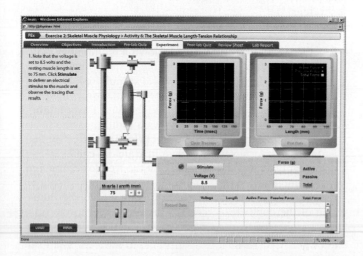

1. Note that the voltage is set to 8.5 volts and the resting muscle length is set to 75 mm. Click **Stimulate** to deliver an electrical stimulus to the muscle and observe the tracing that results.

2. You should see a single muscle twitch tracing on the left oscilloscope display and three data points (representing active, passive, and total force generated during this twitch) plotted on the right display. The yellow box represents the total force, the red dot contained within the yellow box represents the active force, and the green square represents the passive force. Click **Record Data** to display your results in the grid (and record your results in Chart 6).

CHART 6	Skeletal Muscle Length-Tension Relationship		
Length (mm)	Active force (g)	Passive force (g)	Total force (g)

> **? PREDICT Question 1**
> As the resting length of the muscle is changed, what will happen to the amount of total force the muscle generates during the stimulated twitch?
>
> _____
>
> _____

3. You will now gradually shorten the muscle to determine the effect of muscle length on active, passive, and total force.

- Shorten the muscle by 5 mm by clicking the – button beside the muscle length display.

- Click **Stimulate** to deliver an electrical stimulus to the muscle and note the values of the total, active, and passive forces relative to those observed at the original 75 mm.

- Click **Record Data** to display your results in the grid (and record your results in Chart 6).

Repeat these steps until you reach a muscle length of 50 mm.

4. Click **Clear Tracings** to clear the left oscilloscope display.

5. Lengthen the muscle to 80 mm by clicking the + button beside the muscle length display. Click **Stimulate** to deliver an electrical stimulus to the muscle and note the values of the total, active, and passive forces relative to those observed at the original 75 mm.

6. Click **Record Data** to display your results in the grid (and record your results in Chart 6).

7. You will now gradually lengthen the muscle to determine the effect of muscle length on active, passive, and total force.

- Lengthen the muscle by 10 mm by clicking the + button beside the muscle length display.

- Click **Stimulate** to deliver an electrical stimulus to the muscle and note the values of the total, active, and passive forces relative to those observed at the original 75 mm.

- Click **Record Data** to display your results in the grid(and record your results in Chart 6).

  Repeat these steps until you reach a muscle length of 100 mm.

8. Click **Plot Data** to view a summary of your data on a plotted grid. Click **Submit** to record your plot in the lab report.

After you complete the experiment, take the online **Post-lab Quiz** for Activity 6.

Activity Questions

1. Explain what happens in the skeletal muscle sarcomere to result in the changes in active, passive, and total force when the resting muscle length is changed.

_____

_____

2. Explain the dip in the total force curve as the muscle was stretched to longer lengths. (Hint: Keep in mind that you are measuring the sum of active and passive forces.)

_____

_____

▬▬

A C T I V I T Y 7

Isotonic Contractions and the Load-Velocity Relationship

OBJECTIVES

1. To understand the terms _isotonic concentric contraction, load, latent period, shortening velocity,_ and _load-velocity relationship_.

2. To understand the effect that increasing load (that is, weight) has on an isolated skeletal muscle when the muscle is stimulated in an isotonic contraction experiment.

3. To understand the load-velocity relationship in isolated skeletal muscle.

Introduction

Skeletal muscle contractions can be described as either isometric or isotonic. When a muscle attempts to move an object (the **load**) that is equal in weight to the force generated by the muscle, the muscle is observed to contract isometrically. In an isometric contraction, the muscle stays at a fixed length (_isometric_ means "same length").

During an **isotonic contraction,** the skeletal muscle length changes and, thus, the load moves a measurable distance. If the muscle length shortens as the load moves, the contraction is called an **isotonic** _concentric_ **contraction.** An isotonic concentric contraction occurs when a muscle generates a force greater than the load attached to the muscle's end. In this type of contraction, there is a **latent period** during which there is a rise in muscle tension but no observable movement of the weight. After the muscle tension exceeds the weight of the load, an isotonic concentric contraction can begin. Thus, the latent period gets longer as the weight of the load gets larger. When the building muscle force exceeds the load, the muscle shortens and the weight moves. Eventually, the force of the muscle contraction will decrease as the muscle twitch begins the relaxation phase, and the load will therefore start to return to its original position.

An isotonic twitch is not an all-or-nothing event. If the load is increased, the muscle must generate more force to move it and the latent period will therefore get longer because it will take more time for the necessary force to be generated by the muscle. The speed of the contraction (muscle **shortening velocity**) also depends on the load that the muscle is attempting to move. Maximal shortening velocity is attained with minimal load attached to the muscle. Conversely, the heavier the load, the slower the muscle twitch. You can think of lifting an object from the floor as an example. A light object can be lifted quickly (high velocity), whereas a heavier object will be lifted with a slower velocity for a shorter duration.

In an isotonic muscle contraction experiment, one end of the muscle remains free (unlike in an isometric contraction experiment, where both ends of the muscle are held in a fixed position). Different weights (loads) can then be attached to the free end of the isolated muscle, while the other end is held in a fixed position by the force transducer. If the weight (the load) is less than the tension generated by the whole muscle, then the muscle will be able to lift it with a measurable distance, velocity, and duration. In this activity, you will change the weight (load) that the muscle will try to move as it shortens.

> **EQUIPMENT USED** The following equipment will be depicted on-screen: intact, viable skeletal muscle dissected off the leg of a frog; electrical stimulator—delivers the desired amount and duration of stimulating voltage to the muscle via electrodes resting on the muscle; mounting stand—includes a ruler that allows a rapid measurement of the distance (cm) that the weight (load) is lifted by the isolated muscle; several weights (in grams)—can be interchangeably attached to the hook on the free lower tendon of the mounted skeletal muscle; oscilloscope—displays the stimulated isotonic concentric contraction, the duration of the contraction, and the distance that muscle lifts the weight (load).

Experiment Instructions

Go to the home page in the PhysioEx software and click **Exercise 2: Skeletal Muscle Physiology.** Click **Activity 7: Isotonic Contractions and the Load-Velocity Relationship,** and take the online **Pre-lab Quiz** for Activity 7.

After you take the online Pre-lab Quiz, click the **Experiment** tab and begin the experiment. The experiment instructions are reprinted here for your reference. The opening screen for the experiment is shown below.

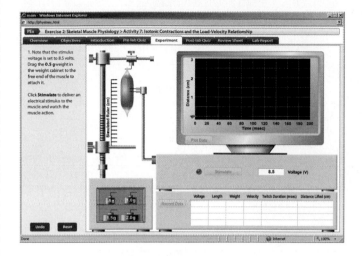

1. Note that the stimulus voltage is set to 8.5 volts. Drag the 0.5-g weight in the weight cabinet to the free end of the muscle to attach it. Click **Stimulate** to deliver an electrical stimulus to the muscle and watch the muscle action.

2. Observe that, as the muscle shortens in length, it lifts the weight off the platform. The muscle then lengthens as it relaxes and lowers the weight back down to the platform. Click **Stimulate** again and try to watch both the muscle and the oscilloscope screen at the same time.

3. Click **Record Data** to display your results in the grid (and record your results in Chart 7).

CHART 7	Isotonic Contraction Results		
Weight (g)	Velocity (cm/sec)	Twitch duration (msec)	Distance lifted (cm)

PREDICT Question 1
As the load on the muscle *increases*, what will happen to the latent period, the shortening velocity, the distance that the weight moved, and the contraction duration?

4. Remove the 0.5-g weight by dragging it back to the weight cabinet. Drag the 1.0-g weight to the free end of the muscle to attach it. Click **Stimulate** and observe the muscle and the oscilloscope screen.

5. Click **Record Data** to display your results in the grid (and record your results in Chart 7).

6. Remove the 1.0-g weight by dragging it back to the weight cabinet. Drag the 1.5-g weight to the free end of the muscle to attach it. Click **Stimulate** and observe the muscle and the oscilloscope screen.

7. Click **Record Data** to display your results in the grid (and record your results in Chart 7).

8. Remove the 1.5-g weight by dragging it back to the weight cabinet. Drag the 2.0-g weight to the free end of the muscle to attach it. Click **Stimulate** and observe the muscle and the oscilloscope screen.

9. Click **Record Data** to display your results in the grid (and record your results in Chart 7).

10. Click **Plot Data** to generate a muscle load-velocity relationship. Watch the display carefully as the program animates the development of a load-velocity relationship for the data you have collected. Click **Submit** to record your plot in the lab report.

After you complete the experiment, take the online **Post-lab Quiz** for Activity 7.

Activity Questions

1. Explain the relationship between the load attached to a skeletal muscle and the initial velocity of skeletal muscle shortening.

2. Explain why it will take you longer to perform ten repetitions lifting a 20-pound weight than it would to perform the same number of repetitions with a 5-pound weight.

NAME _____

LAB TIME/DATE _____

Skeletal Muscle Physiology

ACTIVITY 1 The Muscle Twitch and the Latent Period

1. Define the terms *skeletal muscle fiber, motor unit, skeletal muscle twitch, electrical stimulus,* and *latent period.* _____

   _____

   _____

   _____

   _____

2. What is the role of acetylcholine in a skeletal muscle contraction? _____

   _____

3. Describe the process of excitation-contraction coupling in skeletal muscle fibers. _____

   _____

   _____

4. Describe the three phases of a skeletal muscle twitch. _____

   _____

   _____

   _____

5. Does the duration of the latent period change with different stimulus voltages? How well did the results compare with your

   prediction? _____

   _____

6. At the threshold stimulus, do sodium ions start to move into or out of the cell to bring about the membrane depolarization?

   _____

   _____

ACTIVITY 2 The Effect of Stimulus Voltage on Skeletal Muscle Contraction

1. Describe the effect of increasing stimulus voltage on isolated skeletal muscle. Specifically, what happened to the muscle

    force generated with stronger electrical stimulations and why did this change occur? How well did the results compare with

    your prediction? _____

    _____

2. How is this change in whole-muscle force achieved in vivo? _____

    _____

3. What happened in the isolated skeletal muscle when the maximal voltage was applied? _____

    _____

ACTIVITY 3 The Effect of Stimulus Frequency on Skeletal Muscle Contraction

1. What is the difference between stimulus intensity and stimulus frequency? _____

2. In this experiment you observed the effect of stimulating the isolated skeletal muscle multiple times in a short period with

    complete relaxation between the stimuli. Describe the force of contraction with each subsequent stimulus. Are these results

    called treppe or wave summation? _____

    _____

3. How did the frequency of stimulation affect the amount of force generated by the isolated skeletal muscle when the fre-

    quency of stimulation was increased such that the muscle twitches did not fully relax between subsequent stimuli? Are these

    results called treppe or wave summation? How well did the results compare with your prediction? _____

    _____

    _____

    _____

4. To achieve an active force of 5.2 g, did you have to increase the stimulus voltage above 8.5 volts? If not, how did you

    achieve an active force of 5.2 g? How well did the results compare with your prediction? _____

    _____

    _____

5. Compare and contrast frequency-dependent wave summation with motor unit recruitment (previously observed by increas-

    ing the stimulus voltage). How are they similar? How was each achieved in the experiment? Explain how each is achieved

    in vivo. _____

    _____

ACTIVITY 4 Tetanus in Isolated Skeletal Muscle

1. Describe how increasing the stimulus frequency affected the force developed by the isolated whole skeletal muscle in this activity. How well did the results compare with your prediction? _____

_____

2. Indicate what type of force was developed by the isolated skeletal muscle in this activity at the following stimulus frequencies: at 50 stimuli/sec, at 140 stimuli/sec, and above 146 stimuli/sec. _____

_____

3. Beyond what stimulus frequency is there no further increase in the peak force? What is the muscle tension called at this frequency? _____

ACTIVITY 5 Fatigue in Isolated Skeletal Muscle

1. When a skeletal muscle fatigues, what happens to the contractile force over time? _____

_____

2. What are some proposed causes of skeletal muscle fatigue? _____

_____

3. Turning the stimulator off allows a small measure of muscle recovery. Thus, the muscle will produce more force for a longer time period if the stimulator is briefly turned off than if the stimuli were allowed to continue without interruption. Explain why this might occur. How well did the results compare with your prediction? _____

_____

_____

_____

4. List a few ways that humans could delay the onset of fatigue when they are vigorously using their skeletal muscles _____

_____

ACTIVITY 6 The Skeletal Muscle Length-Tension Relationship

1. What happens to the amount of total force the muscle generates during the stimulated twitch? How well did the results compare with your prediction? _____

_____

_____

2. What is the key variable in an isometric contraction of a skeletal muscle? _____

3. Based on the unique arrangement of myosin and actin in skeletal muscle sarcomeres, explain why active force varies with changes in the muscle's resting length. _____

_____

_____

4. What skeletal muscle lengths generated passive force? (Provide a range.) _____

5. If you were curling a 7-kg dumbbell, when would your bicep muscles be contracting isometrically? _____

ACTIVITY 7 Isotonic Contractions and the Load-Velocity Relationship

1. If you were using your bicep muscles to curl a 7-kg dumbbell, when would your muscles be contracting isotonically?

_____

2. Explain why the latent period became longer as the load became heavier in the experiment. How well did the results compare with your prediction?_____

_____

_____

3. Explain why the shortening velocity became slower as the load became heavier in this experiment. How well did the results compare with your prediction? _____

_____

_____

4. Describe how the shortening distance changed as the load became heavier in this experiment. How well did the results compare with your prediction? _____

_____

_____

5. Explain why it would take you longer to perform 10 repetitions lifting a 10-kg weight than it would to perform the same number of repetitions with a 5-kg weight. _____

_____

6. Describe what would happen in the following experiment: A 2.5-g weight is attached to the end of the isolated whole skeletal muscle used in these experiments. Simultaneously, the muscle is maximally stimulated by 8.5 volts and the platform supporting the weight is removed. Will the muscle generate force? Will the muscle change length? What is the name for this type of contraction? _____

_____

Neurophysiology of Nerve Impulses

Exercise Overview

The nervous system contains two general types of cells: **neurons** and neuroglia (or glial cells). This exercise focuses on neurons. Neurons respond to their local environment by generating an electrical signal. For example, sensory neurons in the nose generate a signal (called a **receptor potential**) when odor molecules interact with receptor proteins on the membrane of these olfactory sensory neurons. Thus, sensory neurons can respond directly to sensory stimuli. The receptor potential can trigger another electrical signal (called an **action potential**), which travels along the membrane of the sensory neuron's axon to the brain—you could say that the action potential is conducted to the brain.

The action potential causes the release of **chemical neurotransmitters** onto neurons in olfactory regions of the brain. These chemical neurotransmitters bind to receptor proteins on the membrane of these brain **interneurons.** In general, interneurons respond to chemical neurotransmitters released by other neurons. In the nose the odor molecules are sensed by sensory neurons. In the brain the odor is perceived by the activity of interneurons responding to neurotransmitters. Any resulting action or behavior is caused by the subsequent activity of **motor neurons,** which can stimulate muscles to contract (see Exercise 2).

In general each neuron has three functional regions for signal transmission: a receiving region, a conducting region, and an output region, or secretory region. Sensory neurons often have a receptive ending specialized to detect a specific sensory stimulus, such as odor, light, sound, or touch. The **cell body** and **dendrites** of interneurons receive stimulation by neurotransmitters at structures called **chemical synapses** and produce **synaptic potentials.** The conducting

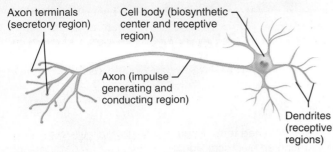

Axon terminals
(secretory region)

Cell body (biosynthetic
center and receptive
region)

Axon (impulse
generating and
conducting region)

Dendrites
(receptive
regions)

**FIGURE 3.1   A neuron with functional areas
identified.**

region is usually an **axon,** which ends in an output region
(the axon terminal) where neurotransmitter is released (view
Figure 3.1).

Although the neuron is a single cell surrounded by a
continuous plasma membrane, each region contains distinct
membrane proteins that provide the basis for the functional
differences. Thus, the receiving end has receptor proteins
and proteins that generate the receptor potential, the conducting region has proteins that generate and conduct action
potentials, and the output region has proteins to package
and release neurotransmitters. Membrane proteins are found
throughout the neuronal membrane—many of these proteins
transport ions (see Exercise 1).

The signals generated and conducted by neurons are
electrical. In ordinary household devices, electric current is
carried by electrons. In biological systems, currents are carried by positively or negatively charged **ions.** Like charges
repel each other and opposite charges attract. In general, ions
cannot easily pass through the lipid bilayer of the plasma
membrane and must pass through **ion channels** formed by
integral membrane proteins. Some channels are usually open
(leak channels) and others are gated, meaning that the channel can be in an open or closed configuration. Channels can
also be selective for which ions are allowed to pass. For
example, sodium channels are mostly permeable to sodium
ions when open, and potassium channels are mostly permeable to potassium ions when open. The term **conductance**
is often used to describe **permeability.** In general, ions will
flow through an open channel from a region of higher concentration to a region of lower concentration (see Exercise 1).
In this exercise you will explore some of these characteristics
applied to neurons.

Although it is possible to measure the ionic currents
through the membrane (even the currents passing through single ion channels), it is more common to measure the potential
difference, or voltage, across the membrane. This membrane
voltage is usually called the **membrane potential,** and the
units are **millivolts (mV).** One can think of the membrane as
a battery, a device that separates and stores charge. A typical household battery has a positive and a negative pole so
that when it is connected, for example through a lightbulb
in a flashlight, current flows through the bulb. Similarly, the
plasma membrane can store charge and has a relatively positive side and a relatively negative side. Thus, the membrane
is said to be **polarized.** When these two sides (intracellular
and extracellular) are connected through open ion channels,
current in the form of ions can flow in or out across the membrane and thus change the membrane voltage.

ACTIVITY 1

The Resting Membrane Potential

OBJECTIVES

1. To define the term *resting membrane potential*.
2. To measure the resting membrane potential in different
   parts of a neuron.
3. To determine how the resting membrane potential
   depends on the concentrations of potassium and sodium.
4. To understand the ion conductances/ion channels
   involved in the resting membrane potential.

Introduction

The receptor potential, synaptic potentials, and action potentials are important signals in the nervous system. These
potentials refer to changes in the membrane potential from its
resting level. In this activity you will explore the nature of the
resting potential. The **resting membrane potential** is really
a potential difference between the inside of the cell (intracellular) and the outside of the cell (extracellular) across
the membrane. It is a steady-state condition that depends on
the resting permeability of the membrane to ions and on the
intracellular and extracellular concentrations of those ions to
which the membrane is permeable.

For many neurons, Na^+ and K^+ are the most important
ions, and the concentrations of these ions are established by
transport proteins, such as the Na^+-K^+ pump, so that the
intracellular Na^+ concentration is low and the intracellular
K^+ concentration is high. Inside a typical cell, the concentration of K^+ is ~150 mM and the concentration of Na^+ is ~5
mM. Outside a typical cell, the concentration of K^+ is ~5 mM
and the concentration of Na^+ is ~150 mM. If the membrane
is permeable to a particular ion, that ion will diffuse down
its concentration gradient from a region of higher concentration to a region of lower concentration. In the generation of
the resting membrane potential, K^+ ions diffuse out across
the membrane, leaving behind a net negative charge—large
anions that cannot cross the membrane.

The membrane potential can be measured with an amplifier. In the experiment the extracellular solution is connected
to a ground (literally, the earth) which is defined as 0 mV. To
record the voltage across the membrane, a microelectrode is
inserted through the membrane without significantly damaging it. Typically, the microelectrode is made by pulling a thin
glass pipette to a fine hollow point and filling the pulled pipette
with a salt solution. The salt solution conducts electricity like
a wire, and the glass insulates it. Only the tip of the microelectrode is inserted through the membrane, and the filled tip of
the microelectrode makes electrical contact with the intracellular solution. A wire connects the microelectrode to the input
of the amplifier so that the amplifier records the membrane
potential, the voltage across the membrane between the intracellular and grounded extracellular solutions.

The membrane potential and the various signals can be
observed on an oscilloscope. An electron beam is pulled up
or down according to the voltage as it sweeps across a phos-
phorescent screen. Voltages below 0 mV are negative and
voltages above 0 mV are positive. For this first activity, the
time of the sweep is set for 1 second per division, and the sensitivity is set to 10 mV per division; a division is the distance
between gridlines on the oscilloscope.

Experiment Instructions

Go to the home page in the PhysioEx software, and click **Exercise 3: Neurophysiology of Nerve Impulses.** Click **Activity 1: The Resting Membrane Potential,** and take the online **Pre-lab Quiz** for Activity 1.

After you take the online Pre-lab Quiz, click the **Experiment** tab and begin the experiment. The experiment instructions are reprinted here for your reference. The opening screen for the experiment is shown below.

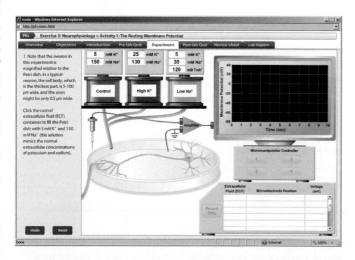

1. Note that the neuron in this experiment is magnified relative to the petri dish. In a typical neuron, the cell body, which is the thickest part, is 5–100 µm wide, and the axon might be only 0.5 µm wide.

Click the **control extracellular fluid (ECF)** container to fill the petri dish with 5 mM K$^+$ and 150 mM Na$^+$ (this solution mimics the normal extracellular concentrations of potassium and sodium).

2. Note that a reference electrode is already positioned in the petri dish. This reference electrode is connected to ground through the amplifier.

Click position **1** on the microelectrode manipulator controller to position the microelectrode tip in the solution, just outside the cell body, and observe the tracing that results on the oscilloscope.

3. Note the oscilloscope tracing of the voltage outside the cell body and click **Record Data** to display your results in the grid (and record your results in Chart 1).

4. Click position **2** on the microelectrode manipulator controller to position the microelectrode tip just inside the cell body and observe the tracing that results.

5. Note the oscilloscope tracing of the voltage inside the cell body and click **Record Data** to display your results in the grid (and record your results in Chart 1). This is the resting membrane potential; that is, the potential difference between intracellular and extracellular membrane voltages. By convention, the extracellular resting membrane voltage is taken to be 0 mV.

6. Click position **3** on the microelectrode manipulator controller to position the microelectrode tip in the solution, just outside the axon, and observe the tracing that results.

7. Note the oscilloscope tracing of the voltage outside the axon and click **Record Data** to display your results in the grid (and record your results in Chart 1).

CHART 1	Resting Membrane Potential	
Extracellular fluid (ECF)	**Microelectrode position**	**Voltage (mV)**

8. Click position **4** on the microelectrode manipulator controller to position the microelectrode tip just inside the axon and observe the tracing that results.

9. Note the oscilloscope tracing of the voltage inside the axon and click **Record Data** to display your results in the grid (and record your results in Chart 1).

> **? PREDICT Question 1**
> Predict what will happen to the resting membrane potential if the extracellular K^+ concentration is increased.
>
> _____
>
> _____

10. You will now change the concentrations of the ions in the extracellular fluid to determine which ions contribute most to the separation of charge across the membrane. The extracellular potassium concentration is normally low, so you will first increase the extracellular potassium concentration.

In the high K^+ ECF solution the K^+ concentration has been increased fivefold, from 5 to 25 mM. To keep the number of positive charges in the extracellular solution constant, the Na^+ concentration has been reduced by 20 mM, from 150 to 130 mM. As you will see, this relatively small decrease in Na^+ will not by itself change the membrane potential. Note that in this activity, the generation of the action potential (which is covered in Activities 3–9) is blocked with a toxin. Click the **high K^+ ECF** container to change the solution in the petri dish to 25 mM K^+ and 130 mM Na^+.

11. Note the voltage inside the axon and click **Record Data** to display your results in the grid (and record your results in Chart 1).

12. Click position **3** on the microelectrode manipulator controller to position the microelectrode tip in the solution, just outside the axon, and observe the tracing that results.

13. Note the voltage outside the axon and click **Record Data** to display your results in the grid (and record your results in Chart 1).

14. Click position **1** on the microelectrode manipulator controller to position the microelectrode tip in the solution, just outside the cell body, and observe the tracing that results.

15. Note the voltage outside the cell body and click **Record Data** to display your results in the grid (and record your results in Chart 1).

16. Click position **2** on the microelectrode manipulator controller to position the microelectrode tip just inside the cell body and observe the tracing that results on the oscilloscope.

17. Note the voltage inside the cell body and click **Record Data** to display your results in the grid (and record your results in Chart 1).

18. Click the *control* ECF container to change back to the normal K^+ concentration and note the change in voltage inside the cell body.

19. You will now decrease the extracellular Na^+ concentration (the extracellular Na^+ concentration is normally high).

The extracellular sodium concentration in the low Na^+ solution has been decreased fivefold, from 150 mM to 30 mM. To keep the number of positive charges constant in the extracellular solution, the Na^+ has been replaced by the same amount of a large monovalent cation. Note that the extracellular Na^+ concentration, even in the low Na^+ ECF, is higher than the intracellular Na^+ concentration. Click the **low Na^+ ECF** container to change the solution in the petri dish to 5 mM K^+ and 30 mM Na^+.

20. Note the voltage inside the cell body and click **Record Data** to display your results in the grid (and record your results in Chart 1).

21. Click position **1** on the microelectrode manipulator controller to position the microelectrode tip in the solution, just outside the cell body, and observe the tracing that results.

22. Note the voltage outside the cell body and click **Record Data** to display your results in the grid (and record your results in Chart 1).

23. Click position **3** on the microelectrode manipulator controller to position the microelectrode tip in the solution, just outside the axon, and observe the tracing that results.

24. Note the voltage outside the axon and click **Record Data** to display your results in the grid (and record your results in Chart 1).

25. Click position **4** on the microelectrode manipulator controller to position the microelectrode tip just inside the axon and observe the tracing that results on the oscilloscope.

26. Note the voltage inside the axon and click **Record Data** to display your results in the grid (and record your results in Chart 1).

After you complete the experiment, take the online **Post-lab Quiz** for Activity 1.

Activity Questions

1. Explain why the resting membrane potential had the same value in the cell body and in the axon.

_____

_____

2. Describe what would happen to a resting membrane potential if the sodium-potassium transport pump was blocked.

_____

_____

3. Describe what would happen to a resting membrane potential if the concentration of large intracellular anions that are unable to cross the membrane is experimentally increased.

_____

_____ ▬

Receptor Potential

OBJECTIVES

1. To define the terms *sensory receptor, receptor potential, sensory transduction, stimulus modality,* and *depolarization.*
2. To determine the *adequate stimulus* for different sensory receptors.
3. To demonstrate that the receptor potential amplitude increases with stimulus intensity.

Introduction

The receiving end of a sensory neuron, the **sensory receptor,** has receptor proteins (as well as other membrane proteins) that can generate a signal called the **receptor potential** when the sensory neuron is stimulated by an appropriate, adequate stimulus. In this activity you will use the same recording instruments and microelectrode that you used in Activity 1. However, in this activity, you will record from the sensory receptor of three different sensory neurons and examine how these neurons respond to sensory stimuli of different modalities.

The sensory region will be shown disconnected from the rest of the neuron so that you can record the receptor potential in isolation. Similar results can sometimes be obtained by treating a whole neuron with chemicals that block the responses generated by the axon. The molecules localized to the sensory receptor ending are able to generate a receptor potential when an adequate stimulus is applied. The energy in the stimulus (for example, chemical, physical, or heat) is changed into an electrical response that involves the opening or closing of membrane ion channels. The general process that produces this change is called **sensory transduction,** which occurs at the receptor ending of the sensory neuron. Sensory transduction can be thought of as a type of signal transduction where the signal is the sensory stimulus.

You will observe that, with an appropriate stimulus, the amplitude of the receptor potential increases with stimulus intensity. Such a response is an example of a potential that is graded with stimulus intensity. These responses are sometimes referred to as *graded potentials,* or *local potentials.* Thus, the receptor potential is a graded, or local, potential. If the response (receptor potential) is a change in membrane potential from the negative resting potential to a less negative level, the membrane becomes less polarized and the change is called **depolarization.**

EQUIPMENT USED The following equipment will be depicted on-screen: three sensory receptors—Pacinian (lamellar) corpuscle, olfactory receptor, and free nerve ending; microelectrode—a probe with a very small tip that can impale a single neuron (In an actual wet lab, a microelectrode manipulator is used to position the microelectrodes. For simplicity, the microelectrode manipulator will not be depicted in this activity.); microelectrode amplifier—used to measure the voltage between the microelectrode and a reference; stimulator—used to select the stimulus modality (pressure, chemical, heat, or light) and intensity (low, moderate, or high); oscilloscope—used to observe voltage changes.

Experiment Instructions

Go to the home page in the PhysioEx software, and click **Exercise 3: Neurophysiology of Nerve Impulses.** Click **Activity 2: Receptor Potential,** and take the online **Pre-lab Quiz** for Activity 2.

After you take the online Pre-lab Quiz, click the **Experiment** tab and begin the experiment. The experiment instructions are reprinted here for your reference. The opening screen for the experiment is shown below.

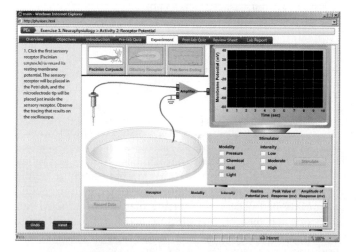

1. Note that the timescale on the oscilloscope has been changed from 1 second per division to 10 milliseconds per division, so that you can observe the responses recorded in the sensory receptors more clearly. Click the first sensory receptor (Pacinian corpuscle) to record its resting membrane potential. The sensory receptor will be placed in the petri dish, and the microelectrode tip will be placed just inside the sensory receptor. Observe the tracing that results on the oscilloscope.

2. Note the voltage inside the sensory receptor and click **Record Data** to display your results in the grid (and record your results in Chart 2).

? PREDICT Question 1
The adequate stimulus for a Pacinian corpuscle is pressure or vibration on the skin. For a Pacinian corpuscle, which modality will induce a receptor potential of the largest amplitude?

CHART 2	Receptor Potential		
		Receptor potential (mV)	
Stimulus modality	Pacinian (lamellar) corpuscle	Olfactory receptor	Free nerve ending
None			
Pressure			
Low			
Moderate			
High			
Chemical			
Low			
Moderate			
High			
Heat			
Low			
Moderate			
High			
Light			
Low			
Moderate			
High			

3. You will now observe how the sensory receptor responds to different sensory stimuli. On the stimulator, click the **Pressure** modality. Click **Low** intensity and then click **Stimulate** to stimulate the sensory receptor and observe the tracing that results. Click **Moderate** intensity and then click **Stimulate** and observe the tracing that results. Click **High** intensity and then click **Stimulate** and observe the tracing that results. Click **Record Data** to display your results in the grid (and record your results in Chart 2).

4. On the stimulator, click the **Chemical** (odor) modality. Click **Low** intensity and then click **Stimulate** to stimulate the sensory receptor and observe the tracing that results. Click **Moderate** intensity and then click **Stimulate** and observe the tracing that results. Click **High** intensity and then click **Stimulate** and observe the tracing that results. Click **Record Data** to display your results in the grid (and record your results in Chart 2).

5. On the stimulator, click the **Heat** modality. Click **Low** intensity and then click **Stimulate** to stimulate the sensory receptor and observe the tracing that results. Click **Moderate** intensity and then click **Stimulate** and observe the tracing that results. Click **High** intensity and then click **Stimulate** and observe the tracing that results. Click **Record Data** to display your results in the grid (and record your results in Chart 2).

6. On the stimulator, click the **Light** modality. Click **Low** intensity and then click **Stimulate** to stimulate the sensory receptor and observe the tracing that results. Click **Moder-ate** intensity and then click **Stimulate** and observe the tracing that results. Click **High** intensity and then click **Stimulate** and observe the tracing that results. Click **Record Data** to display your results in the grid (and record your results in Chart 2).

? PREDICT Question 2
The adequate stimuli for olfactory receptors are chemicals, typically odorant molecules. For an olfactory receptor, which modality will induce a receptor potential of the largest amplitude?

7–12. Repeat steps 1–6 with the next sensory receptor: olfactory receptor.

13–18. Repeat steps 1–6 with the next sensory receptor: free nerve ending.

After you complete the experiment, take the online **Post-lab Quiz** for Activity 2.

Activity Questions

1. Are graded receptor potentials always depolarizing? Do graded receptor potentials always make it easier to induce action potentials?

2. Based on the definition of membrane depolarization in this activity, define membrane *hyperpolarization*.

_____

_____

3. What do you think is the adequate stimulus for sensory receptors in the ear? Can you think of a stimulus that would inappropriately activate the sensory receptors in the ear if the stimulus had enough intensity?

_____

_____ ▬

ACTIVITY 3

The Action Potential: Threshold

OBJECTIVES

1. To define the terms *action potential, nerve, axon hillock, trigger zone,* and *threshold.*
2. To predict how an increase in extracellular K^+ could trigger an action potential.

Introduction

In this activity you will explore changes in potential that occur in the axon. Axons are long, thin structures that conduct a signal called the **action potential.** A **nerve** is a bundle of axons.

Axons are typically studied in a nerve chamber. In this activity the axon will be draped over wires that make electrical contact with the axon and can therefore record the electrical activity in the axon. Because the axon is so thin, it is very difficult to insert an electrode across the membrane into the axon. However, some of the charge (ions) that crosses the membrane to generate the action potential can be recorded from outside the membrane (extracellular recording), as you will do in this activity. The molecular mechanisms underlying the action potential were explored more than 50 years ago with intracellular recording using the giant axons of the squid, which are about 1 millimeter in diameter.

In this activity the axon will be artificially disconnected from the cell body and dendrites. In a typical multipolar neuron (view Figure 3.1 in the Exercise Overview), the axon extends from the cell body at a region called the **axon hillock.** In a myelinated axon, this first region is called the initial segment. An action potential is usually initiated at the junction of the axon hillock and the initial segment; therefore, this region is also referred to as the **trigger zone.**

You will use an electrical stimulator to explore the properties of the action potential. Current passes from the stimulator to one of the stimulation wires, then across the axon, and then back to the stimulator through a second wire. This current will depolarize the axon. Normally, in a sensory neuron, the depolarizing receptor potential spreads passively to the axon hillock and produces the depolarization needed to evoke the action potential. Once an action potential is generated, it is regenerated down the membrane of the axon. In other words, the action potential is **propagated,** or *conducted,* down the axon (see Activity 6).

You will now generate an action potential at one end of the axon by stimulating it electrically and record the action potential that is propagated down the axon. The extracellular action potential that you record is similar to one that would be recorded across the membrane with an intracellular microelectrode, but much smaller. For simplicity, only one axon is depicted in this activity.

> **EQUIPMENT USED** The following equipment will be depicted on-screen: nerve chamber; axon; oscilloscope—used to observe timing of stimuli and voltage changes in the axon; stimulator—used to set the stimulus voltage and to deliver pulses that depolarize the axon; stimulation wires (S); recording electrodes (wires R1 and R2)—used to record voltage changes in the axon. (The first set of recording electrodes, R1, is 2 centimeters from the stimulation wires, and the second set of recording electrodes, R2, is 2 centimeters from R1.)

Experiment Instructions

Go to the home page in the PhysioEx software, and click **Exercise 3: Neurophysiology of Nerve Impulses.** Click **Activity 3: The Action Potential: Threshold,** and take the online **Pre-lab Quiz** for Activity 3.

After you take the online Pre-lab Quiz, click the **Experiment** tab and begin the experiment. The experiment instructions are reprinted here for your reference. The opening screen for the experiment is shown below.

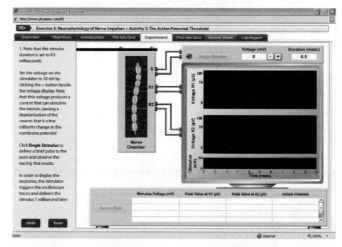

1. Note that the stimulus duration is set to 0.5 milliseconds. Set the voltage on the stimulator to 10 mV by clicking the **+** button beside the voltage display. Note that this voltage produces a current that can stimulate the neuron, causing a depolarization of the neuron that is a change of a few millivolts in the membrane potential.

Click **Single Stimulus** to deliver a brief pulse to the axon and observe the tracing that results. In order to display the response, the stimulator triggers the oscilloscope traces and delivers the stimulus 1 millisecond later.

2. Note that the recording electrodes R1 and R2 record the extracellular voltage, rather than the actual membrane potential. The 10 mV depolarization at the site of stimulation only occurs locally at that site and is not recorded farther down the axon. At this initial stimulus voltage, there was no action potential. Click **Record Data** to display your results in the grid (and record your results in Chart 3, p. PEx-42).

CHART 3	Threshold		
Stimulus voltage (mV)	Peak value at R1 (μV)	Peak value at R2 (μV)	Action potential

3. You will increase the stimulus voltage until you observe an action potential at recording electrode 1 (R1). Increase the voltage by 10 mV by clicking the **+** button beside the voltage display and then click **Single Stimulus.** The voltage at which you first observe an action potential is the **threshold voltage.** Note that the action potential recorded extracellularly is quite small. Intracellularly, the membrane potential would change from −70 mV to about +30 mV. Click **Record Data** to display your results in the grid (and record your results in Chart 3).

> **?** PREDICT Question 1
> How will the action potential at R1 (or R2) change as you continue to increase the stimulus voltage?

4. You will now continue to observe the effects of incremental increases of the stimulus voltage. Increase the voltage by 10 mV by clicking the **+** button beside the voltage display and then click **Single Stimulus.** Repeat this step until you reach the maximum voltage the stimulator can deliver.

Repeat this step until you stimulate the axon at 50 mV and then click **Record Data** to display your results in the grid (and record your results in Chart 3).

After you complete the experiment, take the online **Post-lab Quiz** for Activity 3.

Activity Questions

1. Explain why the threshold voltage is not always the same value (between axons and within an axon).

2. Describe how the action potential is regenerated by local ion flux at each location on the axon.

3. Why doesn't the peak value of the action potential increase with stronger stimuli?

The Action Potential: Importance of Voltage-Gated Na^+ Channels

OBJECTIVES

1. To define the term *voltage-gated channel*.
2. To describe the effect of tetrodotoxin on the voltage-gated Na^+ channel.
3. To describe the effect of lidocaine on the voltage-gated Na^+ channel.
4. To examine the effects of tetrodotoxin and lidocaine on the action potential.
5. To predict the effect of lidocaine on pain perception and to predict the site of action in the sensory neurons (nociceptors) that sense pain.

Introduction

The action potential (as seen in Activity 3) is generated when voltage-gated sodium channels open in sufficient numbers. **Voltage-gated sodium channels** open when the membrane depolarizes. Each sodium channel that opens allows Na^+ ions to diffuse into the cell down their electrochemical gradient. When enough sodium channels open so that the amount of sodium ions that enters via these voltage-gated channels overcomes the leak of potassium ions (recall that the potassium leak via passive channels establishes and maintains the negative resting membrane potential), threshold for the action potential is reached, and an action potential is generated.

In this activity you will observe what happens when these voltage-gated sodium channels are blocked with chemicals. One such chemical is tetrodotoxin (TTX), a toxin found in puffer fish, which is extremely poisonous. Another such chemical is lidocaine, which is typically used to block pain in dentistry and minor surgery.

> **EQUIPMENT USED** The following equipment will be depicted on-screen: nerve chamber; axon; oscilloscope—used to observe timing of stimuli and voltage changes in the axon; stimulator—used to set the stimulus voltage and the interval between stimuli and to deliver pulses that depolarize the axon; stimulation wires (S); recording electrodes (wires R1 and R2)—used to record voltage changes in the axon (The first set of recording electrodes, R1, is 2 centimeters from the stimulation wires, and the second set of recording electrodes, R2, is 2 centimeters from R1.); tetrodotoxin (TTX); lidocaine.

Experiment Instructions

Go to the home page in the PhysioEx software and click **Exercise 3: Neurophysiology of Nerve Impulses.** Click **Activity 4: The Action Potential: Importance of**

Voltage-Gated Na⁺ Channels, and take the online **Pre-lab Quiz** for Activity 4.

After you take the online Pre-lab Quiz, click the **Experiment** tab and begin the experiment. The experiment instructions are reprinted here for your reference. The opening screen for the experiment is shown below.

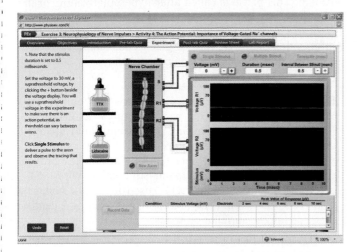

1. Note that the stimulus duration is set to 0.5 milliseconds. Set the voltage to 30 mV, a suprathreshold voltage, by clicking the + button beside the voltage display. You will use a suprathreshold voltage in this experiment to make sure there is an action potential, as threshold can vary between axons. Click **Single Stimulus** to deliver a pulse to the axon and observe the tracing that results.

2. Enter the peak value of the response at R1 and R2 in the field below and then click **Submit** to record your answer in the lab report. _____ μV

3. Click **Timescale** on the stimulator to change the timescale on the oscilloscope from milliseconds to seconds.

4. You will now deliver successive stimuli separated by 2.0-second intervals to observe what the control action potentials look like at this timescale. Set the interval between stimuli to 2.0 seconds by clicking the + button beside the "Interval between Stimuli" display. Click **Multiple Stimuli** to deliver pulses to the axon every 2 seconds. The stimuli will be stopped after 10 seconds.

5. Note the peak values of the responses at R1 and R2 and click **Record Data** to display your results in the grid (and record your results in Chart 4).

? PREDICT Question 1
If you apply TTX between recording electrodes R1 and R2, what effect will the TTX have on the action potentials at R1 and R2?

_____

_____

6. Drag the dropper cap of the TTX bottle to the axon between recording electrodes R1 and R2 to apply a drop of TTX to the axon.

7. Click **Multiple Stimuli** to deliver pulses to the axon every 2 seconds. The stimuli will be stopped after 10 seconds.

8. Note the peak values of the responses at R1 and R2 and click **Record Data** to display your results in the grid (and record your results in Chart 4).

9. Click **New Axon** to select a new axon. TTX is irreversible and there is no known antidote for TTX poisoning.

? PREDICT Question 2
If you apply lidocaine between recording electrodes R1 and R2, what effect will the lidocaine have on the action potentials at R1 and R2?

_____

_____

10. Drag the dropper cap of the lidocaine bottle to the axon between recording electrodes R1 and R2 to apply a drop of lidocaine to the axon.

11. Set the interval between stimuli to 2.0 seconds by clicking the + button beside the "Interval between Stimuli" display. Click **Multiple Stimuli** to deliver pulses to the axon every 2 seconds. The stimuli will be stopped after 10 seconds.

CHART 4	Effects of Tetrodotoxin and Lidocaine						
			Peak value of response (μV)				
Condition	**Stimulus voltage (mV)**	**Electrodes**	**2 sec**	**4 sec**	**6 sec**	**8 sec**	**10 sec**

12. Note the peak values of the responses at R1 and R2. For simplicity, this experiment was performed on a single axon, where the action potential is an "all-or-none" event. If you had treated a bundle of axons (a nerve), each with a slightly different threshold and sensitivity to the drugs, you would likely see the peak values of the action potentials decrease more gradually as more and more axons were blocked. Click **Record Data** to display your results in the grid (and record your results in Chart 4).

After you complete the experiment, take the online **Post-lab Quiz** for Activity 4.

Activity Questions

1. If depolarizing membrane potentials open voltage-gated sodium channels, what closes them?

2. Why must a sushi chef go through years of training to prepare puffer fish for human consumption?

3. For action potential generation and propagation, are there any other cation channels that could substitute for the voltage-gated sodium channels if the sodium channels were blocked?

ACTIVITY 5

The Action Potential: Measuring Its Absolute and Relative Refractory Periods

OBJECTIVES

1. To define *inactivation* as it applies to a voltage-gated sodium channel.
2. To define the *absolute refractory period* and *relative refractory period* of an action potential.
3. To define the relationship between stimulus frequency and the generation of action potentials.

Introduction

Voltage-gated sodium channels in the plasma membrane of an excitable cell open when the membrane depolarizes. About 1–2 milliseconds later, these same channels inactivate, meaning they no longer allow sodium to go through the channel. These inactivated channels cannot be reopened by depolarization for an additional period of time (usually many milliseconds). Thus, during this time, fewer sodium channels can be opened. There are also voltage-gated potassium channels that open during the action potential. These potassium channels open more slowly. They contribute to the repolarization of the action potential from its peak, as more potassium flows out through this second type of potassium channel (recall

there are also passive potassium channels that let potassium leak out, and these leak channels are always open). The flux through extra voltage-gated potassium channels opposes the depolarization of the membrane to threshold, and it also causes the membrane potential to become transiently more negative than the resting potential at the end of an action potential. This phase is called after-hyperpolarization, or the undershoot.

In this activity you will explore what consequences the conformation states of voltage-gated channels have for the generation of subsequent action potentials.

EQUIPMENT USED The following equipment will be depicted on-screen: nerve chamber; axon; oscilloscope—used to observe timing of stimuli and voltage changes in the axon; stimulator—used to set the stimulus voltage and the interval between stimuli and to deliver pulses that depolarize the axon; stimulation wires (S); recording electrode (wires R1)—used to record voltage changes in the axon. (The recording electrode is 2 centimeters from the stimulation wires.)

Experiment Instructions

Go to the home page in the PhysioEx software and click **Exercise 3: Neurophysiology of Nerve Impulses.** Click **Activity 5: The Action Potential: Measuring Its Absolute and Relative Refractory Periods,** and take the online **Pre-lab Quiz** for Activity 5.

After you take the online Pre-lab Quiz, click the **Experiment** tab and begin the experiment. The experiment instructions are reprinted here for your reference. The opening screen for the experiment is shown below.

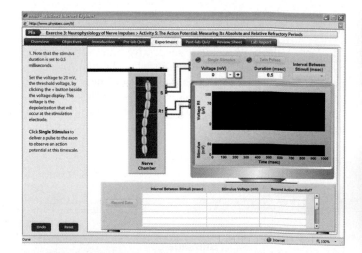

1. Note that the stimulus duration is set to 0.5 milliseconds. Set the voltage to 20 mV, the threshold voltage, by clicking the **+** button beside the voltage display. This voltage is the depolarization that will occur at the stimulation electrode. Click **Single Stimulus** to deliver a pulse to observe an action potential at this timescale.

2. You will now deliver two successive stimuli separated by 250 milliseconds. Set the interval between stimuli to 250 milliseconds by selecting 250 in the "Interval between Stimuli"

pull-down menu. Click **Twin Pulses** to deliver two pulses to the axon and observe the tracing that results. Click **Record Data** to display your results in the grid (and record your results in Chart 5).

CHART 5	Absolute and Relative Refractory Periods	
Interval between stimuli (msec)	Stimulus voltage (mV)	Second action potential?

3. Decrease the interval between stimuli to 125 milliseconds by selecting 125 in the "Interval between Stimuli" pull-down menu. Click **Twin Pulses** to deliver two pulses to the axon and observe the tracing that results. Click **Record Data** to display your results in the grid (and record your results in Chart 5).

4. Decrease the interval between stimuli to 60 milliseconds by selecting 60 in the "Interval between Stimuli" pull-down menu. Click **Twin Pulses** to deliver two pulses to the axon and observe the tracing that results.

Note that, at this stimulus interval, the second stimulus did not generate an action potential. Click **Record Data** to display your results in the grid (and record your results in Chart 5).

5. A second action potential can be generated at this stimulus interval, but the stimulus intensity must be increased. This interval is part of the relative refractory period, the time after an action potential when a second action potential can be generated if the stimulus intensity is increased.

Increase the stimulus intensity by 5 mV by clicking the + button beside the voltage display and then click **Twin Pulses** to deliver two pulses to the axon. Repeat this step until you generate a second action potential. After you generate a second action potential, click **Record Data** to display your results in the grid (and record your results in Chart 5).

If you further decrease the interval between the stimuli, will the threshold for the second action potential change?

_____

_____

6. You will now decrease the interval until the second action potential fails again. (So that you can clearly observe two action potentials at the shorter interval between stimuli, the timescale on the oscilloscope has been set to 10 msec per division.) Decrease the interval between stimuli by 50% and then click **Twin Pulses** to deliver two pulses to the axon. When the second action potential fails, click **Record Data** to display your results in the grid (and record your results in Chart 5).

7. You will now increase the stimulus intensity until a second action potential is generated again. Increase the stimulus intensity by 5 mV by clicking the + button beside the voltage display and then click **Twin Pulses** to deliver two pulses to the axon. Repeat this step until you generate a second action potential. After you generate a second action potential, click **Record Data** to display your results in the grid (and record your results in Chart 5).

8. You will now determine the interval between stimuli at which a second action potential cannot be generated, no matter how intense the stimulus. Increase the stimulus intensity to 60 mV (the highest voltage on the stimulator). Decrease the interval between stimuli by 50% and then click **Twin Pulses** to deliver two pulses to the axon. Repeat this step until the second action potential fails.

The interval at which the second action potential fails is the **absolute refractory period**, the time after an action potential when the neuron cannot fire a second action potential, no matter how intense the stimulus. Click **Record Data** to display your results in the grid (and record your results in Chart 5).

After you complete the experiment, take the online **Post-lab Quiz** for Activity 5.

Activity Questions

1. Explain how the absolute refractory period ensures directionality of action potential propagation.

_____

_____

2. Some tissues (for example, cardiac muscle) have long absolute refractory periods. Why would this be beneficial?

_____

_____

3. What do you think is the benefit of a relative refractory period in an axon of a sensory neuron?

_____

ACTIVITY 6

The Action Potential: Coding for Stimulus Intensity

OBJECTIVES

1. To observe the response of axons to longer periods of stimulation.
2. To examine the relationship between stimulus intensity and the frequency of action potentials.

Introduction

As seen in Activity 3, the action potential has a constant amplitude, regardless of the stimulus intensity—it is an "all-or-none" event. As seen in Activity 5, the absolute refractory period is the time after an action potential when the neuron cannot fire a second action potential, no matter how intense the stimulus, and the relative refractory period is the time after an action potential when a second action potential can be generated if the stimulus intensity is increased.

In this activity you will use these concepts to begin to explore how the axon codes the stimulus intensity as *frequency,* the number of events (in this case, action potentials) per unit time. To demonstrate this phenomenon you will use longer periods of stimulation that are more representative of real-life stimuli. For example, when you encounter an odor, the odor is normally present for seconds (or longer), unlike the very brief stimuli used in Activities 3–5. These longer stimuli allow the axon of the neuron to generate additional action potentials as soon as it has recovered from the first. As seen in Activity 5, the length of this recovery period changes depending on the stimulus intensity. For example, at threshold, a second action potential can occur only after the axon has recovered from the absolute refractory period and the entire relative refractory period.

We will not consider the phenomenon of adaptation, which is a decrease in the response amplitude that often occurs with prolonged stimuli. For example, with most odors, after many seconds, you no longer smell the odor, even though it is still present. This decrease in response is due to adaptation.

EQUIPMENT USED The following equipment will be depicted on-screen: nerve chamber; axon; oscilloscope—used to observe timing of stimuli and voltage changes in the axon; stimulator—used to set the voltage and duration of stimuli and to deliver pulses that depolarize the axon; stimulation wires (S); recording electrode (wires R1)—used to record voltage changes in the axon. (The recording electrode is 2 centimeters from the stimulation wires.)

Experiment Instructions

Go to the home page in the PhysioEx software and click **Exercise 3: Neurophysiology of Nerve Impulses.** Click **Activity 6: The Action Potential: Coding for Stimulus Intensity,** and take the online **Pre-lab Quiz** for Activity 6.

After you take the online Pre-lab Quiz, click the **Experiment** tab and begin the experiment. The experiment instructions are reprinted here for your reference. The opening screen for the experiment is shown below.

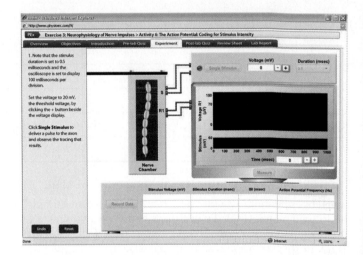

1. Note that the stimulus duration is set to 0.5 milliseconds and the oscilloscope is set to display 100 milliseconds per division. Set the voltage to 20 mV, the threshold voltage, by clicking the + button beside the voltage display. Click **Single Stimulus** to deliver a pulse to the axon and observe the tracing that results.

2. Note how the action potential looks at this timescale and click **Record Data** to display your results in the grid (and record your results in Chart 6).

CHART 6	Frequency of Action Potentials		
Stimulus voltage (mV)	Stimulus duration (msec)	ISI (msec)	Action potential frequency (Hz)

3. Increase the stimulus duration to 500 milliseconds by selecting 500 from the duration pull-down menu. Click **Single Stimulus** to deliver a pulse to the axon and observe the tracing that results. The stimulus is delivered after a delay of 100 milliseconds so that you can easily see the timing of the stimulus.

4. At the site of stimulation, the stimulus keeps the membrane of the axon at threshold for a long time, but this depolarization does not spread to the recording electrode. After one action potential has been generated and the axon has fully recovered from its absolute and relative refractory periods, the stimulus is still present to generate another action potential.

Measure the time (in milliseconds) between action potentials. This interval should be a bit longer than the relative refractory period (measured in Activity 5). Click **Measure** to help determine the time between action potentials. A thin, vertical yellow line appears at the far left side of the oscilloscope screen. You can move the line in 10-millisecond increments by clicking the + and − buttons beside the time display, which shows the time at the line. Click **Submit** to display your answer in the data table (and record your results in Chart 6).

5. The interval between action potentials is sometimes called the interspike interval (ISI). Action potentials are sometimes referred to as spikes because of their rapid time course. From the ISI, you can calculate the action potential frequency. The frequency is the reciprocal of the interval and is usually expressed in hertz (Hz), which is events (action potentials) per second. From the ISI you entered, calculate the frequency of action potentials with a prolonged (500 msec) threshold stimulus intensity. Frequency = 1/ISI. Click **Submit** to display your answer in the data table (and record your results in Chart 6).

6. A stimulus intensity of 30 mV was able to generate a second action potential toward the end of the relative refractory period in Activity 5. With this stronger stimulus, the second action potential can occur after a shorter time. Increase the stimulus intensity to 30 mV by clicking the + button beside the voltage display. Click **Single Stimulus** to deliver this stronger stimulus and observe the tracing that results.

7. Click **Submit** to display your answer in the data table (and record your results in Chart 6). Click **Measure** to help determine the time between action potentials. A thin, vertical yellow line appears at the far left side of the oscilloscope screen. You can move the line in 10-millisecond increments by clicking the + and − buttons beside the time display, which shows the time at the line.

8. From the ISI you entered, calculate the frequency of action potentials with a prolonged (500 msec) 30-mV stimulus intensity. Frequency = 1/ISI. Click **Submit** to display your answer in the data table (and record your results in Chart 6).

9. A stimulus intensity of 45 mV was able to generate a second action potential in the middle of the relative refractory period in Activity 5. With this even stronger stimulus, the second action potential can occur after an even shorter time. Increase the stimulus intensity to 45 mV.

? PREDICT Question 1
What effect will the increased stimulus intensity have on the frequency of action potentials?

_____

_____

10. Click **Single Stimulus** to deliver the stronger, 45-mV stimulus and observe the tracing that results.

11. Click **Submit** to display your answer in the data table (and record your results in Chart 6). Click **Measure** to help determine the time between action potentials. A thin, vertical yellow line appears at the far left side of the oscilloscope screen. You can move the line in 10-millisecond increments by clicking the + and − buttons beside the time display, which shows the time at the line.

12. From the ISI you entered, calculate the frequency of action potentials with a prolonged (500 msec) 45-mV stimulus intensity. Frequency = 1/ISI. Click **Submit** to display your answer in the data table (and record your results in Chart 6).

After you complete the experiment, take the online **Post-lab Quiz** for Activity 6.

Activity Questions

1. Compare the action potential frequency in a temperature-sensitive sensory neuron exposed to warm water and then hot water.

_____

_____

2. When a long-duration stimulus is applied, what two determinants of an action potential refractory period are being overcome?

_____

_____

3. Suggest several ways to pharmacologically overcome a neuron's refractory period and thereby increase the action potential frequency.

_____

_____

ACTIVITY 7

The Action Potential: Conduction Velocity

OBJECTIVES

1. To define and measure *conduction velocity* for an action potential.

2. To examine the effect of myelination on conduction velocity.

3. To examine the effect of axon diameter on conduction velocity.

Introduction

Once generated, the action potential is propagated, or conducted, down the axon. In other words, all-or-none action potentials are regenerated along the entire length of the axon. This propagation ensures that the amplitude of the action potential does not diminish as it is conducted along the axon. In some cases, such as the sensory neuron traveling from your toe to the spinal cord, the axon can be quite long (in this case, up to 1 meter). Propagation/conduction occurs because there are voltage-gated sodium and potassium channels located along the axon and because the large depolarization that constitutes the action potential (once generated at the trigger zone) easily brings the next region of the axon to threshold. The **conduction velocity** can be easily calculated by knowing both the distance the action potential travels and the amount of time it takes. Velocity has the units of distance per time, typically meters/second. An experimental stimulus artifact (see Activity 3) provides a convenient marker of the stimulus time because it travels very quickly (for our purposes, instantaneously) along the axon.

Several parameters influence the conduction velocity in an axon, including the axon diameter and the amount of myelination. **Myelination** refers to a special wrapping of the membrane from glial cells (or neuroglia) around the axon. In the central nervous system, oligodendrocytes are the glia that wrap around the axon. In the peripheral nervous system, the Schwann cells are the glia that wrap around the axon. Many glial cells along the axon contribute a myelin sheath, and the myelin sheaths are separated by gaps called nodes of Ranvier.

In this activity you will compare the conduction velocities of three axons: (1) a large-diameter, heavily myelinated axon, often called an A fiber (the terms axon and fiber are synonymous), (2) a medium-diameter, lightly myelinated axon (called the B fiber), and (3) a thin, unmyelinated fiber (called the C fiber). Examples of these axon types in the body include the axon of the sensory Pacinian corpuscle (an A fiber), the axon of both the olfactory sensory neuron and a free nerve ending (C fibers), and a visceral sensory fiber (a B fiber).

> **EQUIPMENT USED** The following equipment will be depicted on-screen: nerve chamber; three axons— A fiber, B fiber, and C fiber; oscilloscope—used to observe timing of stimuli and voltage changes in the axon; stimulator—used to set the stimulus voltage and to deliver pulses that depolarize the axon; stimulation wires (S); recording electrodes (wires R1 and R2)—used to record voltage changes in the axon. (The first set of recording electrodes, R1, is 2 centimeters from the stimulation wires, and the second set of recording electrodes, R2, is 2 centimeters from R1.)

Experiment Instructions

Go to the home page in the PhysioEx software and click **Exercise 3: Neurophysiology of Nerve Impulses.** Click **Activity 7: The Action Potential: Conduction Velocity,** and take the online **Pre-lab Quiz** for Activity 7.

After you take the online Pre-lab Quiz, click the **Experiment** tab and begin the experiment. The experiment instructions are reprinted here for your reference. The opening screen for the experiment is shown below.

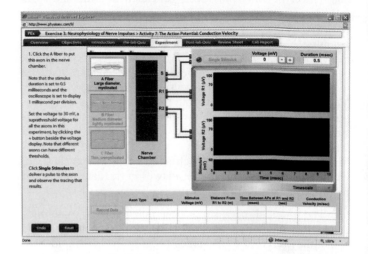

1. Click the A fiber to put this axon in the nerve chamber. Note that the stimulus duration is set to 0.5 milliseconds and the oscilloscope is set to display 1 millisecond per division.

Set the voltage to 30 mV, a suprathreshold voltage for all the axons in this experiment, by clicking the **+** button beside the voltage display. Note that different axons can have different thresholds. Click **Single Stimulus** to deliver a pulse to the axon and observe the tracing that results.

2. Click **Record Data** to display your results in the grid (and record your results in Chart 7).

3. Note the difference in time between the action potential recorded at R1 and the action potential recorded at R2. The distance between these sets of recording electrodes is 10 centimeters (0.1 m). Convert the time from milliseconds to seconds and then click **Submit** to display your results in the grid (and record your results in Chart 7).

4. Calculate the conduction velocity in meters/second by dividing the distance between R1 and R2 (0.1 m) by the time it took for the action potential to travel from R1 to R2. Click

CHART 7	Conduction Velocity					
Axon type	Myelination	Stimulus voltage (mV)	Distance from R1 to R2 (m)	Time between action potentials at R1 and R2		Conduction velocity (m/sec)
				(msec)	(sec)	

Submit to display your results in the grid (and record your results in Chart 7).

> **? PREDICT Question 1**
> How will the conduction velocity in the B fiber compare with that in the A fiber?
>
> _____
>
> _____

5. Click the **B fiber** to put this axon in the nerve chamber. Set the timescale on the oscilloscope to 10 milliseconds per division by selecting 10 in the timescale pull-down menu. Click **Single Stimulus** to deliver a pulse to the axon and observe the tracing that results.

6–8. Repeat steps 2–4 with the B fiber (and record your results in Chart 7).

> **? PREDICT Question 2**
> How will the conduction velocity in the C fiber compare with that in the B fiber?
>
> _____
>
> _____

9. Click the **C fiber** to put this axon in the nerve chamber. Set the timescale on the oscilloscope to 50 milliseconds per division by selecting 50 in the timescale pull-down menu. Click **Single Stimulus** to deliver a pulse to the axon and observe the tracing that results.

10–12. Repeat steps 2–4 with the C fiber (and record your results in Chart 7).

After you complete the experiment, take the online **Post-lab Quiz** for Activity 7.

Activity Questions

1. The squid utilizes a very large-diameter, unmyelinated axon to execute a rapid escape response when it perceives danger. How is this possible, given that the axon is unmyelinated?

_____

_____

2. When you burn your finger on a hot stove, you feel sharp, immediate pain, which later becomes slow, throbbing pain. These two types of pain are carried by different pain axons. Speculate on the axonal diameter and extent of myelination of these axons.

_____

_____

3. Why do humans possess a mixture of axons, some large-diameter, heavily myelinated axons and some small-diameter, relatively unmyelinated axons?

_____

_____

ACTIVITY 8

Chemical Synaptic Transmission and Neurotransmitter Release

OBJECTIVES

1. To define *neurotransmitter, chemical synapse, synaptic vesicle,* and *postsynaptic potential.*
2. To determine the role of calcium ions in neurotransmitter release.

Introduction

A major function of the nervous system is communication. The axon conducts the action potential from one place to another. Often, the axon has branches so that the action potential is conducted to several places at about the same time. At the end of each branch, there is a region called the axon terminal that is specialized to release packets of chemical neurotransmitters from small (~30-nm diameter) intracellular membrane-bound vesicles, called **synaptic vesicles. Neurotransmitters** are extracellular signal molecules that act on local targets as paracrine agents, on the neuron releasing the chemical as autocrine agents, and sometimes as hormones (endocrine agents) that reach their target(s) via the circulation. These chemicals are released by exocytosis and diffuse across a small extracellular space (called the synaptic gap, or synaptic cleft) to the target (most often the receiving end of another neuron or a muscle or gland). The neurotransmitter molecules often bind to membrane receptor proteins on the target, setting in motion a sequence of molecular events that can open or close membrane ion channels and cause the membrane potential in the target cell to change. This region where the neurotransmitter is released from one neuron and binds to a receptor on a target cell is called a **chemical synapse,** and the change in membrane potential of the target is called a synaptic potential, or **postsynaptic potential.**

In this activity you will explore some of the steps in neurotransmitter release from the axon terminal. Exocytosis of synaptic vesicles is normally triggered by an increase in calcium ions in the axon terminal. The calcium enters from outside the cell through membrane calcium channels that are opened by the depolarization of the action potential. The axon terminal has been greatly magnified in this activity so that you can visualize the release of neurotransmitter. Different from the other activities in this exercise, however, this procedure of directly seeing neurotransmitter release is not easily done in the lab; rather, neurotransmitter is usually detected by the postsynaptic potentials it triggers or by collecting and analyzing chemicals at the synapse after robust stimulation of the neurons.

EQUIPMENT USED The following equipment will be depicted on-screen: neuron (in vitro)—a large, dissociated (or cultured) neuron with magnified axon terminals; four extracellular solutions—control Ca^{2+}, no Ca^{2+}, low Ca^{2+}, and Mg^{2+}.

Experiment Instructions

Go to the home page in the PhysioEx software and click **Exercise 3: Neurophysiology of Nerve Impulses.** Click **Activity 8: Chemical Synaptic Transmission and Neurotransmitter Release,** and take the online **Pre-lab Quiz** for Activity 8.

After you take the online Pre-lab Quiz, click the **Experiment** tab and begin the experiment. The experiment instructions are reprinted here for your reference. The opening screen for the experiment is shown below.

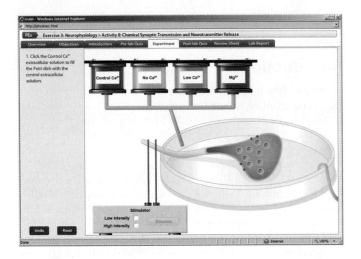

1. Click the **control Ca^{2+}** extracellular solution to fill the petri dish with the control extracellular solution.

2. Click **Low Intensity** on the stimulator and then click **Stimulate** to stimulate the neuron (axon) with a threshold stimulus that generates a low frequency of action potentials. Observe the release of neurotransmitter.

3. Click **High Intensity** on the stimulator and then click **Stimulate** to stimulate the neuron with a longer, more intense stimulus to generate a burst of action potentials. Observe the release of neurotransmitter.

> **? PREDICT Question 1**
> You have just observed that each action potential in a burst can trigger additional neurotransmitter release. If calcium ions are removed from the extracellular solution, what will happen to neurotransmitter release at the nerve terminal?
> _____
> _____

4–6. Repeat steps 1–3 with the *no Ca^{2+}* extracellular solution.

> **? PREDICT Question 2**
> What will happen to the amount of neurotransmitter release when low amounts of calcium are added back to the extracellular solution?
> _____
> _____

7–9. Repeat steps 1–3 with the *low Ca^{2+}* extracellular solution.

> **? PREDICT Question 3**
> What will happen to neurotransmitter release when magnesium is added to the extracellular solution?
> _____
> _____

10–12. Repeat steps 1–3 with the *Mg^{2+}* extracellular solution.

After you complete the experiment, take the online **Post-lab Quiz** for Activity 8.

Activity Questions

1. If you added more sodium to the extracellular solution, could the sodium substitute for the missing calcium?

_____

_____

2. How does botulinum toxin block synaptic transmission? Why is it used for cosmetic procedures?

_____

_____

_____

ACTIVITY 9

The Action Potential: Putting It All Together

OBJECTIVES

1. To identify the functional areas (for example, the sensory ending, axon, and postsynaptic membrane) of a two-neuron circuit.

2. To predict and test the responses in each functional area to a very weak, subthreshold stimulus.

3. To predict and test the responses in each functional area to a moderate stimulus.

4. To predict and test the responses in each functional area to an intense stimulus.

Introduction

In the nervous system, sensory neurons respond to adequate sensory stimuli, generating action potentials in the axon if the stimulus is strong enough to reach threshold (the action potential is an "all-or-nothing" event). Via chemical synapses, these sensory neurons communicate with interneurons that process the information. Interneurons also communicate with motor neurons that stimulate muscles and glands, again, usually via chemical synapses.

After performing Activities 1–8, you should have a better understanding of how neurons function by generating changes from their resting membrane potential. If threshold is reached, an action potential is generated and propagated. If the stimulus is more intense, then action potentials are generated at a higher frequency, causing the release of more neurotransmitter at the next synapse. At an excitatory synapse the chemical neurotransmitter binds to receptors at the receiving end of the next cell (usually the cell body or dendrites of an interneuron), causing ion channels to open, resulting in a depolarization toward threshold for an action potential in the interneuron's axon. This depolarizing synaptic potential (called an excitatory postsynaptic potential) is graded in amplitude, depending on the amount of neurotransmitter and the number of channels that open. In the axon, the amplitude of this synaptic potential is coded as the frequency of action potentials. Neurotransmitters can also cause inhibition, which will not be covered in this activity.

In this activity you will stimulate a sensory neuron, predict the response of that cell and its target, and then test those predictions.

EQUIPMENT USED   The following equipment will be depicted on-screen: neuron (in vitro)—a large, dissociated (or cultured) neuron; interneuron (in vitro)—a large, dissociated (or cultured) interneuron; microelectrodes—small probes with very small tips that can impale a single neuron (In an actual wet lab, a microelectrode manipulator is used to position the microelectrodes. For simplicity, the microelectrode manipulator will not be depicted in this activity.); microelectrode amplifier—used to measure the voltage between the microelectrodes and a reference; oscilloscope—used to observe the changes in voltage across the membrane of the neuron and interneuron; stimulator—used to set the stimulus intensity (low or high) and to deliver pulses to the neuron.

Experiment Instructions

Go to the home page in the PhysioEx software and click **Exercise 3: Neurophysiology of Nerve Impulses.** Click **Activity 9: The Action Potential: Putting It All Together,** and take the online **Pre-lab Quiz** for Activity 9.

After you take the online Pre-lab Quiz, click the **Experiment** tab and begin the experiment. The experiment instructions are reprinted here for your reference. The opening screen for the experiment is shown above.

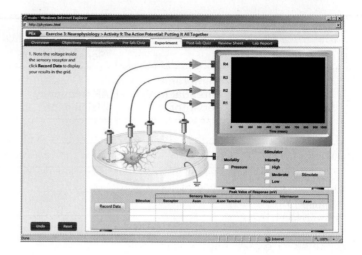

1. Note the membrane potential at the sensory receptor and the receiving end of the interneuron and click **Record Data** to display your results in the grid (and record your results in Chart 9).

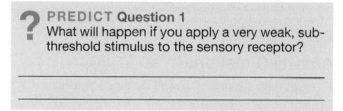

? PREDICT Question 1
What will happen if you apply a very weak, subthreshold stimulus to the sensory receptor?

_____

_____

2. Click **Very Weak** intensity on the stimulator and then click **Stimulate** to stimulate the receiving end of the sensory neuron and observe the tracing that results.

3. Click **Record Data** to display your results in the grid (and record your results in Chart 9). The stimulus lasts 500 msec.

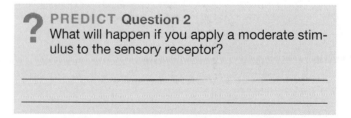

? PREDICT Question 2
What will happen if you apply a moderate stimulus to the sensory receptor?

_____

_____

4. Click **Moderate** intensity on the stimulator and then click **Stimulate** to stimulate the sensory receptor and observe the tracing that results.

5. Click **Record Data** to display your results in the grid (and record your results in Chart 9).

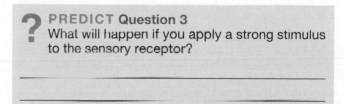

? PREDICT Question 3
What will happen if you apply a strong stimulus to the sensory receptor?

_____

_____

CHART 9	Putting It All Together				
	Sensory neuron			Interneuron	
Stimulus	Membrane Potential (mV) Receptor	AP frequency (Hz) in axon	Vesicles released from axon terminal	Membrane potential (mV) receiving end	AP frequency (Hz) in axon
None					
Weak					
Moderate					
Strong					

6. Click **Strong** intensity on the stimulator and then click **Stimulate** to stimulate the sensory receptor and observe the tracing that results.

7. Click **Record Data** to display your results in the grid (and record your results in Chart 9).

After you complete the experiment, take the online **Post-lab Quiz** for Activity 9.

Activity Questions

1. Why were the peak values of the action potentials at R2 and R4 the same when you applied a strong stimulus?

_____

_____

2. If the axons were unmyelinated, would the peak value of the action potential at R4 change relative to that at R2?

_____

_____

Neurophysiology of Nerve Impulses

NAME _____

LAB TIME/DATE _____

ACTIVITY 1   The Resting Membrane Potential

1. Explain why increasing extracellular K^+ reduces the net diffusion of K^+ out of the neuron through the K^+ leak ~~els.~~

   _____

2. Explain why increasing extracellular K^+ causes the membrane potential to change to a less negative value. How well

   results compare with your prediction? _____

   _____

   _____

3. Explain why a change in extracellular Na^+ did not alter the membrane potential in the resting neuron. _____

   _____

4. Discuss the relative permeability of the membrane to Na^+ and K^+ in a resting neuron. _____

   _____

   _____

5. Discuss how a change in Na^+ or K^+ conductance would affect the resting membrane potential. _____

   _____

ACTIVITY 2   Receptor Potential

1. Sensory neurons have a resting potential based on the efflux of potassium ions (as demonstrated in Activity 1). What passive

   channels are likely found in the membrane of the olfactory receptor, in the membrane of the Pacinian corpuscle, and in the

   membrane of the free nerve ending? _____

   _____

   _____

2. What is meant by the term *graded potential*? _____

   _____

   _____

3. Identify which of the stimulus modalities induced the largest amplitude receptor potential in the Pacinian corpuscle. How well did results compare with your prediction? _____

_____

4. Identify which of the stimulus modalities induced the largest-amplitude receptor potential in the olfactory receptors. How did the results compare with your prediction? _____

_____

The olfactory receptor also contains a membrane protein that recognizes isoamyl acetate and, via several other molecules, transduces the odor stimulus into a receptor potential. Does the Pacinian corpuscle likely have this isoamyl acetate receptor protein? Does the free nerve ending likely have this isoamyl acetate receptor protein? _____

_____

6. What type of sensory neuron would likely respond to a green light? _____

_____

_____

ACTIVITY 3 The Action Potential: Threshold

1. Define the term *threshold* as it applies to an action potential. _____

_____

_____

2. What change in membrane potential (depolarization or hyperpolarization) triggers an action potential? _____

_____

3. How did the action potential at R1 (or R2) change as you increased the stimulus voltage above the threshold voltage? How well did the results compare with your prediction? _____

_____

4. An action potential is an "all-or-nothing" event. Explain what is meant by this phrase. _____

_____

_____

5. What part of a neuron was investigated in this activity? _____

_____

_____

ACTIVITY 4   The Action Potential: Importance of Voltage-Gated Na$^+$ Channels

1. What does TTX do to voltage-gated Na$^+$ channels? _____

_____

_____

2. What does lidocaine do to voltage-gated Na$^+$ channels? How does the effect of lidocaine differ from the effect of TTX?

_____

_____

3. A nerve is a bundle of axons, and some nerves are less sensitive to lidocaine. If a nerve, rather than an axon, had been used in

the lidocaine experiment, the responses recorded at R1 and R2 would be the sum of all the action potentials (called a compound

action potential). Would the response at R2 after lidocaine application necessarily be zero? Why or why not? _____

_____

_____

4. Why are fewer action potentials recorded at R2 when TTX is applied between R1 and R2? How well did the results compare

with your prediction?_____

_____

_____

5. Why are fewer action potentials recorded at R2 when lidocaine is applied between R1 and R2? How well did the results

compare with your prediction? _____

_____

_____

6. Pain-sensitive neurons (called nociceptors) conduct action potentials from the skin or teeth to sites in the brain involved in

pain perception. Where should a dentist inject the lidocaine to block pain perception? _____

_____

_____

ACTIVITY 5   The Action Potential: Measuring Its Absolute and Relative Refractory Periods

1. Define *inactivation* as it applies to a voltage-gated sodium channel. _____

_____

2. Define the *absolute refractory period*. _____

_____

_____

3. How did the threshold for the second action potential change as you further decreased the interval between the stimuli?

How well did the results compare with your prediction? _____

_____

4. Why is it harder to generate a second action potential during the relative refractory period? _____

_____

_____

ACTIVITY 6 The Action Potential: Coding for Stimulus Intensity

1. Why are multiple action potentials generated in response to a long stimulus that is above threshold? _____

_____

_____

2. Why does the frequency of action potentials increase when the stimulus intensity increases? How well did the results

compare with your prediction? _____

_____

3. How does threshold change during the relative refractory period? _____

_____

_____

4. What is the relationship between the interspike interval and the frequency of action potentials? _____

_____

_____

ACTIVITY 7 The Action Potential: Conduction Velocity

1. How did the conduction velocity in the B fiber compare with that in the A fiber? How well did the results compare with your

prediction? _____

_____

_____

2. How did the conduction velocity in the C fiber compare with that in the B fiber? How well did the results compare with your

   prediction? _____

   _____

   _____

3. What is the effect of axon diameter on conduction velocity? _____

   _____

   _____

4. What is the effect of the amount of myelination on conduction velocity? _____

   _____

   _____

5. Why did the time between the stimulation and the action potential at R1 differ for each axon? _____

   _____

6. Why did you need to change the timescale on the oscilloscope for each axon? _____

   _____

   _____

ACTIVITY 8 Chemical Synaptic Transmission and Neurotransmitter Release

1. When the stimulus intensity is increased, what changes: the number of synaptic vesicles released or the amount of

   neurotransmitter per vesicle? _____

   _____

   _____

2. What happened to the amount of neurotransmitter release when you switched from the control extracellular fluid to the

   extracellular fluid with no Ca^{2+}? How well did the results compare with your prediction? _____

   _____

   _____

3. What happened to the amount of neurotransmitter release when you switched from the extracellular fluid with no Ca^{2+} to

   the extracellular fluid with low Ca^{2+}? How well did the results compare with your prediction? _____

   _____

   _____

4. How did neurotransmitter release in the Mg^{2+} extracellular fluid compare to that in the control extracellular fluid? How well did the result compare with your prediction? _____

_____

_____

5. How does Mg^{2+} block the effect of extracellular calcium on neurotransmitter release? _____

_____

_____

ACTIVITY 9   The Action Potential: Putting It All Together

1. Why is the resting membrane potential the same value in both the sensory neuron and the interneuron? _____

_____

_____

2. Describe what happened when you applied a very weak stimulus to the sensory receptor. How well did the results compare with your prediction? _____

_____

_____

3. Describe what happened when you applied a moderate stimulus to the sensory receptor. How well did the results compare with your prediction? _____

_____

_____

4. Identify the type of membrane potential (graded receptor potential or action potential) that occurred at R1, R2, R3, and R4 when you applied a moderate stimulus. (View the response to the stimulus.) _____

_____

_____

5. Describe what happened when you applied a strong stimulus to the sensory receptor. How well did the results compare with your prediction? _____

_____

_____

Endocrine System Physiology

Exercise Overview

In the human body the **endocrine system** (in addition to the nervous system) coordinates and integrates the functions of different physiological systems (view Figure 4.1, p. PEx-60). Thus, the endocrine system plays a critical role in maintaining **homeostasis.** This role begins with chemicals, called **hormones,** secreted from ductless **endocrine glands,** which are tissues that have an epithelial origin. Endocrine glands secrete hormones into the extracellular fluid compartments. More specifically, the blood usually carries hormones (sometimes attached to specific plasma proteins) to their **target cells.** Target cells can be very close to, or very far from, the source of the hormone.

Hormones bind to high-affinity **receptors** located on the target cell's surface, in its cytosol, or in its nucleus. These hormone receptors have remarkable **sensitivity,** as the hormone concentration in the blood can range from 10^{-9} to 10^{-12} molar! A hormone-receptor complex forms and can then exert a **biological action** through signal-transduction cascades and alteration of gene transcription at the target cell. The physiological response to hormones can vary from seconds to hours to days, depending on the chemical nature of the hormone and its receptor location in the target cell.

The chemical structure of the hormone is important in determining how it will interact with target cells. *Peptide* and *catecholamine hormones* are fast-acting hormones that attach to a plasma-membrane receptor and cause a second-messenger cascade in the cytoplasm of the target cell. For example, a chemical called cAMP (cyclic adenosine monophosphate) is synthesized from a molecule of ATP. The

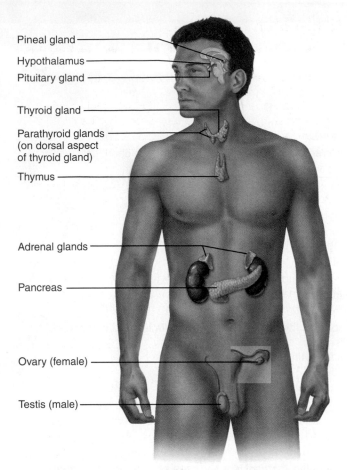

Pineal gland

Hypothalamus

Pituitary gland

Thyroid gland

Parathyroid glands
(on dorsal aspect
of thyroid gland)

Thymus

Adrenal glands

Pancreas

Ovary (female)

Testis (male)

FIGURE 4.1 Selected endocrine organs of the body.

synthesis of this chemical makes the cell more metabolically active and, therefore, more able to respond to a stimulus.

Steroid hormones and *thyroxine* (thyroid hormone) are slow-acting hormones that enter the target cell and interact with the nucleus to affect the transcription of various proteins that the cell can synthesize. The hormones enter the nucleus and attach at specific points on the DNA. Each attachment causes the production of a specific mRNA, which is then moved to the cytoplasm, where ribosomes can translate the mRNA into a protein.

Keep in mind that the organs of the endocrine system do not function independently. The activities of one endocrine gland are often coordinated with the activities of other glands. No one system functions independently of any other system. For this reason, we will be stressing feedback mechanisms and how we can use them to predict, explain, and understand hormone effects.

Given the powerful influence that hormones have on homeostasis, **negative feedback mechanisms** are important in regulating hormone secretion, synthesis, and effectiveness at target cells. Negative feedback ensures that if the body needs a particular hormone, that hormone will be produced until there is too much of it. When there is too much of the hormone, its release will be inhibited.

Rarely, the body regulates hormones via a *positive feedback mechanism.* The release of *oxytocin* from the posterior pituitary is one of these rare instances. Oxytocin is a hormone that causes the muscle layer of the uterus, called the *myometrium,* to contract during childbirth. This contraction of the myometrium causes additional oxytocin to be released,

allowing stronger contractions. Unlike what happens in negative feedback mechanisms, the increase in circulating levels of oxytocin does not inhibit oxytocin secretion.

Many experimental methods can be used to study the functions of an endocrine gland. These methods include removing the gland from an animal and then injecting, implanting, or feeding glandular extracts into a normal animal or an animal deprived of the gland being studied. In this exercise you will use these methods to gain a deeper understanding of the *function* and *regulation* of some of the endocrine glands.

ACTIVITY 1

Metabolism and Thyroid Hormone

OBJECTIVES

1. To understand the terms *basal metabolic rate (BMR), thyroid-stimulating hormone (TSH), thyroxine, goiter, hypothyroidism, hyperthyroidism, thyroidectomized,* and *hypophysectomized.*

2. To observe how negative feedback mechanisms regulate hormone release.

3. To understand thyroxine's role in maintaining the basal metabolic rate.

4. To understand the effect of TSH on the basal metabolic rate.

5. To understand the role of the hypothalamus in regulating the secretion of thyroxine and TSH.

Introduction

Metabolism is the broad range of biochemical reactions occurring in the body. Metabolism includes *anabolism* and *catabolism.* Anabolism is the building up of small molecules into larger, more complex molecules via enzymatic reactions. Energy is stored in the chemical bonds formed when larger, more complex molecules are formed.

Catabolism is the breakdown of large, complex molecules into smaller molecules via enzymatic reactions. The breaking of chemical bonds in catabolism releases energy that the cell can use to perform various activities, such as forming ATP. The cell does not use all the energy released by bond breaking. Much of the energy is released as heat to maintain a fixed body temperature, especially in humans. Humans are *homeothermic* organisms that need to maintain a fixed body temperature to maintain the activity of the various metabolic pathways in the body.

The most important hormone for maintaining metabolism and body heat is **thyroxine** (thyroid hormone), also known as *tetraiodothyronine,* or T_4. Thyroxine is secreted by the thyroid gland, located in the neck.

The production of thyroxine is controlled by the pituitary gland, or hypophysis, which secretes **thyroid-stimulating hormone (TSH).** The blood carries TSH to its target tissue, the thyroid gland. TSH causes the thyroid gland to increase in size and secrete thyroxine into the general circulation. If TSH levels are too high, the thyroid gland enlarges. The resulting glandular swelling in the neck is called a **goiter.**

The **hypothalamus** in the brain is also a vital participant in thyroxine and TSH production. It is a primary endocrine gland that secretes several hormones that affect the pituitary gland, or hypophysis, which is also located in the brain.

Thyrotropin-releasing hormone (TRH) is directly linked to thyroxine and TSH secretion. TRH from the hypothalamus stimulates the anterior pituitary to produce TSH, which then stimulates the thyroid to produce thyroxine.

These events are part of a classic negative feedback mechanism. When circulation levels of thyroxine are low, the hypothalamus secretes more TRH to stimulate the pituitary gland to secrete more TSH. The increase in TSH further stimulates the secretion of thyroxine from the thyroid gland. The increased levels of thyroxine will then influence the hypothalamus to reduce its production of TRH.

TRH travels from the hypothalamus to the pituitary gland via the **hypothalamic-pituitary portal system.** This specialized arrangement of blood vessels consists of a single **portal vein** that connects two capillary beds. The hypothalamic-pituitary portal system transports many other hormones from the hypothalamus to the pituitary gland. The hypothalamus primarily secretes *tropic* hormones, which stimulate the secretion of other hormones. TRH is an example of a tropic hormone because it stimulates the release of TSH from the pituitary gland. TSH itself is also an example of a tropic hormone because it stimulates production of thyroxine.

In this activity you will investigate the effects of thyroxine and TSH on a rat's metabolic rate. The metabolic rate will be indicated by the amount of oxygen the rat consumes per time per body mass. You will perform four experiments on three rats: a normal rat, a thyroidectomized rat (a rat whose thyroid gland has been surgically removed), and a hypophysectomized rat (a rat whose pituitary gland has been surgically removed). You will determine (1) the rat's basal metabolic rate, (2) its metabolic rate after it has been injected with thyroxine, (3) its metabolic rate after it has been injected with TSH, and (4) its metabolic rate after it has been injected with propylthiouracil, a drug that inhibits the production of thyroxine.

EQUIPMENT USED The following equipment will be depicted on-screen: three refillable syringes—used to inject the rats with propylthiouracil (a drug that inhibits the production of thyroxine by blocking the incorporation of iodine into the hormone precursor molecule), thyroid-stimulating hormone (TSH), and thyroxine; airtight, glass animal chamber—provides an isolated, sealed system in which to measure the amount of oxygen consumed by the rat in a specified amount of time (Opening the clamp on the left tube allows outside air into the chamber, and closing the clamp will create a closed, airtight system. The T-connector on the right tube allows you to connect the chamber to the manometer or to connect the fluid-filled manometer to the syringe filled with air.), soda lime (found at the bottom of the glass chamber)—absorbs the carbon dioxide given off by the rat; manometer—U-shaped tube containing fluid (As the rat consumes oxygen in the isolated, sealed system, this fluid will rise in the left side of the U-shaped tube and fall in the right side of the tube.); syringe—used to inject air into the tube and thus measure the amount of air that is needed to return the fluid columns in the manometer to their original levels; animal scale—used to measure body weight; three white rats—a *normal* rat, a *thyroidectomized* (Tx) rat (a rat whose thyroid gland has been surgically removed), and a *hypophysectomized* (Hypox) rat (a rat whose pituitary gland has been surgically removed).

Experiment Instructions

Go to the home page in the PhysioEx software and click **Exercise 4: Endocrine System Physiology.** Click **Activity 1: Metabolism and Thyroid Hormone,** and take the online **Pre-lab Quiz** for Activity 1.

After you take the online Pre-lab Quiz, click the **Experiment** tab and begin the experiment. The experiment instructions are reprinted here for your reference. The opening screen for the experiment is shown below.

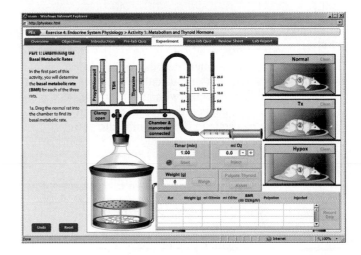

Part 1: Determining the Basal Metabolic Rates

In the first part of this activity, you will determine the basal metabolic rate (BMR) for each of the three rats.

1a. Drag the *normal* rat into the chamber to find its BMR.

1b. Click **Weigh** to determine the rat's weight.

1c. Click the clamp on the left tube (top of the chamber) to close it. This will prevent any outside air from entering the chamber and ensure that the only oxygen the rat is breathing is the oxygen inside the closed system.

1d. Note that the timer is set to one minute. Click **Start** beneath the timer to measure the amount of oxygen consumed by the rat in one minute in the sealed chamber. Note what happens to the water levels in the manometer as time progresses.

1e. Click the T-connector knob to connect the manometer and syringe.

1f. Click the clamp on the left tube (top of the chamber) to open it so the rat can breathe outside air.

1g. Observe the difference between the level in the left and right arms of the manometer. Estimate the volume of O_2 that you will need to inject to make the levels equal by counting the divisions on both sides. This volume is equivalent to the amount of oxygen that the rat consumed during the minute in the sealed chamber. Click the + button under the ml O_2 display until you reach the estimated volume. Then click **Inject** and watch what happens to the fluid in the two arms. When the volume levels are equalized, the word "Level" will appear and stay on the screen.

- If you have not injected enough oxygen, the word "Level" will not appear. Click the + to increase the volume and then click **Inject** again.

- If you have injected too much oxygen, the word "Level" will flash and then disappear. Click the button to decrease the volume and then click **Inject** again. Click **Record Data** when the levels are equalized.

1h. Calculate the oxygen consumption per hour for this rat using the following equation:

$$\frac{\text{ml O}_2 \text{ consumed}}{1 \text{ minute}} \times \frac{60 \text{ minutes}}{1 \text{ hr}} = \text{ml O}_2/\text{hr}$$

Enter the oxygen consumption per hour in the field below and then click **Submit** to record your results in the lab report. _____ ml O$_2$/hr

1i. Now that you have calculated the oxygen consumption per hour for this rat, you can calculate the metabolic rate per kilogram of body weight with the following equation (note that you need to convert the weight data from grams to kilograms to use this equation): Metabolic rate = (ml O$_2$/hr)/ (weight in kg) = ml O$_2$/kg/hr.

$$\text{Metabolic rate} = \frac{\text{ml O}_2/\text{hr}}{\text{weight in kg}} = \text{ml O}_2/\text{kg/hr}$$

Enter the metabolic rate in the field below and then click **Submit** to record your results in the lab report. _____ ml O$_2$/kg/hr

1j. Click **Palpate Thyroid** to manually check the size of the thyroid and, thus, whether a goiter is present. After reviewing the findings, click **Submit** to record your results in the lab report.

1k. Drag the rat from the chamber back to its cage and then click **Restore** (beneath **Palpate Thyroid**) to restore the apparatus to its initial state.

> **? PREDICT Question 1**
> Make a prediction about the basal metabolic rate (BMR) of the remaining rats compared with the BMR of the normal rat you just measured.
> _____
> _____

2a.–2k. Repeat steps 1a–1k for the *thyroidectomized (Tx)* rat.

3a.–3k. Repeat steps 1a–1k for the *hypophysectomized (Hypox)* rat.

> **? PREDICT Question 2**
> What do you think will happen to the metabolic rates of the rats after you inject them with thyroxine?
> _____
> _____

Part 2: Determining the Effect of Thyroxine on Metabolic Rate

In this part of the activity, you will investigate the effects of thyroxine injections on the metabolic rates of all three rats.

4a. Drag the syringe filled with *thyroxine* to the *normal* rat's hindquarters. Release the mouse button to inject thyroxine into the rat. (In this experiment, the effects of the injection are immediate. In a wet lab, you would have to inject the rats daily with thyroxine for 1–2 weeks).

4b. In this part of the activity, the rat's weight, the amount of oxygen consumed by the rat in one minute, the rat's oxygen consumption per hour, the rat's metabolic rate, and the result of the thyroid palpation will be generated automatically after you drag the rat into the chamber.

Drag the injected rat into the chamber and note the results (and record your results in Chart 1).

4c. Drag the rat from the chamber back to its cage and then click **Clean** to clear all traces of thyroxine from the rat and clean the syringe. (In this experiment, the thyroxine is removed instantly. In a wet lab, clearance would take weeks or require that a different rat be used.)

5a.–5c. Repeat steps 4a–4c with the *thyroidectomized (Tx)* rat (and record your results in Chart 1).

6a.–6c. Repeat steps 4a–4c with the *hypophysectomized (Hypox)* rat (and record your results in Chart 1).

> **? PREDICT Question 3**
> What do you think will happen to the metabolic rates of the rats after you inject them with TSH?
> _____
> _____

Part 3: Determining the Effect of TSH on Metabolic Rate

In this part of the activity, you will investigate the effects of TSH injections on the metabolic rates of all three rats.

7a. Drag the syringe filled with *TSH* to the *normal* rat's hindquarters. Release the mouse button to inject TSH into the rat. (In this experiment, the effects of the injection are immediate. In a wet lab, you would have to inject the rats daily with TSH for 1–2 weeks.)

7b. In this part of the activity, the rat's weight, the amount of oxygen consumed by the rat in one minute, the rat's oxygen consumption per hour, the rat's metabolic rate, and the result of the thyroid palpation will be generated automatically after you drag the rat into the chamber.

Drag the injected rat into the chamber and note the results (and record your results in Chart 1).

7c. Drag the rat from the chamber back to its cage and then click **Clean** to clear all traces of TSH from the rat and clean the syringe. (In this experiment, the TSH is removed instantly. In a wet lab, clearance would take weeks or require that a different rat be used.)

8a.–8c. Repeat steps 7a–7c with the *thyroidectomized (Tx)* rat (and record your results in Chart 1).

9a.–9c. Repeat steps 7a–7c with the *hypophysectomized (Hypox)* rat (and record your results in Chart 1).

CHART 1	Effects of Hormones on Metabolic Rate		
	Normal rat	**Thyroidectomized rat**	**Hypophysectomized rat**
Baseline			
Weight	_____ grams	_____ grams	_____ grams
ml O_2 used in 1 minute	_____ ml	_____ ml	_____ ml
ml O_2 used per hour	_____ ml	_____ ml	_____ ml
Metabolic rate	_____ ml O_2/kg/hr	_____ ml O_2/kg/hr	_____ ml O_2/kg/hr
Palpation results	_____	_____	_____
With thyroxine			
Weight	_____ grams	_____ grams	_____ grams
ml O_2 used in 1 minute	_____ ml	_____ ml	_____ ml
ml O_2 used per hour	_____ ml	_____ ml	_____ ml
Metabolic rate	_____ ml O_2/kg/hr	_____ ml O_2/kg/hr	_____ ml O_2/kg/hr
Palpation results	_____	_____	_____
With TSH			
Weight	_____ grams	_____ grams	_____ grams
ml O_2 used in 1 minute	_____ ml	_____ ml	_____ ml
ml O_2 used per hour	_____ ml	_____ ml	_____ ml
Metabolic rate	_____ ml O_2/kg/hr	_____ ml O_2/kg/hr	_____ ml O_2/kg/hr
Palpation results	_____	_____	_____
With propylthiouracil			
Weight	_____ grams	_____ grams	_____ grams
ml O_2 used in 1 minute	_____ ml	_____ ml	_____ ml
ml O_2 used per hour	_____ ml	_____ ml	_____ ml
Metabolic rate	_____ ml O_2/kg/hr	_____ ml O_2/kg/hr	_____ ml O_2/kg/hr
Palpation results	_____	_____	_____

? PREDICT Question 4
Propylthiouracil (PTU) is a drug that inhibits the production of thyroxine by blocking the attachment of iodine to tyrosine residues in the follicle cells of the thyroid gland (iodinated tyrosines are linked together to form thyroxine). What do you think will happen to the metabolic rates of the rats after you inject them with PTU?

_____

_____

Part 4: Determining the Effect of Propylthiouracil on Metabolic Rate

In this part of the activity, you will investigate the effects of propylthiouracil injections on the metabolic rates of all three rats.

10a. Drag the syringe filled with *propylthiouracil* to the *normal* rat's hindquarters. Release the mouse button to inject propylthiouracil into the rat. (In this experiment, the effects of the injection are immediate. In a wet lab, you would have to inject the rats daily with propylthiouracil for 1–2 weeks).

10b. In this part of the activity, the rat's weight, the amount of oxygen consumed by the rat in one minute, the rat's oxygen consumption per hour, the rat's metabolic rate, and the result of the thyroid palpation will be generated automatically after you drag the rat into the chamber.

Drag the injected rat into the chamber and note the results (and record your results in Chart 1).

10c. Drag the rat from the chamber back to its cage and then click **Clean** to clear all traces of propylthiouracil from the rat and clean the syringe. (In this experiment, the propylthiouracil is removed instantly. In a wet lab, clearance would take weeks or require that a different rat be used.)

11a.–11c. Repeat steps 10a–10c with the *thyroidectomized (Tx)* rat (and record your results in Chart 1).

12a.–12c. Repeat steps 10a–10c with the *hypophysectomized (Hypox)* rat (and record your results in Chart 1).

After you complete the experiment, take the online **Post-lab Quiz** for Activity 1.

Activity Questions

1. Using a water-filled manometer, you observed the amount of oxygen consumed by rats in a sealed chamber. What happened to the carbon dioxide the rat produced while in the sealed chamber?

_____

_____

2. What would happen to the fluid levels of the manometer (and, thus, the results of the metabolism experiment) if the rats in the sealed chamber were engaged in physical activity (such as running in a wheel)?

_____

_____

3. Describe the role of the hypothalamus in the production of thyroxine.

_____

_____

4. What does it mean if a hormone is a _tropic_ hormone?

_____

_____

5. How could you treat a thyroidectomized rat so that it functions like a "normal" rat? How would you verify that your treatments were safe and effective?

_____

_____

6. What is the role of the hypothalamus in the production of thyroid-stimulating hormone (TSH)?

_____

_____

7. How does thyrotropin-releasing hormone (TRH) travel from the hypothalamus to the pituitary gland?

_____

_____

8. Why didn't the administration of TSH have any effect on the metabolic rate of the thyroidectomized rat?

_____

_____

9. Why didn't the administration of propylthiouracil have any effect on the metabolic rate of either the thyroidectomized rat or the hypophysectomized rat?

_____

_____

10. Propylthiouracil inhibits the production of thyroxine by blocking the attachment of iodine to the amino acid tyrosine. What naturally occurring problem in some parts of the world does this drug mimic?

_____

_____

ACTIVITY 2

Plasma Glucose, Insulin, and Diabetes Mellitus

OBJECTIVES

1. To understand the use of the terms _insulin, type 1 diabetes mellitus, type 2 diabetes mellitus,_ and _glucose standard curve._
2. To understand how fasting plasma glucose levels are used to diagnose diabetes mellitus.
3. To understand the assay that is used to measure plasma glucose.

Introduction

Insulin is a hormone produced by the beta cells of the endocrine portion of the pancreas. This hormone is vital to the regulation of **plasma glucose** levels, or "blood sugar," because the hormone enables our cells to absorb glucose from the bloodstream. Glucose absorbed from the blood is either used as fuel for metabolism or stored as glycogen (also known as animal starch), which is most notable in liver and muscle cells. About 75% of glucose consumed during a meal is stored as glycogen. As humans do not feed continuously (we are considered "discontinuous feeders"), the production of glycogen from a meal ensures that a supply of glucose will be available for several hours after a meal.

Furthermore, the body has to maintain a certain level of plasma glucose to continuously serve nerve cells because these cell types use only glucose for metabolic fuel. When glucose levels in the plasma fall below a certain value, the alpha cells of the pancreas are stimulated to release the hormone **glucagon.** Glucagon stimulates the breakdown of stored glycogen into glucose, which is then released back into the blood.

When the pancreas does not produce enough insulin, **type 1 diabetes mellitus** results. When the pancreas produces sufficient insulin but the body fails to respond to it, **type 2 diabetes mellitus** results. In either case, glucose remains in the bloodstream, and the body's cells are unable to take it up to serve as the primary fuel for metabolism. The kidneys then filter the excess glucose out of the plasma. Because the reabsorption of filtered glucose involves a finite number of transporters in kidney tubule cells, some of the excess glucose is not reabsorbed into the circulation. Instead, it passes out of the body in urine (hence _sweet urine,_ as the name **diabetes mellitus** suggests).

The inability of body cells to take up glucose from the blood also results in skeletal muscle cells undergoing protein catabolism to free up amino acids to be used in forming glucose in the liver. This action puts the body into a negative nitrogen balance from the resulting protein depletion and tissue wasting. Other associated problems include poor wound healing and poor resistance to infections.

This activity is divided into two parts. In Part 1, you will generate a **glucose standard curve,** which will be explained in the experiment. In Part 2, you will use the glucose standard curve to measure the fasting plasma glucose levels from several patients to diagnose the presence or absence of diabetes mellitus. A patient with FPG values greater than or equal to 126 mg/dl in two FPG tests is diagnosed with diabetes. FPG values between 110 and 126 mg/dl indicate impairment or borderline impairment of insulin-mediated glucose uptake by cells. FPG values less than 110 mg/dl are considered normal.

> **EQUIPMENT USED**   The following equipment will be depicted on-screen: deionized water—used to adjust the volume so that it is the same for each reaction; glucose standard; enzyme color reagent; barium hydroxide; heparin; blood samples from five patients; test tubes—used as reaction vessels for the various tests; test tube incubation unit—used to incubate, mix, and centrifuge the samples; spectrophotometer—used to measure the amount of light absorbed or transmitted by a pigmented solution.

Experiment Instructions

Go to the home page in the PhysioEx software and click **Exercise 4: Endocrine System Physiology.** Click **Activity 2: Plasma Glucose, Insulin, and Diabetes Mellitus,** and take the online **Pre-lab Quiz** for Activity 2.

After you take the online Pre-lab Quiz, click the **Experiment** tab and begin the experiment. The experiment instructions are reprinted here for your reference. The opening screen for the experiment is shown below.

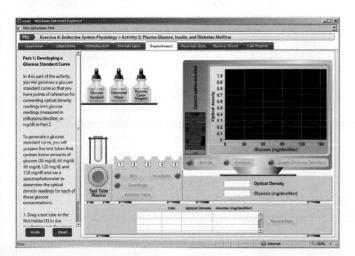

Part 1: Developing a Glucose Standard Curve

In this part of the activity, you will generate a glucose standard curve so that you have points of reference for converting optical density readings into glucose readings (measured in milligrams/deciliter, or mg/dl) in Part 2.

To generate a glucose standard curve, you will prepare five test tubes that contain known amounts of glucose (30 mg/dl, 60 mg/dl, 90 mg/dl, 120 mg/dl, and 150 mg/dl) and use a spectrophotometer to determine the optical density readings for each of these glucose concentrations.

1. Drag a test tube to the first holder (**1**) in the incubation unit. Four more test tubes will automatically be placed in the incubation unit.

2. Drag the dropper cap of the glucose standard bottle to the first tube in the incubation unit to dispense one drop of glucose standard solution into the tube. The dropper will automatically move across and dispense glucose standard to the remaining tubes. Note that each tube receives one additional drop of glucose standard (tube 2 receives 2 drops, tube 3 receives 3 drops, tube 4 receives 4 drops, and tube 5 receives 5 drops).

3. Drag the dropper cap of the deionized water bottle to the first tube in the incubation unit to dispense four drops of deionized water into the tube. The dropper will automatically move across and dispense deionized water to the remaining tubes. Note that each tube receives one less drop of deionized water (tube 2 receives 3 drops, tube 3 receives 2 drops, tube 4 receives 1 drop, and tube 5 does not receive any drops).

4. Click **Mix** to mix the contents of the tubes.

5. Click **Centrifuge** to centrifuge the contents of the tubes. After the centrifugation process, the tubes will automatically rise.

6. Click **Remove Pellet** to remove any pellets formed during the centrifugation process. Pellets can contain reagent precipitates and debris from the laboratory environment.

7. Drag the dropper cap of the enzyme color reagent bottle to the first tube in the incubation unit to dispense five drops of enzyme color reagent into each tube.

8. Click **Incubate** to incubate the contents of the tubes. The incubation unit will gently agitate the test tube rack, evenly mixing the contents of all test tubes throughout the incubation.

9. Click **Set Up** on the spectrophotometer to warm up the instrument and get it ready for your sample readings.

10. Drag tube 1 to the spectrophotometer.

11. Click **Analyze** to analyze the sample. A data point will appear on the monitor to show the optical density and the glucose concentration of the sample. These values will also appear in the optical density and glucose displays.

12. Click **Record Data** to display your results in the grid (and record your results in Chart 2.1). The tube will automatically be placed in the test tube washer.

CHART 2.1	Glucose Standard Curve Results	
Tube	**Optical density**	**Glucose (mg/dl)**
1		
2		
3		
4		
5		

13. You will now analyze the samples in the remaining tubes.

- Drag the next tube into the spectrophotometer.

- Click **Analyze** to analyze the sample. A data point will appear on the monitor to show the optical density and the glucose concentration of the sample. These values will also appear in the optical density and glucose displays.

- Click **Record Data** to display your results in the grid (and record your results in Chart 2.1). The tube will automatically be placed in the test tube washer.

  Repeat this step until you analyze all five tubes.

14. Click **Graph Glucose Standard** to generate the glucose standard curve on the monitor. You will use this graph in Part 2.

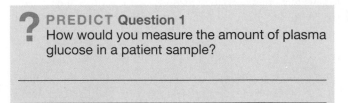

PREDICT Question 1
How would you measure the amount of plasma glucose in a patient sample?

_____

_____

Part 2: Measure Fasting Plasma Glucose Levels

In this part of the activity, you will use the glucose standard curve you generated in Part 1 to measure the fasting plasma glucose levels from five patients to diagnose the presence or absence of diabetes mellitus. Note the addition of two reagent bottles (barium hydroxide and heparin) and blood samples from the five patients. To undergo the fasting plasma glucose (FPG) test, patients must fast for a minimum of 8 hours prior to the blood draw.

A patient with FPG values greater than or equal to 126 mg/dl in two FPG tests is diagnosed with diabetes. FPG values between 110 and 126 mg/dl indicate impairment or borderline impairment of insulin-mediated glucose uptake by cells. FPG values less than 110 mg/dl are considered normal.

15. Drag a test tube to the first holder (**1**) in the incubation unit. Four more test tubes will automatically be placed in the incubation unit.

16. Drag the dropper cap of the first patient blood sample to the first tube in the incubation unit to dispense three drops of the sample. Three drops from each sample will automatically be dispensed into a separate tube.

17. Drag the dropper cap of the deionized water bottle to the first tube in the incubation unit to dispense five drops of deionized water into each tube.

18. Barium hydroxide dissolves and thus clears both proteins and cell membranes (so that clear glucose readings can be obtained). Drag the dropper cap of the barium hydroxide bottle to the first tube in the incubation unit to dispense five drops of barium hydroxide into each tube.

19. Drag the dropper cap of the heparin bottle to the first tube in the incubation unit to dispense a drop of heparin into each tube. Heparin prevents blood clots, which would interfere with clear glucose readings.

20. Click **Mix** to mix the contents of the tubes.

21. Click **Centrifuge** to centrifuge the contents of the tubes. After the centrifugation process, the tubes will automatically rise.

22. Click **Remove Pellet** to remove any pellets formed during the centrifugation process. Pellets can contain reagent precipitates and debris from the laboratory environment.

23. Drag the dropper cap of the enzyme color reagent bottle to the first tube in the incubation unit to dispense five drops of enzyme color reagent into each tube.

24. Click **Incubate** to incubate the contents of the tubes. The incubation unit will gently agitate the test tube rack, evenly mixing the contents of all test tubes throughout the incubation.

25. Click **Set Up** on the spectrophotometer to warm up the instrument and get it ready for your sample readings.

26. Click **Graph Glucose Standard** to display the glucose standard curve you generated in Part 1 on the monitor.

27. Drag tube 1 to the spectrophotometer.

28. Click **Analyze** to analyze the sample. A horizontal line will appear on the monitor to show the optical density of the sample. The optical density will also appear in the optical density display.

29. Drag the movable ruler (the vertical red line on the right side of the monitor) to the intersection of the horizontal yellow line (the optical density of the sample) and the glucose standard curve. Note the change in the glucose display as you move the line. The glucose concentration where the lines intersect is the fasting plasma glucose for this patient. Click **Record Data** to display your results in the grid (and record your results in Chart 2.2). The tube will automatically be placed in the test tube washer, and the monitor will be cleared (except for the glucose standard curve).

CHART 2.2	Fasting Plasma Glucose Results	
Sample	Optical density	Glucose (mg/dl)
1		
2		
3		
4		
5		

30. You will now analyze the samples in the remaining tubes.

- Drag the next tube into the spectrophotometer.

- Click **Analyze** to analyze the sample. A data point will appear on the monitor to show the optical density and the glucose concentration of the sample. These values will also appear in the optical density and glucose displays.

- Click **Record Data** to display your results in the grid. The tube will automatically be placed in the test tube washer (and record your results in Chart 2.2).

    Repeat this step until you analyze all five tubes.

After you complete the experiment, take the online **Post-lab Quiz** for Activity 2.

Activity Questions

1. How would you know if your glucose standard curve was aberrant and thus inappropriate for patient diagnostics?

_____

_____

_____

2. What are potential sources of variability when generating a glucose standard curve?

_____

_____

_____

3. What recommendations would you make to a patient with fasting plasma glucose levels in the impaired/borderline-impaired range who was in the impaired/borderline-impaired range for the oral glucose tolerance test?

_____

_____

_____

4. The amount of corn syrup in the American diet has been described as alarmingly high (especially in the foods that children eat). In the context of this activity, predict the likely trends in the fasting plasma glucose levels of our children as they mature.

_____

_____

ACTIVITY 3

Hormone Replacement Therapy

OBJECTIVES

1. To understand the terms *hormone replacement therapy, follicle-stimulating hormone (FSH), estrogen, calcitonin, osteoporosis, ovariectomized,* and *T score.*

2. To understand how estrogen levels affect bone density.

3. To understand the potential benefits of hormone replacement therapy.

Introduction

Follicle-stimulating hormone (FSH) is an anterior pituitary peptide hormone that stimulates ovarian follicle growth. Developing ovarian follicles then produce and secrete a steroid hormone called **estrogen** into the plasma. Estrogen has numerous effects on the female body and homeostasis, including the stimulation of bone growth and protection against **osteoporosis** (a reduction in the quantity of bone characterized by decreased bone mass and increased susceptibility to fractures).

After menopause, the ovaries stop producing and secreting estrogen. One of the effects and potential health problems of menopause is a loss of bone density that can result in osteoporosis and bone fractures. For this reason, post-menopausal treatments to prevent osteoporosis often include hormone replacement therapy. Estrogen can be administered to increase bone density. Calcitonin (secreted by C cells in the thyroid gland) is another peptide hormone that can be administered to counteract the development of osteoporosis. Calcitonin inhibits osteoclast activity and stimulates calcium uptake and deposition in long bones.

In this activity you will use three **ovariectomized** rats that are no longer producing estrogen because their ovaries have been surgically removed. A **T score** is a quantitative measurement of the mineral content of bone, used as an indicator of the structural strength of the bone and as a screen for osteoporosis. The three rats were chosen because each has a baseline T score of 2.61, indicating osteoporosis. T scores are interpreted as follows: normal = +1 to −0.99; osteopenia (bone thinning) = −1.0 to −2.49; osteoporosis = −2.5 and below.

You will administer either estrogen therapy or calcitonin therapy to these rats, representing two types of **hormone replacement therapy.** The third rat will serve as an untreated control and receive daily injections of saline. The vertebral bone density (VBD) of each rat will be measured with dual X-ray absorptiometry (DXA) to obtain its T score after treatment.

> **EQUIPMENT USED** The following equipment will be depicted on screen: three ovariectomized rats (Note that if this were an actual wet lab, the ovariectomies would have been performed on the rats a month before the experiment to ensure that no residual hormones remained in the rats' systems.); saline; estrogen; calcitonin; reusable syringe—used to inject the rats; anesthesia—used to immobilize the rats for the X-ray scanning; dual X-ray absorptiometry bone-density scanner (DXA)—used to measure vertebral bone density of the rats.

Experiment Instructions

Go to the home page in the PhysioEx software and click **Exercise 4: Endocrine System Physiology.** Click **Activity 3: Hormone Replacement Therapy,** and take the online **Pre-lab Quiz** for Activity 3.

After you take the online Pre-lab Quiz, click the **Experiment** tab and begin the experiment. The experiment instructions are reprinted here for your reference. The opening screen for the experiment is shown on the following page.

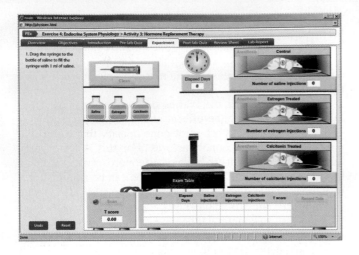

1. Drag the syringe to the bottle of saline to fill the syringe with 1 ml of saline.

2. Drag the syringe to the *control* rat, placing the tip of the needle in the rat's lower abdominal area. Injections into this area are considered *intraperitoneal* and will quickly be circulated by the abdominal blood vessels.

3. Click **Clean** beneath the syringe holder to clean the syringe of all residues.

4. Drag the syringe to the bottle of estrogen to fill the syringe with 1 ml of estrogen.

5. Drag the syringe to the *estrogen-treated* rat, placing the tip of the needle in the rat's lower abdominal area.

6. Click **Clean** beneath the syringe holder to clean the syringe of all residues.

7. Drag the syringe to the bottle of calcitonin to fill the syringe with 1 ml of calcitonin.

8. Drag the syringe to the *calcitonin-treated* rat, placing the tip of the needle in the rat's lower abdominal area.

9. Click **Clean** beneath the syringe holder to clean the syringe of all residues.

10. Click the clock face to advance one day (24 hours).

11. Each rat must receive seven injections over the course of seven days (one injection per day). The remaining injections will be automated. Click the clock face to repeat the series of injections until you have injected each of the rats seven times.

> ? **PREDICT Question 1**
> What effect will the saline injections have on the control rat's vertebral bone density?

> ? **PREDICT Question 2**
> What effect will the estrogen injections have on the estrogen-treated rat's vertebral bone density?

> ? **PREDICT Question 3**
> What effect will the calcitonin injections have on the calcitonin-treated rat's vertebral bone density?

12. Click **Anesthesia** above the *control* rat's cage to immobilize the control rat with a gaseous anesthetic for X-ray scanning.

13. Drag the anesthetized rat to the exam table for X-ray scanning.

14. Click **Scan** to activate the scanner. The T score will appear in the T score display. Click **Record Data** to record your results in the grid (and record your results in Chart 3). The control rat will be automatically returned to its cage.

CHART 3	Hormone Replacement Therapy Results	
Rat	**T score**	

15. You will now obtain the T scores for the remaining rats. Perform these steps to obtain the T score for the *estrogen-treated* rat, then repeat these steps to obtain the T score for the *calcitonin-treated* rat.

- Click **Anesthesia** above the rat's cage to immobilize the rat with a gaseous anesthetic for X-ray scanning.

- Drag the anesthetized rat to the exam table for X-ray scanning.

- Click **Scan** to activate the scanner. The T score will appear in the T score display.

- Click **Record Data** to record your results in the grid (and record your results in Chart 3). The rat will be automatically returned to its cage.

After you complete the experiment, take the online **Post-lab Quiz** for Activity 3.

Activity Questions

1. Recently, hormone replacement therapy has been prominent in the popular press. Describe a hormone replacement therapy that you have seen in the news, and highlight its benefits, its potential risks, the reasons to continue and the reasons to discontinue its use.

2. In hormone replacement therapy, how is the hormone dose determined by the prescribing physician?

_____

_____ ▬

Measuring Cortisol and Adrenocorticotropic Hormone

OBJECTIVES

1. To understand the terms *cortisol, adrenocorticotropic hormone (ACTH), corticotropin-releasing hormone (CRH), Cushing's syndrome, iatrogenic, Cushing's disease,* and *Addison's disease.*

2. To understand how CRH controls ACTH secretion and ACTH controls cortisol secretion.

3. To understand how negative feedback mechanisms influence the levels of tropic CRH and ACTH.

4. To measure the blood levels of cortisol and ACTH in five patients and correlate these readings with symptoms and diagnoses.

5. To distinguish between Cushing's syndrome and Cushing's disease.

Introduction

Cortisol, a hormone secreted by the *adrenal cortex,* is important in the body's response to many kinds of stress. Cortisol release is stimulated by **adrenocorticotropic hormone (ACTH),** a tropic hormone released by the anterior pituitary. A *tropic* hormone stimulates the secretion of another hormone. ACTH release, in turn, is stimulated by **corticotropin-releasing hormone (CRH),** a tropic hormone from the hypothalamus. Increased levels of cortisol negatively feed back to inhibit the release of both ACTH and CRH.

Increased cortisol in the blood, or *hypercortisolism,* is referred to as **Cushing's syndrome** if the increase is caused by an adrenal gland tumor. Cushing's syndrome can also be **iatrogenic** (that is, physician induced). For example, physician-induced Cushing's syndrome can occur when glucocorticoid hormones, such as prednisone, are administered to treat rheumatoid arthritis, asthma, or lupus. Cushing's syndrome is often referred to as "steroid diabetes" because it results in hyperglycemia. In contrast, **Cushing's disease** is hypercortisolism caused by an anterior pituitary tumor. People with Cushing's disease exhibit increased levels of ACTH and cortisol.

Decreased cortisol in the blood, or *hypocortisolism,* can occur because of adrenal insufficiency. In primary adrenal insufficiency, also known as **Addison's disease,** the low cortisol is directly caused by gradual destruction of the adrenal cortex and ACTH levels are typically elevated as a compensatory effect. Secondary adrenal insufficiency also results in low levels of cortisol, usually caused by damage to the anterior pituitary. Therefore, the levels of ACTH are also low in secondary adrenal insufficiency.

As you can see, a variety of endocrine disorders can be related to both high and low levels of cortisol and ACTH. Table 4.1 summarizes these endocrine disorders.

TABLE 4.1	Cortisol and ACTH Disorders	
	Cortisol level	ACTH level
Cushing's syndrome (primary hypercortisolism)	High	Low
Iatrogenic Cushing's syndrome	High	Low
Cushing's disease (secondary hypercortisolism)	High	High
Addison's disease (primary adrenal insufficiency)	Low	High
Secondary adrenal insufficiency (hypopituitarism)	Low	Low

EQUIPMENT USED The following equipment will be depicted on-screen: plasma samples from five patients; HPLC (high-performance liquid chromatography) column—used to quantitatively measure the amount of cortisol and ACTH in the patient samples; HPLC detector—provides the hormone concentration in the patient sample; reusable syringe—used to inject the patient samples into the HPLC injection port; HPLC injection port—used to inject the patient samples into the HPLC column.

Experiment Instructions

Go to the home page in the PhysioEx software and click **Exercise 4: Endocrine System Physiology.** Click **Activity 4: Measuring Cortisol and Adrenocorticotropic Hormone,** and take the online **Pre-lab Quiz** for Activity 4.

After you take the online Pre-lab Quiz, click the **Experiment** tab and begin the experiment. The experiment instructions are reprinted here for your reference. The opening screen for the experiment is shown below.

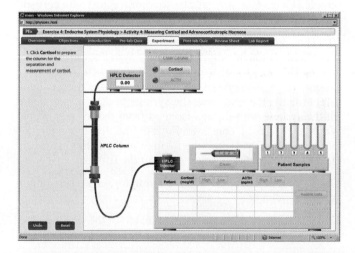

1. Click **Cortisol** to prepare the column for the separation and measurement of cortisol.

2. Drag the syringe to the first tube to fill the syringe with plasma isolated from the first patient.

3. Drag the syringe to the HPLC injector. The sample will enter the tubing and flow through the column. The cortisol concentration in the patient sample will appear in the HPLC detector display.

4. Click **Record Data** to display your results in the grid (and record your results in Chart 4).

CHART 4	Measurement of Cortisol			
Patient	Cortisol (mcg/dl)	Cortisol level	ACTH (pg/ml)	ACTH level
1				
2				
3				
4				
5				

5. Click **Clean** beneath the syringe to prepare it for the next sample. Click **Clean Column** to remove residual cortisol from the column.

6. Drag the syringe to the second tube to fill the syringe with plasma isolated from the second patient.

7. Drag the syringe to the HPLC injector. The sample will enter the tubing and flow through the column. The cortisol concentration in the patient sample will appear in the HPLC detector display.

8. Click **Record Data** to display your results in the grid (and record your results in Chart 4).

9. Click **Clean** beneath the syringe to prepare it for the next sample. Click **Clean Column** to remove residual cortisol from the column.

10. The procedure for the remaining samples will be completed automatically. Drag the syringe to the third tube to fill the syringe with plasma isolated from the third patient. When the cortisol concentration for the third patient is recorded in the grid, drag the syringe to the fourth tube to fill the syringe with plasma isolated from the fourth patient. When the cortisol concentration for the fourth patient is recorded in the grid, drag the syringe to the fifth tube to fill the syringe with plasma isolated from the fifth patient.

11. Click **ACTH** to prepare the column for ACTH separation and measurement.

12. Drag the syringe to the first tube to fill the syringe with plasma isolated from the first patient.

13. Drag the syringe to the HPLC injector. The sample will enter the tubing and flow through the column. The ACTH concentration in the patient sample will appear in the HPLC detector display.

14. Click **Record Data** to display your results in the grid.

15. Click **Clean** beneath the syringe to prepare it for the next sample. Click **Clean Column** to remove residual ACTH from the column.

16. Drag the syringe to the second tube to fill the syringe with plasma isolated from the second patient.

17. Drag the syringe to the HPLC injector. The sample will enter the tubing and flow through the column. The ACTH concentration in the patient sample will appear in the HPLC detector display.

18. Click **Record Data** to display your results in the grid (and record your results in Chart 4).

19. Click **Clean** beneath the syringe to prepare it for the next sample. Click **Clean Column** to remove residual ACTH from the column.

20. The procedure for the remaining samples will be completed automatically. Drag the syringe to the third tube to fill the syringe with plasma isolated from the third patient. When the ACTH concentration for the third patient is recorded in the grid, drag the syringe to the fourth tube to fill the syringe with plasma isolated from the fourth patient. When the ACTH concentration for the fourth patient is recorded in the grid, drag the syringe to the fifth tube to fill the syringe with plasma isolated from the fifth patient.

21. Indicate whether the cortisol and ACTH concentrations (levels) for each patient are high or low using the breakpoints shown in Table 4.2. Click the row of the patient and then click **High** or **Low** next to cortisol and ACTH.

TABLE 4.2	Abnormal Morning Cortisol and ACTH Levels	
ACTH level	High	Low
Cortisol	≥23 mcg/dl	<5 mcg/dl
ACTH	≥80 pg/ml	<20 pg/ml

Note: 1 mcg = 1 µg = 1 microgram

After you complete the experiment, take the online **Post-lab Quiz** for Activity 4.

Activity Questions

1. Discuss the benefits and drawbacks of giving glucocorticoids to young children that have significant allergy-induced asthma.

_____

_____

2. Explain the difference between Cushing's syndrome and Cushing's disease.

_____

_____

NAME _____

LAB TIME/DATE _____

Endocrine System Physiology

ACTIVITY 1 Metabolism and Thyroid Hormone

Part 1

1. Which rat had the fastest basal metabolic rate (BMR)? _____

   _____

2. Why did the metabolic rates differ between the normal rat and the surgically altered rats? How well did the results compare

   with your prediction? _____

   _____

   _____

3. If an animal has been thyroidectomized, what hormone(s) would be missing in its blood? _____

   _____

4. If an animal has been hypophysectomized, what effect would you expect to see in the hormone levels in its body? _____

   _____

   _____

Part 2

5. What was the effect of thyroxine injections on the normal rat's BMR? _____

   _____

   _____

6. What was the effect of thyroxine injections on the thyroidectomized rat's BMR? How does the BMR in this case compare

   with the normal rat's BMR? Was the dose of thyroxine in the syringe too large, too small, or just right? _____

   _____

   _____

7. What was the effect of thyroxine injections on the hypophysectomized rat's BMR? How does the BMR in this case compare with the normal rat's BMR? Was the dose of thyroxine in the syringe too large, too small, or just right? _____

_____

Part 3

8. What was the effect of thyroid-stimulating hormone (TSH) injections on the normal rat's BMR? _____

_____

_____

9. What was the effect of TSH injections on the thyroidectomized rat's BMR? How does the BMR in this case compare with the normal rat's BMR? Why was this effect observed? _____

_____

10. What was the effect of TSH injections on the hypophysectomized rat's BMR? How does the BMR in this case compare with the normal rat's BMR? Was the dose of TSH in the syringe too large, too small, or just right? _____

_____

_____

Part 4

11. What was the effect of propylthiouracil (PTU) injections on the normal rat's BMR? Why did this rat develop a palpable goiter?

_____

_____

12. What was the effect of PTU injections on the thyroidectomized rat's BMR? How does the BMR in this case compare with the normal rat's BMR? Why was this effect observed? _____

_____

13. What was the effect of PTU injections on the hypophysectomized rat's BMR? How does the BMR in this case compare with the normal rat's BMR? Why was this effect observed? _____

_____

_____

ACTIVITY 2 Plasma Glucose, Insulin, and Diabetes Mellitus

1. What is a glucose standard curve, and why did you need to obtain one for this experiment? Did you correctly predict how you would measure the amount of plasma glucose in a patient sample using the glucose standard curve? _____

_____

_____

2. Which patient(s) had glucose reading(s) in the diabetic range? Can you say with certainty whether each of these patients has type 1 or type 2 diabetes? Why or why not? _____

_____

_____

3. Describe the diagnosis for patient 3, who was also pregnant at the time of this assay. _____

_____

_____

4. Which patient(s) had normal glucose reading(s)? _____

_____

5. What are some lifestyle choices these patients with normal plasma glucose readings might recommend to the borderline impaired patients? _____

_____

_____

ACTIVITY 3 Hormone Replacement Therapy

1. Why were ovariectomized rats used in this experiment? How does the fact that the rats are ovariectomized explain their baseline T scores? _____

_____

_____

2. What effect did the administration of saline injections have on the control rat? How well did the results compare with your prediction? _____

_____

3. What effect did the administration of estrogen injections have on the estrogen-treated rat? How well did the results compare with your prediction? _____

_____

_____

4. What effect did the administration of calcitonin injections have on the calcitonin-treated rat? How well did the results compare with your prediction? _____

_____

_____

5. What are some health risks that postmenopausal women must consider when contemplating estrogen hormone replacement therapy? _____

_____

_____

ACTIVITY 4 **Measuring Cortisol and Adrenocorticotropic Hormone**

1. Which patient would most likely be diagnosed with Cushing's disease? Why? _____

_____

_____

2. Which two patients have hormone levels characteristic of Cushing's syndrome? _____

_____

_____

3. Patient 2 is being treated for rheumatoid arthritis with prednisone. How does this information change the diagnosis? _____

_____

_____

4. Which patient would most likely be diagnosed with Addison's disease? Why? _____

_____

_____

Cardiovascular Dynamics

Exercise Overview

The cardiovascular system is composed of a pump—the heart—and blood vessels that distribute blood containing oxygen and nutrients to every cell of the body. The principles governing blood flow are the same physical laws that apply to the flow of liquid through a system of pipes. For example, one very basic law in fluid mechanics is that the flow rate of a liquid through a pipe is directly proportional to the difference between the pressures at the two ends of the pipe (the **pressure gradient**) and inversely proportional to the pipe's **resistance** (a measure of the degree to which the pipe hinders, or resists, the flow of the liquid).

$$\text{Flow} - \text{pressure gradient/resistance} = \Delta P/R$$

This basic law also applies to blood flow. The "liquid" is blood, and the "pipes" are blood vessels. The pressure gradient is the difference between the pressure in arteries and the pressure in veins that results when blood is pumped into arteries. Blood flow rate is directly proportional to the pressure gradient and inversely proportional to resistance.

Blood flow is the amount of blood moving through a body area or the entire cardiovascular system in a given amount of time. Total blood flow is proportional to **cardiac output** (the amount of blood the heart is able to pump per minute). Blood flow to specific body areas can vary dramatically in a given time period. Organs differ in their requirements from moment to moment, and blood vessels have different-sized diameters in their lumen (opening) to regulate local blood flow to various areas in response to the tissues' immediate needs. Consequently, blood flow can increase to some areas and decrease to other areas at the same time.

Resistance is a measure of the degree to which the blood vessel hinders, or resists, the flow of blood. The main factors that affect resistance are (1) blood vessel *radius,* (2) blood vessel *length,* and (3) blood *viscosity.*

Radius

The smaller the blood vessel radius, the greater the resistance, because of frictional drag between the blood and the vessel walls. Contraction of smooth muscle of the blood vessel, or **vasoconstriction,** results in a decrease in the blood vessel radius. Lipid deposits can also cause the radius of an artery to decrease, preventing blood from reaching the coronary tissue, which frequently leads to a heart attack. Alternately, relaxation of smooth muscle of the blood vessel, or **vasodilation,** causes an increase in the blood vessel radius. Blood vessel radius is the single most important factor in determining blood flow resistance.

Length

The longer the vessel length, the greater the resistance—again, because of friction between the blood and vessel walls. The length of a person's blood vessels change only as a person grows. Otherwise, the length generally remains constant.

Viscosity

Viscosity is blood "thickness," determined primarily by **hematocrit**—the fractional contribution of red blood cells to total blood volume. The higher the hematocrit, the greater the viscosity. Under most physiological conditions, hematocrit does not vary much and blood viscosity remains more or less constant.

The Effect of Blood Pressure and Vessel Resistance on Blood Flow

Blood flow is directly proportional to blood pressure because the pressure difference (ΔP) between the two ends of a vessel is the driving force for blood flow. Peripheral resistance is the friction that opposes blood flow through a blood vessel. This relationship is represented in the following equation:

$$\text{Blood flow (ml/min)} = \frac{\Delta P}{\text{peripheral resistance}}$$

Three factors that contribute to peripheral resistance are blood viscosity (η), blood vessel length (L), and the radius of

the blood vessel (r). These relationships are expressed in the following equation:

$$\text{Peripheral resistance} = \frac{8L\eta}{\pi r^4}$$

From this equation you can see that the viscosity of the blood and the length of the blood vessel are directly proportional to peripheral resistance. The peripheral resistance is inversely proportional to the fourth power of the vessel radius. If you combine the two equations, you get the following result:

$$\text{Blood flow (ml/min)} = \frac{\Delta P \pi r^4}{8L\eta}$$

From this combination you can see that blood flow is directly proportional to the fourth power of vessel radius, which means that small changes in vessel radius result in dramatic changes in blood flow.

ACTIVITY 1

Studying the Effect of Blood Vessel Radius on Blood Flow Rate

OBJECTIVES

1. To understand how blood vessel radius affects blood flow rate.
2. To understand how vessel radius is changed in the body.
3. To understand how to interpret a graph of blood vessel radius versus blood flow rate.

Introduction

Controlling **blood vessel radius** (one-half of the diameter) is the principal method of controlling blood flow. Controlling blood vessel radius is accomplished by contracting or relaxing the smooth muscle within the blood vessel walls (vasoconstriction or vasodilation).

To understand why radius has such a pronounced effect on blood flow, consider the physical relationship between blood and the vessel wall. Blood in direct contact with the vessel wall flows relatively slowly because of the friction, or drag, between the blood and the lining of the vessel. In contrast, blood in the center of the vessel flows more freely because it is not rubbing against the vessel wall. The free-flowing blood in the middle of the vessel is called the **laminar flow.** Now picture a fully constricted (small-radius) vessel and a fully dilated (large-radius) vessel. In the fully constricted vessel, proportionately more blood is in contact with the vessel wall and there is less laminar flow, significantly impeding the rate of blood flow in the fully constricted vessel relative to that in the fully dilated vessel.

In this activity you will study the effect of blood vessel radius on blood flow. The experiment includes two glass beakers and a tube connecting them. Imagine that the left beaker is your heart, the tube is an artery, and the right beaker is a destination in your body, such as another organ.

Experiment Instructions

Go to the home page in the PhysioEx software and click **Exercise 5: Cardiovascular Dynamics.** Click **Activity 1: Studying the Effect of Blood Vessel Radius on Blood Flow Rate,** and take the online **Pre-lab Quiz** for Activity 1.

After you take the online Pre-lab Quiz, click the **Experiment** tab and begin the experiment. The experiment instructions are reprinted here for your reference. The opening screen for the experiment is shown below.

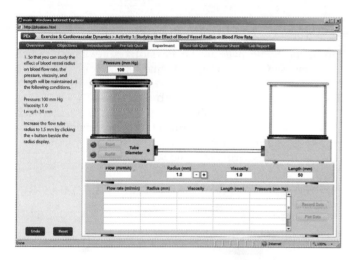

1. So that you can study the effect of blood vessel radius on blood flow rate, the pressure, viscosity, and length will be maintained at the following conditions:

Pressure: 100 mm Hg

Viscosity: 1.0

Length: 50 mm

Increase the flow tube radius to 1.5 mm by clicking the + button beside the radius display.

2. Click **Start** and then watch the fluid move into the right beaker. (Fluid moves slowly under some conditions—be patient!) Pressure propels fluid from the left beaker to the right beaker through the flow tube. The flow rate is shown in the flow rate display after the left beaker has finished draining.

3. Click **Record Data** to display your results in the grid (and record your results in Chart 1).

4. Click **Refill** to replenish the left beaker.

PREDICT Question 1
What do you think will happen to the flow rate if the radius is increased by 0.5 mm?

Flow (ml/min)	Radius (mm)
CHART 1 Effect of Blood Vessel Radius on Blood Flow Rate	

5. Increase the flow tube radius to 2.0 mm by clicking the + button beside the radius display. Click **Start** and watch the fluid move into the right beaker.

6. Click **Record Data** to display your results in the grid (and record your results in Chart 1).

7. Click **Refill** to replenish the left beaker.

8. You will now observe the effect of incremental increases in flow tube radius.

- Increase the flow tube radius by 0.5 mm.

- Click **Start** and then watch the fluid move into the right beaker.

- Click **Record Data** to display your results in the grid (and record your results in Chart 1).

- Click **Refill** to replenish the left beaker.

Repeat this step until you reach a flow tube radius of 5.0 mm.

PREDICT Question 2
Do you think a graph plotted with radius on the X-axis and flow rate on the Y-axis will be linear (a straight line)?

9. Click **Plot Data** to view a summary of your data on a plotted grid. Radius will be displayed on the X-axis and flow rate will be displayed on the Y-axis. Click **Submit** to record your plot in the lab report.

After you complete the experiment, take the online **Post-lab Quiz** for Activity 1.

Activity Questions

1. Describe the relationship between vessel radius and blood flow rate.

2. In this activity you altered the radius of the flow tube by clicking the + and − buttons. Explain how and why the radius of blood vessels is altered in the human body.

_____

_____

_____

3. Describe the appearance of your plot of blood vessel radius versus blood flow rate and relate the plot to the relationship between these two variables.

_____

_____

_____

4. Describe an advantage of slower blood velocity in some areas of the body, for example, in the capillaries of our fingers.

_____

_____

_____

ACTIVITY 2

Studying the Effect of Blood Viscosity on Blood Flow Rate

OBJECTIVES

1. To understand how blood viscosity affects blood flow rate.
2. To list the components in the blood that contribute to blood viscosity.
3. To explain conditions that might lead to viscosity changes in the blood.
4. To understand how to interpret a graph of viscosity versus blood flow.

Introduction

Viscosity is the thickness, or "stickiness," of a fluid. The more viscous a fluid, the more resistance to flow. Therefore, the flow rate will be slower for a more viscous solution. For example, consider how much more slowly maple syrup pours out of a container than milk does.

The viscosity of blood is due to the presence of plasma proteins and formed elements, which include white blood cells (leukocytes), red blood cells (erythrocytes), and platelets (thrombocytes). Formed elements and plasma proteins in the blood slide past one another, increasing the resistance to flow. With a viscosity of 3–5, blood is much more viscous than water (usually given a viscosity value of 1).

A body in homeostatic balance has a relatively stable blood consistency. Nevertheless, it is useful to examine the effects of blood viscosity on blood flow to predict what might occur in the human cardiovascular system when homeostatic imbalances occur. Factors such as dehydration and altered blood cell numbers do alter blood viscosity. For example,

polycythemia is a condition in which excess red blood cells are present, and certain types of anemia result in fewer red blood cells. Increasing the number of red blood cells increases blood viscosity, and decreasing the number of red blood cells decreases blood viscosity.

In this activity you will examine the effects of blood viscosity on blood flow rate. The experiment includes two glass beakers and a tube connecting them. Imagine that the left beaker is your heart, the tube is an artery, and the right beaker is a destination in your body, such as another organ.

> **EQUIPMENT USED** The following equipment will be depicted on-screen: left beaker—simulates blood flowing from the heart; flow tube between the left and right beaker—simulates an artery; right beaker—simulates another organ (for example, the biceps brachii muscle).

Experiment Instructions

Go to the home page in the PhysioEx software and click **Exercise 5: Cardiovascular Dynamics.** Click **Activity 2: Studying the Effect of Blood Viscosity on Blood Flow Rate,** and take the online **Pre-lab Quiz** for Activity 2.

After you take the online Pre-lab Quiz, click the **Experiment** tab and begin the experiment. The experiment instructions are reprinted here for your reference. The opening screen for the experiment is shown below.

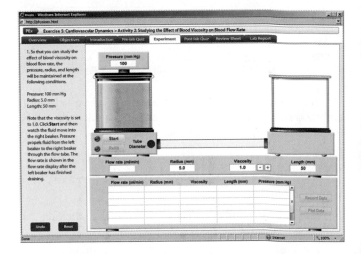

1. So that you can study the effect of blood viscosity on blood flow rate, the pressure, radius, and length will be maintained at the following conditions:

Pressure: 100 mm Hg

Radius: 5.0 mm

Length: 50 mm

Note that the viscosity is set to 1.0. Click **Start** and then watch the fluid move into the right beaker. Pressure propels fluid from the left beaker to the right beaker through the flow tube. The flow rate is shown in the flow rate display after the left beaker has finished draining.

2. Click **Record Data** to display your results in the grid (and record your results in Chart 2).

CHART 2	Effect of Blood Viscosity on Blood Flow Rate
Flow (ml/min)	**Viscosity**

3. Click **Refill** to replenish the left beaker.

> **? PREDICT Question 1**
> What effect do you think increasing the viscosity will have on the fluid flow rate?

4. Increase the fluid viscosity to 2.0 by clicking the + button beside the viscosity display. Click **Start** and then watch the fluid move into the right beaker.

5. Click **Record Data** to display your results in the grid (and record your results in Chart 2).

6. Click **Refill** to replenish the left beaker.

7. You will now observe the effect of incremental increases in viscosity.

- Increase the viscosity by 1.0.
- Click **Start** and then watch the fluid move into the right beaker.
- Click **Record Data** to display your results in the grid (and record your results in Chart 2).
- Click **Refill** to replenish the left beaker.

Repeat this step until you reach a viscosity of 8.0.

8. Click **Plot Data** to view a summary of your data on a plotted grid. Viscosity will be displayed on the X-axis and flow rate will be displayed on the Y-axis. Click **Submit** to record your plot in the lab report.

After you complete the experiment, take the online **Post-lab Quiz** for Activity 2.

Activity Questions

1. Describe the effect on blood flow rate when blood viscosity was increased.

2. Explain why the relationship between viscosity and blood flow rate is inversely proportional.

3. What might happen to blood flow if you increased the number of blood cells?

ACTIVITY 3

Studying the Effect of Blood Vessel Length on Blood Flow Rate
OBJECTIVES

1. To understand how blood vessel length affects blood flow rate.
2. To explain conditions that can lead to blood vessel length changes in the body.
3. To compare the effect of blood vessel length changes with the effect of blood vessel radius changes on blood flow rate.

Introduction

Blood vessel lengths increase as we grow to maturity. The longer the vessel, the greater the resistance to blood flow through the blood vessel because there is a larger surface area in contact with the blood cells. Therefore, when blood vessel length increases, friction increases. Our blood vessel lengths stay fairly constant in adulthood, unless we gain or lose weight. If we gain weight, blood vessel lengths can increase, and if we lose weight, blood vessel lengths can decrease.

In this activity you will study the physical relationship between blood vessel length and blood flow. Specifically, you will study how blood flow changes in blood vessels of constant radius but different lengths. The experiment includes two glass beakers and a tube connecting them. Imagine that the left beaker is your heart, the tube is an artery, and the right beaker is a destination in your body, such as another organ.

> **EQUIPMENT USED** The following equipment will be depicted on-screen: left beaker—simulates blood flowing from the heart; flow tube between the left and right beaker—simulates an artery; right beaker—simulates another organ (for example, the biceps brachii muscle).

Experiment Instructions

Go to the home page in the PhysioEx software and click **Exercise 5: Cardiovascular Dynamics.** Click **Activity 3: Studying the Effect of Blood Vessel Length on Blood Flow Rate,** and take the online **Pre-lab Quiz** for Activity 3.

After you take the online Pre-lab Quiz, click the **Experiment** tab and begin the experiment. The experiment instructions are reprinted here for your reference. The opening screen for the experiment is shown on the following page.

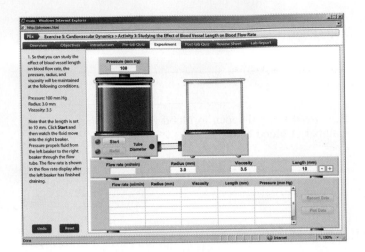

1. So that you can study the effect of blood vessel length on blood flow rate, the pressure, radius, and viscosity will be maintained at the following conditions:

Pressure: 100 mm Hg

Radius: 3.0 mm

Viscosity: 3.5

Note that the length is set to 10 mm. Click **Start** and then watch the fluid move into the right beaker. Pressure propels fluid from the left beaker to the right beaker through the flow tube. The flow rate is shown in the flow rate display after the left beaker has finished draining.

2. Click **Record Data** to display your results in the grid (and record your results in Chart 3).

CHART 3	Effect of Blood Vessel Length on Blood Flow Rate	
Flow (ml/min)	**Flow Tube length (mm)**

3. Click **Refill** to replenish the left beaker.

? PREDICT Question 1
What effect do you think increasing the flow tube length will have on the fluid flow rate?

4. Increase the flow tube length to 15 mm by clicking the + button beside the length display. Click **Start** and then watch the fluid move into the right beaker.

5. Click **Record Data** to display your results in the grid (and record your results in Chart 3).

6. Click **Refill** to replenish the left beaker.

7. You will now observe the effect of incremental increases in flow tube length.

- Increase the flow tube length by 5 mm.
- Click **Start** and then watch the fluid move into the right beaker.
- Click **Record Data** to display your results in the grid (and record your results in Chart 3).
- Click **Refill** to replenish the left beaker.

  Repeat this step until you reach a flow tube length of 40 mm.

8. Click **Plot Data** to view a summary of your data on a plotted grid. Length will be displayed on the X-axis and flow rate will be displayed on the Y-axis. Click **Submit** to record your plot in the lab report.

After you complete the experiment, take the online **Post-lab Quiz** for Activity 3.

Activity Questions

1. Is the relationship between blood vessel length and blood flow rate directly proportional or inversely proportional? Why?

_____

_____

2. Which of the following can vary in size more quickly: blood vessel diameter or blood vessel length?

_____

_____

3. Describe what happens to resistance when blood vessel length increases.

_____

_____

ACTIVITY 4

Studying the Effect of Blood Pressure on Blood Flow Rate

OBJECTIVES

1. To understand how blood pressure affects blood flow rate.
2. To understand what structure produces blood pressure in the human body.
3. To compare the plot generated for pressure versus blood flow to those generated for radius, viscosity, and length.

Introduction

The pressure difference between the two ends of a blood vessel is the driving force behind blood flow. This pressure difference is referred to as a pressure gradient. In the cardiovascular system, the force of contraction of the heart provides the initial pressure and vascular resistance contributes to the pressure gradient. If the heart changes its force of contraction, the blood vessels need to be able to respond to the change in force. Large arteries close to the heart have more elastic tissue in their tunics in order to accommodate these changes.

In this activity you will look at the effect of pressure changes on blood flow (recall from the blood flow equation that a change in blood flow is directly proportional to the pressure gradient). The experiment includes two glass beakers and a tube connecting them. Imagine that the left beaker is your heart, the tube is an artery, and the right beaker is a destination in your body, such as another organ.

> **EQUIPMENT USED** The following equipment will be depicted on-screen: left beaker—simulates blood flowing from the heart; flow tube between the left and right beaker—simulates an artery; right beaker—simulates another organ (for example, the biceps brachii muscle).

Experiment Instructions

Go to the home page in the PhysioEx software and click **Exercise 5: Cardiovascular Dynamics.** Click **Activity 4: Studying the Effect of Blood Pressure on Blood Flow Rate,** and take the online **Pre-lab Quiz** for Activity 4.

After you take the online Pre-lab Quiz, click the **Experiment** tab and begin the experiment. The experiment instructions are reprinted here for your reference. The opening screen for the experiment is shown below.

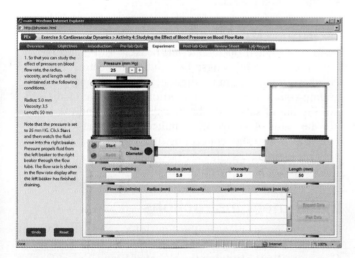

1. So that you can study the effect of pressure on blood flow rate, the radius, viscosity, and length will be maintained at the following conditions:

Radius: 5.0 mm

Viscosity: 3.5

Length: 50 mm

Note that the pressure is set to 25 mm Hg. Click **Start** and then watch the fluid move into the right beaker. Pressure pro-

pels fluid from the left beaker to the right beaker through the flow tube. The flow rate is shown in the flow rate display after the left beaker has finished draining.

2. Click **Record Data** to record your results in the grid (and record your results in Chart 4).

CHART 4	Effect of Blood Pressure on Blood Flow Rate	
Flow (ml/min)	**Pressure (mm Hg)**

3. Click **Refill** to replenish the left beaker.

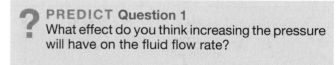

> **PREDICT Question 1**
> What effect do you think increasing the pressure will have on the fluid flow rate?

4. Increase the pressure to 50 mm Hg by clicking the + button beside the pressure display. Click **Start** and then watch the fluid move into the right beaker.

5. Click **Record Data** to record your results in the grid (and record your results in Chart 4).

6. Click **Refill** to replenish the left beaker.

7. You will now observe the effect of incremental increases in pressure.

- Increase the pressure by 25 mm Hg.
- Click **Start** and then watch the fluid move into the right beaker.
- Click **Record Data** to display your results in the grid (and record your results in Chart 4).
- Click **Refill** to replenish the left beaker.

Repeat this step until you reach a pressure of 200 mm Hg.

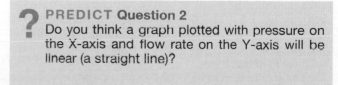

> **PREDICT Question 2**
> Do you think a graph plotted with pressure on the X-axis and flow rate on the Y-axis will be linear (a straight line)?

8. Click **Plot Data** to view a summary of your data on a plotted grid. Pressure will be displayed on the X-axis and flow rate will be displayed on the Y-axis. Click **Submit** to record your plot in the lab report.

After you complete the experiment, take the online **Post-lab Quiz** for Activity 4.

Activity Questions

1. How does increasing the driving pressure affect the blood flow rate?

_____

_____

2. Is the relationship between blood pressure and blood flow rate directly proportional or inversely proportional? Why?

_____

_____

3. How does the cardiovascular system increase pressure?

_____

_____

4. Although changing blood pressure can be used to alter the blood flow rate, this approach causes problems if it continues indefinitely. Explain why.

_____

_____

▬▬

ACTIVITY 5

Studying the Effect of Blood Vessel Radius on Pump Activity

OBJECTIVES

1. To understand the terms *systole* and *diastole*.
2. To predict how a change in blood vessel radius will affect flow rate.
3. To predict how a change in blood vessel radius will affect heart rate.
4. To observe the compensatory mechanisms for maintaining blood pressure.

Introduction

In the human body, the heart beats approximately 70 strokes each minute. Each heartbeat consists of a filling interval, when blood moves into the chambers of the heart, and an ejection period, when blood is actively pumped into the aorta and the pulmonary trunk.

The pumping activity of the heart can be described in terms of the phases of the cardiac cycle. Heart chambers fill

during **diastole** (relaxation of the heart) and pump blood out during **systole** (contraction of the heart). As you can imagine, the length of time the heart is relaxed is one factor that determines the amount of blood within the heart at the end of the filling interval. Up to a point, increasing ventricular filling time results in a corresponding increase in ventricular volume. The volume in the ventricles at the end of diastole, just before cardiac contraction, is called the **end diastolic volume,** or EDV. The volume ejected by a single ventricular contraction is the **stroke volume,** and the volume remaining in the ventricle after contraction is the **end systolic volume,** or **ESV.**

The human heart is a complex, four-chambered organ consisting of two individual pumps (the right and left sides). The right side of the heart pumps blood through the lungs into the left side of the heart. The left side of the heart, in turn, delivers blood to the systems of the body. Blood then returns to the right side of the heart to complete the circuit.

Recall that cardiac output (**CO**) is equal to blood flow. To determine CO, you multiply heart rate (HR) by stroke volume (SV): $CO = HR \times SV$. From the equation for flow (flow $= \Delta P/R$), you can determine the equation for blood pressure: $\Delta P = $ flow $\times R$. Substituting CO in the equation for flow, you get: $\Delta P = HR \times SV \times R$.

Therefore, to maintain blood pressure, the cardiovascular system can alter heart rate, stroke volume, or resistance. For example, if resistance decreases, heart rate can increase to maintain the pressure difference.

In this activity you will explore the operation of a simple, one-chambered pump and apply the physical concepts in the experiment to the operation of either of the two pumps of the human heart. The stroke volume and the difference in pressure will remain constant. You will explore the effect that a change in resistance has on heart rate and the compensatory mechanisms that the cardiovascular system uses to maintain blood pressure.

> **EQUIPMENT USED** The following equipment will be depicted on-screen: left beaker—simulates blood coming from the lungs; flow tube connecting the left beaker and the pump—simulates the pulmonary veins; pump—simulates the left ventricle (the valve to the left of the pump simulates the bicuspid valve, and the valve to the right of the pump simulates the aortic semilunar valve); flow tube connecting the pump and the right beaker—simulates the aorta; right beaker—simulates blood going to the systemic circuit.

Experiment Instructions

Go to the home page in the PhysioEx software and click **Exercise 5: Cardiovascular Dynamics.** Click **Activity 5: Compensation: Studying the Effect of Blood Vessel Radius on Pump Activity** and take the online **Pre-lab Quiz** for Activity 5.

After you take the online Pre-lab Quiz, click the **Experiment** tab and begin the experiment. The experiment instructions are reprinted here for your reference. The opening screen for the experiment is shown on the following page.

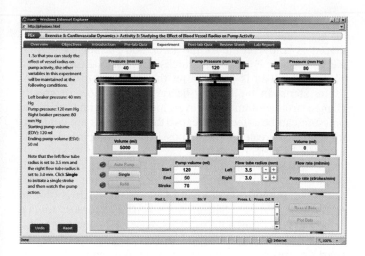

1. So that you can study the effect of vessel radius on pump activity, the other variables in this experiment will be maintained at the following conditions:

Left beaker pressure: 40 mm Hg

Pump pressure: 120 mm Hg

Right beaker pressure: 80 mm Hg

Starting pump volume (EDV): 120 ml

Ending pump volume (ESV): 50 ml

Note that the left flow tube radius is set to 3.5 mm and the right flow tube radius is set to 3.0 mm. Click **Single** to initiate a single stroke and then watch the pump action.

2. Click **Auto Pump** to initiate 10 strokes and then watch the pump action. The flow rate is shown in the flow rate display and the pump rate is shown in the pump rate display after the left beaker has finished draining.

3. Click **Record Data** to display your results in the grid (and record your results in Chart 5).

CHART 5	Effect of Blood Vessel Radius on Pump Activity	
Flow rate (ml/min)	Right radius (mm)	Pump rate (strokes/min)

4. Click **Refill** to replenish the left beaker.

? PREDICT Question 1
If you increase the flow tube radius, what will happen to the pump rate to maintain constant pressure?

5. Increase the right flow tube radius to 3.5 mm by clicking the + button beside the right flow tube radius display. Click **Auto Pump** to initiate 10 strokes and then watch the pump action.

6. Click **Record Data** to display your results in the grid (and record your results in Chart 5).

7. Click **Refill** to replenish the left beaker.

8. You will now observe the effect of incremental increases in the right flow tube radius.

- Increase the right flow tube radius by 0.5 mm.
- Click **Auto Pump** to initiate 10 strokes and then watch the pump action.
- Click **Record Data** to display your results in the grid (and record your results in Chart 5).
- Click **Refill** to replenish the left beaker.

Repeat this step until you reach a right flow tube radius of 5.0 mm.

9. Click **Plot Data** to view a summary of your data on a plotted grid. Right flow tube radius will be displayed on the X-axis and flow rate will be displayed on the Y-axis. Click **Submit** to record your plot in the lab report.

After you complete the experiment, take the online **Post-lab Quiz** for Activity 5.

Activity Questions

1. Describe the position of the pump during diastole.

2. Describe the position of the pump during systole.

3. Describe what happened to the flow rate when the blood vessel radius was increased.

4. Explain what happened to the resistance and the pump rate to maintain pressure when the radius was increased.

ACTIVITY 6

Studying the Effect of Stroke Volume on Pump Activity

OBJECTIVES

1. To understand the effect a change in venous return has on stroke volume.
2. To explain how stroke volume is changed in the heart.
3. To explain the Frank-Starling law of the heart.
4. To define *preload, contractility,* and *afterload.*
5. To distinguish between intrinsic and extrinsic control of contractility of the heart.
6. To explore how heart rate and stroke volume contribute to cardiac output and blood flow.

Introduction

In a normal individual, 60% of the blood contained within the heart is ejected from the heart during ventricular systole, leaving 40% of the blood behind. The blood ejected by the heart—the **stroke volume**—is the difference between the **end diastolic volume (EDV),** the volume in the ventricles at the end of diastole, just before cardiac contraction, and **end systolic volume (ESV),** the volume remaining in the ventricle after contraction. That is, stroke volume = EDV − ESV. Many factors affect stroke volume, the most important of which include *preload, contractility,* and *afterload.* We will look at these defining factors and how they relate to stroke volume.

The Frank-Starling law of the heart states that, when more than the normal volume of blood is returned to the heart by the venous system, the heart muscle will be stretched, resulting in a more forceful contraction of the ventricles. This, in turn, will cause more than normal blood to be ejected by the heart, raising the stroke volume. The degree to which the ventricles are stretched by the end diastolic volume (EDV) is referred to as the **preload.** Thus, the preload results from the amount of ventricular filling between strokes, or the magnitude of the EDV. Ventricular filling could increase when the heart rate is slow because there will be more time for the ventricles to fill. Exercise increases venous return and, therefore, EDV. Factors such as severe blood loss and dehydration decrease venous return and EDV.

The **contractility** of the heart refers to strength of the cardiac muscle contraction (usually the ventricles) and its ability to generate force. A number of extrinsic mechanisms, including the sympathetic nervous system and hormones, control the force of cardiac muscle contraction, but they are not the focus of this activity. The focus of this activity will be the intrinsic controls of contractility (those that reside entirely within the heart). When the end diastolic volume increases, the cardiac muscle fibers of the ventricles stretch and lengthen. As the length of the cardiac sarcomere increases, so does the force of contraction. Cardiac muscle, like skeletal muscle, demonstrates a **length-tension relationship.** At rest, cardiac muscles are at a less than optimum overlap length for maximum tension production in the healthy heart. Therefore, when the heart experiences an increase in stretch with an increase in venous return and, therefore, EDV, it can respond by increasing the force of contraction, yielding a corresponding increase in stroke volume.

Afterload is the back pressure generated by the blood in the aorta and the pulmonary trunk. Afterload is the threshold that must be overcome for the aortic and pulmonary semilunar valves to open. This pressure is referred to as an *after*load because the load is placed after the contraction of the ventricles starts. In the healthy heart, afterload doesn't greatly change stroke volume. However, individuals with high blood pressure can be affected because the ventricles are contracting against a greater pressure, possibly resulting in a decrease in stroke volume.

Cardiac output is equal to the heart rate (HR) multiplied by the stroke volume. Total blood flow is proportional to cardiac output (the amount of blood the heart is able to pump per minute). Therefore, when the stroke volume decreases, the heart rate must increase to maintain cardiac output. Conversely, when the stroke volume increases, the heart rate must decrease to maintain cardiac output.

Even though our simple pump in this experiment does not work exactly like the human heart, you can apply the concepts presented to basic cardiac function. In this activity you will examine how the activity of the pump is affected by changing the starting (EDV) and ending volumes (ESV).

> **EQUIPMENT USED** The following equipment will be depicted on-screen: left beaker—simulates blood coming from the lungs; flow tube connecting the left beaker and the pump—simulates the pulmonary veins; pump—simulates the left ventricle (the valve to the left of the pump simulates the bicuspid valve, and the valve to the right of the pump simulates the aortic semilunar valve); flow tube connecting the pump and the right beaker—simulates the aorta; right beaker—simulates blood going to the systemic circuit.

Experiment Instructions

Go to the home page in the PhysioEx software and click **Exercise 5: Cardiovascular Dynamics.** Click **Activity 6: Studying the Effect of Stroke Volume on Pump Activity** and take the online **Pre-lab Quiz** for Activity 6.

After you take the online Pre-lab Quiz, click the **Experiment** tab and begin the experiment. The experiment instructions are reprinted here for your reference. The opening screen for the experiment is shown below.

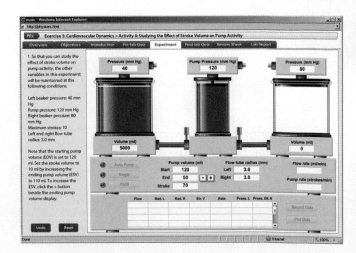

1. So that you can study the effect of stroke volume on pump activity, the other variables in this experiment will be maintained at the following conditions:

Left beaker pressure: 40 mm Hg

Pump pressure: 120 mm Hg

Right beaker pressure: 80 mm Hg

Maximum strokes: 10

Left and right flow tube radius: 3.0 mm

Note that the starting pump volume (EDV) is set to 120 ml. Set the stroke volume to 10 ml by increasing the ending pump volume (ESV) to 110 ml. To increase the ESV, click the + button beside the ending pump volume display.

2. Click **Auto Pump** to initiate 10 strokes and then watch the pump action. The flow rate is shown in the flow rate display and the pump rate is shown in the pump rate display after the left beaker has finished draining.

3. Click **Record Data** to display your results in the grid (and record your results in Chart 6).

CHART 6	Effect of Stroke Volume on Pump Activity	
Flow rate (ml/min)	Stroke volume (ml)	Pump rate (strokes/min)

4. Click **Refill** to replenish the left beaker.

? PREDICT Question 1
If the pump rate is analogous to the heart rate, what do you think will happen to the rate when you increase the stroke volume?

5. Increase the stroke volume to 20 ml by decreasing the ESV. To decrease the ending pump volume, click the − button beside the ending pump volume display.

6. Click **Auto Pump** to initiate 10 strokes and then watch the pump action.

7. Click **Record Data** to display your results in the grid (and record your results in Chart 6).

8. Click **Refill** to replenish the left beaker.

9. You will now observe the effect of incremental increases in the stroke volume.

- Increase the stroke volume by 10 ml by decreasing the ending pump volume (ESV).
- Click **Auto Pump** to initiate 10 strokes and then watch the pump action.
- Click **Record Data** to display your results in the grid (and record your results in Chart 6).
- Click **Replenish** to refill the left beaker.

 Repeat this step until you reach a stroke volume of 60 ml.

10. Increase the stroke volume by 20 ml by decreasing the ending pump volume (ESV). Click **Auto Pump** to initiate 10 strokes and then watch the pump action.

11. Click **Record Data** to display your results in the grid (and record your results in Chart 6).

12. Click **Refill** to replenish the left beaker.

13. Increase the stroke volume by 20 ml by decreasing the ending pump volume (ESV). Click **Auto Pump** to initiate 10 strokes and then watch the pump action.

14. Click **Record Data** to display your results in the grid (and record your results in Chart 6).

15. Click **Plot Data** to view a summary of your data on a plotted grid. Stroke volume will be displayed on the X-axis and flow rate will be displayed on the Y-axis. Click **Submit** to record your plot in the lab report.

After you complete the Experiment, take the online **Post-lab Quiz** for Activity 6.

Activity Questions

1. Describe how the heart responds to an increase in end diastolic volume (include the terms *preload* and *contractility* in your explanation).

_____

_____

2. Explain what happened to the pump rate when the stroke volume increased. Why?

_____

_____

3. Judging from the simulation results, explain why an athlete's resting heart rate might be lower than that of an average person.

_____

_____

ACTIVITY 7

Compensation in Pathological Cardiovascular Conditions

OBJECTIVES

1. To understand how aortic stenosis affects flow of blood through the heart.
2. To explain ways in which the cardiovascular system might compensate for changes in peripheral resistance.
3. To understand how the heart compensates for changes in afterload.
4. To explain how valves affect the flow of blood through the heart.

Introduction

If a blood vessel is compromised, your cardiovascular system can compensate to some degree. Aortic valve stenosis is a condition where there is a partial blockage of the aortic semilunar valve, increasing resistance to blood flow and left ventricular **afterload.** Therefore, the pressure that must be reached to open the aortic valve increases. The heart could compensate for a change in afterload by increasing contractility, the force of contraction. Increasing contractility will increase cardiac output by increasing stroke volume. To increase contractility, the myocardium becomes thicker. Athletes similarly improve their hearts through cardiovascular conditioning. That is, the thickness of the myocardium increases in diseased hearts with aortic valve stenosis and in athletes' hearts (though the chamber volume increases in athletes' hearts and decreases in diseased hearts).

Valves are important in the heart because they ensure that blood flows in one direction through the heart. The valves in the activity will ensure that blood moves in a single direction. Because the right flow tube represents the aorta (which is actually on the left side of the heart), decreasing the right flow tube radius simulates stenosis, or narrowing of the aortic valve.

Plaques in the arteries, known as **atherosclerosis,** can similarly cause an increase in resistance. An increase in peripheral resistance results in a decreased flow rate. Atherosclerosis is a type of **arteriosclerosis** in which the arteries have lost their elasticity. Atherosclerosis is one of the conditions that leads to heart disease.

In this activity you will test three different compensation mechanisms and predict which mechanism will make the best improvement in flow rate. The three mechanisms include (1) increasing the left flow tube radius (that is, increasing preload), (2) increasing the pump's pressure (that is, increasing contractility), and (3) decreasing the pressure in the right beaker (that is, decreasing afterload).

> **EQUIPMENT USED** The following equipment will be depicted on-screen: left beaker—simulates blood coming from the lungs; flow tube connecting the left beaker and the pump—simulates the pulmonary veins; pump—simulates the left ventricle (the valve to the left of the pump simulates the bicuspid valve, and the valve to the right of the pump simulates the aortic semilunar valve); flow tube connecting the pump and the right beaker—simulates the aorta; right beaker—simulates blood going to the systemic circuit.

Experiment Instructions

Go to the home page in the PhysioEx software and click **Exercise 5: Cardiovascular Dynamics.** Click **Activity 7: Compensation in Pathological Cardiovascular Conditions** and take the online **Pre-lab Quiz** for Activity 7.

After you take the online Pre-lab Quiz, click the **Experiment** tab and begin the experiment. The experiment instructions are reprinted here for your reference. The opening screen for the experiment is shown below.

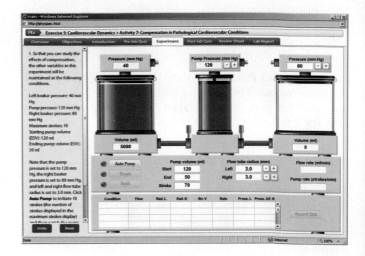

1. So that you can study the effects of compensation, the other variables in this experiment will be maintained at the following conditions:

Left beaker pressure: 40 mm Hg

Maximum strokes: 10

Starting pump volume (EDV): 120 ml

Ending pump volume (ESV): 50 ml

Note that the pump pressure is set to 120 mm Hg, the right beaker pressure is set to 80 mm Hg, and left and right flow tube radius is set to 3.0 mm. Click **Auto Pump** to initiate 10 strokes (the number of strokes displayed in the maximum strokes display) and then watch the pump action. The flow rate is shown in the flow rate display and the pump rate is shown in the pump rate display after the left beaker has finished draining.

2. Click **Record Data** to display your results in the grid (and record your results in Chart 7). This will be your baseline, or "normal," data point for flow rate.

3. Click **Refill** to replenish the left beaker.

4. Decrease the right flow tube radius to 2.5 mm by clicking the − button beside the right flow tube radius display. Click **Auto Pump** to initiate 10 strokes and then watch the pump action.

5. Click **Record Data** to display your results in the grid (and record your results in Chart 7).

6. Click **Refill** to replenish the left beaker.

CHART 7	Compensation Results					
Condition	Flow rate (ml/min)	Left radius (mm)	Right radius (mm)	Pump rate (strokes/ min)	Pump pressure (mm Hg)	Right beaker pressure (mm Hg)

? PREDICT Question 1
You will now test three mechanisms to compensate for the decrease in flow rate caused by the decreased flow tube radius. Which mechanism do you think will have the greatest compensatory effect?

7. Increase the left flow tube radius to 3.5 mm by clicking the + button beside the left flow tube radius display. Click **Auto Pump** to initiate 10 strokes and then watch the pump action.

8. Click **Record Data** to display your results in the grid (and record your results in Chart 7).

9. Click **Refill** to replenish the left beaker.

10. Increase the left flow tube radius to 4.0 mm. Click **Auto Pump** to initiate 10 strokes and then watch the pump action.

11. Click **Record Data** to display your results in the grid (and record your results in Chart 7).

12. Click **Refill** to replenish the left beaker.

13. Increase the left flow tube radius to 4.5 mm. Click **Auto Pump** to initiate 10 strokes and then watch the pump action.

14. Click **Record Data** to display your results in the grid (and record your results in Chart 7).

15. Click **Refill** to replenish the left beaker.

16. Decrease the left flow tube radius to 3.0 mm by clicking the − button beside the left flow tube radius display and increase the pump pressure to 130 mm Hg by clicking the

+ button beside the pump pressure display. Click **Auto Pump** to initiate 10 strokes and then watch the pump action.

17. Click **Record Data** to display your results in the grid (and record your results in Chart 7).

18. Click **Refill** to replenish the left beaker.

19. Increase the pump pressure to 140 mm Hg by clicking the + button beside the pump pressure display. Click **Auto Pump** to initiate 10 strokes and then watch the pump action.

20. Click **Record Data** to display your results in the grid (and record your results in Chart 7).

21. Click **Refill** to replenish the left beaker.

22. Increase the pump pressure to 150 mm Hg. Click **Auto Pump** to initiate 10 strokes and then watch the pump action.

23. Click **Record Data** to display your results in the grid (and record your results in Chart 7).

24. Click **Refill** to replenish the left beaker.

25. Decrease the pump pressure to 120 mm Hg by clicking the − button beside the pump pressure display and decrease the right (destination) beaker pressure to 70 mm Hg by clicking the − button beside the right beaker pressure display. Click **Auto Pump** to initiate 10 strokes and then watch the pump action.

26. Click **Record Data** to display your results in the grid (and record your results in Chart 7).

27. Click **Refill** to replenish the left beaker.

28. Decrease the right (destination) beaker pressure to 60 mm Hg by clicking the − button beside the right beaker pressure display. Click **Auto Pump** to initiate 10 strokes and then watch the pump action.

29. Click **Record Data** to display your results in the grid (and record your results in Chart 7).

30. Click **Refill** to replenish the left beaker.

31. Decrease the right (destination) beaker pressure to 50 mm Hg. Click **Auto Pump** to initiate 10 strokes and then watch the pump action.

32. Click **Record Data** to display your results in the grid (and record your results in Chart 7).

33. Click **Refill** to replenish the left beaker.

> **?** **PREDICT** Question 2
> What do you think will happen if the pump pressure and the beaker pressure are the same?
> _____

34. Increase the right (destination) beaker pressure to 120 mm Hg by clicking the + button beside the right beaker pressure display. Click **Auto Pump** to initiate 10 strokes and then watch the pump action.

After you complete the experiment, take the online **Post-lab Quiz** for Activity 7.

Activity Questions

1. Explain why a thicker myocardium is seen in both the athlete's heart and the diseased heart.

_____

_____

2. Describe what the term *afterload* means.

_____

_____

3. Explain which mechanism in the simulation had the greatest compensatory effect.

_____

_____

4. Describe the mechanism used in the human heart to compensate for aortic stenosis.

_____

_____

NAME _____

LAB TIME/DATE _____

Cardiovascular Dynamics

ACTIVITY 1 Studying the Effect of Blood Vessel Radius on Blood Flow Rate

1. Explain how the body establishes a pressure gradient for fluid flow. _____

2. Explain the effect that the flow tube radius change had on flow rate. How well did the results compare with your prediction?

_____

_____

3. Describe the effect that radius changes have on the laminar flow of a fluid. _____

_____

4. Why do you think the plot was not linear? (Hint: Look at the relationship of the variables in the equation.) How well did the

results compare with your prediction? _____

_____

ACTIVITY 2 Studying the Effect of Blood Viscosity on Blood Flow Rate

1. Describe the components in the blood that affect viscosity. _____

_____

2. Explain the effect that the viscosity change had on flow rate. How well did the results compare with your prediction?

_____

_____

3. Describe the graph of flow versus viscosity. _____

_____

4. Discuss the effect that polycythemia would have on viscosity and on blood flow. _____

_____

_____

ACTIVITY 3 Studying the Effect of Blood Vessel Length on Blood Flow Rate

1. Which is more likely to occur, a change in blood vessel radius or a change in blood vessel length? Explain why.

   _____

   _____

2. Explain the effect that the change in blood vessel length had on flow rate. How well did the results compare with your

   prediction? _____

   _____

3. Explain why you think blood vessel radius can have a larger effect on the body than changes in blood vessel length (use the

   blood flow equation). _____

   _____

4. Describe the effect that obesity would have on blood flow and why. _____

   _____

ACTIVITY 4 Studying the Effect of Blood Pressure on Blood Flow Rate

1. Explain the effect that pressure changes had on flow rate. How well did the results compare with your prediction?

   _____

2. How does the plot differ from the plots for tube radius, viscosity, and tube length? How well did the results compare with

   your prediction? _____

   _____

3. Explain why pressure changes are not the best way to control blood flow. _____

   _____

4. Use your data to calculate the increase in flow rate in ml/min/mm Hg. _____

   _____

ACTIVITY 5 Studying the Effect of Blood Vessel Radius on Pump Activity

1. Explain the effect of increasing the right flow tube radius on the flow rate, resistance, and pump rate. _____

   _____

2. Describe what the left and right beakers in the experiment correspond to in the human heart. _____

_____

3. Briefly describe how the human heart could compensate for flow rate changes to maintain blood pressure. _____

_____

ACTIVITY 6 Studying the Effect of Stroke Volume on Pump Activity

1. Describe the Frank-Starling law in the heart. _____

_____

2. Explain what happened to the pump rate when you increased the stroke volume. Why do you think this occurred? How well

did the results compare with your prediction? _____

_____

3. Describe how the heart alters stroke volume. _____

_____

4. Describe the intrinsic factors that control stroke volume. _____

_____

ACTIVITY 7 Compensation in Pathological Cardiovascular Conditions

1. Explain how the heart could compensate for changes in peripheral resistance. _____

_____

2. Which mechanism had the greatest compensatory effect? How well did the results compare with your prediction? _____

_____

3. Explain what happened when the pump pressure and the beaker pressure were the same. How well did the results compare

with your prediction? _____

_____

4. Explain whether it would be better to adjust heart rate or blood vessel diameter to achieve blood flow changes at a local level

(for example, in just the digestive system). _____

_____

Cardiovascular Physiology

PRE-LAB QUIZ

1. Circle True or False: Cardiac muscle and some types of smooth muscle have the ability to contract without any external stimuli.

2. The total cardiac action potential lasts 250–300 milliseconds and has _____ phases.
   a. two   c. four
   b. three   d. five

3. The _____ nerve carries parasympathetic signals to the heart.
   a. phrenic   c. splanchnic
   b. vagus   d. visceral

4. Circle True or False: Stimulation of the parasympathetic nervous system increases the heart rate and the force of contraction of the heart.

5. What is *vagal escape*?

   _____

6. The frog heart is different from the human heart in that it contains:
   a. one atrium and two ventricles
   b. two atria and a single, incompletely divided ventricle
   c. two atria and two ventricles (it is not different)

7. The _____ node has the fastest rate of depolarization and determines the heart rate.
   a. atrioventricular
   b. atriosinus
   c. sinoatrial
   d. ventriculosino

8. Circle the correct underlined term: Humans are able to maintain an external body temperature of 35.8–38.2°C in spite of environmental conditions and are referred to as <u>poikilotherms</u> / <u>homeotherms</u>.

9. Circle True or False: Sympathetic nerve fibers release epinephrine and acetylcholine at their cardiac synapses.

10. The resting cell membrane in cardiac muscle cells favors the movement of _____ ions.
    a. potassium
    b. calcium
    c. sodium
    d. magnesium

Exercise Overview

Cardiac muscle and some types of smooth muscle contract spontaneously, without any external stimuli. Skeletal muscle is unique in that it requires depolarizing signals from the nervous system to contract. The heart's ability to trigger its own contractions is called **autorhythmicity.**

If you isolate cardiac pacemaker muscle cells, place them into cell culture, and observe them under a microscope, you can see the cells contract. Autorhythmicity occurs because the plasma membrane in cardiac pacemaker muscle cells has reduced permeability to potassium ions but still allows sodium and calcium ions to slowly leak into the cells. This leakage causes the muscle cells to slowly depolarize until the action potential threshold is reached and L-type calcium channels open, allowing Ca^{2+} entry from the extracellular fluid. Shortly thereafter, contraction of the remaining cardiac muscle occurs prior

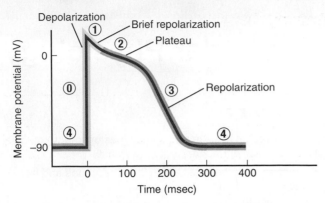

FIGURE 6.1 The cardiac action potential.

to potassium-dependent repolarization. The spontaneous depolarization-repolarization events occur in a regular and continuous manner in cardiac pacemaker muscle cells, leading to **cardiac action potentials** in the majority of cardiac muscle.

There are five main phases of membrane polarization in a cardiac action potential (view Figure 6.1).

- **Phase 0** is similar to depolarization in the neuronal action potential. Depolarization causes voltage-gated sodium channels in the cell membrane to open, increasing the flow of sodium ions into the cell and increasing the membrane potential.

- In **phase 1,** the open sodium channels begin to inactivate, decreasing the flow of sodium ions into the cell and causing the membrane potential to fall slightly. At the same time, voltage-gated potassium channels close and voltage-gated calcium channels open. The subsequent decrease in the flow of potassium out of the cell and increase in the flow of calcium into the cell act to depolarize the membrane and curb the fall in membrane potential caused by the inactivation of sodium channels.

- In **phase 2,** known as the **plateau phase,** the membrane remains in a depolarized state. Potassium channels stay closed, and long-lasting (L-type) calcium channels stay open. This plateau lasts about 0.2 seconds, or 200 milliseconds.

- In **phase 3,** the membrane potential gradually falls to more negative values when a second set of potassium channels that began opening in phases 1 and 2 allows significant amounts of potassium to flow out of the cell. The falling membrane potential causes calcium channels to close, reducing the flow of calcium into the cell and repolarizing the membrane until the resting potential is reached.

- In **phase 4,** the resting membrane potential is again established in cardiac muscle cells and is maintained until the next depolarization arrives from neighboring cardiac pacemaker cells.

The total cardiac action potential lasts 250–300 milliseconds.

Investigating the Refractory Period of Cardiac Muscle

OBJECTIVES

1. To observe the autorhythmicity of the heart.
2. To understand the phases of the cardiac action potential.
3. To induce extrasystoles and observe them on the oscilloscope tracing of contractile activity in the isolated, intact frog heart.
4. To relate the presence or absence of wave summation and tetanus in cardiac muscle to the refractory period of the cardiac action potential.

Introduction

Recall that **wave summation** occurs when a skeletal muscle is stimulated with such frequency that muscle twitches overlap and result in a stronger contraction than a single muscle twitch. When the stimulations are frequent enough, the muscle reaches a state of fused tetanus, during which the individual muscle twitches cannot be distinguished. Tetanus occurs in skeletal muscle because skeletal muscle has a relatively short **absolute refractory period** (a period during which action potentials cannot be generated no matter how strong the stimulus).

Unlike skeletal muscle, cardiac muscle has a relatively long refractory period and is thus incapable of wave summation. In fact, cardiac muscle is incapable of reacting to *any* stimulus before approximately the middle of phase 3, and will not respond to a normal cardiac stimulus before phase 4. The period of time between the beginning of the cardiac action potential and the approximate middle of phase 3 is the **absolute refractory period.** The period of time between the absolute refractory period and phase 4 is the **relative refractory period.** The total refractory period of cardiac muscle is 200–250 milliseconds—almost as long as the contraction of the cardiac muscle.

In this activity you will use external stimulation to better understand the refractory period of cardiac muscle. You will use a frog heart, which is anatomically similar to the human heart. The frog heart has two atria and a single, incompletely divided ventricle.

> **EQUIPMENT USED** The following equipment will be depicted on-screen: oscilloscope display— displays the contractile activity from the frog heart; electrical stimulator—used to apply electrical shocks to the frog heart; electrode holder—locks electrodes in place for stimulation; external stimulation electrode; apparatus for sustaining an isolated frog heart—includes 23°C Ringer's solution; frog heart.

Experiment Instructions

Go to the home page in the PhysioEx software and click **Exercise 6: Cardiovascular Physiology.** Click **Activity 1: Investigating the Refractory Period of Cardiac Muscle,** and take the online **Pre-lab Quiz** for Activity 1.

After you take the online Pre-lab Quiz, click the **Experiment** tab and begin the experiment. The experiment

instructions are reprinted here for your reference. The opening screen for the experiment is shown below.

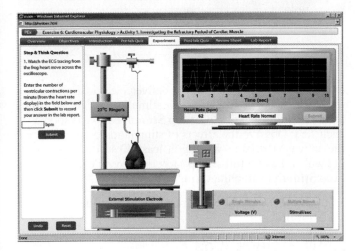

1. Watch the contractile activity from the frog heart on the oscilloscope. Enter the number of ventricular contractions per minute (from the heart rate display) in the field below and then click **Submit** to record your answer in the lab report.

_____ beats/min

2. Drag the external stimulation electrode to the electrode holder to the right of the frog heart. The electrode will touch the ventricular muscle tissue.

> **? PREDICT Question 1**
> When you increase the frequency of the stimulation, what do you think will happen to the amplitude (height) of the ventricular systole wave?

3. Deliver single shocks in succession by clicking **Single Stimulus** rapidly. You might need to practice to acquire the correct technique. You should see a "doublet," or double peak, which contains an **extrasystole,** or extra contraction of the ventricle, and then a compensatory pause, which allows the heart to get back on schedule after the extrasystole. When you see a doublet, click **Submit** to record the tracing in the lab report.

> **? PREDICT Question 2**
> If you deliver multiple stimuli (20 stimuli per second) to the heart, what do you think will happen?

4. Click **Multiple Stimuli** to deliver electrical shocks to the heart at a rate of 20 stimuli/sec. The **Multiple Stimuli** button changes to a **Stop Stimuli** button as soon as it is clicked. Observe the effects of stimulation on the contractile activity and, after a few seconds, click **Stop Stimuli** to stop the stimuli.

After you complete the experiment, take the online **Post-lab Quiz** for Activity 1.

Activity Questions

1. Describe how the frog heart and human heart differ anatomically.

_____

_____

2. What does an extrasystole correspond to? How did you induce an extrasystole?

_____

_____

3. Explain why it is important that wave summation and tetanus do not occur in the cardiac muscle.

_____

_____

_____ ▬

ACTIVITY 2

Examining the Effect of Vagus Nerve Stimulation

OBJECTIVES

1. To understand the role that the sympathetic and parasympathetic nervous systems have on heart activity.
2. To explain the consequences of vagal stimulation and vagal escape.
3. To explain the functionality of the sinoatrial node.

Introduction

The autonomic nervous system has two branches: the **sympathetic** nervous system ("fight or flight") and **parasympathetic** nervous system ("resting and digesting"). At rest both the sympathetic and parasympathetic nervous systems are working but the parasympathetic branch is more active. The sympathetic nervous system becomes more active when needed, for example, during exercise and when confronting danger.

Both the parasympathetic and sympathetic nervous systems supply nerve impulses to the heart. Stimulation of the sympathetic nervous system increases the rate and force of contraction of the heart. Stimulation of the parasympathetic nervous system decreases the heart rate without directly changing the force of contraction. The vagus nerve (cranial nerve X) carries the signal to the heart. If stimulation of the vagus nerve (vagal stimulation) is excessive, the heart will stop beating. After a short time, the ventricles will begin to beat again. The resumption of the heartbeat is referred to as **vagal escape** and can be the result of sympathetic reflexes or initiation of a rhythm by the Purkinje fibers.

The **sinoatrial node (SA node)** is a cluster of autorhythmic cardiac cells found in the right atrial wall in the human heart. The SA node has the fastest rate of spontaneous depolarization, and, for that reason, it determines the heart rate and is therefore referred to as the heart's **"pacemaker."** In the absence of parasympathetic stimulation, sympathetic stimulation, and hormonal controls, the SA node generates action potentials 100 times per minute.

> EQUIPMENT USED The following equipment will be depicted on-screen: oscilloscope display—displays the contractile activity from the frog heart; electrical stimulator—used to apply electrical shocks to the frog heart; electrode holder—locks electrodes in place for stimulation; vagus nerve stimulation electrode; apparatus for sustaining an isolated, intact frog heart—includes 23°C Ringer's solution; frog heart with vagus nerve (thin, white strand to the right).

Experiment Instructions

Go to the home page in the PhysioEx software and click **Exercise 6: Cardiovascular Physiology.** Click **Activity 2: Examining the Effect of Vagus Nerve Stimulation,** and take the online **Pre-lab Quiz** for Activity 2.

After you take the online Pre-lab Quiz, click the **Experiment** tab and begin the experiment. The experiment instructions are reprinted here for your reference. The opening screen for the experiment is shown below.

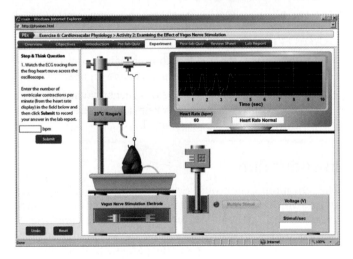

1. Watch the contractile activity from the frog heart on the oscilloscope. Enter the number of ventricular contractions per minute (from the heart rate display) in the field below and then click **Submit** to record your answer in the lab report.

_____ beats/min

2. Drag the vagus nerve stimulation electrode to the electrode holder to the right of the heart. Note that, when the electrode locks in place, the vagus nerve is draped over the electrode. Stimuli will go directly to the vagus nerve and indirectly to the heart.

3. Enter the number of ventricular contractions per minute (from the heart rate display) in the field below and then click **Submit** to record your answer in the lab report.

_____ beats/min

> ? **PREDICT Question 1**
> What do you think will happen if you apply multiple stimuli to the heart by indirectly stimulating the vagus nerve?

_____

4. Click **Multiple Stimuli** to deliver electrical shocks to the vagus nerve at a rate of 50 stimuli/sec. The **Multiple Stimuli** button changes to a **Stop Stimuli** button as soon as it is clicked. Observe the effects of stimulation on the contractile activity and, after waiting at least 20 seconds (the tracing will make two full sweeps across the oscilloscope), click **Stop Stimuli** to stop the stimuli.

After you complete the experiment, take the online **Post-lab Quiz** for Activity 2.

Activity Questions

1. Describe how stimulation of the vagus nerves affects the heart rate.

_____

_____

2. How does the sympathetic nervous system affect heart rate and the force of contraction?

_____

_____

3. Describe the mechanism of vagal escape.

_____

_____

4. What would happen to the heart rate if the vagus nerve were cut?

_____

_____

ACTIVITY 3

Examining the Effect of Temperature on Heart Rate

OBJECTIVES

1. To define the terms *hyperthermia* and *hypothermia*.
2. To contrast the terms *homeothermic* and *poikilothermic*.
3. To understand the effect that temperature has on the frog heart.
4. To understand the effect that temperature could have on the human heart.

Introduction

Humans are **homeothermic,** which means that the human body maintains an internal body temperature within the 35.8–38.2°C range even though the external temperature is changing. When the external temperature is elevated, the hypothalamus is signaled to activate heat-releasing mechanisms, such as sweating and vasodilation, to maintain the body's internal temperature. During extreme external temperature conditions, the body might not be able to maintain homeostasis and either **hyperthermia** (elevated body temperature) or **hypothermia** (low body temperature) could result. In contrast, the frog is a **poikilothermic** animal. Its internal body temperature changes depending on the temperature of its external environment because it lacks internal homeostatic regulatory mechanisms.

Ringer's solution, also known as Ringer's irrigation, consists of essential electrolytes (chloride, sodium, potassium, calcium, and magnesium) in a physiological solution and is required to keep the isolated, intact heart viable. In this activity you will explore the effect of temperature on heart rate using a Ringer's solution incubated at different temperatures.

EQUIPMENT USED The following equipment will be depicted on-screen: oscilloscope display—displays the contractile activity from the frog heart; electrical stimulator—used to apply electrical shocks to the frog heart; electrode holder—locks electrodes in place for stimulation; external stimulation electrode; apparatus for sustaining an isolated, intact frog heart—includes 5°C, 23°C, and 32°C Ringer's solution; frog heart.

Experiment Instructions

Go to the home page in the PhysioEx software and click **Exercise 6: Cardiovascular Physiology.** Click **Activity 3: Examining the Effect of Temperature on Heart Rate,** and take the online **Pre-lab Quiz** for Activity 3.

After you take the online Pre-lab Quiz, click the **Experiment** tab and begin the experiment. The experiment instructions are reprinted here for your reference. The opening screen for the experiment is shown below.

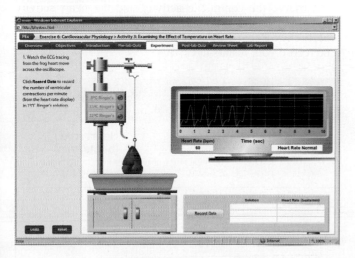

1. Watch the contractile activity from the frog heart on the oscilloscope. Click **Record Data** to record the number of ventricular contractions per minute (from the heart rate display) in 23°C Ringer's solution.

? PREDICT Question 1
What effect will decreasing the temperature of the Ringer's solution have on the heart rate of the frog?

_____

---

2. Click **5°C Ringer's** to observe the effects of lowering the temperature.

3. When the heart activity display reads *Heart Rate Stable,* click **Record Data** to display your results in the grid (and record your results in Chart 3).

CHART 3	Effect of Temperature on Heart Rate
Solution	**Heart rate (beats/min)**

4. Click **23°C Ringer's** to bathe the heart and return it to room temperature. When the heart activity display reads *Heart Rate Normal,* you can proceed.

? PREDICT Question 2
What effect will increasing the temperature of the Ringer's solution have on the heart rate of the frog?

_____

---

5. Click **32°C Ringer's** to observe the effects of increasing the temperature.

6. When the heart activity display reads *Heart Rate Stable,* click **Record Data** to display your results in the grid (and record your results in Chart 3).

After you complete the experiment, take the online **Post-lab Quiz** for Activity 3.

Activity Questions

1. Explain the importance of Ringer's solution (essential electrolytes in physiological saline) in maintaining the auto-rhythmicity of the heart.

_____

_____

2. Describe the effect of lower temperature on heart rate.

_____

_____

3. Explain the effect that fever would have on heart rate. Explain why.

_____

_____

_____ ▬

ACTIVITY 4

Examining the Effects of Chemical Modifiers on Heart Rate

OBJECTIVES

1. To distinguish between cholinergic and adrenergic modifiers of heart rate.
2. To define agonist and antagonist modifiers of heart rate.
3. To observe the effects of epinephrine, pilocarpine, atropine, and digitalis on heart rate.
4. To relate chemical modifiers of the heart rate to sympathetic and parasympathetic activation.

Introduction

Although the heart does not need external stimulation to beat, it can be affected by extrinsic controls, most notably the autonomic nervous system. The sympathetic nervous system is activated in times of "fight or flight," and sympathetic nerve fibers release **norepinephrine** (also known as **noradrenaline**) and **epinephrine** (also known as **adrenaline**) at their cardiac synapses.

Norepinephrine and epinephrine increase the frequency of action potentials by binding to β_1 adrenergic receptors embedded in the plasma membrane of **sinoatrial (SA) node** (pacemaker) cells. Working through a cAMP second-messenger mechanism, binding of the ligand opens sodium and calcium channels, increasing the rate of depolarization and shortening the period of repolarization, thus increasing the heart rate.

The parasympathetic nervous system, our "resting and digesting branch," usually dominates, and parasympathetic nerve fibers release **acetylcholine** at their cardiac synapses. Acetylcholine decreases the frequency of action potentials by binding to muscarinic cholinergic receptors embedded in the plasma membrane of the SA node cells. Acetylcholine indirectly opens potassium channels and closes calcium and sodium channels, decreasing the rate of depolarization and, thus, decreasing heart rate.

Chemical modifiers that inhibit, mimic, or enhance the action of acetylcholine in the body are labeled **cholinergic.** Chemical modifiers that inhibit, mimic, or enhance the action of epinephrine in the body are **adrenergic.** If the modifier works in the same fashion as the neurotransmitter (acetylcholine or norepinephrine), it is an **agonist.** If the modifier works in opposition to the neurotransmitter, it is an **antagonist.** In this activity you will explore the effects of pilocarpine, atropine, epinephrine, and digitalis on heart rate.

> **EQUIPMENT USED** The following equipment will be depicted on-screen: oscilloscope display—displays the contractile activity; apparatus for sustaining an isolated intact frog heart—includes 23°C Ringer's solution; pilocarpine; atropine; epinephrine; digitalis; frog heart.

Experiment Instructions

Go to the home page in the PhysioEx software and click **Exercise 6: Cardiovascular Physiology.** Click **Activity 4: Examining the Effects of Chemical Modifiers on Heart Rate,** and take the online **Pre-lab Quiz** for Activity 4.

After you take the online Pre-lab Quiz, click the **Experiment** tab and begin the experiment. The experiment instructions are reprinted here for your reference. The opening screen for the experiment is shown below.

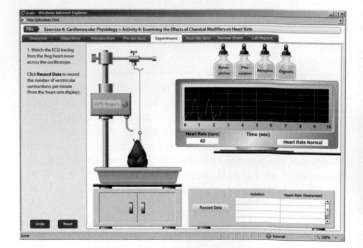

1. Watch the contractile activity from the frog heart on the oscilloscope. Click **Record Data** to record the number of ventricular contractions per minute (from the heart rate display) and record your results in Chart 4.

2. Drag the dropper cap of the epinephrine bottle to the frog heart to release epinephrine onto the heart.

3. Observe the contractile activity and the heart activity display. When the heart activity display reads *Heart Rate Stable,* click **Record Data** to display your results in the grid (and record your results in Chart 4).

CHART 4	Effects of Chemical Modifiers on Heart Rate	
Solution	**Heart rate (beats/min)**

4. Click **23°C Ringer's** (room temperature) to bathe the heart and flush out the epinephrine. When the heart activity display reads *Heart Rate Normal,* you can proceed.

> **? PREDICT Question 1**
> Pilocarpine is a cholinergic drug, an acetylcholine agonist. Predict the effect that pilocarpine will have on heart rate.

5. Drag the dropper cap of the pilocarpine bottle to the frog heart to release pilocarpine onto the heart.

6. Observe the contractile activity and the heart activity display. When the heart activity display reads *Heart Rate Stable,* click **Record Data** to display your results in the grid (and record your results in Chart 4).

7. Click **23°C Ringer's** (room temperature) to bathe the heart and flush out the pilocarpine. When the heart activity display reads *Heart Rate Normal,* you can proceed.

> **? PREDICT Question 2**
> Atropine is another cholinergic drug, an acetylcholine antagonist. Predict the effect that atropine will have on heart rate.

8. Drag the dropper cap of the atropine bottle to the frog heart to release atropine onto the heart.

9. Observe the contractile activity and the heart activity display. When the heart activity display reads *Heart Rate Stable,* click **Record Data** to display your results in the grid (and record your results in Chart 4).

10. Click **23°C Ringer's** (room temperature) to bathe the heart and flush out the atropine. When the heart activity display reads *Heart Rate Normal,* you can proceed.

11. Drag the dropper cap of the digitalis bottle to the frog heart to release digitalis onto the heart.

12. Observe the contractile activity and the heart activity display. When the heart activity display reads *Heart Rate Stable,* click **Record Data** to display your results in the grid (and record your results in Chart 4).

After you complete the experiment, take the online **Post-lab Quiz** for Activity 4.

Activity Questions

1. Define *agonist* and *antagonist.* Clearly distinguish between the two and give examples used in this activity.

2. Describe the effect of epinephrine on heart rate and force of contraction.

3. What is the effect of atropine on heart rate?

4. Describe the effect of digitalis on heart rate and force of contraction.

ACTIVITY 5

Examining the Effects of Various Ions on Heart Rate

OBJECTIVES

1. To understand the movement of ions that occurs during the cardiac action potential.

2. To describe the potential effect of potassium, sodium, and calcium ions on heart rate.

3. To explain how calcium channel blockers might be used pharmaceutically to treat heart patients.

4. To define the terms *inotropic* and *chronotropic.*

Introduction

In cardiac muscle cells, action potentials are caused by changes in permeability to ions due to the opening and closing of ion channels. The permeability changes that occur for the cardiac muscle cell involve potassium, sodium, and calcium ions. The concentration of potassium is greater inside the cardiac muscle cell than outside the cell. Sodium and calcium are present in larger quantities outside the cell than inside the cell.

The resting cell membrane favors the movement of potassium more than sodium or calcium. Therefore, the resting membrane potential of cardiac cells is determined mainly by the ratio of extracellular and intracellular concentrations of potassium. View Table 6.1 on p. PEx-100 for a summary of the phases of the cardiac action potential and ion movement during each phase.

Calcium channel blockers are used to treat high blood pressure and abnormal heart rates. They block the movement of calcium through its channels throughout all phases of the cardiac action potentials. Consequently, because less calcium gets through, both the rate of depolarization and the force of the contraction are reduced. Modifiers that affect heart rate are **chronotropic,** and modifiers that affect the force of contraction are **inotropic.** Modifiers that lower heart rate are negative chronotropic, and modifiers that increase heart rate are positive chronotropic. The same adjectives describe inotropic modifiers. Therefore, negative inotropic drugs decrease the force of contraction of the heart and positive inotropic drugs increase the force of contraction of the heart.

TABLE 6.1

Phase of cardiac action potential	Ion movement
Phase 0 (rapid depolarization)	Sodium moves in
Phase 1 (small repolarization)	Sodium movement decreases
Phase 2 (plateau)	Potassium movement out decreases Calcium moves in
Phase 3 (repolarization)	Potassium moves out Calcium movement decreases
Phase 4 (resting potential)	Potassium moves out Little sodium or calcium moves in

EQUIPMENT USED The following equipment will be depicted on-screen: oscilloscope display—displays the contractile activity from the frog heart; apparatus for sustaining frog heart—includes 23°C Ringer's solution; calcium ions; sodium ions; potassium ions; frog heart.

Experiment Instructions

Go to the home page in the PhysioEx software and click **Exercise 6: Cardiovascular Physiology.** Click **Activity 5: Examining the Effects of Various Ions on Heart Rate,** and take the online **Pre-lab Quiz** for Activity 5.

After you take the online Pre-lab Quiz, click the **Experiment** tab and begin the experiment. The experiment instructions are reprinted here for your reference. The opening screen for the experiment is shown below.

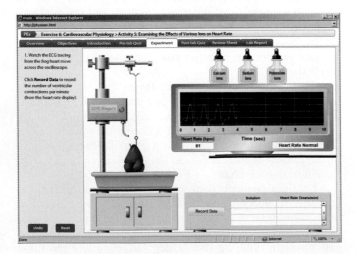

1. Watch the contractile activity from the frog heart move on the oscilloscope. Click **Record Data** to record the number of ventricular contractions per minute (from the heart rate display).

? PREDICT Question 1
Because calcium channel blockers are negative chronotropic and negative inotropic, what effect do you think increasing the concentration of calcium will have on heart rate?

2. Drag the dropper cap of the calcium ions bottle to the frog heart to release calcium ions onto the heart. Note the change in heart rate after you drop the calcium ions onto the heart.

3. When the heart activity display reads *Heart Rate Stable,* click **Record Data** to display your results in the grid (and record your results in Chart 5).

CHART 5	Effects of Various Ions on Heart Rate	
Solution	**Heart rate (beats/min)**

4. **Click 23°C Ringer's** (room temperature) to bathe the heart and flush out the calcium. When the heart activity display reads *Heart Rate Normal,* you can proceed.

5. Drag the dropper cap of the sodium ions bottle to the frog heart to release sodium ions onto the heart. Note the immediate change in the heart rate and the change in heart rate over time after you drop the sodium ions onto the heart.

6. After waiting at least 20 seconds (the tracing will make two full sweeps across the oscilloscope), click **Record Data** to display your results in the grid (and record your results in Chart 5).

7. Click **23°C Ringer's** (room temperature) to bathe the heart and flush out the sodium. When the heart activity display reads *Heart Rate Normal,* you can proceed.

? PREDICT Question 2
Excess potassium outside of the cardiac cell decreases the resting potential of the plasma membrane, thus decreasing the force of contraction. What effect (if any) do you think it will *initially* have on heart rate?

8. Drag the dropper cap of the potassium ions bottle to the frog heart to release potassium ions onto the heart. Note the immediate change in heart rate and the change in heart rate over time after you drop the potassium ions onto the heart.

9. After waiting at least 20 seconds (the tracing will make two full sweeps across the oscilloscope), click **Record Data** to display your results in the grid (and record your results in Chart 5).

After you complete the experiment, take the online **Post-lab Quiz** for Activity 5.

Activity Questions

1. Define chronotropic and inotropic effects on the heart.

_____

_____

2. Describe the effect of adding calcium ions to the frog heart.

_____

_____

3. Calcium channel blockers are often used to treat high blood pressure. Explain how their effects would benefit individuals with high blood pressure.

_____

_____

4. Describe the initial effect of adding potassium ions to the frog heart.

_____

_____

NAME _____

LAB TIME/DATE _____

Cardiovascular Physiology

ACTIVITY 1 Investigating the Refractory Period of Cardiac Muscle

1. Explain why the larger waves seen on the oscilloscope represent ventricular contraction.

   _____

   _____

2. Explain why the amplitude of the wave did not change when you increased the frequency of the stimulation. (Hint: Relate your response to the refractory period of the cardiac action potential.) How well did the results compare with your prediction?

   _____

   _____

3. Why is it only possible to induce an extrasystole during relaxation? _____

   _____

4. Explain why wave summation and tetanus are not possible in cardiac muscle tissue. How well did the results compare with your prediction? _____

   _____

ACTIVITY 2 Examining the Effect of Vagus Nerve Stimulation

1. Explain the effect that extreme vagus nerve stimulation had on the heart. How well did the results compare with your prediction?

   _____

   _____

2. Explain two ways that the heart can overcome excessive vagal stimulation. _____

   _____

3. Describe how the sympathetic and parasympathetic nervous systems work together to regulate heart rate. _____

   _____

4. What do you think would happen to the heart rate if the vagus nerve was cut? _____

   _____

ACTIVITY 3 Examining the Effect of Temperature on Heart Rate

1. Explain the effect that decreasing the temperature had on the frog heart. How do you think the human heart would respond?

   How well did the results compare with your prediction? _____

   _____

2. Describe why Ringer's solution is required to maintain heart contractions. _____

   _____

3. Explain the effect that increasing the temperature had on the frog heart. How do you think the human heart would respond?

   How well did the results compare with your prediction? _____

   _____

ACTIVITY 4 Examining the Effects of Chemical Modifiers on Heart Rate

1. Describe the effect that pilocarpine had on the heart and why it had this effect. How well did the results compare with your

   prediction? _____

   _____

2. Atropine is an acetylcholine antagonist. Does atropine inhibit or enhance the effects of acetylcholine? Describe your results

   and how they correlate with how the drug works. How well did the results compare with your prediction? _____

   _____

3. Describe the benefits of administering digitalis. _____

   _____

4. Distinguish between cholinergic and adrenergic chemical modifiers. Include examples of each in your discussion. _____

   _____

ACTIVITY 5 Examining the Effects of Various Ions on Heart Rate

1. Describe the effect that increasing the calcium ions had on the heart. How well did the results compare with your prediction?

   _____

2. Describe the effect that increasing the potassium ions initially had on the heart in this activity. Relate this to the resting

   membrane potential of the cardiac muscle cell. How well did the results compare with your prediction? _____

   _____

3. Describe how calcium channel blockers are used to treat patients and why. _____

   _____

Respiratory System Mechanics

Exercise Overview

The physiological function of the respiratory system is essential to life. If problems develop in most other physiological systems, we can survive for some time without addressing them. But if a persistent problem develops within the respiratory system (or the circulatory system), death can occur in minutes.

The primary role of the respiratory system is to distribute oxygen to, and remove carbon dioxide from, *all* the cells of the body. The respiratory system works together with the circulatory system to achieve this. **Respiration** includes **ventilation,** or the movement of air into and out of the lungs (breathing), and the transport (via blood) of oxygen and carbon dioxide between the lungs and body cells (view Figure 7.1, p. PEx-106). The heart pumps deoxygenated blood to pulmonary capillaries, where gas exchange occurs between blood and **alveoli** (air sacs in the lungs), thus oxygenating the blood. The heart then pumps the oxygenated blood to body tissues, where oxygen is used for cell metabolism. At the same time, carbon dioxide (a waste product of metabolism) from body tissues diffuses into the blood. This carbon dioxide–enriched, oxygen-reduced blood then returns to the heart, completing the circuit.

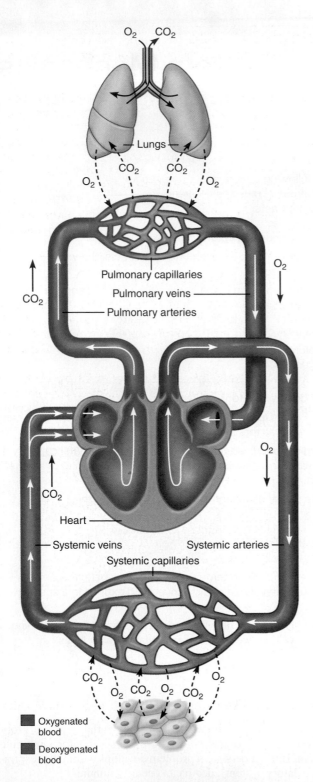

FIGURE 7.1 Relationship between external respiration and internal respiration.

Ventilation is the result of skeletal muscle contraction. When the **diaphragm**—a dome-shaped muscle that divides the thoracic and abdominal cavities—and the **external intercostal muscles** contract, the volume in the thoracic cavity increases. This increase in thoracic volume reduces the pressure in the thoracic cavity, allowing atmospheric gas to enter

the lungs (a process called **inspiration**). When the diaphragm and the external intercostals relax, the pressure in the thoracic cavity increases as the volume decreases, forcing air out of the lungs (a process called **expiration**). Inspiration is considered an *active* process because muscle contraction requires the use of ATP, whereas expiration is usually considered a *passive* process because the muscles relax, rather than contract. When a person is running, however, expiration becomes an active process, resulting from the contraction of **internal intercostal muscles** and **abdominal muscles.** In this case, both inspiration and expiration are considered *active* processes because muscle contraction is needed for both.

The amount of air that flows into and out of the lungs in 1 minute is the pulmonary **minute ventilation,** which is calculated by multiplying the **frequency of breathing** by the volume of each breath (the **tidal volume**). Ventilation must be regulated at all times to maintain oxygen in arterial blood and carbon dioxide in venous blood at their normal levels—that is, at their normal **partial pressures.** The *partial pressure* of a gas is the proportion of pressure that the gas exerts in a mixture. For example, in the atmosphere at sea level, the total pressure is 760 mm Hg. Oxygen makes up 21% of the total atmosphere and, therefore, has a partial pressure (P_{O_2}) of 160 mm Hg (760 mm Hg $\times$ 0.21).

Oxygen and carbon dioxide diffuse down their partial pressure gradients, from high partial pressures to low partial pressures. Oxygen diffuses from the alveoli of the lungs into the blood, where it can dissolve in plasma and attach to hemoglobin, and then diffuses from the blood into the tissues. Carbon dioxide (produced by the metabolic reactions of the tissues) diffuses from the tissues into the blood and then diffuses from the blood into the alveoli for export from the body.

In this exercise you will investigate the basic mechanics and regulation of the respiratory system. The concepts you will explore with a simulated lung will help you understand the operation of the human respiratory system in better detail.

ACTIVITY 1

Measuring Respiratory Volumes and Calculating Capacities

OBJECTIVES

1. To understand the use of the terms *ventilation, inspiration, expiration, diaphragm, external intercostals, internal intercostals, abdominal-wall muscles, expiratory reserve volume (ERV), forced vital capacity (FVC), tidal volume (TV), inspiratory reserve volume (IRV), residual volume (RV),* and *forced expiratory volume in one second (FEV$_1$).*

2. To understand the roles of skeletal muscles in the mechanics of breathing.

3. To understand the volume and pressure changes in the thoracic cavity during ventilation of the lungs.

4. To understand the effects of airway radius and, thus, resistance on airflow.

Introduction

The two phases of **ventilation,** or breathing, are (1) **inspiration,** during which air is taken into the lungs, and (2) **expiration,** during which air is expelled from the lungs. Inspiration occurs as the **external intercostal muscles** and the **diaphragm** contract. The diaphragm, normally a dome-shaped muscle, flattens as it moves inferiorly while the external intercostal muscles, situated between the ribs, lift the rib cage. These cooperative actions increase the thoracic volume. Air rushes into the lungs because this increase in thoracic volume creates a partial vacuum.

During quiet expiration, the inspiratory muscles relax, causing the diaphragm to rise superiorly and the chest wall to move inward. Thus, the **thorax** returns to its normal shape because of the elastic properties of the lung and thoracic wall. As in a deflating balloon, the pressure in the lungs rises, forcing air out of the lungs and airways. Although expiration is normally a *passive* process, **abdominal-wall muscles** and the **internal intercostal muscles** can also contract during expiration to force additional air from the lungs. Such forced expiration occurs, for example, when you exercise, blow up a balloon, cough, or sneeze.

Normal, quiet breathing moves about 500 ml (0.5 liter) of air (the **tidal volume**) into and out of the lungs with each breath, but this amount can vary due to a person's size, sex, age, physical condition, and immediate respiratory needs. In this activity you will measure the following respiratory volumes (the values given for the normal adult male and female are approximate).

Tidal volume (TV): Amount of air inspired and then expired with each breath under resting conditions (500 ml)

Inspiratory reserve volume (IRV): Amount of air that can be forcefully inspired after a normal tidal volume inspiration (male, 3100 ml; female, 1900 ml)

Expiratory reserve volume (ERV): Amount of air that can be forcefully expired after a normal tidal volume expiration (male, 1200 ml; female, 700 ml)

Residual volume (RV): Amount of air remaining in the lungs after forceful and complete expiration (male, 1200 ml; female, 1100 ml)

Respiratory capacities are calculated from the respiratory volumes. In this activity you will calculate the following respiratory capacities.

Total lung capacity (TLC): Maximum amount of air contained in lungs after a maximum inspiratory effort: TLC = TV + IRV + ERV + RV (male, 6000 ml; female, 4200 ml)

Vital capacity (VC): Maximum amount of air that can be inspired and then expired with maximal effort: VC = TV + IRV + ERV (male, 4800 ml; female 3100 ml)

You will also perform two pulmonary function tests in this activity.

Forced vital capacity (FVC): Amount of air that can be expelled when the subject takes the deepest possible inspiration and forcefully expires as completely and rapidly as possible

Forced expiratory volume (FEV$_1$): Measures the percentage of the vital capacity that is expired during 1 second of the FVC test (normally 75%–85% of the vital capacity)

EQUIPMENT USED The following equipment will be depicted on-screen: simulated human lungs suspended in a glass bell jar; rubber diaphragm—used to seal the jar and change the volume and, thus, pressure in the jar (As the diaphragm moves inferiorly, the volume in the bell jar increases and the pressure drops slightly, creating a partial vacuum in the bell jar. This partial vacuum causes air to be sucked into the tube at the top of the bell jar and then into the simulated lungs. As the diaphragm moves up, the decreasing volume and rising pressure within the bell jar forces air out of the lungs.); adjustable air-flow tube—connects the lungs to the atmosphere; oscilloscope; three different breathing patterns: normal tidal volumes, expiratory reserve volume (ERV), and forced vital capacity (FVC).

Experiment Instructions

Go to the home page in the PhysioEx software and click **Exercise 7: Respiratory System Mechanics.** Click **Activity 1: Measuring Respiratory Volumes and Calculating Capacities,** and take the online **Pre-lab Quiz** for Activity 1.

After you take the online Pre-lab Quiz, click the **Experiment** tab and begin the experiment. The experiment instructions are reprinted here for your reference. The opening screen for the experiment is shown below.

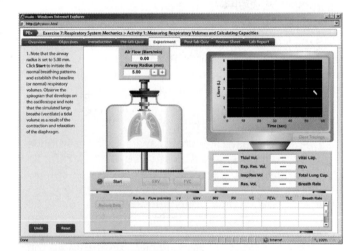

1. Note that the airway radius is set to 5.00 mm. Click **Start** to initiate the normal breathing patterns and establish the baseline (or normal) respiratory volumes. Observe the spirogram that develops on the oscilloscope and note that the simulated lungs breathe (ventilate) a tidal volume as a result of the contraction and relaxation of the diaphragm.

2. Click **Record Data** to display your results in the grid (and record your results in Chart 1, p. PEx-108).

3. Click **Clear Tracings** to clear the spirogram on the oscilloscope.

4. You will now complete the measurement of respiratory volumes and determine the respiratory capacities. First, click **Start** to initiate the normal breathing pattern. After 10 seconds, click **ERV.** Wait another 10 seconds and then click **FVC** to

CHART 1	Respiratory Volumes and Capacities							
Radius (mm)	Flow (ml/min)	TV (ml)	ERV (ml)	IRV (ml)	RV (ml)	VC (ml)	FEV$_1$ (ml)	TLC (ml)

complete the measurement of respiratory volumes. When you click ERV, the program will simulate forced expiration using the contraction of the internal intercostal muscles and abdominal-wall muscles. When you click FVC, the lungs will first inspire maximally and then expire fully to demonstrate forced vital capacity.

5. Note that, in addition to the tidal volume, the expiratory reserve volume, inspiratory reserve volume, and residual volume were measured. The vital capacity and total lung capacity were calculated from those volumes. Click **Record Data** to display your results in the grid (and record your results in Chart 1).

6. Minute ventilation is the amount of air that flows into and then out of the lungs in a minute. Minute ventilation (ml/min) = TV (ml/breath) × BPM (breaths/min). Enter the minute ventilation in the field below and then click **Submit** to record your answer in the lab report. _____ ml/min

PREDICT Question 1
Lung diseases are often classified as obstructive or restrictive. An **obstructive** disease affects *airflow*, and a **restrictive** disease usually reduces *volumes and capacities*. Although they are not diagnostic, pulmonary function tests such as forced expiratory volume (FEV$_1$) can help a clinician determine the difference between obstructive and restrictive diseases. Specifically, an FEV$_1$ is the forced volume expired in 1 second.

In obstructive diseases such as chronic bronchitis and asthma, airway radius is decreased. Thus, FEV$_1$ will:

_____ .

7. You will now explore what effect changing the airway radius has on pulmonary function. Decrease the airway radius to 4.50 mm by clicking the − button beneath the airway radius display.

8. Click **Start** to initiate the normal breathing pattern. After 10 seconds, click **ERV**. Wait another 10 seconds and then click **FVC**. The FEV$_1$ will appear in the FEV$_1$ display beneath the oscilloscope.

9. Click **Record Data** to display your results in the grid (and record your results in Chart 1).

10. You will now gradually decrease the airway radius.

- Decrease the airway radius by 0.50 mm by clicking the − button beneath the airway radius display.

- Click **Start** to initiate the normal breathing pattern. After 10 seconds, click **ERV**. Wait another 10 seconds and then click **FVC**. The FEV$_1$ will appear in the FEV$_1$ display beneath the oscilloscope.

- Click **Record Data** to display your results in the grid (and record your results in Chart 1).

Repeat this step until you reach an airway radius of 3.00 mm.

11. A useful way to express FEV$_1$ is as a percentage of the forced vital capacity (FVC). Using the FEV$_1$ and FVC values from the data grid, calculate the FEV$_1$ (%) by dividing the FEV$_1$ volume by the FVC volume (in this case, the VC is equal to the FVC) and multiply by 100%. Enter the FEV$_1$ (%) for an airway radius of 5.0 mm in the field below and then click **Submit** to record your answer in the lab report.

FEV$_1$ (%) for an airway radius of 5.0 (mm): _____

12. Enter the FEV$_1$ (%) for an airway radius of 3.00 mm in the field below and then click **Submit** to record your answer in the lab report.

FEV$_1$ (%) for an airway radius of 3.00 (mm): _____

After you complete the experiment, take the online **Post-lab Quiz** for Activity 1.

Activity Questions

1. When you forcefully exhale your entire expiratory reserve volume, any air remaining in your lungs is called the residual volume (RV). Why is it impossible to further exhale the RV (that is, *where* is this air volume trapped, and *why* is it trapped)?

_____

_____

2. How do you measure a person's RV in a laboratory?

_____

_____

3. Draw a spirogram that depicts a person's volumes and capacities before and during a significant cough.

Comparative Spirometry

OBJECTIVES

1. To understand the terms *spirometry, spirogram, emphysema, asthma, inhaler, moderate exercise, heavy exercise, tidal volume (TV), expiratory reserve volume (ERV), inspiratory reserve volume (IRV), residual volume (RV), vital capacity (VC), total lung capacity (TLC), forced vital capacity (FVC),* and *forced expiratory volume in one second (FEV$_1$).*

2. To observe and compare spirograms collected from resting, healthy patients to those taken from an emphysema patient.

3. To observe and compare spirograms collected from resting, healthy patients to those taken from a patient suffering an acute asthma attack.

4. To observe and compare the spirogram collected from an asthmatic patient *while* suffering an acute asthma attack to that taken after the patient uses an inhaler for relief.

5. To observe and compare spirograms collected from volunteers engaged in moderate exercise and heavy exercise.

Introduction

In this activity you will explore the changes to normal respiratory volumes and capacities when pathophysiology develops and during aerobic exercise by recruiting volunteers to breathe into a water-filled spirometer. The spirometer is a device that measures the volume of air inspired and expired by the lungs over a specified period of time. Several lung capacities and flow rates can be calculated from this data to assess pulmonary function. With your knowledge of respiratory mechanics, you can predict, document, and explain changes to the volumes and capacities in each state.

Emphysema breathing: With emphysema, there is a significant loss of elastic recoil in the lung tissue. This loss of elastic recoil occurs as the disease destroys the walls of the alveoli. Airway resistance is also increased as the lung tissue

in general becomes more flimsy and exerts less anchoring on the surrounding airways. Thus, the lung becomes overly compliant and expands easily. Conversely, a great effort is required to expire because the lungs can no longer passively recoil and deflate. Each expiration requires a noticeable and exhausting muscular effort, and a person with emphysema expires slowly.

Acute asthma attack breathing: During an acute asthma attack, bronchiole smooth muscle spasms and, thus, the airways become constricted (that is, reduced in diameter). They also become clogged with thick mucus secretions. These changes lead to significantly increased airway resistance.

Underlying these symptoms is an airway inflammatory response brought on by triggers such as allergens (for example, dust and pollen), extreme temperature changes, and even exercise. Like with emphysema, the airways collapse and pinch closed before a forced expiration is completed. Thus, the volumes and peak flow rates are significantly reduced during an asthma attack. Unlike with emphysema, the elastic recoil is not diminished in an acute asthma attack.

When an acute asthma attack occurs, many people seek to relieve symptoms with an inhaler, which atomizes the medication and allows for direct application onto the afflicted airways. Usually, the medication includes a smooth muscle relaxant (for example, a β_2 agonist or an acetylcholine antagonist) that relieves the bronchospasms and induces bronchiole dilation. The medication can also contain an anti-inflammatory agent, such as a corticosteroid, that suppresses the inflammatory response. The use of the inhaler reduces airway resistance.

Breathing during exercise: During *moderate* aerobic exercise, the human body has an increased metabolic demand, which is met, in part, by changes in respiration. Specifically, both the rate of breathing and the tidal volume increase. These two respiratory variables do not increase by the same amount. The increase in the tidal volume is greater than the increase in the rate of breathing. During *heavy* exercise, further changes in respiration are required to meet the extreme metabolic demands of the body. In this case both the rate of breathing and the tidal volume increase to their maximum tolerable limits.

EQUIPMENT USED The following equipment will be depicted on-screen: a classic water-filled spirometer with an attached rotating drum that records the analog spirogram in real time; breathing patterns from a variety of patients: unforced breathing and forced vital capacity for a "normal" patient, a patient with emphysema, and a patient with asthma (during an attack and after using an inhaler); and the breathing patterns from a patient during moderate and heavy exercise.

Experiment Instructions

Go to the home page in the PhysioEx software and click **Exercise 7: Respiratory System Mechanics.** Click **Activity 2: Comparative Spirometry,** and take the online **Pre-lab Quiz** for Activity 2.

After you take the online Pre-lab Quiz, click the **Experiment** tab and begin the experiment. The experiment

instructions are reprinted here for your reference. The opening screen for the experiment is shown below.

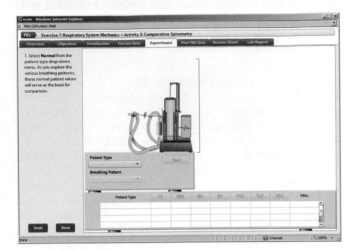

1. Select **Normal** from the patient type drop-down menu. As you explore the various breathing patterns, these normal patient values will serve as the basis for comparison.

2. Select **Unforced Breathing** from the breathing pattern drop-down menu.

3. Click **Start** to record the patient's unforced breathing pattern and watch as the drum starts turning and the spirogram develops on the paper rolling off the drum.

4. Note the volume levels (in milliliters) on the Y-axis of the spirogram. When half the screen is filled with unforced tidal volumes and the spirogram has paused, select **Forced Vital Capacity** from the breathing pattern drop-down menu.

5. Click **Start** to record the patient's forced vital capacity. The spirogram ends as the paper rolls to the right edge of the screen.

6. Click on each of the buttons in the data recorder to measure respiratory volumes and capacities. Start with tidal volume (TV) and work your way to the right. When you measure each volume or capacity, (1) a bracket appears on

the spirogram to indicate where that measurement originates and (2) the value (in milliliters) displays in the grid. After you complete all the measurements, the FEV_1 (%) ratio will automatically be calculated. The FEV_1 (%) = (FEV_1/FVC) × 100%. Record your results in Chart 2.

? PREDICT Question 1
With emphysema, there is a significant loss of elastic recoil in the lung tissue and a noticeable, exhausting muscular effort is required for each expiration. Inspiration actually becomes easier because the lung is now overly compliant. What lung values will change (from those of the normal patient) in the spirogram when the patient with emphysema is selected?

_____

7. Select **Emphysema** from the patient type drop-down menu.

8. Select **Unforced Breathing** from the breathing pattern drop-down menu.

9. Click **Start** to record the patient's unforced breathing pattern and watch as the drum starts turning and the spirogram develops on the paper rolling off the drum.

10. Note the volume levels on the Y-axis of the spirogram. When half the screen is filled with unforced tidal volumes and the spirogram has paused, select **Forced Vital Capacity** from the breathing pattern drop-down menu.

11. Click **Start** to record the patient's forced vital capacity. The spirogram ends as the paper rolls to the right edge of the screen.

12. Click on each of the buttons in the data recorder to measure respiratory volumes and capacities. Start with tidal volume (TV) and work your way to the right. Record your results in Chart 2.

CHART 2	Spirometry Results							
Patient type	**TV (ml)**	**ERV (ml)**	**IRV (ml)**	**RV (ml)**	**FVC (ml)**	**TLC (ml)**	**FEV$_1$ (ml)**	**FEV$_1$ (%)**
Normal								
Emphysema								
Acute asthma attack								
Plus inhaler								
Moderate exercise								
Heavy exercise								

? PREDICT Question 2
During an acute asthma attack, airway resistance is significantly increased by (1) increased thick mucous secretions and (2) airway smooth muscle spasms. What lung values will change (from those of the normal patient) in the spirogram for a patient suffering an acute asthma attack?

13. Select **Acute Asthma Attack** from the patient type drop-down menu.

14. Select **Unforced Breathing** from the breathing pattern drop-down menu.

15. Click **Start** to record the patient's uforced breathing pattern and watch as the drum starts turning and the spirogram develops on the paper rolling off the drum.

16. Note the volume levels on the Y-axis of the spirogram. When half the screen is filled with unforced tidal volumes and the spirogram has paused, select **Forced Vital Capacity** from the breathing pattern drop-down menu.

17. Click **Start** to record the patient's forced vital capacity. The spirogram ends as the paper rolls to the right edge of the screen.

18. Click on each of the buttons in the data recorder to measure respiratory volumes and capacities. Start with tidal volume (TV) and work your way to the right. Record your results in Chart 2.

? PREDICT Question 3
When an acute asthma attack occurs, many people seek relief from the increased airway resistance by using an inhaler. This device atomizes the medication and induces bronchiole dilation (though it can also contain an anti-inflammatory agent). What lung values will change *back* to those of the normal patient in the spirogram after the asthma patient uses an inhaler?

19. Select **Plus Inhaler** from the patient type drop-down menu.

20. Select **Unforced Breathing** from the breathing pattern drop-down menu.

21. Click **Start** to record the patient's unforced breathing pattern and watch as the drum starts turning and the spirogram develops on the paper rolling off the drum.

22. Note the volume levels on the Y-axis of the spirogram. When half the screen is filled with unforced tidal volumes and the spirogram has paused, select **Forced Vital Capacity** from the breathing pattern drop-down menu.

23. Click **Start** to record the patient's forced vital capacity. The spirogram ends as the paper rolls to the right edge of the screen.

24. Click on each of the buttons in the data recorder to measure respiratory volumes and capacities. Start with tidal volume (TV) and work your way to the right. Record your results in Chart 2.

? PREDICT Question 4
During moderate aerobic exercise, the human body will change its respiratory cycle in order to meet increased metabolic demands. During heavy exercise, further changes in respiration are required to meet the extreme metabolic demands of the body. Which lung value will change more during moderate exercise, the ERV or the IRV?

25. Select **Moderate Exercise** from the patient type drop-down menu. Note that the selection of a breathing pattern is not applicable because our central nervous system automatically adjusts and maintains the depth and frequency of breathing to meet the increased metabolic demands while we exercise. We do not normally alter this pattern with conscious intervention.

26. Click **Start** to record the patient's breathing pattern and watch as the drum starts turning and the spirogram develops on the paper rolling off the drum.

27. Click on each of the buttons in the data recorder to measure respiratory volumes and capacities. Start with tidal volume (TV) and work your way to the right. *ND* indicates this measurement or calculation was not done. Record your results in Chart 2.

28. Select **Heavy Exercise** from the patient type drop-down menu.

29. Click **Start** to record the patient's breathing pattern and watch as the drum starts turning and the spirogram develops on the paper rolling off the drum.

30. Click on each of the buttons in the data recorder to measure respiratory volumes and capacities. Start with tidal volume (TV) and work your way to the right. Record your results in Chart 2.

After you complete the experiment, take the online **Post-lab Quiz** for Activity 2.

Activity Questions

1. Why is residual volume (RV) above normal in a patient with emphysema?

2. Why did the asthmatic patient's inhaler medication fail to return all volumes and capacities to normal values right away?

3. Looking at the spirograms generated in this activity, state an easy way to determine whether a person's exercising effort is moderate or heavy.

———————————————————————————

——————————————————————————— ▬

Effect of Surfactant and Intrapleural Pressure on Respiration

OBJECTIVES

1. To understand the terms *surfactant, surface tension, intrapleural space, intrapleural pressure, pneumothorax,* and *atelectasis.*

2. To understand the effect of surfactant on surface tension and lung function.

3. To understand how negative intrapleural pressure prevents lung collapse.

Introduction

At any gas-liquid boundary, the molecules of the liquid are attracted more strongly to each other than they are to the gas molecules. This unequal attraction produces tension at the liquid surface, called **surface tension.** Because surface tension resists any force that tends to increase surface area of the gas-liquid boundary, it acts to decrease the size of hollow spaces, such as the alveoli, or microscopic air spaces within the lungs.

If the film lining the air spaces in the lung were pure water, it would be very difficult, if not impossible, to inflate the lungs. However, the aqueous film covering the alveolar surfaces contains **surfactant,** a detergent-like mixture of lipids and proteins that decreases surface tension by reducing the attraction of water molecules to each other. You will explore the importance of surfactant in this activity.

Between breaths, the pressure in the pleural cavity, the **intrapleural pressure,** is less than the pressure in the alveoli. Two forces cause this negative pressure condition: (1) the tendency of the lung to recoil because of its elastic properties and the surface tension of the alveolar fluid and (2) the tendency of the compressed chest wall to recoil and expand outward. These two forces pull the lungs away from the thoracic wall, creating a partial vacuum in the pleural cavity.

Because the pressure in the intrapleural space is lower than atmospheric pressure, any opening created in the pleural membranes equalizes the intrapleural pressure with atmospheric pressure by allowing air to enter the pleural cavity, a condition called **pneumothorax.** A pneumothorax can then lead to lung collapse, a condition called **atelectasis.** In this activity, the **intrapleural space** is the space between the wall of the glass bell jar and the outer wall of the lung it contains.

EQUIPMENT USED The following equipment will be depicted on-screen: simulated human lungs suspended in a glass bell jar; rubber diaphragm—used to seal the jar and change the volume and, thus, pressure in the jar (As the diaphragm moves inferiorly, the volume in the bell jar increases and the pressure drops slightly, creating a partial vacuum in the bell jar. This partial vacuum causes air to be sucked into the tube at the top of the bell jar and then into the simulated lungs. As the diaphragm moves up, the decreasing volume and rising pressure within the bell jar forces air out of the lungs.); valve—allows intrapleural pressure in the left side of the bell jar to equalize with atmospheric pressure; surfactant—amphipathic lipids (dipalmitoylphosphatidylcholine, phosphatidylglycerol, and palmitic acid) and short, synthetic peptides in a mixture that mimics the surfactant found in human lungs (surfactant molecules reduce surface tension in alveoli by adsorbing to the air-water interface, with their hydrophilic parts in the water and their hydrophobic parts facing toward the air); oscilloscope.

Experiment Instructions

Go to the home page in the PhysioEx software and click **Exercise 7: Respiratory System Mechanics.** Click **Activity 3: Effect of Surfactant and Intrapleural Pressure on Respiration,** and take the online **Pre-lab Quiz** for Activity 3.

After you take the online Pre-lab Quiz, click the **Experiment** tab and begin the experiment. The experiment instructions are reprinted here for your reference. The opening screen for the experiment is shown below.

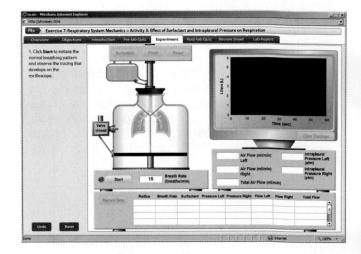

1. Click **Start** to initiate the normal breathing pattern and observe the tracing that develops on the oscilloscope.

2. Click **Record Data** to display your results in the grid (and record your results in Chart 3). This data represents breathing in the absence of surfactant.

3. Click **Surfactant** twice to dispense two aliquots of the synthetic lipids and peptides onto the interior lining of the lungs.

4. Click **Start** to initiate breathing in the presence of surfactant and observe the tracing that develops.

5. Click **Record Data** to display your results in the grid (and record your results in Chart 3).

CHART 3	Effect of Surfactant and Intrapleural Pressure on Respiration				
Surfactant	Intrapleural pressure left (atm)	Intrapleural pressure right (atm)	Airflow left (ml/min)	Airflow right (ml/min)	Total airflow (ml/min)

PREDICT Question 1
What effect will adding more surfactant have on these lungs?

_____

6. Click **Surfactant** twice to dispense two more aliquots of the synthetic lipids and proteins onto the interior lining of the lungs.

7. Click **Start** to initiate breathing in the presence of additional surfactant and observe the tracing that develops.

8. Click **Record Data** to display your results in the grid (and record your results in Chart 3).

9. Click **Clear Tracings** to clear the tracing on the oscilloscope.

10. Click **Flush** to clear the lungs of surfactant from the previous run.

11. Click **Start** to initiate breathing and observe the tracing that develops. Notice the negative pressure condition displayed below the oscilloscope when the lungs inflate.

12. Click **Record Data** to display your results in the grid (and record your results in Chart 3).

13. Click the valve on the left side of the glass bell jar to open it.

14. Click **Start** to initiate breathing and observe the tracing that develops.

15. Click **Record Data** to display your results in the grid (and record your results in Chart 3).

PREDICT Question 2
What will happen to the collapsed lung in the left side of the glass bell jar if you close the valve?

_____

16. Click the valve on the left side of the glass bell jar to close it.

17. Click **Start** to initiate breathing and observe the tracing that develops.

18. Click **Record Data** to display your results in the grid (and record your results in Chart 3).

19. Click the **Reset** button above the glass bell jar to draw the air out of the intrapleural space and return the lung to its normal resting condition.

20. Click **Start** to initiate breathing and observe the tracing that develops.

21. Click **Record Data** to display your results in the grid (and record your results in Chart 3).

After you complete the experiment, take the online **Post-lab Quiz** for Activity 3.

Activity Questions

1. Why is normal quiet breathing so difficult for premature infants?

_____

_____

2. Why does a pneumothorax frequently lead to atelectasis?

_____

_____

NAME _____

LAB TIME/DATE _____

Respiratory System Mechanics

ACTIVITY 1 Measuring Respiratory Volumes and Calculating Capacities

1. What would be an example of an everyday respiratory event the ERV button simulates? _____

2. What additional skeletal muscles are utilized in an ERV activity?

   _____

3. What was the FEV_1 (%) at the initial radius of 5.00 mm?

   _____

4. What happened to the FEV_1 (%) as the radius of the airways decreased? How well did the results compare with your prediction?

   _____

5. Explain why the results from the experiment suggest that there is an obstructive, rather than a restrictive, pulmonary problem.

   _____

ACTIVITY 2 Comparative Spirometry

1. What lung values changed (from those of the normal patient) in the spirogram when the patient with emphysema was selected? Why did these values change as they did? How well did the results compare with your prediction?

   _____

   _____

2. Which of these two parameters changed more for the patient with emphysema, the FVC or the FEV_1? _____

3. What lung values changed (from those of the normal patient) in the spirogram when the patient experiencing an acute asthma attack was selected? Why did these values change as they did? How well did the results compare with your prediction?

   _____

   _____

   _____

4. How is having an acute asthma attack similar to having emphysema? How is it different? _____

   _____

5. Describe the effect that the inhaler medication had on the asthmatic patient. Did all the spirogram values return to "normal"?

   Why do you think some values did not return all the way to normal? How well did the results compare with your prediction?

   _____

   _____

   _____

6. How much of an increase in FEV_1 do you think is required for it to be considered significantly improved by the medication?

   _____

7. With moderate aerobic exercise, which changed more from normal breathing, the ERV or the IRV? How well did the results

   compare with your prediction?

   _____

   _____

8. Compare the breathing rates during normal breathing, moderate exercise, and heavy exercise. _____

   _____

ACTIVITY 3 Effect of Surfactant and Intrapleural Pressure on Respiration

1. What effect does the addition of surfactant have on the airflow? How well did the results compare with your prediction?

   _____

   _____

2. Why does surfactant affect airflow in this manner? _____

   _____

3. What effect did opening the valve have on the left lung? Why does this happen?

   _____

   _____

4. What effect on the collapsed lung in the left side of the glass bell jar did you observe when you closed the valve? How well

   did the results compare with your prediction?

   _____

   _____

5. What emergency medical condition does opening the left valve simulate?

   _____

6. In the last part of this activity, you clicked the Reset button to draw the air out of the intrapleural space and return the lung

    to its normal resting condition. What emergency procedure would be used to achieve this result if these were the lungs in a

    living person? _____

    _____

    _____

7. What do you think would happen when the valve is opened if the two lungs were in a single large cavity rather than separate

    cavities? _____

    _____

Chemical and Physical Processes of Digestion

PRE-LAB QUIZ

1. Circle the correct underlined term: The <u>liver</u> / <u>stomach</u> produces pepsin which, in the presence of hydrochloric acid, digests protein.

2. _____ is a hydrolytic enzyme that breaks starch down to maltose.
   a. Pepsin
   b. Salivary amylase
   c. Pancreatic lipase
   d. Bile

3. A _____ is made in order to compare a known standard to an experimental standard.

4. Circle True or False: A positive IKI test for starch will yield a bright orange to green color.

5. An enzyme has a pocket or _____, which the substrate or substrates must fit into temporarily in order for catalysis to occur.
   a. reagent
   b. hydrolase
   c. active site
   d. polysaccharide

6. During digestion, the chief cells of the stomach secrete _____, which is responsible for the digestion of protein.
   a. peptidase
   b. amylase
   c. lipase
   d. pepsin

7. Circle the correct underlined term: <u>Positive</u> / <u>Negative</u> controls are used to determine whether there are any contaminating substances in the reagents used in an experiment.

8. Circle True or False: At room temperature, both fats and oils are liquid and are soluble in water.

9. A solution containing fatty acids formed by lipase activity will have a _____ than a solution without such fatty acid production.
   a. lower pH
   b. higher pH
   c. lower temperature
   d. higher temperature

Exercise Overview

The **digestive system,** also called the gastrointestinal system, consists of the digestive tract (also called the gastrointestinal tract, or GI tract) and accessory glands that secrete enzymes and fluids needed for digestion. The digestive tract includes the mouth, pharynx, esophagus, stomach, small intestine, colon, rectum, and anus. The major functions of the digestive system are to ingest food, to break food down to its simplest components, to extract nutrients from these components for absorption into the body, and to eliminate wastes.

Most of the food we consume cannot be absorbed into our bloodstream without first being broken down into smaller subunits. **Digestion** is the process of

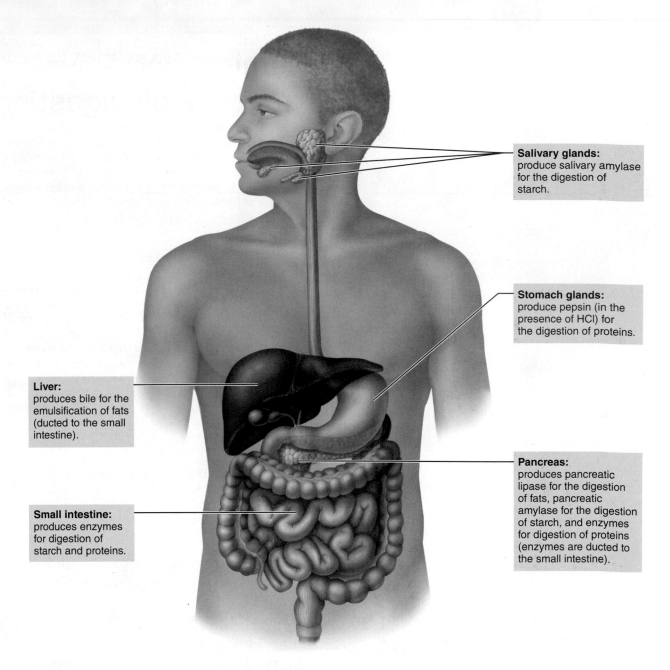

Salivary glands: produce salivary amylase for the digestion of starch.

Stomach glands: produce pepsin (in the presence of HCl) for the digestion of proteins.

Liver: produces bile for the emulsification of fats (ducted to the small intestine).

Pancreas: produces pancreatic lipase for the digestion of fats, pancreatic amylase for the digestion of starch, and enzymes for digestion of proteins (enzymes are ducted to the small intestine).

Small intestine: produces enzymes for digestion of starch and proteins.

FIGURE 8.1 The human digestive system. A few sites of chemical digestion and the organs that produce the enzymes of chemical digestion.

breaking down food molecules into smaller molecules with the aid of enzymes in the digestive tract. **Enzymes** are large protein molecules produced by body cells. They are biological catalysts that increase the rate of a chemical reaction without becoming part of the product. The digestive enzymes are hydrolytic enzymes, or **hydrolases,** which break down organic food molecules, or **substrates,** by adding water to the molecular bonds, thus cleaving the bonds between the subunits, or monomers.

A hydrolytic enzyme is highly specific in its action. Each enzyme hydrolyzes one substrate molecule or, at most, a small group of substrate molecules. Specific environmental conditions are necessary for an enzyme to function optimally. For example, in extreme environments, such as high temperature, an enzyme can unravel, or denature, because of the effect that temperature has on the three-dimensional structure of the protein.

Because digestive enzymes actually function outside the body cells in the digestive tract lumen, their hydrolytic activity can also be studied in vitro in a test tube. Such in vitro studies provide a convenient laboratory environment for investigating the effect of various factors on enzymatic activity. View Figure 8.1 for an overview of chemical digestion sites in the body.

Assessing Starch Digestion by Salivary Amylase

OBJECTIVES

1. To explain how enzyme activity can be assessed with enzyme assays: the IKI assay and the Benedict's assay.

2. To define *enzyme, catalyst, hydrolase, substrate,* and *control.*

3. To understand the specificity of amylase action.

4. To name the end products of carbohydrate digestion.

5. To perform the appropriate chemical tests to determine whether digestion of a particular food has occurred.

6. To discuss the possible effect of temperature and pH on amylase activity.

Introduction

In this activity you will investigate the hydrolysis of starch to maltose by **salivary amylase,** the enzyme produced by the salivary glands and secreted into the mouth. For you to be able to detect whether or not enzymatic action has occurred, you need to be able to identify the presence of the substrate and the product to determine to what extent hydrolysis has occurred. Thus, **controls** must be prepared to provide a known standard against which comparisons can be made. With positive controls, all of the required substances are included and a positive result is expected. Sometimes negative controls are included. With negative controls, a negative result is expected. Negative results with negative controls validate the experiment. Negative controls are used to determine whether there are any contaminating substances in the reagents. So, when a positive result is produced but a negative result is expected, one or more contaminating substances are present to cause the change.

With amylase activity, starch decreases and maltose increases as digestion proceeds according to the following equation.

$$\text{Starch} + \text{water} \xrightarrow{\text{amylase}} \text{maltose}$$

Because the chemical changes that occur as starch is digested to maltose cannot be seen by the naked eye, you need to conduct an **enzyme assay,** the chemical method of detecting the presence of digested substances. You will perform two enzyme assays on each sample. The IKI assay detects the presence of starch, and the Benedict's assay tests for the presence of reducing sugars, such as glucose or maltose, which are the digestion products of starch. Normally a caramel-colored solution, IKI turns blue-black in the presence of starch. Benedict's reagent is a bright blue solution that changes to green to orange to reddish brown with increasing amounts of maltose. It is important to understand that enzyme assays only indicate the presence or absence of substances. It is up to you to analyze the results of the experiments to decide whether enzymatic hydrolysis has occurred.

EQUIPMENT USED The following equipment will be depicted on-screen: amylase—an enzyme that digests starch; starch—a complex carbohydrate substrate; maltose—a disaccharide substrate; pH buffers—solutions used to adjust the pH of the solution; deionized water—used to adjust the volume so that it is the same for each reaction; test tubes—used as reaction vessels for the various tests; incubators—used for temperature treatments (boiling, freezing, and 37°C incubation); IKI—found in the assay cabinet; used to detect the presence of starch; Benedict's reagent—found in the assay cabinet; used to detect the products of starch digestion (this includes the reducing sugars maltose and glucose).

Experiment Instructions

Go to the home page in the PhysioEx software and click **Exercise 8: Chemical and Physical Processes of Digestion.** Click **Activity 1: Assessing Starch Digestion by Salivary Amylase,** and take the online **Pre-lab Quiz** for Activity 1.

After you take the online Pre-lab Quiz, click the **Experiment** tab and begin the experiment. The experiment instructions are reprinted here for your reference. The opening screen for the experiment is shown below.

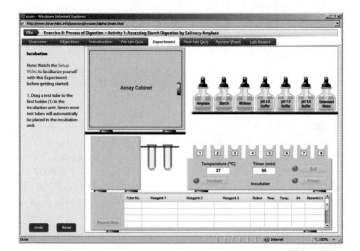

Incubation

1. Drag a test tube to the first holder **(1)** in the incubation unit. Seven more test tubes will automatically be placed in the incubation unit.

2. Add the substances indicated below to tubes 1 through 7.

Tube 1: amylase, starch, pH 7.0 buffer

Tube 2: amylase, starch, pH 7.0 buffer

Tube 3: amylase, starch, pH 7.0 buffer

Tube 4: amylase, deionized water, pH 7.0 buffer

Tube 5: deionized water, starch, pH 7.0 buffer

Tube 6: deionized water, maltose, pH 7.0 buffer

Tube 7: amylase, starch, pH 2.0 buffer

Tube 8: amylase, starch, pH 9.0 buffer

To add a substance to a test tube, drag the dropper cap of the bottle on the solutions shelf to the top of the test tube.

3. Click the number (**1**) under the first test tube. The tube will descend into the incubation unit. All other tubes should remain in the raised position.

4. Click **Boil** to boil tube 1. After boiling for a few moments, the tube will automatically rise.

5. Click the number (**2**) under the second test tube. The tube will descend into the incubation unit. All other tubes should remain in the raised position.

6. Click **Freeze** to freeze tube 2. After freezing for a few moments, the tube will automatically rise.

7. Click **Incubate** to start the run. Note that the incubation temperature is set at 37°C and the timer is set at 60 min. The incubation unit will gently agitate the test tube rack, evenly mixing the contents of all test tubes throughout the incubation. The simulation compresses the 60-minute time period into 10 seconds of real time, so what would be a 60-minute incubation in real time will take only 10 seconds in the simulation. When the incubation time elapses, the test tube rack will automatically rise, and the doors to the assay cabinet will open.

> ? **PREDICT Question 1**
> What effect do you think boiling and freezing will have on the activity of the amylase enzyme?
>
> _____
>
> _____

Assays

After the assay cabinet doors open, notice the two reagents in the assay cabinet. IKI tests for the presence of starch and Benedict's reagent detects the presence of reducing sugars, such as glucose or maltose, which are the digestion products of starch. Below the reagents are eight small assay tubes into which you

will dispense a small amount of test solution from the incubated samples in the incubation unit, plus a drop of IKI.

8. Drag the first tube in the incubation unit to the first small assay tube on the left side of the assay cabinet to decant approximately half of the contents in the test tube into the assay tube. The decanting step will automatically repeat for the remaining tubes in the incubation unit.

9. Drag the IKI dropper cap to the first assay tube to dispense a drop of IKI into the assay tube. The dropper will automatically move across and dispense IKI to the remaining tubes.

10. Inspect the tubes for color change. A blue-black color indicates a positive starch test. If starch is not present, the mixture will look like diluted IKI, a negative starch test. Intermediate starch amounts result in a pale-gray color. Click **Record Data** to display your results in the grid (and record your results in Chart 1).

11. Drag the Benedict's reagent dropper cap to the test tube in the first holder (**1**) in the incubation unit to dispense five drops of Benedict's reagent into the tube. The dropper will automatically move across and dispense Benedict's reagent to the remaining tubes.

12. Click **Boil.** The entire tube rack will descend into the incubation unit and automatically boil the tube contents for a few moments.

13. Inspect the tubes for color change. A green-to-reddish color indicates that a reducing sugar is present; this is a positive sugar test. An orange-colored sample contains more sugar than a green sample. A reddish-brown color indicates even more sugar. A negative sugar test is indicated by no color change from the original bright blue. Click **Record Data** to display your results in the grid (and record your results in Chart 1).

After you complete the experiment, take the online **Post-lab Quiz** for Activity 1.

CHART 1	Salivary Amylase Digestion of Starch							
Tube No.	1	2	3	4	5	6	7	8
Additives	Amylase Starch pH 7.0 buffer	Amylase Starch pH 7.0 buffer	Amylase Starch pH 7.0 buffer	Amylase Deionized water pH 7.0 buffer	Deionized water Starch pH 7.0 buffer	Deionized water Maltose pH 7.0 buffer	Amylase Starch pH 2.0 buffer	Amylase Starch pH 9.0 buffer
Incubation condition	Boil first, then incubate at 37°C for 60 minutes	Freeze first, then incubate at 37°C for 60 minutes	37°C 60 minutes	37°C 60 minutes	37°C 60 minutes	37°C 60 minutes	37°C 60 minutes	37°C 60 minutes
IKI test								
Benedict's test								

Activity Questions

1. Describe the effect that boiling had on the activity of amylase. Why did boiling have this effect? How does the effect of freezing differ from the effect of boiling?

_____

_____

2. What is the purpose for including tube 3 and what can you conclude from the result?

_____

_____

3. Describe how you determined the optimal pH for amylase activity.

_____

_____

4. Judging from what you learned in this activity, suggest a reason why salivary amylase would be much less active in the stomach.

_____

_____

ACTIVITY 2

Exploring Amylase Substrate Specificity

OBJECTIVES

1. Explain how hydrolytic enzyme activity can be assessed with the IKI assay and the Benedict's assay.
2. Understand the specificity that enzymes have for their substrate.
3. Understand the difference between the substrates starch and cellulose.
4. Explain what would be the substrate specificity of peptidase.
5. Explain how bacteria might aid in digestion.

Introduction

In this activity you will investigate the specificity that enzymes have for their substrates. To do this you will hydrolyze starch to maltose and maltotriose using **salivary amylase,** the enzyme produced by the salivary glands and secreted into the mouth. To detect whether or not enzymatic action has occurred, you need to be able to identify the presence of the substrate and the product to determine to what extent hydrolysis has occurred. The **substrate** is the substance that the enzyme acts on. The enzyme has a pocket called the **active site,** which the substrate or substrates must fit into temporarily for catalysis to occur. The substrate is often held in the active site by non-covalent bonds (weak bonds), such as ionic bonds and hydrogen bonds.

With amylase activity, starch decreases and sugar increases as digestion proceeds according to the following equation.

$$\text{Starch + water} \xrightarrow{\text{amylase}} \text{maltose + maltotriose + starch}$$

Because the chemical changes that occur as starch is digested to maltose cannot be seen by the naked eye, you need to conduct an **enzyme assay,** the chemical method of detecting the presence of digested substances. You will perform two enzyme assays on each sample. The IKI assay detects the presence of starch or cellulose and the Benedict's assay tests for the presence of reducing sugars, such as glucose or maltose, which are the digestion products of starch. Normally a caramel-colored solution, IKI turns blue-black in the presence of starch or cellulose. Benedict's reagent is a bright blue solution that changes to green to orange to reddish brown with increasing amounts of maltose. It is important to understand that enzyme assays only indicate the presence or absence of substances. It is up to you to analyze the results of the experiments to decide whether enzymatic hydrolysis has occurred.

Starch is a polysaccharide found in plants, where it is used to store energy. Plants also have the polysaccharide **cellulose,** which provides rigidity to their cell walls. Both polysaccharides are polymers of glucose, but the glucose molecules are linked differently. You will be testing salivary amylase to determine whether it digests cellulose. Also, you will investigate to see whether a bacterial suspension can digest cellulose and whether **peptidase,** a pancreatic enzyme that digests peptides, can break down starch.

> **EQUIPMENT USED** The following equipment will be depicted on-screen: amylase—an enzyme that digests starch; starch—a polysaccharide; pH 7.0 buffer—a solution used to set the pH of the test tube solution; deionized water—used to adjust the test tube solution volume so it is the same for each reaction; glucose—a reducing sugar that is the monosaccharide subunit of both starch and cellulose; cellulose—a complex carbohydrate found in the cell wall of plants; peptidase—a pancreatic enzyme that breaks down peptides; bacteria—a suspension of live bacteria; test tubes—used as reaction vessels for the various tests; incubators—used for temperature treatments (37°C incubation); IKI—found in the assay cabinet; used to detect the presence of starch or cellulose; Benedict's reagent—found in the assay cabinet; used to detect the products of starch and cellulose digestion.

Experiment Instructions

Go to the home page in the PhysioEx software and click **Exercise 8: Chemical and Physical Processes of Digestion.** Click **Activity 2: Exploring Amylase Substrate Specificity,** and take the online **Pre-lab Quiz** for Activity 2.

After you take the online Pre-lab Quiz, click the **Experiment** tab and begin the experiment. The experiment instructions are reprinted here for your reference. The opening screen for the experiment is shown on the following page.

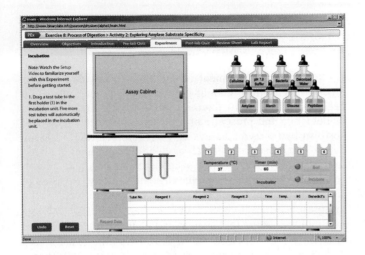

Incubation

1. Drag a test tube to the first holder (**1**) in the incubation unit. Five more test tubes will automatically be placed in the incubation unit.

2. Add the substances indicated below to tubes 1 through 6.

Tube 1: amylase, starch, pH 7.0 buffer

Tube 2: amylase, glucose, pH 7.0 buffer

Tube 3: amylase, cellulose, pH 7.0 buffer

Tube 4: cellulose, pH 7.0 buffer, deionized water

Tube 5: peptidase, starch, pH 7.0 buffer

Tube 6: bacteria, cellulose, pH 7.0 buffer

To add a substance to a test tube, drag the dropper cap of the bottle on the solutions shelf to the test tube.

3. Click **Incubate** to start the run. Note that the incubation temperature is set at 37°C and the timer is set at 60 min. The incubation unit will gently agitate the test tube rack, evenly mixing the contents of all test tubes throughout the incubation. The simulation compresses the 60-minute time period into 10 seconds of real time, so what would be a 60-minute incubation in real life will take only 10 seconds in the simulation. When the incubation time elapses, the test tube rack will automatically rise, and the doors to the assay cabinet will open.

? **PREDICT Question 1**
Do you think test tube 3 will show a positive Benedict's test?

_____

_____

Assays

After the assay cabinet doors open, notice the two reagents in the assay cabinet. IKI tests for the presence of starch and Benedict's reagent detects the presence of reducing sugars, such as glucose or maltose, which are the digestion products of starch. Below the reagents are seven small assay tubes into which you will dispense a portion of the incubated samples, plus a drop of IKI.

4. Drag the first tube in the incubation unit to the first small assay tube on the left side of the assay cabinet to decant approximately half of the contents in the test tube into the assay tube. The decanting step will automatically repeat for the remaining tubes in the incubation unit.

5. Drag the IKI dropper cap to the first assay tube to dispense a drop of IKI into the assay tube. The dropper will automatically dispense IKI into the remaining tubes.

6. Inspect the tubes for color change. A blue-black color indicates a positive starch test. If starch is not present, the mixture will look like diluted IKI, a negative starch test. Intermediate starch amounts result in a pale-gray color. Click **Record Data** to display your results in the grid (and record your results in Chart 2).

7. Drag the Benedict's reagent dropper cap to the test tube in the first holder (**1**) in the incubation unit to dispense five drops of Benedict's reagent into the tube. The dropper will automatically move across and dispense Benedict's reagent to the remaining tubes.

8. Click **Boil.** The entire tube rack will descend into the incubation unit and automatically boil the tube contents for a few moments.

9. Inspect the tubes for color change. A green-to-reddish color indicates that a reducing sugar is present; this is a positive sugar test. An orange-colored sample contains more sugar than a green sample. A reddish-brown color indicates even more sugar. A negative sugar test is indicated by no

CHART 2	Enzyme Digestion of Starch and Cellulose					
Tube No.	**1**	**2**	**3**	**4**	**5**	**6**
Additives	Amylase Starch pH 7.0 buffer	Amylase Glucose pH 7.0 buffer	Amylase Cellulose pH 7.0 buffer	Deionized water Cellulose pH 7.0 buffer	Peptidase Starch pH 7.0 buffer	Bacteria Cellulose pH 7.0 buffer
Incubation condition	37°C 60 minutes	37°C 60 minutes	37°C 60 minutes	37°C 60 minutes	37°C 60 minutes	37°C 60 minutes
IKI test						
Benedict's test						

color change from the original bright blue. Click **Record Data** to display your results in the grid (and record your results in Chart 2).

After you complete the Experiment, take the online **Post-lab Quiz** for Activity 2.

Activity Questions

1. Does amylase use cellulose as a substrate?

_____

_____

2. What effect did the addition of bacteria have on the digestion of cellulose?

_____

_____

_____

3. What effect did the addition of peptidase to the starch have? Why?

_____

_____

_____

4. What is the smallest subunit into which starch can be broken down?

_____

_____

_____

ACTIVITY 3

Assessing Pepsin Digestion of Protein

OBJECTIVES

1. Explain how the enzyme activity of pepsin can be assessed with the BAPNA assay.
2. Identify the substrate specificity of pepsin.
3. Discuss the effects of temperature and pH on pepsin activity.
4. Understand the pH specificity of enzyme activity and how it relates to human physiology.

Introduction

In this activity, you will explore the digestion of protein (**peptides**). Peptides are two or more **amino acids** linked together by a peptide bond. A peptide chain containing 10 to 100 amino acids is typically called a **polypeptide.** Proteins can consist of a large peptide chain (more than 100 amino acids) or even multiple peptide chains.

During digestion, **chief cells** of the stomach glands secrete a protein-digesting enzyme called **pepsin.** Pepsin **hydrolyzes** peptide bonds. This activity breaks up ingested proteins and polypeptides into smaller peptide chains and free amino acids. In this activity, you will use **BAPNA** as a **substrate** to assess pepsin activity. BAPNA is a synthetic "peptide" that releases

a yellow dye **product** when hydrolyzed. BAPNA solutions turn yellow in the presence of an active peptidase, such as pepsin, but otherwise remain colorless.

To quantify the pepsin activity in each test solution, you will use a **spectrophotometer** to measure the amount of yellow dye produced. A spectrophotometer shines light through the sample and then measures how much light is absorbed. The fraction of light absorbed is expressed as the sample's **optical density.** Yellow solutions, where BAPNA has been hydrolyzed, will have optical densities greater than zero. The greater the optical density, the more hydrolysis has occurred. Colorless solutions, in contrast, do not absorb light and will have an optical density near zero.

Some negative controls are included in this activity. With negative controls, a negative result is expected. Negative results with negative controls validate the experiment. Negative controls are used to determine whether there are any contaminating substances in the reagents. So, when a positive result is produced but a negative result is expected, one or more contaminating substances are present to cause the change.

EQUIPMENT USED The following equipment will be depicted on-screen: pepsin—an enzyme that digests peptides; BAPNA—a synthetic "peptide"; pH buffers—solutions used to set the pH of the test tube solution; deionized water—used to adjust the test tube solution volume so it is the same for each reaction; test tubes—used as reaction vessels for the various tests; incubators—used for temperature treatments (boiling and 37°C incubation); spectrophotometer—found in the assay cabinet; used to measure the optical density of solutions.

Experiment Instructions

Go to the home page in the PhysioEx software and click **Exercise 8: Chemical and Physical Processes of Digestion.** Click **Activity 3: Assessing Pepsin Digestion of Protein,** and take the online **Pre-lab Quiz** for Activity 3.

After you take the online Pre-lab Quiz, click the **Experiment** tab and begin the experiment. The experiment instructions are reprinted here for your reference. The opening screen for the experiment is shown below.

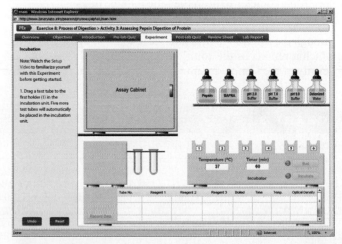

Incubation

1. Drag a test tube to the first holder (**1**) in the incubation unit. Five more test tubes will automatically be placed in the incubation unit.

2. Add the substances indicated below to tubes 1 through 6.

Tube 1: pepsin, BAPNA, pH 2.0 buffer

Tube 2: pepsin, BAPNA, pH 2.0 buffer

Tube 3: pepsin, deionized water, pH 2.0 buffer

Tube 4: deionized water, BAPNA, pH 2.0 buffer

Tube 5: pepsin, BAPNA, pH 7.0 buffer

Tube 6: pepsin, BAPNA, pH 9.0 buffer

     To add a substance to a test tube, drag the dropper cap of the bottle on the solutions shelf to the test tube.

3. Click the number **(1)** under the first test tube. The tube will descend into the incubation unit. All other tubes should remain in the raised position.

4. Click **Boil** to boil tube 1. After boiling for a few moments, the tube will automatically rise.

5. Click **Incubate** to start the run. Note that the incubation temperature is set at 37°C and the timer is set at 60 min. The incubation unit will gently agitate the test tube rack, evenly mixing the contents of all test tubes throughout the incubation. The simulation compresses the 60-minute time period into 10 seconds of real time, so what would be a 60-minute incubation in real life will take only 10 seconds in the simulation. When the incubation time elapses, the test tube rack will automatically rise, and the doors to the assay cabinet will open. The spectrophotometer is in the assay cabinet.

> **? PREDICT Question 1**
> At which pH do you think pepsin will have the highest activity?
> _____
> _____

Assays

6. You will now use the spectrophotometer to measure how much yellow dye was liberated from BAPNA hydrolysis. Drag the first tube in the incubation unit to the holder in the spectrophotometer to drop the tube into the holder.

7. Click **Analyze.** The spectrophotometer will shine light through the solution to measure the amount of light absorbed, which it reports as the solution's optical density. The optical density of the sample is shown in the optical density display.

8. Click **Record Data** to display your results in the grid (and record your results in Chart 3).

9. Drag the tube to its original position in the incubation unit.

10. Analyze the remaining five tubes by repeating the following steps for each tube.

- Drag the tube to the holder in the spectrophotometer to drop the tube into the holder.
- Click **Analyze.**
- Drag the tube to its original position in the incubation unit.

     After you have analyzed all five tubes, click **Record Data** to display your results in the grid (and record your results in Chart 3).

     After you complete the experiment, take the online **Post-lab Quiz** for Activity 3.

Activity Questions

1. Describe the significance of the optimum pH for pepsin observed in the simulation and the secretion of pepsin by the chief cells of the gastric glands.

_____

_____

2. Would pepsin be active in the mouth? Explain your answer.

_____

_____

3. What are the subunit products of peptide digestion?

_____

_____

4 Describe the reason for including control tube 4.

_____

_____

CHART 3	Pepsin Digestion of Protein					
Tube No.	1	2	3	4	5	6
Additives	Pepsin BAPNA pH 2.0 buffer	Pepsin BAPNA pH 2.0 buffer	Pepsin Deionized water pH 2.0 buffer	Deionized water BAPNA pH 2.0 buffer	Pepsin BAPNA pH 7.0 buffer	Pepsin BAPNA pH 9.0 buffer
Incubation condition	Boil first, then incubate at 37°C for 60 minutes	37°C 60 minutes	37°C 60 minutes	37°C 60 minutes	37°C 60 minutes	37°C 60 minutes
Optical density						

ACTIVITY 4

Assessing Lipase Digestion of Fat

OBJECTIVES

1. Explain how the enzyme activity of pancreatic lipase can be assessed with a pH-based measurement.
2. Identify the hydrolysis products of fat digestion.
3. Understand the role that bile plays in fat digestion.
4. Understand the significance of pH specificity of lipase activity and how it relates to human physiology.
5. Discuss the difficulty of using pH to measure digestion when comparing the activity of lipase at various pHs.

Introduction

Fats and oils belong to a diverse class of molecules called lipids. **Triglycerides,** a type of lipid, make up both fats and oils. At room temperature, fats are solid and oils are liquid. Both are poorly soluble in water. This insolubility of triglycerides presents a challenge during digestion because they tend to clump together, leaving only the surface molecules exposed to **lipase** enzymes. To overcome this difficulty, **bile salts** are secreted into the small intestine during digestion to physically emulsify lipids. Bile salts act like a detergent, separating the lipid clumps and increasing the surface area accessible to lipase enzymes.

As a result, two reactions must occur. First,

$$\text{Triglyceride clumps} \xrightarrow[\text{(emulsification)}]{\text{bile}} \text{minute triglyceride droplets}$$

Then,

$$\text{Triglyceride} \xrightarrow{\text{lipase}} \text{monoglyceride} + \text{two fatty acids}$$

Lipase hydrolyzes each triglyceride to a monoglyceride and two fatty acids. In addition to the **pancreatic lipase** secreted into the small intestine, **lingual lipase** and **gastric lipase** are also secreted. Even though bile salts are not secreted in the mouth or the stomach, small amounts of lipids are digested by these other lipases.

Because some of the end products of fat digestion are acidic (that is, fatty acids), lipase activity can be easily measured by monitoring the solution's **pH.** A solution containing fatty acids liberated by lipase activity will have a lower pH than a solution without such fatty acid production. You will record pH in this activity with a **pH meter.**

> **EQUIPMENT USED** The following equipment will be depicted on-screen: lipase—an enzyme that digests triglycerides; vegetable oil—a mixture of triglycerides; bile salts—a solution that physically separates fats into smaller droplets; pH buffers—solutions used to set the pH of the test tube solution; deionized water—used to adjust the test tube solution volume so it is the same for each reaction; test tubes—used as reaction vessels for the various tests; incubators—used for temperature treatments (boiling and 37°C incubation); pH meter—found in the assay cabinet; used to measure pH.

Experiment Instructions

Go to the home page in the PhysioEx software and click **Exercise 8: Chemical and Physical Processes of Digestion.** Click **Activity 4: Assessing Lipase Digestion of Fat,** and take the online **Pre-lab Quiz** for Activity 4.

After you take the online Pre-lab Quiz, click the **Experiment** tab and begin the experiment. The experiment instructions are reprinted here for your reference. The opening screen for the experiment is shown below.

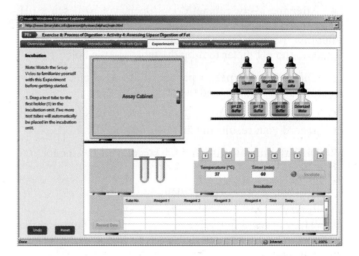

Incubation

1. Drag a test tube to the first holder **(1)** in the incubation unit. Five more test tubes will automatically be placed in the incubation unit.

2. Add the substances indicated below to tubes 1 through 6.

Tube 1: lipase, vegetable oil, bile salts, pH 7.0 buffer

Tube 2: lipase, vegetable oil, deionized water, pH 7.0 buffer

Tube 3: lipase, deionized water, bile salts, pH 9.0 buffer

Tube 4: deionized water, vegetable oil, bile salts, pH 7.0 buffer

Tube 5: lipase, vegetable oil, bile salts, pH 2.0 buffer

Tube 6: lipase, vegetable oil, bile salts, pH 9.0 buffer

To add a substance to a test tube, drag the dropper cap of the bottle on the solutions shelf to the test tube.

3. Click **Incubate** to start the run. Note that the incubation temperature is set at 37°C and the timer is set at 60 min. The incubation unit will gently agitate the test tube rack, evenly mixing the contents of all test tubes throughout the incubation. The simulation compresses the 60 minute time period into 10 seconds of real time, so what would be a 60-minute incubation in real life will take only 10 seconds in the simulation. When the incubation time elapses, the test tube rack will automatically rise, and the doors to the assay cabinet will open.

? PREDICT Question 1
Which tube do you think will have the highest lipase activity?

_____

_____

Assays

4. After the assay cabinet doors open, you will see a pH meter that you will use to measure the final pH of your test solutions. Drag the first tube in the incubation unit to the holder in the pH meter to drop the tube into the holder.

5. Click **Measure pH.** A probe will descend into the sample, take a pH reading, and then retract.

6. Click **Record Data** to display your results in the grid (and record your results in Chart 4).

7. Drag the tube to its original position in the incubation unit.

8. Measure the pH in the remaining five tubes by repeating the following steps for each tube.

- Drag the tube in the incubation unit to the holder in the pH meter to drop the tube into the holder.

- Click **Measure pH.**

- Drag the tube to its original position in the incubation unit.

After you have measured the pH in all five tubes, click **Record Data** to display your results in the grid (and record your results in Chart 4).

After you complete the experiment, take the online **Post-lab Quiz** for Activity 4.

Activity Questions

1. Describe how lipase activity is measured in the simulation.

_____

_____

2. Can you determine if fat hydrolysis occurred in tube 5? Why or why not?

_____

_____

3. Would pancreatic lipase be active in the mouth? Why or why not?

_____

_____

4. Describe the physical separation of fats by bile salts.

_____

_____

CHART 4	**Pancreatic Lipase Digestion of Triglycerides and the Action of Bile**					
Tube No.	1	2	3	4	5	6
Additives	Lipase Vegetable oil Bile salts pH 7.0 buffer	Lipase Vegetable oil Deionized water pH 7.0 buffer	Lipase Deionized water Bile salts pH 9.0 buffer	Deionized water Vegetable oil Bile salts pH 7.0 buffer	Lipase Vegetable oil Bile salts pH 2.0 buffer	Lipase Vegetable oil Bile salts pH 9.0 buffer
Incubation condition	37°C 60 minutes	37°C 60 minutes	37°C 60 minutes	37°C 60 minutes	37°C 60 minutes	37°C 60 minutes
pH						

NAME _____

LAB TIME/DATE _____

Chemical and Physical Processes of Digestion

ACTIVITY 1 Assessing Starch Digestion by Salivary Amylase

1. List the substrate and the subunit product of amylase. _____

   _____

2. What effect did boiling and freezing have on enzyme activity? Why? How well did the results compare with your prediction?

   _____

   _____

3. At what pH was the amylase most active? Describe the significance of this result. _____

   _____

4. Briefly describe the need for controls and give an example used in this activity. _____

   _____

5. Describe the significance of using a 37°C incubation temperature to test salivary amylase activity. _____

   _____

ACTIVITY 2 Exploring Amylase Substrate Specificity

1. Describe why the results in tube 1 and tube 2 are the same. _____

   _____

2. Describe the result in tube 3. How well did the results compare with your prediction? _____

   _____

3. Describe the usual substrate for peptidase. _____

   _____

4. Explain how bacteria can aid in digestion. _____

   _____

ACTIVITY 3 **Assessing Pepsin Digestion of Protein**

1. Describe the effect that boiling had on pepsin and how you could tell that it had that effect. _____

_____

2. Was your prediction correct about the optimal pH for pepsin activity? Discuss the physiological correlation behind your results.

_____

_____

3. What do you think would happen if you reduced the incubation time to 30 minutes for tube 5? _____

_____

ACTIVITY 4 **Assessing Lipase Digestion of Fat**

1. Explain why you can't fully test the lipase activity in tube 5. _____

_____

2. Which tube had the highest lipase activity? How well did the results compare with your prediction? Discuss possible reasons

why it may or may not have matched. _____

_____

_____

3. Explain why pancreatic lipase would be active in both the mouth and the intestine. _____

_____

_____

4. Describe the process of bile emulsification of lipids and how it improves lipase activity. _____

_____

_____

Renal System Physiology

Exercise Overview

The **kidney** is *both* an excretory and a regulatory organ. By filtering the water and solutes in the blood, the kidneys are able to *excrete* excess water, waste products, and even foreign materials from the body. However, the kidneys also *regulate* (1) plasma osmolarity (the concentration of a solution expressed as osmoles of solute per liter of solvent), (2) plasma volume, (3) the body's acid-base balance, and (4) the body's electrolyte balance. All these activities are extremely important for maintaining homeostasis in the body.

The paired kidneys are located between the posterior abdominal wall and the abdominal peritoneum. The right kidney is slightly lower than the left kidney. Each human kidney contains approximately one million **nephrons,** the functional units of the kidney.

Each nephron is composed of a **renal corpuscle** and a **renal tubule.** The renal corpuscle consists of a "ball" of capillaries, called the *glomerulus,* which is enclosed by a fluid-filled capsule, called *Bowman's capsule,* or the glomerular capsule. An **afferent arteriole** supplies blood to the glomerulus. As blood flows through the glomerular capillaries, protein-free plasma filters into the Bowman's capsule, a process called **glomerular filtration.** An **efferent arteriole** then drains the glomerulus of the remaining blood (view Figure 9.1, p. PEx-132).

The filtrate flows from Bowman's capsule into the start of the renal tubule, called the **proximal convoluted tubule,** then into the **loop of Henle,** a U-shaped

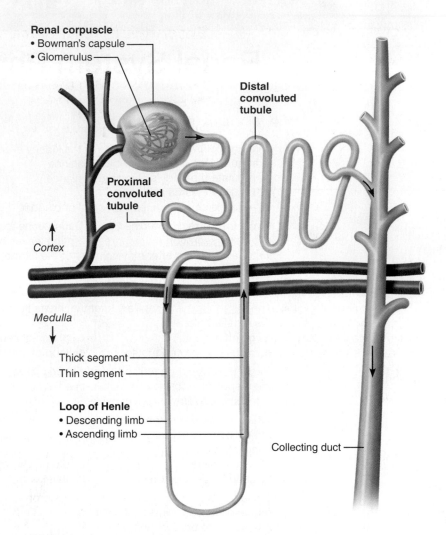

Renal corpuscle
• Bowman's capsule
• Glomerulus

Distal convoluted tubule

Proximal convoluted tubule

Cortex

Medulla

Thick segment
Thin segment

Loop of Henle
• Descending limb
• Ascending limb

Collecting duct

FIGURE 9.1 Location and structure of nephrons.

hairpin loop, and, finally, into the **distal convoluted tubule** before emptying into a **collecting duct.** From the collecting duct, the filtrate flows into, and collects in, the minor calyces.

The nephron performs three important functions that process blood into filtrate and urine: (1) glomerular filtration, (2) tubular reabsorption, and (3) tubular secretion. **Glomerular filtration** is a passive process in which fluid passes from the lumen of the glomerular capillary into the glomerular capsule of the renal tubule. **Tubular reabsorption** moves most of the filtrate back into the blood, leaving mainly salt water and the wastes in the lumen of the tubule. Some of the desirable, or needed, solutes are actively reabsorbed, and others move passively from the lumen of the tubule into the interstitial spaces. **Tubular secretion** is essentially the reverse of tubular reabsorption and is a process by which the kidneys can rid the blood of additional unwanted substances, such as creatinine and ammonia.

The reabsorbed solutes and water that move into the interstitial space between the nephrons need to be returned to the blood, or the kidneys will rapidly swell like balloons. The **peritubular capillaries** surrounding the renal tubule reclaim the reabsorbed substances and return them to general circulation. Peritubular capillaries arise from the efferent arteriole exiting the glomerulus and empty into the renal veins leaving the kidney.

ACTIVITY 1

The Effect of Arteriole Radius on Glomerular Filtration

OBJECTIVES

1. To understand the terms *nephron, glomerulus, glomerular capillaries, renal tubule, filtrate, Bowman's capsule, renal corpuscle, afferent arteriole, efferent arteriole, glomerular capillary pressure,* and *glomerular filtration rate.*

2. To understand how changes in afferent arteriole radius impact glomerular capillary pressure and filtration.

3. To understand how changes in efferent arteriole radius impact glomerular capillary pressure and filtration.

Introduction

Each of the million **nephrons** in each kidney contains two major parts: (1) a tubular component, the **renal tubule,** and (2) a vascular component, the **renal corpuscle** (view Figure 9.1). The **glomerulus** is a tangled capillary knot that filters fluid from the blood into the lumen of the renal tubule. The function of the renal tubule is to process the filtered fluid, also called the **filtrate.** The beginning of the renal tubule is an enlarged end called **Bowman's capsule** (or the glomerular capsule), which surrounds the glomerulus and serves to funnel the filtrate into the rest of the renal tubule. Collectively, the glomerulus and Bowman's capsule are called the renal corpuscle.

Two arterioles are associated with each glomerulus: an **afferent arteriole** feeds the **glomerular capillary** bed and an **efferent arteriole** drains it. These arterioles are responsible for blood flow through the glomerulus. The diameter of the efferent arteriole is smaller than the diameter of the afferent arteriole, restricting blood flow out of the glomerulus. Consequently, the pressure in the glomerular capillaries forces fluid through the endothelium of the capillaries into the lumen of the surrounding Bowman's capsule. In essence, everything in the blood except for the blood cells (red and white) and plasma proteins is filtered through the glomerular wall. From the Bowman's capsule, the filtrate moves into the rest of the renal tubule for processing. The job of the tubule is to reabsorb all the beneficial substances from its lumen and allow the wastes to travel down the tubule for elimination from the body.

During glomerular filtration, blood enters the glomerulus from the afferent arteriole and protein-free plasma flows from the blood across the walls of the glomerular capillaries and into the Bowman's capsule. The **glomerular filtration rate** is an index of kidney function. In humans, the filtration rate ranges from 80 to 140 ml/min, so that, in 24 hours, as much as 180 liters of filtrate is produced by the glomeruli. The filtrate formed is devoid of cellular debris, is essentially protein free, and contains a concentration of salts and organic molecules similar to that in blood.

The glomerular filtration rate can be altered by changing arteriole resistance or arteriole hydrostatic pressure. In this activity, you will explore the effect of arteriole radius on glomerular capillary pressure and filtration in a single nephron. You can apply the concepts you learn by studying a single nephron to understand the function of the kidney as a whole.

EQUIPMENT USED The following equipment will be depicted on-screen: source beaker for blood (first beaker on left side of screen)—simulates blood flow and pressure (mm Hg) from general circulation to the nephron; drain beaker for blood (second beaker on left side of screen)—simulates the renal vein; flow tube with adjustable radius—simulates the afferent arteriole and connects the blood supply to the glomerular capillaries; second flow tube with adjustable radius—simulates the efferent arteriole and drains the glomerular capillaries into the peritubular capillaries, which ultimately drain into the renal vein (drain beaker); simulated nephron (The filtrate forms in Bowman's capsule, flows through the renal tubule—the tubular components—and empties into a collecting duct, which in turn drains into the urinary bladder.); nephron tank; glomerulus—"ball" of capillaries that forms part of the filtration membrane; glomerular (Bowman's) capsule—forms part of the filtration membrane and a capsular space where the filtrate initially forms; proximal convoluted tubule; loop of Henle; distal convoluted tubule; collecting duct; drain beaker for filtrate (beaker on right side of screen)—simulates the urinary bladder.

Experiment Instructions

Go to the home page in the PhysioEx software and click **Exercise 9: Renal System Physiology.** Click **Activity 1: The Effect of Arteriole Radius on Glomerular Filtration,** and take the online **Pre-lab Quiz** for Activity 1.

After you take the online Pre-lab Quiz, click the **Experiment** tab and begin the experiment. The experiment instructions are reprinted here for your reference. The opening screen for the experiment is shown below.

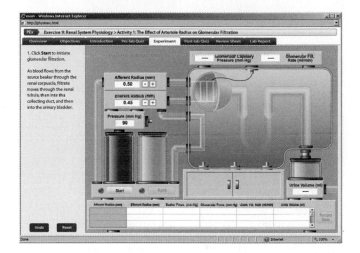

1. Click **Start** to initiate glomerular filtration. As blood flows from the source beaker through the renal corpuscle, filtrate moves through the renal tubule, then into the collecting duct, and then into the urinary bladder.

2. The glomerular capillary pressure display shows the hydrostatic blood pressure in the glomerular capillaries that promotes filtration, and the filtration rate display shows the flow rate of the fluid moving from the lumen of the glomerular capillaries into the lumen of Bowman's capsule. Click **Record Data** to display your results in the grid (and record your results in Chart 1, p. PEx-134).

3. Click **Refill** to replenish the source beaker and prepare the nephron for the next run.

? PREDICT Question 1
What will happen to the glomerular capillary pressure and filtration rate if you decrease the radius of the afferent arteriole?

---

4. Decrease the radius of the afferent arteriole to 0.45 mm by clicking the − button beside the afferent radius display. Click **Start** to initiate glomerular filtration.

5. Note the glomerular capillary pressure and glomerular filtration rate displays and click **Record Data** to display your results in the grid (and record your results in Chart 1).

6. Click **Refill** to replenish the source beaker and prepare the nephron for the next run.

CHART 1	Effect of Arteriole Radius on Glomerular Filtration		
Afferent arteriole radius (mm)	Efferent arteriole radius (mm)	Glomerular capillary pressure (mm Hg)	Glomerular filtration rate (ml/min)

7. You will now observe the effect of incremental decreases in the radius of the afferent arteriole.

- Decrease the radius of the afferent arteriole by 0.05 mm by clicking the − button beside the afferent radius display.
- Click **Start** to initiate glomerular filtration.
- Note the glomerular capillary pressure and glomerular filtration rate displays and click **Record Data** to display your results in the grid (and record your results in Chart 1).
- Click **Refill** to replenish the source beaker and prepare the nephron for the next run.

Repeat this step until you reach an afferent arteriole radius of 0.35 mm.

? PREDICT Question 2
What will happen to the glomerular capillary pressure and filtration rate if you increase the radius of the afferent arteriole?

8. Increase the radius of the afferent arteriole to 0.55 mm by clicking the + button beside the afferent radius display. Click **Start** to initiate glomerular filtration.

9. Note the glomerular capillary pressure and glomerular filtration rate displays and click **Record Data** to display your results in the grid (and record your results in Chart 1).

10. Click **Refill** to replenish the source beaker and prepare the nephron for the next run.

11. Increase the radius of the afferent arteriole to 0.60 mm. Click **Start** to initiate glomerular filtration.

12. Note the glomerular capillary pressure and glomerular filtration rate displays and click **Record Data** to display your results in the grid (and record your results in Chart 1).

13. Click **Refill** to replenish the source beaker and prepare the nephron for the next run.

? PREDICT Question 3
What will happen to the glomerular capillary pressure and filtration rate if you decrease the radius of the efferent arteriole?

14. Decrease the radius of the afferent arteriole to 0.50 mm by clicking the − button beside the afferent radius display. Click **Start** to initiate glomerular filtration.

15. Note the glomerular capillary pressure and glomerular filtration rate displays and click **Record Data** to display your results in the grid (and record your results in Chart 1).

16. Click **Refill** to replenish the source beaker and prepare the nephron for the next run.

17. You will now observe the effect of incremental decreases in the radius of the efferent arteriole.

- Decrease the radius of the efferent arteriole by 0.05 mm by clicking the − button beside the efferent radius display.
- Click **Start** to initiate glomerular filtration.

- Note the glomerular capillary pressure and glomerular filtration rate displays and click **Record Data** to display your results in the grid (and record your results in Chart 1).
- Click **Refill** to replenish the source beaker and prepare the nephron for the next run.

Repeat this step until you reach an efferent arteriole radius of 0.30 mm.

After you complete the experiment, take the online **Post-lab Quiz** for Activity 1.

Activity Questions

1. Activation of sympathetic nerves that innervate the kidney leads to a decreased urine production. Knowing that fact, what do you think the sympathetic nerves do to the afferent arteriole?

_____

_____

2. How is this effect of the sympathetic nervous system beneficial? Could this effect become harmful if it goes on too long?

_____

_____ ▬

The Effect of Pressure on Glomerular Filtration

OBJECTIVES

1. To understand the terms *glomerulus, glomerular capillaries, renal tubule, filtrate, Starling forces, Bowman's capsule, renal corpuscle, afferent arteriole, efferent arteriole, glomerular capillary pressure,* and *glomerular filtration rate.*
2. To understand how changes in glomerular capillary pressure affect glomerular filtration rate.
3. To understand how changes in renal tubule pressure affect glomerular filtration rate.

Introduction

Cellular metabolism produces a complex mixture of waste products that must be eliminated from the body. This excretory function is performed by a combination of organs, most importantly, the paired kidneys. Each kidney consists of approximately one million nephrons, which carry out three crucial processes: (1) glomerular filtration, (2) tubular reabsorption, and (3) tubular secretion.

Both the blood pressure in the **glomerular capillaries** and the **filtrate** pressure in the **renal tubule** can have a significant impact on the **glomerular filtration rate.** During glomerular filtration, blood enters the **glomerulus** from the afferent arteriole. **Starling forces** (hydrostatic and osmotic pressure gradients) drive protein-free fluid between the blood in the glomerular capillaries and the filtrate in **Bowman's capsule.** The glomerular filtration rate is an index of kidney function. In humans, the filtration rate ranges from 80 to 140 ml/min, so that, in 24 hours, as much as 180 liters of filtrate is produced by the glomerular capillaries. The filtrate formed is devoid of blood cells, is essentially protein free, and contains a concentration of salts and organic molecules similar to that in blood.

Approximately 20% of the blood that enters the glomerular capillaries is normally filtered into Bowman's capsule, where it is then referred to as filtrate. The unusually high hydrostatic blood pressure in the glomerular capillaries promotes this filtration. Thus, the glomerular filtration rate can be altered by changing the afferent arteriole resistance (and, therefore, the hydrostatic pressure). In this activity you will explore the effect of blood pressure on the glomerular filtration rate in a single nephron. You can apply the concepts you learn by studying a single nephron to understand the function of the kidney as a whole.

EQUIPMENT USED The following equipment will be depicted on-screen: left source beaker (first beaker on left side of screen)—simulates blood flow and pressure (mm Hg) from general circulation to the nephron; drain beaker for blood (second beaker on left side of screen)—simulates the renal vein; flow tube with adjustable radius—simulates the afferent arteriole and connects the blood supply to the glomerular capillaries; second flow tube with adjustable radius—simulates the efferent arteriole and drains the glomerular capillaries into the peritubular capillaries, which ultimately drain into the renal vein (drain beaker); simulated nephron (The filtrate forms in Bowman's capsule, flows through the renal tubule—the tubular components—and empties into a collecting duct, which in turn drains into the urinary bladder.); nephron tank; glomerulus—"ball" of capillaries that forms part of the filtration membrane; glomerular (Bowman's) capsule—forms part of the filtration membrane and a capsular space where the filtrate initially forms; proximal convoluted tubule; loop of Henle; distal convoluted tubule; collecting duct; one-way valve between end of collecting tube (duct) and urinary bladder—used to restrict the flow of filtrate into the urinary bladder, increasing the volume and pressure in the renal tubule; drain beaker for filtrate (beaker on right side of screen)—simulates the urinary bladder.

Experiment Instructions

Go to the home page in the PhysioEx software and click **Exercise 9: Renal System Physiology.** Click **Activity 2: The Effect of Pressure on Glomerular Filtration,** and take the online **Pre-lab Quiz** for Activity 2.

After you take the online Pre-lab Quiz, click the **Experiment** tab and begin the experiment. The experiment instructions are reprinted here for your reference. The opening screen for the experiment is shown on the following page.

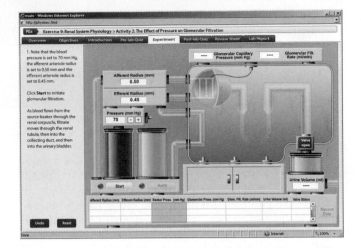

1. Note that the blood pressure is set to 70 mm Hg, the afferent arteriole radius is set to 0.50 mm, and the efferent arteriole radius is set to 0.45 mm. Click **Start** to initiate glomerular filtration. As blood flows from the source beaker through the renal corpuscle, filtrate moves through the renal tubule, then into the collecting duct, and then into the urinary bladder.

2. The glomerular capillary pressure display shows the hydrostatic blood pressure in the glomerular capillaries that promotes filtration, and the filtration rate display shows the flow rate of the fluid moving from the lumen of the glomerular capillaries into the lumen of Bowman's capsule. Click **Record Data** to display your results in the grid (and record your results in Chart 2.)

3. Click **Refill** to replenish the source beaker and prepare the nephron for the next run.

? **PREDICT Question 1**
What will happen to the glomerular capillary pressure and filtration rate if you increase the blood pressure in the left source beaker?

_____

4. Increase the blood pressure to 80 mm Hg by clicking the + button beside the pressure display. Click **Start** to initiate glomerular filtration.

5. Note the glomerular capillary pressure and glomerular filtration rate displays and click **Record Data** to display your results in the grid (and record your results in Chart 2).

6. Click **Refill** to replenish the source beaker and prepare the nephron for the next run.

7. You will now observe the effect of further incremental increases in blood pressure.

- Increase the blood pressure by 10 mm Hg by clicking the + button beside the pressure display.

- Click **Start** to initiate glomerular filtration.

- Note the glomerular capillary pressure and glomerular filtration rate displays and click **Record Data** to display your results in the grid (and record your results in Chart 2).

- Click **Refill** to replenish the source beaker and prepare the nephron for the next run.

Repeat this step until you reach a blood pressure of 100 mm Hg.

? **PREDICT Question 2**
What will happen to the filtrate pressure in Bowman's capsule (not directly measured in this experiment) and the filtration rate if you close the one-way valve between the collecting duct and the urinary bladder?

_____

8. Note that the valve between the collecting duct and the urinary bladder is open. Decrease the blood pressure to 70 mm Hg by clicking the button beside the pressure display. Click **Start** to initiate glomerular filtration.

CHART 2	Effect of Pressure on Glomerular Filtration			
Blood pressure (mm Hg)	Valve (open or closed)	Glomerular capillary pressure (mm Hg)	Glomerular filtration rate (ml/min)	Urine volume (ml)

9. Note the glomerular capillary pressure and glomerular filtration rate displays and click **Record Data** to display your results in the grid (and record your results in Chart 2).

10. Click **Refill** to replenish the source beaker and prepare the nephron for the next run.

11. Click the valve between the collecting duct and the urinary bladder to close it. Click **Start** to initiate glomerular filtration.

12. Note the glomerular capillary pressure and glomerular filtration rate displays and click **Record Data** to display your results in the grid (and record your results in Chart 2).

13. Click **Refill** to replenish the source beaker and prepare the nephron for the next run.

14. Increase the blood pressure to 100 mm Hg. Click **Start** to initiate glomerular filtration.

15. Note the glomerular capillary pressure and glomerular filtration rate displays and click **Record Data** to display your results in the grid (and record your results in Chart 2).

16. Click **Refill** to replenish the source beaker and prepare the nephron for the next run.

17. Click the valve between the collecting duct and the urinary bladder to open it. Click **Start** to initiate glomerular filtration.

18. Note the glomerular capillary pressure and glomerular filtration rate displays and click **Record Data** to display your results in the grid (and record your results in Chart 2).

After you complete the experiment, take the online **Post-lab Quiz** for Activity 2.

Activity Questions

1. Judging from the results in this laboratory activity, what *should be* the effect of blood pressure on glomerular filtration?

_____

_____

2. Persistent high blood pressure with inadequate glomerular filtration is now a frequent problem in Western cultures. Using the concepts in this activity, explain this health problem.

_____

_____

_____ ▬▬

A C T I V I T Y   3

Renal Response to Altered Blood Pressure

OBJECTIVES

1. To understand the terms *nephron, renal tubule, filtrate, Bowman's capsule, blood pressure, afferent arteriole,* *efferent arteriole, glomerulus, glomerular filtration rate,* and *glomerular capillary pressure.*

2. To understand how blood pressure affects glomerular capillary pressure and glomerular filtration.

3. To observe which is more effective: changes in afferent or efferent arteriole radius when changes in blood pressure occur.

Introduction

In humans approximately 180 liters of filtrate flows into the **renal tubules** every day. As demonstrated in Activity 2, the **blood pressure** supplying the **nephron** can have a substantial impact on the **glomerular capillary pressure** and **glomerular filtration.** However, under most circumstances, glomerular capillary pressure and glomerular filtration remain relatively constant despite changes in blood pressure because the nephron has the capacity to alter its **afferent** and **efferent arteriole** radii.

During glomerular filtration, blood enters the **glomerulus** from the afferent arteriole. **Starling forces** (primarily hydrostatic pressure gradients) drive protein-free fluid out of the glomerular capillaries and into **Bowman's capsule.** Importantly for our body's homeostasis, a relatively constant glomerular filtration rate of 125 ml/min is maintained despite a wide range of blood pressures that occur throughout the day for an average human.

Activities 1 and 2 explored the independent effects of arteriole radii and blood pressure on glomerular capillary pressure and glomerular filtration. In the human body, these effects occur simultaneously. Therefore, in this activity, you will alter both variables to explore their combined effects on glomerular filtration and observe how changes in one variable can compensate for changes in the other to maintain an adequate glomerular filtration rate.

EQUIPMENT USED The following equipment will be depicted on-screen: left source beaker (first beaker on left side of screen)—simulates blood flow and pressure (mm Hg) from general circulation to the nephron; drain beaker for blood (second beaker on left side of screen)—simulates the renal vein; flow tube with adjustable radius—simulates the afferent arteriole and connects the blood supply to the glomerular capillaries; second flow tube with adjustable radius—simulates the efferent arteriole and drains the glomerular capillaries into the peritubular capillaries, which ultimately drain into the renal vein (drain beaker); simulated nephron (The filtrate forms in Bowman's capsule, flows through the renal tubule—the tubular components—and empties into a collecting duct, which in turn drains into the urinary bladder.); nephron tank; glomerulus—"ball" of capillaries that forms part of the filtration membrane; glomerular (Bowman's) capsule— forms part of the filtration membrane and a capsular space where the filtrate initially forms; proximal convoluted tubule; loop of Henle; distal convoluted tubule; collecting duct; one-way valve between end of collecting tube (duct) and urinary bladder—used to restrict the flow of filtrate into the urinary bladder, increasing the volume and pressure in the renal tubule; drain beaker for filtrate (beaker on right side of screen)—simulates the urinary bladder.

Experiment Instructions

Go to the home page in the PhysioEx software and click **Exercise 9: Renal System Physiology.** Click **Activity 3: Renal Response to Altered Blood Pressure,** and take the online **Pre-lab Quiz** for Activity 3.

    After you take the online Pre-lab Quiz, click the **Experiment** tab and begin the experiment. The experiment instructions are reprinted here for your reference. The opening screen for the experiment is shown below.

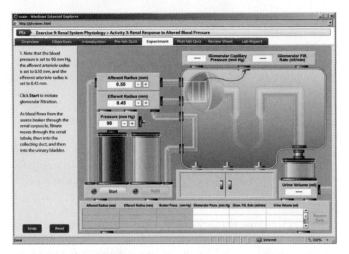

1. Note that the blood pressure is set to 90 mm Hg, the afferent arteriole radius is set to 0.50 mm, and the efferent arteriole radius is set to 0.45 mm. Click **Start** to initiate glomerular filtration. As blood flows from the source beaker through the renal corpuscle, filtrate moves through the renal tubule, then into the collecting duct, and then into the urinary bladder.

2. The glomerular capillary pressure display shows the hydrostatic blood pressure in the glomerular capillaries that promotes filtration, and the filtration rate display shows the flow rate of the fluid moving from the lumen of the glomerular capillaries into the lumen of Bowman's capsule. Click **Record Data** to display your results in the grid (and record your results in Chart 3).

3. Click **Refill** to replenish the source beaker and prepare the nephron for the next run.

4. You will now observe how the nephron might operate to keep the glomerular filtration rate relatively constant despite a large drop in blood pressure. Decrease the blood pressure to 70 mm Hg by clicking the − button beside the pressure display. Click **Start** to initiate glomerular filtration.

5. Note the glomerular capillary pressure and glomerular filtration rate displays and click **Record Data** to display your results in the grid (and record your results in Chart 3).

6. Click **Refill** to replenish the source beaker and prepare the nephron for the next run.

7. Increase the afferent arteriole radius to 0.60 mm by clicking the + button beside the afferent radius display. Click **Start** to initiate glomerular filtration.

8. Note the glomerular capillary pressure and glomerular filtration rate displays and click **Record Data** to display your results in the grid (and record your results in Chart 3).

9. Click **Refill** to replenish the source beaker and prepare the nephron for the next run.

10. Return the afferent arteriole radius to 0.50 mm by clicking the − button beside the afferent radius display and decrease the efferent radius to 0.35 mm by clicking the button beside the efferent radius display. Click **Start** to initiate glomerular filtration.

11. Note the glomerular capillary pressure and glomerular filtration rate displays and click **Record Data** to display your results in the grid (and record your results in Chart 3).

12. Click **Refill** to replenish the source beaker and prepare the nephron for the next run.

> **? PREDICT Question 1**
> What will happen to the glomerular capillary pressure and glomerular filtration rate if both of these arteriole radii changes are implemented simultaneously with the low blood pressure condition?

CHART 3	Renal Response to Altered Blood Pressure			
Afferent arteriole radius (mm)	**Efferent arteriole radius (mm)**	**Blood pressure (mm Hg)**	**Glomerular capillary pressure (mm Hg)**	**Glomerular filtration rate (ml/min)**

13. Set the afferent arteriole radius to 0.60 mm and keep the efferent arteriole radius at 0.35 mm. Click **Start** to initiate glomerular filtration.

14. Note the glomerular capillary pressure and glomerular filtration rate displays and click **Record Data** to display your results in the grid (and record your results in Chart 3).

After you complete the experiment, take the online **Post-lab Quiz** for Activity 3.

Activity Questions

1. How could an increased urine volume be viewed as beneficial to the body?

_____

_____

2. Diuretics are frequently given to people with persistent high blood pressure. Why?

_____

_____

ACTIVITY 4

Solute Gradients and Their Impact on Urine Concentration

OBJECTIVES

1. To understand the terms *antidiuretic hormone (ADH),* *reabsorption, loop of Henle, collecting duct, tubule lumen, interstitial space,* and *peritubular capillaries.*

2. To explain the process of water reabsorption in specific regions of the nephron.

3. To understand the role of ADH in water reabsorption by the nephron.

4. To describe how the kidneys can produce urine that is four times more concentrated than the blood.

Introduction

As filtrate moves through the tubules of a nephron, solutes and water move *from* the **tubule lumen** *into* the **interstitial spaces** of the nephron. This movement of solutes and water relies on the total solute concentration gradient in the interstitial spaces surrounding the tubule lumen. The interstitial fluid is comprised mostly of NaCl and urea. When the nephron is permeable to solutes or water, equilibrium will be reached between the interstitial fluid and the tubular fluid contents.

Antidiuretic hormone (ADH) increases the water permeability of the **collecting duct,** allowing water to flow to areas of higher solute concentration, from the tubule lumen into the surrounding interstitial spaces. **Reabsorption** describes this movement of filtered solutes and water from the lumen of the renal tubules back into the plasma. The reabsorbed solutes and water that move into the interstitial space need to be returned to the blood, or the kidneys will rapidly

swell like balloons. The **peritubular capillaries** surrounding the renal tubule reclaim the reabsorbed substances and return them to general circulation. Peritubular capillaries arise from the efferent arteriole exiting the glomerulus and empty into the renal veins leaving the kidney.

Without reabsorption, we would excrete the solutes and water that our bodies need to maintain homeostasis. In this activity you will examine the process of passive reabsorption that occurs while filtrate travels through a nephron and urine is formed. While completing the experiment, assume that when ADH is present, the conditions favor the formation of the most concentrated urine possible.

EQUIPMENT USED The following equipment will be depicted on-screen: simulated nephron surrounded by interstitial space between the nephron and peritubular capillaries (Reabsorbed solutes, such as glucose, will move from the lumen of the tubule into the interstitial space, and then into the peritubular capillaries that branch out from the efferent arteriole.); drain beaker for filtrate—simulates the urinary bladder; antidiuretic hormone (ADH).

Experiment Instructions

Go to the home page in the PhysioEx software and click **Exercise 9: Renal System Physiology.** Click **Activity 4: Solute Gradients and Their Impact on Urine Concentration,** and take the online **Pre-lab Quiz** for Activity 4.

After you take the online Pre-lab Quiz, click the **Experiment** tab and begin the experiment. The experiment instructions are reprinted here for your reference. The opening screen for the experiment is shown below.

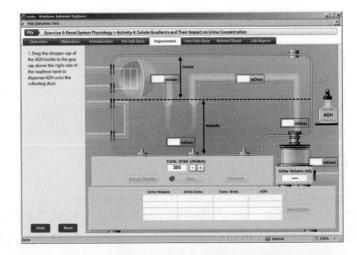

1. Drag the dropper cap of the ADH bottle to the gray cap above the right side of the nephron tank to dispense ADH onto the collecting duct.

2. Click **Dispense** beneath the concentration gradient display to adjust the maximum total solute concentration in the interstitial fluid to 300 mOsm. Because the blood solute concentration is also 300 mOsm, there is no osmotic difference between the lumen of the tubule and the surrounding interstitial fluid.

3. Click **Start** to initiate filtration. Filtrate will flow through the nephron, and solutes and water will move out of the tubules into the interstitial space. Fluid will also move

back into the peritubular capillaries, thus completing the process of reabsorption.

4. Click **Record Data** to display your results in the grid (and record your results in Chart 4).

CHART 4	Solute Gradients and Their Impact on Urine Concentration	
Urine volume (ml)	Urine concentration (mOsm)	Concentration gradient (mOsm)

5. Click **Empty Bladder** to prepare for the next run.

> **? PREDICT Question 1**
> What will happen to the urine volume and concentration as the solute gradient in the interstitial space is increased?
> _____

6. Increase the maximum concentration of the solutes in the interstitial space to 600 mOsm by clicking the + button beside the concentration gradient display. Click **Dispense** to adjust the maximum total solute concentration in the interstitial fluid.

7. Click **Start** to initiate filtration.

8. Click **Record Data** to display your results in the grid (and record your results in Chart 4).

9. Click **Empty Bladder** to prepare for the next run.

10. You will now observe the effect of incremental increases in maximum total solute concentration in the interstitial fluid.

- Increase the maximum concentration of the solutes in the interstitial space by 300 mOsm by clicking the + button beside the concentration gradient display.

- Click **Dispense** to adjust the maximum total solute concentration in the interstitial fluid.

- Click **Start** to initiate filtration.

- Click **Record Data** to display your results in the grid (and record your results in Chart 4).

- Click **Empty Bladder** to prepare for the next run.

Repeat this step until you reach the maximum total solute concentration in the interstitial fluid of 1200 mOsm.

After you complete the experiment, take the online **Post-lab Quiz** for Activity 4.

Activity Questions

1. From what you learned in this activity, speculate on ways that desert rats are able to concentrate their urine significantly more than humans.

_____

_____

2. Judging from this activity, what would be a reasonable mechanism for diuretics?

_____

_____

▬

ACTIVITY 5

Reabsorption of Glucose via Carrier Proteins

OBJECTIVES

1. To understand the terms *reabsorption, carrier proteins, apical membrane, secondary active transport, facilitated diffusion,* and *basolateral membrane.*

2. To understand the role that glucose carrier proteins play in removing glucose from the filtrate.

3. To understand the concept of a glucose carrier transport maximum and why glucose is not normally present in the urine.

Introduction

Reabsorption is the movement of filtered solutes and water from the lumen of the renal tubules back into the plasma. Without reabsorption, we would excrete the solutes and water that our bodies require for homeostasis.

Glucose is not very large and is therefore easily filtered out of the plasma into Bowman's capsule as part of the filtrate. To ensure that glucose is reabsorbed into the body so that it can fuel cellular metabolism, glucose **carrier proteins** are present in the proximal tubule cells of the nephron. There are a finite number of these glucose carriers in each renal tubule cell. Therefore, if too much glucose is present in the filtrate, it will not all be reabsorbed and glucose will be inappropriately excreted into the urine.

Glucose is first absorbed by **secondary active transport** at the **apical membrane** of proximal tubule cells and then it leaves the tubule cell via **facilitated diffusion** along the **basolateral membrane.** Both types of carrier proteins that transport these molecules across the tubule membranes are transmembrane proteins. Because carrier proteins are needed to move glucose from the lumen of the nephron into the interstitial spaces, there is a limit to the amount of glucose that can be reabsorbed. When all glucose carriers are bound with the glucose they are transporting, excess glucose in the filtrate is eliminated in urine.

In this activity, you will examine the effect of varying the number of glucose transport proteins in the *proximal convoluted tubule*. It is important to note that, normally, the number of glucose carriers is constant in a human kidney and that it is the plasma glucose that varies during the day. Plasma glucose will be held constant in this activity, and the number of glucose carriers will be varied.

EQUIPMENT USED The following equipment will be depicted on-screen: simulated nephron surrounded by interstitial space between the nephron and peritubular capillaries (Reabsorbed solutes, such as glucose, will move from the lumen of the tubule into the interstitial space, and then into the peritubular capillaries that branch out from the efferent arteriole.); drain beaker for filtrate—simulates the urinary bladder; glucose carrier protein control box—used to adjust the number of glucose carriers that will be inserted into the proximal tubule.

Experiment Instructions

Go to the home page in the PhysioEx software and click **Exercise 9: Renal System Physiology.** Click **Activity 5: Reabsorption of Glucose via Carrier Proteins,** and take the online **Pre-lab Quiz** for Activity 5.

   After you take the online Pre-lab Quiz, click the **Experiment** tab and begin the experiment. The experiment instructions are reprinted here for your reference. The opening screen for the experiment is shown below.

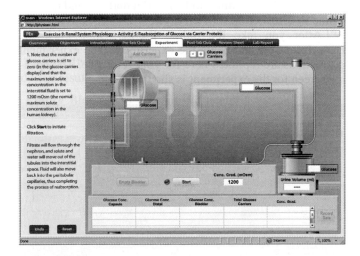

1. Note that the number of glucose carriers is set to zero (in the glucose carriers display) and that the maximum total solute concentration in the interstitial fluid is set to 1200 mOsm (the normal maximum solute concentration in the human kidney). Click **Start** to initiate filtration. Filtrate will flow through the nephron, and solute and water will move out of the tubules into the interstitial space. Fluid will also move back into the peritubular capillaries, thus completing the process of reabsorption.

2. Click **Record Data** to display your results in the grid (and record your results in Chart 5). The concentrations of glucose in Bowman's capsule, the distal convoluted tubule, and the urinary bladder will be displayed in the grid.

CHART 5	Reabsorption of Glucose via Carrier Proteins		
Glucose concentration (m*M*)			
Bowman's capsule	Distal convoluted tubule	Urinary bladder	Glucose carriers

3. Click **Empty Bladder** to prepare the nephron for the next run.

PREDICT Question 1
What will happen to the glucose concentration in the urinary bladder as glucose carriers are added to the proximal tubule?

---

4. Increase the number of glucose carriers to 100 (an arbitrary number) by clicking the + button beside the glucose carriers display. Click **Add Carriers** to insert the specified number of glucose carrier proteins per unit area into the membrane of the proximal tubule.

5. Click **Start** to initiate filtration.

6. Click **Record Data** to display your results in the grid (and record your results in Chart 5).

7. Click **Empty Bladder** to prepare the nephron for the next run.

8. You will now observe the effect of incremental increases in the number of glucose carriers.

   • Increase the number of glucose carriers by 100 by clicking the + button beside the glucose carriers display.

   • Click **Add Carriers** to insert the specified number of glucose carrier proteins per unit area into the membrane of the proximal tubule.

   • Click **Start** to initiate filtration.

   • Click **Record Data** to display your results in the grid (and record your results in Chart 5).

   • Click **Empty Bladder** to prepare the nephron for the next run.

   Repeat this step until you have inserted 400 glucose carrier proteins per unit area into the membrane of the proximal tubule.

After you complete the experiment, take the online **Post-lab Quiz** for Activity 5.

2. Compare the urine volume in your baseline data with the urine volume as you increased the blood pressure. How did the

   urine volume change? _____

   _____

3. How could the change in urine volume with the increase in blood pressure be viewed as being beneficial to the body?

   _____

   _____

4. When the one-way valve between the collecting duct and the urinary bladder was closed, what happened to the filtrate pressure in Bowman's capsule (this is not directly measured in this experiment) and the glomerular filtration rate? How well did

   the results compare with your prediction? _____

   _____

5. How did increasing the blood pressure alter the results when the valve was closed? _____

   _____

ACTIVITY 3   Renal Response to Altered Blood Pressure

1. List the several mechanisms you have explored that change the glomerular filtration rate. How does each mechanism specifi-

   cally alter the glomerular filtration rate? _____

   _____

2. Describe and explain what happened to the glomerular capillary pressure and glomerular filtration rate when *both* arteriole radii changes were implemented simultaneously with the low blood pressure condition. How well did the results compare

   with your prediction? _____

   _____

   _____

3. How could you adjust the afferent or efferent radius to compensate for the effect of reduced blood pressure on the glomerular

   filtration rate? _____

   _____

4. Which arteriole radius adjustment was more effective at compensating for the effect of low blood pressure on the glomerular

   filtration rate? Explain why you think this difference occurs. _____

   _____

5. In the body, how does a nephron maintain a near-constant glomerular filtration rate despite a constantly fluctuating blood

   pressure? _____

   _____

ACTIVITY 4 Solute Gradients and Their Impact on Urine Concentration

1. What happened to the urine concentration as the solute concentration in the interstitial space was increased? How well did

the results compare to your prediction? _____

_____

_____

2. What happened to the volume of urine as the solute concentration in the interstitial space was increased? How well did the

results compare to your prediction? _____

_____

3. What do you think would happen to urine volume if you did not add ADH to the collecting duct? _____

_____

_____

4. Is most of the tubule filtrate reabsorbed into the body or excreted in urine? Explain. _____

_____

_____

5. Can the reabsorption of solutes influence water reabsorption from the tubule fluid? Explain. _____

_____

ACTIVITY 5 Reabsorption of Glucose via Carrier Proteins

1. What happens to the concentration of glucose in the urinary bladder as the number of glucose carriers increases? _____

_____

_____

2. What types of transport are utilized during glucose reabsorption and where do they occur? _____

_____

3. Why does the glucose concentration in the urinary bladder become zero in these experiments? _____

_____

4. A person with type 1 diabetes cannot make insulin in the pancreas, and a person with untreated type 2 diabetes does not
respond to the insulin that is made in the pancreas. In either case, why would you expect to find glucose in the person's urine?

_____

_____

ACTIVITY 6 The Effect of Hormones on Urine Formation

1. How did the addition of aldosterone affect urine volume (compared with baseline)? Can the reabsorption of solutes influence

   water reabsorption in the nephron? Explain. How well did the results compare with your prediction? _____

   _____

   _____

2. How did the addition of ADH affect urine volume (compared with baseline)? How well did the results compare with your
   prediction? Why did the addition of ADH also affect the concentration of potassium in the urine (compared with baseline)?

   _____

   _____

3. What is the principal determinant for the release of aldosterone from the adrenal cortex? _____

   _____

4. How did the addition of both aldosterone and ADH affect urine volume (compared with baseline)? How well did the results

   compare with your prediction? _____

   _____

   _____

5. What is the principal determinant for the release of ADH from the posterior pituitary gland? Does ADH favor the formation

   of dilute or concentrated urine? Explain why. _____

   _____

   _____

6. Which hormone (aldosterone or ADH) has the greater effect on urine volume? Why? _____

   _____

   _____

7. If ADH is not available, can the urine concentration still vary? Explain your answer. _____

   _____

   _____

8. Consider this situation: you want to reabsorb sodium ions but you do not want to increase the volume of the blood by
   reabsorbing large amounts of water from the filtrate. Assuming that aldosterone and ADH are both present, how would you

   adjust the hormones to accomplish the task? _____

   _____

Acid-Base Balance

Exercise Overview

pH denotes the hydrogen ion concentration, [H⁺], in a solution (such as body fluids). The reciprocal relationship between pH and [H⁺] is defined by the following equation.

$$pH = \log(1/[H^+])$$

Because the relationship is reciprocal, [H⁺] is higher at *lower* pH values (indicating higher acid levels) and lower at *higher* pH values (indicating lower acid levels).

The pH of a body's fluid is also referred to as its **acid-base balance.** An **acid** is a substance that releases H⁺ in solution. A **base,** often a hydroxyl ion (OH⁻) or bicarbonate ion (HCO₃⁻), is a substance that binds, or buffers, the H⁺. A **strong acid** completely dissociates in solution, releasing all of its hydrogen ions and, thus, lowering the solution's pH. A **weak acid** dissociates incompletely and does not release all of its hydrogen ions in solution, producing a lesser effect on the solution's pH. A **strong base** has a strong tendency to bind to H⁺, raising the solution's pH. A **weak base** binds less of the H⁺, producing a lesser effect on the solution's pH.

The pH of body fluids is very tightly regulated. Blood and tissue fluids normally have a pH between 7.35 and 7.45. Under pathological conditions, blood pH as low as 6.9 or as high as 7.8 has been recorded, but a higher or lower pH cannot sustain human life. The narrow range from 7.35 to 7.45 is remarkable when you consider the vast number of biochemical reactions that take place in the body. The human body normally produces a large amount of H⁺ as the result of metabolic processes; ingested acids; and the products of fat, sugar, and amino acid metabolism. The regulation of a relatively constant internal pH is one of the major physiological functions of the body's organ systems.

To maintain pH homeostasis, the body utilizes both *chemical* and *physiological* buffering systems. Chemical buffers are composed of a mixture of weak acids and weak bases. They help regulate the body's pH levels by binding H^+ and removing it from solution as its concentration begins to rise or by releasing H^+ into solution as its concentration begins to fall. The body's three major chemical buffering systems are the *bicarbonate, phosphate,* and *protein buffer systems.* We will not focus on chemical buffering systems in this exercise, but keep in mind that chemical buffers are the fastest form of compensation and can return pH to normal within a fraction of a second.

The body's two major physiological buffering systems are the **renal system** and the **respiratory system.** The renal system is the slower of the two, taking hours to days to do its work. The respiratory system usually works within minutes, but cannot handle the amount of pH change that the renal system can. These physiological buffer systems help regulate body pH by controlling the output of acids, bases, or carbon dioxide (CO_2) from the body. For example, if there is too much acid in the body, the renal system may respond by excreting more H^+ from the body in urine. Similarly, if there is too much carbon dioxide in the blood, the respiratory system may respond by increasing ventilation to expel the excess carbon dioxide. Carbon dioxide levels have a direct effect on pH because the addition of carbon dioxide to the blood results in the generation of more H^+. The following equation shows what happens when carbon dioxide combines with water in the blood, producing carbonic acid.

$$H_2O + CO_2 \rightleftarrows \underset{\substack{\text{carbonic} \\ \text{acid}}}{H_2CO_3} \rightleftarrows H^+ + \underset{\substack{\text{bicarbonate} \\ \text{ion}}}{HCO_3^-}$$

ACTIVITY 1

Hyperventilation

OBJECTIVES

1. To introduce pH homeostasis in the body.
2. To understand the normal ranges for pH and P_{CO_2}.
3. To recognize respiratory alkalosis and its causes.
4. To interpret an oscilloscope tracing for hyperventilation and compare it with a tracing for normal breathing.

Introduction

Acid-base imbalances can have respiratory and metabolic causes. When diagnosing these disorders, two key signs are evaluated: the pH and the partial pressure of carbon dioxide in the blood (P_{CO_2}). The normal range for pH is between 7.35 and 7.45, and the normal range for P_{CO_2} is between 35 and 45 mm Hg. When the pH falls below 7.35, the body is said to be in a state of **acidosis.** When the pH rises above 7.45, the body is said to be in a state of **alkalosis.**

Respiratory alkalosis is the condition of too little carbon dioxide in the blood. Respiratory alkalosis commonly results from traveling to high altitude (where the air contains

less oxygen) or hyperventilation, which can be brought on by fever, panic attack, or anxiety. Hyperventilation, defined as an increase in the rate and depth of breathing, removes carbon dioxide from the blood faster than it is being produced by the cells of the body, reducing the amount of H^+ in the blood and, thus, increasing the blood's pH. The following equation shows the shift in the equilibrium that results in the increase in blood pH due to less carbon dioxide in the blood.

$$H_2O + CO_2 \leftarrow \underset{\substack{\text{carbonic} \\ \text{acid}}}{H_2CO_3} \leftarrow H^+ + \underset{\substack{\text{bicarbonate} \\ \text{ion}}}{HCO_3^-}$$

The renal system can compensate for alkalosis by retaining H^+ and excreting bicarbonate ions to lower the blood pH levels back to the normal range.

> **EQUIPMENT USED** The following equipment will be depicted on-screen: simulated lung chamber; pH meter; oscilloscope; two breathing patterns: normal and hyperventilation.

Experiment Instructions

Go to the home page in the PhysioEx software and click **Exercise 10: Acid-Base Balance.** Click **Activity 1: Hyperventilation,** and take the online **Pre-lab Quiz** for Activity 1.

After you take the online Pre-lab Quiz, click the **Experiment** tab and begin the experiment. The experiment instructions are reprinted here for your reference. The opening screen for the experiment is shown below.

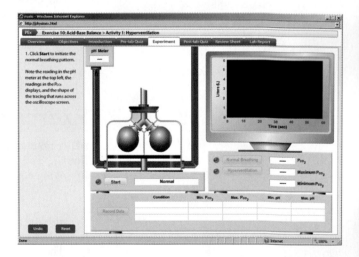

1. Click **Start** to initiate the normal breathing pattern. Note the reading in the pH meter at the top left, the readings in the P_{CO_2} displays, and the shape of the tracing that runs across the oscilloscope screen.

2. Click **Record Data** to display your results in the grid (and record your results in Chart 1).

> **? PREDICT Question 1**
> What do you think will happen to the pH and P_{CO_2} levels with hyperventilation?

CHART 1	Hyperventilation Breathing Patterns			
Condition	Minimum P_{CO_2}	Maximum P_{CO_2}	Minimum pH	Maximum pH

3. Click **Start** to initiate the normal breathing pattern. After the normal breathing tracing runs for 10 seconds, click **Hyperventilation** to initiate the hyperventilation breathing pattern. Note the reading in the pH meter at the top left, the readings in the P_{CO_2} displays, and the shape of the tracing that runs across the oscilloscope screen.

4. Click **Record Data** to display your results in the grid (and record your results in Chart 1).

5. Click **Start** to initiate the normal breathing pattern. After the normal breathing tracing runs for 10 seconds, click **Hyperventilation** to initiate the hyperventilation breathing pattern. After the hyperventilation tracing runs for 10 seconds, click **Normal Breathing** to return to the normal breathing pattern. Note the reading in the pH meter at the top left, the readings in the P_{CO_2} displays, and the shape of the tracing that runs across the oscilloscope screen.

6. Click **Record Data** to display your results in the grid (and record your results in Chart 1).

After you complete the experiment, take the online **Post-lab Quiz** for Activity 1.

Activity Questions

1. At what pH range is the body considered to be in a state of respiratory alkalosis?

_____

_____

2. How can the body compensate for respiratory alkalosis?

_____

_____

3. How did the tidal volume change with hyperventilation?

_____

_____

4. What might cause a person to hyperventilate?

_____

_____

ACTIVITY 2

Rebreathing

OBJECTIVES

1. To understand how rebreathing can simulate hypoventilation.
2. To observe the results of respiratory acidosis.
3. To describe the causes of respiratory acidosis.

Introduction

The body is said to be in a state of **acidosis** when the pH of the blood falls below 7.35 (although a pH of 7.35 is technically not acidic). Respiratory acidosis is the result of impaired respiration, or *hypoventilation*, which leads to the accumulation of too much carbon dioxide in the blood. The causes of impaired respiration include airway obstruction, depression of the respiratory center in the brain stem, lung disease (such as emphysema and chronic bronchitis), and drug overdose.

Recall that carbon dioxide contributes to the formation of carbonic acid when it combines with water through a reversible reaction catalyzed by carbonic anhydrase. The carbonic acid then dissociates into hydrogen ions and bicarbonate ions. Because hypoventilation results in elevated carbon dioxide levels in the blood, the equilibrium shifts, the H^+ levels increase, and the pH value of the blood decreases.

$$H_2O + CO_2 \rightarrow \underset{\substack{\text{carbonic} \\ \text{acid}}}{H_2CO_3} \rightarrow H^+ + \underset{\substack{\text{bicarbonate} \\ \text{ion}}}{HCO_3^-}$$

Rebreathing is the action of breathing in air that was just expelled from the lungs. Rebreathing results in the accumulation of carbon dioxide in the blood. Breathing into a paper bag is an example of rebreathing. (Note that breathing into a paper bag can deplete the body of oxygen and is therefore not the best therapy for hyperventilation because it can mask other life-threatening emergencies, such as a heart attack or asthma.) In this activity, you will observe what happens to pH and carbon dioxide levels in the blood during rebreathing. In the body, the kidneys regulate the acid-base balance by altering the amount of H^+ and HCO_3^- excreted in the urine.

> **EQUIPMENT USED** The following equipment will be depicted on-screen: simulated lung chamber; pH meter; oscilloscope; two breathing patterns: normal and rebreathing.

Experiment Instructions

Go to the home page in the PhysioEx software and click **Exercise 10: Acid-Base Balance.** Click **Activity 2: Rebreathing,** and take the online **Pre-lab Quiz** for Activity 2.

After you take the online Pre-lab Quiz, click the **Experiment** tab and begin the experiment. The experiment instructions are reprinted here for your reference. The opening screen for the experiment is shown below.

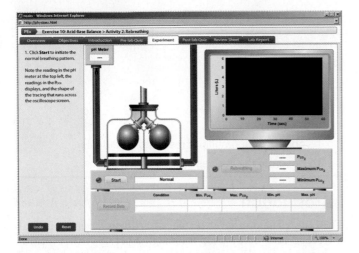

1. Click **Start** to initiate the normal breathing pattern. Note the reading in the pH meter at the top left, the readings in the P_{CO_2} displays, and the shape of the tracing that runs across the oscilloscope screen.

2. Click **Record Data** to display your results in the grid (and record your results in Chart 2).

> **? PREDICT Question 1**
> What do you think will happen to the pH and P_{CO_2} levels during rebreathing?

3. Click **Start** to initiate the normal breathing pattern. After the normal breathing tracing runs for 10 seconds, click **Rebreathing** to initiate the rebreathing pattern. Note the reading in the pH meter at the top left, the readings in the P_{CO_2} displays, and the shape of the tracing that runs across the oscilloscope screen.

4. Click **Record Data** to display your results in the grid (and record your results in Chart 2).

After you complete the experiment, take the online **Post-lab Quiz** for Activity 2.

Activity Questions

1. Did the pH level of the blood change at all with rebreathing? If so, how did it change?

_____

_____

2. What happens to the pH level of the blood when there is too much carbon dioxide remaining in the blood?

_____

_____

3. How did the tidal volumes change with rebreathing?

_____

_____

4. Describe two ways in which too much carbon dioxide might remain in the blood.

_____

_____

_____

Renal Responses to Respiratory Acidosis and Respiratory Alkalosis

OBJECTIVES

1. To understand renal compensation mechanisms for respiratory acidosis and respiratory alkalosis.
2. To explore the functional unit of the kidneys that responds to acid-base balance.
3. To observe the changes in ion concentrations that occur with renal compensation.

Introduction

The kidneys play a major role in maintaining fluid, electrolyte, and acid-base balance in the body's internal environment. By regulating the amount of water lost in the urine, the kidneys defend the body against excessive hydration or dehydration. By regulating the acidity of urine and the rate of electrolyte excretion, the kidneys maintain plasma pH and electrolyte levels within normal limits.

CHART 2	Normal Breathing Patterns			
Condition	**Minimum P_{CO_2}**	**Maximum P_{CO_2}**	**Minimum pH**	**Maximum pH**

Renal compensation is the body's primary method of compensating for conditions of respiratory acidosis or respiratory alkalosis. The kidneys regulate the acid-base balance by altering the amount of H^+ and HCO_3^- excreted in the urine. If we revisit the equation for the dissociation of carbonic acid, a weak acid, we see that the conservation of bicarbonate ion (base) has the same net effect as the loss of acid, H^+.

$$H_2O + CO_2 \rightleftarrows \underset{\substack{\text{carbonic} \\ \text{acid}}}{H_2CO_3} \rightleftarrows H^+ + \underset{\substack{\text{bicarbonate} \\ \text{ion}}}{HCO_3^-}$$

In this activity you will examine how the renal system compensates for respiratory acidosis or respiratory alkalosis. Respiratory acidosis is generally caused by the accumulation of carbon dioxide in the blood from hypoventilation, but it can also be caused by rebreathing. Acidosis results in a lower-than-normal blood pH. Respiratory alkalosis is caused by a depletion of carbon dioxide, often caused by an episode of hyperventilation, and results in an elevated blood pH.

You will primarily be working with the variable P_{CO_2}. Recall that the normal range for pH is between 7.35 and 7.45 and the normal range for P_{CO_2} is between 35 and 45 mm Hg. You will observe how increases and decreases in P_{CO_2} affect the levels of H^+ and HCO_3^- that the kidneys excrete in urine. The functional unit for adjusting the plasma composition is the **nephron.** Remember that although the renal system can partially compensate for pH imbalances with a respiratory cause, the kidneys cannot fully compensate if respirations have not returned to normal because the carbon dioxide levels will still be abnormal.

EQUIPMENT USED The following equipment will be depicted on-screen: source beaker for blood (first beaker on left side of screen); drain beaker for blood (second beaker on left side of screen); simulated nephron (The filtrate forms in Bowman's capsule and flows through the renal tubule—the tubular components, and empties into a collecting duct which, in turn drains into the urinary bladder.); nephron tank; glomerulus—"ball" of capillaries that forms part of the filtration membrane; glomerular (Bowman's) capsule—forms part of the filtration membrane and a capsular space where the filtrate initially forms; proximal convoluted tubule; loop of Henle; distal convoluted tubule; collecting duct; drain beaker for filtrate (beaker on right side of screen)—simulates the urinary bladder.

Experiment Instructions

Go to the home page in the PhysioEx software and click **Exercise 10: Acid-Base Balance.** Click **Activity 3: Renal Responses to Respiratory Acidosis and Respiratory Alkalosis** and take the online **Pre-lab Quiz** for Activity 3.

After you take the online Pre-lab Quiz, click the **Experiment** tab and begin the experiment. The experiment instructions are reprinted here for your reference. The opening screen for the experiment is shown above.

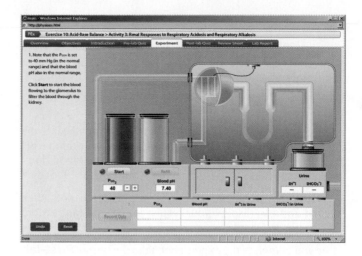

1. Note that the P_{CO_2} is set to 40 mm Hg (in the normal range) and that the blood pH is also in the normal range. Click **Start** to start the blood flowing to the glomerulus to filter the blood through the kidney.

2. Note the $[H^+]$ and $[HCO_3^-]$ in the urine and click **Record Data** to display your results in the grid (and record your results in Chart 3).

CHART 3	Renal Responses to Respiratory Acidosis and Respiratory Alkalosis		
P_{CO_2}	Blood pH	$[H^+]$ in urine	$[HCO_3^-]$ in urine

3. Click **Refill** to replenish the source beaker.

? PREDICT Question 1
What effect do you think lowering the P_{CO_2} will have on $[H^+]$ and $[HCO_3^-]$ in the urine?

---

4. Lower the P_{CO_2} to 30 by clicking the − button beside the P_{CO_2} display. Note the corresponding increase in blood pH (above the normal range). Click **Start** to start the blood flowing to the glomerulus to filter the blood through the kidney.

5. Note the $[H^+]$ and $[HCO_3^-]$ in the urine and click **Record Data** to display your results in the grid (and record your results in Chart 3).

6. Click **Refill** to replenish the source beaker.

> **? PREDICT Question 2**
> What effect do you think raising the P_{CO_2} will have on [H$^+$] and [HCO$_3^-$] in the urine?

7. Raise the P_{CO_2} to 60 by clicking the + button beside the P_{CO_2} display. Note the corresponding decrease in blood pH (below the normal range). Click **Start** to start the blood flowing to the glomerulus to filter the blood through the kidney.

8. Note the [H$^+$] and [HCO$_3^-$] in the urine and click **Record Data** to display your results in the grid (and record your results in Chart 3).

After you complete the experiment, take the online **Post-lab Quiz** for Activity 3.

Activity Questions

1. Describe how the kidneys respond to respiratory acidosis.

_____

_____

2. What P_{CO_2} corresponded to respiratory acidosis?

_____

3. Describe how the kidneys respond to respiratory alkalosis.

_____

_____

4. What P_{CO_2} corresponded to respiratory alkalosis?

_____

ACTIVITY 4

Respiratory Responses to Metabolic Acidosis and Metabolic Alkalosis

OBJECTIVES

1. To understand the causes of metabolic acidosis and metabolic alkalosis.
2. To observe the physiological changes that occur with an increase and decrease in metabolic rate.
3. To explain how the respiratory system compensates for metabolic acidosis and alkalosis.

Introduction

Conditions of acidosis and alkalosis that do not have respiratory causes are termed *metabolic acidosis and metabolic alkalosis*. **Metabolic acidosis** is characterized by low plasma HCO$_3^-$ and pH. The causes of metabolic acidosis include:

- **Ketoacidosis,** a buildup of keto acids that can result from diabetes mellitus
- **Salicylate poisoning,** a toxic condition resulting from ingestion of too much aspirin or oil of wintergreen (a substance often found in laboratories)
- The ingestion of too much alcohol, which metabolizes into acetic acid
- Diarrhea, which results in the loss of bicarbonate with the elimination of intestinal contents
- Strenuous exercise, which can cause a buildup of lactic acid from anaerobic muscle metabolism

Metabolic alkalosis is characterized by elevated plasma HCO$_3^-$ and pH. The causes of metabolic alkalosis include:

- Ingestion of alkali, such as antacids or bicarbonate
- Vomiting, which can result in the loss of too much H$^+$
- Constipation, which may result in significant reabsorption of HCO$_3^-$

Increases or decreases in the body's normal metabolic rate can also result in metabolic acidosis or alkalosis. Recall that carbon dioxide—a waste product of metabolism—mixes with water in plasma to form carbonic acid, which in turn forms H$^+$.

$$H_2O + CO_2 \rightleftarrows \underset{\substack{\text{carbonic} \\ \text{acid}}}{H_2CO_3} \rightleftarrows H^+ + \underset{\substack{\text{bicarbonate} \\ \text{ion}}}{HCO_3^-}$$

An increase in the normal metabolic rate causes more carbon dioxide to form as a metabolic waste product, resulting in the formation of more H$^+$ and, therefore, lower plasma pH, potentially causing acidosis. Other acids that are also normal metabolic waste products (such as ketone bodies and phosphoric, uric, and lactic acids) would likewise accumulate with an increase in metabolic rate.

Conversely, a decrease in the normal metabolic rate causes less carbon dioxide to form as a metabolic waste product, resulting in the formation of less H$^+$ and, therefore, higher plasma pH, potentially causing alkalosis. Many factors can affect the rate of cell metabolism. For example, fever, stress, or the ingestion of food all cause the rate of cell metabolism to *increase*. Conversely, a fall in body temperature or a decrease in food intake causes the rate of cell metabolism to *decrease*.

The respiratory system compensates for metabolic acidosis or alkalosis by expelling or retaining carbon dioxide in the blood. During metabolic acidosis, respiration increases to expel carbon dioxide from the blood, thus decreasing [H$^+$] and raising the pH. During metabolic alkalosis, respiration decreases to promote the accumulation of carbon dioxide in the blood, thus increasing [H$^+$] and decreasing the pH.

The renal system also compensates for metabolic acidosis and alkalosis by conserving or excreting bicarbonate ions. Nevertheless, in this activity, you will focus on respiratory compensation of metabolic acidosis and alkalosis.

Experiment Instructions

Go to the home page in the PhysioEx software and click **Exercise 10: Acid-Base Balance.** Click **Activity 4: Respiratory Responses to Metabolic Acidosis and Metabolic Alkalosis,** and take the online **Pre-lab Quiz** for Activity 4.

After you take the online Pre-lab Quiz, click the **Experiment** tab and begin the experiment. The experiment instructions are reprinted here for your reference. The opening screen for the experiment is shown below.

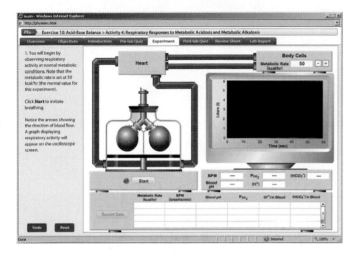

1. You will begin by observing respiratory activity at normal metabolic conditions. Note that the metabolic rate is set at 50 kcal/hr (the normal value for this experiment). Click **Start** to initiate breathing and blood flow. Notice the arrows showing the direction of blood flow. A graph displaying respiratory activity will appear on the oscilloscope screen.

2. Note the data in the displays below the oscilloscope screen and click **Record Data** to display your results in the grid (and record your results in Chart 4).

3. Increase the metabolic rate to 60 kcal/hr by clicking the + button beside the metabolic rate display. Click **Start** to initiate breathing and blood flow.

4. Note the data in the displays below the oscilloscope screen and click **Record Data** to display your results in the grid (and record your results in Chart 4).

5. Click **Clear Tracings** to clear the tracings on the oscilloscope.

? PREDICT Question 1
What do you think will happen when the metabolic rate is increased to 80 kcal/hr?

6. Increase the metabolic rate to 80 kcal/hr by clicking the + button beside the metabolic rate display. Click **Start** to initiate breathing and blood flow.

7. Note the data in the displays below the oscilloscope screen and click **Record Data** to display your results in the grid (and record your results in Chart 4).

8. Click **Clear Tracings** to clear the tracings on the oscilloscope.

9. Decrease the metabolic rate to 40 kcal/hr by clicking the − button beside the metabolic rate display. Click **Start** to initiate breathing and blood flow.

10. Note the data in the displays below the oscilloscope screen and click **Record Data** to display your results in the grid (and record your results in Chart 4).

11. Click **Clear Tracings** to clear the tracings on the oscilloscope.

? PREDICT Question 2
What do you think will happen when the metabolic rate is decreased to 20 kcal/hr?

CHART 4	Respiratory Responses to Metabolic Acidosis and Metabolic Alkalosis				
Metabolic rate	BPM (breaths/min)	Blood pH	P_{CO_2}	$[H^+]$ in blood	$[HCO_3^-]$ in blood

12. Decrease the metabolic rate to 20 kcal/hr by clicking the button beside the metabolic rate display. Click **Start** to initiate breathing and blood flow.

13. Note the data in the displays below the oscilloscope screen and click **Record Data** to display your results in the grid (and record your results in Chart 4).

After you complete the experiment, take the online **Post-lab Quiz** for Activity 4.

Activity Questions

1.   Describe what happens to carbon dioxide and pH with increased metabolism.

_____

_____

_____

2.   Describe the respiratory response to metabolic acidosis.

_____

_____

_____

3.   When the respiratory system compensates for the metabolic acidosis, does the pH increase or decrease in value?

_____

_____

_____

4.   Describe the respiratory response to metabolic alkalosis.

_____

_____

Acid-Base Balance

NAME _____

LAB TIME/DATE _____

ACTIVITY 1 Hyperventilation

1. Describe the normal ranges for pH and carbon dioxide in the blood. _____

_____

_____

2. Describe what happened to the pH and the carbon dioxide levels with hyperventilation. How well did the results compare

with your prediction? _____

_____

_____

3. Explain how returning to normal breathing after hyperventilation differed from hyperventilation without returning to normal

breathing. _____

_____

_____

4. Describe some possible causes of respiratory alkalosis. _____

_____

_____

ACTIVITY 2 Rebreathing

1. Describe what happened to the pH and the carbon dioxide levels during rebreathing. How well did the results compare with

your prediction? _____

_____

_____

2. Describe some possible causes of respiratory acidosis. _____

_____

_____

3. Explain how the renal system would compensate for respiratory acidosis. _____

_____

_____

ACTIVITY 3 Renal Responses to Respiratory Acidosis and Respiratory Alkalosis

1. Describe what happened to the concentration of ions in the urine when the P_{CO_2} was lowered. How well did the results

   compare with your prediction? _____

   _____

   _____

2. What condition was simulated when the P_{CO_2} was lowered? _____

   _____

   _____

3. Describe what happened to the concentration of ions in the urine when the P_{CO_2} was raised. How well did the results compare

   with your prediction? _____

   _____

   _____

4. What condition was simulated when the P_{CO_2} was raised? _____

   _____

   _____

ACTIVITY 4 Respiratory Responses to Metabolic Acidosis and Metabolic Alkalosis

1. Describe what happened to the blood pH when the metabolic rate was increased to 80 kcal/hr. What body system was

   compensating? How well did the results compare with your prediction? _____

   _____

   _____

2. List and describe some possible causes of metabolic acidosis. _____

   _____

   _____

3. Describe what happened to the blood pH when the metabolic rate was decreased to 20 kcal/hr. What body system was

    compensating? How well did the results compare with your prediction? _____

    _____

    _____

4. List and describe some possible causes of metabolic alkalosis. _____

    _____

    _____

Blood Analysis

P R E - L A B   Q U I Z

1. The percentage of erythrocytes in a sample of whole blood is measured by the:
   a. hemoglobin
   b. hematocrit
   c. ABO blood type
   d. erythropoietin

2. Circle the correct underlined term: The protein that transports oxygen from the lungs to the cells of the body is <u>hemoglobin</u> / <u>hematocrit</u>.

3. A lower-than-normal hematocrit is known as _____, in which insufficient oxygen is transported to the body cells.
   a. polycythemia
   b. hemophilia
   c. hemochromatosis
   d. anemia

4. The erythrocyte sedimentation rate (ESR) can be used to follow the progression of all of the following diseases or conditions *except:*
   a. anemia
   b. rheumatoid arthritis
   c. acute appendicitis (within 24 hours)
   d. myocardial infarction

5. Circle the correct underlined term: A <u>rouleaux formation</u> / <u>hemoglobinometer</u> will be used to compare a standard color value to an experimental sample to determine hemoglobin content.

6. Circle the correct underlined term: A person with type <u>AB</u> / <u>O</u> blood has two recessive alleles and has neither type A nor type B antigen.

7. Circle True or False: Blood transfusion reactions are of little consequence and occur when the recipient has antibodies that react with the antigens present on the transfused cells.

Exercise Overview

Blood transports soluble substances to and from all cells of the body. Laboratory analysis of our blood can reveal important information about how well this function is being achieved. The five activities in this exercise simulate common laboratory tests performed on blood: (1) *hematocrit* determination, (2) *erythrocyte sedimentation rate*, (3) *hemoglobin* determination, (4) *blood typing,* and (5) total *cholesterol* determination.

Hematocrit refers to the percentage of red blood cells (RBCs), or erythrocytes, in a sample of whole blood. A hematocrit of 48 means that 48% of the volume of blood consists of RBCs. RBCs transport oxygen to the cells of the body. Therefore, the higher the hematocrit, the more RBCs are present in the blood and the greater the oxygen-carrying potential of the blood. Males usually have higher hematocrit levels than females because males have higher levels of testosterone. In addition to promoting the male sex characteristics, testosterone is responsible for stimulating the release of erythropoietin from the kidneys. Erythropoietin (EPO) is a hormone that stimulates the synthesis of RBCs. Therefore, higher levels of testosterone lead to more EPO secretion and, thus, higher hematocrit levels.

The **erythrocyte sedimentation rate (ESR)** measures the settling of RBCs in a vertical, stationary tube of blood during one hour. In a healthy individual, RBCs do not settle very much in an hour. In some disease conditions, increased production of fibrinogen and immunoglobulins causes the RBCs to clump together, stack up, and form a column (called a *rouleaux formation*). RBCs in a rouleaux formation are heavier and settle faster (that is, they display an increase in the sedimentation rate.)

Hemoglobin (Hb), a protein found in RBCs, is necessary for the transport of oxygen from the lungs to the cells of the body. Four polypeptide chains of amino acids comprise the globin part of the molecule. Each polypeptide chain has a heme unit—a group of atoms that includes an atom of iron to which a molecule of oxygen binds. Each polypeptide chain, if it folds correctly, can bind a molecule of oxygen. Therefore, each hemoglobin molecule can carry four molecules of oxygen. Oxygen combined with hemoglobin forms oxyhemoglobin, which has a bright red color.

All of the cells in the human body, including RBCs, are surrounded by a plasma membrane that contains genetically determined glycoproteins, called antigens. On RBC membranes, there are certain antigens, called **agglutinogens,** that determine a person's blood type. Blood typing is used to identify the **ABO blood groups,** which are determined by the presence or absence of two antigens: **type A** and **type B.** Because these antigens are genetically determined, a person has two copies (alleles) of the gene for these antigens, one copy from each parent.

Cholesterol is a lipid substance that is essential for life—it is an important component of all cell membranes and is the base molecule of steroid hormones, vitamin D, and bile salts. Cholesterol is produced in the human liver and is present in some foods of animal origin, such as milk, meat, and eggs. Because cholesterol is a hydrophobic lipid, it needs to be wrapped in protein packages, called **lipoproteins,** to travel in the blood (which is mostly water) from the liver and digestive organs to the cells of the body.

ACTIVITY 1

Hematocrit Determination

OBJECTIVES

1. To understand the terms *hematocrit, red blood cells, hemoglobin, buffy coat, anemia,* and *polycythemia.*

2. To understand how the hematocrit (packed red blood cell volume) is determined.

3. To understand the implications of elevated or decreased hematocrit.

4. To understand the importance of proper disposal of laboratory material that comes in contact with blood.

Introduction

Hematocrit refers to the percentage of **red blood cells (RBCs),** or erythrocytes, in a sample of whole blood. A hematocrit of 48 means that 48% of the volume of blood consists of RBCs. RBCs transport oxygen to the cells of the body. Therefore, the higher the hematocrit, the more RBCs are present in the blood and the higher the oxygen-carrying potential of the blood. Hematocrit values are determined by spinning a microcapillary tube filled with a sample of whole blood in a special microhematocrit centrifuge. This procedure separates the blood cells from the blood plasma. A **buffy coat** layer

of white blood cells (WBCs) appears as a thin, white layer *between* the heavier RBC layer and the lighter, yellow plasma.

The hematocrit is determined after centrifuging by measuring the height of the RBC layer (in millimeters) and dividing that by the height of the total blood sample (in millimeters). This calculation gives the percentage of the total blood volume consisting of RBCs. The average hematocrit for males is 42–52%, and the average hematocrit for females is 37–47%. A lower-than-normal hematocrit indicates **anemia,** and a higher-than-normal hematocrit indicates **polycythemia.**

Anemia is a condition in which insufficient oxygen is transported to the body's cells. There are many possible causes for anemia, including inadequate numbers of RBCs, a decreased amount of the oxygen-carrying pigment **hemoglobin** in the RBCs, and abnormally shaped hemoglobin. The heme portion of a hemoglobin molecule contains an atom of iron to which a molecule of oxygen can bind. If adequate iron is not available, the body cannot manufacture hemoglobin, resulting in the condition *iron-deficiency anemia. Aplastic anemia* results from the failure of the bone marrow to produce adequate red blood cell numbers. *Sickle cell anemia* is an inherited condition in which the protein portion of hemoglobin molecules folds incorrectly when oxygen levels are low. As a result, oxygen molecules cannot bind to the misshapen hemoglobin, the RBCs develop a sickle shape, and anemia results. Regardless of the underlying cause, anemia causes a reduction in the blood's ability to transport oxygen to the cells of the body.

Polycythemia refers to an increase in RBCs, resulting in a higher-than-normal hematocrit. There are many possible causes of polycythemia, including living at high altitudes, strenuous athletic training, and tumors in the bone marrow. In this activity you will simulate the blood test used to determine hematocrit.

> **EQUIPMENT USED** The following equipment will be depicted on-screen: six heparinized capillary tubes (heparin keeps blood from clotting); blood samples from six individuals: sample 1: a healthy male living in Boston, sample 2: a healthy female living in Boston, sample 3: a healthy male living in Denver, sample 4: a healthy female living in Denver, sample 5: a male with aplastic anemia, sample 6: a female with iron-deficiency anemia; capillary tube sealer—a clay material (shown as an orange-yellow substance) used to seal the capillary tubes on one end so the blood sample can be centrifuged without having the blood spray out of the tube; microhematocrit centrifuge—used to centrifuge the samples (rotates at 14,500 revolutions per minute); metric ruler; biohazardous waste disposal—used to properly dispose of equipment that comes in contact with blood.

Experiment Instructions

Go to the home page in the PhysioEx software and click **Exercise 11: Blood Analysis.** Click **Activity 1: Hematocrit Determination,** and take the online **Pre-lab Quiz** for Activity 1.

After you take the online Pre-lab Quiz, click the **Experiment** tab and begin the experiment. The experiment

instructions are reprinted here for your reference. The opening screen for the experiment is shown below.

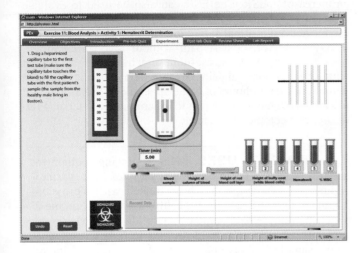

1. Drag a heparinized capillary tube to the first test tube (make sure the capillary tube touches the blood) to fill the capillary tube with the first patient's sample (the sample from the healthy male living in Boston).

2. Drag the capillary tube containing sample 1 to the container of capillary tube sealer to seal one end of the tube.

3. Drag the capillary tube to the microhematocrit centrifuge. The remaining samples will automatically be prepared for centrifugation.

4. Note that the timer is set to 5 minutes. Click **Start** to centrifuge the samples for 5 minutes at 14,500 revolutions per minute. The simulation compresses the 5-minute time period into 5 seconds of real time.

5. Drag capillary tube 1 from the centrifuge to the metric ruler to measure the height of the column of blood and the height of each layer.

6. Click **Record Data** to display your results in the grid (and record your results in Chart 1).

7. Drag capillary tube 1 to the biohazardous waste disposal.

> **? PREDICT Question 1**
> Predict how the hematocrits of the patients living in Denver, Colorado (approximately one mile above sea level), will compare with the hematocrit levels of the patients living in Boston, Massachusetts (at sea level).

8. You will now measure the column and layer heights of the remaining samples.

- Drag the next capillary tube from the centrifuge to the metric ruler.

- Click **Record data** to display your results in the grid (and record your results in Chart 1). The tube will automatically be placed in the biohazardous waste disposal.

Repeat this step for each of the remaining samples.

After you complete the experiment, take the online **Post-lab Quiz** for Activity 1.

CHART 1	Hematocrit Determination				
	Total height of column of blood (mm)	Height of red blood cell layer (mm)	Height of buffy coat (mm)	Hematocrit	% WBC
Sample 1 (healthy male living in Boston)					
Sample 2 (healthy female living in Boston)					
Sample 3 (healthy male living in Denver)					
Sample 4 (healthy female living in Denver)					
Sample 5 (male with aplastic anemia)					
Sample 6 (female with iron-deficiency anemia)					

Activity Questions

1. How do you calculate the hematocrit after you centrifuge the total blood sample? What does the result of this calculation indicate?

_____

_____

2. What is the significance of the "buffy coat" after you centrifuge the total blood sample?

_____

_____

3. As noted in the Exercise Overview, the average hematocrit for males is 42–52%, the average hematocrit for females is 37–47%, and erythropoietin is a hormone that is responsible for the synthesis of RBCs. Given this information, explain how a female could have a consistent hematocrit of 48, large, well-defined skeletal muscles, and an abnormally deep voice.

_____

_____

_____

_____ ▬▬

ACTIVITY 2

Erythrocyte Sedimentation Rate

OBJECTIVES

1. To understand *erythrocyte sedimentation rate (ESR), red blood cells (RBCs),* and *rouleaux formation.*

2. To learn how to perform an erythrocyte sedimentation rate blood test.

3. To understand the results (and their implications) from an erythrocyte sedimentation rate blood test.

4. To understand the importance of proper disposal of laboratory material that comes in contact with blood.

Introduction

The **erythrocyte sedimentation rate (ESR)** measures the settling of **red blood cells (RBCs)** in a vertical, stationary tube of whole blood during one hour. In a healthy individual, red blood cells do not settle very much in an hour. In some disease conditions, increased production of fibrinogen and immunoglobulins cause the RBCs to clump together, stack up, and form a dark red column (called a **rouleaux formation**). RBCs in a rouleaux formation are heavier and settle faster (that is, they exhibit an increase in the settling rate).

The ESR is neither very specific nor diagnostic, but it can be used to follow the progression of certain diseases, including sickle cell anemia, some cancers, and inflammatory diseases, such as rheumatoid arthritis. When the disease worsens, the ESR increases. When the disease improves, the ESR decreases.

The ESR can be elevated in iron-deficiency anemia, and menstruating females sometimes develop anemia and show an increase in ESR. The ESR can also be used to evaluate a patient with chest pains because the ESR is elevated in established myocardial infarction (heart attack) but normal in angina pectoris (chest pain without myocardial infarction). Similarly, it can be useful in screening a female patient with severe abdominal pains because the ESR is not elevated within the first 24 hours of acute appendicitis but is elevated in the early stage of acute pelvic inflammatory disease (PID) or ruptured ectopic pregnancy.

EQUIPMENT USED The following equipment will be depicted on-screen: blood samples from six individuals (each sample has been treated with the anticoagulant heparin): sample 1: healthy individual, sample 2: menstruating female, sample 3: individual with sickle cell anemia, sample 4: individual with iron-deficiency anemia, sample 5: individual suffering a myocardial infarction, sample 6: individual with angina pectoris; sodium citrate—used to bind with calcium and prevent the blood samples from clotting so they can be easily poured into the narrow sedimentation rate tubes; test tubes—used as reaction vessels for the tests; sedimentation tubes (contained in cabinet); magnifying chamber—used to help read the millimeter markings on the sedimentation tubes; biohazardous waste disposal—used to properly dispose of equipment that comes in contact with blood.

Experiment Instructions

Go to the home page in the PhysioEx software and click **Exercise 11: Blood Analysis.** Click **Activity 2: Erythrocyte Sedimentation Rate,** and take the online **Pre-lab Quiz** for Activity 2.

After you take the online Pre-lab Quiz, click the **Experiment** tab and begin the experiment. The experiment instructions are reprinted here for your reference. The opening screen for the experiment is shown below.

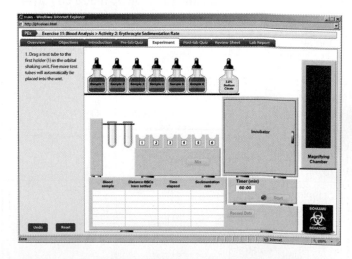

1. Drag a test tube to the first holder (1) in the orbital shaking unit. Five more test tubes will automatically be placed into the unit.

2. Drag the dropper cap of the sample 1 bottle (the sample from the healthy individual) to the first test tube (1) in the orbital shaking unit to dispense one milliliter of blood into the tube. The remaining five samples will be automatically dispensed.

3. Drag the dropper cap of the 3.8% sodium citrate bottle to the first test tube to dispense 0.5 milliliters of sodium citrate into each of the tubes.

4. Click **Mix** to mix the samples.

5. Drag the first test tube to the first sedimentation tube in the incubator to pour the contents of the test tube into the sedimentation tube.

6. Drag the now empty test tube to the biohazardous waste disposal. The contents of the remaining test tubes will automatically be poured into the sedimentation tubes, and the empty tubes will automatically be placed in the biohazardous waste disposal.

7. Note that the timer is set to 60 minutes. Click **Start** to incubate the sedimentation tubes for 60 minutes. The simulation compresses the 60-minute time period into 6 seconds of real time.

8. Drag the first sedimentation tube to the magnifying chamber to examine the tube. The tube is marked in millimeters (the distance between two marks is 5 mm).

9. Click **Record Data** to display your results in the grid (and record your results in Chart 2).

10. Drag the sedimentation tube to the biohazardous waste disposal.

> **? PREDICT Question 1**
> How will the sedimentation rate for sample 6 (unhealthy individual) compare with the sedimentation rate for sample 1 (healthy individual)?

11. You will now measure the sedimentation rate for the remaining samples.

- Drag the next sedimentation tube to the magnifying chamber to examine the tube.

- Click **Record Data** to display your results in the grid (and record your results in Chart 2). The tube will automatically be placed in the biohazardous waste disposal.

Repeat this step for each of the remaining samples.

After you complete the experiment, take the online **Post-lab Quiz** for Activity 2.

Activity Questions

1. Why is ESR useful, even though it is neither specific nor sensitive?

_____

_____

2. Describe the physical process underlying an accelerated erythrocyte sedimentation rate.

_____

_____ ▬

ACTIVITY 3

Hemoglobin Determination

OBJECTIVES

1. To understand the terms *hemoglobin (Hb)*, *anemia*, *heme*, *oxyhemoglobin*, and *hemoglobinometer*.

2. To learn how to determine the amount of hemoglobin in a blood sample.

3. To understand the results and their implications when examining the amounts of hemoglobin present in a blood sample.

4. To understand the importance of proper disposal of laboratory material that comes in contact with blood.

CHART 2	Erythrocyte Sedimentation Rate		
Blood sample	Distance RBCs have settled (mm)	Elapsed time	Sedimentation rate
Sample 1 (healthy individual)			
Sample 2 (menstruating female)			
Sample 3 (individual with sickle cell anemia)			
Sample 4 (individual with iron-deficiency anemia)			
Sample 5 (individual suffering a myocardial infarction)			
Sample 6 (individual with angina pectoris)			

Introduction

Hemoglobin (Hb), a protein found in red blood cells, is necessary for the transport of oxygen from the lungs to the cells of the body. Four polypeptide chains of amino acids comprise the globin part of the molecule. Each polypeptide chain has a **heme** unit—a group of atoms that includes an atom of iron to which a molecule of oxygen binds. Each polypeptide chain, if it folds correctly, can bind a molecule of oxygen. Therefore, each hemoglobin molecule can carry four molecules of oxygen. Oxygen combined with hemoglobin forms **oxyhemoglobin,** which has a bright-red color. Anemia results when insufficient oxygen is carried in the blood.

A quantitative hemoglobin measurement is used to determine the classification and possible causes of anemia and also gives useful information on some other disease conditions. For example, a person can have anemia with a normal red blood cell count if there is inadequate hemoglobin in the red blood cells. Normal blood contains an average of 12–18 grams of hemoglobin per 100 milliliters of blood. A healthy male has 13.5–18 g/100 ml and a healthy female has 12–16 g/100 ml. Hemoglobin levels increase in patients with polycythemia, congestive heart failure, and chronic obstructive pulmonary disease (COPD). Hemoglobin levels also increase when dwelling at high altitudes. Hemoglobin levels decrease in patients with anemia, hyperthyroidism, cirrhosis of the liver, renal disease, systemic lupus erythematosus, and severe hemorrhage.

The hemoglobin level of a blood sample is determined by stirring the blood with a wooden stick to rupture, or lyse, the red blood cells. The color intensity of the hemolyzed blood reflects the amount of hemoglobin present. A **hemoglobinometer** transmits green light through the hemolyzed blood sample and then compares the amount of light that passes through the sample to standard color intensities to determine the hemoglobin content of the sample.

> **EQUIPMENT USED** The following equipment will be depicted on-screen: blood samples from five individuals: sample 1: healthy male, sample 2: healthy female, sample 3: female with iron-deficiency anemia, sample 4: male with polycythemia, sample 5: female Olympic athlete; hemolysis sticks—used to stir the blood samples to lyse the red blood cells, thereby releasing their hemoglobin; blood chamber dispenser—used to dispense a blood chamber slide with a depression for the blood sample; hemoglobinometer—used to analyze the hemoglobin level in each sample; biohazardous waste disposal—used to properly dispose of equipment that comes in contact with blood.

Experiment Instructions

Go to the home page in the PhysioEx software and click **Exercise 11: Blood Analysis.** Click **Activity 3: Hemoglobin Determination,** and take the online **Pre-lab Quiz** for Activity 3.

After you take the online Pre-lab Quiz, click the **Experiment** tab and begin the experiment. The experiment instructions are reprinted here for your reference. The opening screen for the experiment is shown below.

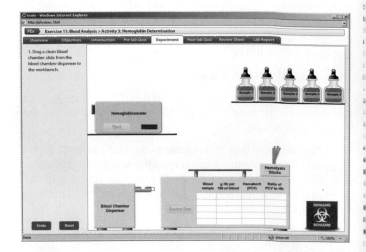

1. Drag a clean blood chamber slide from the blood chamber dispenser to the workbench.

2. Drag the bottle cap from the sample 1 bottle (the sample from the healthy male) to the depression in the blood chamber slide to dispense a drop of blood into the depression.

3. Drag a hemolysis stick to the drop of blood in the chamber to stir the blood sample for 45 seconds, lysing the red blood cells and releasing their hemoglobin.

4. Drag the hemolysis stick to the biohazardous waste disposal.

5. Drag the blood chamber slide to the dark rectangular slot on the hemoglobinometer to analyze the sample. After you insert the blood chamber slide into the hemoglobinometer, you will see a blowup of the inside of the hemoglobinometer.

6. The left half of the circular field shows the intensity of green light transmitted by blood sample 1. The right half of the circular field shows the intensity of green light for known levels of hemoglobin present in blood. Drag the lever on the right side of the hemoglobinometer down until the shade of green in the right half of the field matches the shade of green in the left half of the field and then click **Record Data** to display your results in the grid (and record your results in Chart 3).

7. Click **Eject** to remove the blood chamber slide from the hemoglobinometer.

8. Drag the blood chamber slide from the hemoglobinometer to the biohazardous waste disposal.

> **? PREDICT Question 1**
> How will the hemoglobin levels for the female Olympic athlete (sample 5) compare with the hemoglobin levels for the healthy female (sample 2)?

CHART 3	Hemoglobin Determination		
Blood sample	Hb in grams per 100 ml of blood	Hematocrit (PCV)	Ratio of PCV to Hb
Sample 1 (healthy male)		48	
Sample 2 (healthy female)		44	
Sample 3 (female with iron-deficiency anemia)		40	
Sample 4 (male with polycythemia)		60	
Sample 5 (female Olympic athlete)		60	

9. You will now measure the hemoglobin levels for each of the remaining samples.

- Drag a blood chamber slide to the workbench.
- Drag the bottle cap from the next sample bottle to the depression in the slide.
- Drag a hemolysis stick to the drop of blood in the chamber (after stirring the sample, the hemolysis stick will automatically be placed in the biohazardous waste disposal).
- Drag the blood chamber slide to the dark rectangular slot on the hemoglobinometer.
- Drag the lever on the right side of the hemoglobinometer down until the shade of green in the right half of the field matches the shade of green in the left half of the field and then click **Record Data** to display your results in the grid (and record your results in Chart 3).
- Click **Eject** to remove the blood chamber slide from the hemoglobinometer (the slide will automatically be placed in the biohazardous waste disposal).

Repeat this step until you analyze all five samples.

After you complete the experiment, take the online **Post-lab Quiz** for Activity 3.

Activity Questions

1. As mentioned in the introduction to this activity, hemoglobin levels increase for people living at high altitudes. Given that the atmospheric pressure of oxygen significantly declines as you ascend to higher elevations, why do you think hemoglobin levels would increase for those living at high altitudes?

2. Just by looking at the color of a freshly drawn blood sample, how could you distinguish between blood that is well oxygenated and blood that is poorly oxygenated?

Blood Typing

OBJECTIVES

1. To understand the terms *antigens, agglutinogens, ABO antigens, Rh antigens,* and *agglutinins.*
2. To learn how to perform a blood-typing assay.
3. To understand the results and their implications when examining agglutination reactions.
4. To understand the importance of proper disposal of laboratory material that comes in contact with blood.

Introduction

All of the cells in the human body, including red blood cells, are surrounded by a plasma membrane that contains genetically determined glycoproteins, called **antigens.** On red blood cell membranes, there are certain antigens, called **agglutinogens,** that determine a person's blood type. If a blood transfusion recipient has antibodies (called **agglutinins**) that react with the antigens present on the transfused cells, the red blood cells will become clumped together, or agglutinated, and then lysed, resulting in a potentially life-threatening blood transfusion reaction. It is therefore important to determine an individual's blood type before performing blood transfusions to avoid mixing incompatible blood. Although many different antigens are present on red blood cell membranes, the **ABO** and **Rh antigens** cause the most vigorous and potentially fatal transfusion reactions.

The ABO blood groups are determined by the presence or absence of two antigens: type A and type B. Because these antigens are genetically determined, a person has two copies (alleles) of the gene for these proteins, one copy from each parent. The presence of these antigens is due to a dominant allele, and their absence is due to a recessive allele.

- A person with type A blood can have two alleles for the type A antigen or one allele for the type A antigen and one allele for the absence of either the type A or type B antigen.
- A person with type B blood can have two alleles for the type B antigen or one allele for the type B antigen and one allele for the absence of either the type A or type B antigen.
- A person with type AB blood has one allele for the type A antigen and one allele for the type B antigen.
- A person with type O blood has two recessive alleles and has neither the type A nor type B antigen.

TABLE 11.1	ABO Blood Types	
Blood type	Antigens on RBCs	Antibodies present in plasma
A	A	anti-B
B	B	anti-A
AB	A and B	none
O	none	anti-A and anti-B

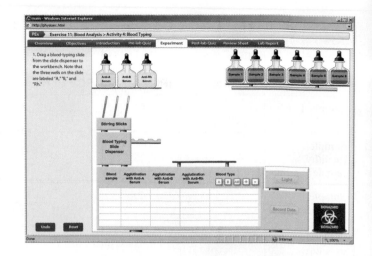

Antibodies against the A and B antigens are found pre-formed in the blood plasma. A person has antibodies only for the antigens not on his or her red blood cells, so a person with type A blood will have anti-B antibodies. View Table 11.1 for a summary of the antigens on red blood cells and the antibodies in the plasma for each blood type.

The Rh factor is another genetically determined protein that can be present on red blood cell membranes. Approximately 85% of the population is Rh positive (Rh$^+$), and their red blood cells have this protein on their surface. Antibodies against the Rh factor are not found preformed in the plasma. They are produced by an Rh negative (Rh$^-$) individual only after exposure to blood cells from someone who is Rh$^+$. Such exposure can occur during pregnancy when Rh$^+$ blood cells from the baby cross the placenta and expose the mother to the antigen.

To determine an individual's blood type, drops of an individual's blood sample are mixed separately with antiserum containing antibodies to either type A antigens, type B antigens, or Rh antigens. An agglutination reaction (showing clumping) indicates the presence of the agglutinogen.

EQUIPMENT USED The following equipment will be depicted on-screen: blood samples from six individuals with different blood types; anti-A serum (blue bottle), anti-B serum (yellow bottle), and anti-Rh serum (white bottle), containing antibodies to the A antigen, B antigen, and Rh antigen, respectively; blood-typing slide dispenser; color-coded stirring sticks—used to mix the blood sample and the serum (blue: used with anti-A serum, yellow: used with the anti-B serum, white: used with the anti-Rh serum); light box—used to view the blood type samples; biohazardous waste disposal—used to properly dispose of equipment that comes in contact with blood.

Experiment Instructions

Go to the home page in the PhysioEx software and click **Exercise 11, Blood Analysis.** Click **Activity 4, Blood Typing,** and take the online **Pre-lab Quiz** for Activity 4.

After you take the online Pre-lab Quiz, click the **Experiment** tab and begin the experiment. The experiment instructions are reprinted here for your reference. The opening screen for the experiment is shown above.

1. Drag a blood-typing slide from the slide dispenser to the workbench. Note that the three wells on the slide are labeled "A," "B," and "Rh."

2. Drag the dropper cap of the sample 1 bottle to well A on the blood-typing slide to dispense a drop of blood into each well.

3. Drag the dropper cap of the anti-A serum bottle to well A on the blood-typing slide to dispense a drop of anti-A serum into the well.

4. Drag the dropper cap of the anti-B serum bottle to well B on the blood-typing slide to dispense a drop of anti-B serum into the well.

5. Drag the dropper cap of the anti-Rh serum bottle to well Rh on the blood-typing slide to dispense a drop of anti-Rh serum into the well.

6. Drag a blue-tipped stirring stick to well A to mix the blood and anti-A serum.

7. Drag the stirring stick to the biohazardous waste disposal.

8. Drag a yellow-tipped stirring stick to well B to mix the blood and anti-B serum.

9. Drag the stirring stick to the biohazardous waste disposal.

10. Drag a white-tipped stirring stick to well Rh to mix the blood and anti-Rh serum.

11. Drag the stirring stick to the biohazardous waste disposal.

12. Drag the blood-typing slide to the light box and then click **Light** to analyze the slide.

13. Under each of the wells, click **Positive** if agglutination occurred (the sample shows clumping) or click **Negative** if agglutination did not occur (the sample looks smooth).

CHART 4	Blood Typing Results			
Blood sample	Agglutination with anti-A serum	Agglutination with anti-B serum	Agglutination with anti-Rh serum	Blood type
1				
2				
3				
4				
5				
6				

14. Click **Record Data** to display your results in the grid (and record your results in Chart 4).

15. Drag the blood-typing slide to the biohazardous waste disposal.

> **? PREDICT Question 1**
> If the patient's blood type is AB⁻, what would be the appearance of the A, B, and Rh samples?

16. You will now analyze the remaining samples.

- Drag a blood-typing slide from the slide dispenser to the workbench. The next sample will be added to each well on the slide, the appropriate antiserum will be added to each well, the sample and antisera will be mixed, and the slide will be placed in the light box.

- Under each of the wells, click **Positive** if agglutination occurred (the sample shows clumping) or click **Negative** if agglutination did not occur (the sample looks smooth).

- Click **Record Data** to display your results in the grid (and record your results in Chart 4).

  Repeat this step until you analyze all six samples.

17. You will now indicate the blood type for each sample and indicate whether the sample is Rh positive or Rh negative.

- Click the row for the sample in the grid (and record your results in Chart 4).

- Click A, B, AB, or O above the blood type column to indicate the blood type.

- Click the − button or the + button above the blood type column to indicate whether the sample is Rh negative or Rh positive.

  Repeat this step for all six samples. Record your results in Chart 4.

After you complete the experiment, take the online **Post-lab Quiz** for Activity 4.

Activity Questions

1. Antibodies against the A and B antigens are found in the plasma, and a person has antibodies only for the antigens that are not present on their red blood cells. Using this information, list the antigens found on red blood cells and the antibodies in the plasma for blood types 1) AB−, 2) O+, 3) B−, and 4) A+.

_____

_____

2. If an individual receives a bone marrow transplant from someone with a different ABO blood type, what happens to the recipient's ABO blood type?

_____

_____

_____

ACTIVITY 5

Blood Cholesterol

OBJECTIVES

1. To understand the terms *cholesterol, lipoproteins, low-density lipoprotein (LDL), hypocholesterolemia, hypercholesterolemia,* and *atherosclerosis*.

2. To learn how to test for total blood cholesterol using a colorimetric assay.

3. To understand the results and their implications when examining total blood cholesterol.

4. To understand the importance of proper disposal of laboratory material that comes in contact with blood.

Introduction

Cholesterol is a lipid substance that is essential for life—it is an important component of all cell membranes and is the base molecule of steroid hormones, vitamin D, and bile salts. Cholesterol is produced in the human liver and is present in some foods of animal origin, such as milk, meat, and eggs. Because cholesterol is a water-insoluble lipid, it needs to be wrapped in protein packages, called **lipoproteins,** to travel in the blood (which is mostly water) from the liver and digestive organs to the cells of the body.

One type of lipoprotein package, called **low-density lipoprotein (LDL),** has been identified as a potential source of damage to the interior of arteries. LDLs can contribute to **atherosclerosis,** the buildup of plaque, in these blood vessels.

A total blood cholesterol determination does not measure the level of LDLs, but it does provide valuable information about the total amount of cholesterol in the blood.

Less than 200 milligrams of total cholesterol per deciliter of blood is considered desirable. Between 200 and 239 mg/dl is considered borderline high cholesterol. Over 240 mg/dl is considered high blood cholesterol (**hypercholesterolemia)** and is associated with an increased risk of cardiovascular disease. Abnormally low blood cholesterol levels (total cholesterol lower than 100 mg/dl) can also suggest a problem. Low levels may indicate hyperthyroidism (overactive thyroid gland), liver disease, inadequate absorption of nutrients from the intestine, or malnutrition. Other reports link **hypocholesterolemia** (low blood cholesterol) to depression, anxiety, and mood disturbances, which are thought to be controlled by the level of available serotonin, a neurotransmitter. There is evidence of a relationship between low levels of blood cholesterol and low levels of serotonin in the brain.

In this test for total blood cholesterol, a sample of blood is mixed with enzymes that produce a colored reaction with cholesterol. The intensity of the color indicates the amount of cholesterol present. The cholesterol tester compares the color of the sample to the colors of known levels of cholesterol (standard values).

EQUIPMENT USED The following equipment will be depicted on-screen: lancets—sharp, needlelike instruments used to prick the finger to obtain a drop of blood; four patients (represented by an extended finger); alcohol wipes—used to cleanse the patient's fingertip before it is punctured with the lancet; color wheel—divided into shades of green that correspond to total cholesterol levels; cholesterol strips—contain chemicals that convert, by a series of reactions, the cholesterol in the blood sample into a green-colored solution; biohazardous waste disposal—used to properly dispose of equipment that comes in contact with blood.

Experiment Instructions

Go to the home page in the PhysioEx software and click **Exercise 11: Blood Analysis.** Click **Activity 5: Blood Cholesterol,** and take the online **Pre-lab Quiz** for Activity 5.

After you take the online Pre-lab Quiz, click the **Experiment** tab and begin the experiment. The experiment instructions are reprinted here for your reference. The opening screen for the experiment is shown below.

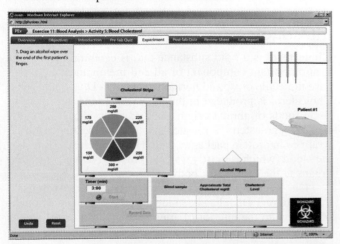

1. Drag an alcohol wipe over the end of the first patient's finger.

2. Drag the alcohol wipe to the biohazardous waste disposal.

3. Drag a lancet to the tip of the patient's finger to prick the finger and obtain a drop of blood.

4. Drag the lancet to the biohazardous waste disposal.

5. Drag a cholesterol strip to the finger to transfer a drop of blood from the patient's finger to the strip.

6. Drag the cholesterol strip to the rectangular box to the right of the color wheel.

7. Click **Start** to start the timer. It takes three minutes for the chemicals in the cholesterol strip to react with the blood. The simulation compresses the 3-minute time period into 3 seconds of real time.

8. Click the color on the color wheel that most closely matches the color on the cholesterol strip.

9. Click **Record Data** to display your results in the grid (and record your results in Chart 5).

CHART 5	Total Cholesterol Determination	
Blood sample	Approximate total cholesterol (mg/dl)	Cholesterol level
1		
2		
3		
4		

10. Drag the cholesterol test strip to the biohazardous waste disposal.

 PREDICT Question 1
Patient 4 prefers to cook all his meat in lard or bacon grease. Knowing this dietary preference, you anticipate his total cholesterol level to be:

_____

11. You will now test the total cholesterol levels for the remaining patients.

- Drag an alcohol wipe over the end of the patient's finger. The alcohol wipe will automatically be placed in the biohazardous waste disposal.

- Drag a lancet to the tip of the patient's finger to prick the finger and obtain a drop of blood. The lancet will automatically be placed in the biohazardous waste disposal.

- Drag a cholesterol strip to the finger to transfer a drop of blood from the patient's finger to the strip.

- Drag the cholesterol strip to the rectangular box to the right of the color wheel. The timer will automatically run for three minutes to allow the chemicals in the cholesterol strip to react with the blood.

- Click the color on the color wheel that most closely matches the color on the cholesterol strip.

- Click **Record Data** to display your results in the grid (and record your results in Chart 5). The cholesterol strip will automatically be placed in the biohazardous waste disposal.

Repeat this step until you determine the total cholesterol levels for all four patients.

After you complete the experiment, take the online **Post-lab Quiz** for Activity 5.

Activity Questions

1. Why do cholesterol plaques occur in arteries and not veins?

_____

_____

2. Phytosterols can alter absorption of certain molecules by the intestinal tract. Why would they be a beneficial dietary supplement for people with high LDL levels?

_____

_____

NAME _____

LAB TIME/DATE _____

Blood Analysis

ACTIVITY 1 Hematocrit Determination

1. List the hematocrits for the healthy male (sample 1) and female (sample 2) living in Boston (at sea level) and indicate whether they are normal or whether they indicate anemia or polycythemia.

   _____

   _____

2. Describe the difference between the hematocrits for the male and female living in Boston. Why does this difference between the sexes exist?

   _____

   _____

3. List the hematocrits for the healthy male and female living in Denver (approximately one mile above sea level) and indicate whether they are normal or whether they indicate anemia or polycythemia.

   _____

   _____

4. How did the hematocrit levels of the Denver residents differ from those of the Boston residents? Why? How well did the results compare with your prediction?

   _____

   _____

5. Describe how the kidneys respond to a chronic decrease in oxygen and what effect this has on hematocrit levels.

   _____

   _____

6. List the hematocrit for the male with aplastic anemia (sample 5) and indicate whether it is normal or abnormal. Explain your response.

   _____

   _____

7. List the hematocrit for the female with iron-deficiency anemia (sample 6) and indicate whether it is normal or abnormal. Explain your response.

_____

_____

ACTIVITY 2 Erythrocyte Sedimentation Rate

1. Describe the effect that sickle cell anemia has on the sedimentation rate (sample 3). Why do you think that it has this effect?

_____

_____

2. How did the sedimentation rate for the menstruating female (sample 2) compare with the sedimentation rate for the healthy individual (sample 1)? Why do you think this occurs?

_____

_____

3. How did the sedimentation rate for the individual with angina pectoris (sample 6) compare with the sedimentation rate for the healthy individual (sample 1)? Why? How well did the results compare with your prediction?

_____

_____

4. What effect does iron-deficiency anemia (sample 4) have on the sedimentation rate?

_____

_____

5. Compare the sedimentation rate for the individual suffering a myocardial infarction (sample 5) with the sedimentation rate for the individual with angina pectoris (sample 6). Explain how you might use this data to monitor heart conditions.

_____

_____

ACTIVITY 3 Hemoglobin Determination

1. Is the male with polycythemia (sample 4) deficient in hemoglobin? Why?

_____

_____

2. How did the hemoglobin levels for the female Olympic athlete (sample 5) compare with the hemoglobin levels for the healthy female (sample 2)? Is either person _deficient_ in hemoglobin? How well did the results compare with your prediction?

_____

3. List conditions in which hemoglobin levels would be expected to decrease. Provide reasons for the change when possible.

_____

_____

4. List conditions in which hemoglobin levels would be expected to increase. Provide reasons for the change when possible.

_____

_____

5. Describe the ratio of hematocrit to hemoglobin for the healthy male (sample 1) and female (sample 2). (A normal ratio of hematocrit to grams of hemoglobin is approximately 3:1.) Discuss any differences between the two individuals.

_____

_____

6. Describe the ratio of hematocrit to hemoglobin for the female with iron-deficiency anemia (sample 3) and the female Olympic athlete (sample 5). (A normal ratio of hematocrit to grams of hemoglobin is approximately 3:1.) Discuss any differences between the two individuals.

_____

_____

ACTIVITY 4 Blood Typing

1. How did the appearance of the A, B, and Rh samples for the patient with AB⁻ blood type compare with your prediction?

_____

_____

2. Which blood sample contained the rarest blood type?

_____

3. Which blood sample contained the universal donor?

_____

4. Which blood sample contained the universal recipient?

_____

5. Which blood sample did not agglutinate with any of the antibodies tested? Why?

_____

_____

6. What antibodies would be found in the plasma of blood sample 1? _____

_____

7. When transfusing an individual with blood that is compatible but not the same type, it is important to separate packed cells from the plasma and administer only the packed cells. Why do you think this is done? (Hint: Think about what is *in plasma* versus what is *on RBCs*.)

_____

_____

8. List the blood samples in this activity that represent people who could donate blood to a person with type B$^+$ blood.

_____

_____

ACTIVITY 5 Blood Cholesterol

1. Which patient(s) had desirable cholesterol level(s)?

_____

2. Which patient(s) had elevated cholesterol level(s)?

_____

3. Describe the risks for the patient(s) you identified in question 2.

_____

_____

4. Was the cholesterol level for patient 4 low, desirable, or high? How well did the results compare with your prediction? What advice about diet and exercise would you give to this patient? Why?

_____

_____

5. Describe some reasons why a patient might have abnormally low blood cholesterol.

_____

_____

Serological Testing

Exercise Overview

Immunology, the study of the immune system, focuses on chemical interactions that are difficult to observe. A number of chemical techniques have been developed to visually represent antibodies and antigens in the **serum,** the fluid portion of the blood with the clotting factors removed. The study and use of these techniques is referred to as **serology.** These techniques are performed in vitro, outside of the body, and are primarily used as diagnostic tools to detect disease. Other applications include pregnancy testing and drug testing. These immunological techniques depend upon the principle that an antibody binds only to specific, corresponding antigens. The tests are relatively expensive to perform, so these activities will allow you to perform them without the sometimes cost-prohibitive supplies.

Antigens and Antibodies

The word **antigen** is derived from two words: *anti*body and *gen*erator. Antigens do not produce antibodies, but early scientists noted that when antigens were present, antibodies appeared. Plasma cells actually produce antibodies.

Antigens include proteins, polysaccharides, and various small molecules that stimulate antibody production. Antigens are often molecules that are described as **nonself,** or foreign to the body. There are also self-antigens that act as identifier tags, such as the proteins found on the surface of red blood cells. Most often,

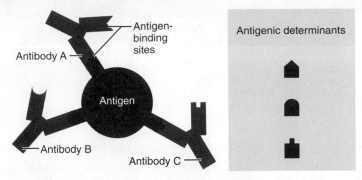

FIGURE 12.1 Antigen-antibody interaction with antigenic determinants.

antigens are a portion of an infectious agent, such as a bacterium or a virus, and the body produces antibodies in response to the presence of the infectious agent.

Antigens are often large and have multiple antigenic sites—locations that can bind to antibodies. We refer to these sites as **antigenic determinants,** or **epitopes.** The antibody has a corresponding antigen-binding site that has a "lock-and-key" recognition for the antigenic determinant on the antigen (view Figure 12.1). All of the simulated tests presented in this exercise take advantage of antigen-antibody specificity. These tests include direct fluorescent antibody technique, Ouchterlony technique, ELISA (enzyme-linked immunosorbent assay), and Western blotting technique.

Nonspecific Binding

The lock-and-key recognition that antigen and antibody have for each other is much like the specificity that an enzyme and its substrate have for one another. However, with antigen and antibody, **nonspecific binding** sometimes occurs. For this reason you will perform a number of washing steps in this exercise to remove any nonspecific binding.

Positive and Negative Controls

You will also use **positive** and **negative controls** to ensure that the test is working accurately. Positive controls include a substance that is known to react positively, thus giving you a standard against which to base your results. Negative controls include substances that should not react. A positive result with a negative control is a "false positive," which would invalidate all other results. Likewise, a negative result with a positive control is a "false negative," which would also invalidate your results.

ACTIVITY 1

Using Direct Fluorescent Antibody Technique to Test for Chlamydia

OBJECTIVES

1. To understand how fluorescent antibodies can be used diagnostically to detect the presence of a specific antigen.

2. To observe how to test for the sexually transmitted disease chlamydia.

3. To distinguish between antigens and antibodies.

4. To understand the terms *epitope* and *antigenic determinant.*

5. To observe nonspecific binding that can result between antigen and antibody.

Introduction

The direct fluorescent antibody technique uses antibodies to directly detect the presence of antigen. A fluorescent dye molecule attached to these antibodies acts as a visual signal for a positive result. This technique is typically used to test for antigens from infectious agents, such as bacteria or viruses. In this activity you will test for the presence of *Chlamydia trachomatis* (a bacterium that invades the cells of its host) using fluorescently labeled antibodies to detect the presence of the antigen and, therefore, the bacterium. *Chlamydia trachomatis* is an important infectious agent because it causes the sexually transmitted disease **chlamydia.** Left untreated, chlamydia can lead to sterility in men and women.

Chlamydia trachomatis is an obligate, intracellular bacterium, which means that it can only survive inside a host cell. The life cycle of the bacterium has two cellular types. The infectious cell type is the small, dense **elementary body,** which is capable of attaching to the host cell. The **reticulate body** is a larger, less-dense cell, which divides actively once inside the host cell. The reticulate body is also referred to as the vegetative form. The life cycle of *Chlamydia* begins when the elementary body enters the host cell and continues as the elementary body changes inside the host cell into a reticulate body. The reticulate body divides into more reticulate bodies and converts back to the elementary body form for release to infect other cells.

In this activity you will test three patient samples and two control samples for the *Chlamydia* infection. An epithelial scraping from the male urethra or from the cervix of the uterus is performed to collect squamous cells from the surface. The elementary bodies are measured by reacting antigen-specific antibodies to infected cells. The fluorescent dye attached to the antigen-specific antibodies makes the complex detectable. The sample is viewed with a fluorescent microscope. The presence of ten or more elementary bodies in a field of view with a diameter of 5 millimeters is considered a positive result. The elementary bodies will be stained green inside red host cells.

EQUIPMENT USED The following equipment will be depicted on-screen: five samples: patient A, patient B, patient C, a positive control, and a negative control; incubator; fluorescent microscope; 95% ethyl alcohol—used for fixing the sample to the microscope slide; chlamydia fluorescent antibody (Chlamydia FA)—antibodies specific for the *Chlamydia* antigen with a fluorescent dye attached; fluorescent antibody mounting media (FA mounting)—used to mount the prepared sample to the slide when ready for viewing under the microscope; phosphate buffered saline (PBS)—used to wash off excess antibodies and prevent nonspecific binding of the antigen and antibody; fluorescent antibody buffer (FA buffer)—used to remove excess ethyl alcohol; petri dishes—used for incubation of the slides to keep them moist; microscope slides—an incubation vessel where the antigen and antibody react; cotton-tipped applicators—used for application and mixing of the antibodies with the samples; filter paper—used to keep the samples moist in the petri dishes; biohazardous waste disposal.

Experiment Instructions

Go to the home page in the PhysioEx software and click **Exercise 12: Serological Testing.** Click **Activity 1: Using Direct Fluorescent Antibody Technique to Test for Chlamydia,** and take the online **Pre-lab Quiz** for Activity 1.

After you take the online Pre-lab Quiz, click the **Experiment** tab and begin the experiment. The experiment instructions are reprinted here for your reference. The opening screen for the experiment is shown below.

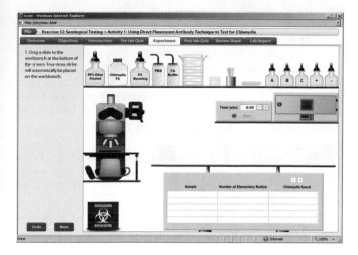

1. Drag a slide to the workbench at the bottom of the screen. Four more slides will automatically be placed on the workbench.

2. The patient samples have been suspended in a small amount of buffer and placed in dropper bottles for ease of dispensing. Drag the dropper cap of the patient A sample bottle to the first slide on the workbench to dispense a drop of the sample onto the slide. A drop from each sample will be placed on a separate slide.

3. Drag the dropper cap of the 95% ethyl alcohol bottle to the first slide on the workbench to dispense three drops of ethyl alcohol onto each slide.

4. Set the timer to 5 minutes by clicking the + button beside the timer display. Click **Start** to start the timer and allow the ethyl alcohol to fix the sample to the slide and prevent the sample from being washed off in the subsequent washing steps. The simulation compresses the 5-minute time period into 5 seconds of real time.

5. Drag the fluorescent antibody (FA) buffer squirt bottle to the first slide to rinse all five slides and remove excess ethyl alcohol.

6. Drag an applicator stick to the chlamydia fluorescent antibody (FA) bottle to soak its cotton tip with antibodies that are specific for *Chlamydia* and labeled with a fluorescent tag.

7. Drag the applicator stick to the first slide to apply the chlamydia fluorescent antibody. Separate applicator sticks will automatically be soaked in chlamydia fluorescent antibody and applied to each slide. Each applicator will automatically be placed in the biohazardous waste disposal.

8. Drag a petri dish to the workbench. A piece of filter paper will be placed into the petri dish. The filter paper has

been moistened with fluorescent antibody buffer to keep the samples from drying out during incubation. Four more petri dishes (and filter paper moistened with fluorescent antibody buffer) will automatically be placed on the workbench.

9. Drag the first slide into the first petri dish. The remaining four slides will automatically be placed into the remaining petri dishes and all five petri dishes will be loaded into the incubator.

10. Set the timer to 20 minutes by clicking the + button next to the timer display. Click **Start** to incubate the samples at 25°C. During incubation the antibodies will react with the corresponding antigens if they are present in the sample. The petri dishes will automatically be removed from the incubator when the time is complete. The simulation compresses the 20-minute incubation time period into 10 seconds of real time.

11. Drag the phosphate buffered saline (PBS) squirt bottle to the first petri dish to wash off excess antibodies and prevent nonspecific binding of the antigen and antibody. The timer will count down 10 minutes for a thorough washing.

12. Click the first petri dish to open the dish and remove the slide. The slides will automatically be removed from the remaining petri dishes.

13. Drag the first petri dish to the biohazardous waste disposal. The remaining petri dishes will automatically be placed in the biohazardous waste disposal.

14. Drag the dropper cap of the fluorescent antibody (FA) mounting media to the first slide to dispense a drop of mounting media onto each slide to mount the sample to the slide.

15. Drag the first slide (patient A) to the fluorescent microscope. Count the number of elementary bodies you see through the microscope (recall that elementary bodies stain green). Click **Submit** to display your results in the grid (and record your results in Chart 1). After you click **Submit,** the slide will automatically be placed in the biohazardous waste disposal.

CHART 1	Direct Fluorescent Antibody Technique Results	
Sample	Number of elementary bodies	Chlamydia result
Patient A		
Patient B		
Patient C		
Positive control		
Negative control		

16. Repeat step 15 for Patient B.

17. Repeat step 15 for Patient C.

18. Repeat step 15 for the Positive Control.

19. Repeat step 15 for the Negative Control.

20. You will now indicate whether each sample is negative or positive for *Chlamydia*. Click the row for the sample in the grid and then click the − button or the + button above the Chlamydia result column to indicate whether the sample is negative or positive for *Chlamydia*. Repeat this step for all five samples. Record your results in Chart 1.

After you complete the experiment, take the online **Post-lab Quiz** for Activity 1.

Activity Questions

1. With this technique, is the antigen or antibody found on the patient sample? Explain how you know this.

_____

_____

2. Explain the difference between an antigen and an epitope (antigenic determinant).

_____

_____

3. When a sample has a small number of elementary bodies but not enough to be a positive result, there appears to have been some nonspecific binding that was not removed by the washing steps. Which sample displayed this property?

_____

_____

Comparing Samples with Ouchterlony Double Diffusion

OBJECTIVES

1. To observe the precipitation reaction between antigen and antibody.
2. To distinguish between *epitope* and *antigen*.
3. To understand the specificity that antibodies have for their epitopes.
4. To observe how related proteins might share epitopes in common.

Introduction

The Ouchterlony technique is also known as double diffusion. In this technique antigen and antibody diffuse toward each other in a semisolid medium made up of clear, clarified agar. When the antigen and antibody are in optimal proportions, cross-linking of the antigen and antibody occurs, forming an insoluble precipitate, called a **precipitin line.** These lines can then be used to visually identify similarities between antigens. If optimum proportions have not been met—for example, if there is excess antigen or excess antibody—then no visible precipitate will form. This technique provides easily visible evidence of the binding between antigen and antibody, and sophisticated equipment is not needed to observe the antigen-antibody reaction.

The Ouchterlony technique is designed to determine whether antigens are identical, related, or unrelated. Antigens have **identity** if they are identical. Identical antigens have all their antigenic determinants, or epitopes, in common. In the case of identity, precipitin lines diffuse into each other to completely fuse and form an arc. Antigens have **partial identity** if they are similar or related. Related antigens have some, but not all, antigenic determinants in common. In the case of partial identity, a spur pointing toward the more similar antigen well forms in addition to the arc. Antigens have **non-identity** if they are unrelated. Unrelated antigens do not have any antigenic determinants in common. In the case of non-identity, the lines intersect to form two spurs that resemble an X.

In the Ouchterlony technique, holes are punched into the agar to form wells. The wells are then loaded with either antigen or antibody, which are allowed to diffuse toward each other. Often, the same antigen is placed in adjacent wells to assess the purity of an antigen preparation. In this case a smooth arc with no spurs should be seen, as the antigens are identical. Multiple antibodies can also be placed in a center well. The antibodies will diffuse out in all directions and react with the antigens that are placed in the surrounding wells.

In this activity you will use human and bovine (from cows) albumin as the antigens, and the antibodies will be made in goats against albumin from either humans or cows. The goals are to identify an unknown antigen and to observe the patterns produced by the various relationships: identity, partial identity, and non-identity.

> **EQUIPMENT USED** The following equipment will be depicted on-screen: goat anti–human albumin (Goat A-H)—an antiserum containing antibodies produced by goats against human albumin; goat anti–bovine albumin (Goat A-B)—an antiserum containing antibodies produced by goats against bovine (cow) albumin; bovine serum albumin (BSA); human serum albumin (HSA); unknown antigen; petri dishes filled with clear agar; well cutter.

Experiment Instructions

Go to the home page in the PhysioEx software and click **Exercise 12: Serological Testing.** Click **Activity 2: Comparing Samples with Ouchterlony Double Diffusion:** and take the online **Pre-lab Quiz** for Activity 2.

After you take the online Pre-lab Quiz, click the **Experiment** tab and begin the experiment. The experiment instructions are reprinted here for your reference. The opening screen for the experiment is shown on the following page.

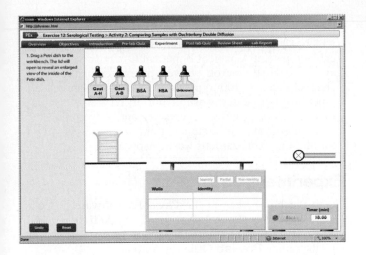

CHART 2	Ouchterlony Double Diffusion Results	
Wells	**Identity**	
2 and 5		
2 and 3		
3 and 4		
4 and 5		

1. Drag a petri dish to the workbench. The lid will open to reveal an enlarged view of the inside of the petri dish.

2. Drag the well cutter to the middle of the enlarged view of the petri dish to punch a hole in the agar in the middle of the petri dish. Drag the well cutter to the upper left, upper right, lower left, and lower right of the petri dish to punch four more holes in the agar. After you punch all five wells into the agar, the wells will be labeled 1–5.

3. Drag the dropper cap of the goat anti–human albumin (Goat A-H) bottle to well 1 to fill it with a sample.

4. Drag the dropper cap of the goat anti–bovine albumin (Goat A-B) bottle to well 1 to fill it with a sample.

5. Drag the dropper cap of the bovine serum albumin (BSA) bottle to well 2 to fill it with a sample.

6. Drag the dropper cap of the bovine serum albumin (BSA) bottle to well 3 to fill it with a sample.

7. Drag the dropper cap of the human serum albumin (HSA) bottle to well 4 to fill it with a sample.

8. Drag the dropper cap of the unknown antigen bottle to well 5 to fill it with a sample.

> **? PREDICT Question 1**
> How do you think human serum albumin and bovine serum albumin will compare?

9. Note that the timer is set to 16 hours. Click **Start** to start the timer. The antigen and antibodies will diffuse toward each other and form a precipitate, detected as a precipitin line. The simulation compresses the 16-hour time period into 10 seconds of real time.

10. You will now examine the precipitin lines that formed and indicate the relationship between each pair of antigens. Click the row for the wells containing the antigens in the grid and then click **Identity, Partial,** or **Non-Identity** above the identity column to indicate whether the antigens have identity, partial identity, or non-identity. Repeat this step for the four pairs of antigens. Record your results in Chart 2.

After you complete the Experiment, take the online **Post-lab Quiz** for Activity 2.

Activity Questions

1. Which type of identity was present between the samples in this activity? Describe this type of identity.

_____

_____

_____

2. Describe the importance of what you place in the center well.

_____

_____

3. Why do you think it is important for the agar to be clear and clarified?

_____

_____

4. Describe the role that albumin plays in the blood.

_____

_____

_____ ▬

Indirect Enzyme-Linked Immunosorbent Assay (ELISA)

OBJECTIVES

1. To understand how the enzyme-linked immunosorbent assay (ELISA) is used as a diagnostic test.

2. To distinguish between the direct and the indirect ELISA.

3. To describe the basic structure of antibodies.

4. To define *seroconversion*.

5. To understand how the indirect ELISA is used to detect antibodies against HIV.

Introduction

The **enzyme-linked immunosorbent assay (ELISA)** is used to test for the presence of an antigen or antibody. The assay is considered enzyme linked because an enzyme is chemically linked to an antibody in both the direct and indirect versions of the test. Immunosorbent refers to the fact that either antigens or antibodies are being adsorbed (stuck) to plastic. If the test is designed to detect an antigen or antigens, it is a **direct ELISA** because it is directly looking for the foreign substance. An **indirect ELISA** is designed to detect antibodies that the patient has made against the antigen. A positive result with the indirect ELISA requires **seroconversion.** Seroconversion occurs when a patient goes from testing negative for a specific antibody to testing positive for the same antibody.

In the direct ELISA, a 96-well microtiter plate is coated with homologous antibodies made against the antigen of interest. The number of wells makes it easy to test many samples at the same time. The patient serum sample is added to the plate to test for the presence of the antigen that binds to the antibody coating on the plate. ELISA takes advantage of the fact that protein sticks well to plastic. A secondary antibody is added to the plate after the patient serum sample is added. If the antigen is present, a "sandwich" of antibody, antigen, and secondary antibody will form. The secondary antibody is chemically linked to an enzyme. When the substrate is added, the enzyme converts the substrate from a colorless compound to a colored compound. The amount of color produced will be proportional to the amount of antigen binding to the antibodies and thus indicates whether the patient is positive for the antigen. If the antigen is not present, the secondary (enzyme-linked) antibodies will be rinsed away with the washing steps and the substrate will not be converted and will remain colorless. A common use of the direct ELISA is a home pregnancy test, which detects human chorionic gonadotropin (hCG), a hormone present in the urine of pregnant women.

In the indirect ELISA, a 96-well microtiter plate is coated with antigens. The patient serum sample is added to test for the presence of antibodies that bind to the antigens on the plate. The secondary antibody that is added has an enzyme linked to it that binds to the **constant region** of the primary antibody if it is present in the patient sample. The constant region of an antibody has the same sequence of amino acids within a class of antibodies (for example, all IgG antibodies have the same constant region). The **variable region** of an antibody provides the diversity of antibodies and is the site to which the antigen binds. The configuration that forms in the indirect ELISA is antigen, primary antibody, and secondary antibody. Just as in the direct ELISA, the addition of substrate is used to determine whether the sample is positive for the presence of antibody.

In this activity you will use the indirect ELISA to test for the presence of antibodies made against human immunodeficiency virus (HIV). You will use positive and negative controls to verify the results. You will note that an indeterminate result can be obtained if there is not enough color produced to warrant a positive result. The cause of an indeterminate result could be either nonspecific binding or that the individual has been recently infected and has not yet produced enough antibodies for a positive result. In either case, the individual would be retested.

EQUIPMENT USED The following equipment will be depicted on-screen: five samples in the samples cabinet: patient A, patient B, patient C, a positive control, and a negative control; 96-well microtiter plate; multichannel pipettor; 100-μl pipettor; microtiter plate reader; pipettor tip dispenser; washing buffer; HIV antigen solution; developing buffer—secondary antibody conjugated with an enzyme; substrate solution; paper towels—used for blotting; biohazardous waste disposal.

Experiment Instructions

Go to the home page in the PhysioEx software and click **Exercise 12: Serological Testing.** Click **Activity 3: Indirect Enzyme-Linked Immunosorbent Assay (ELISA),** and take the online **Pre-lab Quiz** for Activity 3.

After you take the online Pre-lab Quiz, click the **Experiment** tab and begin the experiment. The experiment instructions are reprinted here for your reference. The opening screen for the experiment is shown below.

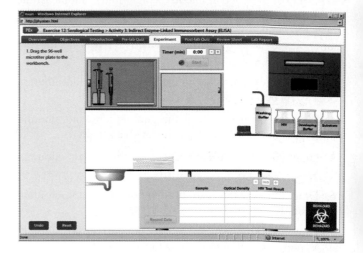

1. Drag the 96-well microtiter plate to the workbench.

2. Drag the multichannel pipettor to the pipette tip dispenser to insert the tips.

3. Drag the multichannel pipettor to the HIV antigens bottle to draw the antigen solution into the tips.

4. Drag the multichannel pipettor directly over the microtiter plate to dispense the liquid into the wells in one row of the plate.

5. Drag the multichannel pipettor to the biohazardous waste disposal for removal and disposal of the tips.

6. Set the timer to 14 hours by clicking the **+** button beside the timer display. This incubation time allows the antigens to stick to the plastic wells of the microtiter plate. Click **Start** to start the timer. The simulation compresses the 14-hour time period into 10 seconds of real time.

7. Drag the washing buffer squeeze bottle to the microtiter plate to remove excess antigens that are not adsorbed (stuck) to the plate.

8. Drag the microtiter plate to the sink to dump the contents of the tray into the sink to remove the washing buffer and excess antigens that are not stuck to the plastic.

9. Drag the microtiter plate to the paper towels. The plate will be pressed to the surface of the paper towels to remove the remaining liquid from the wells. In a typical ELISA, you would perform multiple washing steps to reduce any nonspecific binding. The number of washing steps in this simulation has been reduced for simplicity.

10. Drag the 100-μl pipettor to the tip dispenser to place a tip onto the pipettor.

11. Drag the 100-μl pipettor to the test tube containing the positive control sample (+) to draw the sample into the tip.

12. Drag the 100-μl pipettor to the microtiter plate to dispense the sample into the wells of the plate. The tip will automatically be removed and disposed of in the biohazardous waste disposal. Each of the remaining samples will automatically be dispensed into plate.

13. Set the timer to 1 hour by clicking the + button beside the timer display. This incubation time allows the antigens stuck to the plastic to bind to the antibodies present in the sample. Click **Start** to start the timer. The simulation compresses the 1-hour time period into 10 seconds of real time.

14. Drag the washing buffer squeeze bottle to the microtiter plate to wash off excess antibodies and prevent nonspecific binding of the antigen and antibody.

15. Drag the microtiter plate to the sink to dump washing buffer and unbound antibodies into the sink.

16. Drag the microtiter plate to the paper towels. The plate will be pressed to the surface of the paper towels to remove the remaining liquid from the wells.

17. Drag the multichannel pipettor to the pipette tip dispenser to insert the tips.

18. Drag the multichannel pipettor to the developing buffer bottle to draw the developing buffer into the tips. The developing buffer contains the conjugated secondary antibody.

19. Drag the multichannel pipettor to the microtiter plate to dispense the solution into the wells. The tips will automatically be removed and disposed of in the biohazardous waste disposal.

20. Set the timer to 1 hour and then click **Start** to start the timer and allow the conjugated secondary antibody to bind to the primary antibody if it is present in the sample.

21. Drag the washing buffer squeeze bottle to the microtiter plate to remove any nonspecific binding that occurred.

22. Drag the microtiter plate to the sink to dump the contents of the tray into the sink.

23. Drag the microtiter plate to the paper towels. The plate will be pressed to the surface of the paper towels to remove the remaining liquid from the wells.

24. Drag the multichannel pipettor to the pipette tip dispenser to insert the tips.

25. Drag the multichannel pipettor to the substrate bottle to draw the substrate into the tips.

26. Drag the multichannel pipettor to the microtiter plate to dispense the solution into the wells. The tips will automatically be removed and disposed of in the biohazardous waste disposal.

27. An enlargement of the wells will appear. The development will progress over time. To determine the optical density for each sample (the samples are in the first row, from top to bottom, of the microtiter plate):

- Click the well and the optical density will appear in the window of the microtiter plate reader.
- Click **Record Data** to display your results in the grid (and record your results in Chart 3).

CHART 3	Indirect ELISA Results	
Sample	Optical density	HIV test result
Patient A		
Patient B		
Patient C		
Positive control		
Negative control		

28. You will now indicate whether the result for each sample is negative, indeterminate, or positive for HIV.

- A result of <0.300 is read as negative for HIV-1.
- A result of 0.300–0.499 is read as indeterminate (need to retest).
- A result of >0.500 is read as positive for HIV-1.

Click the row for the sample in the grid and then click the − button, **IND,** or the + button above the HIV test result column to indicate whether the result for the sample is positive, indeterminate, or negative for HIV. Repeat this step for all five samples. Record your results in Chart 3.

After you complete the Experiment, take the online **Postlab Quiz** for Activity 3.

Activity Questions

1. Describe how you can tell that this test is the indirect ELISA rather than the direct ELISA.

_____

_____

2. Describe what the secondary antibody binds to in this activity and why.

_____

_____

3. Define *seroconversion*. How can you tell that a sample has seroconverted?

_____

_____ ▬

Western Blotting Technique

OBJECTIVES

1. To compare the Western blotting technique to the ELISA.
2. To observe the use of the Western blotting technique to test for HIV.
3. To distinguish between antigens and antibodies.

Introduction

Southern blotting was developed by Ed Southern in 1975 to identify DNA. A variation of this technique, developed to identify RNA, was named Northern blotting, thus continuing the directional theme. Western blotting, another variation that identifies proteins, is named by the same convention.

Western blotting uses an electrical current to separate proteins on the basis of their size and charge. This technique uses **gel electrophoresis** to separate the proteins in a gel matrix. Because the resulting gel is fragile and would be difficult to use in further tests, the proteins are then transferred to a **nitrocellulose membrane.** The original Western blotting technique used blotting (diffusion) to transfer the proteins, but electricity is also used now for the transfer of the proteins to nitrocellulose strips. These strips are commercially available, eliminating the need for the electrophoresis and transfer equipment. In this activity you will begin the procedure after the HIV (human immunodeficiency virus) antigens have already been transferred to nitrocellulose and cut into strips.

Western blotting is also known as **immunoblotting** because the proteins that are transferred, or blotted, onto the membrane are later treated with antibodies—the same procedure used in the **indirect enzyme-linked immunosorbent assay (ELISA).** The ELISA is considered enzyme linked because an enzyme is chemically linked to an antibody in both the direct and indirect versions of the test. Immunosorbent refers to the fact that either antigens or antibodies are being adsorbed (stuck) to plastic. If the test is designed to detect an antigen or antigens, it is a **direct ELISA** because it is directly looking for the foreign substance. An **indirect ELISA** is designed to detect antibodies that the patient has made against the antigen.

Similar to the secondary antibodies used in the indirect ELISA technique, the secondary antibodies in the Western blot have an enzyme attached to them, allowing for the use of color to detect a particular protein. The secondary antibody binds to the constant region of the primary antibody found in the patient's sample. The main difference between these techniques is that the ELISA technique uses a well that corresponds to a mixture of antigens, and the Western blot has a discrete protein band that represents the specific antigen that the antibody is recognizing. Like HIV, Lyme disease can also be detected with the Western blot technique.

The initial test for HIV is the ELISA, which is less expensive and easier to perform than the Western blot. The Western blot is used as a confirmatory test after a positive ELISA because the ELISA is prone to false-positive results. The bands from a positive Western blot are from antibodies binding to specific proteins and glycoproteins from the human immunodeficiency virus. A positive result from the Western blot is determined by the presence of particular protein bands (view Table 12.1).

> **EQUIPMENT USED** The following equipment will be depicted on-screen: washing buffer; developing buffer—secondary antibody conjugated with an enzyme; substrate solution; five samples in the samples cabinet: patient A, patient B, patient C, positive control, and negative control; rocking apparatus; nitrocellulose strips; troughs; tray; biohazardous waste disposal.

Experiment Instructions

Go to the home page in the PhysioEx software and click **Exercise 12: Serological Testing.** Click **Activity 4: Western Blotting Technique,** and take the online **Pre-lab Quiz** for Activity 4.

After you take the online Pre-lab Quiz, click the **Experiment** tab and begin the experiment. The experiment instructions are reprinted here for your reference. The opening screen for the experiment is shown below.

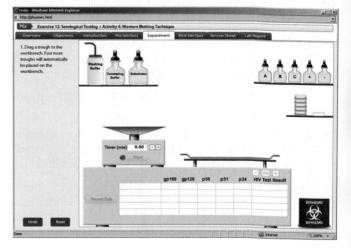

1. Drag a trough to the tray on the workbench. Four more troughs will automatically be placed on the tray.

2. Click the stack of nitrocellulose strips to place a nitrocellulose strip in each trough.

3. Drag the dropper cap of the patient A sample bottle to the first trough to dispense the antiserum from patient A to the nitrocellulose strip. A drop of antiserum for each patient will be dispensed into a separate trough.

4. Drag the tray holding the five troughs to the rocking apparatus.

5. Set the timer to 60 minutes by clicking the **+** button beside the timer display. Click **Start** to gently rock the samples and allow the antibodies to react with the antigens bound to the nitrocellulose. The tray will automatically be returned to the workbench and each trough will be drained into the biohazardous waste disposal when the time is complete. The simulation compresses the 60-minute time period into 10 seconds of real time.

6. Drag the washing buffer squirt bottle to the first trough to dispense washing buffer in each trough. Each trough will be automatically drained into the biohazardous waste container. The washing step removes any nonspecific binding of antibodies that occurred.

7. Drag the dropper cap of the developing buffer bottle to the first trough to dispense developing buffer to each trough.

8. Drag the tray holding the five troughs to the rocking apparatus.

9. Set the timer to 60 minutes by clicking the + button beside the timer display. Click **Start** to gently rock the samples and allow the antibodies to react with the antibodies bound to the nitrocellulose. The tray will automatically be returned to the workbench and each trough will be drained into the biohazardous waste disposal when the time is complete.

10. Drag the washing buffer squirt bottle to the first trough to add washing buffer to each trough. Each trough will be automatically drained into the biohazardous waste container. The washing step removes any nonspecific binding of secondary conjugated antibodies. Excess secondary conjugated antibodies could react erroneously with the substrate and give a false-positive result.

11. Drag the dropper cap of the substrates bottle to the first trough to dispense the substrates (tetramethyl benzidine and hydrogen peroxide) into each trough. The substrates are the chemicals that are being changed by the enzyme that is linked to the antibody.

12. Drag the tray holding the five troughs to the rocking apparatus.

13. Set the timer to 10 minutes by clicking the + button beside the timer display. Click **Start** to gently rock the samples and allow the enzyme to react with the substrates. The tray will automatically be returned to the workbench when the time is complete. The simulation compresses the 10-minute time period into 10 seconds of real time.

14. To determine the antigens present for each sample:

- Click the nitrocellulose strip inside the trough to visualize the results.
- Click **Record Data** to display your results in the grid (and record your results in Chart 4). The bands present

on the nitrocellulose strip represent the antibodies present in the sample that have reacted with the antigens (bands) on the strip (view Table 12.1).

Repeat this step for all five samples.

TABLE 12.1	HIV Antigens
Abbreviation	**Description**
gp160	Glycoprotein 160, a viral envelope precursor
gp120	Glycoprotein 120, a viral envelope protein that binds to CD4
p55	A precursor to the viral core protein p24
gp41	A final envelope glycoprotein
p31	Reverse transcriptase
p24	A viral core protein

15. You will now indicate whether the result for each sample is negative, indeterminate, or positive for HIV. The criteria for reporting a positive result varies slightly from agency to agency. The Centers for Disease Control and Prevention recommend the following criteria:

- If no bands are present, the result is negative.
- If bands are present but they do not match the criteria for a positive result, the result is indeterminate. Patients whose results are deemed indeterminate after multiple tests should be monitored and tested again at a later date.
- If either p31 or p24 is present *and* gp160 or gp120 is present, the result is positive. Click the row for the sample in the grid and then click the – button, **IND,** or the + button above the HIV test result column to indicate whether the result for the sample is positive, indeterminate, or negative for HIV.

Repeat this step for all five samples. Record your results in Chart 4.

After you complete the experiment, take the online **Post-lab Quiz** for Activity 4.

CHART 4	Western Blot Results					
Sample	**gp160**	**gp120**	**p55**	**p31**	**p24**	**HIV test result**
Patient A						
Patient B						
Patient C						
Positive control						
Negative control						

Activity Questions

1   Describe how gel electrophoresis is used to separate proteins.

_____

_____

2.   In a patient sample that is positive for HIV, would antibodies or antigens be present when using the Western blot technique? How do you know?

_____

_____

NAME _____

LAB TIME/DATE _____

Serological Testing

ACTIVITY 1   Using Direct Fluorescent Antibody Technique to Test for Chlamydia

1. Describe the importance of the washing steps in the direct antibody fluorescence test. _____ _____

   _____

   _____

2. Explain where the epitope (antigenic determinant) is located. _____

   _____

   _____

3. Describe how a positive result is detected in this serological test. _____

   _____

   _____

4. How would the results be affected if a negative control gave a positive result? _____

   _____

   _____

ACTIVITY 2   Comparing Samples with Ouchterlony Double Diffusion

1. Describe how you were able to determine what antigen is in the unknown well. _____

   _____

   _____

2. Why does the precipitin line form? _____

   _____

   _____

3. Did you think human serum albumin and bovine serum albumin would have epitopes in common? How well did the results

compare with your prediction? _____

_____

_____

ACTIVITY 3   Indirect Enzyme-Linked Immunosorbent Assay (ELISA)

1. Describe how the direct and indirect ELISA are different. _____

_____

_____

2. Discuss why a patient might test indeterminate. _____

_____

_____

3. How would your results have been affected if your negative control had given an indeterminate result?_____

_____

_____

4. Briefly describe the basic structure of antibodies._____

_____

_____

ACTIVITY 4   Western Blotting Technique

1. Describe why the HIV Western blot is a more specific test than the indirect ELISA for HIV. _____

_____

_____

2. Explain the procedure for a patient with an indeterminate HIV Western blot result. _____

_____

_____

3. Briefly describe how the nitrocellulose strips were prepared before the patient samples were added to them. _____

_____

_____

4. Describe the importance of the washing steps in the procedure._____

_____

_____

Credits

Illustrations

All illustrations are by Imagineering STA Media Services, except for Review Sheet Art and as noted below.

Scissors icon: Vectorpro/Shutterstock. Button icon: justone/Shutterstock. Video (arrow/camera) icons: Yuriy Vlasenko/Shutterstock.

Exercise 1 1.1: Imagineering STA Media Services/Precision Graphics. 1.2, 1.4: Precisions Graphics. 1.9: Source: Adapted from Marieb, Elaine N.; Mallatt, Jon B.; Wilhelm, Patricia Brady, *Human Anatomy*, 5e, F1.10, © 2008. Reprinted and Electronically reproduced by permission of Pearson Education, Inc., Upper Saddle River, New Jersey.

Exercise 3 3.2, 3.3, 3.4, 3.A3: Precision Graphics.

Exercise 4 Opener, 4.3, Table 4.1.3, 4.RS1: Tomo Narashima. 4.2, Table 4.1.4–5: Imagineering STA Media Services/Precision Graphics.

Exercise 5 5.1: Precision Graphics.

Exercise 6 6.2: Precision Graphics.

Exercise 7 Opener, 7.1, 7.2, 7.7: Electronic Publishing Services, Inc.

Exercise 9 9.1–9.3, 9.6, 9.7, 9.9, 9.21: Nadine Sokol.

Exercise 12 12.4: Imagineering STA Media Services/Precision Graphics. 12.5: Electronic Publishing Services, Inc.

Exercise 14 14.1–14.3: Precision Graphics. 14.7–14.9, 14.13–14.18, 14.U1: Biopac Systems.

Exercise 15 15.1: Imagineering STA Media Services/Precision Graphics.

Exercise 16 16.2–16.4: Precision Graphics.

Exercise 17 17.1, 17.5, 17.7–17.11: Electronic Publishing Services, Inc.

Exercise 18 18.4–18.7: Biopac Systems.

Exercise 19 Opener, 19.1b, 19.3b: Electronic Publishing Services, Inc.

Exercise 20 20.7–20.13: Biopac Systems.

Exercise 21 21.1: Electronic Publishing Services, Inc. 21.8, 21.9: Biopac Systems.

Exercise 23 Opener, 23.1, 23.3, 23.4: Electronic Publishing Services, Inc.

Exercise 24 24.1: Shirley Bortoli. 24.4, 24.5: Precision Graphics.

Exercise 25 Opener, 25.2, 25.3, 25.7, 25.9: Electronic Publishing Services, Inc. 25.1: Electronic Publishing Services, Inc./Precision Graphics.

Exercise 26 Opener, 26.1, 26.2: Electronic Publishing Services, Inc.

Exercise 29 29.2, 29.5: Precision Graphics

Exercise 30 Opener, 30.2, 30.3, 30.6: Electronic Publishing Services, Inc. 30.1: Electronic Publishing Services, Inc./Precision Graphics.

Exercise 31 Opener, 31.1: Electronic Publishing Services, Inc. 31.2, 31.4: Precision Graphics. 31.7–31.12: Biopac Systems.

Exercise 32 Opener, 32.3–32.13, 32.15: Electronic Publishing Services, Inc.

Exercise 33 33.2: Precision Graphics. 33.6–33.9: Biopac Systems.

Exercise 34 Opener, 34.6: Electronic Publishing Services, Inc. 34.1, 34.2, 34.4: Precision Graphics. 34.5: Biopac Systems.

Exercise 35 35.8: Precision Graphics.

Exercise 36 Opener, 36.1–36.5, 36.7: Electronic Publishing Services, Inc.

Exercise 37 37.5, 37.6: Precision Graphics. 37.11, 37.13, 37.14: Biopac Systems.

Exercise 38 Opener, 38.1–38.5, 38.7, 38.8, 38.10, 38.15: Electronic Publishing Services, Inc. 38.16: Electronic Publishing Services, Inc./Precision Graphics.

Exercise 39 39.A1–A3: Precision Graphics.

Exercise 40 Opener, 40.1, 40.2: Electronic Publishing Services, Inc.

Exercise 41 41.1: Precision Graphics.

Exercise 42 Opener, 42.1, 42.2, 42.7: Electronic Publishing Services, Inc.

Exercise 43 43.1: Precision Graphics. 43.2: Electronic Publishing Services, Inc.

Exercise 44 44.1: Electronic Publishing Services, Inc.

Exercise 46 46.17: Precision Graphics.

Cat Dissection Exercises 1.1, 3.1, 3.3a, 7.1: Precision Graphics. 2.1a, 2.2a, 3.2, 4.2, 4.4, 6.2, 7.2a, 8.2a, 9.1a: Kristin Mount.

Fetal Pig Dissection Exercises 1.3–1.8, 2.2, 3.1, 3.2, 4.1–4.5, 5.2, 6.1, 6.2, 7.1, 7.2, 8.1, 8.2: Kristin Mount.

PhysioEx Exercises All illustrations by BinaryLabs, Inc. except as noted 2.1: Precision Graphics/Source: Adapted from Stanfield, *Principles of Human Physiology*, 4e. F12.1, (c) 2011. Reprinted and Electronically reproduced by permission of Pearson Education, Inc. Upper Saddle River, NJ. 3.1: Imagineering STA Media Services. 4.1: Electronic Publishing Services, Inc. 6.1: Precision Graphics/Source: Adapted from Stanfield, *Principles of Human Physiology*, 4e. F13.13, (c) 2011. Reprinted and Electronically reproduced by permission of Pearson Education, Inc. Upper Saddle River, NJ. 7.1: Precision Graphics/Source: Adapted from Stanfield, *Principles of Human Physiology*, 4e. F16.1, (c) 2011. Reprinted and Electronically reproduced

by permission of Pearson Education, Inc. Upper Saddle River, NJ. 8.1: Electronic Publishing Services, Inc. 9.1: Electronic Publishing Services, Inc. 12.1: Imagineering STA Media Services.

Photographs

Visual Walkthrough (clockwise from top): michaeljung/Fotolia, Guy Cali/Corbis, Monkey Business/Fotolia, Tyler Olson/Shutterstock.

Exercise 1 Opener, 1.3.1, 1.8a: John Wilson White, Pearson Education. 1.3a: CNRI/Science Photo Library/Science Source. 1.3b: Scott Camazine/Science Source. 1.3c: James Cavallini/Science Source.

Exercise 2 Opener: Arcady/Shutterstock. 2.1a–d, 2.2, 2.3a, 2.4a, 2.5b,c: Elena Dorfman, Pearson Education. 2.3b, 2.4b, 2.5a, 2.6a–c: From *A Stereoscopic Atlas of Human Anatomy* by David L. Bassett, M.D. 2.7: Arcady/Shutterstock.

Exercise 3 Opener, 3.1: Vereshchagin Dmitry/Shutterstock. 3.5: Victor P. Eroschenko, Pearson Education.

Exercise 4 4.1b: Don Fawcett/Science Source. 4.4, 4.RS2: William Karkow, Pearson Education.

Exercise 5 5.2a,b: Richard Megna/Fundamental Photographs. 5.3a–c: David M. Philips/Science Source.

Exercise 6 6.3a,d,g, 6.5a,b,e,g,l, 6.7b,c: William Karkow, Pearson Education. 6.3b,c,f,h, 6.5d,j,k: Alan Bell, Pearson Education. 6.3e, 6.5c,f,i: Nina Zanetti, Pearson Education. 6.5h: Steve Downing, PAL 2.0, Pearson Education. 6.6: Biophoto Associates/Science Source. 6.7a: Nina Zanetti, PAL 1.0, Pearson Education.

Exercise 7 7.2a, 7.6b: William Karkow, Pearson Education. 7.3: Pearson Education. 7.6a, 7.RS2: Marian Rice. 7.7a,b: Lisa Lee, Pearson Education.

Exercise 8 8.4c: William Karkow, Pearson Education. 8.5: Lisa Lee, Pearson Education. 8.RS2: Alan Bell, Pearson Education.

Exercise 9 9.1b, 9.2b, 9.4a,b, 9.21c,d: Larry DeLay, PAL 3.0, Pearson Education. 9.3c, 9.5: Michael Wiley, Univ. of Toronto, Imagineering © Pearson Education. 9.10: Elena Dorfman, Pearson Education. 9.11c: From *A Stereoscopic Atlas of Human Anatomy* by David L. Bassett, M.D. 9.19b: Karen Krabbenhoft, PAL 3.0, Pearson Education. 9.20c: Pearson Education.

Exercise 10 10.5b: CMSP/Newscom. Table 10.1: From *A Stereoscopic Atlas of Human Anatomy* by David L. Bassett, M.D.

Exercise 11 11.5: John Wilson White, Pearson Education. 11.6c: From *A Stereoscopic Atlas of Human Anatomy* by

David L. Bassett, M.D. 11.7d, 11.8c: Karen Krabbenhoft, PAL 3.0, Pearson Education. 11.8a: Mark Neilsen, Pearson Education.

Exercise 12 12.1a: Marian Rice. 12.3, 12.6: Victor P. Eroschenko, Pearson Education. 12.4b: William Karkow, Pearson Education.

Exercise 13 13.4b, 13.5a, 13.9a, 13.11f: Karen Krabbenhoft, PAL 3.0, Pearson Education. 13.8b: William Karkow, Pearson Education. 13.13b: From *A Stereoscopic Atlas of Human Anatomy* by David L. Bassett, M.D.

Exercise 15 15.2b, 15.6b: William Karkow, Pearson Education. 15.3b: Don W. Fawcett/Science Source. 15.4: *Eroschenko's Interactive Histology*. 15.6a: Sercomi/Science Source. 15.6c: Nina Zanetti, Pearson Education. 15.8b: Victor P. Eroschenko, Pearson Education.

Exercise 17 Opener, 17.3, 17.4a, 17.5b, 17.6a,b, 17.10: Karen Krabbenhoft, PAL 3.0, Pearson Education. 17.2c, 17.7c: From *A Stereoscopic Atlas of Human Anatomy* by David L. Bassett, M.D. 17.11a,c: Sharon Cummings, Pearson Education. 17.12–17.14: Elena Dorfman, Pearson Education.

Exercise 18 18.1a: Hank Morgan/Science Source.

Exercise 19 19.2b–d: From *A Stereoscopic Atlas of Human Anatomy* by David L. Bassett, M.D. 19.5: Lisa Lee, Pearson Education. 19.8b: Karen Krabbenhoft, PAL 3.0, Pearson Education.

Exercise 21 21.4–21.6: Richard Tauber, Pearson Education.

Exercise 22 22.1b: Lynn McCutchen. 22.1c,d, 22.2b: Victor P. Eroschenko, Pearson Education.

Exercise 23 23.3b: From *A Stereoscopic Atlas of Human Anatomy* by David L. Bassett, M.D. 23.4b: Lisa Lee, Pearson Education. 23.5: Elena Dorfman, Pearson Education. 23.6b: Stephen Spector, Pearson Education.

Exercise 24 24.6: Dr. Charles Klettke, Pearson Education.

Exercise 25 25.4: Victor P. Eroschenko, Pearson Education. 25.6a–c: Richard Tauber, Pearson Education. 25.8: I. M. Hunter-Duvar, Department of Otolaryngology, The Hospital for Sick Children, Toronto.

Exercise 26 26.1b, 26.3: Victor P. Eroschenko, Pearson Education. 26.2d: Steve Downing, PAL 3.0, Pearson Education.

Exercise 27 27.3a,f: William Karkow, Pearson Education. 27.3b,d: Victor P. Eroschenko, Pearson Education. 27.3c: Lisa Lee, Pearson Education. 27.3e: Benjamin Widrevitz.

Exercise 29 29.3: William Karkow, Pearson Education. 29.4a–e: Nina Zanetti,

Pearson Education. 29.6a–c, 29.7a–d: Elena Dorfman, Pearson Education. 29.8b: Meckes and Ottawa/Science Source. 29.9.1–4, 29.RS1.1 4: Jack Scanlan, Pearson Education.

Exercise 30 30.3b: From *A Stereoscopic Atlas of Human Anatomy* by David L. Bassett, M.D. 30.3c: Philippe Plailly/Look at Sciences/Science Source. 30.3d: Karen Krabbenhoft, PAL 3.0, Pearson Education. 30.6, 30.RS3: William Karkow, Pearson Education. 30.7a,b, 30.8: Wally Cash, Pearson Education.

Exercise 32 32.2: Ed Reschke/Getty Images. 32.4c: From *A Stereoscopic Atlas of Human Anatomy* by David L. Bassett, M.D.

Exercise 33 Opener: John Wilson White, Pearson Education.

Exercise 35 35.2: Eric V. Grave/Science Source. 35.4b, 35.6: William Karkow, Pearson Education. 35.5c: Victor P. Eroschenko, Pearson Education.

Exercise 36 Opener, 36.5a: Richard Tauber, Pearson Education. 36.1a, 36.5b: From *A Stereoscopic Atlas of Human Anatomy* by David L. Bassett, M.D. 36.6b: Victor P. Eroschenko, Pearson Education. 36.7a: William Karkow, Pearson Education. 36.7b: Lisa Lee, Pearson Education.

Exercise 37 37.3, 37.4a,b: Elena Dorfman, Pearson Education.

Exercise 38 38.5b: Karen Krabbenhoft, PAL 3.0, Pearson Education. 38.6a, 38.9a: Nina Zanetti, Pearson Education. 38.6b, 38.13: Victor P. Eroschenko, Pearson Education. 38.6c: Roger C. Wagner. 38.7a,c, 38.14a,b, 38.16b: From *A Stereoscopic Atlas of Human Anatomy* by David L. Bassett, M.D. 38.8d: LUMEN Histology, Loyola University Medical Education Network. 38.9b: Steve Downing, Pearson Education. 38.9c: William Karkow, Pearson Education.

Exercise 40 40.1b: Richard Tauber, Pearson Education. 40.3a: Karen Krabbenhoft, Pearson Education. 40.6a,b, 40.7: Victor P. Eroschenko, Pearson Education.

Exercise 42 42.2b, 42.5, 42.9: Victor P. Eroschenko, Pearson Education. 42.2c: From *A Stereoscopic Atlas of Human Anatomy* by David L. Bassett, M.D. 42.3: Harry H. Plymale. 42.4: Roger C. Wagner. 42.8: vetpathologist/Shutterstock.

Exercise 43 43.2b: Pearson Education. 43.3: William Karkow, Pearson Education. 43.5b: Steve Downing, Pearson Education. 43.6a–c: Victor P. Eroschenko, Pearson Education.

Exercise 45 Opener, 45.1: CNRI/SPL/Science Source. 45.2.1: Ostill/Shutterstock. 45.2.2: Photos.com. 45.2.3–4: Dion Ogust/Image Works. 45.2.5: Boisvieux/Explorer/Science Source. 45.2.6: Anthony Loveday, Pearson Education.

Exercise 46 Opener, 46.1–46.3, 46.5, 46.7, 46.9–46.16, 46.18a–d, 46.19: John Wilson White, Pearson Education.

Cat Dissection Exercises 1.2 1.13, 2.3b, 4.1, 4.3, 4.5b, 6.3, 6.4, 7.3–7.5b, 8.1b, 9.2b: Shawn Miller (dissection) and Mark Nielsen (photography), Pearson Education. 2.1b, 9.1b: Paul Waring, Pearson Education. 2.2b, 7.2b, 8.2b: Elena Dorfman, Pearson Education. 3.3b,c: Yvonne Baptiste-Szymanski, Pearson Education.

Fetal Pig Dissection Exercises 1.1, 1.2: Jack Scanlan, Pearson Education. 1.3b–1.8b, 2.1, 4.1b–4.4b, 5.1, 5.2b, 6.1b, 6.2b, 7.1b, 7.2b, 8.1b. 8.2b: Elena Dorfman, Pearson Education. 3.2b,c: Charles J. Venglarik, Pearson Education.

Trademark Acknowledgments

3M is a trademark of 3M.

Adrenaline is a registered trademark of King Pharmaceuticals.

Albustix, Clinistix, Clinitest, Hemastix, Ictotest, Ketostix, and Multistix are registered trademarks of Bayer.

Betadine is a registered trademark of Purdue Products L. P.

Chemstrip is a registered trademark of Roche Diagnostics.

Coban is a trademark of 3M.

Harleco is a registered trademark of EMD Chemicals Inc.

Landau is a registered trademark of Landau Uniforms.

Lycra is a registered trademark of INVISTA.

Macintosh, Power Macintosh and Mac OS X are registered trademarks of Apple Computer, Inc.

Novocain is a registered trademark of Sterling Drug, Inc.

Parafilm is a registered trademark of Pechiney Incorporated.

Porelon is a registered trademark of IDG, LLC.

Sedi-stain is a registered trademark of Becton, Dickinson and Company.

Speedo is a registered trademark of Speedo International.

VELCRO® is a registered trademark of VELCRO Industries B. V.

Wampole is a registered trademark of Wampole Laboratories.

Windows is either a registered trademark or trademark of Microsoft Corporation in the United States and/or other countries.

Index